# HANDBOOK OF ELECTRONIC PACKAGE DESIGN

# MECHANICAL ENGINEERING

## A Series of Textbooks and Reference Books

**Editor: L.L. FAULKNER**   Columbus Division, Battelle Memorial Institute, and Department of Mechanical Engineering, The Ohio State University, Columbus, Ohio

**Associate Editor: S.B. MENKES**   Department of Mechanical Engineering, The City College of the City University of New York, New York

**Additional Volumes in Preparation**

**Mechanical Engineering Software**

Spring Design with an IBM PC, *by Al Dietrich*

Mechanical Design Failure Analysis: With Failure Analysis System Software for the IBM PC, *by David G. Ullman*

# HANDBOOK OF ELECTRONIC PACKAGE DESIGN

### EDITED BY

## MICHAEL PECHT

*CALCE Center for Electronic Packaging*
*University of Maryland at College Park*
*College Park, Maryland*

MARCEL DEKKER, INC.          NEW YORK · BASEL

**Library of Congress Cataloging-in-Publication Data**

Handbook of electronic package design / edited by Michael Pecht.
     p.   cm. -- (Mechanical engineering ; 76)
    Includes bibliographical references and index.
    ISBN 0-8247-7921-5
    1. Electronic packaging--Handbooks, manuals, etc.  I. Pecht,
Michael.  II. Series: Mechanical engineering (Marcel Dekker, Inc.) ; 76.
TK7870. 15.H36   1991
621.381'046--dc20                                          91-21380
                                                            CIP

This book is printed on acid-free paper

MARCEL DEKKER, INC.
270 Madison Avenue, New York, New York 10016

Current printing (last digit):
10 9 8 7 6

PRINTED IN THE UNITED STATES OF AMERICA

To my parents, George and Dorothy Pecht,
my wife, Judy,
my children, Joann and Jefferson,
and Grandma.

# Preface

This book is intended as a handbook for practitioners and a text for use in teaching electronic packaging concepts, guidelines, and techniques. The treatment begins with an overview of the electronics design process and proceeds to examine the levels of electronic packaging and the fundamental issues in the development of a complete electronic system. The book then addresses the various analysis methods and techniques that can be used during the electronics design process. Emphasis is placed on those engineering tasks and issues essential to the system design for reliability. The book concludes with a chapter on design for reliability and on electronic material properties. A brief description of each of the twelve chapters follows.

Chapter 1 provides an overview of the electronic packaging process and introduces important terminology. The packaging hierarchy of electronic systems is defined and aspects of each level are highlighted. Design trends and considerations for electronic packaging are examined. The roles of electrical and mechanical engineers in design of electronic packages are discussed, and the concepts of concurrent engineering design and life cycle product development are presented and formulated.

Chapter 2 presents the detailed definitions of electronic components with emphasis on microelectronic packages. Issues associated with package design are examined, including package and lead styles. Hybrids, multichip modules, and surface-mounted packages are discussed.

Chapter 3 presents the definitions and design issues associated with printed circuit (PCBs) and printed wiring boards (PWBs). Manufacturing processes and board fabrication steps, from bare substrate formation to machining, metallization, and preparation for component mounting, are discussed for both organic and ceramic PWBs. Design considerations and reliability issues for organic and ceramic boards are discussed. Layout design techniques for meeting specified electrical parameters are given. Developmental board designs, such as direct bond copper to ceramic and thin film polyimide substrates, are also called presented.

Chapter 4 commences with a discussion of elementary subassemblies, which are used as the building blocks for other assemblies and systems. Each level of assembly is presented with details on standard materials, manufacturing processes, interconnection methods, and applications. The applications of elementary assemblies in hybrid assemblies and printed wiring assemblies (PWAs) are reviewed, followed by hybrid applications within PWAs, chassis, and standard electronic modules (SEMs), and PWAs and SEMs utilization within chassis and systems. Design aspects and concurrent engineering considerations for each assembly level are examined. The goal is to meet system requirements by correctly laying out and packaging the devices within the system. Trade-offs between reliability, manufacturability, and costs also play an important part in the design effort. Surface and insertion mounting methods, including various soldering techniques and equipment, are discussed in detail. Electrostatic discharge considerations are reviewed. The chapter concludes with a discussion on how the electronic building blocks are integrated to form the final system.

Chapter 5 presents the definitions and design issues associated with electronic connections and interconnections at six levels of assembly, including chip to lead frame, package to printed circuit board (PCB), PCB to PCB, PCB to cabinets, among cabinets, and among subsystems and systems. The reliability of these interconnections is influenced by the design, construction materials, of construction level of stresses due to temperature, humidity, and environment. Improper design can cause the interconnection reliability to be several orders of magnitude less than the reliability of the systems they connect. Typical interconnection methods at the various levels along with illustrations have been presented. Connector properties and characteristics have been treated in sufficient detail to aid in connector selection.

Chapter 6 deals with two important aspects of layout, placement, and routing. Layout is a transitory process of converting the electrical circuit design into a feasible working process. Partitioning and assignment of the electrical system design is reviewed. Three placement concerns, routability, reliability, and producibility, are then discussed. The placement goal of routability is presented, highlighting both constructive and iterative placement methods. Constructive methods discussed include pair linking and cluster development methods. Iterative placement procedures include Steinberg's algorithm and the pair-wise interchange method. In addition, iterative placement methods, based on force-directed placement and minicut methods, are discussed. A comparison of the various routability placement techniques is also given. Placement for reliability methods provides recent theoretical developments and heuristic placement techniques. Placement is also discussed from a producibility standpoint. Various placement trade-offs are explored by means of examples. Aspects of routing in terms of path determination, layering, path prioritizing, and track layout are then discussed, and the standard Lee's algorithm is highlighted.

Chapter 7 deals with the thermal design and analysis of electronic systems. The objective of the chapter is to introduce the reader to the concepts and tools used to predict and measure the temperatures of electronic systems in service. An overview of thermal considerations that are important in the design of an electronic system is presented, followed by a discussion of thermal conduction with a review of transient three-dimensional thermal conduction fundamentals, the special cases of a small Biot number, steady state, and one-dimensional heat flow and discussions of fins and extended surfaces, contact resistance, and constriction resistance. Natural (free) convection is presented with a review of natural convection fundamentals followed by empirical correlations of natural convection for external flows and for enclosures. A discussion of forced convection concepts followed by empirical correlations of forced convection for external flows and for enclosures, fan design principles, the effects of barriers, missing modules, or odd-sized modules on forced convection, and the use of numerical methods used to solve forced convection problems. A survey of mixed convection (i.e., those flow situations in which both forced convection and natural convection are important), heat transfer processes associated with the condensation or evaporation of a fluid, and radiation is presented.

Chapter 8 deals with the analysis of mechanical stresses generated in electronic assemblies. The objective is to give the reader an understanding of the underlying principles and to present simple, closed-form, quantitative tools for designing against mechanical damage and failures resulting from monotonic and/or cyclic thermomechanical loads in electronic components, assemblies, interconnects, and boards. This chapter develops the required concepts from the basics, with illustrations used throughout the chapter to explain basic concepts. It also provides an overview of the mechanical problems encountered in electronic packages arising from environmental thermal-cycling and/or functional power-cycling. The basic concepts of multidimensional stresses and strains (including thermal strains) at a point in a solid and their thermoelastic interrelationship are first defined. In order to keep the discussion simple, it is intentionally presented from the perspective of strength of materials rather than the mathematical theory of elasticity. Practical examples are presented to illustrate the utility of these concepts in simplified stress analysis of electronic assemblies subject to temperature changes. The concept of fatigue wear-out failures in electronic devices due to cyclic thermomechanical loading is then presented. Empirical quantitative models are given for designing against both high- and low-cycle fatigue phenomena. An alternative approach for fatigue of brittle materials is presented, based on a fracture mechanics model of fatigue crack propagation. The interactions between creep and fatigue in viscoplastic materials at high temperatures are discussed, and empirical methods are referenced for dealing with such phenomena at a practical design level. Topics of current interest in thermomechanical creep and fatigue of electronic assemblies are then discussed. These include common failure mechanisms (including illustrative photographs of actual failed components); physical and mechanical properties of solder materials; thermal and power cycling tests; and experimental reliability tests (including considerations for accelerated test programs) for failure and reliability analysis and predictions.

Chapter 9 presents the design concepts for minimizing failure in electronic systems due to excessive vibration and/or shock. A fundamental review of simple systems is given, describing methods used to estimate the natural frequency of circuit boards. Expressions are then formulated to estimate the useful life of components and boards subjected to vibration and shock loading.

Chapter 10 provides a comprehensive view of environmental failure mechanisms in packages associated with humidity and corrosion. The effects of moisture are presented together with screening procedures. The impact of moisture on plastic integrated circuit (IC) packages is examined by identifying the failure mechanisms and design precautions.

Corrosion in microelectronic packages caused by moisture and contaminates ingress is then discussed. Factors that affect the rate of corrosion are presented, and mathematical models for the corrosion mechanism are discussed.

Chapter 11 presents the issues, tools, and techniques for reliability analysis, assessment, and correction. The chapter begins with reliability concepts and techniques that can be used in the design and assessment of electronic packages and systems. Basic concepts, mathematical models, and failure probability density distributions are discussed. An overview of the physics of failure issues associated with electronic packages are presented, including a discussion of the various stresses that have been known to induce failures. Emphasis is placed on failure mechanisms, common failure sites, and failure modes. Reliability practices and techniques for electronics design and manufacture are then discussed. The goal is to provide the basic understanding of terms and the background information to be able to conduct a detailed reliability program.

Chapter 12 discusses the properties, behavior, and typical applications of electronic materials. Electrical, thermal, mechanical, and applications data have been compiled in 41 tables for over 200 materials. Data are provided for polymers, ceramics, and metals commonly used in the packages or structures that house electronic devices and systems. The discussion and tabulated data are organized into sections corresponding to the various elements of the package system: semiconductors, substrates and chip carriers, laminates, enclosures, joining materials, encapsulants, and conductors.

Thanks to Wendi Kamtman, Thelma Miller, and Sant'ea Byrd, for help with the typing, and Ed Magrab and Dave Weiss of the University of Maryland Engineering Research Center, Jonathan Watts of the CALCE Center for Electronics Packaging, Charlie Harper of Technology Seminars, Inc., and Kenneth LaSala from the U.S. Air Force, for their contributions on reviewing and recommending modifications to the text. Special thanks are given to Dave Harris of Westinghouse for his efforts as liaison, proofreader and terminology coordinator, and to R. M. Windsor of NASA and Stan Ropiak of UNISYS at the Goddard Space Flight Center and E-Systems Management for supporting the effort to compile the data in Chapter 12. Photographs from NASA are made possible through the efforts of Christopher Petrignani of UNISYS. The International Society of Hybrid Microelectronics is appreciated for their contribution to the acronyms, terminology, and standards appendixes. Special thanks to Tammy Young and Tom Topalian of Philips Components, Lawrence Fogel of Signetics Co., Al Ruttner and Kevin Derricotte of North American Philips Corporation, and Jan Van Haaren of N. V. Philips of Eindhoven, The Netherlands, for the photographs supplied for this book, and to Dick Wilcher of Amp, Inc., for the many connector photographs.

*Michael Pecht*

# Contents

# Contributors

**Donald B. Barker** obtained his Ph.D. degree in Engineering Mechanics from UCLA and has published extensively in the general area of experimental mechanics, fracture mechanics, fatigue, dynamic material response, and electronic packaging since joining the faculty at the University of Maryland in 1976. Dr. Barker is an associate professor in the Mechanical Engineering Department and also the Associate Director of the University of Maryland Computer Aided Life Cycle Engineering (CALCE) Center for Electronics Packaging. Dr. Barker's research involves the experimental and numerical modeling of mechanical failures in electronic components and assemblies and the integration into a concurrent engineering design environment. This research is leading to practical approaches in microelectronic package design and the implementation of reliability prediction methodologies, based on physics of failure concepts.

**David Dancer** received a B.S. in Chemical Engineering from the University of Cincinnati and a M.S. in Chemical Engineering and a Ph.D. in Mechanical Engineering from the University of Maryland. Dr. Dancer's areas of interest include heat transfer, fluid mechanics, operations research, and numerical methods in engineering.

**Abhijit Dasgupta,** an Assistant Professor of Mechanical Engineering at the University of Maryland, received his Ph.D. from the University of Illinois in 1989. His dissertations research is in the mechanics of fatigue damage in heterogeneous materials with microstructure, such as fiber and particle reinforced composite materials. He has participated in several sponsored and consulting research projects dealing with the mechanics of fatigue, fracture and damage of metallic, polymeric and ceramic materials and associated composites. He also has wide experience in the use of general-purpose finite element programs and has developed numerous special purpose codes for nonlinear finite element analyses, such as viscoelastic problems, and large deformation stability problems, and for the analysis of mixed-mode fracture in anisotropic materials with the help of conservation integrals. As a faculty member of the CALCE Center, he is involved in research dealing with the failure and reliability of electronic components and systems.

**Jillian Y. Evans** is a Senior Engineer with E-Systems, Melpar Division in Falls Church, Virginia. She is responsible for applied research and development in microelectronic packaging for airborne communications and intelligence systems. Jillian has also been employed with Westinghouse as a thermal analyst and Rockwell International as a research engineer. Jillian received the B.S. in Mechanical Engineering from the University of Iowa in December of 1985. She is pursuing a Master of Science degree in Mechanical Engineering at the University of Maryland. Jillian is also a member of the CALCE Center for Electronics Packaging responsible for research in thermal analysis.

**John W. Evans** has been employed with UNISYS Corporation at the Goddard Space Flight Center in Greenbelt, Maryland since July 1987, as a Senior Engineer. He supports the Office of Flight Assurance at Goddard in the areas of materials, processes and electronic packaging and their impact on spacecraft reliability. John received his undergraduate degree in Mechanical Engineering from the University of Nebraska in May 1983. He earned a Master of Science degree in Materials Engineering from the University of Iowa in December 1985, while working as a materials and processes engineer at Rockwell International. He is currently a Ph.D. candidate in Materials Science and Engineering at Johns Hopkins University in Baltimore, Maryland. John is a member of ASM and has authored 11 technical papers in failure analysis and electronic materials.

**Denise Burkus Harris** received her B.S. and M.S. in Chemical Engineering from the University of Notre Dame. From 1981 to 1985 she worked at Texas Instruments in the areas of photolithography engineering and new process and product development in the Hybrid Microelectronics Lab. In 1985 and 1986 Denise worked in hybrid layout and packaging design at Magnavox. In 1986 she came to Westinghouse where she is now a Senior Engineer in the Mechanical Design and Developmental Engineering Department. Denise has been an active member of the International Society for Hybrid Microelectronics (ISHM) since 1983 and serves as the president of the Capital Chapter of ISHM.

**Dennis K. Karr** received the B.S.E.E. degree from the University of Kentucky in 1977 and the M.S.E.S. degree, Electrical Engineering, from the University of Toledo in 1982. Mr. Karr is an expert in bimetal thermal protectors and has had substantial experience in thermal protector design and application with both bimetal and solid-state temperature sensors in computers and computer peripherals, power supplies, engines, and motors. Mr. Karr is a member of IEEE, ASM, and ASTM, and is currently the editor of the IEEE TC-9 Thermal Management Technical Committee and the SEMI-THERM Symposium newsletters. Mr. Karr is the Product Manager of the Thermal Products Group of the Airpax Corporation and is Vice-President and General Manager of CETAR, Ltd., both North American Philips companies.

**Wing C. Ko** received his Ph.D. in Mechanical Engineering at the University of Maryland, College Park. Dr. Ko's research has concentrated on the corrosion mechanisms of microelectronic packages.

**Pradeep Lall** received a Ph.D. in Mechanical Engineering at the University of Maryland, College Park. Dr. Lall's research has focused on mechanical and electrical failure modes and mechanisms, reliability assessment and development of failure modeling strategies for microelectronic packages. He is a member of ASME, IEEE, and ISHM.

**Michael D. Osterman** received a Ph.D. in Mechanical Engineering from the University of Maryland at College Park. Dr. Osterman is an expert in numerical techniques for the placement of electronic components on printed wiring boards and optimization schemes for component placement based on reliability and routing constraints. Dr. Osterman is a member of ASME, IEEE, and ISHM.

**Milton Palmer, III** received his Ph.D. in 1981 in Mechanical Engineering from the University of Maryland. His expertise is in the areas of thermo-fluid mechanics and numerical methods. He is currently employed at Ramsearch Company in College Park, Maryland.

**Michael Pecht** received a B.S. in Acoustics, M.S. in Electrical and Computer Engineering, M.S. and Ph.D. in Engineering Mechanics from the University of Wisconsin-Madison. Dr. Pecht is a tenured associate professor with a joint appointment in the Mechanical Engineering Department and the Systems Research Center at the University of Maryland. He is director of the CALCE Center for Electronic Packaging research effort. Dr. Pecht is a member of IEEE, IPC, IEPS, SME, ISHM and ASME and is the Editor of the IEEE Transactions on Reliability.

# 1

# Introduction

**Michael D. Osterman and Michael Pecht**  *CALCE Center for Electronic Packaging, University of Maryland, College Park, Maryland*

Electronic packaging is the art and science of creating a physical electronic product—the art being the development of new and unique electronic products whose concepts are born through human activities, and the science representing the sum of human knowledge in the various engineering disciplines employed in the creative process. Electronic packaging covers a broad range of activities, all of which involve grouping electrical and supporting elements into a functional unit. This unit can range from a deceptively simple single-chip package to an entire electronic system as sophisticated as a mission control center for spaceflights.

Regardless of the sophistication of the electronic package, its design depends on the balance of a number of often conflicting factors. These factors include material selection, interconnection and wiring technologies, mechanical support, thermal management, environmental control, and ergonomics (human-machine interaction), to list a few. The complexity of an electronic package is therefore determined by the nature of the trade-offs made between these clashing constraints and the requirements imposed on the performance of the electronic product. For example, consumer electronic products generally emphasize cost factors and hence may tend to sacrifice the factors which most affect product reliability, such as proper material selection and finely tuned manufacturing processes. Military equipment and flight equipment, on the other hand, have strict design codes and standards for reliability and performance. Hence, these types of electronic systems are generally more characteristic of *high-end products* and their costs are correspondingly higher. To perform and conduct design trade-offs successfully, the design of electronic packaging must integrate the skills of electrical, mechanical, manufacturing, industrial, material, and systems engineers.

## 1.1  PACKAGING HIERARCHY

Electronic packages provide the medium for electrical bridges or interconnections, as well as the mechanical support and protection of the delicate electronic circuitry. For this reason, electronic packages are typically classified into *levels* based on the number and sophistication of the electronic elements of which they are comprised. This classification scheme, or *hierarchy*, is illustrated in Figure 1.1. A brief description of the electronic packaging hierarchy is provided here to familiarize the reader with electronic structures, as well as some important terminology.

The lowest, or *zeroth*, level of packaging is generally considered to be the semiconductor chip, although discrete passive devices such as resistors and capacitors may also be included. The zeroth packaging level consists of the logic gates, transistors, and gate-to-gate interconnections which are formed directly on the chip. The chip itself is created from a thin circular slice (wafer) of a crystalline semiconductor material, usually silicon, gallium arsenide, or indium phosphide. In the chip fabrication process, the wafer is modified by repeated photolithographic and etching processes and the diffusion and/or ion implantation of impurities through a process known as *doping*. After forming the various regions on the wafer, the electrical circuits are completed by establishing inter-

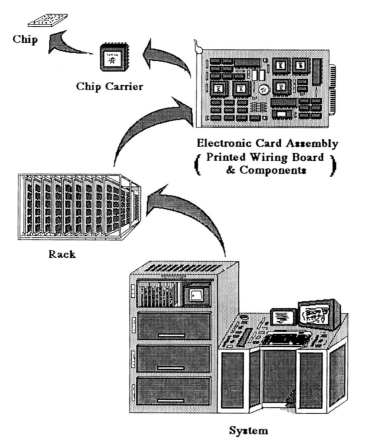

**Figure 1.1**  Packaging hierarchy of an electronic system.

connections which are usually formed by selectively opening regions on the wafer and depositing a fine aluminum layer directly onto the wafer. In general, several chips (dies) are formed simultaneously on a single wafer. They are then cut from the wafer in a process known as *dicing*.

The term *level of integration* is used to describe the number of microcircuit components placed on the chip and is thus a representation of the complexity and sophistication of the zeroth package level. When the first integrated circuits were developed in the early 1960s, they generally consisted of less than 50 components per chip, referred to as small-scale integration (SSI). The next level of integration, called medium-scale integration (MSI), consisted of chips containing as many as 1000 components. In the 1970s, technological advances made it possible to place thousands and tens of thousands of components on a single chip and large-scale integration (LSI) was achieved. The 1980s heralded the introduction of very large-scale integration (VLSI), wherein hundreds of thousands to millions of components were placed on a single chip. With further advances in the art of optical, electron beam (E-beam), and X-ray lithography, the achievement of ultra large-scale integration (ULSI), $10^7$ to $10^9$ components per chip, and even giga-scale integration (greater than a billion components per chip) may become possible. Figure 1.2 illustrates this increasing trend in chip complexity.

In addition to single-chip integration, wafer scale integration (WSI) is being utilized in an effort to increase the level of integration by interconnecting the individual chip areas on the wafer, thus avoiding the need to separate the wafer into individual chips (dies). Because the wafer remains intact as a single integrated microsystem, WSI is particularly suited for the development of array-based systems, such as memory arrays.

The packaging of a chip or a set of chips in a functional and protective chip carrier

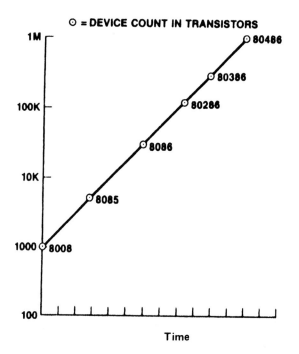

**Figure 1.2**   Trends in chip complexity. (Courtesy of Intel Corp.)

is referred to as *first-level packaging*. Chip carriers can range from single-chip (monolithic) carriers such as the common *dual in-line package* (DIP) to very sophisticated multichip modules such as the IBM thermal conduction module (TCM), which contains over a hundred individually cooled chips. To allow for connections between the chip and the outside world, single chip carriers usually contain a metallized pattern which is commonly referred to as a *lead frame*. Electrical interconnections are then established between conductor pads formed on the chip and the lead frame of the chip carrier through either wire bonding, direct solder bonding, or tape-automated bonding (TAB) processes. In multichip packages, several chips are mounted directly onto a multilayer substrate which contains metallized patterns which interconnect the chips and connect to external leads.

Wire bonds connect each conductor pad on the chip to the corresponding tip on the lead frame with a fine wire, typically composed of an aluminum alloy. The joints between the wire and the conductor pads on the chip and the tips of the lead frame are made through either a compression weld, a thermocompression weld, or a vibratory weld. In rare instances, direct solder bonds are made between the chip pads and the lead frame of the chip carrier.

In TAB, the connection pads on the chip are connected en masse using a thermo-compression welding head to a metallized pattern formed on a polymer tape. The TAB pattern may serve as the lead frame or may be connected to the chip carrier lead frame through a similar thermocompression welding process. For both wire bonding and TAB, the chip is attached directly to the chip carrier by either eutectic solder alloy (typically AuSi) or a polymer adhesive.

The extension of the metal lead frame in the chip carrier to the outside of the package serves to connect the chip circuitry to the second level of packaging. The external con-nections of a chip carrier serve to classify the component into one of the two major technological categories: *through hole* components (THC) and *surface mount* components. For through hole components, the leads (external connectors) or *pins* are inserted through a common mounting surface. For surface mount components, the chip carrier is connected directly to the mounting surface.

The *second level of packaging* is sometimes referred to as the electrical circuit assembly (ECA). At this level, the individual chip carriers are mounted on a common base, usually a printed wiring board (PWB). The board, which may be composed of an organic material, a polymer- or glass-coated metal composite, a flex film, or an injection-molded polymer, provides a mounting surface on which the chip carriers may be positioned, a medium for chip carrier-to-chip carrier interconnections, electrical test sites, and off-board power and signal connections. To facilitate interconnections, metallized conductor paths for signal and power transmission, *footprints* for mounting the chip carriers, and *vias* for signal propagation and heat transfer between the (various) board surfaces are formed on the PWB. Metallized conductor paths on the *substrate* are formed through a photoimaging process; this is followed by the chemical deposition of palladium ion *seeds* and electro-plating of copper in the case of an additive process or chemical etching of a copper-clad board in a subtractive process.

The *third level of packaging* typically involves the interconnection of circuit boards and power supplies to a physical interface, such as a chassis, control, and/or electro-mechanical device or system. The third level of packaging may also involve the connection of several boards within a supporting or protective structure such as a cabinet. ECA-to-ECA interconnections are commonly made on another, larger PWB, which is commonly referred to as the *backplane*. In other designs, a backplane may be omitted as several ECAs are mounted in a *rack* and *cabled* together. Several racks may then be mounted and cabled together in a single cabinet. When several cabinets are then joined together, a *fourth-level package* is created.

## 1.2  TRENDS IN ELECTRONICS

The development of the first integrated circuit in the late 1950s marked the beginning of the electronics revolution. Since then, major advances in electronic technologies and fabrication processes have resulted in the rapid growth and diversity of electronic products. Automated teller machines, satellite telecommunications systems, televisions, radios, telephone systems, automotive ignition and control systems, and computer systems are examples of how electronics have permeated the consumer market. Communication, data processing, information storage, and control systems are additional examples of how sophisticated electronic systems provide important modern services. Future trends in electronics will lead to products whose applications are limited only by the designer's imagination.

The sophistication of electronic packaging has been driven primarily by the technological advances made at the zeroth packaging level. For instance, with the invention of the integrated circuit (IC) at the close of the 1950s, electronic components had evolved from vacuum tubes to transistors. Today, electronic components have matured into single-chip and multichip devices that may contain more than a million components on a single $5 \times 5$ mm chip.

Because of the high reliability of on-chip electrical connections and the advances made in photographic imaging processes, the industry trend has been to push as much as possible of the electrical circuitry onto the chip while simultaneously reducing the cell area. For instance, advances in optical lithography made since the early 1970s have reduced chip feature size by an order of magnitude to approximately 0.8 $\mu$m in 1990, as illustrated in Figure 1.3. With further advances, optical lithography may be able to support feature sizes smaller than 0.5 $\mu$m.

Optical lithography is, however, approaching the limit of its capabilities. For this reason, E-beam and X-ray lithography techniques are being explored as possible alternatives which can generate feature sizes as small as 0.1 $\mu$m. The trend toward increasing the level of integration while simultaneously reducing chip cell sizes is illustrated in Figure 1.4. Even with the reduction in feature and cell sizes, chip areas have increased in order to allow for higher levels of circuit integration, as depicted in Figure 1.5. It is interesting to note that the increase in the number of active devices on the chip has been driven primarily by the decrease in feature size rather than the increase in chip size.

The increased levels of chip integration have produced a number of higher-level electronic packaging challenges. For example, *pinout*, the number of input and output (I/O) connections on the chip, has dramatically increased with the advances made in chip integration, as illustrated in Figure 1.6. As a result, the capabilities of wire bonding are gradually being surpassed and TAB is becoming the favored interconnection technology.

Another challenge resulting from the increase in chip integration is the increased heat flux to be dissipated from the chip surface area, despite the trend to reduce chip power. The thermal management challenge is best illustrated in *hybrids*, components composed of multiple chips with chip-to-chip interconnections contained within a chip carrier, which regularly generate hundreds of watts of heat flux per cubic centimeter. As a result, thermal management considerations have required new system architectures, advances in material sciences, and innovative cooling schemes such as liquid and impingement cooling techniques.

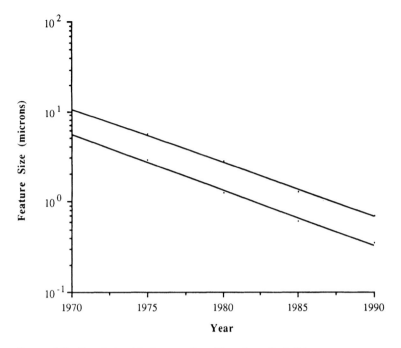

**Figure 1.3**   Trends in chip feature size. (Data from Ref. 9.)

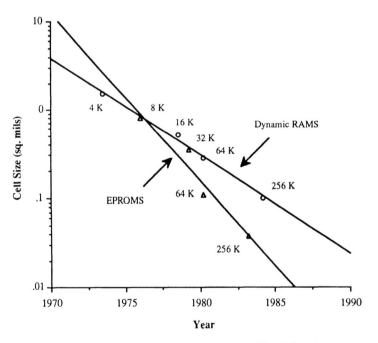

**Figure 1.4**   Trends in chip cell sizes. (Courtesy of Intel Corp.)

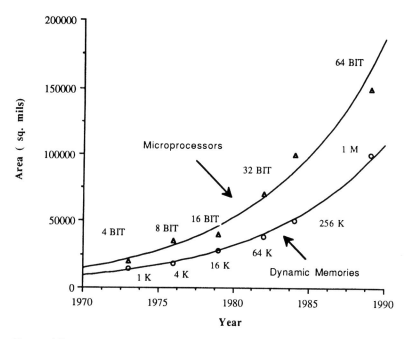

**Figure 1.5** Trends in chip size. (Data from Ref. 9.)

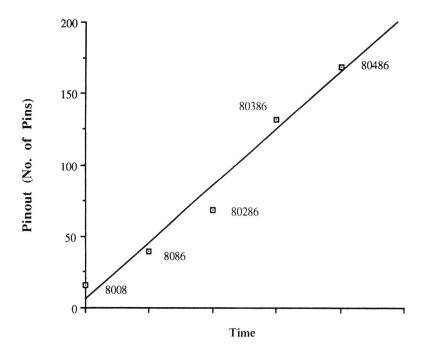

**Figure 1.6** Trend in pinout versus chip complexity. (Data from Intel Corp., Ref. 11.)

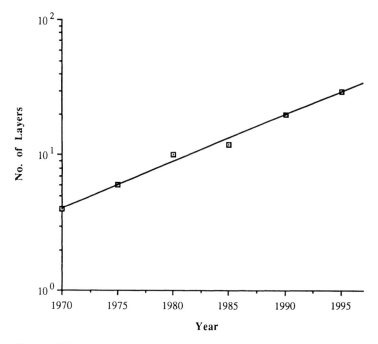

**Figure 1.7**  Trends in number of board layers. (Data from Ref. 5.)

**Table 1.1**  Printed Wiring Board
Technology of 1960

**Devices**
Discrete transistors
Small scale integrated (SSI)
  Components

**Materials**
Epoxy/glass laminate

**Packages**
T-cans
SOIC
Dual in-line packages (DIP)

**Methods: tolerances (in)**
Manual artwork:  ±0.010 features
Mechanical drill:  ±0.007 location
Plating:          ±0.005 thickness
Hole size:        ±0.050 ±0.010

**Interconnection**
Double-sided boards

Changes in chip carrier geometries have spawned changes in the design requirements of the PWB. In order to handle increasing numbers of signal, power, and ground connections, *multilayer boards* were developed. The average number of board layers has thus moved from 4 layers in 1970 to 20 layers in 1990, and boards of up to 42 layers have been successfully achieved. This trend in increasing board layers is illustrated in Figure 1.7. Tables 1.1 through 1.4 and Figures 1.8 through 1.12 summarize some additional changes in printed wiring board technology that have occurred since 1960.

New techniques for interconnecting chip carriers to PWB are also being developed. One area of development is the *fuzz button*, a solderless connection which makes use of pressure connections at the juncture of the chip carrier leads and the component's footprint on the PWB. These pressure connections rely on externally applied loads, e.g., from an elastomer or spring, to maintain the necessary force for proper component-to-footprint contact and have been applied in limited areas, such as flexible substrates.

Some of the most exciting work in electronic packaging has been based on *optical technologies*. Optical and *optoelectronic devices* as well as optical interconnections are now being developed as alternatives in electrical packaging techniques. For instance, optical device and transmission techniques are currently being used in communication systems and attempts are being made to develop completely optical computers. Optical

**Table 1.2** Printed Wiring Board
Technology of 1970

**Devices**
Medium scale integration
(MSI) components
Memory components

**Materials**
Epoxy/glass laminate
Teflon
Kapton
Arcylic

**Packages**
Dual in-line package (DIP)
Hybrid

**Methods: tolerances (in)**

| | |
|---|---|
| Gerber plotter: | ±0.010 features |
| Mechanical tooling: | ±0.007 location |
| Mechanical drill | ±0.005 location |
| Plating: | ±0.0005 thickness |
| Circuit size: | ±0.013 ±0.003 |
| Hole size: | ±0.040 ±0.005 |

**Interconnection**
Multilayer (single channel)
Microwave
Flex and rigid flex stripline

**Table 1.3**   Printed Wiring
Board Technology of 1980

**Devices**
Large scale integration (LSI)
Very large scale integration
   (VLSI)
Very high speed integrated
   circuits (VHSIC)
Application specific integrated
   circuits (ASIC)

**Materials**
Epoxy/glass laminate
High reliability epoxy
Teflon
Kapton
Acrylic
Polyimides
Quartz
Kelvar
Copper foils—direct bonded
Clad metals

**Packages**
DIP
Hybrid
Chip carrier
Pin grid arrays

**Methods: tolerances (in)**
Plotted artwork: $\pm 0.001$ features
Optical tooling: $\pm 0.001$ location
Laser drill:        $\pm 0.001$
Plating:            $\pm 0.0003$ thickness
Circuit size:      $\pm 0.005$ $\pm 0.002$
Hole size:         $\pm 0.006$ $\pm 0.002$

**Interconnection**
Polymer on metal (POM) multilayer
   (two channel)
Advanced polymer on metal
(APOM) (eight channel)
Microwave
Stripline
Flex, rigid flex
Optical (e.g., LEDs and detectors)

**Table 1.4**   Printed Wiring
Board Technology of 1990s

---

**Devices**
VLSI
VHSIC
ASIC
Gate arrays
Wafer scale integration (WSI)

**Packages**
DIP
Hybrid
Chip carrier
Pin grid arrays
Pad area arrays
SMD
TAB

**Materials**
High reliabiity epoxy
Flexible adhesive films
Polyimide
Organic reinforced laminate
Teflon variations
Adhesiveless flux
Composite multilayer heat sinks
Liquid/photographic dielectrics
Programmable silicon

**Methods**
Design/fabrication/assembly Integration
Thick film, thin film, and PWB Integration
''Hands off'' manufacturing
Interactive CAD/CAM
Advanced chemistry sensors
Automated inspection
X-Ray aligned drill
Plasma hole generation
Laser direct imaging
Electron beam patterning

**Interconnection**
Conservative technology insertion
Multilayer (four channel)
POM/APOM combinations
Molded microwave
Stripline
Flex, rigid flex
Solderless
Optical
Fiber optics

---

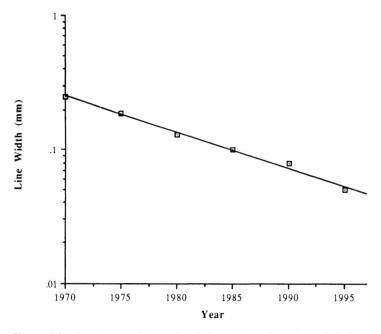

**Figure 1.8**   Trends in multilayer board line widths. (Data from Ref. 5.)

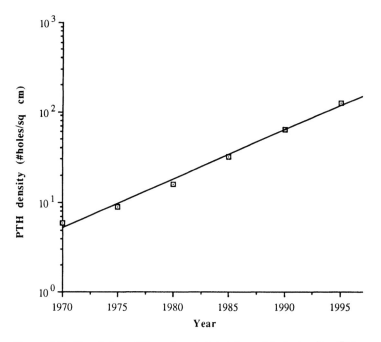

**Figure 1.9**   Trends in multilayer board plated through hole density. (Data from Ref. 5.)

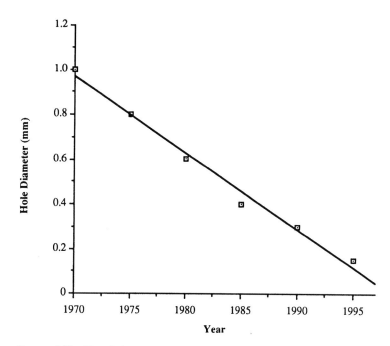

**Figure 1.10**   Trends in plated through hole diameters. (Data from Ref. 5.)

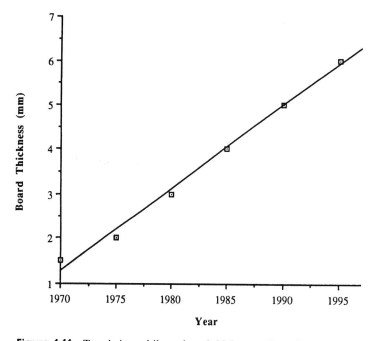

**Figure 1.11**   Trends in multilayer board thickness. (Data from Ref. 5.)

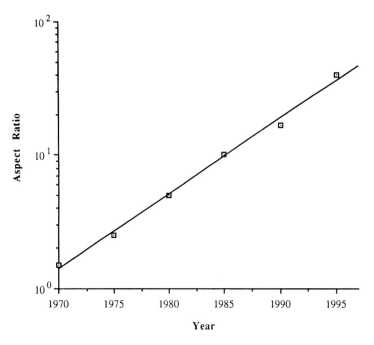

**Figure 1.12**  Trends in multilayer board aspect ratios. (Data from Ref. 5.)

signals operate at higher speeds and do not suffer from signal distortion and interference, which are common problems in electrical transmissions. In addition, the optical signal requires less energy and can handle higher signal densities than electrical transmissions.

## 1.3   THE ELECTRONIC DESIGN AND PRODUCTION PROCESS

In the conventional design environment of the 1980s, most electronic products were designed exclusively by electrical engineers and technicians, the driving forces in their designs being market requirements and technological innovation. Manufacturing and mechanical engineers were then required to fabricate and assemble the design with acceptable yields. Current industry trends in electronic design, however, have demonstrated that the influences from systems, quality, reliability, mechanical, and manufacturing engineers must be integrated as an active force in the design process, as shown in Figure 1.13.

As with all processes and actions, an initiating or driving force is required to start the design and development process for electronic systems. In general, market research anticipates the need for a new electronic product or product modification and initiates the design process. In effect, marketing sells the services of the electronics producer to the customer and supplies required production and design information to the *program or-ganization* of the electronics producer. This information may be derived from contractual specifications with the customer or projected needs of prospective users and may include product specifications and requirements, schedule limitations, and cost limitations. Design specifications include restrictions on the choices of usable devices or device types; production processes; system weight, size, speed, and reliability; maintenance intervals; and environmental considerations such as vibrational and temperature requirements.

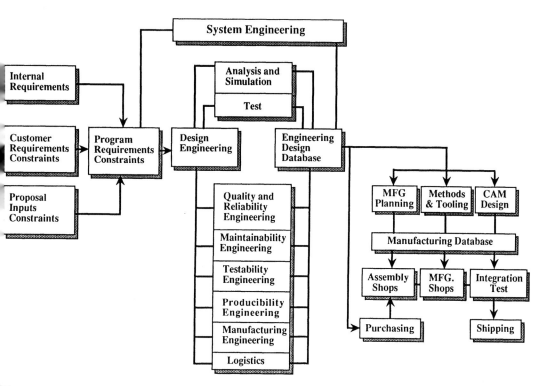

**Figure 1.13** Functional design flow diagram with parallel processes.

The program organization governs the project and is responsible for maintaining communications with the customer and furnishing progress reports to the electronics producer. In addition, the program organization is responsible for breaking the project down into a series of subprojects which may be better handled within the design and production facilities. For example, if the electronics producer was consigned to develop a satellite, the program organization would oversee the design, fabrication, and assembly of the entire satellite. To facilitate the completion of the project, however, the program organization would break the satellite down into various subsystems, such as the telemetry system, which could be handled as a subproject. Subsystems, with their corresponding work orders, design requirements, and system constraints, would then be passed down to the *functional organization* level.

At each functional organization level, system design requirements are refined into a series of work orders encompassing the design, fabrication, and assembly of the subsystem. The functional organization then develops a production schedule, which remains compatible with the global program production schedule, and a production process plan compatible with the production cost estimates of the global program cost forecast. The functional organization then assigns a *design team* to the subsystem project.

The design team is responsible for the success or failure of an electronics product. Subsequently, the design team must ensure that the product complies with the design requirements as specified by the program organization and ultimately the customer. Some typical design requirements for an electronic system are illustrated in Figure 1.14.

In the initial stages of the design process, the design team generates the logic equations

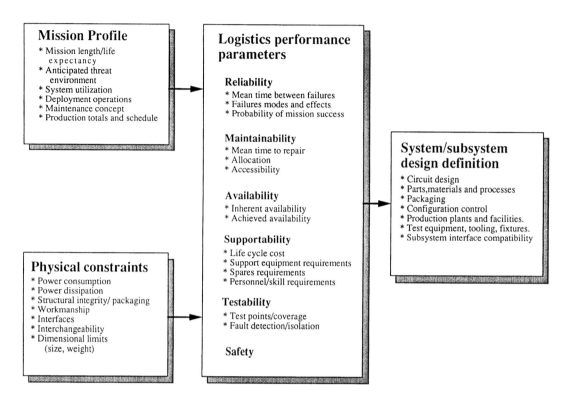

**Figure 1.14**   Design requirements.

and functional block diagrams. Once a design is approved, the functional block diagrams are partitioned to form standard packageable functions and the logic net lists for the circuits are derived. The schematic of the particular unit is then established, and the devices which perform the logical functions are identified and selected on the basis of their adherence to the design requirements and their suitability for use in the selected production processes. Packaging issues, such as those represented in Table 1.5, are then addressed to predict the reliability and maintainability of the system. Some of the interactions and design trade-offs conducted among the systems, electrical, packaging, and manufacturing engineering disciplines through these steps in the design process are illustrated in Figures 1.15 through 1.17.

Throughout the design process, design reviews are conducted in order to evaluate the design on the basis of its compliance with the company's and the customer's requirements. Reviews provide feedback to the design team on the acceptability of the design as the design process continues. Typically, the review process consists of preliminary, critical, and customer evaluations. If, by chance, the design fails any of these design reviews, it is returned to the design team with the appropriate design change recommendations and subsequent work orders.

Once the schematic and parts list have passed the design review process, the manufacturing and any circuit-timing constraints are addressed, and a layout for the assembly design is generated. The components are then placed in accordance with this layout, their orientation being dictated by interconnection, timing, assembly, and thermal considerations. In addition, test points on the individual boards are determined.

**Table 1.5** Packaging
Considerations

---

**System considerations**
  -Size
  -Weight
  -Hardware/Interfaces
  -Life Cycle Cost
  -Power requirements

**Lowest replaceable
unit (LRU)
considerations**
  -Size
  -Weight
  -Hardware
  -Interfaces
  -Partitioning
  -Power requirements

**Board considerations**
  -Size
  -Weight
  -Number of boards
  -Hardware
  -Connector I/O
  -Components placement
  -Routing
  -Power requirements
  -Critical circuitry

**Components considerations**
  -Speed
  -Power consumption
  -Power dissipation
  -Fan-out
  -Noise margin
  -Interfaces

---

Following the generation of the component layout, the design is subjected to detailed thermal and reliability modeling to ensure that the design does not generate harmful thermal stresses due to the formation of extreme hot spots or severe thermal gradients. In addition, the thermal modeling ensures that the components are operating within their nominal operating temperature ranges and thus conform to the required reliability criterion. Should the design not pass either the reliability or the thermal analysis, the components may be replaced, reoriented, or repositioned, or an alternative system architecture may be proposed.

If the design is deemed acceptable, the component positions and placement coordinates are finalized through a superposition of manufacturability considerations; the power supply and signal traces are then routed. Sample manufacturability considerations may include

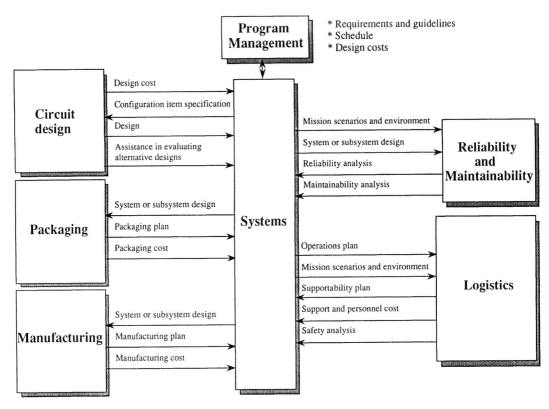

**Figure 1.15**  Systems engineering interfaces.

placement tool space requirements, placement tolerances, registration tolerances, component sequencing considerations, and required manufacturing footprint shifts to facilitate product assembly. Following the finalization of the layout, the circuit drawing package is *released*.

Classically, the release cycle in an electronics production facility serves as a means of locking a design in a set form. In this way, the design team is forced to curtail the design process and focus on the finite transition point between the design and manufacturing stages. In its most rudimentary form, a drawing release consists of the design description; the complete circuit, chassis, cabinet, cabling and connection, and fixturing drawing sets; the *photomasters* for PWB fabrication; the finalized parts, net, and materials lists; the production and assembly work orders; the parts and materials purchase orders; and the departmental and design team sign-off sheets. As each piece of information is completed and *signed off*, it is considered final and not subject to revision without the expressed approval of the program organization.

Following the design release, three procedures are conducted simultaneously. First, the design schematics are reviewed by the manufacturing engineers so that a global process plan may be developed governing the fabrication of the system cabinets, chassis, and fixtures and the assemblage of the electronic system. Once developed, the process plans are subjected to a quality review to ensure that the procedures conform to in-house quality guidelines and a further manufacturing review to ensure that the process plans can indeed be followed within the production facility.

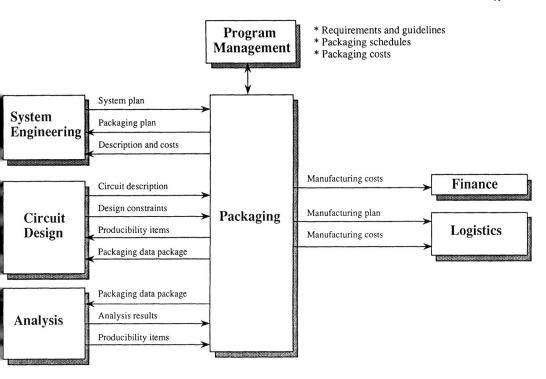

**Figure 1.16** Packaging engineering interfaces.

Should the plans be rejected by either review, they are returned to the manufacturing group for revision. If they are accepted, the plans are *kitted* in inventory, by function, with the appropriate control documentation and design schematics. The kit remains in inventory until completed. It is then distributed to the appropriate production facilities and assembly lines at the start of the production schedule.

The second procedure begins with the purchase orders being submitted to the parts and materials vendors. As the parts and materials are received from the vendor, they are subjected to a statistical inspection, catalogued, and kitted with the appropriate process plans, control documentation, and design schematics.

The third procedure begins with the design photomasters and PWB production work orders being sent to the PWB fabrication facility. At the fabrication facility, the photomasters are inspected and used to create the required *phototools* for PWB fabrication. The bare PWBs are then produced and the work order completed. The actual PWB fabrication process will be discussed in greater detail in Chapter 3.

Upon completion of the work order, the bare PWBs are inspected for fabrication flaws and design compliance. Should a board fail inspection, it may be either *reworked* (repaired) or returned to the PWB fabricator for replacement. If the boards pass inspection, they are cleaned, usually through a vapor degreasing and spray solvent cleaning operation, and then baked to remove any trapped water and solvent vapor. Upon completion of the cleaning and baking procedure, the boards are placed in inventory and kitted with the assembly process plans, control documentation, and parts.

The board assembly process often begins with the tinning and cleaning of the parts, wherein the components are *fluxed*, *tinned*, cleaned, and replaced in the kit. Once com-

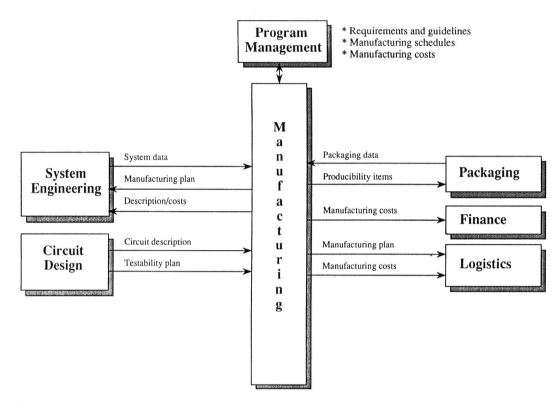

**Figure 1.17**  Manufacturing engineering interfaces.

pleted, the kit is delivered to the assembly area and assembled according to the assembly process plan. Typically, the assembly operation will be divided into *primary side* surface-mounted device (SMD) placements, primary side through hole component insertions, *secondary side* SMD placements, and *en masse* soldering processes. In higher-volume production, the assembly tasks are generally completed through the use of automated production equipment and the components are fed into the assembly line on sequenced spools, or presented in component tubes or trays. Although this manner of manufacturing and assembly tends to remove the steps of kitting the components with the bare PWB, the total assembly and production processes for high- and low-volume production operations differ only slightly.

Once the board is assembled, it undergoes an aqueous cleaning and bake-out operation, followed by a visual inspection. During the inspection process, manufacturing defects such as solder balls, lead bridging, and missing or misaligned components and board defects like measling and blistering may also be detected. If a defect is detected, the board may be reworked or scrapped. The decision to rework or scrap is dependent on flaw type, volume of production, contractual requirements, and/or operating practices. In some cases, components from scrapped boards may be salvaged.

Following the visual inspection, the board is electrically tested in a simulated operating environment to verify that the board does indeed function as required. In some instances, the board is subjected to a burn-in cycle in order to uncover manufacturing and workmanship defects that may lead to premature failures. The issues of reliability and burn-in failures are covered in Chapter 11.

Should a board fail the electrical tests, it is again inspected to determine whether the failure was the result of a manufacturing or component defect or of a design flaw. If the failure is the result of a manufacturing or component defect, the board may be scrapped or reworked, cleaned, visually inspected, and retested. If the failure resulted from a design flaw, the problem is presented to the *material review board*, of which the customer is a member, and approval to repair and/or modify the design is sought. Should approval not be granted, the design is returned to the design team for revision and the design process is reinitiated. Instances of this are, however, quite rare due to the extreme cost of scrapping a design after the completion of a production run.

Once the board passes electrical simulation and testing, it is again cleaned through a vapor degreasing and spray solvent cleaning process, subjected to a cleanliness test, and then baked. Following baking, the board may be conformally coated, visually inspected, assembled at the chassis level, and placed in inventory to await final assembly at the project level.

The design and manufacturing described herein serve only as an indication of the sequencing of operations and are not specific to any particular design or system house. In addition, some portions of this process description may not be utilized by some specialized production facilities.

## 1.4 DESIGN CONSIDERATIONS

The complexity of modern electronic design is governed by the trade-offs which inevitably occur. These trade-offs are based on design considerations which tend to vary with individual design group perceptions. For successful designs, each design group must recognize the overall design objectives and how individual design decisions are carried through the product development and operation stages. In the following sections several major design considerations are reviewed. These include material selection, thermal considerations, cost, performance, and design for *ilities*.

### 1.4.1 Material Selection

With the increased demands on the electrical performance and the increased level of integration, material performance has become a critical design consideration. The materials used in electronic packaging affect practically all of the important factors of electronic products including performance, cost, functionality, manufacturability, and reliability. For this reason, the material selection process must consider the demands of the electrical engineer for improved electrical properties; the mechanical engineer for improved thermal and structural properties; and the manufacturer for compliance with practical processing, such as cutting, drilling, laminating, and gluing. Some of the material parameters which play a large role in the design of electronic systems include the dielectric constant, material permeability, coefficient of thermal expansion, thermal coefficient of conductance, mechanical strength and stiffness, ductility, glass transition temperature, flammability, propensity to outgas, chemical susceptibility, solderability, and machinability.

One of the chief electrical material parameters is the *dielectric constant*. Dielectrics are the insulating materials which are used to support, separate, and maintain the conductive electrical configurations that make up electrical designs. The dielectric constant is the ratio of an electronic configuration's capacitance based on the separating material to ground to the capacitance of the same configuration in the absence of the material (i.e., vacuum). The dielectric constant determines the transmission speed and is based on the geometry of the electrical system. In addition to the dielectric constant, *dielectric loss*, the amount of electrical energy transformed into heat in a dielectric subjected to a charging electric field; *dielectric absorption*, the property of an imperfect dielectric material that holds a portion of the electrical charge generated by an electrical field; and *dielectric strength*, the maximum voltage that a dielectric material can withstand without resulting in a voltage breakdown are all important electrical considerations.

A number of material parameters and considerations are critical to the mechanical aspects of an electronic product. One of the chief mechanical concerns is the development of stresses in electronic packages. Subsequently, the *coefficient of thermal expansion* (CTE) of electronic packaging materials is a particularly important material parameter in terms of the development of stress at the interfaces between the various packaging materials employed in the creation of an electronic product. Thermal expansion has been linked to failure on leadless chip carriers (LCCs) as well as die fractures, plated through holes, and pad lifting.

In addition to thermally induced stresses, the ability of electronic products to withstand shock and vibrational environments is critical to product reliability. Material properties which represent the strength of materials include Young's modulus (or elastic modulus), $E$, the modulus of rigidity, $G$, and the yield strength. The strength of the attachment between the dielectric base material and the metallized conductor pads, sometimes referred to as *peel strength*, is another important material consideration. The ductility, which is a measure of a material's ability to undergo plastic deformations, is particularly important in plated through hole (PTH) design due to the thermal expansion problems in the $z$ axis. The further increase in aspect ratios of PTHs aggravates the problem.

Along with strength and ductility, material weight and thermal cooling aspects of materials are important considerations. The material density is important in determining weight. Weight considerations are particularly important is avionics and space application. The thermal cooling rates of electronic assemblies are governed by the *thermal conductivity* of the packaging material and *convection heat transfer coefficients* of the cooling fluids. Thus, the thermal properties of electronic packaging material must be well understood.

Permeability is the property of a material that allows penetration by a gas or liquid. Certain applications, particularly military ones, require devices which are *hermetically* (gastight) sealed. One of the chief concerns is the propensity for moisture to enter microelectronic devices. By itself, moisture is relatively harmless, since it is a poor conductor. However, the combination of moisture with the chemical remnants of the fabrication processes can cause catastrophic failure due to corrosion and electrical breakdown. Most military and commercial products must also meet flammability requirements set by government regulations. As a result, flame-retardant agents are generally added to certain materials such as plastic encapsulates and organic PWBs to prevent flaming at high temperatures. The propensity of a material to outgas or to release chemical gases is another important material parameter. When combined with moisture, the gases released can create corrosive environments. In addition, outgassing is a particularly important concern in space applications, where the physical properties of material are significantly altered as a result of outgassing. Chemical susceptibility or the ability of a material to withstand chemical attacks is also important when considering the many chemical processes involved in the formation of electronic packages.

   The *solderability* of the metallized conductor pads and other metal materials used in electronic packaging is an important manufacturing issue. The ability of a metal surface to accept a solder bond (wet) is critical to the formation of proper electrical and mechanical joints in the electronic assembly. Solderability is affected by the solder type, flux, and soldered material. In addition, the material must be compliant with the machining and chemical processes to which it will be exposed in the creation of the final electronic product.

## 1.4.2   Thermal Considerations

The primary purpose of thermal control is to ensure reliable operation of the electronic system by preventing thermal stresses from changing signal values or damaging the fragile circuitry. The advent of higher-density digital and analog circuit designs has substantially increased the power dissipation required per unit volume. In addition, power supplies that generated only 1 or 2 watts per cubic inch in the past may now generate 20 to 30 times that amount. Because temperature accounts for 50% or more of the stress sources in electronic packages, as depicted in Figure 1.18, the thermal design process, at all packaging levels, must attempt to equalize the thermal profile such that hot spots with potential damaging effects do not occur. In order to constrain temperature changes and the rate of temperature change to allowable limits, heat transfer paths utilizing conduction, convection, and/or radiation processes must be designed into the system.

   Improvements in second-level packaging thermal control can be achieved by modified packaging architecture and increased conductivity of the board material. Increasing the horizontal and vertical conductivity of the ECA is one of the more easily achieved alternatives and requires less change to existing systems. The conductivity through the board is sometimes improved by the use of thermal vias—plated through holes used to conduct heat away from a chip carrier as opposed to carrying an electrical signal. Higher-con-

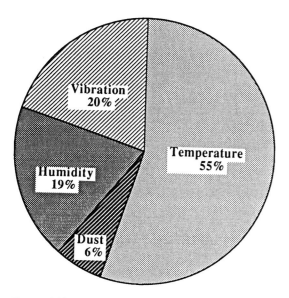

**Figure 1.18**   Sources of stress in electronic equipment.

ductivity substrate materials and/or conduction cooling planes, such as copper, aluminum, or alloy, imbedded in the board are also used to improve lateral conduction.

In particularly high dissipation applications, the board may be mounted to a *cold plate*, a flat convectively cooled heat exchanger. The fin structure in a cold plate is typically staggered to induce greater mixing and improved convective heat transfer. The working fluid in these systems is typically air, but the separation of the board and the cooling fluid allows for liquids which can dramatically increase the convection heat transfer coefficient. Cold plates are typically used in military applications where the electronic devices must be separated from the cooling fluid.

Some of the most interesting and innovative cooling schemes have been developed for high-end multichip packages. Cooling techniques include sophisticated conduction and convection cooling, such as jet impingement cooling; heat pipes; two-phase convection cooling, such as boiling; and immersion cooling. These specialized cooling techniques are dependent on the volumetric heating densities and temperature requirements of the systems.

### 1.4.3  Design for Cost

Cost is a somewhat relative term which is used with various meanings in a variety of situations. Cost may be applied to financial expenditures, time spent on a particular activity, or a penalty for following a particular path. In general, the cost of any product should be determined based on all the relevant life cycle costs of the product. As a result, it is often a complex task to assign and recognize a system cost.

Design to cost establishes cost as an active parameter in the design process. Cost targets are established for the product and the designer must attempt to establish a viable design which meets the target goals. The ability to establish realistic goals and to meet deadlines is critical for successful designs. In this endeavor, cost trade-offs invariably occur and designers must have a clear understanding of how these various trade-offs are translated to the life cycle cost of the product. For example, production costs are a direct result of the selection of parts, materials, and manufacturing processes dictated by the designer and manufacturer based on requirements, constraints, contractual agreements, and personal preferences.

### 1.4.4  Design for Performance

Performance is a measure of how well a product carries out the specific task(s) for which it was designed. Performance is established by determining functional parameters of a product based on similar products or identified by the designer and users. With electronic packages, performance measures depend on the specific function of the product and are generally concerned with how well or how fast the product can achieve its desired function. For example, the memory packages are judged on storage capacity and access speeds. Computer processing units are judged on computational speed—i.e., million instructions per second (MIPS) or billion floating-point operations per second (GFLOPS). The rise and fall times of chip carrier packages are other important performance parameters with which designers of high-speed circuitry are concerned.

Electrical performance characteristics of electronic packages are based on the level of packaging and can be fairly complex. Circuit densities, interconnection, terminals, and power are all factors which govern electrical performance. Thermal performance is characterized by the ability of an electronic package to efficiently remove the heat generated by the electrical elements and maintain controlled operating temperatures. The applied cooling technology for the electronic product directly affects the thermal performance. For example, convection cooling may include natural convection, which has a relatively low contribution to thermal management, and/or force convection, which can remove a greater amount of heat. Mechanical performance indicates how well the product stands up to environmental and internal stresses, and it is governed by the strength of the packaging materials and the ruggedness of the design.

## 1.4.5   Design for Reliability

Reliability is the probability that an object (in this case an electronic package) will perform its expected function(s) in the field application for which it was intended without failure. Hence, the reliability task is intended concurrently to assess a design and correct possible omissions and shortcomings that may affect its ability to perform its intended function. Reliability determination involves predicting the environmental impact and generated stresses on the equipment and its various components, including the electrical and mechanical stresses that may occur during the manufacturing process, transportation, testing and product screening, and projected operation of the system. Thus, reliability assessment provides valuable information for determining what course of action or design revisions need be made to guarantee survival of the system.

In electronic packaging, reliability predictions were initially prepared after a design was completed and were based on available life data from prior designs and field experience. In this approach, reliability was assessed in a design, build, and test methodology. Reliability information was then determined based on the test results and/or field data obtained from customers. The design modifications were based on these results. Due to the complexity of modern electronics, redesign or modifications were generally costly and time-consuming. In addition, consumer confidence in the product was lost.

As electronic systems were employed in areas of human and financial risk, reliability could no longer remain a background issue. Subsequently, an active reliability assessment was adopted that involved an understanding of the failure mechanisms and the systematic elimination or reduction of the risk of failure during the design phase. With increased recognition of the factors affecting reliability, additional requirements based on thermomechanical, vibrational, and electrical stress analyses were needed. These analyses were to establish the fact that stresses imposed on the design were within bounds to meet certain reliability goals. Figure 1.18 shows the typical sources of stress on electronic equipment. Computer simulation has significantly reduced the need for physical testing and plays a significant role in design and reliability verification.

## 1.4.6   Design for Maintainability

Maintainability refers to the ability to keep an existing level of system performance through preventive or corrective repairs. Due to the expense of labor, design for maintainability is based on a system cost versus repair cost basis. As a result, certain systems or units may be disposable while others are repairable. Thus, design for maintainability is generally reserved for high cost modular systems.

The purpose of design for maintainability is to facilitate easy maintenance and repair. The ease of maintenance depends on the mechanical design of the electronic product and thus is a determining factor in the design of most electronic systems. Maintainability and logistics activities are generally concerned with field support of the equipment, including overall maintenance concepts, spare parts provisioning, and in many cases safety analysis.

In designing for maintainability, the product should be fashioned in such a way that first-level maintenance primarily consists of unit replacements (i.e., line replaceable units, LRUs). Design for maintainability also dictates that these replaceable units should be easily accessible and easily removable. Such a doctrine is important for any number of reasons but can be best illustrated through an example. Take, for instance, a frequently replaced LRU in a large system. If the LRU is not easily accessible, the maintenance process is hindered, dramatically increasing the repair time, the system maintenance costs, and the system downtime. These increased costs and inconveniences are then passed on to the customer, thereby reducing the perceived quality and desirability of the system and subsequently reducing future market demands. Similarly, a unit designed for socket insertion and manual latch-lock retention is preferable to one requiring the use of screws and fasteners or the reflowing and reformation of solder joints because it reduces the design complexity, reduces the service time requirements, reduces the service interval sophistication, reduces the service technician training requirements, and ultimately reduces the costs transferred to the customer.

Ergonomics, the interaction between machine and human, dictates that the operating panel layout should be optimized for visual comfort and accessibility of all the controls. A product which is confusing to view, uncomfortable to handle, and cumbersome to operate will reduce operator productivity and interest, thereby again increasing the costs associated with using the system and ultimately causing its demise. Thus, ergomonics dictates that the design should maximize operator efficiency.

### 1.4.7  Design for Manufacturability

Increasing demands on electronics producers to supply a variety of reliable electronics products at a reasonable cost have affected all aspects of electronics production. To meet these changing demands, the design process must incorporate production issues. In this way, many production difficulties and corresponding hidden costs are addressed and resolved early in the design process.

The philosophy of design for manufacturability describes a methodology by which a product is designed to be compatible with the most capable manufacturing processes, equipment, and production practices available. Thus, one must examine the techniques and technologies being employed in the manufacturing process and tailor the complexity of the design to optimize the abilities of the selected equipment and the skills of the operators employed. The result of such a practice is a product with the highest possible yields, throughput, and quality with minimal production costs.

The industry drive toward higher-density electronics designs has triggered the conversion from the current industry standard of insertion technology to surface mount technology (SMT) with the expectation that 30 to 50% of new and revised circuit designs will incorporate SMT by 1992. Surface mount technology is a process-intensive manufacturing technology in which many factors can affect the yield, quality, cost of production, and corresponding costs associated with the system. With SMT, discrete component leads are terminated on the surface of the PWB instead of being inserted into and terminated within holes drilled through the PWB. In this way, SMT simplifies the fabrication of the PWB by reducing the number of holes that must be drilled, as plated through holes are only required for conductive cooling and interlayer connections in multilayer board designs.

As electronics producers convert to SMT, mixed-technology board designs comprising both SMDs and through hole components (THCs) are being developed. In mixed-technology designs, portions of a currently used through hole design are redesigned to alter the function of the assembly. In the component selection process, however, SMDs are used to replace the through hole components in the original design. Mixed-technology designs may also result from the replacement of discrete through hole components with SMDs for the purpose of reducing the size and weight of the completed assembly.

The presence of both THCs and SMDs on the same PWB poses two significant problems with respect to the manufacturability issue. These are the selection of the production equipment to be used and the determination of an applicable global process plan. As the two component types require significantly different handling, tinning, cleaning, placement, soldering, and inspection techniques, electronics producers are compelled to consign the design of all phases of the manufacturing and assembly processes to a design team consisting of electrical, mechanical, and manufacturing engineers. Therefore, an effort must be made to optimize the design to satisfy conflicting constraints by interactions between systems, manufacturing, mechanical, electrical, and design engineering.

With these often conflicting constraints in mind, design for manufacturability requires that the design be physically produced in a cost effective manner, that there are ways to test the end product with a guaranteed level of confidence, and that product quality can be maintained throughout the required life of the product. Table 1.6 shows the manufacturing design activities from the design review to the fabrication of the electronics product.

## 1.5   ELECTRONICS DESIGN AND THE ELECTRICAL ENGINEER

Satisfactory design of electronic systems depends on the acceptable performance of electrical functions at all levels of the packaging hierarchy. The electrical engineer's role in electronic design is therefore generally limited to the areas of logic design, partitioning and assignment, circuit analysis, performance analysis and verification, and establishment of electrical requirements. The logic design defines the actual functions which the electronic systems will perform. The development of digital logic involves the combination of basic logic building blocks (e.g., INVERT, AND, OR). At a higher level, existing electronic devices which perform the basic functions are combined to create a device which is capable of more sophisticated logical functions.

The timing and sequence at which the various logical units combine and operate establish the functional performance of the product and hence are critical to the performance of the product. For this reason, accurate modeling of component packages into equivalent electrical circuits is a critical function of the electrical engineer. Because of the complexity of modern electronic systems, the electrical engineer generally requires computer-aided modeling and simulation for circuit and logic analysis. SPICE (Simulation Program for Integrated Circuit Emphasis) is one such commonly used circuit simulation program. Simulation includes logic testing and time analysis. While logic tests are geared to verify correct responses, timing tests are necessary to ensure the correct sequencing of logic events.

In addition to circuit modeling and simulation, the electrical engineer is responsible for evaluating and reducing negative and potential failure-inducing electrical interactions such as reflections and cross talk. Reflections are typically produced by impedance dif-

**Table 1.6** Manufacturing Design Activities

| | | |
|---|---|---|
| **Testability recommendations** | **Manufacturing plan** | **QC requirements** |
| —Initialization | —Automation | —Configuration control |
| —Test control | —Proceses | —Cost and schedule |
| —Test point availability | —Skills | |
| —Fan-in/fan-out | —Make/buy | **Fabrication** |
| —Bit | —Quality control | —Execute manufacturing |
| —Partitioning | —Schedule | design |
| —Software | **Factory requirements** | —Maintain cost and |
| **Producibility trade studies** | —Manpower | schedule status |
| —Processes | —Skills | —Develop work arounds |
| —Assembly/installation | —Capital equipment | as requirements |
| —Quality assurance | —Space | —Initiate additional |
| —Facilities | —Shop load | producibility requirements |
| —Materials | —Cost and schedule | —Monitor subcontractors |
| —Equipment compatibility | **Tooling requirements** | —Verify shop orders |
| —Cost and schedule | —Types | |
| **Manufacturing costs** | —Quantities | |
| —Material | —Make/buy | |
| —Labor | —Cost and schedule | |
| **Schedule** | **Test requirements** | |
| **Manufacturing operations resources** | —Equipment | |
| —Factory | —Software | |
| —Tooling | —Cost and schedule | |
| —Quality control and test | **Part and Material Requirements** | |
| | —Parts | |
| | —Materials | |
| | —Process qualifications | |
| | —Make/buy processes | |
| | —Cost and schedule | |

ferences, usually between the driver impedance, the characteristic line impedance, and the terminal impedance of the receiving circuit. Reflections can, however, be reduced significantly through control of the conductor path impedance by altering the signal line geometry and the PWB and signal trace material characteristics. Cross talk, on the other hand, is produced by the mutual capacitance and inductance generated between neighboring signal traces and can initiate unwanted currents and voltages in the system. This phenomenon is referred to as *signal noise* and is especially evident in parallel signal traces. Noise is any unwanted electrical signal occurring in the system produced by normal operation of electrical devices, such as switching in digital electronics and electromagnetic radiation.

Power and ground distribution is another important task which the electrical engineer must address. The conversion of power (e.g., AC to DC) must be considered based on the need of the electronic system. Power supplied to the individual electronic devices must be regulated and maintained within specified tolerances. Power deprivation must be eliminated and signal spikes must be filtered. In addition, proper grounding of electronic devices must be maintained for reliable performance and safe system operation.

## 1.6 ELECTRONICS DESIGN AND THE PACKAGING ENGINEER

Because the packaging engineer is responsible for the design of thermal and environmental controls for the electronic system, his or her role in electronics design lies primarily in the areas of mechanical analysis, packaging, manufacturing, reliability, and maintainability. The packaging engineer must, however, perform this role concurrently with the systems engineer, the electrical engineer, and the logistics staff. Thus the packaging engineer plays a significant role in the electronics design process which involves a continuous interaction with all of the other participants in the system design process.

The thermal design of an electronic system generally consists of specifying the principal heat removal paths, strategically orienting and locating components to generate a uniform or optimum temperature profile across the ECA, and verifying the design through a computer simulation. Because excessive heat can produce a number of structural and performance problems, the packaging engineer must also model and design the ECA chassis, the chassis drawers, and the cabinets to supply the required heat removal paths and the volumetric heat transfer capability while also meeting the required structural and vibrational design specifications.

The structural integrity of each of the interconnections and the physical packaging of the chip carriers and hybrid assemblies are other important aspects of packaging which concern the mechanical engineer. In the case of individual electric components, the interconnections may serve as the supporting structure. If so, physical testing and computer simulation of the system and subsystems in shock and vibrational modes are necessary.

## 1.7 CONCURRENT ENGINEERING

Increased demand for sophisticated electronics and the competitive pressure to meet market windows have placed a heavy burden on design engineers. The continuing growth in electronic technologies and the ever-increasing numbers of electronic components compound the problem. In many cases, the traditional design methodology has broken down under these increased pressures. In particular, the development of electronic equipment has traditionally followed a sequential process from design to test to manufacturing. The sequential process often fragmented the design into a series of isolated engineering tasks and hence was cyclic in nature, as each task phase generally required one or more redesign and retest cycles. As additional considerations, constraints, and requirements have been added to the design process, new specialty areas have been developed to deal with each critical issue as depicted in Figure 1.19.

In order to meet the new demands imposed on the designers, a fundamental change has occurred in the design process in the form of a design methodology called concurrent engineering, simultaneous engineering, or concurrent design. The idea behind concurrent engineering has been around for many years and is, in fact, a common sense approach to the difficulties facing designers and manufacturers. Concurrent engineering replaces the test feedback cycle with the interactive passing of knowledge gained and created during the development process. It has been demonstrated that the total cost of a product, regardless of the quantity produced, is highly dependent on early development decisions. In particular, military products, which are subject to long service lives, can have total life cycle costs that far exceed the initial acquisition costs. Early decisions are generally concerned with design costs, functionality, and performance, while later considerations include production

Design Cycle Time

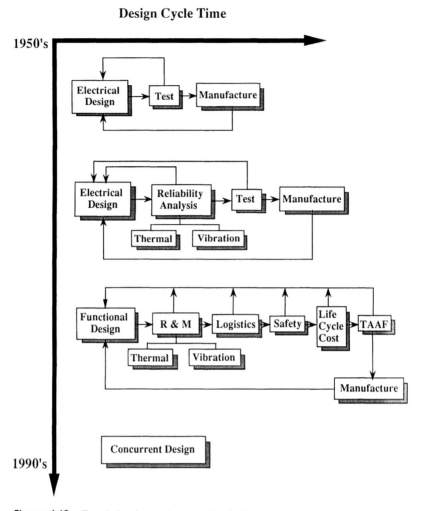

**Figure 1.19**   Trends in electronic assembly design.

cost, quality, reliability, maintainability, manufacturability, logistics, and supportability. Concurrent engineering seeks to accommodate all considerations and helps to ensure design compliance with manufacturing resources and processes while meeting functional and performance goals. The intent is to cause designers to consider all aspects of the product's life from design to disposal at the onset of the development.

Concurrent engineering is achieved through the cooperation of the multiple engineering disciplines, part suppliers, and users early in product development, as depicted in Figure 1.20. As a direct result of this interaction, simultaneous tasks are mapped out and common goals are established. Development time is decreased by performing simultaneous tasks and having a clear understanding of the various design viewpoints. Common goals ensure cooperation between the various design groups. One of the most important functions of concurrent engineering is to provide insight into the cause and effects of each proposed design change as viewed by the different design groups.

The development of easily accessible information and smooth information flow is

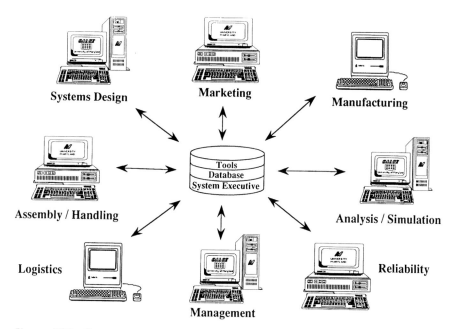

**Figure 1.20**  Concurrent product development concept.

critical to concurrent engineering. In addition, design tools that aid in analysis and monitoring of the design are also important. Design tools are needed to predict outcomes and
keep the specialty designers abreast of other design parameters which are affected by their
actions. Thus, decision support must be provided in such a way as to convey a global
picture of the design to the various specialty designers.

Studies of Japanese manufacturing techniques have demonstrated the effectiveness
of concurrent engineering practices [6]. Similarly, U.S. companies that have adopted
concurrent engineering practices have seen benefits such as improved quality, shortened
production schedules, and lower costs [6]. Further benefits of concurrent engineering can
be observed from Tables 1.7–1.9. Despite the advantages of concurrent engineering,
industry has been slow to apply the practice due to the radical changes which inevitably
occur within the hierarchical structure of many companies. Top management must make

**Table 1.7**  Design decisions costs[a]

| Development process | Percent of total cost | |
| --- | --- | --- |
| | Incurred | Committed |
| Conception | 3–5 | 40–60 |
| Design engineering | 5–8 | 60–80 |
| Testing | 8–10 | 80–90 |
| Process planning | 10–15 | 90–95 |
| Production | 15–100 | 95–100 |

[a] Data: Computer-Aided Manufacturing-International Inc.
[8].

**Table 1.8** Traditional Engineering
Design Change Cost[a]

| Design change | Cost |
|---|---|
| Design | $1,000 |
| Design test | $10,000 |
| Process planning | $100,000 |
| Test production | $1,000,000 |
| Final production | $10,000,000 |

[a] Data: Dataquest Inc. [8].

**Table 1.9** Benefits of Concurrent Engineering

| Benefits | Percent[a] |
|---|---|
| Development time | 30–70% less |
| Engineering changes | 65–90% fewer |
| Time to market | 20–90% less |
| Overall quality | 200–600% higher |
| White-collar productivity | 20–110% higher |
| Dollar sales | 5–50% higher |
| Return on assets | 20–120% higher |

[a] Data: National Institute of Standards and Technology,
Thomas Group Inc., Institute for Defense Analysis [8].

a total commitment to the concurrent design approach, which entails a certain loss by management in the control of the development process.

## 1.8 SOURCES OF INFORMATION

In the previous sections, electronic packaging hierarchy has been discussed and trends in electronic packaging have been highlighted. The electronic design process was presented to illustrate the factors confronting the electronics design team when attempting to develop a new electronic system. Design considerations of material selection, performance, reliability, maintainability, and manufacturability were examined. As part of this discussion, the tasks facing electrical and mechanical engineers were described and the trend toward concurrent engineering was discussed. The rest of the book conveys information concerning electronic packaging which will promote a more thorough understanding of the mechanical and packaging engineering issues in the design process.

The electronics field is so complex and versatile that students and practitioners must be aware of all current trends and places where information may be obtained. The International Society of Electrical and Electronic Engineers (IEEE) provides a full range of publications which offer current research results. The American Society of Mechanical Engineers (ASME) also offers publications which deal directly and indirectly with problems associated with the electronic packaging process. In particular, the *Transactions of ASME*, *Journal of Electronic Packaging* provides a forum for technology exchange. Other societies which are concerned with the electronics industry include the International Society for

Hybrid Microelectronics (ISHM), the International Electronics Packaging Society (IEPS), the Institute for Interconnecting and Packaging Electronic Circuits (IPC), Semiconductor Equipment and Materials International (SEMI), and the Printed Circuit and Interconnection Federation (PCIF). In addition, commercial publications which serve the community of electronic packaging include *Electronic Packaging and Productions*, *Printed Circuit Fabrication*, *Circuit Design*, and *Circuit World*. A comprehensive list of sources of information for electronic packaging is provided in Table 1.10.

**Table 1.10**  Sources of Information

| Source | Language[a] |
| --- | --- |
| *Adhesives Age* | EN |
| *Adhaesion* | GE |
| *AEG Technik Magazin* | GE |
| *Assembly Engineering* | EN |
| *Aviation Week & Space Technology* | EN |
| *Blick durch die Wirtschaft* | GE |
| *Brazing & Soldering*, now: *Soldering & Surface Mount Technology* | EN |
| *CAD/CAM* | GE |
| *CADS* | GE |
| *CAE-Journal* | GE |
| *Ceramic Industry* | EN |
| *Circuit Design*, formerly: *Printed Circuit Design* | EN |
| *CircuiTree* | EN |
| *Circuits Fabrication* | EN |
| *Circuits Manufacturing* | EN |
| *Circuit World* | EN |
| *Cogito* | GE |
| *Computer Aided Engineering* | EN |
| *Computer & Elektronik* | GE |
| *Computer Design* | EN |
| *Computerwoche* | GE |
| *Connection Technology* | EN |
| *Der Elektriker—Der Energieelektroniker* | GE |
| *Design & Elektronik* | GE |
| *Dinero* | SP |
| *Du Pont Magazin* | GE |
| *"e" (Elektronik-Technologie/-Anwendung/-Marketing)* | GE |
| *EDN* | EN |
| *EDV & Recht* | GE |
| *Electri-Onics*, now: *Electronic Manufacturing* | EN |
| *Electric Products* | EN |
| *Electrical Design News* | EN |
| *Electromagnetics* | EN |
| *Electronic Business* | EN |
| *Electronic Design* | EN |
| *Electronic Engineering Times* | EN |
| *Electronic Manufacturing*, formerly: *Electri-Onics* | EN |
| *Electronic Packaging & Production* | EN |

**Table 1.10**   Continued

| Source | Language[a] |
|---|---|
| *Electronic Production* | EN |
| *Electronic Products* | EN |
| *Electronic Purchasing* | EN |
| *Electronics* | EN |
| *Electronics Manufacture & Test* | EN |
| *Electronic System Design* | EN |
| *Electronic Week* | EN |
| *Elektronik* | GE |
| *Elektronik Entwicklung* | GE |
| *Elektroniker* | GE |
| *Elektronik Industrie* | GE |
| *Elektronik Informationen* | GE |
| *Elektronik Journal* | GE |
| *Elektronikpraxis* | GE |
| *Elektronik Produktion & Prueftechnik* | GE |
| *Elektronik-Report* | GE |
| *Elektronikschau* | GE |
| *Elektrotechnik* | GE |
| *Elektrotechnische Zeitschrift (ETZ)* | GE |
| *Elettron Oggi* | IT |
| *European Chemical News* | EN |
| *Evaluation Engineering* | EN |
| *Feingeraetetechnik* | GE |
| *Feinwerktechnik & Messtechnik* | GE |
| *Finishing* | EN |
| *Finomechanika-Mikrotechnika* | HU |
| *Fujitsu Scientific and Technical Journal* | EN |
| *Funk-Technik* | GE |
| *Galvano Organo* | FR |
| *Galvanotechnik* | GE |
| *Gazeta Mercantil* | IT |
| *Gold Bulletin* | EN |
| *Handelsblatt* | GE |
| *HighTech* | GE |
| *Hybrid Circuits* | EN |
| *Hybrid Circuit Technology* | EN |
| *IBM Journal of Research and Development* | EN |
| *IBM Technical Bulletin* | EN |
| *IEE Review (London)* | EN |
| *IEEE Circuits and Devices Magazine* | EN |
| *IEEE Control Systems Magazine* | EN |
| *IEEE Spectrum* | EN |
| *IEEE Transactions on Antennas & Propagation* | EN |
| *IEEE Transactions on CAD* | EN |
| *IEEE Transactions on CHMT* | EN |
| *IEEE Transactions on Computers* | EN |
| *IEEE Transactions on Education* | EN |

**Table 1.10** Continued

| Source | Language[a] |
|---|---|
| *IEEE Transactions on Electromagnetic Compatibility* | EN |
| *IEEE Transactions on Pattern Analysis & Machine Intelligence* | EN |
| *IEEE Transactions on Reliability* | EN |
| *Industrial Engineering* | EN |
| *Industrie Anzeiger* | GE |
| *Industrie Diamanten Rundschau* | GE |
| *Industrie Elektrik und Elektronik* | GE |
| *Info Circuits* | FR |
| *Ingenieur-Werkstoffe* | GE |
| *IS + L* | GE |
| *International Journal of Electrical Engineering Education* | EN |
| *Journal of Manufacturing Systems* | EN |
| *Journal of Surface Mount Technology* | EN |
| *Journal of the Electrochemical Society* | EN |
| *Kontrolle* | GE |
| *Konstruktion und Elektronik* | GE |
| *Laser (Landsberg)* | GE |
| *Laser-Praxis (Munich)* | GE |
| *Leiterplatten* | GE |
| *Machine Design* | EN |
| *Manufacturing Systems* | EN |
| *Markt & Technik* | GE |
| *Messen, Pruefen, Automatisieren* | GE |
| *Mesures* | FR |
| *Metal Finishing* | EN |
| *Metall* | GE |
| *Metalloberflaeche* | GE |
| *Microelectronic Manufacturing and Testing* | EN |
| *Microprocessors and Microsystems* | EN |
| *Microwaves & RF* | EN |
| *Mitsubishi Denki Giho* | JA |
| *Nachrichtentechnik Elektronik* | GE |
| *NEC Research & Development* | EN |
| *New Electronics* | EN |
| *Oberflaeche* | GE |
| *Oberflaeche Surface* | GE |
| *Paper Film Foil Converter* | EN |
| *PCB Magazine* | IT |
| *Plaste und Kautschuk* | GE |
| *Plastics Technology* | EN |
| *Plating Surface Finishing* | EN |
| *Printed Circuit Assembly* | EN |
| *Printed Circuit Design*, now: *Circuit Design* | EN |
| *Printed Circuit Fabrication* | EN |
| *Printed Circuit Network* | EN |
| *Product Finishing* | EN |

**Table 1.10** Continued

| Source | Language[a] |
|--------|-------------|
| *Productronic* | GE |
| *Produktion* | GE |
| *PRONIC* | GE |
| *Qualitaet und Zuverlaessigkeit* | GE |
| *Radio Fernsehen Elecktronik* | GE |
| *Radioelectronics & Communication Systems* | EN |
| *Reinraumtechnik* | GE |
| *Review of Scientific Instruments* | GE |
| *RF Design* | EN |
| *Robot* | JA |
| *Schweizer Maschinenmarkt* | GE |
| *Schweissen und Schneiden* | GE |
| *Scientific and Technical Aerospace Reports* | EN |
| *Semiconductor International* | EN |
| *Sensors Instrumentation News* | EN |
| *Siemens Components* | GE |
| *SIP Siebdruck-Infopost* | GE |
| *SMD Magazin* | GE |
| *Soldering & Surface Mount Technology*, formerly: *Brazing & Soldering* | EN |
| *Surface Mount Technology* (USA) | EN |
| *Surface Mount Technology* (UK) | EN |
| *Surface Mount Technology* (West Germany) | GE |
| *Surface Engineering* | EN |
| *Svecia News Bulletin* | GE |
| *Technical Disclosure Bulletin* | EN |
| *Technical Review IPC* | EN |
| *Technische Mitteilungen Schweizerische Post* | GE |
| *Technische Rundschau* | GE |
| *The Independent* | EN |
| *Toute l'Electronique* | FR |
| *Umwelt* | GE |
| *VDI Nachrichten* | GE |
| *Verbindungstechnik in der Elektrotechnik* | GE |
| *Welding Journal* | EN |
| *Werkstattstechnik* | GE |
| *Wissenschaftliche Zeitschrift der TH Ilmenau* | GE |
| *ZWF/CIM*, formerly: *Zeitschrift fuer wirtschaftliche Fertigung* | GE |

[a] These newspapers/journals are written in the following languages: EN, English; GE, German; IT, Italian; FR, French; SP, Spanish; HU, Hungarian; JA, Japanese.

## EXERCISES

1. Discuss the packaging hierarchy of a personal computer.
2. Compare the advantages and disadvantages of using tape-automated bonding and wire bonding.
3. Discuss the performance, cost, reliability, maintainability, and manufacturability issues associated with an automatic teller machine (ATM).
4. Discuss the concerns in the design of an electronic system which will be used in a space application, such as a telemetry system.
5. What are the chief concerns of the mechanical engineer in the design of an electronic product?
6. Find a recent article on a particular trend in electronics packaging and summarize the article. (Note: The reader should also find other articles which indicate the need for the trend or dispute the article. State your opinion based on the information you have accumulated.)

## REFERENCES

1. Tummala, R. R. and Rymaszewski, E. J., eds., *Microelectronics Packaging Handbook*, Van Nostrand Reinhold, New York, 1989.
2. Edosomwan, J. A. and Ballakur, A., eds., *Productivity and Quality Improvement in Electronics Assembly*, McGraw-Hill, New York, 1989.
3. *Electronic Material Handbook, Packaging*, Vol. 1, ASM International, Materials Park, Ohio, 1989.
4. Mahalinggam, M., "Design Considerations," in *Electronic Material Handbook, Packaging*, Vol. 1, ASM International, Materials Park, Ohio, 1989.
5. Spitz, S. L., "Guide to Plating High Aspect Ratio Holes," *Electronic Packaging and Production 30*(3): 54–56 (1990).
6. Evanczuk, S., "Concurrent Engineering: The New Look of Design," *High Performance Systems 11*(4):16–27 (1990).
7. Barrett, C., "Negotiating Roadblocks," Vol. 27, No. 1, *IEEE Spectrum 27*(1): 42–43 (1990).
8. Woodruff, D. and Phillips, S., "A Smarter Way to Manufacture," *Business Week*, April 30, 1990, pp. 110–117.
9. *Component Quality/Reliability Handbook*, Intel Corp., Santa Clara, CA, 1984.
10. *Component Quality and Reliability*, Intel Corp., Mt. Prospect, IL, 1990.
11. *Military Products Handbook*, Vol. 2, Intel Corp., Santa Clara, CA, 1989.

# 2

# Electronic Components

**Denise Burkus Harris**    *Westinghouse Electric Corporation, Baltimore, Maryland*

**Michael Pecht and Pradeep Lall**    *CALCE Center for Electronic Packaging, University of Maryland, College Park, Maryland*

The complexity and assortment of components which are being employed in electronic systems have grown at an alarming rate. In fact, the growth of the electronics industry can be directly tied to developments made in the fundamental building blocks, the electronic components. Table 2.1 shows the historical trends of the dominating technology advances in electronic components. This growth makes the selection of components increasingly difficult. Lists of components, their functions, and their performance characteristics are published in parts application manuals and catalogs, and technical information data sheets are available from component manufacturers. Lists of preferred parts are generally defined by company policies based on precedent or are generated and published by the government. These approved government lists, referred to as Qualified Materials Lists (QMLs) and Qualified Product Lists (QPLs) give approved vendors and products that have passed military standard qualification testing. These lists and component information pertaining to reliability and availability must be available to the design engineers.

Electronic systems are composed of electronic devices, such as resistors, capacitors, inductors, switches, relays, semiconductors (diodes, transistors, integrated circuits), and hybrids, as well as auxiliary components which provide structural support and aid in reliability and maintainability. Such auxiliary components include heat sinks, fasteners, wedge locks, mounting clips, and connectors.

Electrical components can generally be categorized into active and passive types. Active components require a power supply to perform their functions, whereas passive components do not. Passive components include resistors and capacitors. Active com-

**Table 2.1**  Technology Changes

| | |
|---|---|
| 1945–1955 | Vacuum tube electronics |
| 1955–1965 | Transistors |
| 1965–1975 | Integrated circuits (ICs) |
| 1975–1985 | Large-scale integration (LSI) > 1000 gates |
| 1985– | Very large scale integration (VLSI) and ultra large scale integration (ULSI) |

ponents include transistors, diodes, and integrated circuits. Active components may also incorporate passive components.

## 2.1 RESISTORS

This section will discuss various aspects of resistors. The basic definition of the resistor and its electrical characteristics will be given. Examples of the role a resistor can play in an electronic assembly will be given. The tools needed for designing a resistor will be presented, along with some examples. The section will end by discussing the fabrication methods for manufacturing some different types of resistors.

### 2.1.1 Electrical Characteristics of a Resistor

The resistor is a device which regulates current. Resistors are characterized by resistances, resistance tolerances, power ratings, voltage ratings, thermal coefficients, frequency ratings, noise, and stability. Resistance is a function of applied voltage versus generated current. This relationship is very simple:

$$R = V/I \qquad (2.1)$$

Since power is simply the voltage times the current, we can see that power is related to resistance as follows:

$$P = VI = RI^2 \qquad (2.2)$$

where $R$ is resistance (in ohms), $V$ is voltage (in volts), $I$ is current (in amperes), and $P$ is power (in watts) [1, p. 11]. As the resistance is increased, the current is decreased, less current is allowed through the resistor, and thus the resistor "resists" the flow of the current.

A network of resistors in series add to give the equivalent resistor of the network. Resistors in parallel must be added reciprocally to give the equivalent resistance of the network [1, pp. 42–43].

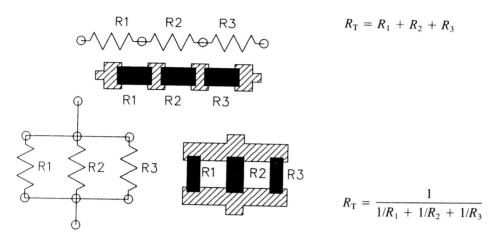

$$R_T = R_1 + R_2 + R_3$$

$$R_T = \frac{1}{1/R_1 + 1/R_2 + 1/R_3}$$

For example, if $R_1 = 10\ \Omega$, $R_2 = 100\ \Omega$, and $R_3 = 1000\ \Omega$, and they are in series, the total resistance would be $1110\ \Omega$. However, if the same three resistors are in parallel, the total resistance would be only $9.009\ \Omega$.

## 2.1.2 Resistor Design

The geometry of the resistor and the resistor and substrate materials—e.g., the resistivity (ohms per square) of the resistor material and the power capabilities of the resistor material and substrate—determine the resistance, tolerance, and stability of the resistor. This relationship is based on the following simple formula:

$$R = \rho \frac{l}{A_s} \qquad (2.3)$$

where $R$ is resistance in ohms ($\Omega$), $\rho$ is volume resistivity in ohm-cm, $l$ is the track length in cm, and $A_s$ is the cross-sectional area of the track (cm²) [2, p. F-89]. If the sheet resistivity is used, a standard sheet thickness is assumed and factored into the resistivity. Thus, the sheet resistivity is given in $\Omega$/square. Typically, resistors are rectangular in shape; therefore, the length ($l$) divided by the width ($w$) gives the number of squares within the resistor (see Figure 2.1). This multiplied by the resistivity will give the resistance:

$$R = \rho \frac{l}{w} \qquad (2.4)$$

In the above equation $R$ is the design or desired value of the resistor and $\rho$ is the sheet resistivity of the resistor material. Dividing $R$ by $\rho$ gives the ratio of the desired resistance over the resistivity. This value is referred to as the aspect ratio ($n$). Applying the above equation, it can be seen that

$$n = \frac{R}{\rho} = \frac{l}{w} \qquad (2.5)$$

Thus the aspect ratio ($n$) is equal to the number of squares in the resistor.

The surface area ($A$) of the resistor determines its stability and power dissipation

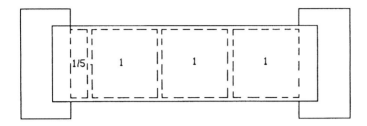

THE ABOVE RESISTOR IS 3.2 SQUARES
IF $\rho = 100\,\Omega/\square$ , THEN R = 3.2$\square$ X 100$\Omega/\square$ = 320$\Omega$

**Figure 2.1**  Number of squares within a rectangular resistor.

capabilities. The power ($P$) from the resistor is dissipated in the form of heat. Heat transfer is directly proportional to the area through which the heat is being transferred.

$$P = QA \tag{2.6}$$

where $Q$ is the power density [3, p. 119]. The power density capability, or maximum $Q$, of the resistor is a function of the resistor material, the substrate material and its thermal properties, the temperature the resistor will see, and the geometry of the resistor. This power density is found empirically. Its value will depend on the factors just listed and the process capabilities and control of the resistor manufacturer. The minimum area needed to dissipate the required power is [3, p. 119]

$$A_{\min} \geq \frac{P}{Q_{\max}} \tag{2.7}$$

Thus, as the area increases, the ability to dissipate power also increases. In other words, as the area increases, the temperature rise the resistor will see during operation will be decreased. The values of $Q_{\max}$ are determined by the resistor manufacturer. Some companies set the value at a constant for all resistors; others have it as a function of the desired stability and/or the substrate material used. For example, as the stability requirement is tightened, $Q_{\max}$ is decreased—the more stability required the more area needed for power dissipation. Or, the $Q_{\max}$ for a resistor designed on beryllia might be five times greater than the $Q_{\max}$ for a resistor designed on alumina, since beryllia has a thermal conductivity approximately five to six times greater than that of alumina. Thus, a resistor designed on beryllia can be much smaller yet able to dissipate much more power.

For a simple rectangular resistor, the area is

$$A = lw \tag{2.8}$$

Rearranging Eq. (2.4) and multiplying both sides by $w$, we have

$$lw = \frac{R}{\rho} w^2 \tag{2.9}$$

Combining Eqs. (2.5), (2.7), (2.8), and (2.9) and solving for $w$, we find that

$$w = \left(\frac{P}{nQ_{\max}}\right)^{1/2} \tag{2.10}$$

This equation is used to determine the minimum value of $w$, or the minimum width of the resistor. Once the value of $w$ is found, it can be substituted into Eq. (2.5) to solve for $l$.

Design a 150-$\Omega$ resistor that can dissipate 5 W. Assume that the resistor is thick film on alumina so that $Q_{max} = 100$, and 100 $\Omega$/square paste is to be used.

$$n = \frac{R}{\rho} = \frac{l}{w} = \frac{150}{100} = 1.5$$

$$w = \left(\frac{P}{nQ_{max}}\right)^{1/2} = \frac{5}{1.5 \times 100}^{1/2} = 0.1826$$

$$w = 0.183 \text{ in.}$$

Substituting this value of $w$ back into our first equation, we can solve for $l$:

$$1.5 = \frac{l}{0.183} \quad \text{or} \quad l = 0.274 \text{ in.}$$

To verify that the designed values will yield the stability required, we can check them by using Eq. (2.7):

$$A_{min} \geq \frac{P}{Q_{max}} \quad \text{or} \quad lw \geq \frac{P}{Q_{max}}$$

$$(0.183)(0.274) = 0.0501 \text{ in.}^2 > \frac{5}{100} = 0.05$$

Therefore, the design values are correct.

## Serpentine Resistors

Figure 2.2 gives an example of a meander resistor, often called a serpentine resistor because of its snake- or serpent-like shape. The figure also furnishes the equations for calculating the aspect ratio ($n$) in terms of the number of squares within the resistor. Figure 2.3 displays various resistor shapes with their effective number of squares, including their correction factors. Utilizing the information given in Figures 2.2 and 2.3 and combining it with the resistor design equations already given, a serpentine resistor may be designed.

What is the resistance of a five-meander serpentine resistor taking up the same area and using the same resistivity as the rectangular resistor given in the previous example? Assume no end connection effects ($k' = 0$), and set $w' = 0.005$ in.
From the previous example, $L = 0.274$ and $W = 0.183$. From Figure 2.2, the number of meanders $= (L - Ws)/(Ws + w')$; thus

$$5 = \frac{(0.274 - Ws)}{(Ws + 0.005)}$$

Solving for $Ws$ yields $Ws = 0.042$ in. From Figure 2.3, $K = 0.559$. From Figure 2.2,

$$n = \left(\frac{W - 2w' + Ws}{w'} + 2K\right)\left(\frac{L - Ws}{Ws + w'}\right) + \frac{Ws}{w'} + 2k'$$

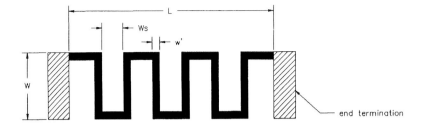

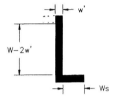

Length of one meander = W−2w' + Ws + 2 bends
The number of squares in one meander = the length
of one meander/w' = (W−2w'+Ws)/w' + 2K,
where K is the correction factor for the bend

**Figure 2.2** Meander resistor. The number of meanders = the length available for the meanders divided by the width of one meander. The length available for meanders = the resistor length minus the end connection length = $L - Ws$. Therefore, the number of meanders = $(L - Ws)/W' + Ws)$. Thus, the total number of squares within the resistor (n) = the number of squares in one meander times the number of meanders + the end connections = $[(W - 2w' + Ws)/w' + 2K] \times [(L - Ws)/(w' + Ws)] + Ws/w' + 2k'$ where Ws/w' accounts for the length of resistor for the one end connection that is not included within a meander and k' is the correction factor for the termination or end connection.

Substituting all known and calculate values thus far yields

$$n = \left(\frac{0.183 - 2(0.005) + 0.042}{0.005} + 2(0.559)\right)\left(\frac{0.274 - 0.042}{0.042 + 0.005}\right) + \frac{0.042}{0.005} = 228.89$$

From Eq. (2.5):

$$R = n\rho = 228.89(100) = 22,889\ \Omega \text{ or } 22.9\ \text{k}\Omega$$

The dissipative area is the same as for the rectangular resistor. Therefore, a serpentine resistor, capable of dissipating the same power and taking up the same area as an equivalent rectangular resistor, can have a resistance value two orders of magnitude higher.

### 2.1.3 Resistor Tolerance
### Resistor Trimming

Once a resistor is printed, whether thick film or thin film, the value can be adjusted by trimming the resistor. Trimming is accomplished with a laser. As shown in Figure 2.4, the laser cuts into the resistor. Cutting into the resistor changes the effective resistor width,

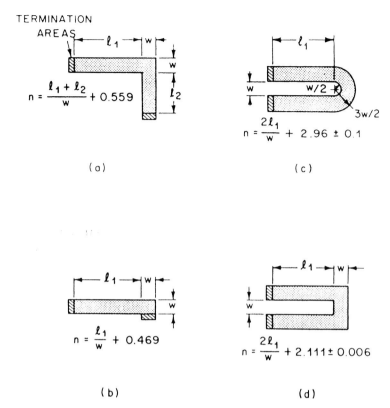

**Figure 2.3** Effective number of squares and correction factors for various geometries. (R. W. Berry, *Thin Film Technology*, as quoted in ref. 3, p. 118).

the width available for carrying the current. This change in width simultaneously changes the number of squares, which, in turn, changes the aspect ratio, which defines the resistance for the given material resistivity. As can be seen in Figure 2.4, trimming the resistor can only result in increasing the number of squares. Thus, a resistor can only have its resistance brought up in value by trimming. For this reason, the designer will typically design the resistor to be 90 to 95% the actual desired value, so that the resistor can be brought up to the appropriate value by laser trimming. Chapter 3 describes the laser trimming process in detail in the sections entitled Substrate Formation and Thick-Film Metallization.

## Temperature Coefficient of Resistance (TCR)

The resistivity ($\rho$) of the resistor material is a function of temperature:

$$\rho = \rho_0 [1 + \alpha(T - T_0)] \tag{2.11}$$

where $\rho_0$ is the sheet resistivity at $T_0$, $T$ is the operating temperature, $T_0$ is the ambient temperature, and $\alpha$ is the temperature coefficient of resistance [3, p. 120].

The TCR is given by [3, p.121]

$$TCR = \alpha = \frac{1}{R_0} \frac{dR}{dT} \tag{2.12}$$

(a) UNTRIMMED RESISTOR

(b) "L" CUT AND SINGLE PLUNGE CUT

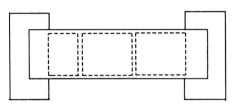

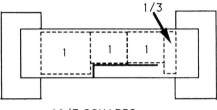

n = 1/w = 8/3 SQUARES
IF ρ = 100 Ω/□ , R = 267 Ω

n = 10/3 SQUARES
IF ρ = 100 Ω/□, R = 333 Ω

(c) DOUBLE PLUNGE CUT

(d) SERPENTINE CUT

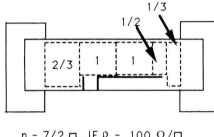

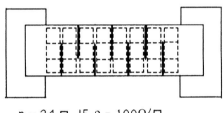

n = 7/2 □, IF ρ = 100 Ω/□

n = 24 □, IF ρ = 100Ω/□
R = 2400 Ω

**Figure 2.4**  Laser trimming of resistors.

■ RESISTOR

▨ CONDUCTOR / TERMINATION

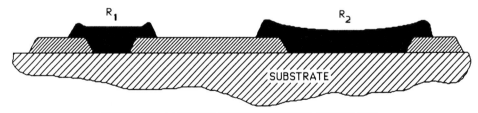

$R_1$  MAINTAINS A UNIFORM PRINT THICKNESS ACROSS THE
MAIN RESISTOR BODY

$R_2$  HAS A LARGE LENGTH WHICH ALLOWS THE SQUEEGE
PRESSURE TO PUSH OR FORCE OUT THE PASTE FROM
THE CENTER OF THE RESISTOR BODY; RESULTING IN
LOW PRINT THICKNESS AT THE CENTER

**Figure 2.5**  Uniformity of printing.

The less the temperature changes, the less effect the TCR will have. Therefore, if the area of the resistor is increased, the resistor can dissipate more heat. Thus a larger area does not allow the temperature to increase, resulting in smaller TCR effects, so the stability is increased with increased area.

This directly proportional relationship between area and stability is true to a point. If the area of the resistor is increased too much, the print uniformity is compromised. As the print uniformity (the print thickness uniformity) is reduced, so is the stability. Figure 2.5 shows the difference in a uniform print versus a nonuniform print. With the nonuniform print, there exist low or thin spots. These thin spots cannot adequately transfer the current traveling through the resistor. They will, therefore, heat up faster than the surrounding resistor area, forming a hot spot which can lead to an increase in TCR effects and in the potential for thermally overstressing the resistor. The existence of these hot spots lowers the stability and reliability of the resistor. TCR values are typically supplied in the form of a graph of resistance versus temperature by the resistor paste manufacturers. An example of the TCR effects for a Du Pont thick-film resistor material is given in Figure 2.6. If the resistor will be required to perform over a large temperature range or at extreme temperatures, the effects of TCR must be taken into account by the designer.

## 2.1.4 Resistor Types
### Insertion Mountable: Leaded Resistors

Resistors can be categorized into two major groups, fixed and variable. The most common type of fixed resistor is a cylindrical carbon resistor with leads extending from both ends. The resistance value and tolerance can be found from a color coding, which is represented by various colored bands located around the periphery of the resistor. The standard color code is given in Figure 2.7. Figure 2.8 shows a insertion mount metal film resistor. It has the same shape as the color-coded carbon resistors. Other types of fixed resistors include metal oxide, wire wound, noninductive, and trimmers.

Variable resistors or potentiometers are used mainly as current-controlling devices. An example of a potentiometer is a volume control on a radio. When the volume is low, the resistance is high and thus a limited amount of current flows through the amplifier. As the volume control is set to a higher volume, the resistance decreases and more current flows through the amplifier. Potentiometers are also useful for attenuation and fine circuit adjustment.

## Chip Resistors

For surface mount applications, leadless chip resistors are used. Figure 2.9 gives an example of a metal electrode face (MELF) thin-film resistor. As shown in the figure, chip resistors are available in tape and reel. The tape is fed into a pick-and-place machine, which then picks up the component and mounts it onto the PWB. Another common form of chip resistor is a ceramic tab with thick-film resistors printed on it. (Chapter 3 will cover the screen printing process in detail.) Typically the ceramic used is alumina. The first step is to print on the conductor metallization to form the terminations of the resistor. The conductor pastes commonly used for these terminations are palladium silver, platinum gold, and gold. After the terminations have been printed and fired, the resistor paste is printed. Resistor pastes are commonly made up of metal oxides. The resistor pastes have different formulations depending on the resistivity desired. Low-ohmage pastes have more

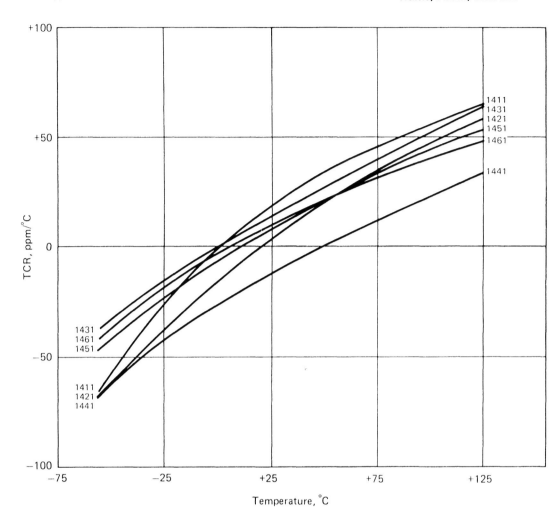

**Figure 2.6** Temperature coefficient of resistance. The specified TCR of BIROX 1400-series compositions is $0 \pm 100$ ppm/°C between $-55$ and $+25$°C and between $+25$ and $+125$°C. The typical TCRs are actually much better, particularly over limited temperature ranges. Between 0 and $+70$°C TCRs of $0 \pm 50$ ppm/°C are generally attainable. TCRs can be affected by variations in resistor geometry and conductor type. Data shown are for 5 mm long × 2.5 mm wide (200 mil long × 100 mil wide) resistors terminated with palladium/gold conductor composition 8651. (Courtesy of Du Pont Electronics.)

metals so that they are more conductive (less resistive); high-ohmage pastes are mostly oxides or ceramics. For example, if a high-power resistor with a value of less than 1 Ω is needed, the resistor can be printed using a platinum gold paste or a palladium silver-based resistor paste. Most resistor pastes, however, need higher resistivities and have compositions containing ruthenium oxide. These pastes are often referred to as "cermet," or a combination of *cer*amic and *met*al.

Thin-film resistors are very similar to thick-film resistors except that they are fabricated using thin-film metallization processes. (See Section 2.8 for more details on thin-film metallization processing.) These resistors can be built up on silicon, gallium arsenide

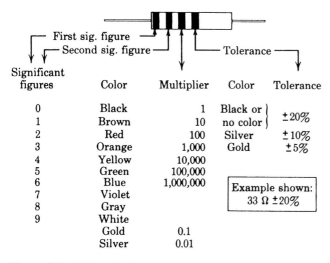

Figure 2.7   Color code for resistors. (From ref. 1, p. 748.)

(GaAs), and a number of different ceramics (alumina, beryllia, aluminum nitride, etc.). The thin-film processing allows for finer lines and smaller resistor dimensions. In the case of thin-film resistors, the resistor material is deposited and patterned prior to applying the metallization for the end terminations (see Figure 2.10).

Both the thick- and thin-film resistors just described are typically mounted inside a

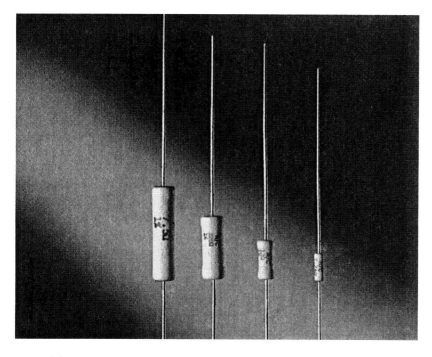

Figure 2.8   Insertion mount/leaded resistors. (Courtesy of Philips Components.)

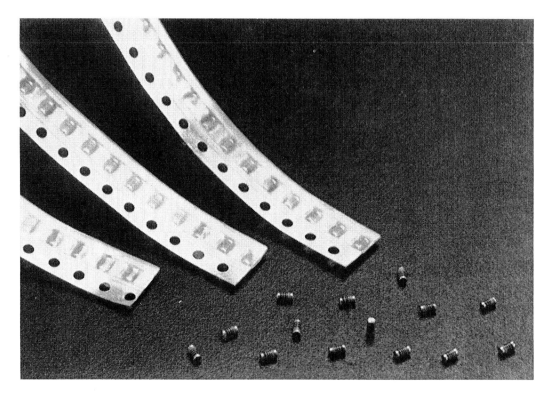

**Figure 2.9**  MELF resistor (Courtesy of Philips Components)

hermetically sealed device. Interconnection to the resistor is made by wire bonding to the end terminations on the top of the chip, as shown in Figure 2.11. To convert these chip resistors to surface-mountable devices, a way to simultaneously mount and electrically interconnect to the chip is needed. This dual functionality is accomplished by wrapping the terminations around to the bottom of the chip. Figure 2.12 shows different methods of forming the wraparounds. The first method is to extend the metallization around the two edges where the resistor terminations are located. The other method of edge wrap involves attaching a piece of conductive material to the ends of the chip. These end caps are epoxied (with a conductive adhesive) or soldered to the metallized terminations. These edge-wrapped chip resistors can then be surface mounted as show in Figure 2.13.

## Resistor Networks

Resistor networks are patterns or networks of different-value resistors printed on a single substrate. They can be fabricated using thin- or thick-film techniques. The designer can control the value of the resistance along a given connection by how the design bonds to the network. Figure 2.14 gives an example of a resistor network's schematic and the corresponding substrate. The table in Figure 2.14 indicates the different resistances obtainable with each different connection for three different designs. For example, taking design 1, if a probe is placed on pin 1 and a second probe is placed on pin 2, the resulting resistance is equal to R1 + R2 (or 10 + 20 = 30 $\Omega$). However, if the second probe is moved to pin 3, the resulting resistance is equal to R1 and R3 in series, or R1 + R3 = 10 + 30 = 40 Ω. If the probes are moved again, this time to pin 7 and pin 10, the resistance is

$$R7 + R8 + \left(\frac{1}{R9} + \frac{1}{R10}\right)^{1/2} = 70 + 80 + \left(\frac{1}{90} + \frac{1}{100}\right)^{1/2} = 197.37 \ \Omega$$

Thus one small chip or die, approximately 0.100 to 0.250 in. square, can provide numerous resistor values by interconnecting to various combinations of the few resistors printed on to the chip. Resistor networks come in different sizes, with various pin or I/O counts and abundant resistor combinations. These networks can also be custom made to contain the exact resistor values and ways of combining them that are needed for the given design.

Resistor networks can be purchased in chip or bare die form so that they can be mounted inside a package and wire bonded per design, as in commonly done within hybrid packages. They can also be purchased as surface-mountable components. In this case, the supplier has already mounted the chip inside a standard package (e.g., a dual in-line package, or DIP, for insertion mounting or a chip carrier for surface mounting). All I/O on the resistor network chip has been wire bonded or tape assisted bonded (TAB) to the I/O of the package. The resistance needed is chosen by how the designer routes the I/O connections of the DIP or chip carrier on the PWB.

## 2.2   CAPACITORS

### 2.2.1   Electrical Characteristics of Capacitors

Capacitors provide impedance to current inversely proportional to the frequency of an applied voltage. "Capacitance is a measure of the ability of a device to store energy in the form of separated charge or in the form of an electric field" [1, p. 19]. Capacitance is defined as

$$C = \frac{Q}{V} \tag{2.13}$$

where $C$ is capacitance in farads, $Q$ is charge in coulombs, and $V$ is potential in volts [2, p. F-68]. Capacitance is the measure of the charge required to change the potential of the device.

The change in charge in relationship to time is the definition of current [1, p.13]:

$$I = \frac{dQ}{dt} \tag{2.14}$$

Combining Eqs. (2.13) and (2.14) yields the following relationship between current and capacitance:

$$I \sim C \frac{dV}{dt} \tag{2.15}$$

Capacitors are used for blocking direct-current and alternating-current devices, for bypassing unwanted alternating-current components to ground, and as frequency-determining devices in resonant circuits. A useful property of capacitors is that their impedance is greatly reduced when they encounter a rapid increase in voltage. Capacitors are often connected

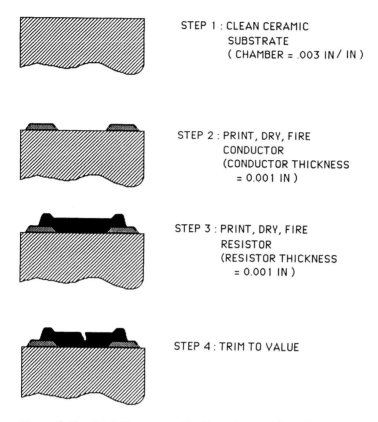

STEP 1 : CLEAN CERAMIC
SUBSTRATE
( CHAMBER = .003 IN / IN )

STEP 2 : PRINT, DRY, FIRE
CONDUCTOR
(CONDUCTOR THICKNESS
= 0.001 IN )

STEP 3 : PRINT, DRY, FIRE
RESISTOR
(RESISTOR THICKNESS
= 0.001 IN )

STEP 4 : TRIM TO VALUE

**Figure 2.10**   Thick film versus thin film resistor configuration.

in parallel with more important components. If an unexpected voltage surge should occur, the capacitor will short and draw most of the current, thus protecting the more important component.

Capacitors are classified by dielectric materials, capacitances, voltage ratings, temperature variations, power factor (the ratio of equivalent series resistances to impedance at specific frequencies), the ratio of capacitive reactivity to the equivalent series resistance, dissipation factor, noise level, and stability.

A network of capacitors in parallel add to give the equivalent capacitance of the network. Capacitors in series must be added reciprocally to give the equivalent capacitance of the network [1, pp. 42–43].

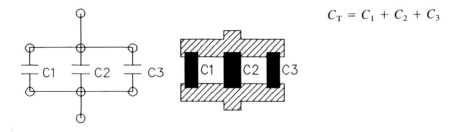

$$C_T = C_1 + C_2 + C_3$$

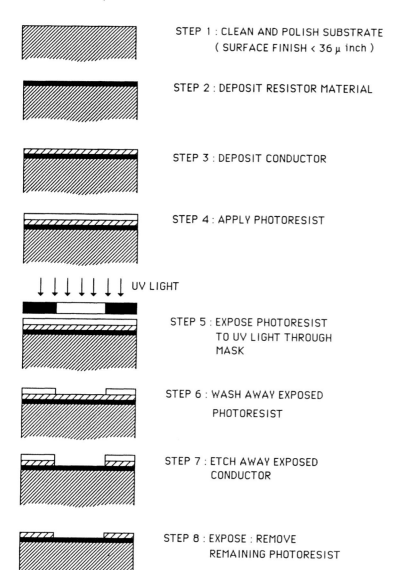

STEP 1 : CLEAN AND POLISH SUBSTRATE
( SURFACE FINISH < 36 μ inch )

STEP 2 : DEPOSIT RESISTOR MATERIAL

STEP 3 : DEPOSIT CONDUCTOR

STEP 4 : APPLY PHOTORESIST

UV LIGHT

STEP 5 : EXPOSE PHOTORESIST
TO UV LIGHT THROUGH
MASK

STEP 6 : WASH AWAY EXPOSED
PHOTORESIST

STEP 7 : ETCH AWAY EXPOSED
CONDUCTOR

STEP 8 : EXPOSE : REMOVE
REMAINING PHOTORESIST

**Figure 2.10** Continued

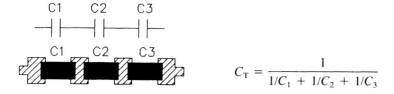

$$C_T = \frac{1}{1/C_1 + 1/C_2 + 1/C_3}$$

For example, if $C_1 = 10$ μF, $C_2 = 100$ μF, and $C_3 = 1000$ μF and they are in parallel, the total capacitance is 1110 μF. However, if the same three capacitors are in series, the total capacitance is only 9.009 μF.

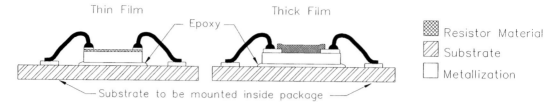

**Figure 2.11** Chip resistor attach and interconnect within a hermetic package.

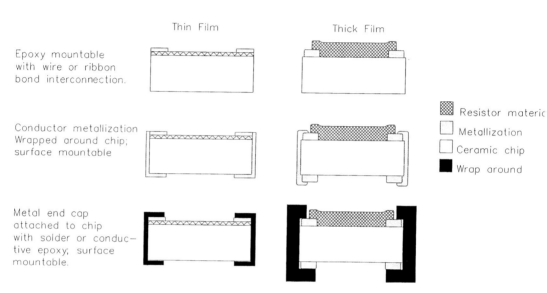

**Figure 2.12** Surface-mountable resistor chips.

## 2.2.2   Capacitor Design
## Parallel-Plate Capacitor

A parallel-plate capacitor consists of a dielectric material sandwiched between two parallel conductor plates. This type of capacitor can be in chip form, where a film of dielectric material is layered between two pieces of metal foil or metal tabs, as shown in Figure 2.15a. Also shown in Figure 2.15 are other parallel-plate capacitors. These types of parallel-plate capacitors are made an integral part of the integrated circuit/die or the printed circuit board. In these latter cases, the capacitor is built up on the board or substrate by depositing (or printing in the case of a thick-film circuit) the first metal pad or plate. The

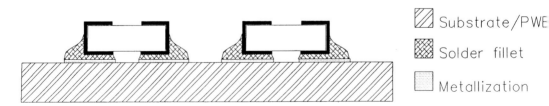

**Figure 2.13** Surface mounting of chip components.

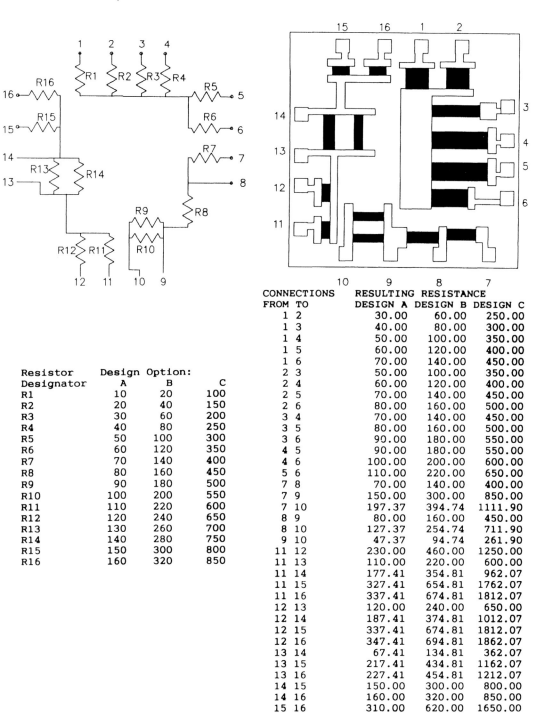

| CONNECTIONS | | RESULTING RESISTANCE | | |
|---|---|---|---|---|
| FROM | TO | DESIGN A | DESIGN B | DESIGN C |
| 1 | 2 | 30.00 | 60.00 | 250.00 |
| 1 | 3 | 40.00 | 80.00 | 300.00 |
| 1 | 4 | 50.00 | 100.00 | 350.00 |
| 1 | 5 | 60.00 | 120.00 | 400.00 |
| 1 | 6 | 70.00 | 140.00 | 450.00 |
| 2 | 3 | 50.00 | 100.00 | 350.00 |
| 2 | 4 | 60.00 | 120.00 | 400.00 |
| 2 | 5 | 70.00 | 140.00 | 450.00 |
| 2 | 6 | 80.00 | 160.00 | 500.00 |
| 3 | 4 | 70.00 | 140.00 | 450.00 |
| 3 | 5 | 80.00 | 160.00 | 500.00 |
| 3 | 6 | 90.00 | 180.00 | 550.00 |
| 4 | 5 | 90.00 | 180.00 | 550.00 |
| 4 | 6 | 100.00 | 200.00 | 600.00 |
| 5 | 6 | 110.00 | 220.00 | 650.00 |
| 7 | 8 | 70.00 | 140.00 | 400.00 |
| 7 | 9 | 150.00 | 300.00 | 850.00 |
| 7 | 10 | 197.37 | 394.74 | 1111.90 |
| 8 | 9 | 80.00 | 160.00 | 450.00 |
| 8 | 10 | 127.37 | 254.74 | 711.90 |
| 9 | 10 | 47.37 | 94.74 | 261.90 |
| 11 | 12 | 230.00 | 460.00 | 1250.00 |
| 11 | 13 | 110.00 | 220.00 | 600.00 |
| 11 | 14 | 177.41 | 354.81 | 962.07 |
| 11 | 15 | 327.41 | 654.81 | 1762.07 |
| 11 | 16 | 337.41 | 674.81 | 1812.07 |
| 12 | 13 | 120.00 | 240.00 | 650.00 |
| 12 | 14 | 187.41 | 374.81 | 1012.07 |
| 12 | 15 | 337.41 | 674.81 | 1812.07 |
| 12 | 16 | 347.41 | 694.81 | 1862.07 |
| 13 | 14 | 67.41 | 134.81 | 362.07 |
| 13 | 15 | 217.41 | 434.81 | 1162.07 |
| 13 | 16 | 227.41 | 454.81 | 1212.07 |
| 14 | 15 | 150.00 | 300.00 | 800.00 |
| 14 | 16 | 160.00 | 320.00 | 850.00 |
| 15 | 16 | 310.00 | 620.00 | 1650.00 |

| Resistor | Design Option: | | |
|---|---|---|---|
| Designator | A | B | C |
| R1 | 10 | 20 | 100 |
| R2 | 20 | 40 | 150 |
| R3 | 30 | 60 | 200 |
| R4 | 40 | 80 | 250 |
| R5 | 50 | 100 | 300 |
| R6 | 60 | 120 | 350 |
| R7 | 70 | 140 | 400 |
| R8 | 80 | 160 | 450 |
| R9 | 90 | 180 | 500 |
| R10 | 100 | 200 | 550 |
| R11 | 110 | 220 | 600 |
| R12 | 120 | 240 | 650 |
| R13 | 130 | 260 | 700 |
| R14 | 140 | 280 | 750 |
| R15 | 150 | 300 | 800 |
| R16 | 160 | 320 | 850 |

**Figure 2.14** Resistor network: schematic, die, and design and resulting values.

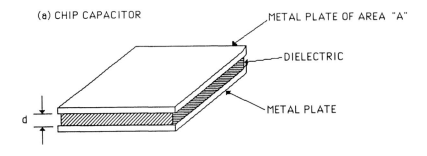

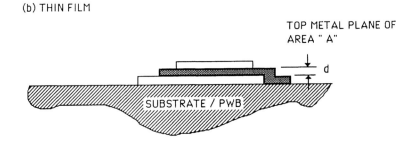

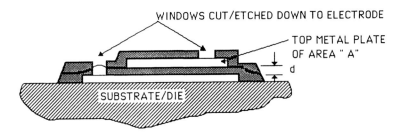

**Figure 2.15**  Parallel-plate capacitor

dielectric is then deposited over the first metal plate, followed by the top metal layer. The first metal plate is larger than the dielectric pad or top metal plate. This larger size allows a window to the first plate to be etched away to permit contact to the base electrode of the capacitor.

The capacitance of a parallel-plate capacitor can be calculated by

$$C = \frac{K_0 \epsilon_0 A}{d} \tag{2.16}$$

where $C$ is the capacitance in farads (F); $K_0$ is the *relative* dielectric constant of the dielectric or oxide; $\epsilon_0$ is the permittivity of free space; $A$ is the area of the top, smaller plate in cm$^2$; and $d$ is the dielectric thickness in cm [4, p. 14]. The permittivity of free space, $\epsilon_0$, is equal to $1/(36\pi) \times 10^{-9}$ F/m [1, p. 739].

Substituting the value of $\epsilon_0$ into Eq. (2.16) and converting to electrostatic units (1 F $= 9 \times 10^{11}$ electrostatic units) for the capacitance and cm for units of length yields

$$C = \frac{KA}{4\pi d} \tag{2.17}$$

where $K$ is the medium's dielectric constant [2, p. F-68].

Design a 30-pF parallel-plate capacitor for an integrated circuit. Assume that the wafer material is silicon.

The standard dielectric material used in silicon wafer fabrication is silicon dioxide ($SiO_2$) with a relative dielectric constant of 3.9, at a thickness of 500 angstroms.

From Eq. (2.16):

$$A = \frac{Cd}{K_0\epsilon_0} = \frac{(30 \times 10^{-12}\ \text{F})(5 \times 10^{-8}\ \text{m})}{3.9(8.86 \times 10^{-12}\ \text{F/m})}$$
$$= 4.34 \times 10^{-8}\ \text{m}^2 = 6.73 \times 10^{-5}\ \text{in.}^2$$

The capacitor will have a top plate which is 8.2 mils square.

As can be seen by examining Eq. (2.16) and the above example, to obtain a small-area capacitor the dielectric thickness must remain as small as possible. For this reason, parallel-plate capacitors are typically not integrated within thick-film systems; they would require too much substrate area. This type of capacitor normally is incorporated within integrated circuits (ICs) or semiconductor dice. Chip forms of these capacitors are often used within microwave assemblies, but due to their configuration require interconnection typically in the form of a ribbon or wire bond to the top plate.

## Interdigital Capacitor Design

There are two kinds of interdigital capacitors, as shown in Figure 2.16. The first has interlacing fingers patterned on top of the dielectric layer that covers the ground plane; the second type, referred to as a bipolar interdigital capacitor, has the fingers placed over a strip of dielectric that encapsulates a metal strip acting as a floating electrode. A bipolar interdigital capacitor isolates each finger as an individual cell, so that the effective capacitance approximates a parallel combination of all the individual cells.

The total capacitance of an interdigital capacitor is given by

$$C = \frac{(K_0)\epsilon_0 L}{W}[(N - 3)A_1 + A_2] \tag{2.18}$$

where $C$ is given in F/length, $N$ is the number of fingers, $L$ is the length of finger overlap, $W$ is the width of the capacitor, and $A_1$ and $A_2$ are correction factors for the interior and two exterior fingers, respectively. These correction factors are a function of the ratio of dielectric thickness to the fingers' line width [3, p. 144].

Examination of Eq. (2.18) reveals that the capacitance increases with decreasing capacitor width and with decreasing finger line width. Thus the most effective interdigital capacitors are those manufactured using thin-film methodology. For this reason, these capacitors are typically integral parts of ICs or thin-film substrates—making these substrates

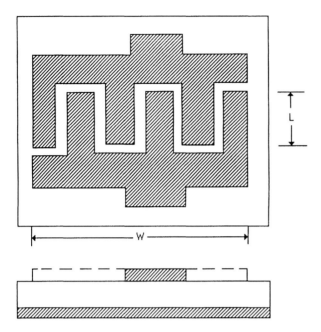

**Figure 2.16** Interdigital capacitor.

printed circuit boards (PCBs). The connection to the electrodes of these capacitors is accomplished through the routing of the PCB. Chip or component forms of these capacitors are used within packaged hybrid units but are not typically used for surface mounting due to the necessity of bonding to the top surface—to the fingers—of the chip.

### 2.2.3 Capacitor Types

Capacitors come in a variety of forms. Electrolytic capacitors are the largest common type of capacitor. They range in size from one to several thousand microfarads. Tantalum capacitors have the highest capacitance values for their size. Tantalum capacitors typically range from about 0.05 μF up to approximately 200 μF. The most commonly used type of capacitor is the ceramic disk, which is physically about the same size as the tantalum capacitors but has capacitance values ranging from approximately 1 pF to 1 μF. Other types of capacitors are metal film; nonpolarized; slow discharge, which can be used in place of batteries in certain applications; and variable capacitors, which are used for fine circuit tuning.

### Chip Capacitors

Insertion mount capacitors consist of dielectric material formed in the shape of a disk or cylinder with leads extending from them. Figure 2.17 shows a typical monoaxial capacitor in tape and reel form. These devices are fed into placement equipment which cuts and forms the leads prior to insertion into a PWB.

For surface-mounted assemblies, chip capacitors are typically used. As can be seen in Figure 2.18, chip capacitors are very similar to chip resistors in appearance. They are

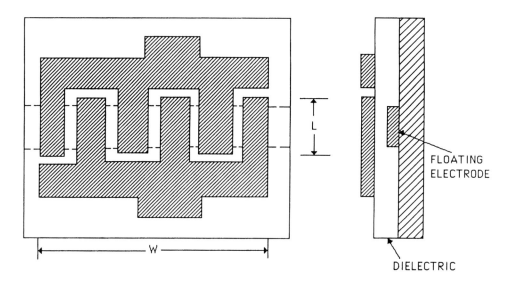

**Figure 2.16** Continued.

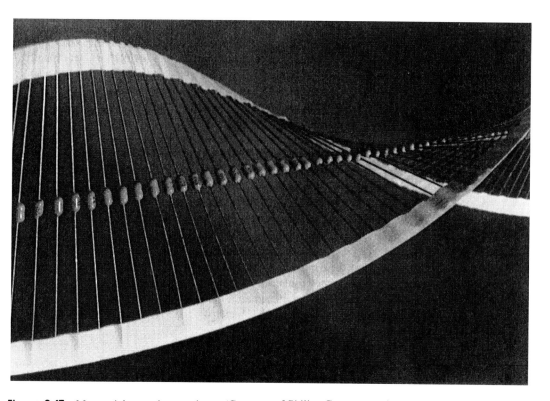

**Figure 2.17** Monoaxial ceramic capacitors. (Courtesy of Philips Components.)

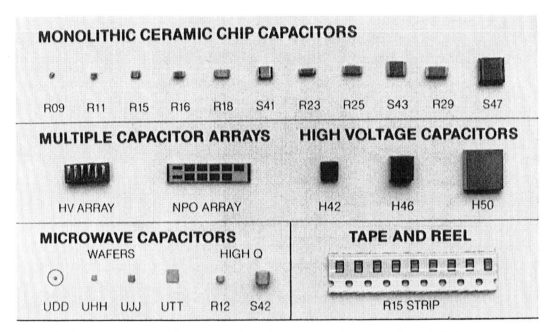

**Figure 2.18**   Chip capacitors. (Courtesy of Westinghouse Electronic Systems Group [ESG].)

composed of a dielectric material sandwiched between two metal end terminations. The dielectric material is typically a ceramic. The end terminations are normally solder dipped or solder or tin plated for solderability during the surface mount process. They are also available with nickel/gold-plated termination for wire bondability when epoxy mounted inside a hermetically sealed package.

Chip capacitors are fabricated in the same manner as chip resistors. They are also mounted in the same fashion as previously shown (Figure 2.13). Chip components, resistors and capacitors alike, can be purchased as individual chips in boxes or bags of hundreds or in a tape and reel strip for feeding into automatic placement equipment, as seen in Figure 2.18.

Adjustable capacitors or trimmer capacitors allow the capacitance to be slightly adjusted by a simple turn of the screw. An example of a trimmer capacitor is given in Figure 2.19.

Capacitor arrays, like resistor networks, are small substrates on which an array of capacitors is patterned. How the array is interconnected, in series or parallel, will determine the resulting capacitance. Capacitor banks are assemblies of multiple capacitors connected together to obtain a desired effective capacitance not available in a single-chip form. Capacitor banks will be covered in detail in Chapter 4.

## 2.3   INDUCTORS

Inductance is the constant of proportionality which relates the voltage to the rate of change in the current:

**Figure 2.19**   Trimmer capacitor. (Courtesy of Philips Components).

$$V \sim L\frac{di}{dt} \tag{2.19}$$

where $V$ is in volts, $i$ is the current in amperes, and $L$ is the inductance in henrys [1, p. 13].

A small amount of voltage is needed to maintain a steady current through an inductor. However, a large voltage is required to change the current flow rapidly. Thus, inductors resist current surges. Connecting an inductor in series with an important component will help prevent the component from failing due to an unexpected current surge. Inductors store energy in the form of moving charges or magnetic fields.

Inductors, like capacitors, are useful for eliminating unwanted frequencies. The resistance to an alternating current is referred to as the impedance. The impedance of an inductor is proportional to the frequency of the applied voltage:

$$Z_{\text{L}} = jvL = vL \angle 90° \tag{2.20}$$

where $Z$ is the impedance in ohms, $v$ is the frequency, and $L$ is the inductance in henrys. The phase angle of 90° is indicated by $j$ [1, p.124]. Thus, an inductor's impedance increases with frequency. Therefore, inductors attenuate or reduce high frequencies while allowing low frequencies to pass with little attenuation.

Inductors are wires turned into a coil to induce a magnetic flux, or are metal spirals patterned on dielectric material. Spiral inductors are those that are patterned using thin-film methods on PCBs or ICs. Surface- or insertion-mountable inductors are actual coils that are mounted on the PWB. These coils can be small springs with extended tails or terminations at both ends. Inductor coils can also be surface mounted in chip form. These inductor chips are coils that are encapsulated within a dielectric with the coil ends connected to the end caps or terminations of the chip. These devices are similar in appearance to chip capacitors.

## 2.4  DIODES

Semiconductor devices are subdivided into several device categories, including diodes, transistors, and integrated circuits. Diodes are voltage-controlling devices. They allow current to pass in one direction with a low applied voltage. In the other direction, a much higher voltage (reverse bias) is required for current to flow. Diodes are commonly used in bride rectifiers which convert AC signals to DC signals. Diodes are also used in silicon controlled rectifiers, which are used as current-controlling devices. Zener diodes or avalanche diodes have the characteristic of virtually short-circuiting when a high enough reverse bias is applied. This is especially useful for safety purposes, because if the voltage is dangerously high, the zener diodes will short and draw all the current to ground. Other types of diodes include light-emitting diodes (LEDs), which are useful for visual displays, and limiters, which have a limit on the maximum output voltage. Diodes are usually rated in terms of maximum voltage and maximum power. An example of a surface mount MELF diode is given in Figure 2.20.

**Figure 2.20**  Cross-sectional view of MELF diode.
(Courtesy of Philips Components.)

## 2.5 TRANSISTORS

The transistor is the device that replaced vacuum tubes as switches and amplifiers. Fundamentally, the transistor permits electronic control (amplification) of current flow. There are two types of transistors: bipolar junction transistors (BJTs) and field effect transistors (FETs). The BJT has wide applications as an amplifier and a switch. The FET is a three-terminal solid-state device that, like a junction transistor, can function as an amplifier or switch. Unlike the BJT, which is current controlled, the FET is a voltage-controlled device. This gives the FET three advantages over the BJT. First, the FET requires only voltage as a control element and virtually zero current; thus, it possesses a high input impedance. Second, the FET produces much higher current gains. Third, the FET is simpler and smaller than the BJT. Bipolar transistors switch faster than field effect transistors but require more power to operate. Transistors are usually rated in terms of rise time, power dissipation, maximum allowable voltage, and propagation delay. Figures 2.21 and 2.22 give examples of packaged, surface-mountable transistors. Figure 2.22 shows how the transistor chip is mounted and bonded inside the package.

## 2.6 INTEGRATED CIRCUITS

Integrated circuits can be broadly classified into analog (or linear) and digital circuits. In digital circuits the active elements are used primarily as switches and are operated in either a turned-on (1) or turned-off (0) mode. With analog integrated circuits there is a continuous and often linear relationship between input and output. The basic difference between analog and digital circuits is in the ways the information is coded into the input and output signals. In analog the information resides in the instantaneous magnitudes of the signal. The analog circuit reshapes the signal in a characteristic manner. The information in a digital signal resides not in its shape but in its levels at discrete intervals of time. A specific period of time must elapse before this information can be retrieved. The signal

**Figure 2.21**   Surface-mountable transistor. (Courtesy of Philips Components.)

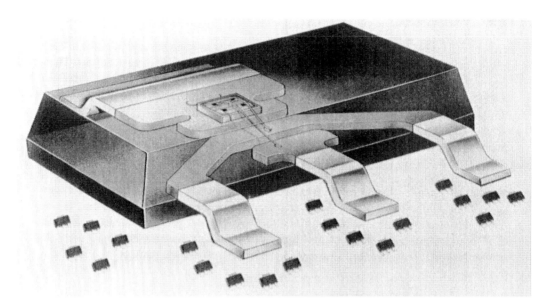

**Figure 2.22**  Surface mount transistor showing internal die attach and wire bonding. (Courtesy of Philips Components.)

levels represent sequences of zeroes and ones decoded into a stream of numbers. The digital circuit acts mathematically or logically on this stream. The output is also encoded in a stream of numbers.

Because the signal period of time must elapse before any information can be retrieved or processed, digital components are not capable of the radio frequencies (RFs) or microwave frequencies associated with some analog devices. Very high speed integrated circuits (VHSICs), the fastest digital circuits, run around 100–300 megahertz. RF devices, by definition, run at over 1 gigahertz.

### 2.6.1  Digital Integrated Circuits

Digital integrated circuits consist of transistors, capacitors, and resistors. Devices such as memory, logic, microprocessors, multiplexers, and operational amplifiers are all integrated circuits. The components are made on the surface of a semiconducting material using photolithography. The most commonly used semiconducting material is silicon. Geranium and indium phosphide are now seldom used in integrated circuits. Gallium arsenide, once used only in high-frequency communication devices and LEDs, is gaining acceptance throughout the semiconductor industry because of its low dielectric constant and potential for high-density integration, which will result in increased speed.

Digital integrated circuits are categorized by the number of devices on a chip. They are subdivided as follows: 1 ≤ small scale integration (SSI) ≤ 100 devices/chip, 100 ≤ medium scale integration (MSI) ≤ 1000 devices/chip, 1000 ≤ large scale integration (LSI) ≤ 10,000 devices/chip, very large scale integration (VSLI) ≥ 10,000 devices/chip. Table 2.2 shows some of the changes in the characteristics of integrated circuit components over the years.

**Table 2.2**   Typical Extremes in IC Technology Trends

|                              | 1980 | 1988 | 1996 |
|------------------------------|------|------|------|
| Relative density             | 1    | 200  | 1250 |
| Max. chip power (W)          | 2    | 10   | 40   |
| Chip power density (W/cm$^2$)| 8    | 17   | 25   |
| Chip area (cm$^2$)           | 0.25 | 1.0  | 1.6  |
| Pins/chip                    | 64   | 200  | 400  |
| System clock (MHz)           | 5    | 25   | 125  |

In a large-scale digital system, such as a computer or a data processor, there are a few basic operations which must be performed repeatedly. The five circuits most commonly employed in such systems: OR, NOR, NOT, AND, and FLIP-FLOP. The first four are logic circuits and the FLIP-FLOP is a memory circuit. These circuits can be used to implement extremely complex logic and mathematical expressions, such as random access memory components (RAMs), read-only memory components (ROMs), programmable read-only memory components (PROMs), central processing units (CPUs), resistors, buffers, arithmetic logical units (ALUs), and entire microprocessors.

Integrated circuits are used for RAMs, ROMs, PROMs, CPUs, resistors, buffers, ALUs, and entire microprocessors. The major requirements and desired features of integrated circuits include the following:

1. Logic flexibility
2. Speed
3. Availability of complex functions
4. High noise immunity
5. Wide operating temperature range
6. Low power dissipation
7. Low noise generation
8. Low cost

Logic flexibility describes the capability and the variety of uses that can be obtained from a logic family. Logic flexibility can be compared between families on the basis of wired logic capabilities, complement outputs, line driving capability, indicator driving, I/O interfacing, driving other logic forms, and multiple gates. Wired logic refers to the capability to tie outputs of gates or functions together to perform additional logic without extra hardware or components. Complement outputs are useful when both a variable and its complement are required in a logic system, so that the need for an inverter is avoided. Line drawing refers to the capability to drive nonstandard loads such as lamps and indicator tubes. Finally, in the design of logic systems the gate count is minimized if all gates are available in the family, though this is not always the case. Sometimes more gates may be included to facilitate design implementation.

In applications where high packing density and large flexibility combined with high volume are needed, gate array carriers are often used whereby prefabricated gates are connected according to the needs of the customer. Gate arrays are a class of integrated circuits which provide integration of standard logic circuits. Gate arrays are used when more complex integrated circuits are not available or a customized circuit is too expensive.

The gate arrays offer high packaging density, no system redesign, high reliability, and short turnaround time.

Speed refers to the process time in terms of the number of computations per unit time. The speed may be increased by increasing the logic and information access rates. The speed of a logic circuit is indicated by its propagation delay and pair delay. The propagation delay is the time for a signal to be transferred from the input to the output pin of the device. Pair delay is the average of the propagation times for a positive and a negative output transmission, which can be unequal.

As the sophistication of design has increased, complex functions have become more common. A complex function is a grouping of basic gates which perform a calculation or map information. Gate-to-pin ratios which increase with complexity give the benefit of decreasing assembly costs per gate and increasing reliability per gate.

In order to prevent the occurrence of false logic signals, high noise immunity is desired. Switching transients, excessive coupling between signal leads, external sources as relays, circuit breakers, and power line transients may cause these erroneous signals. The higher the noise immunity of a circuit, the fewer precautions are required to prevent such an occurrence. Voltage noise immunity is specified in millivolts.

A wide operating temperature range is desired in most electronic equipment. The military has a required operating temperature range from $-55$ to $125°C$. Commercial products typically have temperature limits which range from 0 to $100°C$. Degradation due to temperature extremes may still occur even when circuits are within specifications. Increased temperature ranges are generally associated with increased IC cost.

Low power dissipation is desired in large systems because it increases reliability, lowers cooling costs and cooling distribution costs, and thus reduces mechanical design problems. As chips increase in complexity, the power dissipation has increased on a per gate basis. The complexity is then limited by heat dissipation requirements arising from system design and allowable semiconductor temperature.

Logic family cost evaluations are made by comparing the pack count for a given system speed and the layout and shielding costs for systems with two or more different families. An effort is then made to optimize the cost against design restrictions.

## 2.7 DIGITAL TECHNOLOGY FAMILIES

Most semiconductor digital circuits are centered around a few major technologies. Here, the dominant technologies, TTL, ECL, MOS, and CMOS, will be discussed. Table 2.3 presents a comparison of these families.

The transistor-transistor logic (TTL) family is composed of a combination of transistors which form a logic circuit. The basic circuit for this family is a NAND gate. It consists of a multiemitter transistor AND gate followed by an inverted transistor which results in the NOT-AND, or NAND.

Advantages of TTL include its ability to provide a large number of devices and perform complex functions. With wired logic, TTL offers an extra level of logic without an increase in power dissipation. Low output impedance results in a superior drive capability. TTL has a good immunity to externally generated noise and also has better speed than other forms of logic.

Disadvantages of TTL include the high rate of change of voltage and current with time, which require more attention in layout and mechanical design. TTL devices are

**Table 2.3**  Comparison of the Major Digital Logic Families

| Parameter/logic | TTL | ECL | MOS | CMOS |
|---|---|---|---|---|
| Basic gates | NAND | OR-NOR | NAND | NOR-NAND |
| Propagation delay per gate (ns) | 12–6 | 4–1 | 300–200 | 70–40 |
| Power dissipated per gate (mW) | 10–22 | 40–55 | 0.2–10 | 0.01 mW at static; 1mW at 10 MHz |
| Clock rate (MHz) | >0 to11 | 60–400 | 2–10 | 5–20 |
| Fan-out | 10 | 20 | 20 | >50 |
| immunity to external noise | Very good | Good | Nominal | Very good |
| No. of functions; family growth rate | Very high | High; growing | Moderate; growing | Moderate: growing |

also known to generate glitches during switching and thus require additional capacitors for bypassing. Finally, the implied AND function is not available by tying outputs together.

The emitter-coupled logic (ECL) family is sometimes referred to as current mode logic. This is different from standard saturating logic in that the circuit operation is analogous to that of some linear devices. The basic ECL circuit is the OR/NOR gate. The output is thus not high if one or more of the inputs is high.

ECL gate design permits operation over a wide power supply voltage range. Complementary output of ECL circuits adds greatly to the flexibility of design and usually eliminates requirements for extra gates used simply as inverters. ECL exhibits the industry's best in speed and is used for high-speed logic systems. It has the shortest propagation delay of any logic system. It is also possible to form a transistor tree building up from a current source. This permits even better speed since the same current source may be used three times per logic decision. Other advantages of ECL include high fan-out capability, constant current supply against frequency, low noise generation, low cross talk between signal leads, the ability to tie outputs together giving implied OR functions, and minimum degradation of parameters with temperature. Disadvantages include high power dissipation, lower external noise immunity than some saturated logic, and the need for transistors to interface with saturated logic.

ECL is used for instrumentation, high-speed counters, military systems, aerospace and communications satellite systems, ground support systems, digital communication systems, data transmission frequency synthesizers, phase array radars, and high-speed memories.

The p-channel MOS (PMOS) family acts like a variable resistor modulated in value by the gate voltage. The basic four-input P-MOS gate performs the NAND logic function. The input impedance of P-MOS is essentially capacitive and output is basically resistive.

P-MOS design results in small LSI complexity devices since it permits high-density circuitry. P-MOS devices have tended to be unique to a given application rather than of wide general use, and thus production of a given device is low and unit costs are high. P-MOS is used in small computers, high-density and solid-state memories, calculators, multiplexers, frequency dividers, and communications.

Advantages of P-MOS include fast turnover times because of simple processing, high circuit density, more logic per given area than other semiconductor techniques, and high input impedance, allowing high fan-out. Being bidirectional, P-MOS gives designers more flexibility.

The greatest disadvantage is that P-MOS needs more care than a bipolar device because it is surface sensitive. It has high output impedance, which limits its driving capabilities. Furthermore, the operating speed is limited to a maximum of a few megahertz.

A complementary metal oxide semiconductor (C-MOS) uses *p*-channel and *n*-channel devices on the same substrate. The *n*-channel and *p*-channel devices are connected in series. Only one device is turned on at a time, resulting in extremely low power dissipation. Dissipation is largely a result of switching of devices through the active region and charging and discharging capacitances.

Advantages of C-MOS include low power dissipation, good noise immunity, wide range of power supply variations, high fan-out, large logic swings, and shorter propagation delay. Disadvantages are a relatively larger chip size than P-MOS and high input impedance, which restricts interfacing capabilities. Moreover, more complex processing is required for C-MOS.

C-MOS is used in battery-operated equipment, noisy environments, aerospace logic systems, portable digital communications, and many commerical electronic devices.

## 2.8  MANUFACTURING OF SEMICONDUCTOR DEVICES

The electronic revolution has now permitted thousands of transistors and other electronic components to be mass produced in a chip a quarter of an inch square at low cost. Interconnections are made using microlines etched on the chip. These chips are called integrated circuits. Such integrated circuits are combined to create systems by mounting them on a printed circuit board on which the interconnecting wiring has been etched. Thus, in a sense, a PCB is a macro version of an IC (or vice versa). This chapter describes the basic steps employed in the manufacture and assembly of ICs and printed circuit boards, an understanding of which is very important to the PCB designer.

Processing techniques for ICs are concerned with modifying properties on the surface of the semiconductor material, which is usually silicon, gallium arsenide, or indium phosphide. The process of fabrication is called a planar process because the devices are fabricated on one surface of the material. The substrate is a single crystal slice, called a wafer, cut from a large crystal cylinder. Since the electronic properties of semiconductors are sensitive to impurities in amounts undetected by normal chemical tests, the first steps are devoted to obtaining very pure material for crystal growth. For the following discussion, the material is assumed to be silicon, the most commonly used semiconductor material. The stages in the process are as follows:

1. Production of the silicon wafer
2. Epitaxial growth
3. Oxidation
4. Patterning the surface
5. Impurity introduction and redistribution (doping)
6. Contact deposition
7. Chip testing
8. Packaging
9. Final testing

## 2.8.1  Silicon Wafer Production

The starting point for semiconductor devices is the silicon wafer. Silicon is one of the basic materials used in the manufacture of the semiconductor devices. It is one of the most abundant materials on the surface of the earth but has a great affinity toward oxygen and other compounds. It therefore needs to be separated before it can be used in the fabrication process.

There are three basic steps in the production of pure silicon for wafer manufacture:

· Production of polycrystalline silicon
· Crystal growth
· Manufacture of wafer

## Production of Polycrystalline Silicon

Raw materials containing silicon in the form of quartzite are made to react with carbon at a high temperature in a furnace. In this furnace the raw materials are loaded into a crucible containing an electrode. The electrode produces an arc, melting the materials and eventually submerging them in the melt. As the melt sinks toward the bottom of the crucible, various reactions take place. The gases produced as a result of the reaction process escape and the liquid silicon sinks to form a layer at the bottom of the crucible. The metallurgical silicon (Si) produced by the reaction is about 98% pure. The reaction results in the formation of silicon and carbon monoxide (CO). The reaction is

$$SiO_2 + 2C \xrightarrow{\text{heat}} Si + 2CO$$

This reaction produces silicon along with a gaseous by-product. This metallurgical grade silicon is further purified by converting it into trichlorosilane ($SiHCl_3$). The metallurgical silicon is pulverized and reacted with hydrochloric acid (HCl) in a fluidized bed at 1260°C, forming a liquid chloride of silicon ($SiHCl_3$). The reaction is

$$Si + 3HCl \xrightarrow{1260°C} SiHCl_3 + H_2$$

Trichlorosilane ($SiHCl_3$) is a liquid at room temperature and is purified to required standards by fractionation. Once trichlorosilane is purified, it is reduced in hydrogen at high temperature to form polycrystalline silicon. The reduction reaction that takes place is

$$SiHCl_3 + H_2 \xrightarrow{\text{heat}} Si + 3HCl$$

These three steps take several hours. The resulting silicon in the form of polycrystalline rods is 100% pure.

## Crystal Growth

One of the techniques of crystal growth is called zone refining. In this technique impure silicon from the reduction process in the form of a rod is heated to produce a small molten zone. Impurities in the zone tend to float to the zone boundaries. If the molten region is slowly moved from one end of the rod to the other the impurities collect at the ends and can be removed (Figure 2.23). Usually several passes are required to achieve a desired degree of purity.

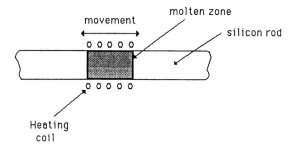

**Figure 2.23**   Impurity removal from silicon rod.


Another technique for crystal growth is the Czochraslki technique (or the CZ process). This process begins with a large quartz crucible of high purity into which electronic grade silicon rods are broken and melted. A previously produced piece of single-crystal silicon is used as a seed crystal, which is touched on the surface of the melt and then rotated and withdrawn. The material in the melt makes a transition to the solid-phase crystal at the solid-liquid interface, so the newly created material accurately replicates the crystal structure of the seed crystal. The pull rate of the seed crystal varies during the growth cycle. During the entire growth cycle the crucible is rotated in one direction at 12–14 rpm while the seed holder is rotated at 6–8 rpm in the opposite direction. This constant stirring prevents the production of local hot or cold regions. Rotation speed of the rod and the crucible, pull rate, and temperature can be accurately controlled to obtain precise control over the rod diameter. As the crystal is grown the melt level in the crucible drops relative to the hot zone area of the furnace. To maintain the crystal in the hot zone area of the furnace, the crucible is continuously elevated during growth.

Eventually most of the melt is condensed and a very large bologna-shaped rod of single-crystal silicon up to 6 in. in diameter and several feet long has been formed. Dopant is introduced into the crystal during the growth process by adding small amounts of heavily doped silicon in the melt. The dopant creates silicon which has either a surplus of free electrons (*n* type) or a shortage of free electrons or holes (*p* type).

The electrical properties of silicon crystals depend on their chemical purity and crystal quality. An important source of impurities is the crystal grower. The crucible is made of quartz and introduces some amount of oxygen into the silicon. This impurity can affect both electrical properties and distribution, and results can be beneficial or adverse. Carbon is another common impurity transmitted from graphite components, such as the susceptor used to support the crucible. Vacuum or argon atmospheres are used during the crystal-growing process to prevent oxygen contamination from the atmosphere.

## Manufacture of the Wafer

The wafer is now manufactured from the grown crystal. First the crystal is ground perfectly round and its rotational orientation is determined. One or more flats are ground along the crystal prior to sawing into wafers. The flats are used as a reference during all the subsequent processing steps. After grinding, the crystal is chemically etched to remove any surface damage. The silicon crystal is then sliced into thin slices called wafers (Figure 2.24).

A silicon etchant is then used to remove saw marks produced as a result of the slicing process, on both sides of the wafer. Both sides of the slice are then lapped to achieve

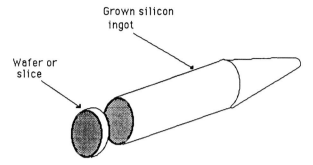

**Figure 2.24**  Wafer slicing.

the required flatness and to provide a uniform finish. After lapping, the wafer is processed through an edge-rounding machine which removes the square corners. These corners are loaded with microcracks which are potential sites for failure. Any damage resulting from lapping is removed by chemical etching. After the wafer has been inspected it is polished using a polishing solution which simultaneously etches and mechanically polishes the wafers.

## 2.8.2  Epitaxial Growth

The wafer forms the substrate for the integrated circuit. In some cases the circuits are formed directly on the wafer. In other cases a thin silicon layer is grown on the polished surface of the wafer to provide a near-perfect crystal structure for the active region of the devices and also to enable the conductivity of the region to be accurately controlled. A flawless structure is essential to prevent the electrical properties from being masked or degraded by impurities in the crystal. The process of deposition of a single crystal layer on the substrate, such that the crystal structure of the layer is an extension of the crystal structure of the substrate, is called epitaxial growth. The deposition sequence in the epitaxial process is as follows:

1. Substrate cleaning
2. Wafer load
3. Heat-up
4. HCl etch
5. Deposition

The substrate is degreased and cleaned using etches (mild acids) and/or deionized wafer (DI water). The usual sequence is sulfuric acid followed by nitric acid, hydrochloric acid, and hydrofluoric acid. The wafer is then dried. The cleaning operation is of great importance, since any residual particles may give rise to imperfections in the deposited layer.

After the cleaning operation, great care is exercised to ensure that the front side of the wafer is not touched. The wafer is then placed on a wafer holder, with care taken to ensure that no particles are transferred from the holder to the substrate. Since the dimensions of pattern features are just below 1 $\mu$m, particles of similar size can be a cause of contamination or even failure. If a 2-$\mu$m particle is located in a wafer, it can prevent a continuous line from being deposited, thus causing an electrical open. For this reason, all wafer fabrication is done within "clean rooms." These clean rooms have different classes: 5, 10, 100, 1000 or 1K, 10K, and 100K. The classes indicate the number of particles of size equal to or greater than 2 μm within 1 cubic foot of air.

The system is then sealed and purged with hot nitrogen to remove any residual gas which might be present. This is continued until a temperature of approximately 500°C is reached, after which the nitrogen is replaced by hydrogen. This replacement is made because of the tendency of nitrogen to etch silicon.

Once the heat-up cycle has been completed, an HCl etch is used to remove a thin layer of damaged silicon from the surface. The process is carried out with great care to guarantee that the properties of the devices being fabricated are not adversely affected.

A silicon epitaxial layer can be grown using liquid-phase epitaxy or gas-phase epitaxy. In gas-phase epitaxy, the epitaxial layer is grown on the wafer surface by passing silicon tetrachloride gas and hydrogen over the wafers heated to about 100°C in a silicon glass tube. The deposited atoms take up a regular array established by the atoms at the polished wafer surface. The grown layer can be made $p$ or $n$ type by introducing a dopant gas such as phosphine as a source of phosphorus dopant or diborane as a source of boron dopant. This process of deposition results in an epitaxial layer with a desired thickness and resistivity. The dopant concentration is controlled by flow rate of the dopant gas. Gas-phase epitaxy is most widely used. Crystalline epitaxial layers can also be grown by liquid-phase epitaxy. This process uses the principle that mixtures of a semiconductor and another element melt at a lower temperature than the semiconductor on its own. Thus a semiconductor crystal can grow in a melt of a mixture containing the semiconductor.

### 2.8.3 Oxidation of the Silicon Wafer

Since a silicon wafer is a chemically reactive substance, a chemically stable layer of silicon dioxide is grown on the surface to act as an insulating layer in addition to providing a mask for the fabrication sequence. This silicon dioxide layer is grown in an atmosphere containing either oxygen or water vapor at a temperature of 900 to 1300°C. Oxidation takes place when the oxygen or the water vapor reacts with silicon. The chemical reactions are

$$Si + O_2 \longrightarrow SiO_2$$
$$Si + 2H_2O \longrightarrow SiO_2 + 2H_2$$

### 2.8.4 Development of Surface Patterns

The geometric pattern that is required to produce the required electrical behavior is transferred to the surface of the wafer using a photolithographic process, as shown in Figure 2.25. A working mask is used to remove areas of the silicon dioxide surface layer on the wafer and allow dopant to enter the epilayer. The silicon dioxide layer is coated with a thin layer of photosensitive emulsion known as photoresist, which is subsequently baked. The photoresist may be positive or negative.

1. If the photoresist is positive, the light from the ultraviolet source increases the solubility of the resist in the developing solution.
2. If the photoresist is negative, the exposure to ultraviolet light causes polymerization of the exposed area of the resist.

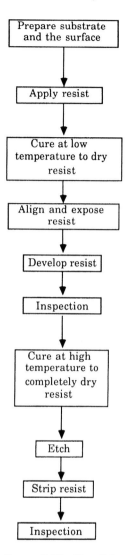

**Figure 2.25**  Photolithography processing.

The emulsion is exposed to a specific wavelength of ultraviolet light through a photographic mask. If the photoresist is negative, the windows in the mask allow the ultraviolet light to polymerize the exposed area of the emulsion, rendering it insoluble in trichloroethylene, a substance in which the wafer is washed after exposure to UV light. Regions in the photoresist which remain unexposed are unpolymerized. These regions are dissolved in the washing/developing process, resulting in the production of windows in the resist layer. The wafer is then etched in a solution such as hydrogen fluoride. The windows in the photoresist allow the etching solution to remove a region in the layer of silicon dioxide, forming a window through which the dopant can be diffused or implanted in the epilayer below. The polymerized photoemulsion is then removed by washing in hot sulfuric acid.

The windows etched through the silicon dioxide layer yield positions where dopant is to be introduced into the epilayer, thus forming the working mask for the doping process.

Positive photoresist follows the same basic steps: deposition of photoresist on wafer, exposure to UV light, development, etching removal of photoresist. However, in the case of positive photoresist, the exposed photoresist is softened by the UV light. It can then be removed with a developer, leaving windows down to the silicon dioxide layer. Windows in the silicon dioxide layer can be etched away to allow the dopant to be diffused or implanted in the epilayer. Once the doping process is complete, the remaining positive photoresist is exposed to the correct frequency of UV light. This softens the photoresist, allowing it to be washed away by the developer.

### 2.8.5   Impurity Introduction and Redistribution

Impurities (i.e., dopants) are introduced into specific regions of the semiconductor wafer to become the transistors, diodes, and other devices that form integrated circuits. Diffusion is used to introduce a controlled amount of impurities into selected regions of the semiconductor crystal. This can be accomplished by two distinct processes—thermal predeposition and ion implantation.

*Thermal predeposition.* A controlled amount of a desired impurity is introduced into the semiconductor at a high temperature. Wafers are first cleaned to remove any contamination from the previous steps. During the process the semiconductor substrates are heated to a particular temperature and an excess amount of dopant is introduced at the surface of the wafer. The dopants enter the material until a maximum concentration called the solid solubility is reached.

*Ion implantation.* This process selects ions of the desired dopant, accelerates them using an electric field, and scans them across the wafer to obtain uniform predeposition. The energy imparted to the dopant ion determines the ion implantation depth. The actual process uses a hot filament to generate ions from a gaseous source. The correct ions are separated from any others by bending them through a preset angle using an electromagnetic field. The selected ions are then accelerated using an electric field and strike a target wafer, penetrating the crystal lattice. The regions implanted with the accelerated ions are selected by using a patterned layer of material such as silicon dioxide or photoresist as a mask. After implantation and the subsequent cleaning steps, the implanted wafers are put through a high-temperature furnace to activate any ions that may not have come to rest in electrically active regions in the crystal structure. Figure 2.26 shows the setup for the process.

The advantages of ion implantation include negligible disturbance of the previously doped regions (as ion implantation is a low-temperature process), negligible lateral effects due to the directional nature of the beam, use of dopants which cannot be diffused, and the possibility of a dopant density greater than its solid solubility. One disadvantage is that crystal damage due to bombardment and channeling in certain directions through the crystal may cause ions to penetrate deeper in some directions than in others.

### 2.8.6   Drive-In

Once the impurity has been introduced, it is redistributed to obtain the final profile in a process called "drive-in." In this step no additional dopant is introduced into the semiconductor. The process is carried out in an oxidizing atmosphere to regrow a protective

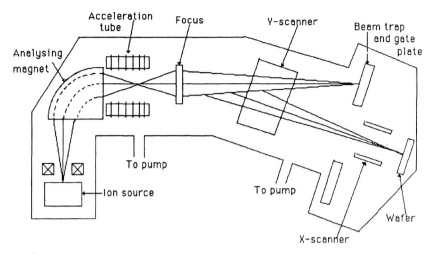

**Figure 2.26** Ion implantation.

layer of silicon dioxide on the surface of the freshly diffused region. During drive-in, the variables of time, temperature, and ambient gases are controlled to obtain desired values of

- The final junction depth of the diffusion
- The final oxide thickness of the newly doped regions
- The exact profile of the dopant in the semiconductor

## 2.8.7  Contact Deposition

The electrical contact of the doped regions is made by forming ohmic metal-semiconductor contacts on the surface of the chip. Windows are opened in the silicon dioxide layer to expose the semiconductor where contact is to be made, using the oxidation, photomasking, and etching processes. The surface of the entire wafer is then coated with a layer of aluminum or gold by vapor deposition or sputtering. This process is called the metallization. The unwanted metal is then removed using a further photomasking and etching stage, leaving the device contact pads.

## 2.8.8  Chip Testing

After formation of the electrical contacts (i.e., the aluminum bonding/probing pads), the individual chips on the wafer are tested for correct operation. The tests check the basic operation and performance parameters and are performed automatically with a probe which makes contact with the contact pads of the device. This test operates on a step-and-repeat sequence over the entire wafer. The test results are recorded with the type and geometric pattern of the failures, enabling rapid identification of failure sites in defective chips. The defective chips are typically marked with a dot of ink by the tester/prober. After testing is completed, the bottom of the wafer is lapped or polished off to the desired final thickness. If the back plane is active or the chips are to be solder or eutectically bonded to the assembly during the packaging process, the back side of the wafer will be metallized.

This back-side metallization is typically a diffused gold. The wafer is then diamond sawed into individual chips or dice.

## 2.9  PACKAGING STYLES

The transistor, one of the first semiconductor devices, was developed to replace the bulky vacuum tube. Semiconductors were initially packaged in cans, called to-cans, which are round capped packages with leads extending from the base. To-cans became difficult to work with as the number of I/O pins required by semiconductor devices began to rise. As a result, new forms of packages for semiconductor integrated circuits were developed. Some of the common package outlines are shown in Figure 2.27.

### 2.9.1  Dual In-Line Package

### Plastic Dual In-line Packages

One of the earliest packaging standards is the rectangular dual in-line package (DIP). The DIP has I/O leads which extend from two opposite sides of the package and are bent downward (Figure 2.28). DIPs have been a main stream of the microelectronics industry since 1968 and still account for some 80% of all packaging of integrated circuits. However, the application of DIPs has been reduced because of newly emerging packages such as chips carriers, small-outline integrated circuits (SOICs) and pin-grid arrays (PGAs).

DIPs were invented to overcome the difficulty associated with handling and inserting packages onto mounting boards, typically called printed wiring boards (PWBs). The leads of DIPS are inserted through holes in the PWB and soldered to the back of the board. With DIPS, it has become possible to service these boards in the field with standard hand tools and soldering irons.

Plastic DIPs consist of a lead frame typically made of kovar. The lead frame is punched out of sheet metal in the desired shape, which normally includes a small platform or pad at the center with the leads fanning out from this pad. The leads are formed so that they bend down for insertion onto the PWB. This lead frame is then plated with nickel as an underplate, or corrosion barrier, and tin or gold surface plating for wire bondability and/or solderability. The semiconductor die is then mounted to the center pad of the lead frame and the I/O pins of the die are wire bonded to the appropriate leads. The entire assembly is then molded in plastic, forming the plastic DIP body.

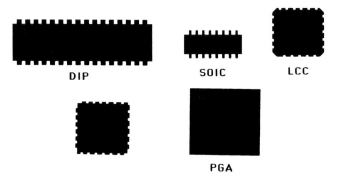

**Figure 2.27**  Common package outlines.

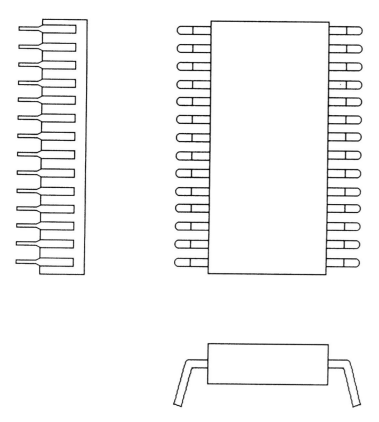

**Figure 2.28**  Dual in-line package.

Since 1972 most plastic packages have been made of Epoxy B or Novalac. Plastic packages are available with up to 68 leads. Above 68 leads, molding and lead frame problems limit the production of larger DIPs because of excessive costs. Up to 68 leads, the manufacture of chips in plastic packages is easily automated, which is one of the prime reasons for their popularity.

Environmental tests on a plastic-encased 4096-bit random access memory have been conducted by Texas Instruments. The test showed that most failures occurred in the pressure-cooker test, which is equivalent to putting the ICs in steam. Although the conditions are rarely found in actual operation, this test indicates that plastic DIPS are most vulnerable to moisture. Still, in the real world, plastic DIPs function reliably in most consumer, computer, and industrial applications where coupled high temperature and high humidity is not encountered.

The advantages of plastic DIPS are low cost and a predictable uniform factor. Furthermore, they have the fewest infant mortality failures of any packaging type. The cofired or multilayer alumina ceramic DIPs are the most reliable after screening and the most expensive insertion mount components.

Plastic-encapsulated ICs use an injection-molded epoxy or silicone encapsulant and are not suitable for high-temperature operation. Because most plastic encapsulants are susceptible to moisture ingress via permeation, they have not been recommended for use in high-temperature/moisture environments. Moisture provides a medium for electrolytic corrosion at wire bonds or conductor tracks through any gaps or holes in the glassivation layer.

## Ceramic Dual In-Line Packages

Plastic DIPs are typically used in commerical applications, whereas military systems typically utilize ceramic (alumina) DIP packages. In ceramic packages, the integrated circuit die is sandwiched between the two ceramic plate elements. The element on the bottom half of the sandwich holds the integrated circuit die. The ceramic section on the top of the sandwich protects the integrated circuit die from mechanical stress during sealing operations. Each of the ceramic elements is coated with low-melting-temperature glass for subsequent joining and sealing.

Ceramic packages are preferred for high-quality electronic systems, where increased reliability takes precedence over increased cost. Ceramic packages are typically made from either beryllia or alumina. Beryllia has better thermal characteristics than alumina; however, beryllia can be toxic if inhaled in a powdered form. For this reason, alumina is the most widely used ceramic material in packaging. Alumina is superior to epoxy in terms of heat transfer and hermeticity. Alumina packages are more expensive than epoxy because of higher material costs and higher manufacturing costs.

The leads of ceramic DIPs, like the leads on plastic DIPS typically have a pitch of 100 mils (1 mil equals 0.001 in.). The materials of the DIP lead frame are mainly kovar (high- nickel steel) or copper. The lead frame is stamped and bent into its final shape. The excessive material is intended to preserve pin alignment. The holes at each end are for the keying jig used in the final sealing operation. The lower half of the ceramic package is inserted into the lead frame. The die is mounted in the well and leads are attached. The top ceramic element is bonded to the bottom element, and the excess material is removed from the package.

The manufacture of multilayer ceramic DIPs, often called cer paks, is an elaborate process. The basic element is uncured (green) alumina tape. Screened onto the top of the tape are metalized patterns, which where necessary include vias (square holes filled with the screened-on metal) for connection to other layers. These metallized layers are then laminated together and fired in a furnace, creating a strong, monolithic structure. A lead frame is brazed to the top or side of the package, which can now be shipped to the user along with the metal lid. The user then mounts the desired die into the cavity of the cer pak and wire bonds the I/O pads of the die to the appropriate wire bond pads of the cer pak. These pads are connected through the multilayers of the package to the brazed-on leads. The first cer paks were design to be one-for-one replacements for DIPs. Thus, these first units were rectangular and had 100-mil centered leads extending from the two long sides of the package. However, the cofired process allowed for finer line resolution, so cer paks are now available in much smaller sizes and lead pitches. Development of the chip carrier soon followed that of the cer pak, so the use of cer paks never really caught on. For more information on the cofired ceramic process, see Chapter 3.

The cofired DIP can handle as many as 68 leads successfully and is at present the largest cofired DIP available. Since this unit is about 3.2 in. long by 0.9 in. wide, it is doubtful whether there will be too many packages made that are large. Other advantages of cofired ceramic DIPs are true hermeticity and superior heat dissipation. The ceramic DIPs are superior to plastic DIPs in lead strength, because the lead frames are brazed on.

Designing a new package typically requires 10 to 18 weeks and $5000 to $35,000 for tooling costs. Ceramic is fragile, making it harder to ship and handle automatically than plastic packages.

DIPs have been used by the military and in industry for a long time; therefore, problems associated with DIPS and the failure mechanisms of through hole leads are well known and most of the problems have been solved. DIPs, however, have the limitation that as the number of the leads is increased, the DIPS take too much space on a printed circuit board. In particular, the smallest mounting hole that can be drilled on a circuit board is 40 mils. As the lead count of the die increases the area for mounting the DIPs increases at an even greater rate.

With the introduction of VLSI, the lower available pin counts of a rectangular DIP have become a limiting factor. With pins spaced 100 mils apart on only two sides of the package, the size and weight of the DIPs have become too great. Consequently, the cost of drilling holes for DIP insertion has also become a problem. For example, a DIP with 64 pins is 0.9 by 3.2 in. with 100-mil pin spacing. On the other hand, the size of the circuit elements has been reduced, which in turn has led to reduced power consumption and chip delay time due to greater circuit integration, thus reducing the cost per circuit function. As a result, DIP packages have become up to 50 times as big as the chips themselves, which defeats the objective of shrinking the size of the integrated circuits. Thus DIPS are not an effective packaging method for VLSI.

## 2.9.2 Flatpack Packages

With the increase in IC pin count, other methods of packaging and mounting have appeared. The emergence of surface- or planar-mounted components has overcome some of the space problems associated with through hole mounting and the high-pin-count DIP packages.

One of the earliest forms of surface-mounted (as opposed to through hole-mounted) components is the flatpack. The flatpack is similar to the DIP with the exception that the leads are bent outward to form a flat surface and are mounted on pads rather than inserted into holes in the board. Flatpacks are found in both dual in-line configurations called small outline integrated circuits (SOIC) and with leads which protrude on all four sides of the component package. SOICs typically have gull wing-type leads and are preferred for lead counts up to 20 pins.

## 2.9.3 Chip Carrier Package

The chip carrier is a package (often square) with I/O connections on all four sides. There are three main types of chip carriers: ceramic and plastic leaded chip carriers and leadless ceramic chip carriers (LCCCs often just called LCCs). Figures 2.29 shows a typical leadless chip carrier geometry. Figure 2.30 shows a leaded ceramic chip carrier before and after lead forming. In this case the leads are top brazed and are formed in a gull-winged (stepped) fashion for surface mounting. The chip carriers have notches in three corners to ensure proper mounting orientation.

Chip carriers offer higher packaging densities than their corresponding DIP packages. For comparison, a 28-pin DIP is 3.725 cm long and 1.475 cm wide and has a 0.5-cm$^2$ chip-mounting area. A 28-lead chip carrier is 1.125 cm on each side with a slightly larger chip capacity (Figure 2.31).

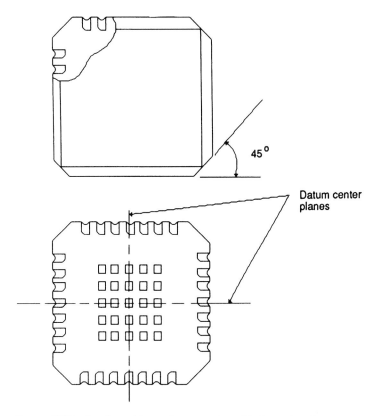

**Figure 2.29**   Leadless chip carrier geometry with thermal pads.

Chip carriers having 16 to 84 terminals meet the requirements of the Joint Electronic Device Engineering Council (JEDEC). The JEDEC standards include two families, 50 mils and 40 mils terminal pitch. Above 84 terminals there are as yet no generally accepted standards, although JEDEC has proposed certain terminal counts and configurations with up to 264 terminals and 25 and 20 mils pitch. Non-JEDEC chip carriers are available in single layer, multilayer, cavity up, cavity down, rectangular, and square configurations. Chip carriers are available in plastic, alumina, beryllia, and aluminum nitride, with thick-film, thin-film, and cofired metallizations. They are open tooled (meaning the tooling is open to the public for purchase) with lead counts over 300.

## Plastic Chip Carriers

The plastic leaded chip carriers were developed in the early 1980s by Texas Instruments as a low-cost alternative to the leadless chip carriers. Plastic chip carriers are fabricated in the same manner as plastic DIPs. The leads on a plastic chip carrier can have the same configuration as the DIPs for insertion mounting. They can also be formed in a gull-winged DIP configuration or a J-lead configuration for surface mounting. Gull-winged chip carriers have the same lead configuration as flatpacks. J leads start off similar to DIP leads, but are bent up just below the package in the shape of a J.

## Ceramic Chip Carriers

Single-layer leadless chip carriers, or SLAMs, have a conductor pad at the center of the chip carrier for mounting the die. Surrounding the die are wire bond pads. These pads are normally on 0.008-in. centers which match the bond pads of most dice. Conductor tracks fan out from these wire bond pads, connecting them to the I/O of the package. The dice are mounted and wire bonded prior to sealing the package by attaching a dome-shaped lid with a glass sealant.

Multilayered chip carriers (see Figure 2.32 for examples) are fabricated using cofired ceramic technology. They do not have leads, just conductors routed through the layers from the edges of the carrier to the wire bond pads. The die is secured in the package's central cavity. The most common LCCs are designed by JEDEC. JEDEC type A has the contacts on the top side. Type B has the contacts on the edges as well as the top and bottom. Type C is different from type B in the way the corners are shaped. Figures 2.33 through 2.38 give examples of the different JEDEC chip carriers.

All ceramic chip carriers can be configured to be leaded or leadless. For leaded chip carriers, leads are brazed onto the I/O pads along the perimeter of the package. Type A chip carriers would have the leads brazed to the top of the case, hence the term top-brazed chip carrier. These top-brazed leads can be formed into gull-winged packages for surface mounting (see Figure 2.30). They are formed in a step-down shape to the PWB. These steps offer enough stress relief to handle the mismatch in coefficient of thermal expansion (CTE) of the chip carrier and the PWB. Figure 2.39 shows a leaded chip carrier mounted to a PWB. Types B and C can have the leads brazed to the top of the carrier, to the bottom, or along the side. The side braze configuration is similar to DIPs and is used for insertion mounting.

**Figure 2.30(a)**   Leaded ceramic chip carrier.

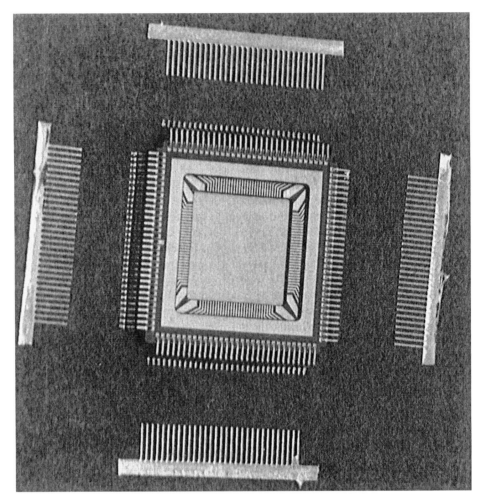

**Figure 2.30  Continued**  (b) leaded chip carrier with gull-wing formed leads for surface mount. (Courtesy of Westinghouse ESG.)

Leadless chip carriers have "castellations" instead of leads. A semicylindrical column is cut out of the side of the chip carrier at every I/O. The insides of these columns are metallized the same way that vias between the layers are metallized. They are then plated as is all external metallization of the package. The plating used is typically nickel followed by gold. To attach LCCs to the board, solder in paste form is applied, components are placed, and joining is accomplished by heating (e.g., vapor phase or infrared reflow), which melts and solidifies the solder. During the reflow process, the molten solder wicks up the metallized columns to form the solder joints, as shown in Figure 2.40. Prevention of solder joint cracking under extremes of temperature requires closely matching the CTEs) of the LCCs to the mounting boards. (See Chapter 4 for more details on the surface mount process and design considerations.) Leadless chip carriers can be reworked by reflowing

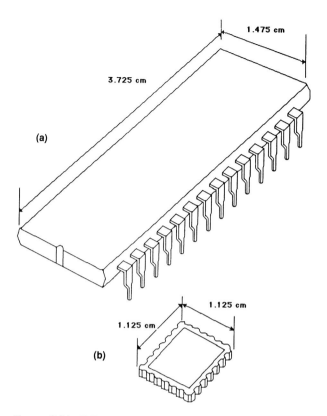

**Figure 2.31**  DIP versus LCC comparison.

the solder joints with the aid of an extractor, as shown in Figure 2.41. An extractor is nothing more than a soldering wand with an end effector in the shape of the chip carrier to be removed.

Inserting a type A leaded chip carrier face down in sockets puts the primary heat-dissipating surface away from the mounting surface, resulting in more effective cooling. Type B has notches for sockets, but the chip carriers can also be soldered directly to boards. Type C LCCs have beveled corners instead of notches, and they are meant to be soldered directly to boards.

LCCs with a typical pitch of 40 or 50 mils take up as little as 30% of the room required by DIPs that have the same lead count. Alumina LCCs typically have a much smaller coefficient of thermal expansion than the standard organic printed circuit boards they are soldered to. As a result, heat cycling in a system, which can cause repeated strain on the solder joint, will eventually cause fatigue failures. Therefore, it is recom-mended that only leaded chip carriers be mounted on organic PWBs. However, if a ceramic PWB is to be used, leadless ceramic chip carriers can be surface mounted with very high reliability.

Figure 2.42 shows the relationship between the wiring length of the package and the number of pins for various package types. TAB is the most efficient packaging technology with respect to wiring length. This is an important consideration, since the wire length determines the time for the signal to propagate from one chip to the next.

**Figure 2.32** Leadless chip carriers. (Courtesy of Westinghouse ESG.)

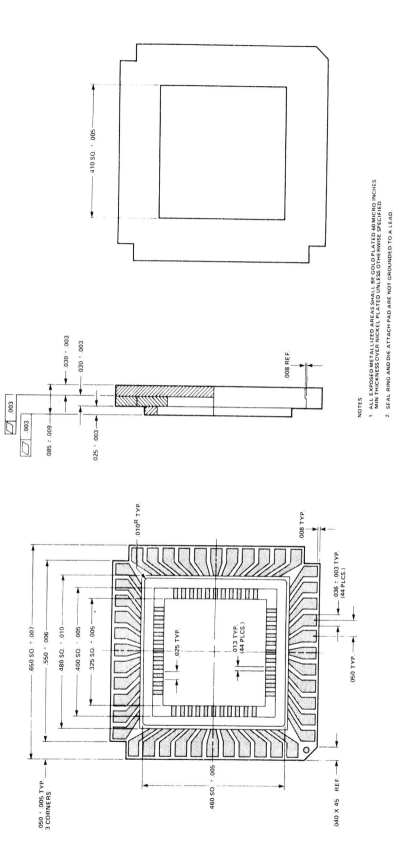

**ALL DIMENSIONS IN INCHES**

**Figure 2.33** JEDEC type A, LCC example. (Courtesy of Kyocera America.)

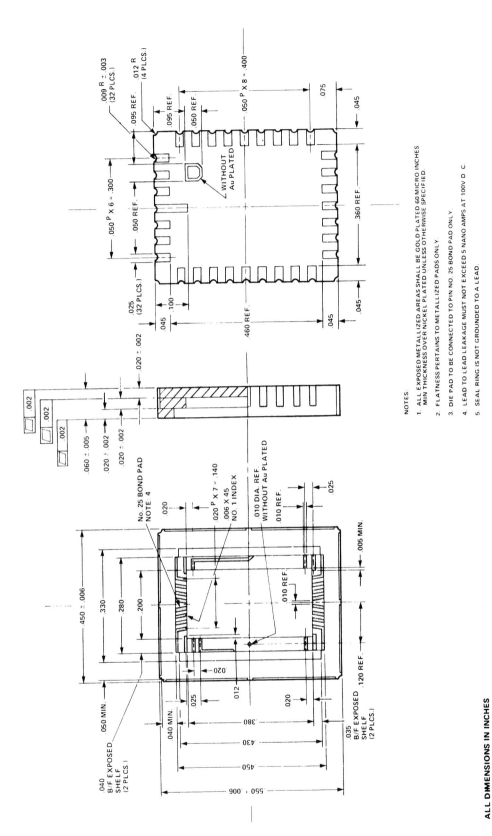

ALL DIMENSIONS IN INCHES

**Figure 2.34** JEDEC type B LCC example. (Courtesy of Kyocera America.)

NOTES:

1. ALL EXPOSED METALLIZED AREAS SHALL BE GOLD PLATED 60 MICRO INCHES MIN THICKNESS OVER NICKEL PLATED UNLESS OTHERWISE SPECIFIED.

2. FLATNESS PERTAINS TO METALLIZED PADS ONLY

3. DIE PAD TO BE CONNECTED TO PIN NO. 25 BOND PAD ONLY.

4. LEAD TO LEAD LEAKAGE MUST NOT EXCEED 5 NANO AMPS AT 100V D C.

5. SEAL RING IS NOT GROUNDED TO A LEAD.

86

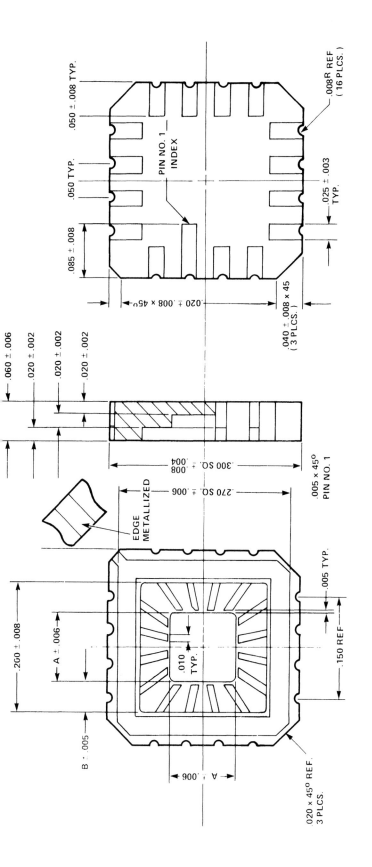

NOTES:

1. ALL EXPOSED METALLIZED AREAS SHALL BE GOLD PLATED 60 MICRO INCHES MIN THICKNESS OVER NICKEL PLATED UNLESS OTHERWISE SPECIFIED.

2. SEAL RING FLATNESS .003 IN./IN. N.L.T. ±.002 T.I.R.

3. SEAL RING AND DIE ATTACH PAD ARE NOT GROUNDED TO A LEAD.

|  | A | B |
|---|---|---|
| OPTION 1 | .100 | .050 |
| OPTION 2 | .120 | .040 |

ALL DIMENSIONS IN INCHES

**Figure 2.35** JEDEC type C, LCC example. (Courtesy of Kyocera America.)

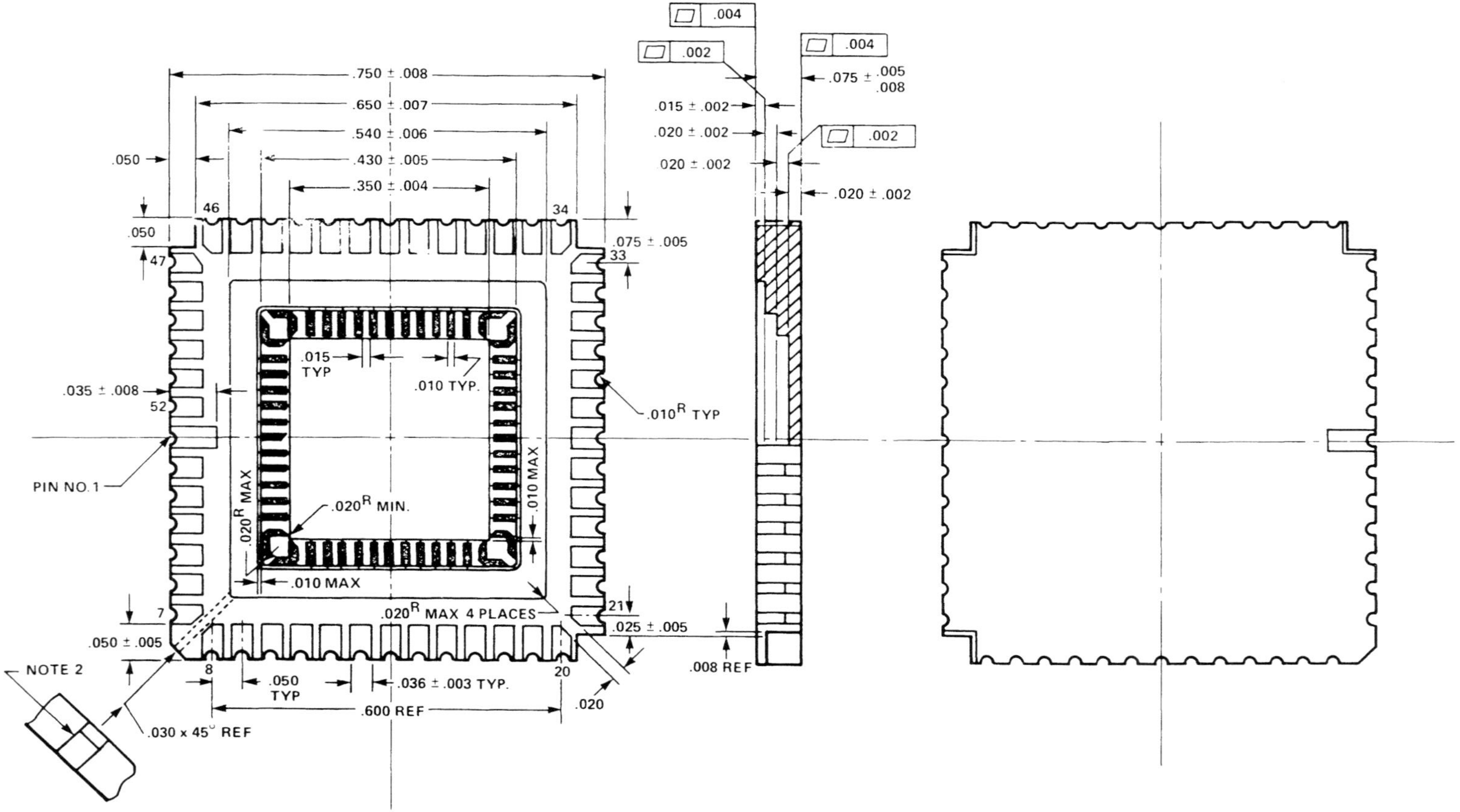

**Figure 2.36** JEDEC type D, LCC example. (Courtesy of Kyocera America.)

ALL DIMENSIONS IN INCHES

NOTES:

1. ALL EXPOSED METALLIZED AREAS SHALL BE GOLD PLATED 60 MICRO INCHES MIN THICKNESS OVER NICKEL PLATED UNLESS OTHERWISE SPECIFIED.

2. THIS METALLIZED AREA ELECTRICALLY CONNECTED TO SEAL AREA.

88

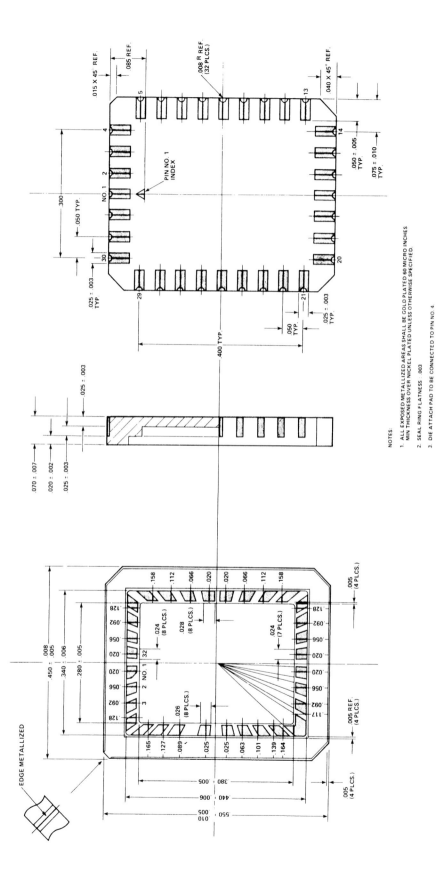

**ALL DIMENSIONS IN INCHES**

**Figure 2.37** JEDEC type E, LCC example. (Courtesy of Kyocera America.)

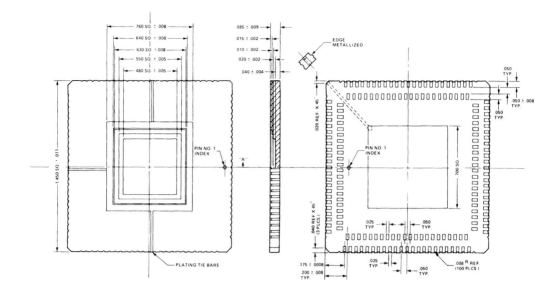

(a)

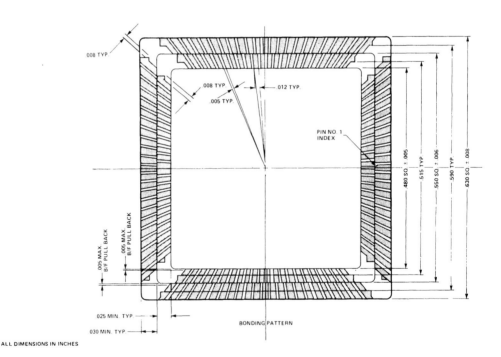

(b)

**Figure 2.38** Non-JEDEC, high I/O, LCC example. (Courtesy of Kyocera America.)

**Figure 2.39**   Leaded chip carrier surface mounted in PWB. (Courtesy of Westinghouse ESG.)

### 2.9.4   Pin-Grid Array Package

Another type of package is the pin-grid array, which is typically used with VLSI chips with lead counts greater than 100. The PGA is often composed of a square, multilayered, ceramic chip carrier with a matrix of butt-brazed, through hole pins on its under surface. Figure 2.43 shows the pin configuration for a PGA, and Figure 2.44 shows a 132-pin PGA with test points.

The pins on the PGA are typically spaced on 100-mil centers over the complete component area. Since the leads are in an array over the entire area, the obtainable lead counts far exceed those for components with peripheral leads only. PGAs are available in cavity-up and cavity-down versions. The cavity-up package has the chip-mounting cavity on the substrate opposite the leads. With this configuration, the entire lead side can be occupied with the pin grid, resulting in very high I/O density. The cavity-down version has the chip cavity face on the same side as the pin grid. This configuration is necessary for chips with high heat loads where heat exchangers or radiators are placed on the top of the package.

**Figure 2.40**   Leadless chip carrier mounted to a PWB. (Courtesy of Westinghouse ESG.)

PGAs typically have 100-mil center-to-center spacing of the array leads. This spacing allows for insertion mounting and wave soldering. Among the obvious disadvantages of the PGA is the extra cost associated with insertion mounting. PWB routing needed to handle the pin density can involve extra layering. Mounting of PGAs directly into the PWB can also increase the drilling cost of the PWB fabrication and can create process control difficulties with the mounting. All pins must be perpendicular to the PGA, which in turn must be parallel to the PWB. Both the PGA and PWB must have tight tolerances on their planarity or flatness. The leads on the PGA must also remain straight and not be bent or skewed. If any of these parameters is not kept within control, insertion mounting of the PGA can be a manufacturing nightmare.

To eliminate the process control problems of insertion mounting the PGA into the PWB, a socket can be used. Here a socket, or female match to the PGA, is mounted to the PWB. The holes in the socket are initially larger than the leads on the PGA. These larger hole openings guide the pins. These holes then taper down to hold the pins snugly in place. However, if a socket is employed, there is an additional level of interconnection, which decreases the gain in packaging density by increasing the volume. This additional interconnection layer also decrease the reliability by impeding the thermal path and increasing the sites for potential failures.

The cost of pin-grid arrays is higher than that of other packages because they are made of ceramic and are generally multilayered. Efforts are under way to economize by use of plastics or single-layer ceramics. The current developmental step for pin-grid arrays is to tighten the array spacing to a 50-mil grid, resulting in four times the lead density. Also, PGAs generally require more board interconnection layers because of their high I/O count.

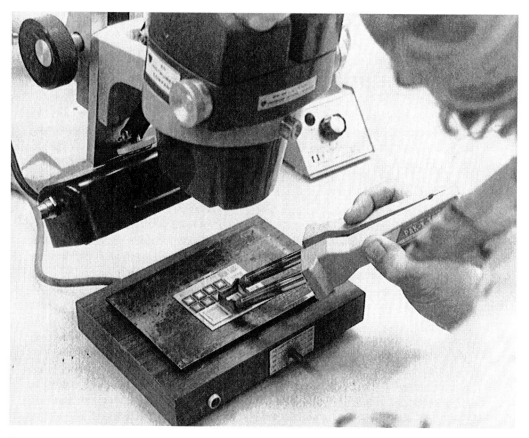

**Figure 2.41**  Rework of a LCC with an extractor. (Courtesy of Westinghouse ESG.)

### 2.9.5  Hybrid Package

Microelectronic hybird packages, referred to as hybrids, are often purchased as components. They are self-contained units that are insertion- or surface-mounted to a PWB or within a module. They are typically leaded, with the leads formed for either insertion mounting, as with DIPs, or in a gull-winged configuration similar to flatpacks.

A hybrid, by definition, contains multiple components within one device. Resistors and capacitors can be incorporated into the substrate or can be chip components surface mounted on conductor pads on the substrate. Bare dice are mounted to the substrate using eutectic, solder, or epoxy methods and the wire bonded to establish interconnection. The assembled substrate is then either encapsulated or mounted inside a package. This package typically has a flatpack configuration.

Hybrids, although purchased and mounted as components onto the PWBs, are actually assemblies themselves. They can vary in design and construction and involve a vast choice of processing techniques. The design and processing associated with hybrids will be discussed at great length in Chapter 4.

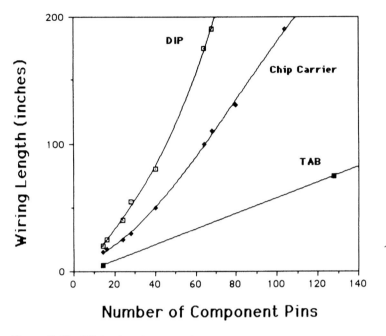

**Figure 2.42**   Wiring length versus pin count.

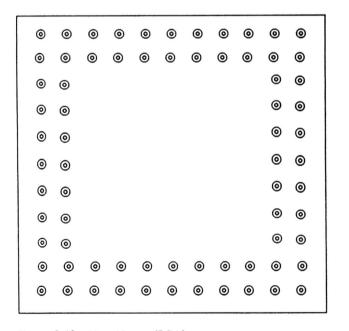 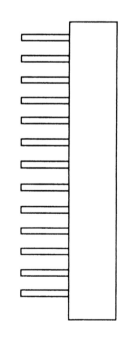

**Figure 2.43**   Pin-grid array (PGA).

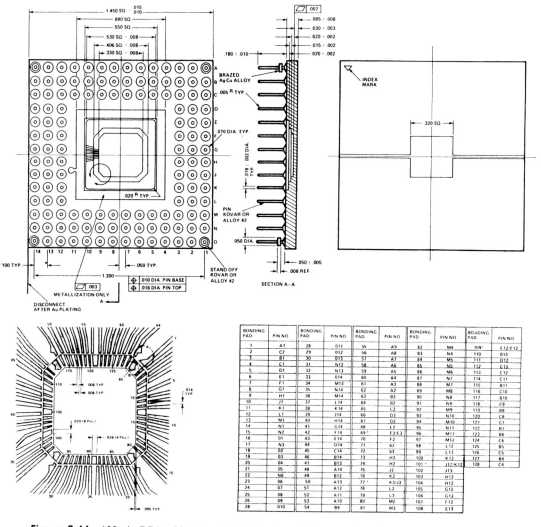

| BONDING PAD | PIN NO | BONDING PAD | PIN NO | BONDING PAD | PIN NO | BONDING PAD | PIN NO | BONDING PAD | PIN NO |
|---|---|---|---|---|---|---|---|---|---|
| 1 | A1 | 28 | O11 | 55 | A9 | 82 | M4 | 109 | E12/F12 |
| 2 | C2 | 29 | O12 | 56 | A8 | 83 | N4 | 110 | D13 |
| 3 | B1 | 30 | O13 | 57 | A7 | 84 | M5 | 111 | D12 |
| 4 | C1 | 31 | N12 | 58 | A6 | 85 | N5 | 112 | C13 |
| 5 | D1 | 32 | N13 | 59 | A5 | 86 | M6 | 113 | C12 |
| 6 | E1 | 33 | O14 | 60 | A4 | 87 | N7 | 114 | C11 |
| 7 | F1 | 34 | M13 | 61 | A3 | 88 | M7 | 115 | B11 |
| 8 | G1 | 35 | N14 | 62 | A2 | 89 | M8 | 116 | C10 |
| 9 | H1 | 36 | M14 | 63 | B3 | 90 | N8 | 117 | B10 |
| 10 | J1 | 37 | L14 | 64 | B2 | 91 | N9 | 118 | C9 |
| 11 | K1 | 38 | K14 | 65 | C3 | 92 | M9 | 119 | B8 |
| 12 | L1 | 39 | J14 | 66 | D3 | 93 | N10 | 120 | C8 |
| 13 | M1 | 40 | H14 | 67 | D2 | 94 | M10 | 121 | C7 |
| 14 | N1 | 41 | G14 | 68 | E2 | 95 | N11 | 122 | B7 |
| 15 | N2 | 42 | F14 | 69 | F3/E3 | 96 | M11 | 123 | B6 |
| 16 | O1 | 43 | E14 | 70 | F2 | 97 | M12 | 124 | C6 |
| 17 | N3 | 44 | D14 | 71 | G2 | 98 | L12 | 125 | B5 |
| 18 | O2 | 45 | C14 | 72 | G3 | 99 | L13 | 126 | C5 |
| 19 | O3 | 46 | B14 | 73 | H3 | 100 | K13 | 127 | B4 |
| 20 | O4 | 47 | B13 | 74 | H2 | 101 | J12/K12 | 128 | C4 |
| 21 | O5 | 48 | A14 | 75 | J2 | 102 | J13 | | |
| 22 | N6 | 49 | B12 | 76 | K2 | 103 | H13 | | |
| 23 | O6 | 50 | A13 | 77 | K3/J3 | 104 | H12 | | |
| 24 | O7 | 51 | A12 | 78 | L2 | 105 | G12 | | |
| 25 | O8 | 52 | A11 | 79 | L3 | 106 | G13 | | |
| 26 | O9 | 53 | A10 | 80 | M2 | 107 | F13 | | |
| 28 | O10 | 54 | B9 | 81 | M3 | 108 | E13 | | |

**Figure 2.44** 132-pin PGA with test points. (Courtesy of Kyocera America.)

## 2.10 LEAD STYLES AND PACKAGE-TO-BOARD INTERCONNECTIONS
### 2.10.1 Lead Configuration

The lead style is important in the development and selection of components. Figure 2.45 shows some commonly used lead styles. The DIP leads are inserted into a socket or board. DIP leads are rectangular in shape and tapered to a cross-sectional size of 18 by 11 mils. The leads are typically on 100-mil centers. Surface-mounted components typically have leads on 50-mil centers. Newer packages are available with leads as close as 20 mils from center to center (the Japanese are working with 8-mil pitch leads). This greatly increases packing density, which results in increased speed and performance.

With DIPs the leads are typically inserted into plated through holes and soldered to the back side of the PWB. Often a wave soldering process is used, in which liquid solder

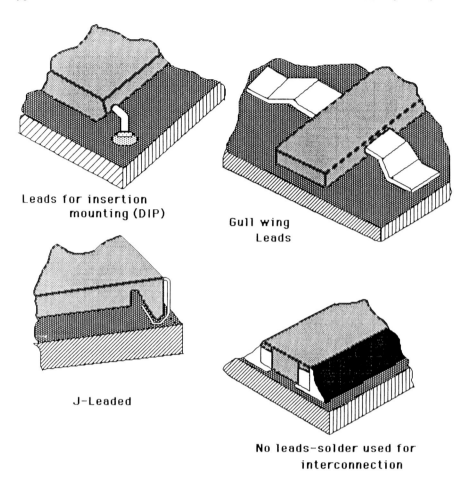

Leads for insertion
mounting (DIP)

Gull wing
Leads

J-Leaded

No leads-solder used for
interconnection

**Figure 2.45**  Lead styles.

flows across the bottom of the board and adheres to the leads and the walls of the through holes. This method confines the leaded components to a single side of the board. Chip components can be mounted to the bottom of the board by epoxy attaching them prior to solder reflow, so that their terminations are connected to the PWB by the wave solder; or they are hand soldered after the wave solder process. DIPs can also be surface mounted. This requires leads formed, or bent, into an L shape under the component. The short leg of the L is then soldered to the mounting pads on the PWB. This method does allow for surface mounting of the DIPs; however, with the 100-mil pitch of the DIP package, it also requires a large increase in the needed surface mounting area on the PWB. Other package styles (e.g., chip carriers) offer a much more efficient surface mounting design.

Surface mount leads have received considerable attention as their usage has increased. There are three basic forms of surface mount lead structures: J, gull-winged, and butt. The J lead takes its name from the lead shape, which forms a J curve under the chip package. The gull-winged lead, with flat or round leads, was the earliest form of surface mount lead. Gull-wing leads have many advantages and disadvantages compared to

J-type leads. The advantages include a low profile, a visible solder joint, a top side electrical probe, and a stress-relieving stepped shape. The disadvantages of the gull-wing leads include extended leads that require more PCB areas, leads that are easily bent, difficulty in cleaning under packages (if the package body is not attached to the PWB), and socketing difficulties. Butt leads are those used for PGAs. Here the pedestal-shaped leads are mounted perpendicular to the case bottom and are butt brazed to the case in an array pattern.

J-type leads have advantages over gull-wing leads. These include ease of socketing, minimum PCB area, protected leads, and available space under the body for cleaning. Since the J leads are bent up under the package, they do not protrude from the package sides. This eliminates potential lead damage during shipping. Gull-wing packages have difficulty ensuring that the leads do not get bent or torqued during shipping. The disadvantages of J-type leads are a higher profile, difficulty in probe testing and solder joints that are only marginally visible. The higher profile increases the thermal resistance of the mounted component. The only available thermal path is through the leads and solder joints. The heat cannot be dissipated over the entire body of the case, which can greatly reduce reliability for high heat-dissipating devices. The higher profile also can greatly increase the volume of a system (see Chapter 4, PWB Assemblies). With the leads bent up under the case, the majority of the solder joint is under the package, thus limiting the visibility of the solder joints. This can greatly hinder the ability to inspect the joints for cracking and proper solder wetting (adhesion).

The lack of standards among component vendors producing SMT devices with regard to widths, heights, and lengths complicates board layouts. In some cases, lead finishes and the thickness of the metals used in the finish vary. Another major problem is that the coplanarity (flatness with respect to their board counting plane) is often outside acceptable tolerances, which makes achieving strong package-to-board contacts difficult.

## 2.10.2   Interconnection Methods

In addition to standard insertion mounting, microelectronic components can either be plugged or soldered into IC sockets or soldered directly onto a board. Plugging ICs into sockets provides advantages for testing and maintenance. The failed component can easily be replaced with less danger of damaging the board or other components, the testing and diagnosis is made much easier, and components which are subject to modifications can easily be exchanged. Disadvantages include extra weight and height and reduced heat transfer.

Surface mounting technology (SMT) was developed in the Swiss watch industry in the late 1960s as a way to decrease the volume of electronics in watches. Surface mount components have leads that attach directly to the surface of the PWB or are leadless, as with LCCs which have castellations instead of leads. With SMT, only one side of the PWB is required for mounting a given component; leads no longer extend out the other side. Thus, packages can be placed on both sides of the board. The rule of thumb is that 35 to 60% more surface mount components can be placed than components requiring insertion mounting. Furthermore, in SMT there is no need for the through holes or the large, round solder contact areas surrounding the holes. Thus component leads can be closer together, which means either smaller packages or many more leads for the same area.

There are many factors that will drive acceptance of the surface mount manufacturing processes. Factors which promote the acceptance of the SMT include:

1. The ability of the technology and the component packaging formats to provide improvements in system performance though increased circuit density, faster switching speeds, and increased transfer rates and access times as a result of adoption of SMT.
2. The growing demand for active devices with high I/O lead counts in small, lighter-weight packages with improved operating characteristics.
3. The growing belief that adoption of the surface mount manufacturing process will result in cost leveling in the present and cost savings in the future.
4. The increasing availability of surface mount formats for a broad range of active and passive component products.
5. The increasing availability of automatic insertion equipment.

Potential users and component suppliers must evaluate the complex issues associated with adoption of the surface mount manufacturing process. The major drawbacks include:

1. Lack of general availability of all types of surface-mounted components.
2. Lack of industry-wide standards.

Surface mount assemblies can be categorized into three different groups. The first type consists only of surface mount components. The assembly can have components on one side or both sides of the PWB. The second type is characterized by the mixed technology used in the assembly. It enables the designer to use components available in through hole when they are not available in surface mount technology or the cost is too high. The third type appears at first glance to be a through hole assembly. This type allows the designer to use the bottom side of the PWB, even though there are leads protruding from it, for mounting small passive or active components. The surface mount components are glued in place on the bottom prior to the wave solder operation. The processes used for assembling the different surface mount types will be covered in detail in Chapter 4.

## 2.11  COST CONSIDERATIONS

In terms of production, automated placement machines have been developed to handle tape and reel, tube, and trays of components. Purchase of an electronic device requires understanding the functions, speed, cost, and form of the packaged component for smooth production.

Figure 2.46 shows the trend between cost per function and number of functions per chip. The electronics industry has now reached a situation where it has been able to reduce the overall costs in such a manner that the minimum cost per function could be reached while doubling the amount of functions every 2 years. The semiconductor industry has been able to furnish the market with the necessary types of components as long as the volume was high enough to justify the developmental costs and the availability of technology was given. For cases where the technology is not available or the integration is too costly, hybrid circuits are used. These hybrids combine available components in one package that will meet the required integration and functionality.

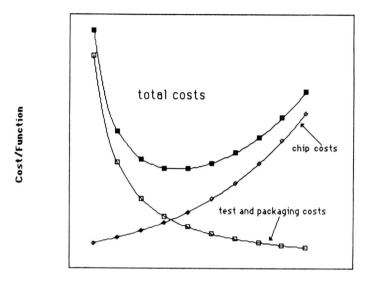

**Figure 2.46**   Cost/function versus functions/chip.

## EXERCISES

1. Describe the different types of components and their basic functions.
2. Describe the difference between passive and active devices.
3. Describe the different configurations and mounting/interconnection methods used for passive components.
4. List the different packaging styles available for semiconductor dice.
5. Sketch the answers found in question 4. Be sure to point out all needed levels of interconnection in each case.
6. How are resistor values adjusted after fabrication?
7. Design a 300 ohm, 3000 ohm, and 300K ohm resistor (each dissipates 1 W).
8. Design a 650 ohm, 400 mW resistor (a) on alumina, (b) on beryllia, (c) on aluminum nitride. (List all assumptions used.)
9. Design a 1.7K ohm, 5 W serpentine resistor. List all assumptions.
10. Design a 1.0 microfarad parallel plate capacitor. List all assumptions.
11. Design a 300 picofarad interdigital capacitor. Assume that $A_1 = 5$ and $A_2 = 1$.
12. Derive the design equation for a top hat resistor. (Hint: A top hat resistor is a single meander that has no spacing within the track loop.)
13. Design a resistor network, with less than 10 I/O, which can provide the following resistances: 100Ω, 400Ω, 750Ω, 1 KΩ, 1.7 KΩ, 30 KΩ, 30 KΩ, 1.1 KΩ, 10 KΩ, 14 KΩ, and 25 KΩ.

# REFERENCES

1.  Smith, R. J., *Circuits, Devices, and Systems—A First Course in Electrical Engineering*, 3rd ed., Wiley, New York, 1976.
2.  Weast, R. C., et al., eds., *Handbook of Chemistry and Physics*, 52nd ed., Chemical Rubber Company, Cleveland, 1971–1972.
3.  Glaser, A. B. and Subak-Sharpe, G. E., *Integrated Circuit Engineering—Design, Fabrication, and Applications*, Addison-Wesley, Reading, Mass., 1979.
4.  Hamilton, D. J. and Howard, W. G., *Basic Integrated Circuit Engineering*, McGraw-Hill, New York, 1975.

# SUGGESTED READINGS

Amerasekera, E. A., et al., *Failure Mechanisms in Semiconductor Devices*, Wiley, New York, 1987.

Clark, R. H., *Handbook of Printed Circuit Manufacturing*, Van Nostrand Reinhold, New York, 1985.

Coombs, C. F., *Printed Circuits Handbook*, McGraw-Hill, New York.

Dally, J. W., *Electronic Packaging—A Mechanical Engineering Perspective*, New York, 1989.

Grodge, M. E., *Semiconductor Device Technology*, Howard W. Same & Co., New York, 1983.

Grove, A. S., *Physics and Technology of Semiconductor Devices*, Wiley, New York, 1967.

Kear, F. W., *Printed Circuit Assembly Manufacturing*, Marcel Dekker, New York, 1987.

Ruska, W. S., *Microelectronic Processing—An Introduction to the Manufacture of Integrated Circuits*, McGraw-Hill, New York, 1987.

# 3

# Printed Wiring Board Design and Fabrication

**Denise Burkus Harris**    *Westinghouse Electric Corporation, Baltimore, Maryland*

**Pradeep Lall**    *CALCE Center for Electronic Packaging, University of Maryland, College Park, Maryland*

The basic building block of any electronic system or assembly is the printed wiring board (PWB). Actually the PWB is the base of most electronic assemblies and provides the support or "board" on which the individual components are mounted. When the PWB is patterned or "printed" on conductor lines or tracks, it provides the routing or "wiring" required to interconnect the mounted components.

In addition to printing the necessary interconnections for the mounted components, discrete function devices—resistors and capacitors—can be incorporated into the metallized pattern within the layers of the board. When this is done, the board is no longer just a printed wiring board, providing only the wiring needed for interconnection, but is a printed circuit board (PCB) containing the wiring and functional circuitry of its own. For example, a ceramic PWB can be made into a PCB by printing a thick-film resistor onto its surface (see Chapter 2 for more information on thick-film resistor processing).

PWBs are made of a dielectric material and are patterned with a conductor. The most common PWB materials are epoxy glass and polyimide glass, referred to as organic PWBs and alumina, aluminum nitride, and beryllia, referred to as ceramic PWBs. This chapter will discuss these types of PWBs along with layout design considerations and basic processing techniques.

## 3.1  ORGANIC PRINTED WIRING BOARDS

The dielectric board materials used to fabricate organic PWBs consist of glass fibers suspended within an organic medium. The organic medium most commonly used is either an epoxy resin or a polymer such as polyimide. The conductor used for these boards is copper. The individual layers consist of sheets of the epoxy glass or polyimide glass with copper foil attached to one or both sides of the sheet. The foil is attached with an epoxy resin.

### 3.1.1  Fabrication of Printed Wiring Boards

The most convenient method for mounting and interconnecting electronic components is by using a PWB. The conducting paths for the interconnection are provided by thin lines of a conductor and the component support by a sheet of insulating material that also serves a variety of other functions. Multilayer boards are used where component density is such that a single layer is inadequate to provide all the required interconnections. Multilayer boards can be single- or double-sided boards with conducting layers sandwiched together by layers of partially cured epoxy glass sheets. A common multilayer configuration consists of a five-layer circuit board (i.e., one ground plane, one power plane, the top mounting pad layer, and two routing layers), but up to 42 layers have been used.

The manufacturing process for PWBs involves combinations of a few basic operations. The conducting layer for interconnection can be constructed by an additive or subtractive process. In the additive process, the conductor is selectively plated over the insulator to create the conducting paths. In the subtractive process, a conducting foil is selectively etched leaving the conducting paths untouched. The subtractive process is more commonly used on organic boards. This section outlines the fabrication process for a double-sided organic board and briefly discusses testing and assembly of such a board. Since a multilayer board is fabricated by sandwiching single- or double-sided boards with layers of partially cured epoxy glass sheets, the manufacturing steps for the manufacture of a multilayer board are discussed at the end of the section.

A copper-clad laminate is the basic material for the manufacture of a printed wiring board. It basically consists of partially cured epoxy, or polyimide, glass sheets with copper foil on each side. The typical fabrication process for a double-sided laminate consists of the following operations:

- Preparation of artwork
- Material preparation
     Shearing
     Drilling and piercing
     Deburring and cleaning
- Electroless copper plating
- Imaging
     Dry/liquid resist application
     Silk screening
- Electroplating
- Resist removal
- Etching
- Making electrical connections to board

- Reflow
- Inspection
- Sealing board surface
- Separation of the boards from the panel
- Testing of the printed wiring board
- Component assembly
- Final inspection

Each of these steps is discussed below in greater detail.

## Preparation of Artwork

The artwork includes all circuit traces, solder pads, component identification numbers, outlines, part numbers, polarity, symbols, dates, and other useful manufacturing information. The artwork may be manufactured by manual or automated methods.

Manual preparation of the artwork is usually generated at four times the original size, although scales of up to 20× have been used to improve the accuracy and tolerance of the artwork. Manual preparation involves using adhesive tape to lay down circuit traces and solder pads on a polyester film. Polyester film is chosen for dimensional stability. The accuracy or precision of the enlarged artwork depends on the skill of the operator and usually ranges from 4 to 12 mils. The final copy of the artwork is photographically reduced to the actual required size. This reduction of size also reduces the tolerance level to 1 to 3 mils. Copies of this final artwork are made by contact printing, again on a polyester film such as Mylar. These copies are subsequently used for the manufacturing process. The original artwork is stored at controlled humidity and temperature for future production of additional photocopies. Environmental controls are maintained to prevent the artwork film from expanding or contracting with changes in the temperature, absorbing moisture and swelling and/or warping, and to prevent contamination.

Automated artwork production provides a digital record of all feature sizes, locations, and dimensions. The system automatically routes the circuit traces and records all vias. A via is a plated hole in one or more layers, used to connect circuits from one layer to another. The digital output of the system is usually in the form of a tape. One tape contains information about the drilling of the holes and goes to the drilling machines. The second tape has information about the circuit layout and the solder pads. This information is fed to the photoplotter (e.g., laser artwork generator, LAG). See Chapter 6 for more information on automatic component placement and routing.

The technologies for the photoplotters are *vector* and *raster*. Vector photoplotters have a flat bed to hold and move the film in the *x–y* plane under a beam of light. Only one feature is imaged at a time. The efficiency is proportional to the speed of the bed and the number of features being imaged.

Raster photoplotters form images of connecting pixels. As the size of the pixels decreases, the image quality improves. To form images of acceptable quality the feature size is usually about six times the size of the pixels. The raster system does not image one feature at a time; instead, all the features are imaged by forming pixels simultaneously along the sweep line as the bed is swept through. The time required to cover the bed depends on the number and the complexity of the features.

Automatically generated artwork is still typically produced on a scale of 4:1. Since most computer-aided design (CAD) workstations are capable of accuracy within 2 mils, the photographically reduced artwork can have line tolerances as low as ±0.5 mil. This improvement of accuracy and the speed of the CAD system have made manual artwork generation obsolete.

Manual artwork preparation is now rarely used in a production environment. The manual generation of artwork is too time consuming, has too great a potential for human error, and does not offer the accuracy of automatic methods. Furthermore, any design change, even a change as simple as moving one track a few mils on one layer, can result in the need to reconstruct the entire layer. If the change of moving one track also results in the need to move a via, more than one layer is involved resulting in a need to regenerate multiple layers on clean polyester film. On most CAD systems, the lines are digitally connected to their associated vias. By instructing the computer to move one line, all connecting vias and lines are also shifted. Thus executing one simple command can accomplish the required redesign, without the need to regenerate all the artwork. The updated digital data of the CAD are downloaded through a digital tape as before to the artwork generator. Such ease of redesign can save a facility a great deal of time and money.

## Material Preparation

The copper-clad laminate material is supplied in sheets that are much larger than required by PWB fabrication. The laminate consists of an insulating material with metal foil bonded on both its faces, as shown in Figure 3.1.

The large laminate sheets permit layout of more than one printed wiring board. The panel size is determined by the tooling hole locations and the PWB size (Figure 3.2). The tooling holes are placed on the PWB for the purpose of alignment for subsequent drilling, piercing, contouring, and automatic component insertion. These holes are larger than functional circuit holes. There may be as few as three tooling holes per panel, or each circuit may have its own tooling holes.

The next operation is to drill or pierce all the functional circuit holes. Drilling generally provides cleaner holes than piercing. Holes are drilled with single- or multiple-spindle drills. More than one panel may be stacked for the drilling operation.

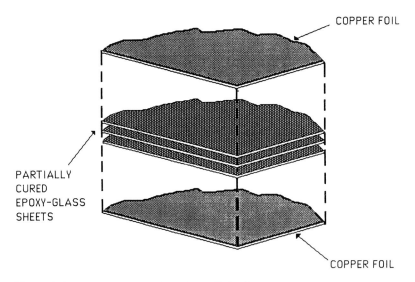

**Figure 3.1**  General construction of a PWB laminate.

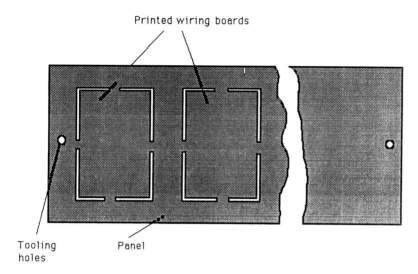

Figure 3.2 PWB panel.

When tooling holes and circuit holes have been completed, the panel is deburred and cleaned. Burrs on holes can seriously affect the quality and functioning of the holes. Subsequent plating would not be successful in the presence of contaminants; therefore, contaminants on the surface of the holes due to machining operations are also unacceptable. An abrasive-type cleaner is used to remove burrs and surface contaminants.

## Electroless Copper Plating

After the panel has been drilled and cleaned, it is processed through an electroless copper bath, which produces a thin conducting coating of copper in the drilled holes (vias) to connect the circuits on the two sides. Electroless copper plating is a chemical deposition process that makes use of special activators (e.g., colloidal solutions of stannous and palladium ions) to sensitize the holes. The board is dipped into the solution and the stannous ions are deposited on the surface to be catalyzed. The palladium ions are then deposited, creating metallic sites for the electroless plating process. After the board has been activated, it is transferred to a copper bath. The copper ions are reduced to metallic copper. This deposition process proceeds at a very low rate (a few microinches per minute). In order to reduce the process time, only a very thin layer is deposited by the process. The printed wiring board is then transferred to an electrolytic bath where the required thickness of copper is deposited. The electroless copper plating process steps are described in Figure 3.3.

## Imaging

Imaging is the process of placing the circuit image on the electroplated copper. The imaging process is normally accomplished by (1) photoresist (dry or liquid) or (2) silk screening. Photoresists are thin coatings produced from organic compounds which, when exposed to light of the proper wavelength, are chemically changed in their solubility to certain solvents. The two common types of photoresists are negative and positive acting.

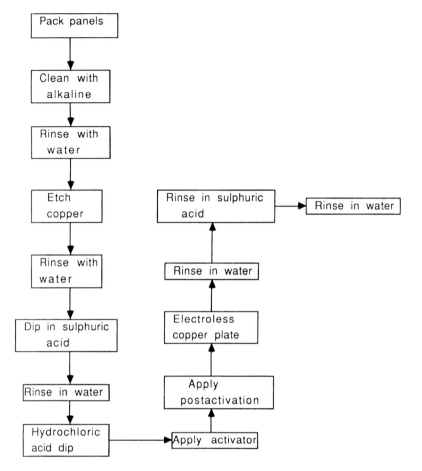

**Figure 3.3** Electroless copper plating process.

Negative-acting photoresist is initially soluble in the developer, but after exposure it becomes polymerized and insoluble in the developer. Exposure to ultraviolet light is made through the film pattern or *mask*. The unexposed resist is selectively dissolved, softened, or washed away, leaving the desired resist pattern on the copper-clad laminate. Positive-acting resists work in the opposite fashion. Exposure to light makes the polymer mixture soluble in the developer. The resist pattern that remains after development is insoluble and chemically resistant to the cleaning, plating, and etching solutions used in the production of printed circuit boards.

Another technique of image application is screen or stencil printing, which uses a stencil image of the circuit or pattern. The photoresist is printed through the screen or stencil, transferring the resist to the copper-clad laminate. (See later under Thick-Film Metallization for more details on screen printing.) After the image has been applied, the panel is taken to the plating area for electroplating. A thin coating of tin-lead electroplate is deposited on the panel. The resist acts as a negative mask (Figure 3.4), allowing metal to be deposited on the circuit areas.

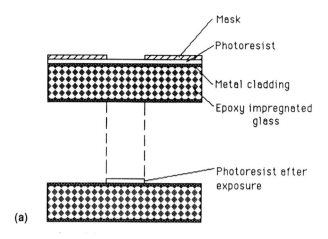

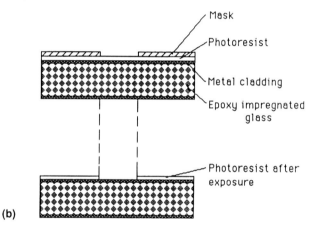

**Figure 3.4** Photoresists actions: (a) actions of the negative-acting photoresist; (b) action of the positive-acting photoresist.

## Electroplating

Electroplating is the process of depositing an adherent metallic coating on a negatively charged electrode by the passage of electric current in a conducting medium. Normally the medium is water to which metal salts have been added as a source of metal ions. Electroplating is employed in several steps in the manufacture of a printed circuit board, e.g., in providing an etch-resistant metal surface, plating through holes, applying gold coating for connector contacts, and thick solder plating. The process is closely monitored for the plating thickness for the following reasons:

1. The thickness of copper and the metallic resist affect the diameter of the plated through hole, which is of primary concern in the assembly of the PWBs.
2. The thickness of the plated metal may be critical in cases of contact finger plating.
3. Some metals are more expensive and hence additional thickness increases the manufacturing cost.

4. Some metals are plated as an etch resist and additional plated thickness would result in an increased time for the etching process.
5. The metal thickness determines the thermal and current-carrying capabilities.

The panel to be plated is placed in an electroplating bath to build up a metal thickness of approximately 0.001 to 0.002 in. in the holes. If copper buildup is not wanted in the tooling holes, they are masked for this operation. The deposition rate of the electroplated metal is a function of current density and current distribution. Current density is not distributed uniformly across the surface of the panel. It is greatest at the corners and the edges of the panel and isolated holes and is least at the center of the panel and the land areas. This makes it necessary to check the holes in numerous locations on the panel and within individual circuits.

The bath chemistry affects the ability of the plating bath to deposit metal at a high or an optimum current density. Every plating bath for printed circuit manufacture has an optimum range for plating bath constituents. Should the metal fall below that range, the throwing power of the bath decreases. The *throwing power* is the ability of the plating bath to "throw" the metal ions onto the surface that is being plated. Further, if the metal content is too high above that range, the conductivity of the bath decreases. This has an effect on the ability of the plating bath to operate at its optimum current density. The presence of organic contamination reduces the range of current densities at which the plating bath can operate. Plating baths generally have an optimum temperature below which the throwing power of the bath is greatly reduced.

Good plating distribution requires agitation of the plating bath. Lack of agitation can lead to depletion of metal ions at the surface of the panels, leading to burning of the plating. Agitation replenishes the metal ions so that the plating may continue in a uniform manner.

The distance between the cathode and the anode is vital for a uniform plating distribution. If the printed circuit panel is too near the anode, the portion of the panel directly opposite the anode will be very heavily plated.

The sequence of plating operations during manufacture may follow two paths: (1) panel plating and (2) pattern plating. Panel plating is independent of the image transfer process. It consists of copper plating the entire board area (including the holes) immediately after the electroless copper process. Figure 3.5 shows the sequence of steps for the panel plating process. In pattern plating, an image is applied using a photoresist or screen printing process. A layer of tin-lead solder is plated on the areas of the board that are free from the resist to protect the circuit details and the plated-through holes as excess copper cladding is etched away. It also protects the copper surfaces on pads and traces from oxidation during the storage period and prior to assembly. After solder plating, the image is stripped off and the circuit is then etched of additional copper. The solder serves as an etch-resistant coating that protects the copper used to form the surface features from chemical attack by the etchant. When the board is ready to be soldered, the solder plating is reflowed. In the liquid state, solder acts to wet the surfaces to be joined.

Pattern plating is completed in a six-step sequence, as depicted in Figure 3.6. The first step is application of the photoresist after the board has been electroless copper plated. The image is transferred on to the photoresist and the board is then electroplated with copper. Solder is then deposited selectively on the copper to protect it from the etchant. The resist is stripped and the exposed copper is etched away to provide the circuit pattern on the board. The sequence of operations is pictorially displayed in Figure 3.7.

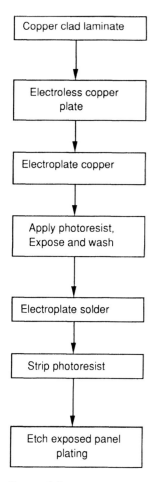

**Figure 3.5** Panel plating process.

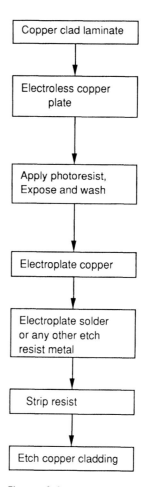

**Figure 3.6** Pattern plating process.

## Resist Removal

After electroplating, the resist is removed with a suitable solvent. Usually chlorinated hydrocarbons are suitable for resist removal. Even after resist has been removed there is a solid metal surface on each side of the panel, as shown in Figure 3.8.

## Etching

The removal of metal to isolate printed conductors is called etching. It is a chemical machining process used to strip cladding away from regions of the board not protected by the etch-resistant coating. This etching transforms an image into a circuit. The copper that remains after etching is used to define the board features such as circuit traces and copper pads. The etching process involves (1) photoresist coating to protect copper cladding from the etchant and (2) solder plating as a protective coating from the etchant.

Etching is done in automated etching lines that spray ammonium hydroxide on both surfaces of the board. The ammonium hydroxide combines with copper ions and forms

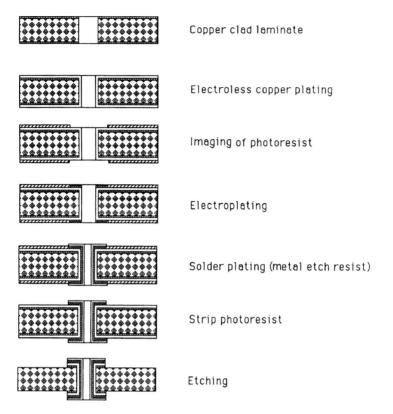

**Figure 3.7**  A cross-sectional view of the print and etch process (using pattern plating).

the chemical compound $Cu(NH_3)_4^{2+}$. This dissolved copper is recovered by recycling the etchant fluid. Recycling also helps avoid the trouble of disposing of the toxic wastes. After etching, the board is completely washed to remove the etchant.

The etching process is not selective and results in an equal rate of etching in all directions. The spraying of the etchant results in a certain degree of selectivity, since fresh etchant is continuously being delivered to the bottom of the area being etched. The nonselective nature of the etching process results in undercutting. Undercutting is the removal of copper from the side wall of the conductor as copper foil is etched, as shown

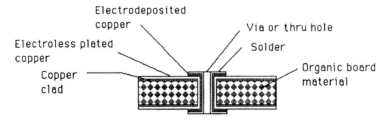

**Figure 3.8**  Plating configuration after resist removal.

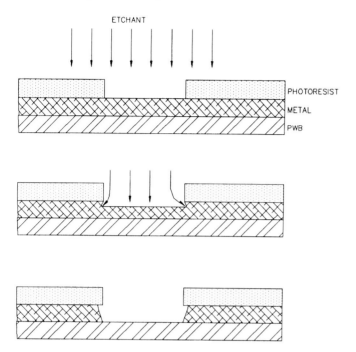

**Figure 3.9** Etch undercutting.

in Figure 3.9. In general, the longer the etching time, the larger is the undercut. An effort is made to reduce the undercut during the etching process.

The lack of selectivity in the etching process causes an undercut of the etch-resistant coating. This undercut is extremely important because the features of the board as designed before undercut have to be different from the final dimensions. The top surface of the cladding is undersized by an amount $u$ as a result of the undercut (Figure 3.10). The magnitude of this undersizing is expressed as an etch factor, $F_e$, defined as

$$F_e = \frac{u}{t}$$

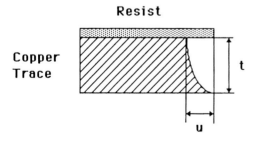

Etch Factor = u/t

**Figure 3.10** Undercut etch factor.

where $u$ is the undercut and $t$ is the thickness of the copper foil. A design that did not take into account the undercut in the cladding would result in an undersized trace. The expression used to calculate the size of the trace prior to etching is given by

$$X_L = X_F + 2F_e t$$

where $X_L$ is the width at the layout stage, $X_F$ is the width after etching, $F_e$ is the etch factor, and $t$ is the thickness of the copper foil.

The speed of etching is a vital design consideration. It depends on the etchant and the conditions prevailing. The lower the speed, the more undercut is produced. When the circuit has been etched and thoroughly cleaned, the nonplated holes are drilled. Figure 3.7 shows the various steps in the print and etch process.

## Making Electrical Connections to Board

Electrical connections to printed circuit boards can be made through directly soldered connections on contact fingers. A contact finger is a row of evenly spaced pads along one edge of the PWB. Contact fingers supply the input/output (I/O) or the interconnection sights of the PWB to the outside world. (The outside world is the chassis or system of which the PWB is a subassembly.) One of the advantages of contact fingers is the ease with which a printed circuit may be installed and removed from the electronic equipment. The circuit board can be simply pushed in or pulled out. The requirements of simplicity, ease of use, and reliability have placed many restrictions on the fabrication of the contact finger. The contact finger must be free of oxidation and corrosion and remain so for the expected life of the equipment. It must also be wear resistant. These needs are generally satisfied by using gold plating on nickel. The nickel provides a barrier metal that prevents copper/gold intermetallics from forming. Both the nickel and the gold act as protection from corrosion. Contact fingers are plated after the printed circuit has been pattern plated, resist stripped, and copper foil etched off the epoxy laminate.

Contact fingers are found in most commercial applications. By using the fingers as the means to interconnect, the need to purchase and mount a connector is eliminated, thus reducing costs. In military applications, however, connectors are permanently mounted to the PWB assembly. The use of connectors enables higher I/O counts by having multiple rows of leads or pins in the connector. Standard pin count connectors can be employed, even if the pins are not all necessary. Even if additional I/O is added to later revisions of the PWB design, the connector can still be "plugged" into the same chassis or system. By using a two-part connector, higher I/O density, through the use of multiple rows, can be obtained without limiting the ease of connecting the board to the higher-level assembly. The board would contain the male part of the connector, while the chassis would have the mating female part. (See Chapter 6 for more details on connectors.)

## Reflow

The process of etching is accompanied by undercutting; i.e., removal of copper from the side wall of the conductor as copper foil is etched (see Figure 3.9). This results in an overhang of the plated etch resist, whether tin-lead, tin, tin-nickel, or nickel, which can create problems of short-circuiting. If the overhanging conductor breaks off, it can form electrical bridges between two points on the circuit. These metal slivers are a leading cause of electrical short circuits. Reducing overhang can help increase overall board yields. In order to reduce the overhang of the metal etch resist, the solder is reflowed. Tin-lead is a dull gray metal, very porous and easily oxidized. The reflow operation melts the tin-lead for a few seconds. This fuses the tin and lead together into a bright, shiny corrosion-resistant alloy known as solder. Reflow is accomplished by two methods: infrared (IR) and hot oil. Chapter 4 discusses IR reflow in detail. After being reflowed, the panels will exhibit shiny traces and holes, with gold contact fingers.

## Sealing Board Surface

Circuits and insulating materials have to meet high quality standards. In order to meet these standards throughout the life of the product it is necessary to ensure that electrical parameters do not degrade. The surface of the board is therefore sealed with material having excellent insulating or dielectric properties and excellent stability. This helps prevent the formation of solder bridges or contamination during the assembly soldering operation.

One sealing material used to solder mask is a dielectric epoxy that is applied to one or both surfaces of the board by film lamination or screen printing. The film lamination is similar to the photoresist process. The polymer is applied in the form of a thin lamination on the surface and imaged using a phototool or mask. The resist is then stripped, exposing only the junctions to be accessed for soldering. The accuracy achieved by this process is excellent, usually less than ±0.001 in. Nevertheless, it is more expensive than the screen printing process.

In addition to the functions mentioned above, the solder mask greatly reduces the potential for high voltage–induced dendritic growth of the copper crystals, which could lead to short circuits of closely placed leads. The section on biomedical hybrids Chapter 4 discusses the ability of a polymer coating to protect a surface from moisture. Figure 3.11 shows a cross-sectional view of a board with an applied solder mask.

The design of solder masks is important not only to the function of PWBs but also to their manufacture. The solder mask design evolves from two considerations:

1. Material to be used
2. Geometric configuration to be used

The objective of solder masks is to provide a permanent protective cover for certain areas of the wiring board and to expose other areas. The mask layout consists of a wiring board image that covers the entire board except for areas to be soldered or accessed for electrical connection. The electrical contact areas may be required for testing the board or for connecting other circuits and/or connectors.

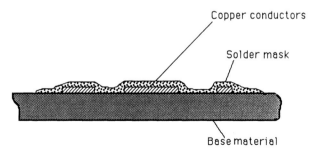

Copper conductors

Solder mask

Base material

**Figure 3.11** Solder mask.

## Separation of the Boards from the Panel and Finishing

A single panel usually consists of many boards. Once the panel has been processed, the individual boards must be separated. This step consists of cutting the boards from the panel. Beveling of the contact fingers is done to facilitate loading of the printed circuit board interconnects. The holes on the PWB which are not plated may be made in two ways:

1. Drill the holes in the beginning and plug during electroplating.
2. Drill the holes at the end after the electroplating operation. The holes that were not drilled in the beginning are drilled in this step.

Product nomenclature to identify locations and print names can be applied using the screen printing, ink stamping, or laser writing process.

## Inspection and Testing of the Printed Wiring Board

In this step the broken traces (i.e., *opens*) or tracks bridged by loose conductor particles (i.e., short circuits or *shorts*) are detected by visual and/or electrical inspection. This process isolates the defective boards, assesses quality, and highlights areas needing attention.

Visual inspection and electrical testing for continuity and isolation of all PWBs are required to provide a measure of the functional quality of the bare boards. This testing seeks to locate unwanted opens, shorts, and capacitive coupling. Methods of bare board testing depend on the end use of the board. A printed wiring board used in low-frequency applications does not have critical capacitive coupling requirements. Open circuits, a well as shorts, are not acceptable under any circumstances. Shorts produce a finite amount of resistance between two conductors that are supposed to show infinite resistance. This is very important in applications in which leakage current is a critical performance parameter.

Continuity/isolation testing verifies that no opens exist (that the circuitry is continuous) and that no shorts exist (that the needed isolation between certain signals is in place). There are various methods for conducting these tests. An operator can manually probe all the different points within the net list with an ohmmeter. One probe can be placed on the first probe site—an I/O of the board or a mounting pad for a certain pin of a given component—while the other probe is moved from one probe site to the next, until all other sites have been probed. The operator must keep track of the ohmage of each site. The ohmage for an isolation site should approach infinity (usually the order of $10^6\Omega$), and the ohmage for continuity should approach zero (or be very low, on the order of milliohms). The process is then repeated for all other probe sites until all combinations of probe sites have been measured. This process is very time consuming and tedious. It has a very high potential for human error. The two more common methods are automated. One such method is conducted with a ''flying probe.'' This equipment carries out the probing in the same manner as the manual method. One probe is placed while the other probe ''flies'' around the board, probing all other sites. The first probe is then moved to the second probe site and the process continues until all probe combinations have been tested. The beauty of a flying probe is that the net list and the board layout—all probe site locations—are automatically downloaded from the CAD system to the computer on the flying probe. Thus, once the board is placed on the probe station and the orientation is confirmed, the computer automatically probes and measures all points in seconds. All the data are automatically processed and any violations are recorded or printed out.

The other common method for circuit check is to probe the board with a "bed of nails." This method first requires the design and fabrication of a panel or fixture with all the probes needed to touch all probes sites simultaneously. Once completed, this fixture typically has a full array of probing pins, which resembles a miniature bed of nails. Each probe is wired to a computer testing station. The computer is then programmed to activate or measure the resistance between all the different combinations of probe sites. Some bed-of-nails testing equipment will accept layout coordinates and net list information from the CAD system. Other equipment requires manual input and programming of the testing sequence.

When the boards have passed these inspections and testing, they are ready to be sent to assembly. In assembly they will have all components and connectors mounted. Chapter 4 will discuss the assembly processes in detail.

### 3.1.2 Multilayer Board Fabrication

Multilayer boards (MLBs) are manufactured in the same way as double-sided boards. When the inner layers have been laminated into the panel, the process consists of the following steps. Thin epoxy-impregnated fiberglass and copper laminate is blanked to panel size. Special tooling holes are punched or drilled into the artwork and the laminate to align layers during lamination into panel form. These ensure proper registration and orientation throughout the process. Inner laminate is imaged with dry film resist. The only difference in the manufacture of double-sided and multilayer boards is that dry screen printing is used for imaging in MLB manufacture and photoresist application is used in double-sided board manufacture. The layers are etched to remove the unwanted copper foil, and the resist is stripped, leaving the copper circuitry on the epoxy-impregnated fiberglass substrate. The etched layers are immersed in a hot caustic oxidizing bath that forms dendritic crystals of black oxide, which helps in bonding the layers during the fabrication of the multilayer board. The layers are carefully sequenced and registered during the stack-up operation. The sheets of prepreg are placed between the interlayers to bond them together. The interlayers and prepreg are pinned together by caul plates and pressed in a multilayer press. Pressure and temperature are applied to bond and cure the prepreg. After the panels are removed from the press, they are drilled and processed like the double-sided boards. The vias are then placed for interconnection of layers. Vias are categorized as through, blind, and buried. Through (sometimes abbreviated as thru) vias connect one external layer to the internal layers and to the other external layer. The hole passes through every layer, and conductor pads connect to traces on the appropriate layers as shown in Figure 3.12.

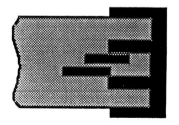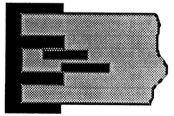

**Figure 3.12** Through holes.

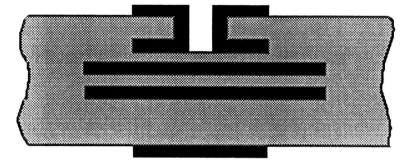

**Figure 3.13**   Blind vias.

Blind vias connect one outside layer to one or more internal layers. Since the blind via does not pass through every layer, the conductors may pass through the same plane on other internal layers as shown in Figure 3.13.

Buried vias connect multiple internal layers without connecting to outer layers. This allows conductors to pass through the same plane on any remaining internal layers as shown in Figure 3.14. Buried or blind vias can be drilled using either mechanical or laser techniques. Blind vias, required for interconnecting to other components, can eliminate the need for one or more additional layers. Figures 3.15 to 3.17 show the process flow for producing through and blind vias.

The drilling process is accompanied by a large amount of heat, which smears the epoxy on the inside of the holes. This could prevent the electrical connection between the layers. The smear is therefore removed by chemical means, by application of sulfuric acid or chromic acid. For example, a HF–$H_2SO_4$ etch may be used to remove smeared epoxy glass and etch back additional epoxy glass. Figure 3.18 gives an overview of a fabrication for a multilayer board.

### 3.1.3   Registration Errors in Assembly

The holes in PWBs are formed by drilling or punching. Punching is typically restricted to single-sided boards fabricated from a copper-reinforced phenolic material. The holes for insertion mount devices are normally larger in diameter to allow for lead clearance. Vias used to connect one signal to another may be smaller. The minimum size of a hole on the board is established by the drilling technology. As the size of the hole to be drilled

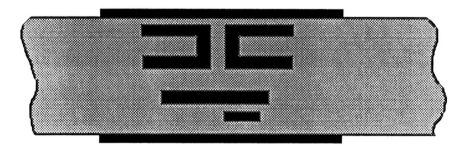

**Figure 3.14**   Buried vias.

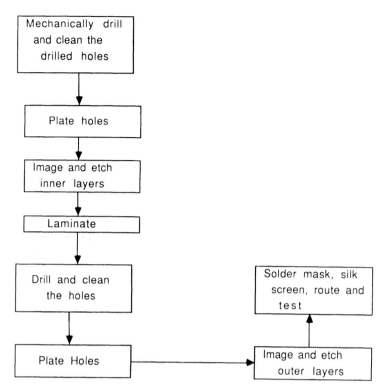

**Figure 3.15**  Blind hole processing.

decreases, the demands on positional accuracy increase. Errors in positional accuracy accumulate from four major sources: (1) the lead diameter, (2) the position of the table on which the board is mounted for the purpose of automated assembly, (3) the position of the insertion head holding the component over the assembly table, and (4) the accuracy of the position of the drilled hole. These errors are accounted for in the manufacturing process by increasing the diameter of the hole on the board. The minimum diameter of a hole that can be handled by any manufacturing process, subject to the constraints of dimensional accuracy that can be achieved in spite of the cumulative error from the sources outlined previously, can be calculated from

$$D_f = D_0 + E_{r1} + E_{r2}$$

where $D_f$ is the minimum diameter of the hole that can be achieved by the manufacturing process, $D_0$ is the diameter of the lead wire, $E_{r1}$ is the positional error due to the insertion machine, and $E_{r2}$ is the positional error due to the process of drilling the hole.

## 3.1.4  Solder Methods

Solder methods in manufacturing systems are designed to maximize efficiency and minimize defects. The cost of defective solder joint escalates as a function of time after the soldering process is complete. After the board is placed in the system, the defective board must be located, and the defective joint must be found before repair can be initiated. The

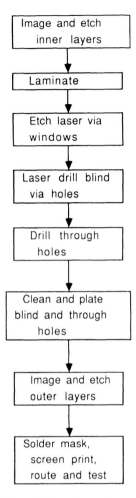

**Figure 3.16** Lasered machined blind vias.

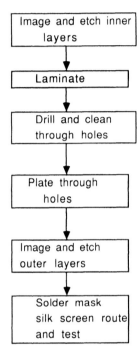

**Figure 3.17** Through
holes processing.

available soldering methods include solder pots, wave soldering, vapor phase, and IR reflow for pretinning and component attach, and thick solder plating for board pretinning.

Solder pot is the simplest form of soldering. The pot is fabricated from cast iron and is heated by electrical elements arranged along the bottom and sides of the pot to maintain a uniform temperature distribution in the molten solder bath. They vary in size from a few pounds to several hundred pounds of solder capacity. The heaters are thermostatically controlled to maintain a steady temperature during the manufacturing process. The molten solder oxidizes with time, forming a film called dross, which floats on the surface of the solder pot and deters the soldering process. Rosin flux is added to the solder pot to reduce the formation of dross.

Another approach is to remove the dross as it accumulates during the soldering process, by skimming the surface. The components are soldered or pretinned by dipping them in the solder pot and holding them in the molten solder until the surface is wetted and coated with solder. In the case of leadless ceramic chip carriers, the carriers are placed on top

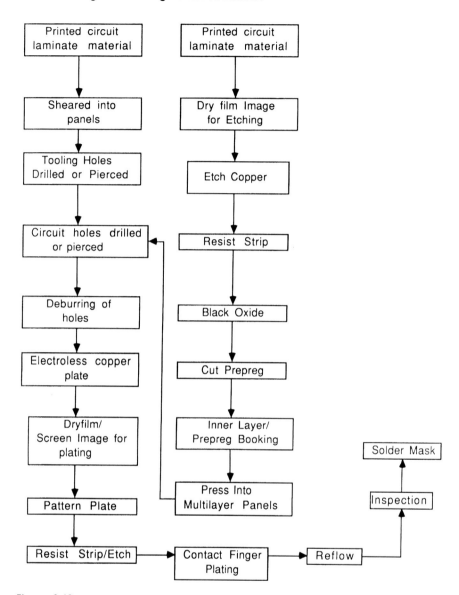

**Figure 3.18** Multilayered organic PWB process flow.

of the solder and allowed to float on the molten solder. As they float, the molten solder wicks up the castellations, pretinning the I/O of the chip carrier. For producing solder joints in larger applications, the pot is equipped with a pump to generate a wave of solder on the underside of the circuit (Figure 3.19). The solder wave is adjusted to impinge on the underside of the circuit board. This wave simultaneously supplies solder to all the solder joints on the board.

Wave soldering, though effective for through hole components, cannot be used for surface mount devices. Vapor-phase soldering and IR soldering are common methods for surface mount devices. The components are mounted and temporarily held in place with

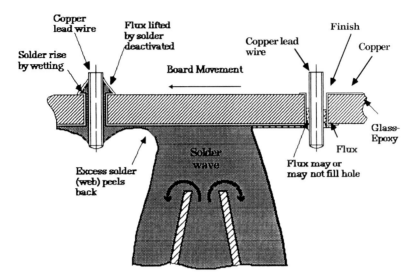

**Figure 3.19**   Wave soldering.

solder paste, which acts as a tacky substance. When the solder paste melts it forms a solder joint, which solidifies when the board is cooled to room temperature. The soldering process requires that the temperature of the joint be increased until the solder melts and flows. The surface temperature of the mating parts must be above the melting temperature of the solder for the solder to wet the surfaces properly. Flux is used to free all the oxide surfaces that accumulate during the manufacturing process. The process may involve preheating the circuit board to about 100°C before moving it into the vacuum chamber to avoid component damage due to thermal shock. This prebake of the solder paste can also reduce the amount of nonactive volatiles (e.g., solvents added to the paste to yield a printable rheology), which could cause solder slumping or bridging during the reflow process. However, some solder pastes (e.g., Cermalloy brand) contain active organics whose effectiveness would be degraded by prebaking. Such solder paste vendors will recommend that the prebake be eliminated. This elimination of the prebake does not eliminate the ability to reduce thermal stressing. Thermal stressing can be reduced by programming the vapor-phase process to allow for a slow descent of the boards and components into the vapors. This will allow the boards and components to heat up slowly enough to avoid unnecessary thermal stressing, yet fast enough so that active solvents within the solder paste will not be outgassed before accomplishing their task of increasing wettability. Upon entry into the chamber, the PWB or PCB is subjected to vapor from Freon FC-70 or Freon FC-71, which is at a temperature of 215 or 240°C, respectively. The vaporized Freon gives up latent heat of vaporization as it condenses on the board, components, and solder joints. The latent heat produces a rapid increase in the temperature of the entire board and component surfaces. The circuit board is then removed from the chamber and allowed to cool. Some vapor phases are equipped with fans or solvent sprays to permit the cooling rate to be controlled. The solder solidifies, simultaneously forming all the joints on the board. The process achieves excellent temperature control over the board. However, this control deteriorates with further processing as flux and the compounds formed by the combination of flux and the oxides the flux removes accumulate in the vapor phase. For this reason, the Freon in the vapor phase must be replaced on a regular basis.

These and other soldering process will be covered in more detail in Chapter 4.

## 3.1.5 Cleaning

The board is thoroughly cleaned after the soldering process to remove the surface residues that accumulate during the solder reflow process. The surface contaminants fall into two categories, polar contaminants and nonpolar contaminants. Polar contaminants include the plating residues and solder flux activators. Nonpolar contaminants include oil, grease, and rosin fluxes. Vapor degreasing is the most commonly employed process for removing both types of surface contaminants. A solvent cleaning solution is heated in a sump until it boils. Vapors from the solvent rise and condense on the circuit board assembly, which is positioned above the sump. The solvent that condenses on the board washes the board, and rinsing with the pure solvent continues until the board reaches the boiling point of the solvent and the condensation process terminates. The impurities that collect in the sump are removed periodically.

To further enhance the removal of the polar contaminants, the degreasing process is sometimes followed by a rinse in deionized water (DI water). DI water is water that is fine filtered (down to 2 µm) and has all ions removed. DI water has a typical value of 18 MΩ. This highly resistive solvent is "hungry" for ions and will remove any polarized contamination from the board assembly.

Other aspects of the cleaning process will be covered in more detail in Chapter 4.

## 3.2 CERAMIC PRINTED WIRING BOARDS
## 3.2.1 Material

Several ceramic materials are used to fabricate ceramic PWBs. The most common is alumina or $Al_2O_3$. Another material that has been in use for many years is beryllia or BeO. This material's thermal conductivity is roughly 4.5 times greater than that of $Al_2O_3$. However, BeO is used only when thermal requirements demand. BeO raw material is two or three times as expensive as alumina. Furthermore, BeO dust, generated whenever the material is fabricated or machined, is highly toxic. The EPA imposes several restrictions on the fabrication and machining of this material. Over $1 million worth of capital equipment is required to meet these restrictions and to ensure a safe working environment. This further drives up the cost of purchasing and processing this material and increases the life cycle time needed to purchase and assembly.

Other ceramics being used are aluminum nitride (AlN), boron nitride (BN), silicon carbide (SiC), and silicon nitride (SiN). These have better thermal conductivities than BeO, without the toxicity. They are also being formulated to yield more desirable co-efficients of thermal expansion (CTEs). Boron nitride has excellent thermal and electrical properties; however, metallization of this material has thus far proved to be very difficult, with minimal adhesion of the metals to the BN. Silicon nitride formulations in which this material is combined with glasses and other ceramics to obtain the desired properties and a producible package are under investigation. Silicon carbide has been successfully used in some military applications; however, its processing is still immature. Aluminum nitride has been the most promising. Thick- and thin-film metal compositions and metallization techniques have been developed which have passed initial military standards testing. Therefore, most of the development work currently in progress has been with AlN. Limited production of AlN substrates and packaging is being done within the electronics industry. Much research has been done with thick-film metallization, both single layer and multilayer. Currently, AlN is as expensive as BeO. However, as recent trends have already indicated, the cost of this material and its processing will decrease as the technology matures.

Porcelainized metals also fall into the category of ceramic PWBs. The process and material used in fabricating these substrates are very similar to those used in making porcelainized metal cookware. This material is very inexpensive when purchased in quantity. However, multilayering is not as common because of the lack of dielectric materials capable of withstanding the thermal expansions that occur within the porcelanized metal. The use of these materials is also limited by their poor thermal properties. Their CTE is not well matched to standard electronic components. These boards also have poor thermal conductivity despite their metal cores; the porcelain, which totally encapsulates the metal, inflicts high thermal resistance. Lastly, these boards are extremely heavy compared with other organic or ceramic boards. For these reasons, the applications utilizing this type of board material are small and involve low packaging density and high volume (e.g., a small automotive ignition switch containing less than four components, or a temperature sensor for a coffee pot).

Glass boards are sometimes metallized to provide signal tracks or a interconnect board. By definition, such an apparatus is a printed wiring board. Nevertheless, these substrates are not normally included in discussions of PWBs because they can only provide single-layer metallization and cannot support the mounting of components. Neither of these aspects associated with PWBs is possible because of the mismatch of the CTE of the glass to that of a multilayer dielectric or common electronic components. Applications of this technology are usually limited to thermal print heads.

Table 3.1 lists some of the mechanical and electrical properties of various ceramic materials commonly used in the electronics field. This list includes the ceramic materials used for both PWB and packaging applications.

## 3.2.2 Substrate Formation

Ceramic substrates can be formed in three different ways: pressed, fired, and roller pressed. In the first process, ceramic powder of uniform size and shape is suspended in solvents to form a slurry and then pressed into sheets between smooth, flat metal plates under extreme pressures. The slurries used for pressed ceramic boards normally contain approximately 4% volatiles. These volatiles are burned off or outgassed during the heat treatment of the ceramic. Roller pressing the substrates is done in a similar manner, except that the ceramic powder mixture is pressed, under high temperature, between two rollers.

The method most commonly used to form ceramic substrates is firing or curing. The first step in manufacturing a fired substrate is to mix the ceramic powders with solvents, binders, and emulsifiers to form a slurry. This slurry, which contains approximately 16% volatiles, must be carefully mixed to provide a homogeneous fluid that does not contain any air bubbles. This is accomplished by slowly rolling the slurry. This rolling or mixing must be slow enough so that the flow within the roller is laminar. Turbulent flow would mix up the ingredients much more rapidly but would allow bubbles to form within the slurry [1].

When the materials have been homogeneously mixed to the proper viscosity and composition, the slurry is fed onto a moving tape. The tape is a polymer material that is inert to the ceramic slurry (e.g., Mylar). It provides the support needed to form the ceramic tape. The Mylar is connected to a rotating spool, which rolls up the Mylar and ceramic tape. As the slurry is poured onto the Mylar tape, a blade skims over the top of the slurry to smooth the surface. The height of the blade, flow rate of the slurry, and rotational speed of the spool determine the thickness of the ceramic tape. This ceramic tape is considered to be in the "green" or unfired state. It is very flexible and fragile, tearing easily. The tape commonly comes in 0.005 to 0.125 in. thicknesses.

The rolls of tape can be stored in a controlled environment for several months until it is time to fabricate the substrates. Substrates are created by first unrolling the tape and cutting it into squares larger than the desired end product. These squares are removed from their Mylar support and stacked on top of each other to obtain the desired prefire thickness, or the tape is originally rolled to the desired prefire thickness. The stack-up is then laminated together under controlled pressure and temperatures. At this time, the blanks are cut or punched into the desired shape and size. The tape will shrink approximately 2% in length and width and 0.7% in thickness during the firing stage. Therefore, the size to which the laminate stack is cut must take into consideration the expected shrinkage. The laminated and cut piece is placed on a flat carrier. This carrier is usually made of graphite, so that the carrier will not melt or deform during the firing and the ceramic material will not adhere to the carrier. The stack-up is fired in a belt furnace at a peak temperature slightly higher than the reflow temperature of the ceramic material (approximately 2000°C). At this temperature the ceramic reflows together, forming a lattice structure. It takes on the shape of the carrier it sits on; thus, the flatness of the carrier will determine the flatness of the finished blank substrate. The top side of the fired substrate, the side that was facing up while being fired, has a slightly more porous surface than the bottom, due to the outgassing of the solvents during the firing process. This surface can be left in its "as-fired" state or can be polished to a smoother surface. This polishing is normally done for thin-film substrates.

There are four different methodologies for metallizing ceramic: thick film, thin film, cofired, and direct bonded. Each of these technologies will be discussed separately in the following sections.

## 3.2.3  Metallization of Ceramic PWBs

### Thick-Film Metallization

By definition, thick-film metallization has conductor thicknesses of 0.0005 to 0.001 in., dielectric thicknesses of 0.0015 to 0.0025 in., and printed resistor thicknesses of 0.001 to 0.0015 in. The layers are built up on the substrate by printing methods similar to silk screen printing (used to print everything from street signs to T-shirts).

The first step in the design of a thick-film circuit is to generate the artwork. The artwork is generated in the same manner as for organic PWBs. When 1:1 scale artwork film has been generated for each layer, a screen or stencil for each layer can be fabricated.

A screen consists of a wire mesh stretched over a metal frame. The mesh comes in various wire diameters and mesh counts. The mesh count is the number of wires contained in a linear inch of mesh. The finer the wire diameter and the higher the mesh count, the

**Table 3.1** Typical Properties of Ceramic Materials Supplied by Frenchtown Ceramics[a]

| Material designation | Alumina | | | | | | | | | | Beryllia Thermalox 995[b] | Cordierite 9954 |
|---|---|---|---|---|---|---|---|---|---|---|---|---|
| | FA-85 | 6096 | 7231 | FA-94 | 4462 | 2082 | FA-96 | FA-98 | FA-995 | FA-999 | | |
| Alumina (% nominal) | 85 | 86 | 92 | 94 | 94 | 94 | 96 | 98 | 99.5 | 99.9+ | — | — |
| Color | White | White | Black | White | Brown | White | White | White | White | White | White | Tan |
| Specific gravity (g/cm$^3$) | 3.42 | 3.44 | 3.71 | 3.63 | 3.70 | 3.60 | 3.73 | 3.74 | 3.93 | 3.96 | 2.80 | 1.91 |
| Pore volume (%) | — | — | — | — | — | — | — | — | — | — | — | 26.7 |
| Water absorption (%) | 0 | 0 | 0 | 0 | 0 | 0 | 0 | 0 | 0 | 0 | 0 | 12.1 |
| Maximum use temperature (°C) | 1400 | 1400 | 1400 | 1550 | 1400 | 1550 | 1650 | 1650 | 1705 | 1910 | 1700 | 1300 |
| Thermal conductivity at 25° (cal/cm$^2$/cm/sec/°C) | 0.040 | 0.040 | 0.041 | 0.052 | 0.047 | 0.051 | 0.054 | 0.057 | 0.103 | 0.108 | 0.600 | — |
| Thermal expansion coefficient (per °C) 25–100°C ($\times 10^{-6}$) | 5.34 | 5.33 | 5.26 | 5.60 | 3.60 | 5.56 | 4.80 | 5.10 | 5.26 | 5.25 | 5.07 | 0.34 |
| 25–400°C ($\times 10^{-6}$) | — | — | — | — | — | — | — | — | — | — | — | — |
| 25–700°C ($\times 10^{-6}$) | 6.86 | 6.85 | 6.25 | 6.20 | 6.11 | 6.21 | 7.00 | 7.00 | 6.39 | 6.35 | 6.80 | 0.87 |
| | 7.44 | 7.42 | 7.20 | 6.83 | 7.29 | 6.84 | 8.30 | 8.05 | 6.95 | 6.90 | 8.43 | 1.26 |

| Property | | | | | | | | | | | | |
|---|---|---|---|---|---|---|---|---|---|---|---|---|
| Flexural strength (psi) | 35,000 | 35,000 | 46,000 | 53,000 | 55,000 | 51,000 | 55,000 | 41,000 | 50,200 | 75,000 | 35,000 | 2580 |
| Compressive strength (psi) | 335,000 | 330,000 | 400,000 | 410,000 | 420,000 | 410,000 | 410,000 | 310,000 | 450,000 | 500,000 | 225,000 | — |
| Tensile strength (psi) | 24,000 | 23,000 | 28,000 | 29,000 | 29,500 | 28,500 | 29,000 | 45,000 | 35,000 | 50,000 | 22,000 | — |
| Impact resistance (Charpy in.-lb) | 7 | 6 | 7 | 7 | 7 | 7 | 7 | 7 | 6.5 | 6.5 | — | — |
| Hardness (Rockwell 45N) | 76 | 76 | 78 | 80 | 78 | 80 | 80 | 76 | 80 | 90 | 60 | — |
| Dielectric strength (V/mil) |  | 250 | 250 | 250 | 250 | 250 | 250 | 250 | 270 | 280 | 240 | — |
| Dielectric constant at 1 MHz | 8.40 | 8.40 | 9.80 | 9.80 | 9.28 | 9.76 | 9.0 | 9.75 | 9.80 | 9.90 | 6.5 | — |
| Dissipation factor at 1 MHz | 0.00082 | 0.00086 | 0.00028 | 0.00015 | 0.00030 | 0.00017 | 0.00030 | 0.00015 | 0.00010 | 0.0003 | 0.0004 | — |
| Loss factor at 1 MHz | 0.00689 | 0.00722 | 0.00270 | 0.00146 | 0.00278 | 0.00166 | 0.0010 | 0.0015 | 0.0010 | 0.003 | 0.004 | — |
| $T_c$ (°C) | 850 | 810 | 540 | 900 | 790 | 900 | 875 | >1000 | >1000 | >1000 | >1000 | — |
| Volume resistivity (ohm-cm) at 25°C | $6.8 \times 10^{13}$ | $6.8 \times 10^{13}$ | $2.6 \times 10^{12}$ |  | $1.6 \times 10^{14}$ | $1.9 \times 10^{14}$ | $2.0 \times 10^{14}$ | $2 \times 10^{14}$ | $1.0 \times 10^{14}$ | $1.0 \times 10^{15}$ | $>1 \times 10^{15}$ | — |
| at 500°C | $6.7 \times 10^{7}$ | $4.4 \times 10^{7}$ | $1.1 \times 10^{6}$ | $8.3 \times 10^{8}$ | $8.5 \times 10^{7}$ | $8.0 \times 10^{8}$ | $1.0 \times 10^{10}$ | $1.8 \times 10^{9}$ | $1.3 \times 10^{9}$ | $3.0 \times 10^{12}$ | $1 \times 10^{10}$ | — |
| at 1000°C | $2.1 \times 10^{5}$ | $1.8 \times 10^{5}$ | $1.3 \times 10^{4}$ | $2.7 \times 10^{5}$ | $3.8 \times 10^{5}$ | $2.9 \times 10^{5}$ | $7.0 \times 10^{5}$ | $7.0 \times 10^{6}$ | $2.1 \times 10^{6}$ | $1.0 \times 10^{7}$ | $1 \times 10^{8}$ | — |

[a] These Frenchtown materials find more use than other materials which Frenchtown has developed for special applications. Other materials are available.

[b] Thermalox 995 is a registered trademark of Brush Wellman.

Courtesy of Frenchtown Ceramics, Inc., Frenchtown, N.J.

a) Mesh opening in a screen.

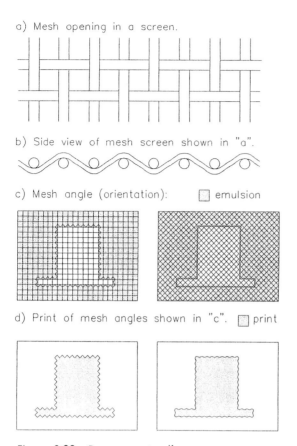

b) Side view of mesh screen shown in "a".

c) Mesh angle (orientation):          ☐ emulsion

d) Print of mesh angles shown in "c".     ☐ print

**Figure 3.20**   Screens vs. stencils.

finer the obtainable print resolution. Fine wire also, unfortunately, means less durability or usable life of the screen. The mesh count is limited by the paste's particle sizes. For example, thick-film pastes consist of very fine particles, while solder pastes and silver-filled epoxy pastes contain much larger particles. Typically, a mesh count of 80 with 0.9 to 1.1 mil wire diameter is used for epoxy and solder pastes. Thick-film pastes can be printed with mesh counts of 200 to 400, again with 0.9 to 1.1 mil wire diameter. For gross patterns, i.e., lines and spaces greater than 0.010 in., or for thick prints, e.g., printing of thick conductor tracks for current bussing, the lower 200 mesh count is sometimes used. However, most ceramic PWBs require fine lines and spaces of 0.004 to 0.008 in. These screens would use 325 to 400 mesh counts.

The wire used for fabricating thick-film screens is stainless steel. The stainless steel provides a durable material that can tolerate the chemicals within the pastes. It is also clean and prevents contamination of the pastes with screen particles. Typically, a stainless steel screen will last for approximately 4 months of continuous use (8-hour shifts each day, 5 days a week). Nylon and polyester mesh can also be used for thick-film screens. They have lower costs but are not as durable and can stretch or be affected by temperature and humidity. These tendencies can lead to the screen losing the tension needed for fine line resolution. Therefore, stainless steel is the norm. Stainless steel screens can also be recycled. If a specific layer pattern is no longer needed, the emulsion on the screen can be removed and a new pattern can be built up on the screen.

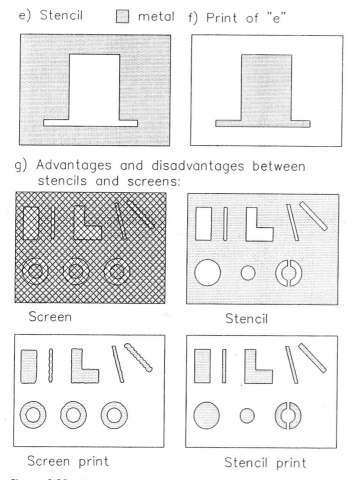

e) Stencil    ▨ metal  f) Print of "e"

g) Advantages and disadvantages between stencils and screens:

Screen

Stencil

Screen print

Stencil print

**Figure 3.20** Continued.

After the wire diameter and mesh count are chosen, the emulsion thickness must be defined. The emulsion thickness will determine the wet print thickness, which in turn determines the final print thickness. Figure 3.20 shows a close-up of the mesh making up the screen. The wire diameter typically used is 0.9 to 1.1 mils. The reciprocal of the mesh count gives the spacing, $S$, or pitch of the wire:

$$S = \frac{1}{\text{mesh count}}$$

The opening, $O$, of the mesh is the spacing minus the wire diameter, $D$:

$$O = S - 2\left(\frac{D}{2}\right) = S - D$$

The percent open area, $P$, of the mesh is the total area minus the area of the wire, or the area of the opening divided by the spacing area:

$$P = \frac{Q^2}{S^2}$$

A typical conductor screen has 1.1 mil wire diameter and a mesh count of 325. This yields a percent open area of

$$P = \frac{[(1/325) - 0.0011]^2}{(1/325)^2} = 0.413 = 41.3\%$$

Knowing the percent open area of the mesh, we can estimate the print thickness for a given emulsion thickness by first calculating the volume of the screen opening. By definition, volume is

$$V = A \times T$$

where $V$ is volume, $A$ area, and $T$ thickness. However, when calculating the open volume of a screen, the volume taken up by the wire mesh must be considered. Thus the open volume of the screen is the sum of the open volume in the emulsion and the open volume of the mesh:

$$V_{total} = V_{emulsion} + V_{mesh}$$

The thickness of the mesh is roughly equal to the wire diameter:

$$T_{mesh} = D$$

Thus the open volume of the mesh is the percent open area times the area times the diameter:

$$V_{mesh} = 0.413 \times A \times D$$

Thus,

$$V_{total} = A \times T_{emulsion} + 0.413 \times A \times D$$

Taking the basic volume equation, we can calculate the wet print thickness:

$$T_{print} = \frac{V_{total}}{A}$$

$$T_{print} = \frac{A \times T_{emulsion} + A \times 0.413 \times D}{A} = T_{emulsion} + 0.413 \times D$$

Thus if an emulsion thickness of 0.0005 in. is used, the wet print thickness will be approximately

$$T_{print} = 0.0005 + 0.413(0.0011) = 0.00095 \text{ in.}$$

or approximately 1 mil.

The emulsion is applied to the screen at the desired thickness. The emulsion that is applied is typically a photosensitive polyvinyl or a polyimide. The emulsion is then patterned using the photolithography methods discussed under Printed Wiring Board Design and Fabrication. Figure 3.21 shows the mask or artwork being placed in contact with the emulsion on the screen. After the screen with mask is exposed to the specified frequency of ultraviolet light and the unpolymerized emulsion is washed away, the pattern is opened up in the screen as shown in Figure 3.22. As Figure 3.22 also shows, the identification

**Figure 3.21**   Photoimaging a screen. (Courtesy of Westinghouse Electronics Systems Group [ESG].)

and alignment markings of the artwork, which are not to be printed on the actual substrate, are filled in with emulsion. This same "blackout" emulsion can be used to make corrections on the screens (e.g., fill in a short between electrically uncommon tracks).

An alternative to printing with a screen is to print with a stencil. A stencil is similar to a screen, except that there is no mesh to deal with. Using the same frame as a screen, a metal foil is attached. The foil is typically nickel or brass. In some cases, a mesh is still attached to the frame and the metal foil is built up on top of the mesh. The mesh directly under the area to be patterned is then etched away, leaving foil stretched over a perimeter of mesh that is in turn stretched over the frame. This makes the stencil more elastic and allows for longer wear. The stencil is patterned using photolithography methods, as discussed earlier in the section on organic PWB fabrication. A photosensitive polymer or photoresist is applied to the foil. A mask is placed over the resist while exposing the surface to ultraviolet light. The exposed positive, or unexposed negative, photoresist is then washed away, leaving windows down to the foil. The stencil is then subjected to an etch, which removes the metal exposed by the photoresist. After the etching process is completed, the remaining photoresist is removed (e.g., by exposing it to the ultraviolet light if it is a positive resist) and the stencil is cleaned.

**Figure 3.22**   Finished screen. (Courtesy of Westinghouse ESG.)

Stencils offer greater resolution, reduced dimensioning capabilities, more uniform print thicknesses, and easier process control. As shown in Figure 3.20, the pattern in the stencil has clean and crisp lines, whereas the screen's lines can have sawtooth edges due to the effect of the mesh. As can be seen in the figure, the artwork pattern should always be oriented along a 45° angle to minimize this sawtooth effect. Furthermore, when the paste is printed or pushed through the screen, it is forced through the mesh. This can result in the deposited material having the mesh pattern embedded in it. A screen print has peaks and valleys, while the stencil print has a smooth top surface yielding a uniform print thickness. In addition, the paste can stick to the mesh during the print and be lifted with the screen after printing. This results in nonuniform deposits and/or smearing of the print. For this reason, the bottom of the screens must be periodically wiped to remove any pastes that might be sticking or even hanging from the mesh. Furthermore, to keep the paste from remaining on the mesh, rather than depositing on the substrate, screen printing must be done with a "snap-off." In other words, as the squeegee forces the paste through the mesh it also pushes down on the screen, placing it in contact with the surface. As the squeegee continues to move across the screen, the screen lifts behind the squeegee and peels away from the surface, so that the paste remains on the surface instead of lifting up with the screen. The snap-off is the distance from the top of the substrate surface to the bottom of the screen mesh. Screen printing is typically done with a 0.025 to 0.060 in. snap-off. As a screen is used and the mesh is repeatedly stretched in this manner, the screen can lose its tension. This can result in poor peeling, smeared prints, and poor alignment. Therefore, the screen tension must be periodically monitored if the printing process is to be controlled.

As discussed earlier, the print thickness with a screen can be estimated but the actual thickness cannot be accurately calculated, nor is it uniform. With a stencil, the paste is flowed through a window or opening, not a mesh. Therefore, the paste has nothing to stick to or be lifted with when the stencil is lifted off the substrate. Therefore, no snap-off is needed with stencil printing. Actually, stencil printing is done with the stencil in contact with the substrate. This contact printing further improves the print resolution by allowing the metal foil of the stencil to seal to the substrate surface. In this way the potential for paste to be forced under the foil around the edges of the pattern is reduced, thus reducing the potential for print smearing. Furthermore, with contact printing through a stencil the wet print thickness is equal to the stencil foil thickness. The foil thickness of the stencil ranges from 0.001 to 0.020 in. Thus solder paste can be printed 0.010 in. thick, epoxy can easily be printed to the desired 0.003 to 0.005 in., and thick-film pastes can be printed to their exact wet print thickness.

The vendor-recommended print thickness should be followed. Typically conductors are printed at 0.001 in. and resistors at 0.0015 in., and dielectric needs a final wet thickness of 0.002 to 0.003 in., depending on the required voltage breakdown limits (the higher the voltage the thicker the dielectric must be). With a stencil it would be possible to print the desired final wet thickness with one stencil print. However, to avoid the possibility of pinholes or pores forming within the dielectric layer, which could allow arcing between signal layers, the dielectric layers are typically printed as two consecutive print, dry, fire steps. This allows the second print/fire to flow dielectric paste into any pinholes or pores left after the first print.

Stencil printing is more easily controlled. By using the stencil, emulsion thickness, wire diameter, mesh count, percent open area, mesh tension, and snap-off parameters are eliminated. There are fewer things to try to control. The remaining screen printing parameters must still be controlled for both screen and stencil printing: squeegee wear and hardness, squeegee pressure and speed, screen or stencil flatness, pattern alignment, and paste viscosity.

The disadvantages of stencil printing are the limitations in possible patterns. Any pattern containing an isolated track within an opening (e.g., a donut) is not possible with a stencil. Since there is no mesh to support the centered, isolated pattern (e.g., the donut hole), it would simply fall out of the stencil. Stencils start to lose their superior print resolution on multilayered designs. As the layers are built up the thick film starts to develop a topography. As the topography grows, the surface flatness is decreased. With the decreased flatness the stencil can no longer seal to the surface. Thus, the paste can start to be forced under the edges of the pattern and smear. Furthermore, stencils are not as durable as screens. Thy lack the elasticity of the mesh. Greater handling care must be used so that the foil is not deformed: punctured, edges rolled or folded, or dimpled. Thus stencils must be more frequently replaced on the production line. This can add to tooling costs and slow down deliveries by having to stop production while the stencil is being replaced.

Once the screens or stencils have been generated from all the artwork layers, the substrate can be screen printed. The substrate blank is first cut to size. This machining

**Figure 3.23**  Laser machining. (Courtesy of Westinghouse ESG.)

of the substrate can be done in three different ways. The most common and versatile method is to use a laser. A carbon dioxide ($CO_2$) laser is used for ceramic machining. YAG lasers are not typically used because of the burned residue they can leave on the ceramic. This residue greatly hinders the adhesion of the thick-film paste to the ceramic. The power needed will depend on the ceramic thickness. If standard 0.025- to 0.040-in. thick alumina is used for the substrate, the power of the laser need only be 100 to 150 W. However, if ceramic window frames are being machined and the ceramic thickness is 0.125 in., then 400 or more watts might be required.

The substrate is loaded into the laser and the laser is programmed to cut the desired pattern and dimensions. The laser then cuts through the ceramic with a high-power beam of laser light as shown in Figure 3.23. This beam actually melts through the ceramic, and the "cut" is a series of pulsed points. The spacing of these pulses will determine the tolerances and smoothness of the cut. The laser beam is typically 0.003 in. As it cuts or melts through the ceramic, the laser leaves a cone-shaped hole as shown in Figure 3.24. The top of the hole is 0.003 in. and the bottom ranges from 0.0005 to 0.0025 in., depending on the thickness of the material being machined. If these holes are overlapped during the laser drilling, then the cut surface is smooth and can yield tolerances of ±0.001 in. If the holes have slight spacings between them, the finished product can have a perforated

(a)                                                                         (b)

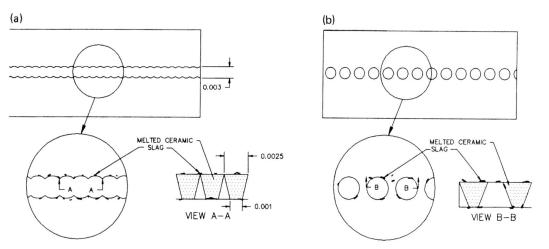

**Figure 3.24**   Laser cut cross section, (a) laser drilling (b) laser scribing.

appearance, with tolerances of approximately 0.003 in. (see Figure 3.24). The process of making spaced holes is referred to as laser scribing. This is typically done when machining a "snapstrate," a substrate that has been divided into an array of substrates that can be processed (i.e., printed) simultaneously. After the layering has been completed the individual substrates can be separated by "snapping" the snapstrate apart along the perforation.

After the substrate has been laser drilled, it must be cleaned. This cleaning can include a deslagging step. The slag is the melted ceramic or glass residue that is left after laser machining (see Figure 3.24). Chemically cleaning to remove the slag would require etching off the glass. Unfortunately, the etch would also attack the substrate surface, resulting in pitting. An early method was to sandblast the slag by forcing a small stream of alumina slurry through the holes at very high pressure. This process effectively removes the slag but can result in microcracking of the ceramic along the machined edges. Furthermore, if the stream is misdirected it can damage the flat surface of the substrate, roughening up the finish. The sandblasting is followed by a degreasing or cleaning to remove surface contaminants. This cleaning may involve sending the substrates through a Freon solvent degreaser or cleaning them in hot isopropyl alcohol (IPA). The solvent 1,1,1-tricholor-ethane was used in the past, but its use has decreased because of its carcinogenic properties. Most facilities conduct a further cleaning step just prior to printing the first layer. This is a heat cleaning step in which the substrate is sent through a furnace profiled with a peak temperature of 800 to 925°C (the same as the profile used to fire the thick-film materials). This "prefire" will burn off any organic contaminates on the surface of the substrate. The laser machining company Laserage has developed a high-temperature cleaning process that is patented under the name LaTite. In this process the substrate is heat treated at temperatures approaching the ceramic reflow temperature. This process burns off the slag left by laser machining and any organic contaminant. It also leaves the ceramic with an as-fired surface, which eliminates any microcracks that may have formed during the machining due to the thermal stress imposed by the laser [2].

The other ceramic machining processes are diamond scribing/cutting and ultrasonic milling. The former process can be used only on thinner substrates (less than 0.040 in.)

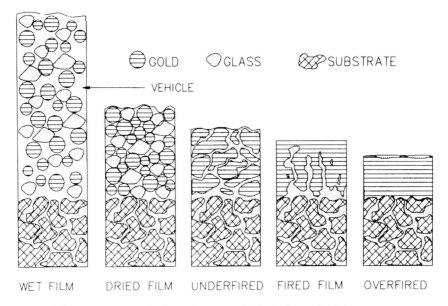

**Figure 3.25**   Thick-film adhesion. (Courtesy of Westinghouse ESG.)

and only to cut straight lines through the entire length of the substrate. First, a line is scribed into the surface of the ceramic with a diamond-tipped cutting tool. This fine line imposes a weakness in the lattice structure of the ceramic. When the substrate is snapped apart, the snap follows the lattice line, leaving a very clean and smooth cut surface. Minimal microcracking results. But, as mentioned before, this process can be used only if a straight-line cut through the entire length is desired—it can yield only a square-shaped substrate. Furthermore, the right angle corners of the substrate have high stress concentration, which can result in the corners chipping during subsequent processing and handling.

Ultrasonic milling, like laser drilling, can be used to machine the ceramic into any desired shape or dimension. This process involves machining a carbide or stainless steel milling tool for every substrate design. The tool is vibrated against the substrate ultrasonically, while an aqueous solution is continually supplied at the milling surface. This solution allows for smooth cuts and reduces the potential for microcracking due to the thermal stress imposed by the temperature gradient created by the heat of friction. Microcracking, however, can still result from the vibrational stress of the process. This process is very slow—one 0.125-in.-thick alumina substrate can take over 1 hour to machine—compared with laser machining several substrates in less than a minute. Furthermore, the process demands the use of a milling tool, which must be replaced periodically as its cutting edges dull. The laser program can be stored and recalled at any time to cut a specific substrate design. All three machining methods must be followed by a cleaning process (e.g., solvent cleaning and/or prefiring) prior to screen printing.

With the substrate cut to size and cleaned and all stencils or screens in place, the metallization process can begin. Thick-film metallization is done on unpolished, fired or pressed blanks. The as-fired surface is smooth enough to allow uniform deposits, yet is rough enough to yield good adhesion. Polishing the substrates could actually decrease the adhesion of the conductors. The polished surface would not yield the needed roughness for the glass frit (glass particles) within the thick-film paste to "grab" the surface of the substrate. Figure 3.25 shows the methodology of thick-film paste adhesion to a ceramic substrate. The paste contains the desired metals or conductors, the glass frit for bonding to the ceramic, and the needed solvents for the rheology of a printable paste. After the paste is printed or deposited on the substrate, it is dried to volatize the solvents. The dried substrate is then fired at the reflow temperature of the paste. Here, the metals start to fuse together on the surface to form the conductor tracks, while the glass melts into the substrate, forming the thick-film bonding method.

If the thick-film paste is underfired, not maintained at the peak temperature for the proper amount of time, the conductor particles cannot melt or fuse together to leave a smooth solid metal surface for component mounting or wire bonding. Likewise, the glass cannot reflow and bond as well with the ceramic, thus reducing the adhesion strength of the film. Overfiring, however, can also cause degradation. If the paste is overfired, all the conductor particles fuse together. This can force glass to rise to the surface of the film, reducing the bondability of component mounting and wire bonding at the surface. It can also reduce the adhesion between the conductor and glass by forming an interface between these two materials. For these reasons, it is crucial that the furnace profile be optimized and controlled during the firing process.

Figure 3.26 shows a substrate being printed. Here the substrate is pushed up against

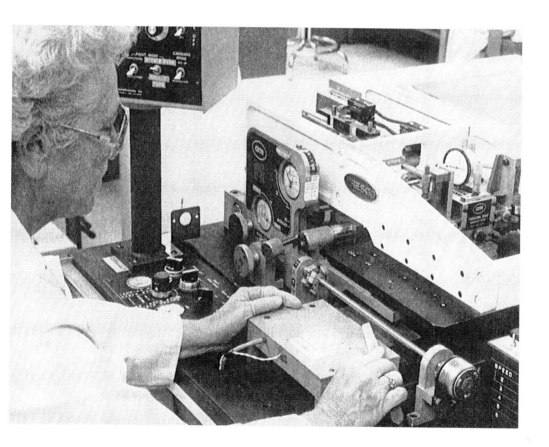

**Figure 3.26**  Screen printing. (Courtesy of Westinghouse ESG.)

**Figure 3.27**  Thick-film drying. (Courtesy of Westinghouse ESG.)

alignment pins on the print station platform. The substrate orientation is verified and the alignment of the screen/stencil to the substrate is verified under a microscope with $10 \times$ to $30 \times$ magnification. After the substrates are printed they are dried as shown in Figure 3.27. Drying the paste reduces the voltatiles within the paste and allows the paste to "set up" on the substrate. At this point the print is inspected. If a problem is found, the dried paste can be washed off with alcohol and the layer reprinted. Figure 3.28 shows the dried substrates being sent through a furnace profiled to fire the paste. Figure 3.29 shows a batch of substrate exiting the furnace. These steps are repeated for each layer. Figure 3.30 shows a substrate at various steps within the multilayering.

When printing resistors, each resistor layer is printed and dried and then all resistors are fired together. The cofired resistors are the last high-temperature firing. Subsequent high-temperature firings can further oxidize the resistor paste, which can greatly increase the resistance of the material. After the resistors are printed, a low-temperature (425 to 525°C) glass encapsulant can be printed and fired over selected resistors and tracks as a protective overcoat or solder mask. Resistor encapsulants must be composed of the proper refractory glasses to allow a laser beam to pass through them and trim the resistors to value. Figure 3.31 shows a substrate ready to be laser trimmed. In this process, the substrate and all the resistors on it are probed so that the laser equipment can monitor

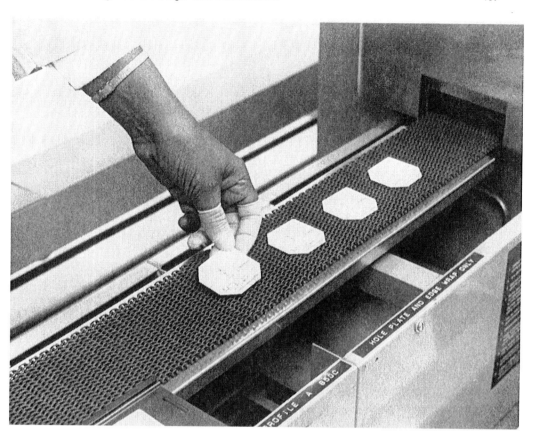

**Figure 3.28**   Thick-film firing. (Courtesy of Westinghouse ESG.)

the resistor values during the trimming process. Chapter 2 discusses thick-film resistor design and how the trimming process alters the resistor values. The laser used to trim resistors has a lower power than those used to drill or machine substrates. Here the laser need only penetrate the resistor and not the ceramic below it. Care must be taken when programming the laser so that the laser beam will not cut through the dielectric layer below the resistors (on multilayer substrates).

The camber of thick-film boards can be greatly affected by metallization processing. The thick-film pastes are applied over the entire surface of one or both sides of the substrates and fired at 800 to 950°C. The final thickness of the thick film can be equivalent to over half the blank substrate thickness. For example a 0.025-in. blank substrate with 10 conductor layers would have 9 dielectric layers with a final fired thickness of 0.0015 in. minimum each. Thus the substrate would have a minimum thick-film thickness of 9 × 0.0015 + 0.0005 (for the top conductor) or 0.014 in., a little over half the thickness of the blank substrate. The dielectric and conductor materials, having different CTEs than the ceramic, would contract less than the ceramic upon postfire cooling, restricting the contraction of the top surface of the substrate. The effect would be to warp the substrate so that the metallized surface would bow up. The effects of this warpage can change the initial camber of 0.003 in./in. to 0.009 in./in. This warpage can be minimized with the

**Figure 3.29** Firing furnace. (Courtesy of Westinghouse ESG.)

proper process control and design. Since the warpage occurs during firing, reducing the number of firings can reduce the warpage. This can be accomplished by optimizing the routing to eliminate unnecessary layers or by cofiring layers (e.g., instead of a print, dry, fire, print, dry, fire dielectric layer sequence, the dielectric layers can be processed by print, dry, print, dry, fire). Another way to minimize the warpage effects is to balance the firing and material deposits on both sides of the substrate. This can be accomplished by alternating prints/firing on both sides (i.e., print and fire layers 1–3 of side A, then print layers 1–3 of side B, then print layers 4–6 of side A, etc.) or by printing one layer of material with a large CTE mismatch on the bottom side for every three or four top-side layers.

## Thin-Film Metallization

Thin-film metallization requires that the ceramic substrate surface be very flat and smooth. For these reasons ceramics of higher percent purity are used (e.g., 99% alumina, 99.9% beryllia). The surfaces of these substrates might even have to be polished to yield finer finishes.

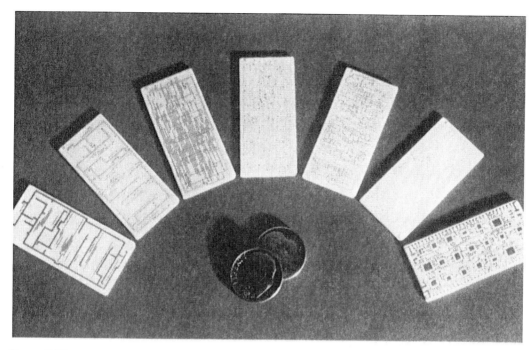

**Figure 3.30** Multilayered ceramic substrate. (Courtesy of Westinghouse ESG.)

Thin-film metallized ceramic PWBs are not commonly used. Thin-film ceramics are difficult to multilayer and are more expensive than their thick-film counterparts. For these reasons, thin film is normally restricted to single-layer applications or specialized designs. If the PWB design can be done in single layer and requires high power dissipation, then a thin-film beryllia substrate would be used in preference to a thick-film one because of the better adhesion offered by thin-film metallization on beryllia. Other applications might include microstrips or similar microwave devices. Here the thin-film metallization offers improved electrical performance and frequencies over thick-film substrates.

Interconnection from one layer to the next is done with buried vias, just as is the case with all ceramic PWBs. However, through holes—metallized holes through the ceramic—can be used to enable connection of the top and bottom metallization of the substrate or to provide a more direct thermal path, a path of less thermal resistance. Through holes or thermal vias in cofired substrates are discussed under Cofired Metallization. Thick-film substrates can have through holes. These must be laser drilled and deslagged, as discussed under Thick-Film Metallization. The holes must then be filled or coated with the thick-film paste. This normally requires additional processing steps. One

**Figure 3.31**  Resistor trimming. (Courtesy of Westinghouse ESG.)

advantage that thin-film metallizing offers is the case of metallizing all surfaces. The side walls of the substrate and/or of a laser-drilled hole in the substrate can be electroless plated along with the top surface of the substrate. Figure 3.32 shows a thin-film ceramic substrate with metallized through holes.

Thin-film ceramics are metallized and patterned using plating, photolithography, etching, vapor depositions, and sputtering methods. These methods are discussed in detail in Chapter 2 under Integrated Circuits and in the present chapter under Organic Printed Wiring Boards.

### Cofired Metallization

In this process the metallization is printed onto unfired ceramic tape and then cofired with the ceramic. The metals used in this process are refractory metals. Noble metals (e.g., gold, silver, copper) cannot be used because their vaporization temperatures are lower than the ceramic reflow temperature. The refractory metals are molybdenum, tungsten copper, and tungsten, the last being the norm. These refractory metals have higher vaporization temperatures and lower thermal and electrical conductivities than their noble

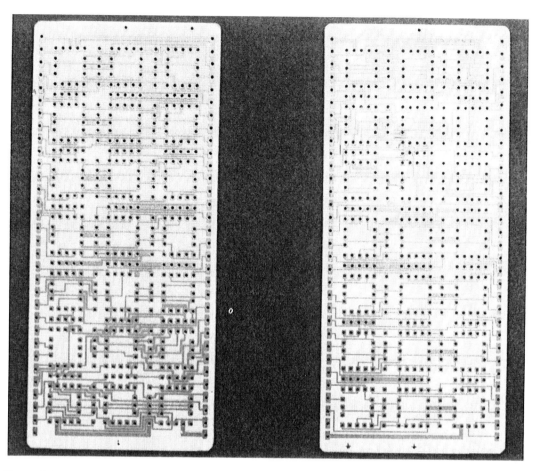

**Figure 3.32**   Thin film ceramic PWB with through holes. (Courtesy of Westinghouse ESG.)

metal counterparts. Because of the lower conductivities, cofired PWBs have typically been limited to digital applications. Table 3.2 lists the properties of metals commonly used in the cofired process.

The metal is applied in paste form, being printed in the same fashion as thick-film pastes on already fired substrates. Interconnection between the layers is done with buried vias. Prior to printing the metal paste onto the tape, vias or holes are punched into the tape. When the paste is printed onto the tape these holes are filled with the metal. The different layers are stacked, laminated, and fired as described earlier when discussing fired ceramic under Substrate Formation.

After the metal and ceramic have been cofired, the exposed refractory metal is electroplated with nickel and gold. The nickel provides a barrier metal that will not form degrading intermetallics and will not cause an electrolytic reaction between the surface gold and tungsten base metal. This nickel plate is typically 0.000080 to 0.000350 in. (as required in the military standards for a corrosion barrier). The gold is plated on top of

**Table 3.2**  Material Thermal Properties

| Materials | K (W/in.-°C) | CTE (10⁻⁶/°C) | Young's modulus (10⁶ psi) | Specific gravity (g/cm³) | Specific heat (cal/g-°C) | Max use temperature (°C) |
|---|---|---|---|---|---|---|
| AL 6061-T6 | 4.5 (0–50°C)<br>4.26 (50–100°C) | | | | | |
| AL 1100 | 5.66 (0–50°C)<br>5.35 (50–100°C) | | | | | |
| Alumina | | | | | | |
| 92% | 0.42 (at 20°C)<br>0.54 | 6.9 | 39 | 3.6 | 0.19 | 1500 |
| 94% | 0.56 (25–75°C)<br>0.48 (75–125°C) | 6.2 | 45 | 3.67 | 0.19 | 1550 |
| 96% | | 7.0 | | 3.73 | | 1650 |
| 99.5% | 0.81 (25–75°C)<br>0.61 (75–125°C) | 6.39 | | 3.93 | | 1705 |
| BeO 99.9% | 7.2 (25–75°C)<br>5.8 (75–125°C) | 7.0 | 21 | 2.8 | 0.25 | 1700 |
| AlN | 4.25 | 4.6 | | | | |
| Glass 7052 | 0.425 | 5.63 | 8.2 | 2.27 | | 436 (strain point) |
| Kovar | 2.17 | 15 | 20 | 8.4 | 0.104 | |
| Nickel | | 4.95–5.6 | 30 | 8.8 | 0.13 | |
| Molybdenum | 4.12 (at 20°C)<br>3.57 (at 100°C) | | 50 | | | |
| Steel | 1.49 (at 20°C)<br>1.70 (at 100°C) | 14 | 28 | 8.7 | | |

| | | | | | |
|---|---|---|---|---|---|
| SS 304 | 3.51 (at 20°C) | 19 | 28 | 7.8 | 0.12 |
| | 4.15 (at 100°C) | | | | |
| Alloy 42 | 0.38 (at 20°C) | 7.3 | 20 | 8.2 | 0.12 |
| | 0.44 (at 100°C) | | | | |
| Alloy 52 | 0.35 | 9.8 | | | |
| OFHC copper | 10.1 | 17.5 | 17 | 8.9 | 0.12 |
| CuFe | 6.77 | 16.3 | | | 0.092 |
| WCu10 | 5.96 (at 20°C) | 6.0 | 37 | 17.2 | 0.039 |
| | 5.47 (at 100°C) | | | | |
| WCu15 | 5.96 (at 20°C) | 6.5 | 36 | 16.5 | 0.041 |
| | 5.74 (at 100°C) | | | | |
| WCu20 | 6.39 (at 20°C) | 7.0 | 35 | 15.6 | 0.044 |
| | 6.28 (at 100°C) | | | | |
| WCu25 | 6.71 (at 20°C) | 7.5 | 34 | 14.8 | 0.047 |
| | 6.82 (at 100°C) | | | | |
| GaAs | 1.125 | | | | |
| SiC | 6.75 | | | | |
| AuGe | 6.4 | | | | |
| AuSn (80/20) | 8.8 | | | | |
| Sn62 | 1.27 | | | | |
| InPb (70/30) | 1.62 | | | | |
| AuSi | 6.91 | | | | |
| Thick film | | | | | |
| Conductor | 7.55 | | | | |
| Dielectric | 0.053 | | | | |

the nickel to provide a solderable and/or wire-bondable surface. The minimum plating thickness required for solderability is 0.000050 in., while wire bondability would require a minimum thickness of 0.000075 to 0.000100 in. for gold and aluminum wire bonds, respectively. The military standards set a maximum thickness of 0.000275 in. This reduces the possibility of solder joint embrittlement due to the formation of gold-tin intermetallics. It also reduces product costs by reducing the amount of gold used in the system.

In addition to cofired ceramic tape systems, many companies have come out with dielectric tape systems. The dielectric tapes are formed in the same manner as the ceramic tapes (see Substrate Formation), but instead of the high-reflow-temperature ceramic material, these tapes are composed of other dielectric materials that reflow at much lower temperatures. These materials include glasses similar to those found in dielectric thick-film pastes. Since these tapes can be fired at much lower temperatures, noble metal pastes, similar to those used in thick-film processing, can be printed and cofired on the tape. These tapes offer a cofired substrate that can be used in microwave and high-speed applications. Furthermore, no addition postfiring plating is needed for these dielectric tape systems.

Both cofired methods offer certain advantages over thin- or thick-film processing. Multilayering capabilities of cofired boards are limited only by the board thickness limitations. Boards of 20 or more layers are not uncommon in advanced technological applications such as avionics. Some military contractors have even fabricated boards approaching 100 layers. Each layer of a cofired board will have a 0.003 to 0.012 in. fired thickness. A 20-layer board will have a board thickness of 0.060 to 0.240 in., depending on what prefired thickness of tape was used. Thus, if the board has a specified maximum thickness of 0.040 in., the maximum number of layers the board can have is 13.

Another advantage of cofired processing is the ability to incorporate thermal vias in the design. An array of vias or holes can be punched and subsequently filled with conductors in each layer. These vias can be stacked up through all the layers in such a way that they will provide a thermal path from directly under the component to the heat sink on which the substrates are mounted. Therefore, even though the dielectric tape systems do not offer high material conductivity, they can be designed to offer a very conductive path through which the heat can dissipate.

Cofired boards can also be designed to contain cavities. The top layers may have windows cut out of them so that when the layers are stacked up and laminated together a cavity will be formed, as is done with multilayer ceramic chip carriers. This would allow a high-dissipating and/or tall component to be mounted in a cavity so that its thermal path/resistance is reduced and/or the board profile (i.e., assembled height) is minimized. In another application of integral windows/cavities, a resistor can be buried in a inner layer with a window above it to the surface to allow for laser trimming the resistor to value.

Cofired designs, unlike thick- or thin-film ceramic designs but similar to organic ones, allow all layers to be patterned before they are laminated together. Thus each layer can be inspected prior to their final stack-up, and if one layer has a defect it can be reworked or reprinted and substituted into the stack. For example, if layer 3 has an open or a short (due to an overlooked defect in the artwork or screen, or a smeared or incomplete print), it can be manually fixed by hand painting on the missing piece of track or wiping off the shorting smear of conductor. Even if the defect cannot be fixed, only the tapes of layer 3 have to be reprocessed and fit into the stack-up. If the same problem is found after firing the fourth layer in a thick-film board, nothing can be done to rework the defect and the entire lot of boards must be scrapped and a new lot started. All the materials and labor that went into the first four layers are lost.

The disadvantages of cofired systems is the longer life cycle time they require. Each layer requires a screen or stencil and punch tooling. The via punch can be accomplished with "soft" or "hard" tooling. Soft tooling is actually a program input and stored on automatic punch equipment. The tape is first punched with alignment holes, which fit over the alignment pins on all the punching, printing, and laminating workstations. The punching equipment has a robotic arm that picks up the desired punching tool (the punch with the desired diameter) and moves it to the preprogrammed $X$, $Y$ coordinates on the workstation. This tooling can be easily modified if a design change is made in the layout, i.e., if vias are added, deleted, or moved. However, this method punches only one via at a time. A complicated design consisting of many layers can have thousands of vias. Thus this method can be costly in terms of labor and turnaround times. For this reason, soft tooling is typically restricted to low-volume prototyping.

Production labor costs mandate the use of hard tooling. Hard tooling consists of a plate in which all needed punches have been incorporated. Thus all vias within a given layer are punched in one step. These tools cannot be easily modified. Design changes in the layout and via positioning would require new hard tooling to be designed and fabricated. Furthermore, these hard tools would have to be replaced periodically as the punches within them dulled.

The need for each layer to have its own set of tooling increases the design-to-fabrication transition time. It also adds to the nonrecurring costs of the designs. Furthermore, ceramic cofired systems require high-temperature processing, which dictates the need for expensive capital equipment. However, once the design is mature and the tooling and equipment are in place, the costs and time to produce cofired boards are very low. For these reasons, suppliers of cofired boards have a rule of thumb that cofired boards are cost effective if the design has four or more layers and is to be produced at medium-volume levels. Figure 3.33 shows a small cofired ceramic PWB with footprints for surface mounting leadless ceramic chip carriers (LCCs).

## Direct-Bonded Copper

A direct-bonded copper (DBCu) substrate consists of copper foil bonded directly to a ceramic substrate. As with the other ceramic PWBs already discussed, the ceramic most commonly used is alumina. Direct bonding has also been successfully accomplished with beryllia and aluminum nitride, which offer great improvements in thermal conductivity. However, the structure of direct-bonded copper to alumina also offers great thermal conductivity and structural improvements over conventional thick- or thin-film metallized alumina.

In this process a copper foil is treated so that a copper oxide is grown on its outer surface. The foil is then placed against a ceramic substrate. The two pieces are sent through a furnace profiled to reflow the copper oxide. This causes the oxides within the ceramic to fuse together with the copper oxide, thus directly bonding the two materials together. Care must be taken that the unit does not reach the reflow temperature of the copper, so that the foil will not lose its shape.

The copper foil is available in the range of 0.001 to 0.080 in. The ceramic thickness that can be used is any currently available thickness (e.g., 0.005 to 0.125 in.). It is recommended that the ceramic thickness be equal to or greater than the foil thickness to be attached, to prevent the copper from causing the substrate to warp or crack.

**Figure 3.33**  Co-Fired PWB. (Courtesy of Westinghouse ESG.)

The two materials bond together at a temperature of approximately 1000°C. As the new composite material cools, the copper, which has a much higher CTE than the ceramic, will contract much faster than the ceramic. This places the ceramic in compression, thus increasing its tensile strength by an order of magnitude. Therefore, DBCu is much stronger than either material is individually. Because of this property, very thin ceramic substrates can be used to reduce the PWB assembly height and, more important, vastly reduce the thermal resistance of the PWB. Coupling a decrease in thermal path (thickness) with the ability of the copper to spread the heat rapidly over its area, a DBCu substrate made with alumina can offer reductions in thermal resistance comparable to a thick film BeO substrate design, as shown in Table 3.3.

The copper foil can be patterned by punching prior to attachment to the ceramic, or it can be attached and patterned by photolithography and etching methods after bonding to the ceramic. (See organic PWB fabrication for details of the photolithography and etching processes.) The later process enables fine line resolution, provided that the copper is very thin. If a thick copper foil (over 0.001 in.) is used, undercutting can be a problem and can greatly limit the dimensioning. Prepunching the foil also limits the dimensioning. Typically, DBCu subsurates have minimum lines and spaces of 0.015 in.

**Table 3.3** Thermal Resistance of a Minimal-Thickness Two-Layer DBCu (Alumina) Substrate Versus a Minimal-Thickness Two-Layer BeO/Thick-Film Substrate

| Material | Thickness (in.) | Conductance (W/in.-°C) | Thermal resistance (in.$^2$-°C/W) |
|---|---|---|---|
| **DBCu** | | | |
| Copper | 0.001 | 10 | 0.001/10 = 0.0001 |
| Alumina | 0.005 | 0.9 | 0.005/0.9 = 0.0055 |
| Copper | 0.001 | 10 | 0.001/10 = 0.0001 |
| Alumina | 0.005 | 0.9 | 0.005/0.9 = 0.0055 |
| | | | 0.0112 |
| **BeO** | | | |
| TF gold | 0.0005 | 7.55 | 0.0005/7.55 = 0.00006 |
| TF diel. | 0.0015 | 0.05 | 0.0015/0.05 = 0.030 |
| TF gold | 0.0005 | 7.55 | 0.0005/7.55 = 0.00006 |
| BeO | 0.020 | 6.35 | 0.020/6.35  = 0.0032 |
| | | | 0.03332 |

Multilayering is accomplished by stacking up alternate layers of copper and ceramic. Interconnection from one layer to the next is accomplished with buried vias, made by filling windows in the ceramic with copper spheres or particles before placing the next copper foil on top of the stack. Thus far, multilayering of a DBCu substrate has been limited to three conductor layers.

The advantages of DBCu are improved thermal management and structural strength. DBCu also has excellent current-carrying capabilities. Furthermore, the top copper metallization offers excellent solderability, and bondability when nickel and gold plated. The disadvantages are the limitations in dimensioning and multilayering.

## 3.2.4 Advantages of Ceramic PWBs

Ceramic boards have many advantages over organic ones. They are rigid by nature and do not require the support of a carrier plate or framing. This rigidity is only slightly affected by processing. Cofired boards have the same camber (flatness) as the fixture used to hold them during firing. Therefore, cofired substrates can be very flat, with cambers typically 0.002 to 0.003 in./in. and even as low as 0.001 in./in. Component mounting, wire bonding, or package sealing is done at a relatively low temperature compared with the firing temperature of the ceramic. Epoxy component mounting is typically done at 150°C, solder mounting at 183 to 240°C, and wire bonding at approximately 300°C—all of which are much lower than the 1600°C needed to fire ceramic. Furthermore, none of these processes add significant deposits of materials. The volume and area of the component attachment material or the wire bond are insignificant compared with the overall size and mass of the ceramic PWB. Thus these additions do not affect the tension or compression of the PWB, and the camber of a cofired board is not significantly affected by this processing. As discussed under Thick-Film Metallization, the camber of a thick-film ceramic board can also be controlled to yield a rigid and flat PWB.

Ceramic boards can offer improved thermal management. The ceramic materials have higher thermal conductivities than their organic counterparts. They can be many times more conductive if more advanced ceramics such as aluminum nitride are used. Both organic and ceramic PWBs can have thermal vias incorporated in their layouts to improve their thermal performance. However, a ceramic board, with its smaller buried thermal vias, can provide a low-resistance thermal path while sacrificing less routing area.

As later chapters will point out, thermal stress is the main cause of failure in electronic assemblies. Leadless ceramic components, such as ceramic chip carriers and chip capacitors and resistors, can be directly mounted to a ceramic PWB. These components have a CTE matched to that of the PWB. Thus thermal stresses due to different expansion and contraction rates are eliminated. The only potential thermal stress is that imposed by the formation of temperature gradients during processing. Chapter 4 will discuss the formation of such gradients and how they can be minimized. With this reduction of potential thermal stress, ceramic PWBs can offer highly reliable leadless surface-mounted assemblies. By eliminating the leads on all of the components, the area and volume of ceramic PWB assemblies can be greatly reduced. Chapter 4 will give examples of the potential decreases. These reductions can be very crucial in military and avionics applications, where size and weight limitations are specified.

The disadvantages of ceramic PWBs are the increased costs and life cycle or design time. The ceramic and noble metal (e.g., gold alloy) materials are more costly than the polyimide or epoxy glass and copper materials used in organic PWB designs. Ceramic board fabrication requires additional tooling, further increasing design time and costs. Furthermore, most commercial applications do not require the improved thermal management or reduced board sizes, so ceramic boards are not in as much demand as organic ones. Ceramic PWBs have typically been limited to low-volume military or avionics applications. This lower demand has limited the emergence of large automated ceramic PWB facilities and has also helped to maintain the higher costs of ceramic designs.

## 3.3  DEVELOPMENTAL SUBSTRATES

Advances in printed circuit technology are ongoing. Many new methods and applications for printed circuits are currently being developed. Some of these methods involve new materials and techniques (e.g., programmable silicon), while others are new combinations of standard materials and methods that have been in use for years (e.g., thin-film polyimide on ceramic or silicon).

Programmable substrates are made of a noncrystalline silicon alloy. The material is structured so that the nonconductive, noncrystalline silicon alloy is sandwiched between an array of points of conductive material. Connections from one layer to the next can be obtained by applying a specified voltage at these points. This voltage causes the silicon to crystallize and become conductive.

Several companies are now marketing thin-film multilayered silicon or ceramic substrates. These substrates have polyimide applied over their surface by spinning or screen printing on the polyimide. The polyimide is then metallized and patterned using thin-film processes. Thus these substrates combine standard thick-film screening and thin-film metallizing techniques. The difference is that instead of thick-film glass dielectric layers, the substrate has polyimide dielectric layers. These substrates offer very fine line resolution due to the thin-film/photolithographic metallizing processes (e.g., 0.001–0.002 in. lines and spaces). These boards are being put to use in multichip packaging (MCP), where bare dice are mounted and wire bonded directly to the substrate.

## 3.4  ELECTRICAL PARAMETERS: LAYOUT DESIGN

The design of integral components within a PCB is the responsibility of the mechanical/layout designer. Thick- and thin-film resistors and capacitors within ceramic PCBs are designed in the same manner discussed in Chapter 2. In addition to the design of these components, certain electrical properties must be designed into the circuitry. The most common is line resistance. Every component, track, and part of the circuit will have a resistance. Certain connections within the net list require that the resistance be kept under specified levels, or in multiple-channel devices the resistance of parallel signals must be matched.

The resistance $(R)$ in ohms of a conductor is defined as

$$R = \rho \frac{l}{A}$$

where $\rho$ is the specific resistance or resistivity (ohm-cm); $l$ the length of the conductor (cm), and $A$ the area (cm) [3, p. F-89].

If the resistance of a signal between two mounting pads must be kept under 0.1 $\Omega$; how far apart can the two pads be placed on an organic PWB or a ceramic thick-film PWB?

For a ceramic thick-film PWB, if we assume that the standard line thickness of 0.008 in. is used and that the metallization is the standard multilayer thick-film gold, we can calculate the maximum length the track can be to meet the maximum resistance of 0.100 $\Omega$.

Using the vendor literature for multilayer gold No. 8880 of Electro-Science Laboratories, Inc., we find that the conductor's recommended fired thickness is 0.005 in. and it has a resistivity of 0.002 to 0.004 $\Omega$/square.

The number of squares within a given track of uniform width is the number of equal-sided units laid side by side within the track. For a uniform track this is equal to the length divided by the width:

$$\text{Number of squares} = \frac{l}{w}$$

Knowing the resistivity per square, we can find the resistance:

$$R = \rho \times \text{number of squares}$$

Combining the two previous equations and substituting the given maximum values, we can calculate for $R$:

$$0.1\Omega = \frac{0.004\ \Omega}{\text{squares}} \times \frac{l}{0.008\ \text{inches}}$$

$$l = \frac{0.1\ \Omega}{0.004\ \Omega} \times 0.008\ \text{in.} = 0.200\ \text{in.}$$

The two pads must be within 200 mils of each other.

For an organic PWB, the standard metallization is 1 ounce copper printed 0.008 in. wide. Given that the specific gravity of copper is 8.89 and the resistivity is 1.7241 microhm-cm [3, p. E-72], we can calculate the print thickness and the maximum length of the track.

"One ounce copper" implies that 1 ounce of copper was placed over 1 square foot of polyimide prior to patterning. Dividing this number by the density of the copper yields the thickness of the copper metallization:

$$\frac{1 \text{ oz/ft}^2}{8.89 \text{ g/cm}^3} = \frac{1 \text{ oz}}{\text{ft}^2} \times \frac{\text{cm}^3}{8.89 \text{ g}} \times \frac{1 \text{ lb}}{16 \text{ oz}} \times \frac{454 \text{ g}}{1 \text{ lb}}$$

$$\times \frac{1 \text{ ft}^2}{144 \text{ in.}^2} \times \frac{1 \text{ in.}^3}{2.54^3 \text{ cm}^3} = 0.00135 \text{ in.}$$

Thus the print thickness is 0.00135 in. This times the print width will yield the cross-sectional area of the conductor:

$$0.00135 \text{ in.} \times 0.008 \text{ in.} = 0.0000108 \text{ in.}^2 = 1.08 \times 10^{-5} \text{ in.}^2$$

Substituting all these values into the definition of resistance of a conductor, we have

$$0.1 \, \Omega = 1.724 \times 10^{-6} \, \Omega\text{-cm} \times \frac{l}{1.08 \times 10^{-5} \text{ in.}^2}$$

Solving for the length,

$$l = 1.59 \text{ in.}$$

Another common characteristic that must be designed into the layout is current capacity. We know that

$$I = \frac{V}{R}$$

where $I$ = current (amperes) $V$ voltage (volts), and $R$ resistance ($\Omega$) [3, p. F-71]. Substituting the equation for resistance of a conductor, we have

$$I = \frac{V}{\rho(l/A)}$$

How thick would a track have to be to be able to carry 60 amps on an organic PWB operating at 5 V?

From the above example we know that an organic PWB uses 0.00135-in.-thick copper tracks and has a resistivity of $1.724 \times 10^{-6}$ $\Omega$-cm. We also know that the cross-sectional area is

$$A = t \times w$$

where $t$ is the thickness and $w$ the width of the track.

Substituting in the above equations and values, we have

$$60 \text{ A} = 5 \frac{V}{1.724 \times 10^{-6} \, \Omega\text{-cm} \times (l/0.00135 \text{ in.} \times w)}$$

Solving for $w$,

$$w = 0.006l$$

Thus the width would have to be a minimum of 0.006 times the maximum length.

## EXERCISES

1. What is the difference between a PWB and a PCB?
2. What is the difference between insertion and surface mounting?
3. What are the benefits of surface mounting?
4. Describe the different ways to interconnect signals on different layers?
5. List 4 metallization techniques for ceramic PWBs.
6. Compare organic PWBs to ceramic PWBs.
7. Provide a flow chart for the fabrication process of (a) an multilayer organic PWB, (b) a thick film multilayer ceramic PWB, (c) a cofired multilayer ceramic PWB.
8. Compare the thermal resistance of a DBCu, a polyimide PWB, and a thick film ceramic PWB. List all assumptions.
10. Describe 3 different ways to reduce the resistance of a signal.
11. Design the necessary track configuration on a 3 ounce copper PWB for a signal capable of carrying 3 amps and requiring a 13 ohm maximum resistance.
12. Design the same track given in question 11 for (a) a cofired ceramic PWB, (b) a thick film ceramic PWB, and (c) a DBCu PWB.

## REFERENCES

1. Interview with Keith Mason, Senior Ceramist, Thick Film Laboratory, Westinghouse Electronics Systems Group, Baltimore, November 22, 1989.
2. Capp, L, and Luther, R. R., "Analysis of the Effect of Laser Machining on 96% Alumina Ceramic Substrates and the Advantages of New LaTITE Finish," in *Proceedings of the 1985 International Symposium on Microelectronics*, Reston, Va., International Society of Hybrid Microelectronics, 1985.
3. Weast, R. C., et al., eds., *Handbook of Chemistry and Physics*, 52nd ed., Chemical Rubber Co., Cleveland, 1971–1972.

## SUGGESTED READING

Brzozowski, V. J., Lee, K., and Harris, D. B., "Printed Circuit," in *Encyclopedia of Science and Technology*, McGraw-Hill, New York, in press.

Clark, R. H., *Handbook of Printed Circuit Manufacturing*, Van Nostrand Reinhold, New York, 1985.

Coombs, C. F., *Printed Circuits Handbook*, McGraw-Hill, New York,

Dally, J. W., *Electronic Packaging—A Mechanical Engineering Perspective*, New York, 1989.

Foley, E., and Rees, G., "Preliminary Investigation of Potential Beryllium Exposure While Laser Trimming Resistors on Beryllia Substrates," Brush Wellman, Inc., Elmore, Ohio, 1984.

Grove, A. S., *Physics and Technology of Semiconductor Devices*, Wiley, New York, 1967.

Iwase, N., and Tsuge, A., "Development of High Thermal Conductive A/N Ceramic Substrate Technology," *Int. J. Hybrid Microelectron.* 7(4): 49–53 (1984).

Iwase, N., et al., "Thick Film and Direct Bond Copper Forming Technologies for Aluminum Nitride Substrate," Toshiba R & D Center, Kawasaki, Japan, 1985.

Kear, F. W., *Printed Circuit Assembly Manufacturing*, Marcel Dekker, New York, 1987.

Povolny, H. R., ''Thick Film Base-Metal Conductors on Beryllia,'' Brush Wellman, Inc., Elmore, Ohio, 1983.

Powers, M. B., ''Potential Beryllium Exposure While Processing Beryllia Ceramics for Electronic Applications,'' Brush Wellman, Inc., 1985.

''Predicting Thermal Resistance in Ceramic Integrated Circuit Packages,'' Brush Wellman, Inc., Cleveland, 1984.

# 4

# Electronic Assemblies

**Denise Burkus Harris** *Westinghouse Electric Corporation, Baltimore,*
*Maryland*

Electronic assemblies consist of any number of electronic and nonelectronic devices (screws, heat sinks, carrier plates, etc.) that are grouped together to perform a function. The assembly is a self-contained production item. A production item is built within a particular facility in which each product follows its own process flow. For example, an insertion-mounted printed wiring board (PWB) can be built in the same facility as a surface-mounted PWB. However, different equipment and processes are used. Both boards can be populated with the same pick-and-place machine, but solder reflow is commonly done with a wave solder system for insertion boards, whereas surface mount boards use solder paste or thick solder plate that is reflowed in a vapor phase or infrared (IR) furnace.

There can be many levels of assembly within a given system. The final system, or top assembly, can contain many subassemblies. Each of these subassemblies can contain its own subassemblies, and so on. Most subassemblies are manufactured as deliverable, self-contained units, built at various facilities. The system house obtains all of these subassemblies and integrates them into the final system.

The parts used within a system are commonly listed according to the subassemblies they make up. This list starts with the final system at the top, with the major subassemblies as branches below. These subassemblies then branch out further with their own subassemblies and components. This listing is referred to as a system parts list, system breakdown, or Xmas tree. Figure 4.1 gives an example of a common way to document a system Xmas tree. In this case a subassembly is denoted by indenting in the list.

This chapter describes the basic subassemblies commonly used within deliverable electronic units and the assembly processing associated with each subassembly. It then demonstrates how these subassemblies are grouped together to form new assemblies,

**153**

Final System: Transmitter for a Radar
    20 #6 Screws
    1 Rack
    1 Cover
    1 Power supply module
        REF 234553 Test Spec
        1 Module housing
        1 Cover
        1 Power supply PWB assembly
            1 PWB
            25 $3\mu$ capacitors
            10 $0.1\mu$ capacitors
            1 Transformer
            1 Power hybrid
              1 Case
              1 Cover
              1 Solder preform
              2 FET assemblies
                1 FET
                1 BeO tab, thin-film metallized
                AR 0.010-in.-diameter Al wire
              1 Logic substrate assembly
                1 Substrate
                4 2N2222 transistors
                2 Rectifier chips (ICs)
                3 $0.01\mu$ capacitors
                AR conductive epoxy
                AR 0.001-in.diameter Au wire
            6 LM139 DIPs
            8 Com-04 cer paks
    1 Matrix plate
        32 8-pin connectors
        1 10-pin connector
        15 ft $\frac{1}{4}$-in. tubing
        33 Cables
    32 Transmitter modules
        1 Module housing
        1 Cover
        1 T/R PWB assembly
            1 PWB
            15 $2\mu$ capacitors
            1 Gate array
              1 Gate array die
              1 84-pin leaded chip carrier
            1 Stripline
            1 Combiner
            4 Regulator hybrids
              1 Case
              1 Cover
              1 Epoxy preform
              1 Substrate assembly
                1 Substrate
                2 10-k$\Omega$ resistor
                3 1N3600 diodes
                3 $0.033\mu$ capacitors
                AR conductive epoxy
                AR 0.001-in.-diameter Au wire
        1 Epoxy preform
        1 Cold plate
    64 Cables

**Figure 4.1**   System parts list or Xmas tree.

which are then used as the building blocks for larger assemblies. Finally, the integration of these assemblies to form the final system will be discussed.

In the example in Figure 4.1, each indentation indicates a lower level of assembly, or a subassembly within the given heading. The quantities given are for one assembly. For example, there are 32 transmitter modules in the radar. Each transmitter module has one module housing, one cover, one PWB assembly, and so on. In reality, when ordering all the piece parts for the radar, 32 module housings, 32 covers, 32 PWB assemblies, and so on must be ordered. AR stands for "as required." This term is used for materials purchased in bulk quantities. REF stands for reference.

## 4.1 ELEMENTARY SUBASSEMBLIES

### 4.1.1 Die Assemblies

The simplest subassemblies are known as die assemblies. They consist of a bare die, such as a transistor or diode, mounted on a tab having a conductive surface. Figure 4.2 shows examples of typical die assemblies. These die assemblies are then mounted on top of a substrate which is protectively overcoated (in commercial applications) or within a hermetically sealed package or hybrid (in military applications). There are three reasons for building these assemblies:

1. These assemblies need only be slightly bigger then the dice themselves, making them much smaller then a discretely packaged or "canned" diode or transistor. Why not just mount the dice directly on the higher assembly's substrate to further decrease the size needed by the dice? This is, indeed, the common practice used in hybrid assemblies. However, by first mounting the dice on tabs and not directly on the substrate, it is possible to test the dice functionally prior to committing them to the more expensive higher assembly.

2. With the ability to pretest these die assemblies, they are utilized for low-yield dice, such as application-specific integrated circuits (ASICs); high-power dice, such as field effect transistors (FETs); or when it is essential to "match" pairs of dice for functional reasons.

3. The tab allows for easier rework. When a die is determined to be inoperative, it must be replaced. To replace the die, the substrate assembly must be heated to reflow the solder or soften the epoxy with which the die is mounted. The die is then extracted from the substrate. Since most dice are only 0.010-in.-thick silicon, they tend to come off in pieces. These pieces must be carefully pried off one at a time. Extreme care must be taken to not damage the neighboring components in the process. Die assemblies with the support of the tab lift up in one piece. This greatly eases the reworkability of the assembly.

### Die Assembly Materials

The tabs used for die assemblies are metallized ceramic or metal that has been plated to allow for proper mounting and wire bonding. Metal tabs provide for the back plane of the die assembly to have the same electrical function as the back plane of the die. If a transistor is mounted to a metal tab, the tab becomes the collector. The back plane input or output (I/O), the collector in the case of a transistor, can now be accessed by connecting to the back plane of the die assembly or by probing or connecting to the top of the tab. The most common metal tabs are kovar or molybdenum. These metals are chosen because

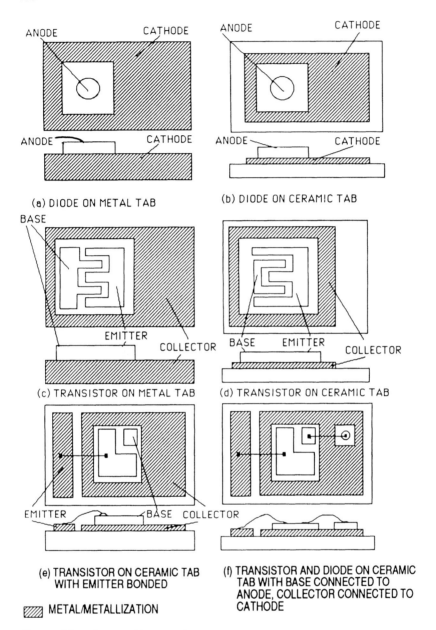

ANODE          CATHODE          ANODE          CATHODE

ANODE          CATHODE          ANODE          CATHODE

(a) DIODE ON METAL TAB          (b) DIODE ON CERAMIC TAB

BASE

EMITTER
          COLLECTOR     BASE     EMITTER
                                            COLLECTOR

(c) TRANSISTOR ON METAL TAB     (d) TRANSISTOR ON CERAMIC TAB

EMITTER          BASE  COLLECTOR

(e) TRANSISTOR ON CERAMIC TAB          (f) TRANSISTOR AND DIODE ON CERAMIC
    WITH EMITTER BONDED                     TAB WITH BASE CONNECTED TO
                                            ANODE, COLLECTOR CONNECTED TO
▨▨ METAL/METALLIZATION                      CATHODE

**Figure 4.2**  Die assembly examples.

of their high thermal conductivities and because their coefficients of thermal expansion (CTEs) are between that of the silicon dice mounted on top of them and that of the ceramic substrate to which they are mounted. The metal tabs are then plated, first with 0.0001 to 0.0003 in. of nickel to provide a corrosion barrier metal and then with 0.000050 to 0.000150 in. of gold to allow for wire bonding and solderability. If aluminum wire bonding is used, the minimum thickness of gold that is recommended is 0.0001 in.

Ceramic tabs are employed when it is not desired to have the backplane of the die assembly active. By mounting the die on a ceramic tab, the surface metallization of the tab becomes the collector, in the case of our transistor. The ceramic tabs are metallized with thick-film gold alloys or thin-film metallization. If 96% alumina is used, the ceramic substrate is laser drilled, dividing the substrate into an array of the desired tab dimensions. This perforated substrate is called a snapstrate. The thick film is printed and fired onto the snapstrate in an array of bonding pads within the perforation. The snapstrate is then snapped apart, forming all of the individual tabs. The back side of the snapstrate may be left as bare ceramic if the die assemblies are to be dielectric epoxy mounted, or it may be metallized in the same manner as the top surface to allow for solder mounting of the die assemblies. If thin-film metallization is desired, 99% alumina is used. These substrates then have a base metal, such as Cr, Ti, W, or Cu, deposited on them. They are then Ni-Au plated in the same manner as metal tabs. After plating is completed, the substrates are diamond sawed into the desired dimensions for the tabs. If BeO tabs are required to provide high thermal conductivity, they are metallized using thin-film methods. Thick film does not provide good adhesion to the BeO. Figure 4.3a and b show the different methodologies of bonding thick-film metals to alumina and BeO, respectively.

## Die Assembly Processing

*Die Mounting*

The dice are mounted to the tabs using solder or eutectic mounting methods. Figure 4.4 shows a typical phase diagram for a solder. The $x$ axis of the diagram gives the composition of the alloy. The $y$ axis is temperature. The liquidus lines indicate the temperature at which a given composition will completely melt, or be in a liquid phase. The solidus lines indicate the temperature at which a given composition will crystallize, or become completely solid. The areas between the liquidus and solidus lines represent the plastic zones of a given composition. The plastic zone is one in which both liquid and solid phases are present. In other words, as the temperature is increased for a given composition, that composition starts to melt when the temperature matches that of the solidus and has completely melted when the temperature matches that of the liquidus. The eutectic is the point in which the solidus and liquidus intersect. A eutectic alloy is one whose composition yields the eutectic point; it is the composition with the lowest melting point, going from solid to liquid at one temperature (i.e., a eutectic alloy has no plastic zone). Eutectic die mounting is the method for mounting die which uses a eutectic alloy as the means of attachment.

Eutectic mounting can be accomplished with or without the use of a solder preform. In the latter case, a silicon die is placed on the gold-plated tab. The tab in turn is placed on the hot plate of a die bonding station. Dry nitrogen is blown over the die mounting area, or the entire process is done in an environmental chamber that is filled with dry nitrogen. The temperatures of the tab and die are brought up to approximately 400 to 420°C, about 20 to 40°C above the gold silicon eutectic point. The die is gently rubbed back and forth or "scrubbed" into the gold plating until the gold and silicon melt together to form the eutectic alloy. At this point the die is eutectically bonded to the tab. This method is very labor intensive and has a high potential for damaging the die during the scrubbing process.

The most common die mounting method is to mount the die to the tab using a preform of the desired eutectic alloy. This method is referred to as eutectic mounting or solder preform mounting. The choice of alloy depends on the metallization of the substrate and the back plane of the die and on the desired reflow temperature. Other considerations

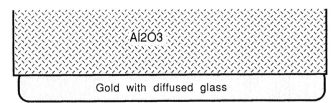

1. Thick film gold is screened onto the alumina.

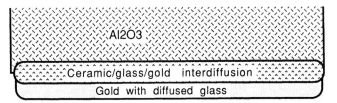

2. Gold is fired onto the alumina causing an interface area of interdiffused alumina/glass with the gold/glass. This forms a very strong metallurgical bond.

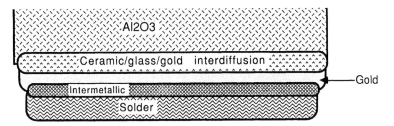

3. Solder reflow forms an intermetallic between the gold and solder.

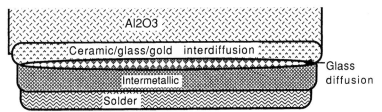

4. The intermetallic grows with exposure to heat until all the gold has gone into solution. Possibly at this point, glass diffuses into the intermetallic.

(a)

**Figure 4.3**   Thick-film metallization to (a) alumina ($Al_2O_3$) and (b) beryllia (BeO).

include the formation of intermetallics, which can affect the long-term reliability of the unit. For example 80% gold 20% tin has a eutectic point of 280°C. This temperature is high enough that surface mounting of the substrate with eutectic lead-tin (63% lead 37% tin, which reflows at 183°C) can be done without risk of reflowing the die mount, but it is lower than some of the other commonly used gold eutectics, thus limiting the exposure of the die to extreme temperature. This alloy also has good wetting properties to both the gold metallization commonly found on the substrate and the diffused gold on the die backplane. However, gold-tin intermetallics are readily formed at relatively low tem-

BeO fabrication and assembly:

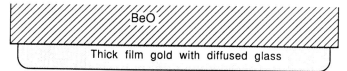

1. Gold is fired onto the BeO.

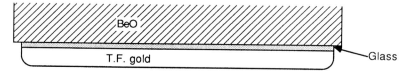

2. Subsequent firings causes the glass in the gold to migrate to the BeO surface.

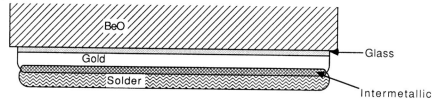

3. Solder is reflowed onto the gold, an intermatallic of gold/solder is formed.

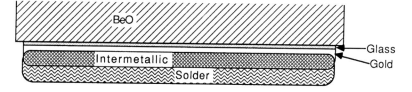

4. Exposure to temperatures as low as 100ºC. causes growth of the intermetallic.

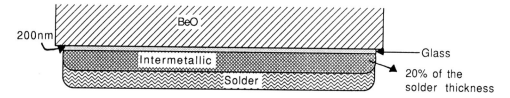

5. Intermetallic growth causes complete solution of the gold into the solder.

**(b)**

peratures. These Au-Sn intermetallics are very brittle and can lead to mechanical fracture failures. Silver eutectics are not brittle but silver can migrate, leading to small changes in resistivity over time. Silver is also notorious for corrosion problems. Table 4.1 lists some commonly used solder alloys.

Once the alloy is determined, the die can be mounted by a method similar to the Au-Si scrubbing method discussed earlier (however, the preform, which is punched with a

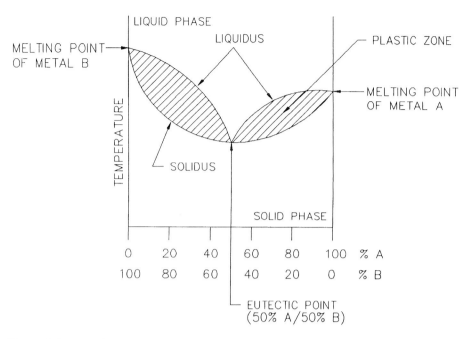

**Figure 4.4** Eutectic alloy phase diagram.

**Table 4.1** Solder Alloys

| Temperature (°C) | | | Elements (%) | | | | | | | | |
|---|---|---|---|---|---|---|---|---|---|---|---|
| Solidus | Liquidus | Eutectic | Pb | Sn | Ag | Ln | Sb | Bi | Cd | Au | Zn |
| | | 93 | | 42.0 | | 44.0 | | | 14.0 | | |
| | | 95 | 32.0 | 15.5 | | | | 52.5 | | | |
| | | 117 | | 48.0 | | 52.0 | | | | | |
| 118 | 125 | | | 50.0 | | 50.0 | | | | | |
| | | 125 | 43.5 | | | | | 56.5 | | | |
| | | 125 | 9.6 | 15.0 | | 70.0 | | | 5.4 | | |
| | | 138 | | | | 42.0 | | 58.0 | | | |
| | | 139 | | 43.0 | | | | 57.0 | | | |
| | | 143 | | | 3.0 | 97.0 | | | | | |
| | | 145 | 30.6 | 51.2 | | | | | 18.2 | | |
| 155 | 149 | | 15.0 | | 5.0 | 80.0 | | | | | |
| | | 146 | 32.0 | 50.0 | | | | | 18.0 | | |
| | | 162 | 18.0 | 70.0 | | 12.0 | | | | | |
| 144 | 163 | | 43.0 | 43.0 | | | | 14.0 | | | |
| 156 | 165 | | 25.0 | | | 75.0 | | | | | |
| 160 | 174 | | 30.0 | | | 70.0 | | | | | |
| | | 177 | | 67.8 | | | | | 32.2 | | |
| | | 179 | 36.1 | 62.5 | 1.4 | | | | | | |
| | | 183 | 38.1 | 61.9 | | | | | | | |

**Table 4.1** Continued.

| Temperature (°C) | | | Elements (%) | | | | | | | | |
|---|---|---|---|---|---|---|---|---|---|---|---|
| Solidus | Liquidus | Eutectic | Pb | Sn | Ag | Ln | Sb | Bi | Cd | Au | Zn |
| 183 | 186 | | 35.0 | 65.0 | | | | | | | |
| 179 | 189 | | 36.0 | 62.0 | 2.0 | | | | | | |
| 183 | 190 | | 40.0 | 60.0 | | | | | | | |
| 183 | 192 | | 30.0 | 70.0 | | | | | | | |
| 183 | 195 | | 25.0 | 75.0 | | | | | | | |
| 183 | 195 | | 42.0 | 58.0 | | | | | | | |
| | | 198 | | 91.0 | | | | | | | 9.0 |
| 183 | 202 | | 20.0 | 80.0 | | | | | | | |
| 183 | 203 | | 45.0 | 55.0 | | | | | | | |
| 180 | 209 | | 50.0 | | | 50.0 | | | | | |
| 183 | 209 | | 15.0 | 85.0 | | | | | | | |
| 183 | 214 | | 50.0 | 50.0 | | | | | | | |
| 183 | 215 | | 10.0 | 90.0 | | | | | | | |
| | | 215 | 85.0 | | 15.0 | | | | | | |
| 183 | 218 | | 48.0 | 52.0 | | | | | | | |
| | | 221 | | 96.5 | 3.5 | | | | | | |
| 183 | 224 | | 5.0 | 95.0 | | | | | | | |
| 183 | 225 | | 55.0 | 45.0 | | | | | | | |
| 179 | 227 | | 35.5 | 61.5 | 3.0 | | | | | | |
| 221 | 229 | | | 96.0 | 4.0 | | | | | | |
| 183 | 238 | | 60.0 | 40.0 | | | | | | | |
| 233 | 240 | | | 95.0 | | | 5.0 | | | | |
| 183 | 242 | | 62.0 | 38.0 | | | | | | | |
| 221 | 245 | | | 95.0 | 5.0 | | | | | | |
| 179 | 246 | | 36.0 | 60.0 | 4.0 | | | | | | |
| | | 246 | | 89.5 | | | 10.5 | | | | |
| 183 | 247 | | 65.0 | 35.0 | | | | | | | |
| | | 248 | 82.6 | | | | | | 17.4 | | |
| | | 252 | 88.9 | | | | 11.1 | | | | |
| 183 | 258 | | 70.0 | 30.0 | | | | | | | |
| 250 | 264 | | 75.0 | | | 25.0 | | | | | |
| 183 | 268 | | 75.0 | 25.0 | | | | | | | |
| | | 280 | | 20.0 | | | | | | 80.0 | |
| 183 | 280 | | 80.0 | 20.0 | | | | | | | |
| 225 | 290 | | 85.0 | 15.0 | | | | | | | |
| 268 | 299 | | 88.0 | 10.0 | 2.0 | | | | | | |
| 296 | 301 | | 93.5 | 5.0 | 1.5 | | | | | | |
| 268 | 302 | | 90.0 | 10.0 | | | | | | | |
| | | 304 | 97.5 | | 2.5 | | | | | | |
| | | 309 | 97.5 | 1.0 | 1.5 | | | | | | |
| 301 | 314 | | 95.0 | 5.0 | | | | | | | |
| | | 318 | 99.5 | | | | | | | | 0.5 |
| 316 | 322 | | 98.0 | 2.0 | | | | | | | |
| 304 | 365 | | 94.5 | | 5.5 | | | | | | |

Courtesy of the International Society for Hybrid Microelectronics.

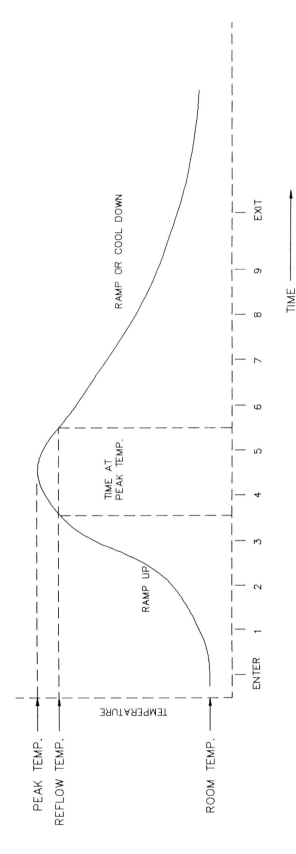

**Figure 4.5**  A typical furnace profile. This example shows a typical 9 zone furnace profile for thick film paste. The parts enter at room temperature. The first 3 zones are programmed so that they will ramp up the temperature quickly enough for the paste to reach its reflow temperature before the active solvents totally volatilize. Zone 4, set at the peak temperature (approximately 30–50 C higher than the reflow temperature) guarantees that the paste will reach and maintain the reflow temperature for the proper duration to accomplish total reflow, without overfiring. The remaining zones are programmed to cool down the parts fast enough to establish proper grain growth, but slow enough to prevent high thermal stressing. The parts exit the furnace at a slightly elevated temperature, but cool enough so that hitting room temperature air will not thermally stress them.

"cookie cutter" die or simply cut out of the alloy foil with scissors, is placed between the die and the mounting surface). The other method used in high-volume production is furnace mounting or fluxless solder reflow. In this method the die and preform are placed on the tab or substrate and sent through a furnace filled with forming gas. Forming gas is approximately 5% hydrogen and 95% nitrogen. This gas gives a reducing atmosphere, which prevents corrosion and contamination while allowing the wetting to occur. Industrial furnaces normally have three to nine zones. Each zone is programmed to maintain a specified gas flow and temperature. The parts are placed on a belt that travels at a programmed speed through the furnace. The combination of gas flow, belt speed, and zone temperatures yields a profile of the temperature that the part reaches. A profile is determined by attaching a thermocouple to a load of parts and running it through the furnace. A typical profile is shown in Figure 4.5.

*Interconnection*

Interconnection to the die assembly is most commonly accomplished with wire bonds. The standard wire used is 0.0007 or 0.001-in. gold. For higher currents, larger-diameter gold wires, up to 0.005 in., can be used. However, because of the price of gold, this means of interconnection can be costly. Therefore, when better current-carrying capacity than that of 0.001 in. gold wire is required, 0.005- to 0.010-in.-diameter aluminum wire is employed. Table 4.2 lists the current-carrying capabilities and fusing limits of some standard-size wires.

Gold wire bonds are made using thermocompression or thermosonic bonding methods. In both cases the wire is fed through a capillary. Heat is applied to melt the wire and form a "ball" of material at the end of the wire. This ball is then "pressed" (for thermocompression bonds) or "scrubbed" into place on the die. These bonds are called ball bonds. The latter technique involves applying heat and pressure while vibrating the wire at ultrasonic frequencies, hence the term thermosonic bonding. The wire is then pulled through the capillary and looped over to the bond pad on the substrate, where the capillary presses or scrubs the wire into the gold-metallized bonding pad on the substrate, forming the "stitch." The wire is then broken, leaving a "tail." Figure 4.6a shows the steps of forming a ball bond.

When gold (substrate metallization) and aluminum (wire) are placed together and sufficient heat is applied, gold-aluminum intermetallics can form. These intermetallics are very brittle and can greatly weaken the bonding strength of the wire bond. For this reason, thermal bonding methods cannot be used to form aluminum wire bonds. Aluminum wire bonds are formed using ultrasonic bonding techniques. Ultrasonic techniques are similar to thermosonic techniques but are performed at room temperature. Because no heat is applied, no ball is formed. Both ends of the wire bond are stitched. These stitches

**Table 4.2**  Wire Bond Fusing Currents

| Wire diameter (in.) | Nonfusing current | Fusing current | Maximum wire length (in.) |
|---|---|---|---|
| 0.0007 (gold) | 160 mA | — | 0.070 |
| 0.001 (gold) | 250 mA | 1 A | 0.090 |
| 0.005 (aluminum) | 3 A | 5–7 A | 0.400 |
| 0.010 (aluminum) | 10 A | 17 A | 0.500 |

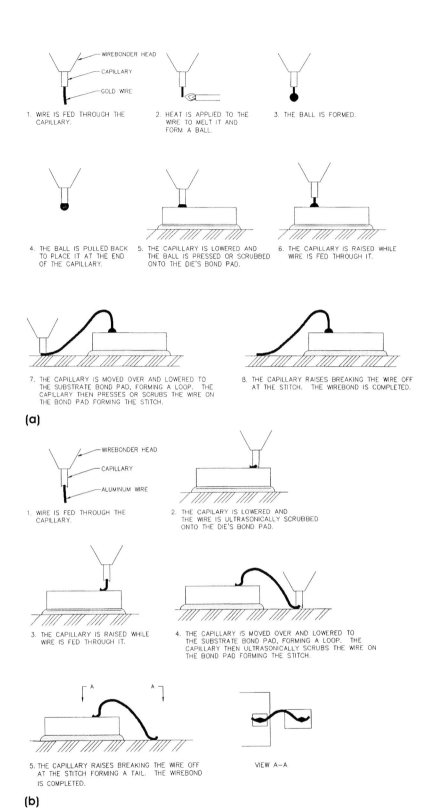

**Figure 4.6**  Formation of (a) a ball wire bond (thermosonic or thermocompression gold) and (b) a wedge wire bond (ultrasonic aluminum).

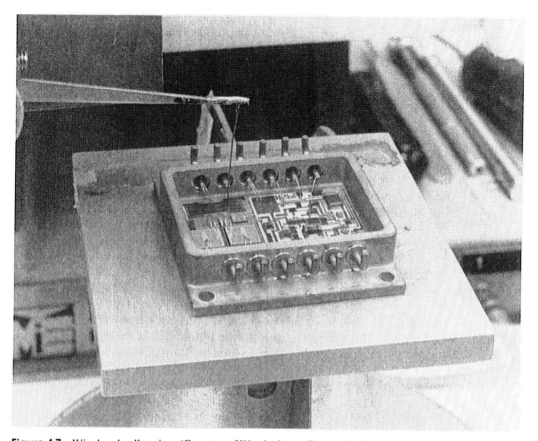

**Figure 4.7**   Wire bond pull testing. (Courtesy of Westinghouse Electronics Systems Group [ESG].)

are also called wedges, and the bonds are known as wedge bonds. Figure 4.6b depicts the formation of a wedge bond.

Military and space applications specify that the bonds used be tested. The standard test used is called "pull testing" and is a military standard test. It can be done as a sample destructive test, when the wires are pulled until they fail. The failure is recorded by documenting the type of failure (ball lift, break in wire loop, break at stitch, etc.) and the force at which the failure occurred. Nondestructive pull testing pulls each wire to a specified force. If the wire bond does not break or fail, the unit passes the test. Figure 4.7 shows a wire bond being pull tested.

## Applications

Military and space applications typically require the maximum reductions in weight, area, and volume, along with the highest reliability possible. Such applications also often require the use of state-of-the-art or leading-edge technology. For example, a military electronic system might be a radar for a fighter jet. This radar must have the most advanced capabilities, yet be small and lightweight so that it can fit into the nose of the aircraft. Use of discrete packages is very uncommon because of the extra volume and weight they impose. Despite their higher cost, bare dice assemblies (i.e., hybrids) are commonly used as a means of meeting the sizing limitations of the system. In addition to increased packing densities, military applications generally have higher power densities. These higher densities often require that certain transistors be tested at high temperatures prior to mounting within the sealed unit, or that these transistors be mounted on a high-spreading tab (e.g., a metallized BeO tab) to prevent thermal stressing of the dice during high-temperature operations and thermal cycling tests. Furthermore, the advanced functionality of the military system often requires the use of a new die that is still immature in its design. The yields of such units are usually low, so it is necessary to test them functionally prior to placement within the hybrid. This testing can be accomplished only when the die is mounted to a conductive surface because of the die's active backplane.

Die assemblies are not commonly used in commercial applications. Most commercial applications do not need to restrict their sizes to merit the extra expense of additional subassemblies and, therefore, use discretely packaged dice. Furthermore, most commercial applications use products that are mature and produced in high volume. These discretely packaged devices are very inexpensive (transistors can be as little as $0.30 each when bought in large quantities) and have high yield rates because of their simplicity and maturity. In other words, these devices have been produced for so long in such high volumes that the processes used to produce them and the products themselves have been fully debugged and optimized.

Die assemblies are used in commercial products that have high-temperature environments. For example, car ignitions require the use of FET transistors, which must be capable of withstanding the high-temperature environment of the car engine. Discretely packaged FETs, with the extra interfaces of the discrete package, its mounting material, and its board material, would not yield adequate heat dissipation to maintain a proper junction temperature during operation. Therefore, a die assembly consisting of the necessary FET eutectically mounted to a highly thermally conductive tab such as molybdenum is used. This die assembly is then encapsulated with a high-temperature protective overcoat.

Another commercial application that requires the use of die assemblies is electronic sensing devices used in oil drilling. These are miniature electronic devices located at the tip of the drill. These assemblies are used to sense the conditions of the drill site, e.g., the pressure, temperature, and viscosity. This information is then transmitted through the drill rod to the surface, where it is analyzed to determine how close the drill is to hitting the oil deposit. The environment here can exceed 200°C due to the thermal heat of the earth and to the friction experienced by the drill.

## 4.1.2  Capacitor Banks

Capacitors, as discussed in Chapter 3, are commonly used in electronic systems. They are used to regulate the current and to store energy for the electronic assembly. In high-power applications, higher farad values are needed to meet the functionality of the assembly. In most cases, several capacitors must be placed in parallel to obtain the necessary capacitance.

For capacitors in series the total capacitance ($C_T$) is the reciprocal of the sum of the reciprocal capacitances of the individual capacitors:

$$C_T = \frac{1}{1/(C_1 + 1/C_2 + 1/C_3 + \cdots + 1/C_n}$$

For capacitors in parallel, the total capacitance is the sum of the individual capacitances:

$$\stackrel{|}{\underset{|}{=}} \; C_T \,(5 \text{ F}) \quad = \quad \stackrel{|}{\underset{|}{=}} \; C_1 \,(1 \text{ F}) \quad \stackrel{|}{=} \; C_2 \,(1 \text{ F}) \quad \stackrel{|}{=} \; C_3 \,(1 \text{ F}) \quad \stackrel{|}{=} \; C_4 \,(2 \text{ F})$$

where $C_T$ is the total capacitance; $C_1$, $C_2$, $C_3$, . . . , $C_n$ are the values of the individual capacitors; F is the symbol for farad, the standard measure of capacitance; and the standard symbol for a capacitor is:

$$\stackrel{|}{\underset{|}{=}}$$

Capacitors can constitute the major portion of the parts list for a power supply. Power supplies, in turn, are needed in almost all systems, to provide and regulate the necessary power for the system.

Placing multiple capacitors flat on a board so that they are hooked up in parallel can take up a great deal of surface area or board real estate. For this reason, capacitor (cap) bank subassemblies are sometimes employed. In these subassemblies the capacitors are stood up on end and placed side by side. The bodies of the caps are bonded together with dielectric epoxy to prevent shorting the end terminations of any one cap. The end terminations of one side of the bank are bonded together with conductive epoxy or a low-temperature solder (e.g., 60/40 SnPb or eutectic SnPb). In some applications electrical isolation is required between end terminations of different caps.

Figure 4.8 shows an example of a cap bank. All the negative terminations are grounded to the floor of the module housing with conductive epoxy. To obtain electrical isolation between the positive terminations, small pieces of kapton tape are placed between each cap with dielectric epoxy. If the caps were ceramic with gold-plated terminations, the connections to the positive terminations could be accomplished with wire bonds or by attaching a gold ribbon with conductive epoxy. If the caps were tinned or had solder-coated terminations, a jumper wire could be hand soldered to the appropriate termination. In the example shown in Figure 4.8, a small ceramic board with a metallized pattern is placed over the caps. Tolerancing was calculated to allow for the cap leads to fit into the drilled holes of the board. The metallized pattern is only on the top surface of the board so that the ceramic material of the board, in direct contact with the caps, will act as an insulator. The leads are connected to the ring of metal around the holes with solder or conductive epoxy. Attached to these rings are the bonding pads needed for making the connections to the appropriate I/O on the associate substrate. Fuse wires can also be added to this design, as needed.

## 4.1.3  Microwave and Radio-Frequency Subassemblies

Microwave functionality is governed by mechanical design as much as by electrical design. The placement, orientation, and proximity of components to one another can greatly influence the electromagnetic fields of the devices and thus affect the functionality of the system when operating at the frequencies associated with microwave applications. At radio frequencies (RFs), signal speeds are very fast. The signal or current travels along the metal surface. If the path length is too long, it can slow down the signal. For this reason, in cases where signal speed is the top priority, the signal path must be shortened

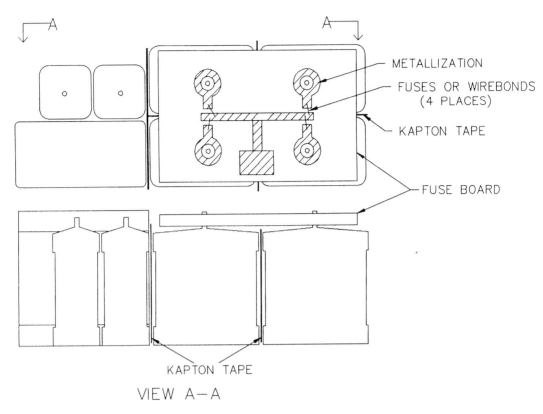

**Figure 4.8** Capacitor bank with interconnecting board and fuses.

by reducing the resistance. Resistance can be reduced by having the signal path made of a high electrically conductive material, by using highly conductive die attach materials, and by shortening the length of the paths as much as possible. In some extreme cases, the only component placement that will meet the operating rise times required by the design is to stack the components on top of each other. Figure 4.9 displays the differences in resistance between components placed side by side and those stacked on top of each other. In other cases, electromagnetic interference (EMI) shielding is required. To provide this shielding, metal barriers are added to the dice assembly or made an integral part of the module housing.

Just as component placement affects the electromagnetic fields, so does the configuration of the wire bonds. When wire bonds are formed, the wire is looped between the connection points. The loop of the wire bond acts as a loop within an inductor coil. This loop with its imposed inductance and capacitance can change the functionality of the circuit design. For this reason, ribbon bonds are commonly used in microwave applications. A ribbon bond does not "loop" up. Therefore, the ribbon does not add unwanted inductance to the connection, as does a wire bond. The potential inductance can be further limited by tacking down the ribbon along the connection path, either by welding it to the track at multiple locations or by simply tacking it down to the substrate surface with conductive epoxy. Figure 4.10 demonstrates the induced inductance of wire bonds versus ribbon bonds [1]. The ribbon is connected to the die and the substrate metallization by split-tip

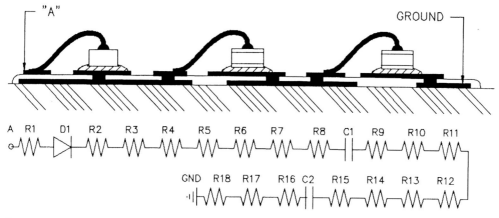

R1 = RESISTANCE OF WIREBOND TO DIODE
D1 = DIODE
R2 = RESISTANCE OF DIODE EPOXY ATTACH
R3 = RESISTANCE OF DIODE MOUNTING PAD
R4 = RESISTANCE OF VIA DOWN FROM MOUNTING PAD DOWN TO BURIED LAYER
R5 = RESISTANCE OF BURIED LAYER
R6 = RESISTANCE OF VIA UP TO C1 WIREBOND PAD.
R7 = RESISTANCE OF C1 WIREBOND PAD
R8 = RESISTANCE OF C1 WIREBOND
C1 = FIRST PARALLEL PLATE CAPACITOR
R9 = RESISTANCE OF C1 EPOXY ATTACH
R10 = RESISTANCE OF C1 MOUNTING PAD
R11 = RESISTANCE OF VIA FROM C1 MOUNTING PAD TO BURIED LAYER
R12 = RESISTANCE OF BURIED LAYER
R13 = RESISTANCE OF VIA UP TO C2 WIREBOND PAD
R14 = RESISTANCE OF C2 WIREBOND PAD
R15 = RESISTANCE OF WIREBOND TO C2
C2 = SECOND CAPACITOR
R16 = RESISTANCE OF C2 EPOXY ATTACH
R17 = RESISTANCE OF C2 MOUNTING PAD
R18 = RESISTANCE OF VIA DOWN TO BURIED GROUND PLANE

(a)

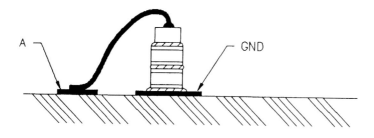

R1 = RESISTANCE OF WIREBOND
D1 = DIODE
R2 = RESISTANCE OF EPOXY BETWEEN D1 AND C1
C1 = FIRST CAPACITOR
R3 = RESISTANCE OF EPOXY BETWEEN C1 AND C2
C2 = SECOND CAPACITOR
R4 = RESISATNCE OF EPOXY BETWEEN C2 AND GROUND PAD

(b)

**Figure 4.9** Component placement effects on signal resistance; (a) standard mounting of one diode and two capacitors; (b) stacked mounting of same diode and capacitors.

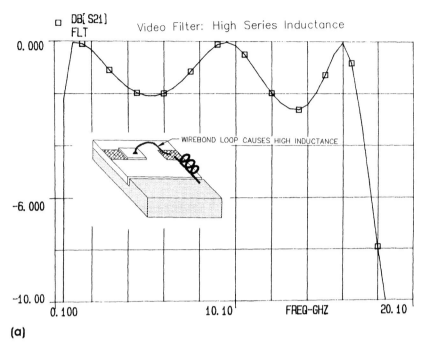

**(a)**

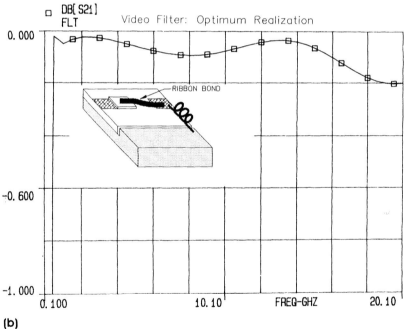

**(b)**

**Figure 4.10**   Inductance comparison: (a) Wire bond effect on noise and (b) ribbon bond effect on noise. (From Ref. 1.)

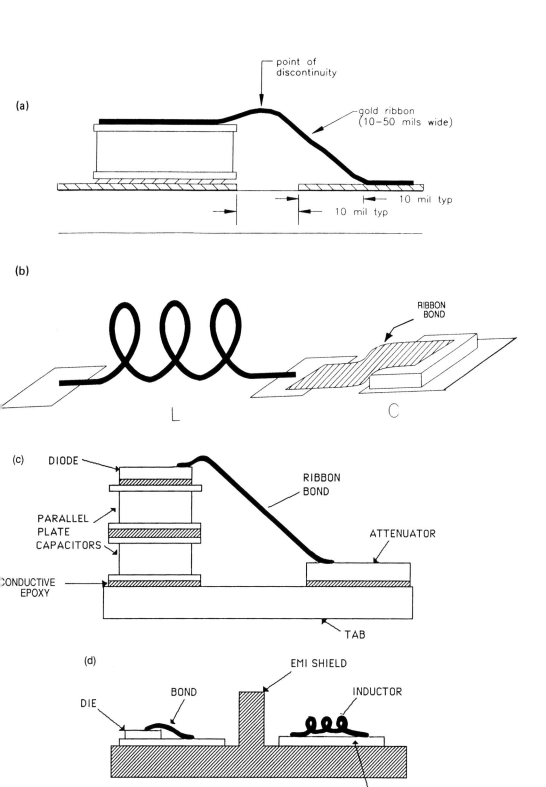

**Figure 4.11** Examples of microwave subassemblies; (a) parallel plate capacitor, (b) inductor and capacitor in series, (c) stacked assembly, (d) EMI shielded assembly.

171

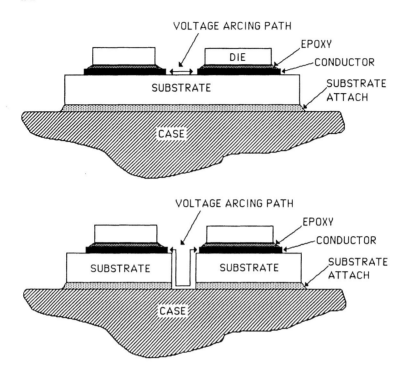

**Figure 4.12**  Elimination of voltage arcing by proper mechanical design. In the top figure the path the voltage has to travel to arc or short the two dice is short, straight and on a single plane. In the bottom figure the voltage would have to travel a further distance and around 4 corners, yet the dice placement and the devices area and volume remain the same.

welding. In this process a welding tip that is split in half is placed with one tip at eacn end of the width of the ribbon. Power is applied and the ribbon is welded across its width to the surface metallization.

As can be seen, microwave subassemblies are highly customized and are unique to the electrical requirements of the device in which they will be utilized. Figure 4.11 gives some examples of microwave subassemblies.

### 4.1.4  Summary

Elementary subassemblies are used when area restrictions are critical or when special pretesting is required. They are typically used in military and space equipment, specifically avionics or satellite applications, where weight and size are of the utmost importance. Elementary subassemblies are typically composed of simple discrete functions: a die assembly is typically only a diode or transistor, and a cap bank is several caps assembled together to form one subassembly functionally equivalent to one large farad-valued capacitor. The need for elementary subassemblies is very design dependent. If the design requires a means of spreading heat under a high-power FET, a die assembly utilizing a BeO tab might be employed. If fast rise times are required, stacking of components may be the answer. These subassemblies can be built or purchased ahead of time as individual products, or they can be built as an integral part of a higher assembly, such as a hybrid.

In high-voltage applications, cross talk and arcing potentials are critical. To prevent such problems, adequate spacing must be designed into the assembly. However, the real estate for such spacing may not be available. Since the current or charge travels on the surface, one way to increase spacing between two tracks without increasing the area needed for the design is to break up the substrate into different subassemblies. When this is done, the path of potential arcing must travel down the side of one substrate, across the floor of the case, and up the side of another substrate, as shown in Figure 4.12. Thus, without increasing the needed surface area, the spacing between two high voltage potentials is increased threefold.

## 4.2    CHIP CARRIER ASSEMBLIES

Chip carriers are plastic or ceramic packages that carry a single chip or die. Chapter 2 discussed the outline configuration of various chip carriers. The assembly processes used to make chip carriers are very similar to those used to make dual in-line packages (DIPs) for plastic chip carriers or ceramic packages (cer paks) for ceramic chip carriers.

### 4.2.1    Plastic Chip Carrier Fabrication and Design

The first step in fabricating a plastic chip carrier is to punch or etch a lead frame pattern into sheet metal, typically kovar that has been plated with nickel and gold. The die is epoxied to the center of the lead frame and then wire-bonded accordingly. The lead frame with die and wire bonds is then encapsulated within a plastic mold. In the case of plastic chip carriers, the leads that extend out of the molded body can have a gull-winged or DIP configuration or be shaped like the letter J, hence the name J-leaded chip carriers. (Plastic chip carriers are always leaded.)

### 4.2.2    Ceramic Chip Carrier Fabrication and Design

Ceramic chip carriers are made of cofired ceramic with tungsten metallization. See Chapter 3 for a review of the cofired process. Any exposed tungsten is nickel and gold plated. The ceramic body is available in two styles, slam or flatpack, and multilayered with a cavity for the die. The slam chip carrier can be single or multilayered. The die mount pad is the center of the slam and is surrounded by the wire bond pads. These wire bond pads via down to the buried layer and/or fan out toward the edge of the chip carrier, where they either via up to the top of the castellation or via down to the solder mount pad/ castellation, in the case of a multilayered design. On the surface of the chip carrier surrounding the wire bond pads is a ring frame of metallization used to solder mount a dome-shaped lid or has a domed lid glass mounted in the case of a single layer design. Figure 4.13a shows the configuration of a slam chip carrier.

Multilayered ceramic chip carriers are cofired ceramic layers with windows cut out in the top layers. When these layers are fired together the windows form a cavity. The bottom layers of the chip carrier form the bottom of the package. The next layers have small windows that form the lower cavity where the die is mounted. The third set of layers, with a slightly larger window, form the wire-bonding ledge within the chip carrier. The top layer has the metallization ring surrounding the window, which is used for solder sealing the lid onto the chip carrier. Figures 4.13b & 4.14 give examples of multilayered chip carriers.

Chip carriers are available in a cavity-up or cavity-down configuration. In a cavity-up design, the cavity faces up from the surface mount pads on the bottom of the package.

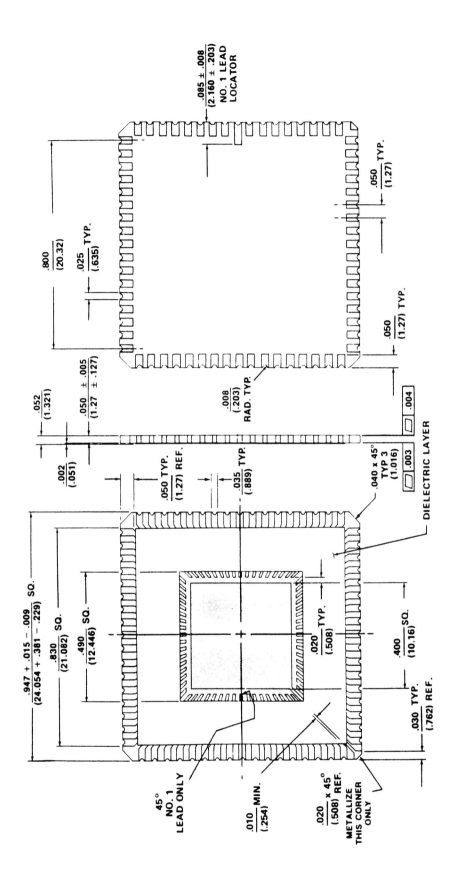

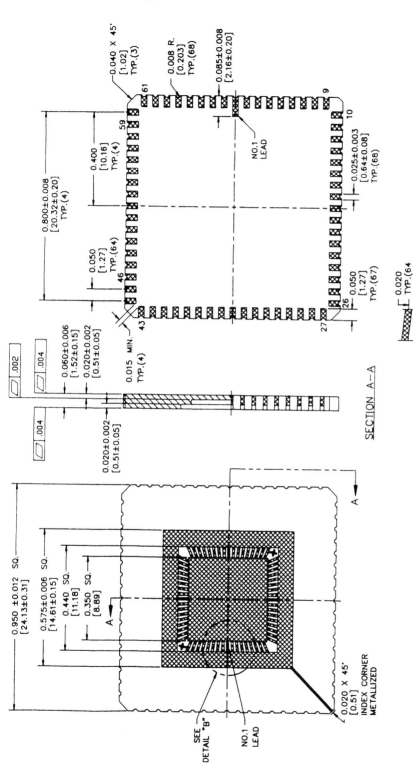

**Figure 4.13** (Top) Slam ceramic chip carrier. Single-layer chip carrier requires domed lid. Example shown is an open tool 68 pin LCC. (Bottom) Multilayer ceramic chip carrier. Example shown is an open tool 68 pin LCC. (Courtesy of Kyocera America.)

**Figure 4.14**  Examples of multilayered chip carriers. (Courtesy of Westinghouse ESG.)

In other words, the lid goes on the top of the package and the bottom of the package is mounted down to the PWB. In a cavity-down package the lid is on the bottom of the package and is placed down on the PWB. A cavity-up package is used in traditional mounting configurations in which both routing and thermal path are directed down through the PWB. A cavity-down package allows the routing to be directed down to the PWB, but the thermal path is directed toward the top of the package, away from the PWB. This thermal design is sometimes used in convection-cooled systems.

Ceramic chip carriers are available in a variety of lead configurations, as discussed in Chapter 2. Leadless chip carriers have castellations. When the chip carriers are fabricated, the green tape used to make them is larger than the desired package, even with consideration of the shrinkage of the tape during firing. Holes are punched into all the tape layers around the outer perimeter of the chip carrier. The edges of these holes are metallized with the tungsten. After the layers are laminated together, the prefired stack-up is cut to size. The outer ring of green tape is cut off. This cut runs through the holes, leaving evenly spaced semicircular columns around the outer perimeter of the chip carrier. These metallized half-circles form the castellations. The castellations provide a column of metal that allows the solder to flow up, forming the solder fillet during the surface-mounting process. Figure 4.15 shows a leadless chip carrier mounted to a board. Note the solder fillets that have formed in the castellations. Figure 4.16 shows a leaded chip carrier mounted to a board. Here the leads have been top brazed to the package to give the standard gull-winged

**Figure 4.15**   Leadless chip carrier (LCC) mounted to a PWB. (Courtesy of Westinghouse ESG.)

**Figure 4.16**   Gull winged chip carrier mounted to a PWB. (Courtesy of Westinghouse ESG.)

leaded style. The leads have been formed to step down to the board. This step or bend acts as a shock absorber. As the board sees fluctuations in temperature, the chip carrier will expand at a different rate than the board to which it is mounted. The bent lead will "give," or absorb the thermal stress between the carrier and the board.

Leaded chip carriers are fabricated in the same manner as leadless ones, with an additional step of brazing on leads after the plating of the exposed tungsten metallization. The leads are kovar (with nickle and gold plating). The leads are attached using a high-temperature braze, such as a eutetic copper-silver. As discussed in Chapter 2, the leads can be brazed to the top of the package, in a gull-winged configuration; to the bottom of the package; to the side of the package, for insertion mounting; or to the side of the package in a J-lead form for surface mounting.

### 4.2.3   Pin Grid Array Packaging

Pin grid array (PGA) packages are high-lead-count chip carriers. To accommodate the high lead count without increasing the perimeter of the chip carrier, the I/O is designed into an array of pads on the bottom of the package instead of a single row along the package perimeter. The leads are pedestals that are brazed onto the array of pads, yielding an array of pins for insertion mounting to the PWB. Likewise, to accommodate the high I/O count dual ledges for wire bonding are typically found inside a PGA. PGAs are available in both cavity-up and cavity-down configurations.

### 4.2.4   Chip Carrier Assembly Processing

Chip carrier processing is very simple. The die is mounted with epoxy or a solder preform. The die is then wire bonded by thermocompression or thermosonically. (These mounting and bonding methods are discussed in detail under Die Assembly Processing.) After a bake out within a dry nitrogen or vacuum oven to outgas all potential contaminants, the lid with preform is clamped on top of the seal ring of the chip carrier. The lid is ceramic with a metallized ring along the perimeter that matches the seal ring on the chip carrier, or is nickel- and gold-plated kovar with a ring-shaped solder preform tacked along its perimeter. The solder preform is eutectic gold-tin. The chip carriers with their clamped-on lids are run through a furnace programmed to reflow the gold-tin solder. After the chip carriers are removed from the furnace, they are tested. For military and space applications, this testing will include burn-in. The final product is available with the nickle/gold-plated finish or with a pretinned finish. Pretinning involves dipping the chip carrier into a solder pot of 60% tin–40% lead solder.

Dice can also be mounted inside the chip carrier with tape-assisted bonding (TAB) techniques. These techniques will be discussed in detail under Interconnection.

Chip carriers are sometimes customized during the assembly process. In this case, a pattern of separate mounting pads is laser scribed into the cavity. This permits mounting multiple components without shorting all the terminations and/or back planes together. The components are then mounted with epoxy, wire bonded, and sealed as before. These customized chip carrier assemblies are referred to as chip carrier hybrids.

## 4.3   HYBRID MICROELECTRONICS ASSEMBLIES
### 4.3.1   Definition of a Hybrid

A hybrid microelectronics device, often referred to as just a hybrid, is a combination of two or more electronic components mounted and interconnected via a substrate. The substrate is made of a dielectric material that is patterned with metallized signals or tracks. The hybrid is a customized electronic function, packaged as a single device.

Hybrids are known by many names, including multicircuit package, multichip package (MCP), multicircuit hybrid package (MHP), and power hybrid package (PHP) for hybrids with power densities greater than 10 W/in$^2$. Some microwave hybrids are also known as integrated microwave assemblies (IMAs). However, it is important to remember that IMA is the name used for any microwave assembly in the sense that all microwave hybrids are IMAs, but not all IMAs are hybrids. From all the names just given, it can be discerned that the essential definition of a hybrid is that it contain multiple components.

### 4.3.2   Hybrid Design

A hybrid differs from a PWB assembly in that all the components are packaged together in a single hermetic case. Unlike PWBs, in which all components are individually packaged and then mounted to the board, hybrids use bare, unpackaged, dice. In other words, a given schematic, with a parts list to define the components needed, can be built with individually packaged and leaded components mounted on an organic PWB, which provides the necessary routing between all the individual components. The same device can be built using bare dice and chip components (e.g., capacitors) and thick-film resistors, all mounted to a multilayer ceramic substrate. The substrate provides all of the internal routing. The substrate assembly is then mounted inside a metal package that is hermetically sealed and forms the hybrid.

A hybrid can be as simple as mounting two diodes into one chip carrier, or can be as complicated as a multiple channel amplifier, regulator, or analog-to-digital (A/D) converter. Hybrids can be digital, analog, or a combination of both.

The most common hybrid design has bare dice and chip components mounted on a multilayer ceramic substrate. This substrate is designed and fabricated in much the same manner as discussed in Chapter 4 for ceramic surface mount boards. A hybrid substrate is typically 96% alumina with thick-film metallization. However, as discussed in Chapter 4, green tape, cofired ceramic, and thin-film multilayering using polyimide as the dielectric material may be employed depending on the particular requirements of the design. The main difference from a surface-mounted board is that bare dice, not individually packaged dice, are mounted. It is necessary to package and/or seal the substrate assembly to provide necessary protection to the bare dice. The substrate assembly is either mounted inside a leaded case or encapsulated within a protective overcoat, as shown in Figure 4.17.

The advantages of hybrids are the vast reduction of volume, area, and weight; improved thermal management; increase functional densities; increased frequency capabilities and improved electrical performance. The disadvantage of hybrids is the increased cost over that of an equivalent PWB assembly.

The elimination of the individual packaging of the components is the most obvious contribution to the size reduction. By this elimination, the bare dice can be mounted

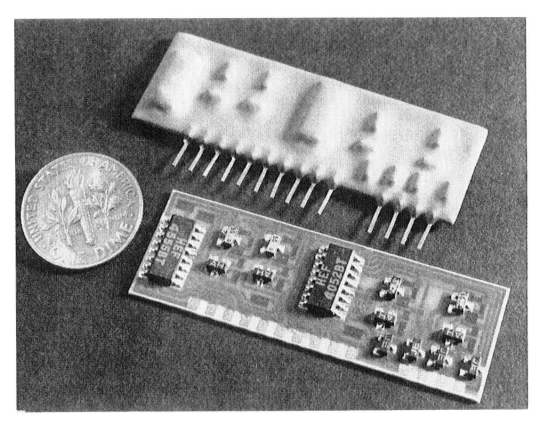

**Figure 4.17**   Encapsulated hybrid. (Courtesy of Westinghouse ESG.)

directly onto the substrate. The component placement is much closer, and interconnections between the components are more direct. Instead of wire bonding from the bare die to the die package's internal leads and having the signal travel through the feedthrough to the external leads, which then have to be solder connected to the PWB, which then routes the signal to the solder joint of another packaged die etc., the bare die on a hybrid can be wire bonded to the substrate, which directly connects to the wire bond pad of the other die in question. Thus, electrical performance and frequencies are improved. Likewise, thermal interfaces are eliminated. On an organic PWB, the heat must travel from the die through the die-attaching media, through the die's package and the package leads, through the solder, and then finally through the PWB to some heat sink. In a hybrid, the heat travels from the die through the die-attaching material to the substrate and through the hybrid case to the heat sink. Not only is the number of thermal interfaces reduced, but also the thermal conductivities of the materials used in the hybrid are higher and capable of dissipating more heat. Thus, thermal management is improved.

The disadvantage of using a hybrid is the increase in cost and turnaround time, or the time it takes to go from electrical design to finished product. Organic PWBs use polyimide with copper metallization. Hybrids utilize ceramics with gold, silver, or gold alloy (e.g., PtPdAu) metallization, all of which are higher in cost.

Dice, whether discretes such as diodes and transistors or integrated circuits (ICs), can be purchased in two forms: already packaged or bare. Packaged dice come in many varieties: leadless chip carriers (LCCs) for surface mounting, DIPs for insertion mounting, plastic J-leaded or gull-winged chip carriers for surface mounting, or "canned" such as the three-leaded metal can packages commonly used for transistors. All of these packaged dice are designed for mounting on PWBs. The suppliers package the dice and then test and sort them. For example, in the case of a diode, all the bare dice are probed while still in wafer form. This probe or test indicates only that the dice act like diodes or do not work at all. This type of testing is often referred to as a "go–no go" test. All diodes that pass the go–no go test are packaged. After being packaged, the diodes are fully tested under full power and operating conditions. They are then sorted into bins (e.g., there might be three different bins for a particular diode: bin 1 for diodes that operate only under 3 V, bin 2 for 3–7 V, and bin 3 for 7–12 V). These units can even be fully screened per the military standards, including burn-in to weed out any infant failures. Packaged dice are in greater demand than bare dice and are manufactured in very large quantities using automated techniques. Therefore, the cost of these packaged units is only pennies more than that of the bare dice. In some cases the cost of purchasing bare dice is even higher than that of their packaged counterparts because of the increase in handling difficulties, electrostatic discharge (ESD) damage, and decreased yields associated with the bare dice, which cannot be completely tested and screened prior to packaging. For some ICs, especially those utilized in microwave applications, packaging increases yield due to the EMI and ESD protection the packaging offers. Bare dice are more susceptible to damage than their packaged counterparts. Their surface metallization can be scratched or contaminated, damaged by ESD, and chipped when picked up by tweezers, or they can simply be lost because of their tiny size (e.g., a transistor can be as small as $0.011 \times 0.007$ in.). Packaged dice have high postmounted yields, typically over 98%. Bare dice have lower postmounted yields because of their vulnerability to handling or ESD damage during the mounting process. Bare die yields can be as low as 60% depending on the complexity, sensitivity, and process control used in their mounting. Typically, under proper conditions, a bare die yield of 90 to 97% can be expected.

The yield of a PWB assembly is normally quite high because all elements and components used in the assembly are fully tested prior to commitment to the PWB. Hybrids, however, use the bare dice, which have lower yields. Hybrids utilize several bare dice, thus compounding the overall yield problem of the end product.

If a hybrid uses 12 diodes, each with an expected yield of 95%, what is the best expected yield of the hybrid? (Assume that the substrate has been fully screened and tested and that no rework is allowed.)

The yield of the hybrid would be $(0.95)^{12} = 0.54$ or 54%.

Hybrids can also be more costly in terms of time to design. As with PWBs, the layout and routing are done on the computer. Once this layout has been checked, the information from the design station is downloaded into a laser artwork generator (LAG). The LAG draws out the layers on the film (e.g., Mylar). For a PWB, this artwork is used to pattern the metal on the board. For hybrids, this artwork must first be reduced (in scale) and then converted into screens and stencils, which will then be used to pattern the metallization onto the substrate [2]. After board fabrication, the PWB can be assembled by mounting the components and reflowing the solder. Hybrids must have the components mounted and wire bonded. The substrate assembly must then be packaged. Thus, hybrids can be more labor intensive than PWBs.

Lead time—the time it takes to order and receive components—is typically longer for hybrids. Since hybrids are used only when size limitations are extreme, as in space and military applications, they are typically not a high-volume production item. For this reason, hybrid piece parts such as bare die and chip components are not in as much demand as surface mount components for PWBs. Thus, component costs and delivery times are often greater.

Extra manufacturing steps, component availability and lead times, material and component costs, turnaround times, and yield issues can all lead to higher costs for hybridization of a design. However, actions can be taken to reduce these cost impacts. Die placement and wire bonding can be automated, with the right capital expenditures. Overall yields can be improved by conducting full functional testing of the dice prior to mounting. This testing would involve the use of very sophisticated and expensive universal die testing equipment. Such equipment can cost approximately $2 million. Proper precautions can be employed to reduce possible ESD damage. Process control and operator training can be implemented to improve the yields through the various processing steps and reduce the possible handling damages. Rework procedures can be developed and improved to increase final yields.

Design guidelines and checklists can be developed to reduce design cycle time. Design engineers should keep themselves educated about the processing capabilities of the facility being used for manufacture. Likewise, manufacturing must maintain awareness of the designer's needs. Both hybrid design and manufacturing engineers must stay abreast of the system trends and design demands. This mutual knowledge can greatly enhance the producibility of the design. Proper planning of the overall tasks to design and fabricate a hybrid can be the most effective way to reduce the cost of the hybridization. For example, defining the piece parts and components in the first phase of the design and having an expeditious purchasing activity would eliminate bottlenecks during manufacturing. Proper planning can also prevent the electrical design from going to layout prior to electrical simulation and design debugging. This would eliminate time-consuming false starts during layout efforts. Concurrent engineering is having all fields of engineering involved from conceptual design through actual production. System, electrical, and mechanical design must work with manufacturing, reliability, and production engineering to design a producible product. Engineering must, in turn, work with purchasing and management to plan and coordinate efficient design and build efforts. Concurrent engineering enables a company to stay at the leading edge of the technology while maintaining cost-effectiveness.

### 4.3.3  Hybrid Processing
### Component Mounting

The most common method used for mounting the components on the substrate is to attach them with epoxy. In the 1960s and 1970s dielectric (i.e., nonconductive) epoxies were used. However, with advances in polymer engineering the adhesion of epoxies was greatly improved, along with techniques for "filling" the epoxies with metals to make them electrically conductive. To obtain a low enough resistance to allow for proper conduction, the epoxies must be 70 to 80% metal. The metals used for filling the epoxies are gold and silver. Copper, although an excellent electrical conductor, is not typically used because of its highly corrosive characteristics. Metals other than those mentioned are either not conductive enough, too expensive, or not conducive to filling. Because of the price of

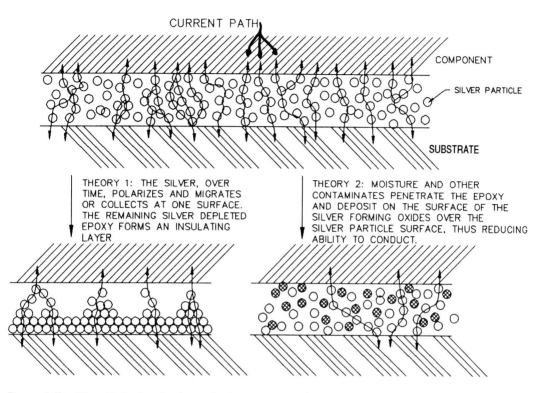

**Figure 4.18**   Silver (Ag) migration in conductive epoxy over time.

gold and silver's lower resistivity, silver is the material more commonly used to fill epoxies.

In recent years, there has been much controversy over the use of silver-filled epoxies in military applications. It was feared that the silver in the epoxies would oxidize and thus greatly reduce the conductivity of the material with time. It has also been feared that the silver would migrate over long periods of time, collecting in one area of the epoxy joint as shown in Figure 4.18. As the silver migrates, the area it migrates away from forms an insulating layer. Data showing changes in resistance of the epoxy over time would seem to support these two fears. However, more recent studies have shown that if the epoxy is properly processed and is contained in a hermetic environment, these changes in resistance over time are insignificant. The theory is that if potential contaminants and oxidizing agents such as air and moisture are kept away from the epoxy, the silver cannot oxidize and its resistance is not greatly effected. Silver migration still occurs, but its effect is small and noticeable only over very long periods of time or at very high frequencies.

The organics within the epoxies can "spoil" if not properly stored. Care must be taken to maintain the integrity of the material. Premixed epoxies must be kept frozen or refrigerated. The shelf life of epoxy kept frozen is 1 year; refrigerated, 6 months; and stored at room temperature, 1–3 months. When the epoxy is to be used, it must be taken out of the freezer and allowed to thaw before opening the container. If the container is opened before the material reaches room temperature, humidity in the air can condense into the material and contaminate it. Further protection of the material must be employed:

only clean stainless steel utensils should be used to mix or apply the epoxy. Wood "tongue depressors" can leave wood fibers that can react with the epoxy organics. Old material left on the screen after printing should never be mixed back in with the fresh epoxy. This material should be gathered in its own container and recertified prior to use on deliverable product or set aside for engineering or prototyping use only.

If the device is mounted with conductive epoxy, the epoxy acts as the electrical connection. For example, a chip capacitor mounted with conductive epoxy is mounted with the epoxy only under the two end terminations. No epoxy is under the body of the capacitor, for it would cause a short. In high-current or microwave applications in which the resistivity of the conductive epoxy or the minute changes of resistance over time (due to the silver migration) do not meet the electrical performance criteria of the design, the capacitors are mounted with dielectric epoxy under the entire length of the capacitor body and the end terminations are wire bonded to provide the electrical connection. Both epoxies, dielectric and conductive, employ the same component mounting techniques. However, the use of dielectric epoxy dictates the need for additional interconnections, usually in the form of wire bonds.

Epoxy can be applied as either a film or a paste. Epoxy films can contain a fiberglass mesh support as found in the prepreg or B stage used to laminate the layers of an organic PWB (see Chapter 3). Epoxy films can also be formed by printing a layer of epoxy on a nonstick, inert surface (e.g., Mylar tape) and partially curing it. This partial cure leaves it flexible but no longer fluid. Film adhesives can be purchased as sheets of approximately $12 \times 12$ in. or preformed by cutting or punching into the desired size. Film adhesives come in thicknesses from 0.003 to 0.012 in. Most film adhesives require the application of pressure during the cure cycle.

Film adhesives are used for mounting large-area devices (e.g., mounting LCCs approximately 0.300 in. square to larger die assemblies or chip carriers with higher I/O counts). Film adhesives are also used for mounting the substrates to the case floor, or mounting the case to the PWB, or attaching the cold plate to the module housing (e.g., areas that can exceed 10 in. square). Preforms are not practical for mounting small dice or chip components. Pastes are used for this purpose. Epoxy pastes can be hand applied, using a stainless steel spatula or dental picks. Wooden utensils should never be used with epoxies, as the wood fibers can contaminate the material. Hand application is normally done only when building prototypes. Automatic dispensing is used by some facilities. The epoxy paste is placed in the dispenser gun and pressure is used to push the epoxy out. In a manual dispenser, the operator controls the pressure by a trigger or foot pedal. In more advanced equipment, a set pressure pulse is automatically applied, giving uniform amounts of epoxy. Some automatic dispensers can be programmed to dispense the epoxy in predetermined locations, much as pick-and-place machines are programmed to place components. The most commonly used method for epoxy application is to print the epoxy. The epoxy is screen printed in the same manner as thick-film paste is screen printed. The differences are that a larger mesh, typically 80 count, and a thicker emulsion, 0.002 to 0.010 in, is used. The larger mesh is needed to accommodate the epoxy's larger particle size and higher viscosity. The thicker emulsion is used to obtain the thickness of the epoxy deposit needed. Stencil printing can also be used, as used with solder paste printing. Advantages of stencils will be discussed later in Surface Mounting.

Other component mounting methods use solders. As discussed earlier under Die Assembly Processing, the die can be scrubbed down by using eutectic alloy formation between the die and the substrate metallization or by using eutectic solder alloy preforms.

Furnace mounting with the use of eutectic alloy preforms is another process used with hybrids.

Chip components can also be mounted using solder paste or solder wire. The solders used for these applications are not always eutectic, although they can be . These solders are normally lead/tin alloys, rather than the gold alloys used for die attaching. These solders reflow at lower temperatures and are not necessarily eutectic (i.e., do not have a single reflow temperature). To get good uniform adhesion, or "wetting," of the solder to the surface metal, oxides must be removed. Removal of oxides can be accomplished in two ways. In one method the solder is reflowed in a reducing atmosphere in a furnace filled with forming gas. Forming gas is a mixture of inert nitrogen and approximately 5% hydrogen. The nitrogen prevents the introduction of any new oxides and the hydrogen, at the reflow temperature, reduces the oxides that have formed on the solder. This method is normally used for higher-temperature or eutectic solders. The other method is to utilize flux to increase the wetting of the material. The flux chemically reduces any oxides on the metal surface or in the solder. Flux can be applied to the metal surface or on the component prior to mounting. For example, flux can be printed on a board that has been pretinned or has been thick solder plated. The flux can then act as the tacky substance that temporarily holds the components in place prior to solder reflow. Another means of introducing the flux is to dip the already pretinned components into flux before placing them down on the board surface This method, however, is labor intensive and not frequently used.

Solder pastes and most solder wires have flux in them. Solder pastes have particles of the solder alloy suspended in a solution containing the flux, emulsifiers, and volatile solvents, which burn off during reflow. Solder wires have a rosin core that contains the flux.

Solder paste or wire can be reflowed by applying heat locally with a soldering wand (sometimes called a soldering gun) or a hot-air gun. Solder paste can be applied to the substrate by hand or by a screening method. The components are then placed in the paste and the solder is reflowed.

The reflowing of the solder paste can be accomplished by a variety of methods. As already mentioned, the paste can be reflowed locally, or the entire substrate can be submitted to the heating cycle needed to reflow the paste. Reflow of the solder can be accomplished by placing the substrate on top of a hot plate or on a belt reflow system. The latter system has a belt that travels down the length of the reflow equipment through heated zones. Zones are heated conductively by heating coils under the belt and/or convectively by hot air or nitrogen blown down on top of the belt.

Furnace reflow can involve fluxless solder reflow, as utilized in eutectic die attach (see Die Assembly Processing), or fluxed solder. In the case of fluxed solder the reflow is performed in a nitrogen atmosphere; a reducing atmosphere is not needed because the flux provides the reducing or wetting agents. Another furnace reflow method is to provide the energy needed to reflow the solder by exposing the substrate within the furnace to infrared (IR) radiation, hence the name IR furnace. Finally, solder can be reflowed using a vapor phase. In this case the substrate is lowered into a chamber filled with the vapor of a Freon. The Freon and its boiling point determine the temperature of the vapor. As the part is lowered into the vapor, the board surface, and thus the solder paste, is exposed to the temperature of the vapor. The solder is reflowed as the vapor condenses on the surface of the substrate and transfers its latent energy of condensation to the solder. Figure 4.19 shows the conceptional construction of typical vapor-phase systems. Whenever these

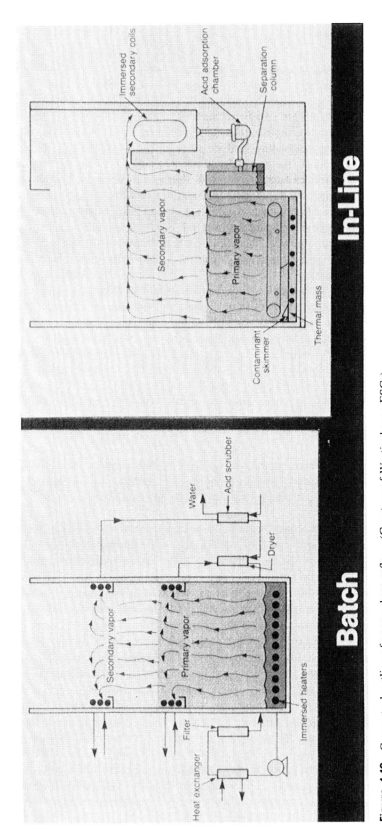

**Figure 4.19** Conceptual outline of vapor phase reflow. (Courtesy of Westinghouse ESG.)

186

methods are used, the substrates must be defluxed or cleaned prior to mounting any bare dice. It is best to deflux the unit before it has a chance to cool down fully. In the cool-down process the flux is baked onto the board and penetrates into the surface of the board. If cleaning is conducted while the board is still warm (i.e., while the flux is still tacky, prior to baking on the flux), the flux residue left on the board will be minimized. Any flux residue left on the board can outgas later and damage the bare die.

Bare dice cannot be mounted using methods involving flux because of the damage the flux does to them. Even though dice are passivated, or given a protective overcoat of a glass or an oxide, this passivation is performed at the wafer level. When the wafer is sawed into the individual dice, the passivation is also cut. Thus, only the surface of the dice, not the sides is protected. The flux and flux fumes can deposit on the dice and migrate under the passivation at the sides of the dice. The flux can then attack the oxides and metallization of the die, causing loss of functionality over time (as the flux eats away at the IC layers).

## Interconnection

The most common means of interconnection in a hybrid is by use of wire bonds. The methods of wire bonding are covered in detail under Die Assembly Processing.

In high-power and/or microwave applications, a single wire bond connection does not always meet the thermal or electrical properties. In high-power FETs, the current and thus the heat are dissipated through the gate connection. If a single wire bond is used for this connection, even if it has a large diameter (0.005 in.), it might not be able to handle the thermal transfer. When the heat travels from the bond pad surface of the gate to the much smaller surface area of the wire bond, the power is bottlenecked and cannot transfer fast enough to the wire. This power loss should not be confused with the power loss in the silicon due to the FET switching. This can result in a burned-out gate. For this reason, multiple wire bonds are often used, although the wire itself might be rated for the current being carried. Sometimes it is better to use three or four smaller-diameter gold wire bonds instead of a single larger-diameter aluminum wire bond. However the increased surface area of multiple bonds, as shown in Figure 4.20, coupled with the higher thermal conductivity of the gold wire, can greatly decrease the potential for a thermal failure under high-power conditions.

Other interconnection methods that improve current (and heat) carrying capabilities include bonding the die to the substrate with tape-automated bonding and ribbon bonding. TAB is a process in which copper metallization is built up on top of a sheet or "tape" of kapton. The copper is patterned using photolithography or etching methods. If the design requires multilayering, another layer of kapton is placed on top of the patterned copper and the etching process for the second layer is repeated. The copper between the layers of kapton is sealed and thus protected from corrosion. The exposed copper, including the leads, is nickel and gold plated to generate a corrosion barrier and a bondable surface, respectively. Figure 4.21a displays unetched or unpatterned tape. Figure 4.21b shows a close-up of tape that has been patterned and plated. At the end of the leads, a "bump" of copper and/or plating exists, as shown in Figure 4.22. It is at these bumps that the interconnection to the die and the substrate or chip carrier is made, using a process similar to resistance welding. After the die has been bonded, it can be probed and tested while still in tape form. Figure 4.23 shows a tape-automated bonded die. Note that the leads fan out to larger probing pads, which enable the die to be functionally tested. These probe pads are cut off when the die and its bonds are punched out of the tape as shown in Figure

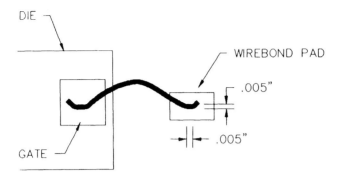

0.005 " DIAMETER WIRE HAS A WEDGE BOND AREA OF APPROX.
$(0.005)^2 = 2.5 \times 10^{-5} in^2$

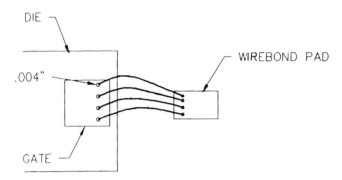

0.001" WIREBONDS HAVE 0.004" BALLBONDS, GIVING A WIREBOND
SURFACE AREA OF : 4 BONDS X $\pi(D/2)^2$ AREA/BOND $\Rightarrow$ 4 $\pi D^2/4$
= $\pi(0.004)^2$ = 5.03 X $10^{-5} in^2$ OR TWICE AS MUCH AS A SINGLE
0.005" WIREBOND.

**Figure 4.20**   Surface area comparison of a single 0.005″ diameter wire
bond and multiple 0.001″-in.-diameter wire bonds.

4.24. After removal from the tape, the leads must be formed or bent to be mounted and
bonded to the chip carrier or substrates displayed in Figure 4.25.

The TAB bonding area is much larger than that of a ball or wedge bond. The lead
itself provides a larger cross-sectional area for carrying the current and power. The copper
also gives one of the highest thermal conductivities and current capacities. The disadvantages of TAB are the time and cost of designing and fabricating the tape and the capital
expense of the TAB bonding equipment. Each die must have its own tape patterned for
its bonding configuration. Likewise, each TAB design requires its own equipment program
and setup. For this reason, TAB has typically been limited to high-volume production
applications.

Ribbon bonding is a process in which a gold ribbon is split-tip welded to the die and
substrate. (The split-tip welding process is described in Section 4.1.3.) Ribbons range
from 0.005 to 0.050 in. wide and 0.001 to 0.005 in. thick. The cross-sectional area of
a ribbon can offer more current-carrying capacity than a wire bond. The advantage of the
ribbon is that the larger surface area can allow for power transfer from the die to the

(a)

(b)

**Figure 4.21** (a) Unpatterned tape for tape-automated bonding (TAB); (b) patterned tape for TAB. (Courtesy of Westinghouse ESG.)

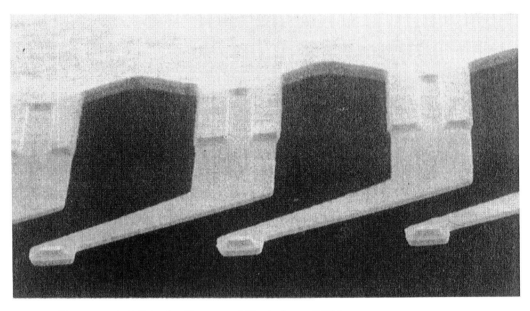

**Figure 4.22**  Bumped TAB leads. (Courtesy of Westinghouse ESG.)

**Figure 4.23**  Tape-automated bonded die. (Courtesy of Westinghouse ESG.)

**Figure 4.24**   TAB excisor. (Courtesy of Westinghouse ESG.)

ribbon. An example comparison of cross-sectional and bonding surface area is given below.

|                              | Cross-sectional area $(10^{-5}$ in.$^2)$ | Bond surface area $(10^{-5}$ in.$^2)$ |
| ---------------------------- | :-: | :-: |
| 0.005-in.-diameter wire      | 1.9 | 2.0 |
| 0.010 × 0.002 in. ribbon     | 2.0 | 5.0 |

As can be seen in this tabulation, a ribbon gives 2.5 times the heat-transferring area for approximately the same cross-sectional area. This factor can be further improved by tacking down the ribbon multiple times on the pad site.

## Substrate Mounting

There are various techniques for mounting a substrate assembly. The first is to mount it in place mechanically. This requires that the substrate assembly be attached to a carrier plate that supports screws used to mount the substrate mechanically to the case. The screw support is usually a flange that extends out from under the substrate. Because a ceramic cannot withstand the torque or stress a screw would impose on it, the ceramic must be

**Figure 4.25**   TAB lead forming. (Courtesy of Westinghouse ESG.)

attached to a metal carrier plate, which can then be screwed down into the case. The ceramic can be formed or machined into the required shape. The forming can be a rounded notch out of a corner or a hole in the center to provide the area needed for the screw. (See Chapter 3 for more details of ceramic machining.) This method is not commonly used for military hybrids, for if the screw was designed to penetrate the case floor it would mean a lost of hermeticity. Furthermore, if the floor was thick enough to support the screws, it would defeat the purpose of going to the weight and volume reduction offered by hybridization of the design. Besides, if the ceramic substrate must be bonded to a carrier plate to allow mechanical mounting, why not just attach the substrate directly to the case? In commercial applications, substrates are mounted to carrier plates, protectively overcoated, and then mechanically mounted within the systems. This mounting method offers quick and easy replacement of the substrate assembly, at the cost of increased weight and volume with decreased hermeticity.

The most common methods for substrate mounting are epoxy and solder mounting. Epoxy mounting can utilize epoxy paste, but this requires extensive process control of the printing and curing. If the deposit of epoxy is not smooth and uniform, air pockets or voids can form when the substrate is mounted. When cured, epoxy outgases the volatile solvents used to make it fluid enough for printing. As outgassing occurs under the larger area of a substrate, as compared with dice, pockets of gas can accumulate, forming voids. Voids result in a decreased of adhesion and of heat-transferring area. If a void is located directly under a "hot" component, the thermal resistance is greatly increased. This, in turn, could result in a thermal failure. For this reason, epoxy preforms are the norm. The preform can be cut or punched to the desired size and shape and comes in a very uniform thickness. The preform is placed in the case with the substrate on top of it. Pressure is applied by placing a weight (a few hundred grams) on top of the substrate, if component spacing allows it, or by clamping the ends of the substrate down. The unit is then placed in a vacuum or nitrogen-purged oven and cured. Vacuum ovens are recommended for their ability to assist in pulling out the outgassed material. The door to the oven must not be opened at any time during the cure because this would greatly affect the profile or epoxy cure schedule. If the epoxy is not allowed to reach cure temperature in the right amount of time, complete curing or molecular interlocking cannot be obtained. If the ramp up to the cure temperature is too long (because of heat escaping from the oven when the door is opened), the organics that promote the interlocking can volatilize before the interlocking can be completed. The results in a reduction of the adhesive strength of the epoxy.

Another way to mount substrates is with a solder preform. Typically, a low-temperature solder, such as a lead-tin composition, is used. The case is typically gold plated to allow for wetting of the solder to its surface. The back side of the substrate is also metallized for solderability. If the substrate is thick-film alumina, the metallization can be a platinum-gold (PtAu), platinum-palladium-gold (PtPdAu), or silver (Ag) thick-film paste. Gold paste is not used because of the tendency of gold to "leach" off the ceramic surface. Leaching is a process in which gold diffuses into the solder to form intermetallics. This mass transfer accelerates at elevated temperatures (upper operating conditions) until all of the gold has diffused into the solder and none is left to maintain adhesion to the ceramic. The addition of Pt or Pd to the gold alloy will prevent all of the material from leaching off the substrate surface. If the substrate is BeO, thin-film metallization must be used (see Die Assembly Materials).

The actual process flow for solder mounting of a substrate varies for different companies and applications. One approach is to reflow the solder on the floor of the case with the aid of flux and a hot plate. The case is then defluxed, typically in a degreaser. The substrate backplane is burnished to increase the surface tension or wetting by roughening up the surface. The substrate is then placed on top of the solder and scrubbed into place while the solder is reflowed on a hot plate under a dry nitrogen flow or within a environmental glove box, as shown in Figure 4.26. The substrate can also be reflowed in place by clamping the preform and substrate into the case and sending the unit through a hydrogen (i.e., forming gas) furnace. This last reflow method cannot be used if thick-film resistors are on the substrate being mounted. Exposure of these thick-film resistors to a reducing atmosphere can cause up to 600% changes in the resistor value because of reduction of the oxides making up the resistor paste.

BIOMEDICAL HYBRIDS

Biomedical hybrids are not typically mounted to any carrier plate or case; they are implanted. Thus the need for a carrier plate is nonexistent. Mounting within a metal case would increase the volume and produce package corners, which could irritate the unit's surroundings. Although metal cases may provide a hermetic seal from air and moisture, they would not protect the unit from the hostile environment of body fluids (e.g., acids, enzymes, and other organics). Biomedical hybrids must be protectively overcoated with a polymeric material that will be inert in its environment, yet withstand the chemical hostility of its surroundings.

Biomedical companies have found that by eliminating any potential contaminants prior to and during the application of the protective overcoat and carefully controlling the curing process, complete and total adhesion is obtained. In other words, the entire surface is coated; no voids or pockets are formed along the substrate assembly surface. Therefore, although the coating will absorb moisture from its surroundings, the moisture that is absorbed has no place to go. The moisture penetrates to a certain depth of the coating and stops, There is no void to which the moisture can mass-transfer and condense. Therefore, the moisture is trapped in the overcoat material and cannot reach the hybrid. Once the surface of the protective overcoat is saturated with moisture, no more moisture transfers into it [3].

Thermal stress in a biomedical hybrid is not as great as in military and most commercial applications. The temperature is slightly elevated, 98.6°F, but fairly constant. The maximum temperature gradient that a biomedical device might see is only a few degrees. Thus thermal management is not a crucial issue for these designs.

After the substrate is in place, the substrate I/O is wire bonded to the case leads. At this time a final internal visual and full functional electrical test is performed. Figure 4.27 shows examples of delidded hybrid assemblies.

## Package Sealing

Once the hybrid passes precap inspection and electrical testing, it is sealed. The most common sealing methods are welding methods, although solder sealing can be used, as is done with chip carriers. In resistance welding, an electrode is rolled over the edges of the lid while current and pressure are applied. The current heats up the lid and case because of the resistance of the case material. Resistance welding works only on cases and lids made of materials with higher resistivities, such as kovar or stainless steel. If the case itself is made of a copper alloy for thermal and grounding purposes, then a kovar or stainless steel seal ring can be brazed to the top of the case side walls to provide the resistance needed for this type of welding.

Another welding method for sealing involves using a laser to weld the lid and case materials together to form a hermetic seal. Laser welding can be accomplished on a wide variety of materials (e.g., aluminum-silicon alloys, ferrous alloys, nickel alloys). Figure 4.28 shows the different lid configurations used for solder and welding methods.

Lid deflection must be taken into account when designing the packaging for an electronic device, whether it be for a hybrid or a module. To enable resistance welding, the metal thickness must be small enough to allow the lid and seal ring surface to heat up to a welding temperature without heating up the entire case. For this reason, the lids used for resistance welding have a outer perimeter thickness of 0.005 in. The center metal is usually 0.010 to 0.015 in. thick, as shown in Figure 4.28. This added center thickness

**Figure 4.26** Substrate mounting in a glove box. (Courtesy of Westinghouse ESG.)

gives the lid more rigidity. However, this thickness still allows lid deflection to occur during operation and testing conditions. This deflection must be considered during the package design and the substrate layout.

For example, in military applications, hybrids must pass various environmental tests, including leak testing, centrifuging, and vibrational testing. All these test impose stresses on the package which can cause lid deflection. For example, when a hydrid seal is tested (i.e., the unit is leak tested), the sealed package is expose to a differential pressure of 2 atmospheres (29.4 psi). If a standard 1 by 2 in. hybrid is leak tested, what would be the expected lid deflection?

Total deflection for a uniformly distributed load can be expressed as

$$y = k_1 \frac{\omega r^4}{E t^3}$$

where $y$ is the total deflection, $E$ the Young's modulus of the lid material, $r$ the width of the lid, $R$ the length of the lid, $t$ the lid thickness, $\omega$ a uniformly distributed load in psi, and $k_1$ a coefficient dependent on the ratio of $R$ to $r$ and the way the lid is supported [4, p. 5–52].

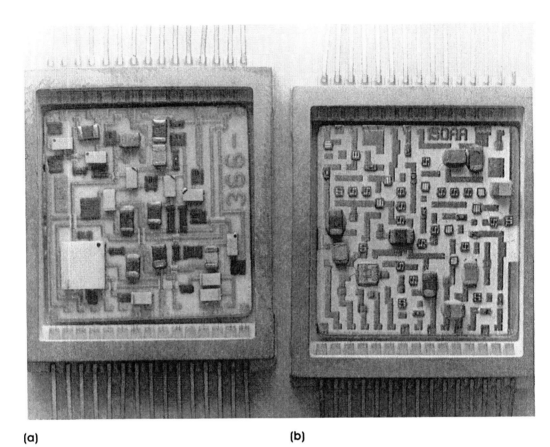

(a)                                                           (b)

**Figure 4.27**  Hybrid assemblies: (a) single layer with crossover thick film substrate in a ceramic case; (b) single layer thick film substrate in a ceramic case.

A standard $1 \times 2$ in. hybrid would have a $1 \times 2 \times 0.015$ in. kovar lid. Thus $r = 1$ in., $R = 2$ in, $t = 0.015$ in. and $E = 20 \times 10^6$ psi (see Table 4.3). This lid is welded or attached to the case, which would imply that the lid is fixed on all edges.

Using a fixed supported model and knowing that $R/r = 2$, we know that $k_1 = 0.0277$ [4, p. 5–54]. Substituting the given information into the equation yields the following:

$$y = 0.0277 \frac{29.4 \times 1^4}{20,000,000 \times 0.015^3}$$

$y = 0.012$ in.

Checking this number against an actually measured deflection during leak testing, we find our calculated valve to be much smaller than the measured value of 0.030 in. Further investigation of the case shows that the case walls are not totally rigid; they can flex at the floor of the case during leak testing. This suggests that the lid is not totally fixed but rather somewhere between fixed and simply supported.

(c)

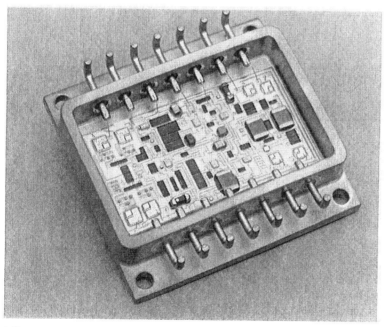

(d)

**Figure 4.27** Continued (c) power hybrid, multiple substrate assembly; (d) power hybrid, single substrate with laser-trimmed thick film resistors as integral part of substrate. (Courtesy of Westinghouse ESG.)

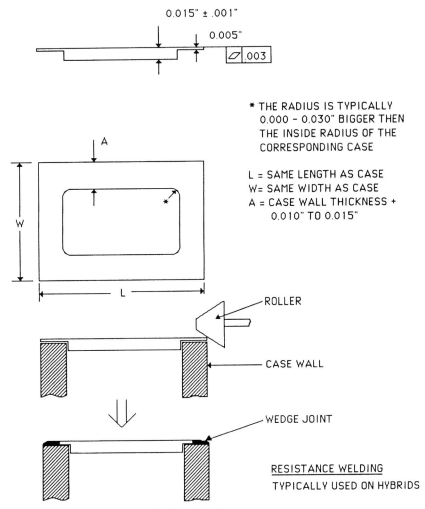

**Figure 4.28**   (a) Sealing lids for hybrid packages.

Recalculating for a simply supported model changes $k_1$ from 0.0277 to 0.1106 [4, p. 5–54]. This in turn gives a deflection of 0.048 in. This does indeed indicate that the actual lid deflection (0.030 in.) is between that of a fixed lid (0.012 in.) and that of a simply supported lid (0.048 in.). For future calculations of this hybrid's deflections a simply supported model will be used. This will ensure that the calculated value always gives the worst-case situation.

From the foregoing example it is obvious that lid deflection must be accounted for when designing a hybrid. To ensure that the lid does not come into contact with any components, potentially shorting components or damaging wire bonds, the bottom of the lid must be 0.048 in. above the highest component or wire bond in the hybrid. (The height of said

(CONTD.)

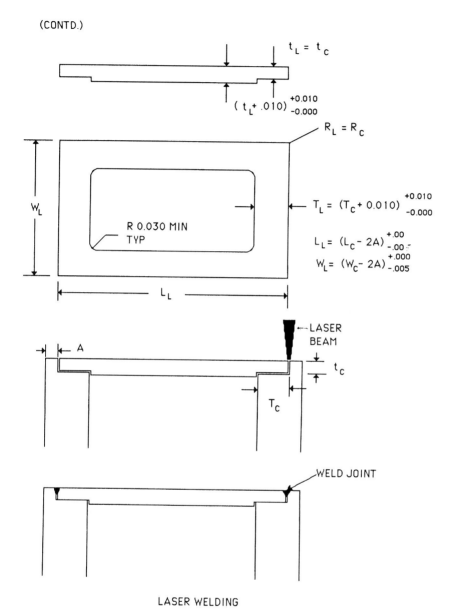

$t_L = t_C$

$(t_L + .010)\,^{+0.010}_{-0.000}$

$R_L = R_C$

R 0.030 MIN
TYP

$T_L = (T_C + 0.010)\,^{+0.010}_{-0.000}$

$L_L = (L_C - 2A)\,^{+.00}_{-.00}$

$W_L = (W_C - 2A)\,^{+.000}_{-.005}$

$W_L$

$L_L$

←LASER
BEAM

A

$t_C$

$T_C$

WELD JOINT

LASER WELDING

**Figure 4.28** Continued (b) solder sealing.

component must be the maximum component stack-up.) Figure 4.29 shows an example of how to calculate the minimum cavity depth needed for a given hybrid design.

If the additional height needed to allow for lid deflection is not available because of system height limitations, the design must be modified in another way to compensate for lid deflection. One approach might be to go to a solder seal or laser weld instead of a resistance weld so that the lid can be made thicker and thus decrease the magnitude of

**Table 4.3** Typical Properties of Metallic and Ceramic Materials Commonly Used in Frenchtown Ceramic-Metal Brazed Assembled[a]

| Property | Metallic materials | | | | | | | | | | | Ceramic materials | | |
|---|---|---|---|---|---|---|---|---|---|---|---|---|---|---|
| | Steel | Stainless steel 304 | Kovar | Alloy 42 | Nickel | Molybdenum | OFHC (copper) | WCu10 | WCu15 | WCu20 | WCu25 | Alumina FA-94 | Alumina 2082 | Beryllia thermal BeO[b] |
| Composition by weight % | Fe 99+ | Fe 67 Cr 20 Ni 10 | Fe 54 Co 17 Ni 29 | Fe 58 Fe 42 | Ni 99.0 min. | MO 99.9 | CU 99.96 | Cu 10 W 90 | Cu 15 W 85 | Cu 20 W 80 | Cu 25 W 75 | Al₂O₃ 94 | Al₂O₃ 94 | BeO 99.5 |
| Specific gravity (g/cm$^3$) | 8.7 | 7.8 | 8.4 | 8.2 | 8.8 | 10.2 | 8.9 | 17.2 | 16.5 | 15.6 | 14.8 | 3.63 | 3.60 | 2.80 |
| Hardness Rockell | 50 | 90 | 70–85 | 60–80 | 50–80 | 55–60 | 10–45 | 30 | 25 | 105 | 100 | 80 | 80 | 60 |
| (B, C, or 45N scale) | B | B | B | B | B | B | B | C | C | B | B | 45N | 45N | 45N |
| Tensile strength (× 10$^3$ psi) | 55 | 85 | 70 | 60 | 60 | 170 | 30 | 125 | 120 | 115 | 100 | 35 | 33 | 22 |
| Compressive strength × 10$^3$ psi | — | — | — | — | — | — | — | — | — | — | — | 410 | 410 | 225 |
| Young's modulus of elasticity × 10$^6$ psi | 28 | 28 | 20 | 20 | 30 | 50 | 17 | 37 | 36 | 35 | 34 | 45 | 45 | 21 |
| Coefficient of linear thermal expansion (× 10$^{-7}$/°C, 20–500°C) | 140 | 190 | 53 | 73 | 150 | 56 | 180 | 60 | 65 | 70 | 75 | 71 | 70 | 73 |
| Thermal conductivity (cal cm/cm$^2$ sec°C) 20°C 100°C | 0.14 0.16 | 0.33 0.39 | 0.040 0.043 | 0.036 0.041 | 0.204 0.200 | 0.380 0.330 | 0.940 0.940 | 0.50 0.49 | 0.55 0.53 | 0.59 0.58 | 0.62 0.63 | 0.05 0.04 | 0.05 0.04 | 0.55 0.40 |
| Specific heat (cal/g °C) | — | 0.12 | 0.104 | 0.120 | 0.130 | 0.060 | 0.092 | 0.039 | 0.041 | 0.044 | 0.047 | 0.19 | 0.19 | 0.25 |
| Electrical conductivity | — | — | — | — | — | — | 100 | 40 | 45 | 50 | 55 | — | — | — |

a These materials are most often used by Frenchtown Ceramics to produce ceramic-metal brazed assemblies. Other materials are available.
b Registered trademark, Brush Wellman.
Courtesy of Frenchtown Ceramics, Inc., Frenchtown, N.J.

Given:  Standard 1" X 2" case with 0.030" base,
        standard 0.025" substrate with 5 conductor
        layers, standard lid, and tallest component
        having a height of 0.045".

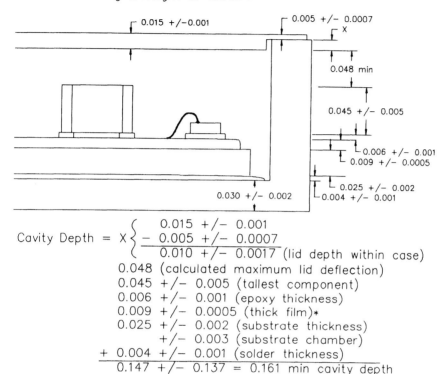

$$
\text{Cavity Depth} = X \begin{cases} & 0.015 +/- 0.001 \\ - & 0.005 +/- 0.0007 \\ \hline & 0.010 +/- 0.0017 \end{cases} \text{(lid depth within case)}
$$

0.048 (calculated maximum lid deflection)
0.045 +/- 0.005 (tallest component)
0.006 +/- 0.001 (epoxy thickness)
0.009 +/- 0.0005 (thick film)*
0.025 +/- 0.002 (substrate thickness)
     +/- 0.003 (substrate chamber)
+ 0.004 +/- 0.001 (solder thickness)
─────────────────────────────────────────
0.147 +/- 0.137 = 0.161 min cavity depth

* 4 Dielectric layers @ 0.0015 each + 5 top and 1 bottom
  conductor layers @ 0.0005 each.

**Figure 4.29**  Package cavity depth calculation.

the deflection. Another approach might be to reinforce the lid by adding ridges or "dimples" to it. The most common solution is to incorporate standoff posts, or spacers in the assembly. This could be done by mounting posts on the surface of the substrate assembly that would support the lid and limit its deflection. Care must be taken when designing these posts. They must support the lid without deforming it (posts that are too pointed could dent or even puncture the lid during leak testing). They must be designed to account for the stack-up tolerances, so that they will adequately limit the deflection without placing a force on the lid during the sealing process.

Another consideration in standoff post design is knowing all forces and resulting deflections to which the case will be exposed. For example, a 1 × 2 in. hybrid tested per MIL-H-38534 will be exposed to a 2-atmosphere pressure differential during leak testing, 5000 $G$'s during centrifuging, and vibration at 200 $G$'s rms during particle impact noise detection (PIND) testing. From the previous example we know that the leak testing results is a worst-case (simply supported) deflection of 0.048 in. Similar calculations yield expected worst-case deflections of 0.037 in. and 0.0015 in. for centrifuging and PIND, respectively. From these calculations we know that the maximum expected deflection will be 0.048 in. during leak testing. We also know that we can expect the lid to vibrate 0.0015 in. during PIND. Therefore, the posts must be designed to prevent the lid from deflecting and coming into contact with the components. Yet the designer cannot forget the 1.5-mil vibration potential. This design will involve a complex structural analysis to determine what number of posts at what height and placement will meet this requirement. In this case the posts must also be short enough to remain below the vibration range, so that the lid will not tap against the posts during PIND and cause a noise detection failure.

### 4.3.4 Hybrid Packaging

Once the substrate assembly, component mounting, and wire bonding are completed, the substrate is ready to be mounted in the package or case. Typical hybrid cases are metal with glass or ceramic feedthroughs or ceramic with brazed-on leads.

### Metal Packaging

Metal cases are fabricated in three different ways. The first is a two piece construction with a kovar plate brazed to a window frame of kovar. Both pieces are nickel and gold plated prior to brazing. The window frame is typically machined out of a solid piece of metal. The second technique for fabricating a metal package is to machine a "bathtub" out of a solid piece of metal. The third metal case fabrication method consists of having the metal molded in the shape of a bathtub.

However the side walls of a gull-winged leaded case are formed—by brazing a ring frame to a plate, machining out a "bathtub," or molding the bathtub shape—the side walls must have the holes for the feedthroughs drilled out of them prior to plating. For insertion-mounted hybrids the holes for the feedthroughs are drilled in the bottom or floor of the case, along the inside perimeter, again prior to plating. The unplated holes in the window frame or case have a donut of glass or a glass bead placed into them. In turn, an unplated kovar lead is placed in the center of the glass donut. The case with glass beads and leads in place is sent through a furnace profiled to allow the glass to melt. The glass reflows and fuses with the oxide on the surface of the kovar. A mild etch or cleaning method is then applied to prepare the remaining kovar surface for plating. The glass used in this process has a CTE close, or matched, to that of the kovar. These glass-to-metal feedthroughs are referred to as matched seals. Figure 4.30 shows a side view of a matched glass seal.

Another glass seal commonly used is a compression glass seal. In this case, the glass is reflowed to a kovar eyelet or ferrule. The lead and ferrule are then plated as usual. Finished feedthroughs are brazed into the side walls of the case. Figure 4.31 shows the case machined to accept the feedthroughs, the feedthroughs, the solder preforms, and the fixturing used to braze the feedthroughs into the case. Finally, a finished metal case with compression feedthroughs is shown in the figure. The fixturing used for such brazing processing is typically graphite. Graphite can withstand the braze temperatures without losing its shape, and the braze material will not adhere to the graphite. The case wall material can be kovar or in power applications, a copper alloy. The ferrule and the glass are put in compression because the expansion rate of the surrounding metal is greater than that of the glass. The glasses typically used in feedthroughs have a design tensile strength of only 1000 psi. Tensile strengths of approximately 2000 psi may be obtained with certain glass compositions. This strength can be further increased to 3000 psi by annealing the glass or to 20,000 psi by tempering the glass [5]. Unfortunately, these strengthening methods cannot be employed to feedthroughs. However, the strength of the glass can be greatly increased by placing the glass in compression. Actual compression strength measurements are difficult to obtain. The glass usually exhibits a tensile failure before the compression strength is gauged, because of slight bending or torquing during the measurement. It is theorized that the glass has a compression strength in the 100-Ksi range. So although this case configuration involves more processing, the glass seals are much stronger.

Glass-to-metal cases have been used in the industry for decades. They are readily available, use proven technology, have relatively low cost, and offer hermeticity. However, glass forms a meniscus during the reflow process (see Figure 4.30). The meniscus is thin and brittle and is not in compression, so the glass within the meniscus has lower strength. For this reason, glass-to-metal seals are plagued with meniscus cracking and chip-out problems due to the thermal and mechanical stresses applied during temperature cycling and the lead forming or centrifugal forces, respectively. When the glass chips out in the meniscus area, the lead base metal, typically kovar, is exposed. As discussed earlier, the glass is reflowed to the unplated kovar; therefore, the metal exposed when a glass chip-out occurs does not have an anticorrosive nickel plate to protect it. Thus exposed base metal becomes a corrosion site. If the chip-out occurs within the interior of the hermetically sealed package, corrosion formation is much slower because of the dry, hermetic environment. However, on the exterior of the case, it is only a matter of time before a chip-out does corrode. Cracks can lead to chip-outs and in the matched glass configuration can propagate, theoretically to the point of a loss of hermeticity. For these reasons the military has established visual inspection criteria specifying cracks and chip-outs as rejectable.

Some methods for repairing glass chip-outs have been tried. The most common is to cover the chipped-out area with an epoxy or overcoat of some sort. One argument against this practice is that the epoxy or overcoat would not be sealed within a hermetic atmosphere and would absorb moisture that would eventually migrate and reach the kovar surface, thus causing a corrosion site or a potential site for dendritic growth. The second argument is that the epoxy would just cover up the chip-out problem, along with any other problem that might exist. The epoxy would make it impossible to see the extent of the chip-outs and/or cracks or whether any other problem existed (e.g., foreign material, embedded material in the glass, glass overrun). Because of these two arguments, the military has rejected this repair process. Another method sometimes attempted is to brush plate the exposed base metal with gold. This again covers up the exposed base metal, giving the appearance of the correct surface finish. Unfortunately, without the nickel plating under the gold, the gold by itself does not provide a corrosion barrier. Thus this method only postpones the development of a corrosion site. Furthermore, the gold in direct contact with the current-carrying kovar lead sets up an electrolytic reaction. In other words, the gold/kovar interface acts like a tiny battery and can actually result in more extensive corrosion.

In an effort to eliminate crack glass yield problems, Department of Defense (DoD) contractors are incorporating ceramic feedthroughs. Ceramic packages and metal packages with ceramic feedthroughs have been used in space applications for many years. Ceramic is much stronger than glass: ceramic (95% alumina, for example) has a tensile strength of 30,000 psi and a compression strength of 300,000 psi, making a ceramic feedthrough an order of magnitude stronger than a glass feedthrough [5]. Furthermore, ceramic feed-

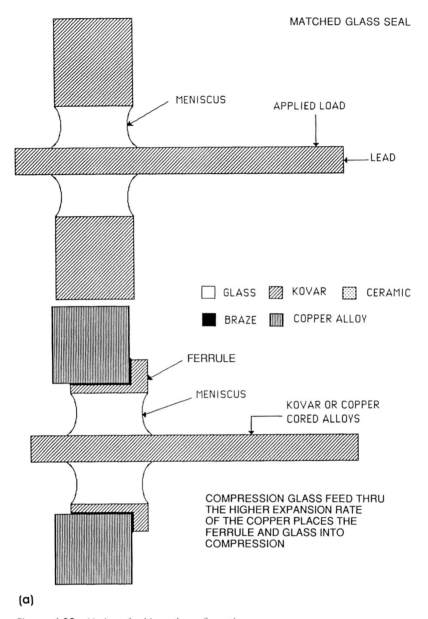

MATCHED GLASS SEAL

MENISCUS

APPLIED LOAD

LEAD

GLASS    KOVAR    CERAMIC

BRAZE    COPPER ALLOY

FERRULE

MENISCUS

KOVAR OR COPPER
CORED ALLOYS

COMPRESSION GLASS FEED THRU
THE HIGHER EXPANSION RATE
OF THE COPPER PLACES THE
FERRULE AND GLASS INTO
COMPRESSION

(a)

**Figure 4.30** Various feedthrough configurations.

throughs do not have a meniscus. Ceramic feedthroughs are formed by metallizing the
inside hole and outer ring of a ceramic donut and then brazing in the lead. These feedthroughs can then be brazed into the case wall with a lower-temperature braze. Figure
4.30 depicts the differences between glass and ceramic feedthroughs. Ceramic feedthroughs
offer greater strength and the elimination of a meniscus and its associated cracking and
chip-out problems. Because of the increased strength of the material, less metal is needed
to surround the feedthrough to ensure mechanical integrity of the package. With glass

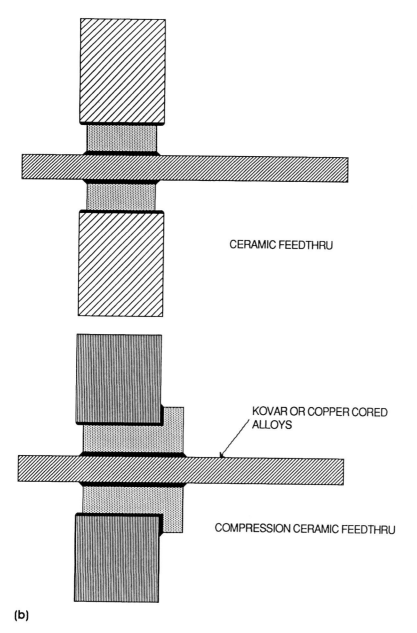

CERAMIC FEEDTHRU

KOVAR OR COPPER CORED
ALLOYS

COMPRESSION CERAMIC FEEDTHRU

(b)

seals, the military standard MIL-H-38534 requires that there be at least 0.040 in. of metal between the glass and the seal surface, as shown in Figure 4.32. This distance of metal will protect the glass from the stress applied during the resistance welding process (discussed earlier under Package Sealing). No standard for ceramic is given in MIL-H-38534; however, ceramic feedthrough packages used for space applications typically have less than 0.040 in. from the ceramic to the seal surface. Of course, any package used for military or

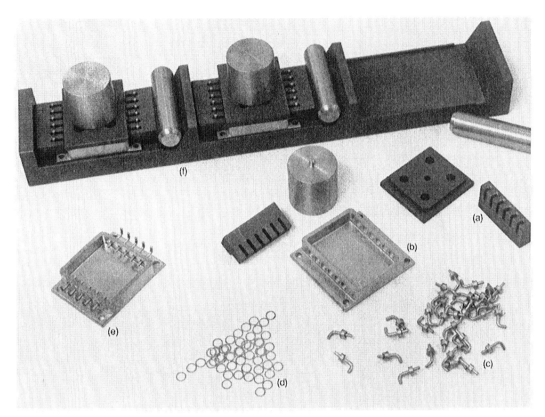

**Figure 4.31** Compression feedthrough case's components and fixturing; (a) feedthru fixture, (b) case blank, (c) feedthrus, (d) solder preforms, (e) finished case, (f) reflow fixture. (Courtesy of Westinghouse ESG.)

space application must be qualified prior to use. This qualification involves passing very stringent environments and testing levels.

Although ceramic is stronger than glass and more capable of withstanding the stresses imposed during processing, handling, operation, and testing, ceramic feedthrough packages have disadvantages. They are more expensive than glass feedthrough packages. With glass, the feedthrough and the lead are put into the case with one step, in a furnace programmed to reflow the glass. Ceramic feedthroughs are much more labor intensive. They require that the ceramic be formed into donuts. This can be done by machining a sheet of fired ceramic. The sheet thickness is the same as the desired donut height. The machining is accomplished by laser drilling or ultrasonically milling out the donut shapes. These machined parts, especially if laser machined, must be deslagged. As the laser cuts through the ceramic, it actually melts the material. This melted material splatters onto the surrounding surface and is referred to as slag. Slag forms burrs with very smooth glasslike surfaces. Metallization will not adhere properly to the slag. So, to prepare the ceramic donut for metallization, the surfaces must be cleaned to eliminate any slag. This cleaning can involve sandblasting, in which abrasive slurry is forced through the holes under high pressure. An etching process is sometimes used, but it can lead to pitting of the surface. Another method developed by Laserage, a laser machining company, involves heat treating the laser-machined ceramic [6]. This heat treatment leaves the machined

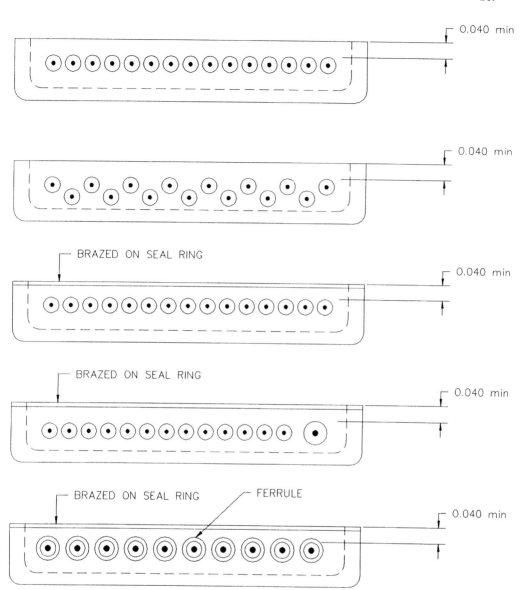

**Figure 4.32** Military standard requirement for insulation to seal surface distance.

surfaces with a finish similar to that of fired ceramic. Once the ceramic is prepared, it must be metallized both inside the hole and on the outer perimeter. This can be done by painting on a thick-film metal paste and firing the metal to the ceramic.

Another method of forming the ceramic donuts needed for the feedthroughs is to mold them into the donut shape while the ceramic is still in green, or nonfired, state. Before firing the ceramic, a refractory metal is applied to the donut hole and perimeter. The metal and ceramic are then cofired. The refractory metal must then be nickel plated to provide a corrosion barrier, and the nickel must be plated with a solderable or brazable metal, typically gold.

Once the ceramic donut is formed and metallized, the leads or pins must be brazed into the donut holes. Then the ceramic feedthrough must be brazed or soldered into the metal package. In some cases both brazing steps are done together, but this requires special tooling to hold the leads and ceramic donuts in place during the brazing process.

Ceramic feedthrough cases involve more processing and are more labor intensive than glass feedthrough cases. Ceramic feedthru case manufacturing also requiries higher-temperature processing. Because the ceramic firing and brazing temperatures are much higher than the glass reflow temperature, more sophisticated furnaces are required for the processing. The additional labor, the higher cost of the ceramic material compared to the glass, and the equipment needed for the ceramic processing all lead to ceramic feedthrough cases being more expensive than their glass counterparts.

Ceramic also has different electrical properties than the standard glass used in electronic packaging. These differences may or may not affect the package design, depending on the electrical requirements of the unit. For example, ceramic may be desirable in high-voltage packages because the more porous surface of the ceramic will hinder the voltage from traveling across the ceramic insulator and shorting to the case. However, in analog or microwave applications, ceramic feedthroughs may not be desirable because of the ceramic's higher dielectric constant.

The dielectric constant of an insulating material in a feedthrough will determine the impedance of that feedthrough. Figure 4.33 gives the definition of impedance. The following equation represents the impedance of a coaxial cable or a feedthrough:

$$Z = \frac{60}{\sqrt{\epsilon}} \ln \frac{b}{a}$$

where $Z$ is the characteristic impedance in ohms ($\Omega$), $\epsilon$ is the dielectric constant of the insulating material (i.e., the glass or ceramic), $b$ is the radius of the insulating material, and $a$ is the radius of the lead [1, Sec. 2].

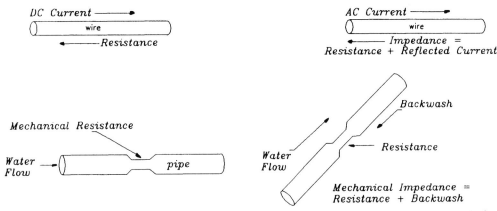

IMPEDANCE: "A measure of the total opposition to current flow in an alternating-current (ac) circuit, equal to the ratio of the rms electromotive force in the circuit to the rms current produced by it, and usually represented in complex notation as $Z = R + iX$, where R is the ohmic resistance and X is the reactance."

**Figure 4.33** Definition of impedance. (From Ref. 1. Definition text from *The American Heritage Dictionary of the English Language*, New York: Houghton Mifflin [1982].)

The dielectric constant of the glass material used for feedthrough manufacture is approximately 5, and dielectric constants of ceramics are higher (e.g., 96 to 99% alumina ceramics have dielectric constants of approximately 9 to 10, respectively [1, Sec. 2]). Thus, if a glass feedthrough were converted to ceramic, with no change in dimensions, the impedance would be decreased.

Given a feedthrough design with an insulator diameter of 0.116 in. and a pin diameter of 0.018 in. what would the impedance be for a glass and a ceramic feedthrough?

$$\text{For glass,} \quad Z = \frac{60}{\sqrt{5}} \ln \frac{0.058}{0.009} = 50 \ \Omega$$

$$\text{For ceramic,} \quad Z = \frac{60}{\sqrt{9}} \ln \frac{0.058}{0.009} = 37 \ \Omega$$

To maintain the same impedance of 50 $\Omega$ and lead diameter of 0.018 in., how would the ceramic feedthrough design change?

$$Z = \frac{60}{\sqrt{\epsilon}} \ln \frac{b}{a}$$

$$\frac{Z\sqrt{\epsilon}}{60} = \ln \frac{b}{a}$$

$$\frac{50\sqrt{9}}{60} = \ln \frac{b}{0.009}$$

$$2.5 = \ln \frac{b}{0.009}$$

$$\exp 2.5 = \frac{b}{0.009}$$

$$b = 0.110 \text{ in.}$$

The diameter of the insulator (i.e., the ceramic) would have to be increased to 0.220 in. Figure 4.34 shows the resulting changes in the case design.

As can be seen in the foregoing example, converting to ceramic feedthroughs for certain analog applications may not be possible, depending on the dimensional limitations of the design. In this example, to have a ceramic feedthrough with a 50-$\Omega$ impedance, the minimum height of the case would have to be 0.300 in. (if the military standard 0.040-in. minimum distance from insulator to seal surface is met). If the system design requiries low-profile packaging with a maximum height of less than 0.300 in. the conversion to ceramic feedthroughs would not be possible.

Industry is developing ceramic-filled glasses that will yield the best properties of both materials. Some of these filled glasses can be processed with lower temperatures, like the glass, but are stronger. Others might require the higher-temperature processing but would have lower dielectric constants, making them very appealing for analog applications. As these materials are developed and become available to the industry at reasonable prices, they will enable electronic packaging to be further miniaturized with improved electrical, specifically analog, capabilities.

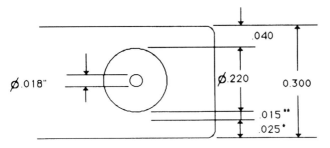

( ALL DIMENSIONS GIVEN ARE MINIMUMS )

\* PACKAGE BASE
\*\* CLEARANCE FROM BASE TO FEEDTHRU

**Figure 4.34**   Ceramic feedthrough case with 50 Ω impedance design. (\* = Package base, \*\* = clearance from base to feed-through.)

## Ceramic Packaging

Cofired ceramic is an alternative case material that is frequently used. These hybrid cases are manufactured in the same manner as ceramic chip carriers. The difference is that they are larger and usually come with top brazed leads on two sides (i.e., gull-winged lead configuration). Figure 4.35 shows some examples of ceramic cases used for hybrids.

Ceramic cases can have the substrate mounted inside or can have the base of the case act as the multilayered substrate. This customized configuration is referred to as an integral substrate package, an example of which is shown in Figure 4.36. In medium- to high-volume applications (e.g., orders of approximately 1000+ cases), integral substrate packaging can greatly reduce manufacturing costs. Once the electrical design is fixed and the packaging vendor has tooled up for such a case, these cases can be produced at very low prices, lower than those of metal cases in the same quantities.

In addition to reducing the cost of the purchased case, an integral substrate case eliminates the need to fabricate a substrate. This reduces the hybrid's material and labor costs. The integral substrate also eliminates the need to mount the substrate in a case and

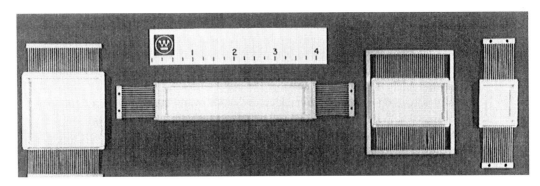

**Figure 4.35**   Ceramic hybrid cases. (Courtesy of Westinghouse ESG.)

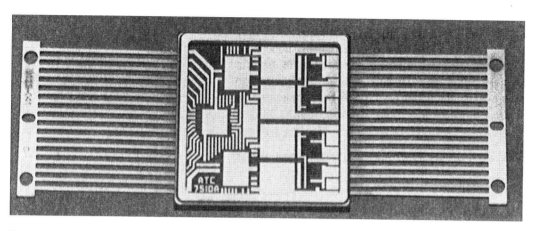

**Figure 4.36**   Integral substrate ceramic hybrid case. (Courtesy of Kyocera America.)

interconnect from the substrate via wire bonding to the package leads. This interconnection is internal within the multilayers of the case, further reducing the hybrid's labor costs.

Other advantages of ceramic packaging are improved yields and reliability. Ceramic cases eliminate yield problems associated with glass feedthrough cracking and chip-outs. The ceramic package is appealing in space and military applications because of its lower weight and noncorrosive nature compared with metal packaging.

As with ceramic feedthroughs, conversion from metal cases to all-ceramic cases is not practical in all applications. Again, in analog or microwave devices, the refractory metals used to fabricate cofired ceramic packages (see Section 4.2.2) may not carry the currents fast enough, so the unit cannot operate at the frequencies needed. Furthermore, the dielectric constant of the ceramic package may affect the electrical performance of a microwave device (e.g., the ceramic might cause an impedance mismatch). Care must be taken when designing microwave packaging. Microwave packaging design is an example of the critical need for concurrent mechanical, electrical, and material engineering.

## Customization of Electronic Packaging

Many designs will dictate specific requirements that cannot be met with the standard metal or ceramic cases. High-power hybrids might utilize a case with a heat-spreading BeO bottom and weldable kovar window frame. Others might have a metal base, ceramic side walls, special grounding requirements, etc. Ceramic and metal combinations are frequently utilized as the design requirements demand. Just as the hybrid itself is a customized electronic function, hybrid packaging often requires customized cases to meet the unit's specifications.

The ceramics most commonly used in production today are 92 to 99% alumina and 99% beryllia. Table 4.3 lists the properties of industrial ceramics and metals commonly used in electronics packaging. Common packaging metals include kovar, copper alloys, molybdenum, copper- or nickel-clad molybdenum, copper-clad invar, copper tungsten, and aluminum. These ceramics and metals are being combined through special techniques and advanced materials to meet the packaging requirements. Special compliant adhesives have been developed that can absorb the mismatch in thermal expansion between the metals and the ceramic. Other methods of attaching the metals to the ceramic have been developed, such as the direct bonding methods discussed in Chapter 3.

Much development work is being conducted with materials that offer even higher thermal conductivities, increased strength, better matched CTEs, lighter weight, etc. The number of these developmental materials is ever growing. Today's leading-edge materials include aluminum nitride, silicon carbide, A40 ( silicon-aluminum material), and other metal matrix and composite materials.

## 4.4  PRINTED WIRING BOARD ASSEMBLIES

As discussed in Chapter 3, there are various kinds of printed wiring or printed circuit boards. The oldest and most common are the organic boards. Ceramic boards, offering better thermal properties with closer CTE matching to the components, have been around for more than 20 years, yet have recently had an increase in use in high-power applications, such as standard electronic module (SEM) power supplies. The combination of polyimide and integrated circuit technology has resulted in high-density surface mount boards with finer line definition, less than 0.001 in. Another board fabrication method that involves directly bonding copper foil to ceramic, appropriately called direct bond copper (DBCu), will be discussed. Directly bonded copper enables vast improvements in power dissipation.

As defined in Chapters 2 and 3, insertion mounting is mounting a leaded component by inserting or plugging it into the PWB. Surface mounting is mounting the component to the surface, or on top of the board. PWBs can be mixed technology; they can have both insertion-mounted and surface-mounted components. However, a single component must be either inserted or surface mounted. Many people interchange this terminology, calling both insertion mounting and mounting to the top of the board ''surface mounting.'' For clarity in this text, insertion mounting will refer only to components mounted by insertion and surface mounting will refer only to components mounted to the surface of the boards.

All PWBs have similar assembly capabilities. However, the type of board will limit the type of assembly. An organic board with through hole intralayer routing can be assembled by insertion mounting or surface mounting, depending on the components and the layout design used. Ceramic boards are not insertion mounted, only surface mounted. However, ceramic boards can easily be double sided by metalizing both sides or by sandwiching two boards to a carrier plate or heat sink. Direct-bond copper offers strong, lightweight boards for high-power applications but has limited layering and line definition. Finally, thin-film polyimide boards are capable of supporting bare die applications. This section will discuss the various types of assembly processing associated with the different types of board designs.

### 4.4.1  Organic PWB Assembly

Organic boards are composite structures; e.g., epoxy glass boards have glass fibers suspended in an epoxy resin. Polyimide boards, like the epoxy glass boards, are composed of glass fabric with polyimide used as the dielectric material. These are the most common combinations that are available within industry. The individual layers are metallized with copper, have etched patterns, and then are laminated together. Interconnection between the layers is accomplished with through holes, blind vias, or buried vias (see Chapter 3).

The method of intralayer connection and the design of the mounting pads determine the type of assembly used. If only insertion-mounted components are to be used, then all connections from the components to the board are done with through holes. The mounting pads are actually holes in the board. Routing of one component's signal to another, from one through hole to another, can be accomplished by through hole, blind via, or buried via. Since the assembled board will have holes with leads extending through them, insertion boards typically do not have components mounted to both sides. If any components are mounted to the "bottom" of the board, they are typically small, hermetic chip components, mounted in position with epoxy. The epoxy holds them in place as they travel through the wave solder.

Boards routed with buried vias, thus limiting the number of blind vias the through holes, have limited reworkability and lower yields. However, they offer increased surface mounting area while decreasing the number of layers needed for routing. These surface mount designs can also be routed to enable double-sided mounting. Surface mount boards must be laid out to include the mounting pads of the right size, dimensioned to allow for component tolerances and proper solder fillets. Surface mount boards often have both leadless and leaded components. The leaded components require surface mount lead forming.

As discussed earlier under the subassembly summary, electrical parameters must be taken into account during the board layout. Components with high voltage potential between them should be separated to prevent voltage arcing between the components or tracks. When a high-voltage arc occurs, the voltage travels across the board material. As this voltage moves along the polymer surface, it can burn a trail into the board surface. This trail is commonly referred to as carbon tracking. If the arcing itself does not short out the circuit or damage the components in question, the carbon track left behind can cause shorts later on. Carbon tracking can be avoided by having additional spacing between all tracks and components, especially the critical ones. This solution, however, is not normally available because of size limitations of the system. The more common method of carbon tracking prevention is to give the board assemblies a protective overcoat. The overcoating reduces the available surface path for the voltage.

Boards can also be designed for mixed technology, in which some of the components are inserted and others surface mounted. The layout of the mixed-technology boards greatly influences the assembly processing. If all inserted components are mounted on one side, the surface-mounted components, provided they are low-profile, hermetic, chip components, can be tacked down, or epoxied, to the "bottom" side and reflow can be accomplished with a wave solder. Therefore, the processing for this mixed-technology board is similar to that for an insertion-mounted board. If inserted- and surface-mounted components are placed on the same side, the processing is similar to surface mounting methodology.

The following sections will discuss the main assembly steps: thermal plane attachment, connector attachment, and component mounting. The design's relationship to the processing used for assembly will be detailed.

## Heat Sink Attachment

In low-power commercial applications, components are mounted on the PWB directly. In high-power or military applications, the first step in the assembly process is to attach the PWB to the heat sink/carrier plate. In the case of an insertion-mounted board, the carrier is either an edge support or a "thermal ladder" (i.e., a slotted plate). The edge

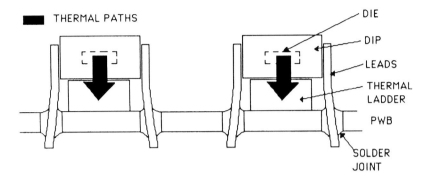

(a) INSERTION MOUNT ORGANIC PWB WITH THERMAL LADDER.  HEAT IS
    DISSIPATED FROM THE DIE THROUGH THE DIP PACKAGE FLOOR TO
    THE HEAT SINK

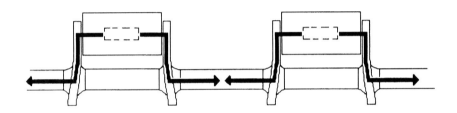

(b) INSERTION MOUNTED ORGANIC PWB.  SINCE THE DIP STANDS OFF
    THE PWB, THE HEAT MUST BE TRANSFERRED FROM THE DIE, THROUGH
    THE DIP PACKAGE, THROUGH THE LEADS, THROUGH THE SOLDER,
    TO THE PWB, OUT TO AN EDGE SUPPORT HEAT SINK.

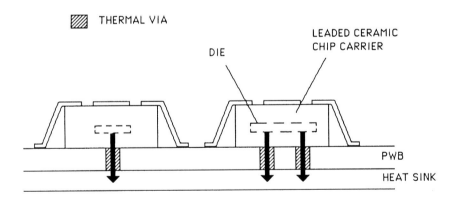

(c) LEADED SURFACE MOUNTED ORGANIC PWB WITH CAVITY UP CHIP
    CARRIERS

**Figure 4.37**   Thermal paths.

support, as its name applies, supports the edge of the PWB, leaving the center of the board free for the device placement. An edge support can be attached to the bottom or top of the PWB or wrapped around the edges of the PWB. A thermal ladder is a heat sink that is attached to the top of the PWB with windows or slots cut out of it. The component (e.g., a plastic DIP, ceramic DIP, or cer pak) straddles the material between the slots such that the leads go through the slots and insert into the PWB, as shown in Figure 4.37. This design provides a thermal path directly from the ceramic component to the heat sink, instead of having to go through the leads of the plastic component (e.g., DIP).

Surface mount boards typically have the same edge supports attached to them as do the insertion mount boards. Thermal ladders are not as effective for surface mounting. Surface mount components are leadless and thus incapable of straddling the thermal ladder, or they have a leaded configuration, such as gull winged, that permits both leads and package body to be mounted on the plane of the board, again making straddling the thermal plane difficult.

Organic surface mount PWBs can also have heat sinks that are attached to the bottom of the PWB with a B-stage epoxy. The PWB, epoxy film, and carrier plate are cured together under controlled pressure and temperature. In some cases the PWB assembly is a double-sided board in which two boards are sandwiched over the heat sink. This ''cored'' board doubles the design's surface-mountable area with a minimal increase in board assembly height or volume. This design forces the thermal path to travel through the PWB, but at least the components' bodies are in direct contact with the surface of the PWB, not standing off the surface leaving only the leads for primary cooling, as depicted in Figure 4.37. The thermal management of this type of design is usually enhanced by placement of vias under the packages. These vias, in addition to being used for routing, now act as thermal vias, giving the assembly a more efficient and direct thermal path from component to heat sink. However, these sandwiched or cored boards involve particular design considerations, which will be discussed in the section on leadless ceramic chip carriers on organic PWBs. Boards mounted to aluminum heat sinks are often used (e.g., SEM modules).

Heat sink or core materials are those offering improved thermal conductivity over the PWB material. Examples of heat sink materials include aluminum, kovar, copper-clad invar, copper-clad nickel, graphite composites, copper-clad molybdenum, and metal matrix composites such as silicon-aluminum materials.

Aluminum has high thermal conductivity and is lightweight, less corrosive than most metals, inexpensive, easy to machine, and readily available. Copper has the same properties as aluminum except that it is highly corrosive and heavy. Both aluminum and copper have excessive CTEs compared to those of the organic or surface mount components. Copper-clad invar offers the superior thermal and electrical conductivity of copper, while the center invar gives the composite material a more desirable CTE. Copper-clad invar has a CTE smaller than that of copper but larger than that of invar. When the board is bonded to the core material, it gives the cored assembly a CTE that approaches that of the leadless ceramic chip carriers (LCCCs). However, the copper-clad invar is many times heavier and is more expensive than aluminum, and it is available only from Texas Instruments (TI) or sales representatives contracted by TI. Graphite fiber composites offer high thermal conductivity, approaching that of the copper-clad material, yet are lightweight like aluminum. However, the graphite is difficult to machine. Composites, like graphite, and metal matrix composite heat sinks can have their thermal conductivity and CTE tailored to the assembly. The direction or orientation of the fibers within a graphite composite will direct the thermal path. Thus the thermal resistance along the path between the components and heat sink can be reduced by having the fibers oriented in the same direction as the desired thermal path. Likewise, by controlling the composition of the metal matrix or composite material, the conductivity and CTE can be controlled to a degree. For example, if an aluminum-silicon composite is to be used, the thermal conductivity can be increased by having a greater percentage of aluminum in the composition of the material, aluminum being the better conductor. The cost of improved conductivity is an increase in CTE. As the amount of aluminum is increased, both the conductivity and CTE move toward that of aluminum. Thus these composite materials must be designed from a metallurgical and mechanical standpoint to accomplish heat sink design optimization. Ongoing research and development of new composite materials address the thermal and CTE design issues.

All of these heat sinks are usually attached with epoxy films. These films are usually thermally conductive B-stage or prepreg sheets. Heat sinks can also be attached with other film adhesives. The heat sink material and adhesive choices depend on the requirements of the system or PWB assembly. If cost and weight are more critical than thermal management (i.e., if the PWB assembly will not be subjected to severe temperatures or rapid temperature changes), then aluminum attached with standard B stage would give a lower cost and one of the lightest weight combinations. If thermal management or power dissipation is critical but there are no restrictions on weight or outgassing, then copper invar copper attached with a flexible silicon adhesive to absorb the CTE mismatch might be the desired design. If thermal management and weight were both critical, the answer could be a graphite heat sink with silicone adhesive. If the same requirements just discussed were imposed along with military specifications, the components of the PWB would determine the heat sink and adhesive design; if the PWB has only hermetically sealed devices mounted to it, and the PWB card is mounted in an open air chassis, then the design given in the previous example would meet the requirements. However, if the PWB assembly was to be mounted inside a hermetic module that also contained nonhermetic devices or bare chip devices, the silicon adhesive could not be used because of its inability to meet the military outgassing limitations. In this case, the heat sink would have to be designed not only to dissipate the heat so that the maximum junction temperature would not be exceeded but also to have a CTE closely matching that of the PWB. In this design, the more expensive developmental composite materials would have to be considered. If a heat sink with the right combination of conductivity and CTE was designed, a less flexible film adhesive that would meet the military outgassing limitations could be used. Furthermore, in such a design, the components being mounted would have to be designed or packaged so that they could withstand their CTE mismatch to the PWB. All the components would have to be small enough so that the CTE would not significantly affect them, or they would have to be leaded and have the leads formed so that they could absorb the strain and resulting stresses caused by the CTE mismatch.

## Insertion Mounting

In an insertion-mounted board all components and subassemblies must be leaded. The leads must be configured or mechanically formed to be inserted into the mounting holes of the board. For PGAs the leads are already in the correct configuration for insertion. They are perpendicular to the bottom of the package. DIPs also are designed for insertion mounting. The leads are formed along the side of the package and point down. The military

**Figure 4.38**   Automatic component placement. (Courtesy of Westinghouse ESG.)

standard configuration has the leads at a 6° angle out from perpendicular. Components such as resistors and capacitors are the leaded round cylinders. These components normally have their leads formed or bent into a right angle for insertion by pick-and-place equipment. The placement machine sends a robotic arm to the bin, tape, or feeder to select the component. The robotic arm has a specifically designed end effector that picks up the component by vacuum or by grabbing it as with a pair of tweezers. The arm then takes the components requiring lead forming to the lead former. The lead former might be a fixture similar to a trough into which the arm pushes the component, causing the leads to be bent as they are forced against the side walls of the trough. Lead forming can also be done by the robotic arm locking into a set position while two other arms come over, grab the leads, and turn to bend them appropriately. The robotic arm then carries the lead-formed part over to the board, where it inserts or mounts the part in a preprogrammed location as shown in Figure 4.38. Of course, there are many other ways of accomplishing the lead forming. This is all done within tight tolerances and within seconds. DIPs are normally supplied in tubes, or feeders for the placement equipment, with the leads already formed. Chip carriers and hybrids normally require preforming of the leads before they are put into the magazines that act as feeders to the pick-and-place equipment. This lead forming is done with a die or fixture designed to cradle the package while arms come down on the leaded sides, grab the leads, and force them into the L shape needed for insertion. Figure 4.39 shows an example of lead forming. For high-volume applications,

(a)

**Figure 4.39** Lead forming; (a) lead forming die, (b) chip carrier before forming, (c) chip carrier after forming (Courtesy of Westinghouse ESG.)

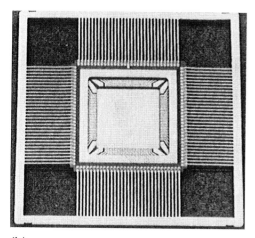

**(b)**

**(c)**

the feeders are designed to form the leads. Here, the feeders form the leads in the manner just discussed prior to feeding the lead-formed part to the placement equipment.

After the components have been inserted, the leads coming out of the bottom of the board must be dealt with. Most equipment today has automatic lead trimming. As the leads are inserted into the board from the top with the appropriate end effector, the associated lead-trimming fixture trims the leads and flattens them on the bottom of the board, as shown in Figure 4.40.

To prepare the boards for component placement, the PWB is tin or lead-tin plated, as covered in Chapter 3. This plating can be a tin flash used as corrosion protection for the copper metallization, or it can be a thick plate of solder used to pretin the boards. Once all the components are inserted, the board is sent through a wave solder. The board is attached to a belt that pulls it across a wave of flux foam and then a wave or "waterfall" of liquid solder. The bottom of the board floats over these waves. The flux foam deposits flux over the entire board surface. The flux is activated with temperature and removes the oxides on the surfaces to be soldered. This deposit of flux allows the solder to wet these surfaces and wick up the leads and into the holes to form the solder joints. In most cases the components are mounted only on the top side of the board. If it is desired to mount components on the bottom side, these components must be small enough to ride in the waves without causing the bottom surface of the board to lose contact with the waves, and they must be attached or tacked down to the bottom surface of the board with an epoxy so that they will not fall off before the wave solder has a chance to form the solder fillet interconnections to the leads. After being soldered, the boards are cleaned and inspected prior to final electrical testing. Figure 4.41 shows an example of an insertion-mounted PWB assembly.

## Surface Mounting

As just discussed, insertion-mounted PWBs require, by definition, that all components be leaded and that the PWB be routed with through holes or plated through vias (at least all connections to the components are done with through holes). Surface mount components

(a)

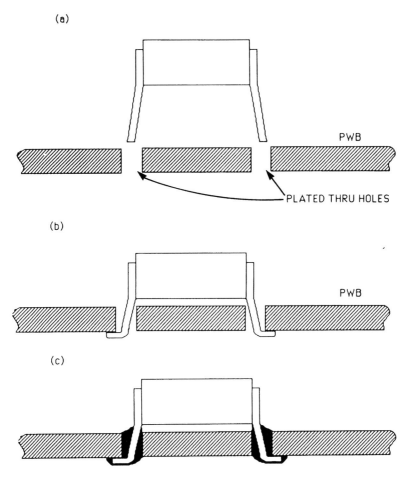

PWB

PLATED THRU HOLES

(b)

PWB

(c)

**Figure 4.40** Insertion mounting and lead trimming, (a) component and board, (b) component inserted in a board with leads cut and clenched, (c) after wave solder.

can be either leaded or leadless, such as chip capacitors, chip resistors, or LCCs. Surface mount organic boards can be routed with any combination of through holes, blind vias, or buried vias. Furthermore, surface mount PWBs can have heat sinks attached directly to the back side of the board, since there is no need to leave this area clear for lead protrusions.

The advantageous aspects of surface mount technology over insertion mounting are the reductions in routing layers, in the assembly's profile, and in board area. A surface mount design does not require through holes for component mounting. The components can be mounted on pads on the board's surface. In addition, the board routing can be accomplished with blind or buried vias. These two factors eliminate the need for through holes, which in turn eliminates the need for extra layering to allow for routing around the holes. Surface-mounted components can also be reworked (i.e., replaced) without heating/reflowing the entire assembly. This rework is accomplished with the use of an extractor, which heats up only the component to be replaced. Figure 4.42 shows a component being removed with an extractor.

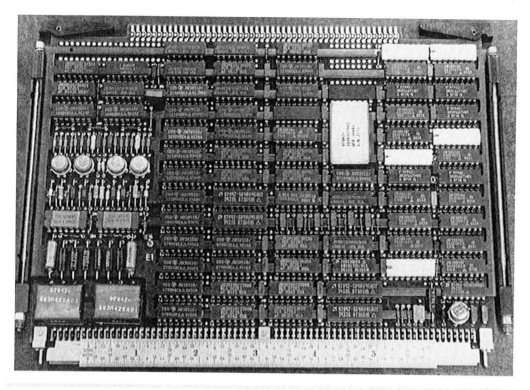

**Figure 4.41**  Insertion-mounted organic PWB assembly. (Courtesy of Westinghouse ESG.)

Surface mount components usually have less height than their insertion mount counterparts. They also sit directly on the board surface, not slightly above the surface as shown in Figure 4.37. Moreover, surface mount chip carriers offer great reductions in board area compared with comparable DIPs or cer paks. This decrease in area is further enhanced if leadless chip carriers are employed. For example, a 20-pin DIP has 0.100-in. center-to-center lead spacing (i.e., "pitch"), making it at least 1.1 in. long by approximately 0.250 in. wide. The area of the component itself is 0.275 in.$^2$. The footprint for this DIP would add an additional 0.150 in. to each side. Thus its required mounting area would be 0.605 in.$^2$. The body of the DIP would be approximately 0.250 in. thick and would sit approximately 0.050 in. off the board, making its height or profile 0.300 in. overall. A 20-pin chip carrier has a body that is 0.300 in. square, with a height as low as 0.060 in. This component's I/O is along the perimeter on all four sides. If the chip carrier is leaded, the footprint would extend 0.150 in. out from the body of the carrier, making the overall mounting area 0.600$^2$ or 0.360 in.$^2$. If this component were an LCC, the footprint would extend only 0.050 in. out from its body, or require a mounting area of only 0.160 in.$^2$. Either leaded or leadless, the chip carrier would sit only 0.005 in. above the board. The leaded device would have an epoxy preform under it, taking up 0.003 to 0.005 in. of height, while the LCC would have 0.005-in. solder joint standoff. The example just discussed, along with a second example with a higher pin count, is tabulated on the following page. The magnitude of the reductions increases with pin count, as can be seen in the 68-pin example given in the table.

|                             | Component area (in.$^2$) | Footprint area (in.$^2$) | Profile (in.) |
|-----------------------------|-------------------------|--------------------------|---------------|
| 20-Pin DIP                  | 0.275                   | 0.605                    | 0.300         |
| 20-Pin leaded chip carrier  | 0.090                   | 0.360                    | 0.065         |
| 20-Pin LCC                  | 0.090                   | 0.160                    | 0.065         |
| 64-Pin DIP                  | 2.880                   | 3.870                    | 0.300         |
| 68-Pin chip carrier         | 0.903                   | 1.560                    | 0.065         |
| 68-Pin LCC                  | 0.903                   | 1.103                    | 0.065         |

With the system trends of increased complexity, decreased size, and cost-effectiveness, it is obvious why surface mount designs are desirable. Surface mount components are comparable in cost to their insertion mount counterparts. However, the surface mount versions offer great reductions in both height and area. Combining these factors can greatly reduce the system volume. For example, consider a system requiring 30 PWB assemblies. If each assembly uses 10 large-scale integrated (LSI) circuits with I/O counts of 60 to 68, and assuming that the footprints account for 60% of the needed board area (the other 40% is required for spacings, small discrete components, and connector attachment), an

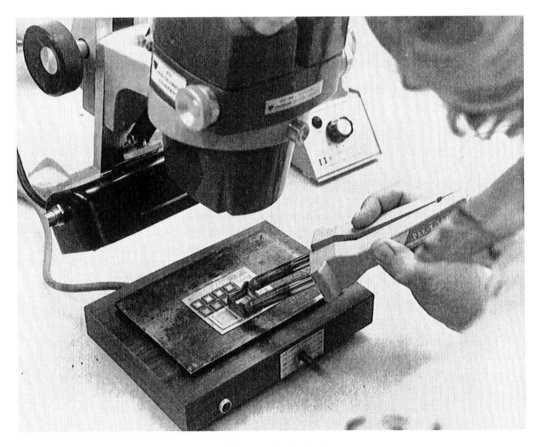

**Figure 4.42**  Extractor rework equipment. (Courtesy of Westinghouse ESG.)

**Figure 4.43**   Leaded surface-mounted organic PWB. (Courtesy of Westinghouse ESG.)

insertion PWB version would require a minimum area of 64.5 in.$^2$. Thus each board would be approximately 8 by 8 in. The insertion mount boards would also require center-to-center spacings of 0.450 in. (0.300 in. component height + 0.040 in. board thickness + 0.010 in. clearance for bottom side lead extension + 0.050 in. spacing between boards). This same design in leaded surface mount technology would require boards of approximately 6 by 4 in. or 5 in. square with center-to-center spacings of 0.165 in. Thus an insertion mount system of 30 such assemblies would require a minimum volume of 14 by 8.5 by 8.5 in., or 7.024 ft$^2$, while the same system in a leaded surface mount design would require a minimum volume of only 5.1 by 6.5 by 4.5 in., or 1.036 ft$^2$. This conversion to surface mounting would result in a 6.8:1 volume reduction. Figure 4.43 shows a leaded-component, surface-mounted PWB assembly.

This reduction in volume for comparable cost can be further improved by converting to a leadless surface mount design. Considering the previous example, leadless board assemblies would require the same spacing of 0.165 in. but would have to be only 4 in. square. Thus, the system would need a volume of only 0.717 ft$^2$, giving an overall reduction of 9.8:1. The general rule of thumb used to estimate the area reductions for conversion from insertion mounting to leadless surface mounting is 8:1. Furthermore, the price of leadless components is comparable to or less than that of their leaded counterparts. A leaded chip carrier is typically slightly more expensive then the leadless version, since it requires taking the leadless version and brazing on the leads. Moreover, leadless components eliminate the need for lead forming while allowing for larger placement tolerancing due to the LCCs' ability to self-align. Therefore, a leadless surface mount design offers not only volume reductions but also decreased labor and tooling costs by eliminating the need for lead forming.

*Leadless Ceramic Chip Carrier Surface Mount on Organic PWB*

The feasibility of surface mounting leadless ceramic chip carriers on organic PWBs has been continuously debated over the years. As discussed in the previous section, such designs can solve many sizing and spacing problems. However, without the leads to act as strain relief for the CTE mismatch between the polyimide or epoxy glass PWBs and the ceramic chip carriers, initial attempts at LCCC mounting to organic PWBs were discouraging. Small LCCCs, approximately 0.400 in. square, with 28 pins or less, mounted to organic PWBs had some limited success. However, as the LCCCs increase in size, the effect of the CTE mismatch between the ceramic components and the polyimide or epoxy glass PWB also increases. If the PWB assemblies are subjected to extreme temperature changes, as required in military applications, the thermal stress imposed by the CTE mismatch can be great enough to cause catastrophic solder joint failures. For this reason, surface mounting LCCCs to organic PWBs has been limited to the smaller LCCCs in military applications, if allowed at all.

*Cored Boards*

Attempts have been made to improve the CTE match between the LCCCs and the PWBs by mounting the PWBs to a heat sink material, or core. PWBs are laminated to the cores with B-stage or film epoxy. This process involves application of heat and pressure to the sandwich of PWB-core-PWB. The resulting cored boards have a composite CTE, between those of the PWB material and the core material, and this composite CTE more closely matches that of the LCCCs. However, the cored PWB artwork dimensions are slightly changed. The core material, having a higher CTE than the PWB, would expand more than the PWB during the heat application of the lamination process. The lamination would be completed at the peak temperature. The now-bonded cored board is allowed to cool. As it cooled, the core would contract more than the PWB would normally, causing some shrinkage of the PWB. The artwork layer for the solder mask has to be modified, as a one-to-one image of solder pads will be slightly bigger than the solder pads on the cored boards. This makes alignment of the solder stencil to the PWB impossible. The artwork has to be adjusted depending on the core material. An aluminum or copper-clad invar cored board has to be adjusted to approximately 96 to 98% of its original size. The amount of adjustment will vary with the dimensions of the board and the CTEs of the board and the core material.

*Components to Boards: Solder Interconnection*

Typically, Pb-Sn solders with RMA (rosin mildly active) flux are used. Eutectic Pb-Sn or 60Pb-40Sn alloys are used because of their lower reflow temperature of approximately 183°C. These solders can withstand the typical commercial and military operating temperature ranges, yet reflow at substantially lower temperature than most die attach solder alloys. Thus, there is no potential for die attach reflow during the surface mount reflow process.

*Solder Deposition*

The application of the solder is accomplished in one of two ways. The first is to print a solder paste onto a tin or solder-flashed board. (The copper-metallized boards have these flashes or coatings to prevent corrosion.) Solder paste is normally printed 0.009 to 0.011 in. thick. These wet print thicknesses yield reflowed solder joint heights of 0.005 to 0.006 in., respectively. This is the desired solder joint height, allowing the solder joint to provide stress relief during temperature cycling. Any additional print thickness would make processing more difficult and cause more solder balling and bridging, examples of which are shown in Figure 4.44, while only slightly increasing the stress relief benefits.

The solder pastes can be screen or stencil printed. An 80-mesh-count screen with 9 mils of emulsion will give the desired wet print thickness; however, the print will give nonuniform paste coverage on the mounting pads. Screen printing can leave too much of the solder behind on the screen, and the wire mesh will leave peaks and valleys in the solder paste surface after printing. This nonuniform paste distribution can result in solder bridging between two adjacent pads with higher volumes of deposited paste and in starved solder joints where too little paste has been deposited. In extreme cases, after reflow, LCCCs will not sit parallel to the board or may be skewed in regard to the footprint.

Self-alignment occurs during the solder reflow provided the LCCC has not been bonded to the board with a thermally conductive adhesive for heat transfer enhancement. As the solder melts, its surface tension increases, thus increasing its wetting (or adhesion) to the PWB's mounting pads and the LCCCs' castellations. These increases in the surface tension cause the solder joint to pull the LCCC into alignment over its footprint. Self-alignment allows the process control placement window to open up, improving manufacturability. However, with nonuniform paste distribution, nonuniform surface tensions result. This can lead to inadequate self-alignment, leaving some of the larger/heavier LCCCs skewed on the board. In some cases the LCCC can be pulled by solder that has bridge together several mounting pads to the point of total misalignment (i.e., pin 1 of the LCCC would be sitting on pin 2 of the PWB footprint).

Mounting pads with excessive amount of solder paste could slump during prebake or reflow. In other words, the paste could spread beyond the footprint metallization. During reflow, as the solder melts, not all of the solder wets to the surface; thus not all of the solder is pulled by surface tension onto the metallized pad. The solder not pulled onto the mounting pad does not wet or adhere to the polyimide or epoxy glass surface of the

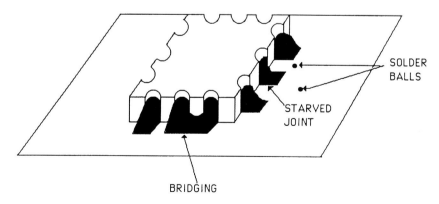

**Figure 4.44** Solder joint defects.

PWB. The surface tension within this displaced amount of solder causes it to form a ball. Solder balls come in various sizes. If not removed from the board, they could potentially roll around and become situated between two electrically uncommon tracks. If large enough, such a solder ball could directly cause a short or could be a potential stepping stone for dendritic growth. Dendritic growth occurs when moisture, which condenses on the PWB, becomes polarized (by the electromagnetic fields). These polarized particles or ions will start to attract each other and form a chain. This chain grows between conductive surfaces that attract the ions. If the ion chain grows enough to bridge the conductive pads, it can carry enough low current to cause intermittent or complete shorting.

For all the reasons just discussed, ''screen'' printed boards require substantial rework. This rework involves having a trained operator manually remove or add solder to individual joints as needed. The operator also has to remove any solder balls manually. This process is very labor intensive. Therefore, an alternative method is recommended: replace the wire mesh screen with a brass stencil. Chapter 3 describes the differences between screen and stencil fabrication and screen and stencil printing. The stencil thickness should be the desired wet print thickness. A 0.010-in. stencil will yield a uniform 0.010-in.-thick wet print. A more uniform wet print will yield more uniform solder joints. Thus a stencil will give better print resolution, which makes printing alignment easier and more accurate. This improves alignment, further decreasing the formation of solder balls and bridges. Alignment is achieved in the manner described in Chapter 3 under ceramic PWBs.

To further reduce solder balling and solder bridging problems, the solder paste can be subjected to a prebake prior to reflow. The prebake volatilizes some of the paste's solvents needed to provide a printable viscosity. With these additional solvents baked out, the paste becomes more tacky and sets up. The prebake must be controlled to verify that only viscosity-lowering solvents are baked out and that the flux and other reactive solvents needed for proper reflow and wetting are not volatilized. For the same reason, time limitations between printing, baking, and reflow must be observed. Best results are obtained if all soldering steps occur within 4 hours of each other. It is worth noting that not all solder pastes require a prebake. For example, Cermalloy solder pastes use organic wetting agents that cannot be exposed to elevated temperature prior to reflow. Therefore, with Cermalloy, reflow immediately following printing is recommended.

The second solder deposit method is to ''thick plate'' the solder onto the boards. As implied, the solder is plated to the desired thickness. This requires that all top-layer metal that is not to be plated be masked. These surfaces can be masked in a variety of ways. A permanent mask can be applied. This may be a photosensitive polymer that is deposited over the entire surface. The coated surface is then exposed to a specific frequency of ultraviolet light through a photomask. The mask will expose either the areas to be soldered or the areas to be solder masked, depending on whether a positive or negative photoresist is used. The board is then sent through a developer solution that washes away the unwanted polymer, leaving the polymer over the areas to be masked. Another way to apply a permanent solder mask is to cover the areas to be masked with a masking film cut into the desired shapes and sizes. In either method, the solder mask must be capable of withstanding the plating and soldering processes.

A temporary mask can also be applied. This involves printing on a masking material with a stencil or screen. This material is a dielectric so that the solder will not wet to it. It also has to be capable of withstanding the soldering temperature and exposure to the flux. Finally it must be capable of being washed away after reflow by sending the board through a degreaser or some other simple cleaning process. Such temporary masking has been accomplished by printing on dielectric or glass thick-film paste. These materials can be printed on the boards but left unfired. After the mask is printed and dried, the solder paste can be printed. The mask can withstand the solder temperatures without any potential of curing. Although these materials typically contain oxides that will be attacked by the flux, they are thick-film printed, so only the surface of the deposit is affected. Finally, they can be washed off with an isopropyl alcohol (IPA) rinse. An added advantage of a temporary mask is that as it is washed away, so is the flux that was left on its surface. The disadvantages are that this method can be used only when the solder is printed on the boards in paste form and that any unsoldered top metallization must be protected from corrosion. Thus, the copper must be tin or nickel plated prior to masking, or the top conductor layer must be designed to contain only solder pads and/or tracks that can be solder coated. Digital tracks or signal lines can be solder coated without any significant effect on functionality. However, at RF and microwave frequencies, at which the current travels on the conductor surface and the speed of the signals is highly critical, a solder-coated track may significantly slow the signal. Solder alloys typically have higher resistivities than the copper metallization of the boards, so even though the solder coating increases the cross-sectional area, it does not necessarily improve the signal speeds for RF designs.

The best design for thick solder plate boards is to lay the boards out so that only the solder mounting pads are on the top layer, as required for military designs. In other words, all signal tracks are buried. This eliminates the need for any solder masking. However, the accomplishment of such a layout may require additional layering.

### Solder Reflow

The most common reflow methods are vapor phase and infrared (IR) reflow. Belt reflow, as discussed earlier, is sometimes used for prototype or small-volume products. Belt reflow, however, is not practical for production applications.

Both vapor phase and IR furnace reflow require product-specific programming of the equipment. The thermal mass and configuration of the product will determine the program needed to provide the product with the proper temperature profile. The profile must be such that the product consistently sees the correct ramp-up of temperature, is held at the proper peak temperature for an adequate time interval, and is cooled down to allow proper solder joint formation without overstressing the parts thermally. If the product has a large thermal mass (e.g., has several large ceramic chip carriers), it will require more time or more energy input to heat up the components and their solder joints. Thus small units may require a faster descent into the vapor, or a faster belt speed into the furnace, whereas larger devices may require a slower decent and longer dwell time in the vapors or a slower belt speed and the higher zone temperatures in the furnace.

The first step in optimizing a vapor phase reflow process is to develop a profile that would steadily ramp up the board temperature rapidly enough to prevent volatilization of reactants prior to reaching the melting point of the solder alloy, thus accomplishing complete reflow, but slowly enough to allow the flux to work, thus enabling complete wetting, and slowly enough to allow self-alignment. The profile must also permit the part to sit in the vapors at peak temperature long enough to allow complete reflow, but not long enough to allow board degradation or overformation of intermetallics or to bake on the flux. Overexposure to the vapors can cause polyimide or epoxy glass boards to discolor or possibly delaminate. A certain amount of intermetallics must form to allow wetting. It is the formation of these intermetallics that allows the metals in the solders to combine and adhere to the metals on the board surface. However, if all the surface metals are diffused into intermetallics, adhesion of the metallization to the board is compromised. Overreflowing, or dwelling at the peak temperature too long, can result in overformation of intermetallics. Finally, the profile has to enable rapid cooling to provide proper grain growth, without overstressing the boards thermally. Such a profile can be accomplished by slowly lowering the parts into the vapors, letting them sit in the vapors for the correct peak time duration, and then pulling them out of the vapor chamber more rapidly. Once out of the chamber, the boards can be cooled to approximately 90°C. The boards should then be immediately sent through a degreaser to clean and deflux them before the flux has a chance to bake and harden onto the board.

Adjustment of the vapor phase profile depends on the component with the largest thermal mass. The heat is transferred from the vapor to the component, and the vapor hits all surfaces simultaneously. Thus the temperature at the surface of the component and its solder joints is independent of surface area or location. The rate of heat transfer into the solder joints is dependent on the rate of heat transfer into the component, which in turn depends on the thermal mass of the component. Therefore, if the vapor phase is programmed to allow for complete reflow of the component with the largest thermal mass, it will completely reflow all components.

Even with an optimized printing process and vapor phase profile, an unforeseen problem may be encountered. As the vapor comes into contact with the PWB, it transfers its latent heat to the PWB and condenses. The condensed liquid then accumulates on the board surface until it eventually runs off the sides. The thermal mass of the larger LCCCs causes them to reflow at a slower rate than their neighboring smaller LCCCs. This later reflow timing allows the liquid to accumulate under the LCCCs before the solder fully melts, thus preventing the fluid solder's surface tension from reaching a level that would hold the LCCCs in place. In other words, the condensed vapor is allowed to lift up the LCCCs, washing away the solder paste and floating the LCCCs off the board. This problem, however, can be quickly remedied by placing the board on a slight angle while it is in the vapor phase. This angle has to be just large enough to cause the condensed fluids to run off without a chance to accumulate under the LCCCs, but small enough that gravity will not cause any of the LCCCs to move during reflow. An angle of approximately 10° meets the requirements.

Reflow in multiple-zone furnaces, although successful for small assemblies such as solder sealing chip carriers or solder chip mounting or sealing of hybrids, has not been very successful for solder mounting components onto PWBs. The main reason here is not the limitations of the equipment but rather the limitations of the modes of heat transfer. In a standard multiple-zone furnace, heat is transferred by convection and conduction. In both cases, the transfer is dependent on surface area. The increase in temperature of a given component is dependent on its area, its mass, its specific heat, the area of the overall assembly, the component's location on the assembly. In other words, the components and the boards are both being heated up. The larger the component's thermal mass, the longer it will take to heat up. Likewise, the larger the board, the longer it will take to heat up. In both cases, the outer edges of the unit will be the first to heat up. As the heat continues to transfer into the unit, a temperature gradient will form. As a component is being heated, the board under it acts as a heat sink and conductively pulls heat away from the component. Thus, components at the center of the board will thermally lag behind those along the edges. In some cases, the board layout can be designed to compensate for this phenomenon. The larger components can be placed on the perimeter of the board, with increasingly smaller devices being placed closer to the center. This placement approach will typically work only for smaller PWB assemblies, because temperature gradients will still form even with such placement. Furthermore, this placement is typically not electrically conducive and would cause unnecessary layering or routing difficulties.

By introducing the heat transfer mode of radiation, furnace reflow has been made a viable reflow process. In an IR furnace, in addition to convective and conductive heat transfer, energy is transferred into the components and their solder joints by exposing them to IR radiation. This radiation is transferred to all surfaces—all components and their solder joints—simultaneously. It is the primary mode of heat transfer. There are still thermal gradients and heat transfer is still dependent on component area and location, but to a much lesser degree. For practical purposes, by proper equipment programming, most PWB assemblies can be reflowed using an IR furnace.

The advantages of IR furnaces are improved long-term process control and operating costs and elimination of the use of carcinogenic solvents. The Freon used in the vapor phase can be toxic if inhaled in large doses. It will also chemically react with the organics (e.g., flux) of the solder paste being reflowed. The flux and other chemical by-products accumulate with usage and can degrade the performance of the vapor phase. These accumulations can cause a shift in the vapor temperature or introduce contaminants on the PWBs. For these reasons, lot-to-lot process control can be difficult, and periodic replacement of the solvents is necessary. The solvents used in vapor-phase equipment are very costly and require special safety precautions, along with costly waste removal procedures. They also represent a danger to the ozone layer. The use of IR reflow eliminates all of these problems.

In both vapor phase and IR reflow, once an optimized profile window is found for one product, it cannot be used for all subsequent products. The program—i.e., the decent speed, dwell time, and accent speed for vapor phase or the belt speed and energy levels or zone settings for furnace—must be adjusted for each product to yield the established profile. A product of higher thermal mass will require lower processing speeds and higher energy settings to have the same profile as a product of lower thermal mass. Thus the established profile is the program's goal.

Single versus double reflow was an issue during the early stages of surface mount technology process development. Double reflow involves stenciling on the solder paste, reflowing it, mounting the components, and reflowing the solder again. In other words, both components and the boards are pretinned prior to assembly. One theory was that by reflowing the solder twice, voiding, caused by incomplete outgassing of the paste volatiles, would be reduced. It was hypothesized that the second reflow would force the gases trapped in the solder to outgas. It was thought that the presence of the voids within the solder joints would weaken the joints or cause cracks to form and propagate. However, the risks of double reflow are increased thermal stressing of the assemblies and increased embrittlement of the solder joints by further intermetallic formation. Furthermore, double reflow would also mean increased assembly labor costs. Thus, there was a second theory that the gains of double reflow would be canceled out by the risks. Still another theory held that voiding acted as a relieve stress—that cracks did not propagate from the voids, but rather their propagation would be stopped upon reaching a void.

By comparing the amount of voiding after one reflow versus two reflows through X-ray and microsectioning, it can be determined that there is no correlation between the number of reflows and the amount of voiding, nor is there a correlation between voiding and solder joint failures. A second reflow has mixed effects on the voiding. Solder joints with small voids near their surface and not located under the LCCCs can completely outgas during the second reflow. In these cases a second reflow can decrease the percent of voiding. However, multiple voids, especially if they are located under the LCCCs, can join together to form larger single voids after the second reflow. In both cases, the number of voids is decreased; however, in the later case the percent of voiding remains the same. These larger voids formed after a second reflow are usually directly under the LCCCs, where the amount of solder (the solder thickness) is smallest while the amount of thermal stress (the CTE mismatch) is greatest. Furthermore, such voiding directly under a self-aligned LCCC, or an LCCC with solder-mounted thermal pads under the dice, greatly hinders the thermal path of power dissipation from the dice inside the LCCCs, through the solder, into the PWB, and finally to the heat sink or core of the PWB assembly. (This thermal solder voiding issue can be addressed by mounting the body of the LCCC with a thermally conductive adhesive. The adhesive would then become the primary thermal path.) It has also been determined that voiding within the solder joints is not a function of the size or location of the LCCCs. The amount and location of voids are totally random.

Studies in which both single and double reflowed boards were subjected to over 1000 temperature cycles of $-55$ to 125°C have shown no correlation between solder joint failures and voiding within the solder joints. A failure is typically defined as a specified percent increase in resistance. These failures are normally located randomly—they are not a function of LCCC size or location. Failure analysis, through microsectioning of the board assemblies and/or delidding of the LCCCs, showed that the decreases in resistance were not due to solder joint voiding or cracks. The source of the majority of failures was not the solder joint, nor was it wire bond failures. Further analysis showed that the stress concentration was located within the plated through holes. In organic boards cored to aluminum, copper-clad invar, and graphite, the CTE mismatch between the LCCCs and the PWB is overshadowed by the mismatch between the epoxy glass or polyimide and the core material. The blind vias and plated through holes of the PWB fail before the solder joints. Redesigning the cored PWBs to reduce the thermal stress between the PWB and the heat sink can reduce the thermal failures within the plated through holes and blind vias. This can be accomplished by designing the core material to have a CTE closer to that of the organic PWB and by laminating the PWB to the core with a flexible adhesive capable of absorbing the thermal stress. However, when this is done, solder joint reliability is once again compromised. The solder responds to the applied strains and resulting stresses by time-dependent plastic deformation. The result is solder joint failures due to accumulating fatigue damage [7].

*Special Mounting Techniques*

In production applications mounting of the components is typically done with pick-and-place equipment. If solder is applied by printing solder paste onto the boards, then placement immediately follows. The tackiness of the paste holds the parts in place until the solder is reflowed and self-alignment or the surface tension of the liquid solder takes over. Therefore, single-sided solder paste PWBs have no need for additional placement or mounting techniques. For high heat dissipation and improved conductive cooling, thermally conductive adhesives are used between the components and the boards. Just as insertion of the components' leads in an insertion-mounted board acts as the components' alignment and holds them in place until the solder joints are formed, the solder paste and the liquid solder's self-alignment accomplishes the same for surface-mounted boards. Again, like their insertion mount counterparts, solder paste-mounted boards can have pretinned or nonpretinned components placed into the solder paste. The solder paste, thanks to the flux within it, will wick up and wet to the component castellations/terminations; thus pretinning of the components is not absolutely necessary. Pretinning is, however, recommended to ensure complete and uniform wetting to all I/O. The solder paste's initial tackiness and its later surface tension, while in liquid form, will not totally overcome gravity as a wave-soldered, inserted, and clenched component will. In addition, the solder paste must be processed within the time limitations of the paste's solvents. Thus a solder paste surface-mounted board does require extra process control and handling.

Thick solder-plated boards do require some means of holding the parts in place prior to solder joint formation. This involves adhesion of the parts to the boards with an epoxy or at least "tacking" the parts in place prior to reflow. In the former case, epoxy preforms can be manually or automatically placed (via pick-and-place equipment) under the components. This is typically done with larger chip carrier or hybrid components. The epoxy is partially or totally cured. Partial curing will enable the epoxy to hold the components in place while solder reflow forms the solder joints and finishes the curing of the epoxy. The same technique can be used with epoxy pastes. The paste application can be manual or printed. The latter process can be accomplished by printing a viscous flux over the thick plated solder pads. The flux serves two purposes: it provides a tacky surface that holds the components in place prior to reflow, and it provides the cleaning/reducing agents needed to rid the solder surfaces of the pretinned board and parts of any oxides that may have formed during their storage.

Double-sided mounting also requires special processing. The parts mounted to the bottom side of the boards must be held in place both prior to and during reflow. The tackiness of the solder paste or the printed flux on thick solder plating may hold the parts in place prior to reflow, but if the parts are heavy enough, the surface tension of the liquid solder cannot be counted on to hold them in place on the bottom of the board until the solder joint is formed. Therefore, components mounted to the bottom side of the boards must be epoxied in place so that they will remain in place during the reflow of the solder.

## 4.4.2 Ceramic PWB Assembly

Ceramic PWBs can have cofired, thick-film, or thin-film metallization (see Chapter 3). All interlayer connections or routing is accomplished with buried vias. The vias are windows left out or punched out in the printed dielectric layer or in the green tape, respectively. All components are surface mounted; nothing is insertion mounted. Figure 4.45 gives an example of a ceramic surface mount assembly. However, these substrates may have through holes. In a cofired substrate these are stacked vias with metallization which give a finished via or hole through the substrate similar to an organic PWB plated through hole. In a thick-film substrate, a hole is machined—punched out in the green stage, or laser drilled or sonic milled out of the fired ceramic. This hole is then cleaned and metallized by coating the walls with a thick-film conductor paste. The result is a plated through hole that can be used to route a signal from one side of the board to the other, or as a thermal via to increase the thermal path at a specific location on the substrate.

A patterned or metallized substrate is the starting point for any ceramic PWB assembly. These substrates can be mounted to a heat sink or left unmounted. A ceramic PWB by

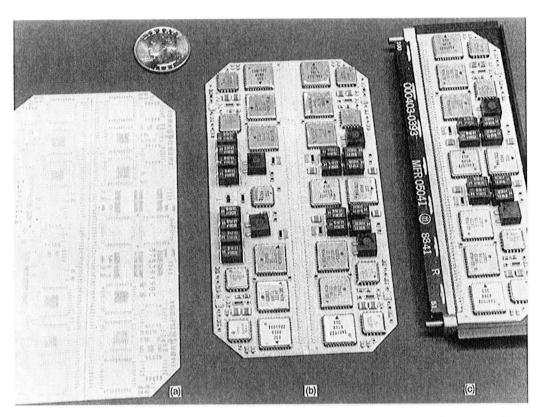

**Figure 4.45**  Ceramic PWB assembly, (a) multilayer ceramic snap strate or PWB, (b) substrate assembly, (c) final assembly with connector and heat sink. (Courtesy of Westinghouse ESG.)

definition is a rigid flat board that maintains its shape. The camber of such boards is typically 0.002 to 0.003 in./in. This camber is approximately maintained throughout processing and operation. These boards will not warp during solder processing as much as organic boards can. Also by definition, the ceramic boards's CTE matches that of the ceramic components that are mounted to it. Thus a CTE mismatch no longer exists. However, thermal stress is still generated by temperature gradients forming between the components and the board. These gradients form because of the difference in the rate of heating due to the differences in the thermal masses. Thus, even though there is no CTE mismatch, there is still a difference in expansion rates due to the temperature gradients. However, the need for ceramic chip carriers to be leaded to allow for stress relief is greatly reduced. Leadless chip carriers can be mounted to the ceramic PWBs with minimized degradation of the long-term reliability of the assembly. Furthermore, the ceramic material from which the boards are fabricated offers much better thermal conductivity than its organic PWB counterparts. Use of more advanced ceramic materials, such as AlN, and incorporation of thermal through holes in the substrates can further enhance the thermal performance of a ceramic PWB.

For the reasons just discussed, ceramic PWBs are greatly beneficial in military and space applications. They offer reductions in thermal stress; reductions in volume because of their ability to incorporate LCCCs in their design; improved thermal management because of their superior conductivity and thermal viaing; and improved long-term reliability because of a CTE matched to that of their components, the reduction of thermal stress, and their rigidity. However, as always, these improved properties typically mean higher costs and longer design times.

## Heat Sink Attachment

Although the ceramic boards are rigid and can stand on their own, heat sinks are commonly attached for thermal reasons. The heat sink increases the ability of the assembly to dissipate heat. The heat sink can also act as the carrier plate, providing a flanged support that can be readily bolted or inserted into a chassis. In standard electronic modules (SEMs) the module housing is the heat sink/carrier plate for the ceramic PWBs.

The materials commonly used are kovar, tungsten-copper alloys, composites such as graphic fiber plates or silicon-aluminum alloys, and molybdenum. All of these materials offer thermal conductivities similar to or better than those of the ceramic and CTEs similar to those of the ceramic. The CTE matching, thermal environmental requirements of the assembly, outgassing restrictions (if any), and linear size of the PWB dictate the attachment method.

Ceramic PWBs can be attached to their heat sinks with epoxy or solder. The attachment methods used for ceramic PWBs are the same as those for ceramic hybrid substrates. The mismatch in CTE is more critical for a ceramic PWB because of the increased size as compared to a hybrid substrate. A flexible, stress-absorbing adhesive such as a silicone adhesive is typically used, provided that no bare dice are mounted on the ceramic PWB. The section Substrate Mounting details the mounting methods and thermal design considerations that apply to a ceramic PWB.

The back of the ceramic PWB is metallized according to the attachment method. For example, if the PWB is to be mounted to the heat sink with solder, its back is metallized with a platinum-palladium-gold (PtPdAu) thick film to allow proper wetting while preventing total leaching of the metallization into the solder. If the PWB is a thin film or cofired ceramic, the base metal is nickel and gold plated. The gold plating provides a good wetting surface, while the nickel acts as a barrier metal that provides corrosion protection while preventing total leaching. Other metallization and attachment considerations were discussed earlier under Substrate Mounting.

Ceramic PWBs can also be directly bonded to a heat sink during the conductive patterning step. Chapter 3 covered the processing and design aspects of direct-bonded copper PWBs in more detail.

## Surface Mounting

As already discussed, ceramic PWBs do not utilize insertion mounting. All components are surface mounted. These components can have leads formed for surface mounting; however, this practice is not necessary and lessens the area and volume reduction benefits of using a surface mount ceramic PWB design.

Leadless components can be used with high reliability because they have CTEs very close to the CTE of the ceramic PWB. These components include ceramic chip carriers, ceramic chip capacitors, and ceramic thin-film and/or thick-film resistor chips. Resistors can easily be incorporated into the layout of the board, making it a PCB (not just a PWB).

The most common mounting method for ceramic boards is to print solder paste on the boards, mount the components, and reflow the boards in a vapor phase or IR furnace.

Thick solder plating of the mounting pads can be done to a thin-film or cofired board; however, solder paste printing is still commonly used for these boards as well as for thick-film boards. See earlier under Surface Mounting for details on the processing associated with this technique.

### 4.4.3  Connector Attachment

In commercial applications connectors need not be mounted to the PWB assembly. Instead, the PWB assembly or card has pads representing the I/O of the assembly along one edge. The card is then inserted or plugged into an end card connector. Therefore, the PWB itself acts as the connector for the PWB assembly. The connector half that the boards are plugged into is typically an integral part of the back plane or mother board of the system.

In some commercial and most military applications, connectors are usually mated pairs. Half of the pair is attached to the PWB assembly while the mating half is mounted to the back plane or mother board of the system. (See Chapter 5 for more details on connectors.) The connector can be attached before, during, or after component mounting. Attachment of the connector before component mounting is not commonly done. Having the connector in place before component mounting can interfere with the solder paste printing and component placement. The connector would prevent the PWB from lying flat on the printer stage and on the placement equipment. Assembly of a PWB with connector in place would require special fixturing and handling. Thus it is most common to attach the connector at the same time as or after the component mounting.

Connectors are typically placed along the edge of the PWB or cored PWB. Some connectors can be mounted and solder reflowed at the same time as the components. This is the common practice with thick-plated organic PWB assemblies. The connector is snapped onto the edge of the assembly and is typically aligned with rivets at each end of the connector. The assembly with connector and components in place is then run through the wave solder for insertion mount boards or through a vapor phase or IR furnace for surface mount boards. To decrease the possibility of dislodging components while snapping on the connector, the connector is sometimes attached after component mounting and solder reflow. The I/O pads to which the connector pins are attached are pretinned. This is accomplished by thick plating these pads at the same time as thick plating the component mounting pads, or while the board is traveling through the wave solder, or by printing solder paste on these connector pads and letting the paste reflow during component mounting/reflow. After the components are mounted and reflowed into place and the boards are cleaned, the connector is positioned with the spring-loaded pins over the pretinned pads. The connector pads are then locally reflowed by exposing this area to a hot-air gun (while shielding the rest of the board with a metal plate) or by placing a hot bar over the connector pads.

Another method of connector attachment involves the use of Raychem's Solder Kwik product. With this product the connector pads can or cannot be pretinned. After component mounting and reflow, the connector is positioned. The Solder Kwik preform is then placed next to the pads, aligning the pads with the capillaries through which the solder will flow. A hot bar is placed over the reservoir of solder. The solder melts and flows through the capillaries of the Solder Kwik preform onto the connector pads, where it wets to the pads while wicking up over the pins.

In addition to edge connectors, flex cables are sometimes surface mounted to the PWB assembly in microwave or power modules (SEMs). Since the PWB is typically mounted to the floor (or central web) of the module, it cannot have a connector snapped over its edge. Therefore a flex cable is surface mounted to the board. The other end of the flex usually has a plug-in type of connector that is then inserted into the module wall to provide interconnection from the module to the system in which it is mounted.

## 4.5  ELECTROSTATIC DISCHARGE

Electrostatic discharge (ESD) is very commonplace. One example that everyone is familiar with is the discharge of electricity that occurs when you touch a metal object after rubbing your feet on a rug. Rubbing your feet on the rug builds up a voltage potential in your body, and this potential is released when you "ground" your body by touching an electrically conductive surface. As you touch the conductive surface, the voltage can arc. Thus if you touch a metal object after rubbing your feet on a rug, you will see a spark. This discharge has very low current, usually microamperes, but can generate very large voltages, up to 40,000 V.

Most semiconductor devices (i.e., ICs) can withstand only 100 to 300 V. Any voltage higher than the ESD limit of the device can burn out a signal track. Exposure of voltages lower than the ESD limit yet higher than the normal operating voltage can degrade a device. The exposure to ESD levels higher than the operating voltage damages the IC. If repeated, the damage is accumulative. For example, if a 15-V diode has an ESD limit of 100 V, then one exposure to 100 V can kill the device. However, an exposure to 30 V will not kill the device but can degrade its ability to withstand subsequent ESD exposures. So, although one or two 30-V exposures will not kill the device, repeated exposures will.

Many steps must be followed to eliminate ESD damage. All ESD-sensitive devices should be handled only by properly grounded individuals and equipment. All equipment, workstations, chairs, and work surfaces must be grounded. These devices should always be stored and carried in ESD containers. The humidity of the work area must be controlled to minimize static electricity in the air. All operators should be trained in ESD handling.

There are simple rules to follow for ESD-sensitive devices. When removing the device from its ESD protective container, first place the container on to a grounded surface (a workstation with a grounded ESD mat on its surface); stand on an ESD floor mat that is properly grounded or sit on a chair on an ESD floor mat and have both feet flat on the mat; and ground yourself by wearing an ESD wrist strap that is properly grounded (e.g., attached to the workstation ground). The metal portion of the wrist strap must be in contact with your skin. Never open an ESD container without grounding it. The ESD container may have a voltage potential build up on its outer surface. By carrying the container through a hallway or area that is dry (that contains static electricity) or by handling the container, a voltage potential can build up on the outer surface of the container. This potential can be transferred to the device inside if the container is not grounded before it is opened. Another ESD handling control is always to have the device on an ESD mat of a grounding surface. Never carry the devices on an open tray or even in your hands unless you and the tray are properly grounded throughout the transport. (Note: electronic assemblies, especially bare-die assemblies, should never be handled by ungloved hands because of the potential for contamination by body oils and salt.)

By following these simple rules and maintaining an awareness of the sensitivity of the devices or assemblies being handled, ESD damage can be greatly reduced. ESD damage can occur at any time—while the supplier is placing the dice into a waffle pack, while the device is being shipped if it is improperly packaged, during die mounting, or even while installing the system. The damage can reach a failure point at any time—during final testing, after burn-in, during system testing, or during system operation. Remember that ESD damage is cumulative. A device can be damaged such that it will still pass the final test but will fail with the next ESD exposure.

## 4.6 SYSTEM INTEGRATION

This chapter has shown how components such as bare dice and capacitors are the building blocks of many electronic assemblies. Subassemblies in addition to components, such as chip carriers, PGAs, and hybrid assemblies, are the building blocks for PWB assemblies. In turn, PWBs are the building blocks for the final systems. The PWB assemblies can be the components inserted into the mother board or back plane of the system (see Figure 4.46). They can also be mounted inside SEMs, which in turn are plugged into the system chassis. How these building blocks become a part of the final system will dictate their design.

The designer must weigh all the requirements against all the options. Often the best design for the system is actually the best compromise between thermal management, cost, weight, volume, electrical performance, long-term reliability, and time. The time to design the PWB assembly can greatly affect the cost of the unit. Making corrections to a design

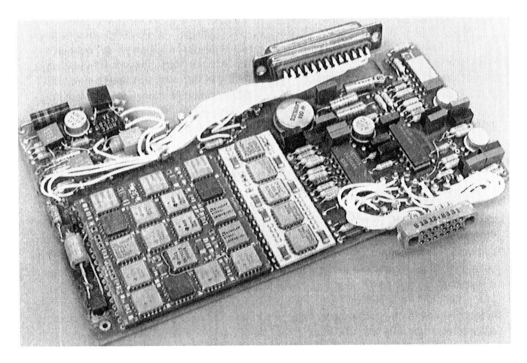

**Figure 4.46**  Insertion mount mother board assembly containing two connectors, jumper wires, one ceramic daughter board assembly, one organic surface mount assembly and discrete insertion mount components. (Courtesy of Westinghouse ESG.)

in progress is the most time-consuming part of the design phase. Part availability and shipping schedules are typically the bottlenecks in manufacturing. Process development is essentially to state-of-the-art technology. Without the development of new processes and materials, the product would not be state of the art. Yet, having to develop a process for a design already in production is time consuming, costly, and often affects the reliability of the product or requires redesign for the process once developed. The best way a designer can control the product's cost is through self-education. The designer must be fully aware of all system requirements. The designer must take into consideration all the specifications, test levels, operational conditions (environmental, electrical, and mechanical), cost limitations, and delivery schedules and educate himself or herself about the materials and processes available. This education should include not only the technical aspects of the material but also the methodology of its use and its availability, machinability, tooling requirements, approximate cost, etc. The designer must create a producible design, not finish the design on paper and hand it to manufacturing to figure out how to build it. The system designer can set a goal for the final size and functionally of the system, but the feasibility of that goal can be realized only through a joint effort of electrical, mechanical, and manufacturing engineering.

## EXERCISES

1. Give 5 examples of assemblies within assemblies (e.g., a light bulb is an assembly of a filament within a glass vacuum tube, but is a subassembly of a lamp.)
2. How can the potential of arcing be reduced within an electronic assembly?
3. What are the differences in material, processing, and electrical performance of wedge and ball bonds?
4. Describe the 3 different die mounting methods.
5. What are the advantages and disadvantages of using conductive epoxy for component mounting?
6. Describe the cost and reliability trade offs between insertion mounted PWBs and leadless ceramic surface mounted PWBs.
7. Calculate the area needed for a circuit containing the components listed below if designed on (a) an organic insertion mount board, (b) an organic surface mount board, (c) a ceramic surface mount board, and (d) a hybrid. Give all assumptions.

| Quantity | Description |
|----------|-------------|
| 3 | 68 I/O dice |
| 5 | 16 I/O dice |
| 2 | $30\Omega$, .5W resistors |
| 2 | $1K\Omega$, 2W resistors |
| 8 | 0.1 $\mu F$ capacitors |

8. Give the definition of hermeticity.
9. Describe 3 different sealing methods.
10. Calculate the expected deflection for a $5'' \times 6'' \times 1.040''$ aluminum lid that is laser welded on to an aluminum module.
11. Design a $60\Omega$ impedance glass feedthru capable of carrying 10 amps.
12. Design a $50\Omega$ impedance ceramic feedthru capable of carrying 50 amps.

13. Describe 3 different solder reflow methods. Compare these methods in terms of process control, cost, and design limitations.
14. Generate a process flow chart for (a) a die assembly, (b) a hermetic hybrid, (c) an encapsulated hybrid, (d) a circuit card assembly.

## REFERENCES

1. Leahy, K. "Microwave Hybrid Design Tutorial," Sect. 1–3, Capital chapter of International Society for Hybrid Microelectronics (ISHM), 20th Annual May Symposium, Baltimore, May 1989.
2. Lee, K., Brzozowski, V., and Harris, D., "Printed Circuit," in *Encyclopedia of Science and Technology*, McGraw-Hill, New York, in print.
3. Troyk, P., "Encapsulants as Packaging for Implanted Electronics," presentation given at Capital chapter of ISHM Symposium, May 1987, Baltimore, MD.
4. Avallone, E. A., and Baumeister, T., III, *Marks' Standard Handbook for Mechanica Engineers*, 9th ed., McGraw-Hill, New York, 1987.
5. Interview with Jerry Stamps, Sr., Mechanical Engineer, Structural Analysis Group, Westinghouse Electronics Systems Group, Baltimore, February 27, 1990.
6. Capp, M., and Luther R., "Analysis of the Effect of Laser Machining on 96% Alumina Ceramic Substrates and the Advantages of New LaTITE Finish."
7. Engelmaier, E., "Surface Mount Solder Joint Long-Term Reliability: Design, Testing, Prediction," *Soldering Surface Mount Technol.* 14–22.

## SUGGESTED READING

Barber, C. S., et al., "High Performance Conductive Thermosets: II," in *Proceedings of the Society of Plastics Engineers*, Society of Plastics Engineers, Washington, D.C., May 1985.
Hunadi, R., et al., "New Ultra-High Purity, Electrically Conductive Epoxy Die Attach Adhesive for Advanced Microelectronic Applications," in *Proceedings of the 1985 International Symposium on Microelectronics*, Anaheim, Calif., International Society for Hybrid Microelectronics, November 1985.
Johnson, R. R., "Multichip Modules: Next-Generation Packages," *IEEE Spectrum 27*(3): (1990).
Oscilowski, A., and Sorrells, D. L., "Use of Thermogravimetric Analysis (TGA) in Predicting Outgassing Characteristics of Electrically Conductive Adhesives," in *Proceedings of the Technical Conference—IEPS, Fourth Annual International Electronics Packaging Society*, Baltimore, IEPS, October 1984.
Pandiri, S. M., "Behavior of Silver Flakes in Conductive Epoxy Adhesives," *Adhesives Age*, *30*(11):31–35 (1987).
Sorrells, D. L., et al., "Selection and Cure Optimization of Conductive Adhesives for Use in Au/ Sn Sealed Microelectronic Packages," in *Proceedings of the 1984 International Symposium on Microelectronics*, Dallas, International Society for Hybrid Microelectronics, September 1984.
Weast, R. C., et al., *Handbook of Chemistry and Physics*, 52nd ed., Chemical Rubber Company, Cleveland, 1971–1972.

# 5

# Interconnections and Connectors

**Pradeep Lall and Michael Pecht**    *CALCE Center for Electronic Packaging, University of Maryland, College Park, Maryland*

## 5.1  INTRODUCTION

Early airborne electronic equipment was packaged by individually mounting discrete passive components, such as capacitors, inductors, and resistors, on a sheet metal chassis. Electron tubes were installed using plug-in sockets, and the circuit was interconnected with insulated wires soldered to component and socket terminals. Increasing aircraft performance requirements drove available volume and allowable weight downward for electronic equipment, leading to the development of smaller and lighter components and innovative methods of interconnecting them.

A major breakthrough in interconnection technology was achieved with the development of the printed wiring board (PWB) with copper conductor patterns photoetched on plastic sheets. Holes were drilled through enlarged conductor areas known as pads, and lead wires emanating from component bodies were inserted into the holes and soldered to the pads.

A major revolution in both electronic and interconnection technology occurred with the development of the integrated circuit by J. S. Kilby and R. Noyce in 1958 and the planar transistor by J. Hoerni in 1959. The first crude integrated circuits were connected by soldered metallic conductors between conductor pads on the die and external pins in the package.

The complex design and maintainability requirements of many modern electronic systems necessitate construction from relatively less complex, often functionally independent, building blocks, which are then assembled and interconnected. The advent of computer-aided design techniques has eased the bookkeeping aspects of interconnection management. Yet the problem still remains of ensuring that the connections provide good contact and are made as short as possible to maximize speed and reliability, especially with increasing I/O requirements.

## 5.2  INTERCONNECTIONS: CLASSIFICATION

Interconnections are involved at all levels of packaging. In advanced electronic systems, the interconnections can be defined at six distinct levels. At the lowest level, the silicon chip supports connections between the chip and the lead frame. At the second level, the leads of the individual components are connected to each other by such methods as printed wiring. At the third level, printed wiring boards are interconnected to each other. The interconnection method used at this level depends on the system's complexity. For small systems, board edge connectors may be used. For large systems, back planes with multilayered wiring patterns provide a compact design for interconnection. At the fourth level of interconnection, the back panels are housed in a drawer or gate in order to interconnect subsystems to each other. The fifth level of interconnection involves wiring of gates or drawers to I/O connections on the cabinet. Cable harnesses are used for signal I/O and bus bars for power distribution. The sixth level of interconnection involves large systems and connections between cabinets and stations.

## 5.3  CHIP-TO-LEAD-FRAME INTERCONNECTIONS

The need for miniaturization along with a less labor-intensive interconnection method led to the concept of bonding small-diameter uninsulated malleable metallic wires by mechanical compression in various combinations with thermal and sonic energy. The wires served to connect the small pads on the die to the external package leads (Figure 5.1).

Wire bonding is a process that is accomplished by bringing two conductors to be joined into intimate contact such that the atoms of the materials interdiffused. The common materials used for wires include aluminum and gold. The materials used for pads are aluminum, gold, nickel, copper, and chromium.

Wire bonds in microelectronic packages connect the bond pad and associated die metallization to wires connected to the I/O leads, or connect two bond pads in hybrid packages. These connections are susceptible to fatigue-type failures as a result of various mechanical and thermal stresses generated during operating life. In fact, wire and wire bonds constitute a large portion of the package-related failures even when wire and wire bond failures as a result of poor manufacturing conditions are eliminated by screening. Most of these failures are a result of various fatigue mechanisms.

Large compressive stresses along with ultrasonic vibrations or thermal energy are applied at the wire-pad interface during the bonding process. Cracks in the bond pad or in the substrate may result if the bonding stress is excessively high [1]. Cracks are difficult to detect visually without detaching the bond. Proper control of the bonding parameters is required to ensure reliable wire bonds.

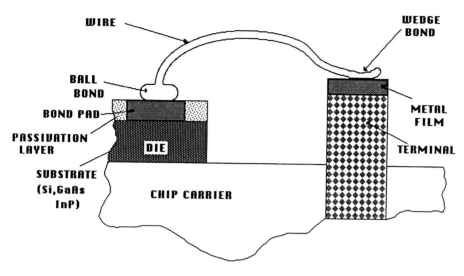

**Figure 5.1**   The wire-wire bond assembly.

## 5.3.1  The Wire Bond
### Wire Materials

Wire materials in common use include gold and aluminum. The characteristics of the wire materials are of vital importance to the strength of the wire bond. These include wire dimensions, tensile strength, elongation, and contamination.

It is important to have a wire of constant dimensions and known cross-sectional area because the bond quality depends on the mass of the wire involved in making the bond. The tensile strength is an important specification for both ultrasonic and thermocompression bonds. Typically, the bonding process consists of formation of the first bond on the chip followed by the second bond on the lead frame. The wire is intentionally made to break after bond formation is completed at the lead frame to prepare the wire for first bond of the next segment. On the other hand, a wire of lower tensile strength would reduce the range of bonding schedules for satisfactory bonding. A softer wire is more difficult to align under the tool. Limited elongation is important so that the wire may be broken off after the second end of the wire is bonded. A large elongation would result in an excessively long tail in the second bond and undue deformation of the wire used in the first bond of the next wire. This would cause the bond to be inferior. Contamination control is critical during the bonding process because water or ionic contaminants that may be included in the lubricants could cause corrosion of the aluminum metallization or subsequent device degradation.

### Bonding Surface

Materials used for pads on the semiconductor die include gold, aluminum, nickel, copper, and chromium. Aluminum is the most commonly used material in the semiconductor die, though gold is used sometimes to avoid problems in the formation of the gold-aluminum intermetallics. Because gold does not adhere well to silicon dioxide, other metals are used to form a multilayer metallization to avoid direct contact between gold and silicon.

Bonding to the die and the terminal is affected by many film-related factors, including surface finish, film hardness and thickness, film preparation, and surface contamination. Howell and Slemmons [2] indicated that for thermocompression bonds the uniformity, composition, and thickness of metallization are important and that surface irregularities could prevent adequate diffusion across the wire-metallization interface and hence interfere with the bonding process. Hill [3] reported that by improving uniformity of the metal-lization, the reliability of the bond was increased. The hardness of the aluminum metal-lization is also said to be important. The metallization should be somewhat softer than the wire so that the surface irregularities may easily be smeared out to better conform to the wire. The thickness of the metallization can have an effect on the bondability and the subsequent reliability of the bond. An excessively thick metallization may be soft, which would be very difficult to bond. To avoid subsequent bond failure due to intermetallic compound growth and Kirkendall voids at the interface, Philofsky [4] suggested that the thickness of the metal film be minimized, consistent with good bonding and device design. This suggestion applies both to bonding gold wire to aluminum and to bonding aluminum wire to gold-plated terminals.

When the device is subjected to thermal or power cycling, the bond wire flexes at the heel. Philofsky [4] suggested that under these circumstances the thickness of the aluminum metallization should be less than one sixth of the wire thickness at the heel of the gold wire wedge or stitch bonds, to avoid growth of the intermetallic compounds in this region and consequent failure of the wire at the heel of the bond. In the case of aluminum wires bonded to gold-plated terminals it was suggested that the plating thickness be less than one third of the wire thickness at the heel of the bond.

Excessive roughness of the bonding surfaces has been found to influence the quality of the wire bonds. The bonding surface roughness should be such that the area of the bond is large compared to the peak-to-peak variations in the surface [5].

Contamination of the wire bonding surface should be avoided. For thermocompression bonding, contaminants can interfere with the intimate contact and interdiffusion of the wire and the metal film and result in poor bonding. The problem of contamination is less significant in ultrasonic bonding because of the ultrasonic agitation. Contaminants of concern include residues of chemicals used in the photoresist and packaging plating operations, water spots, silicon monoxide, silicon dust, and aluminum oxide [5].

## Bond Types

There are several different types of wire bonds, including the wedge bond, ball bond, and stitch bond [6]. Wedge bonds (see Figure 5.2) are made with a wedge or chisel-shaped tool. The end of this tool is rounded with a radius one to four times that of the wire being bonded and is made of sapphire or similar hard materials. This tool is used to apply pressure to the lead wire located on the bonding pad, which has been heated to the bonding temperature. Different methods are used for precise coalignment of the bonding pad, wire, and wedge. Difficulties with wedge bonding include imprecise temperature control, poor wire quality, inadequately mounted silicon chips, and a poorly finished bonding tool.

Ball bonding is a process in which a small ball is formed on the end of the wire by severing the wire with a flame and deformed under pressure against the pad area on the silicon chip (Figure 5.3). The lead wire is perpendicular to the silicon chip as it leaves the bond area. The number of steps in this bonding operation is small and the strength of the bond obtained is strong. Aluminum wire cannot be used because of its inability to form a ball when severed with a flame. However, gold wire is an excellent electric conductor, is more ductile than aluminum, and is chemically inert. For ball bonding, hard gold wire may be used since the balling process determines the ductility of the gold to be deformed. A disadvantage of ball bonding is that a relatively large bond pad is required.

Stitch bonding combines some of the advantages of both wedge and ball bonding (see Figure 5.2c). The wire is fed through the bonding capillary, the bonding area is smaller than for ball bonds, and no hydrogen flame is required. Both gold and aluminum wires can be bonded at a high rate.

## Gold-Aluminum Intermetallics

A gold-aluminum interface exists at the wire-pad interface. Gold-aluminum intermetallic compounds form at this interface at a rate that increases with temperature. Above a temperature of about 125 to 150°C the growth becomes significant with respect to long-term reliability of the wire bonds [5].

The compounds are formed by diffusion of gold and aluminum across the interface. Gold has a greater diffusion rate and as a result vacancies are formed in the gold. This process, known as Kirkendall void formation, can lead to two types of failure: a mechanical stress–induced failure along the locus of the voids and an electrical open circuit caused by coalescence of the voids. Five different intermetallic compounds appear in the interfacial region, depending on the relative concentrations of Au and Al. These include $Au_4Al$, $Au_5Al_2$, $Au_2Al$, $AuAl$, and $AuAl_2$ [5].

The types of reliability problems that result from gold-aluminum interactions depend on the wire bond type and whether a direct or expanded contact is used. Electrical failure occurs when gold wire is ball bonded to aluminum bond pads because of the formation of an annular Kirkendall opening around the bond. The development of voids at the perimeter of the bond is accompanied by increases in the electrical resistance of the bond with time. The rate of increase of resistance with exposure to elevated temperature is larger for thinner aluminum metallization. Bond adherence of these ball bonds is unimpaired by intermetallic compounds that reach to the oxide. The intermetallic compounds adhere well to the silicon dioxide and, though brittle, can sustain a greater tensile stress than either gold or aluminum.

Mechanical failure can occur when gold is ball bonded to thick aluminum films because the supply of aluminum for reaction with the gold is practically unlimited. The process of void formation therefore continues uninterrupted. In this case, the void formation at the interface results in a mechanically fragile bond after high-temperature storage. Similar degradation can occur if an aluminum wire bond is made to a gold-plated terminal where the gold plating is too thick [5].

Thus, to minimize the degradation effects due to gold-aluminum interactions, bonding to thick films and excessive bond deformation should be avoided.

## 5.3.2  Wire Bond Manufacturing Methods

There are several manufacturing methods for wire bonding to thick-film circuits. The most common are thermocompression bonding, ultrasonic bonding, and a combination of both. All these lead-bonding techniques depend on achieving intimate contact between the materials to obtain an atomic contact at the interface [7].

Gold wire is typically bonded using thermocompression, and aluminum wire is typically bonded using an ultrasonic process. Gold tends to age in the amorphous state with

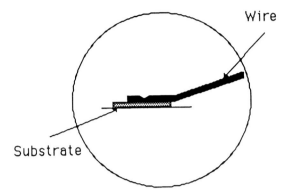

**Figure 5.2(a)**   Wedge bond.

a consequent decrease in tensile strength. Hard drawn gold wire ages significantly at room temperature, so a softer and relatively more stable stress-relieved wire is recommended. Pure aluminum cannot be hardened sufficiently to allow it to be drawn to a diameter of 1 mil. Therefore aluminum is usually hardened by adding about 1% of an impurity such as silicon or (less frequently) magnesium.

## Thermocompression Wire Bonds

Thermocompression wire bonding, as the name indicates, depends on heat and pressure. In general, the bonding equipment includes a microscope, a heated stage, and a heated wedge or capillary that will apply pressure to the wire at the interface of the bonding surface as shown in Figure 5.4. In addition, a wire feed mechanism is required, as is some method for manipulation and control. The thermocompression techniques can produce bond strengths in excess of the ultimate strength of the wire; i.e., instead of the bond breaking, the wire will break during a pull test.

Three primary parameters affecting thermocompression bonding are force, temperature, and time. These parameters are interdependent and are affected by other conditions and factors. Minor changes in these variables can cause significant differences in the bond characteristics. Hence it is necessary to optimize the primary parameters to obtain a satisfactory bond. Short bonding time is desirable for production purposes. Low bonding temperature is desirable to avoid degradation of the wire bonds due to gold-aluminum interactions. Low pressures are desirable to avoid fracturing or otherwise damaging the silicon beneath the bond. Too large a force may damage the semiconductor substrate or deform the wire excessively, and too small a force may produce inadequate bonding. The wire in some cases may be weaker than the bond. In the ball bond, the weakest link occurs in the annealed wire leading to the bond. In the stitch and wedge bonds it occurs in the region of the wire in which the cross section has been reduced by the bonding tool. The bonding tool used in the process may be of tungsten carbide, titanium carbide, sapphire, and ceramics [5].

## Ultrasonic Wire Bonds

Ultrasonic wire bonding also involves heat and pressure, but the heat is supplied by ultrasonic energy rather than by a heated stage or capillaries as shown in Figure 5.5. In addition, with aluminum wire, the ultrasonic energy and the acoustical high-frequency

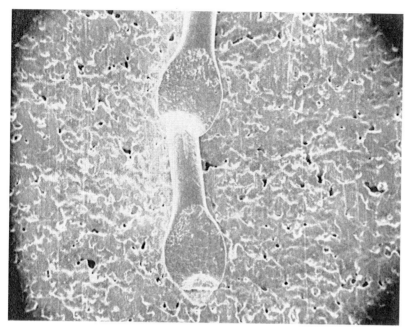

**Figure 5.2(b)**  Thermocompression wedge bond (300×). (Courtesy Westinghouse Electric Corp.)

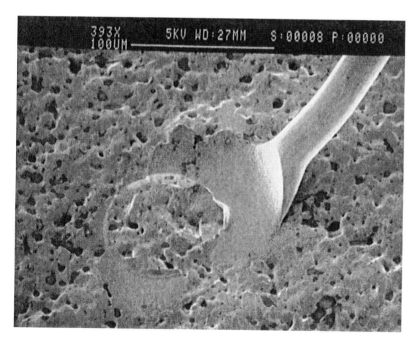

**Figure 5.2(c)**  Thermosonic stitch bond (393×). (Courtesy Westinghouse Electric Corp.)

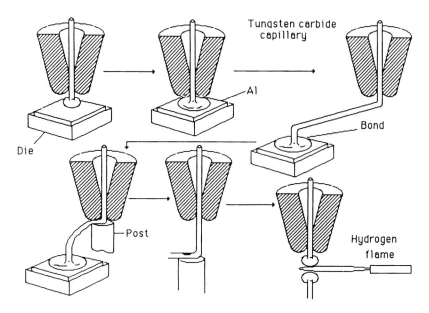

**Figure 5.3(a)**   Ball bonding.

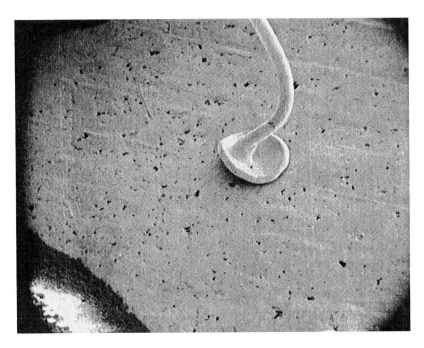

**Figure 5.3(b)**   Poorly adhered ball bond (~100×). (Courtesy Westinghouse Electric Corp.)

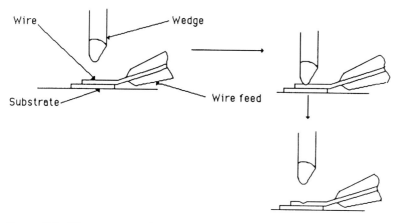

**Figure 5.4** Thermocompression wedge bonding.

movement of the wire against the conductor pad break the refracting oxides surrounding the aluminum wire. Pressure is also used but is incidental to the effect of the ultrasonic energy. The ultrasonic vibratory energy causes a temperature rise at the wire-conductor interface that can approach 30 to 50% of the melting point of the metal. One of the advantages of ultrasonic aluminum wire bonding is the absolute avoidance of ''purple plague.'' Purple plague, which results in embrittlement of the bond, has been found to be a result of the combination of aluminum, gold, silicon, and heat. It is avoided by eliminating gold and heat.

The three primary factors affecting ultrasonic wire bonds are force, time, and ultrasonic power. The ultrasonic power available to make the bond is dependent on the power setting of the oscillator power supply and the frequency adjustment of the tool. The force used is generally of the order of tens of grams and is large enough to hold the wire in place without slipping and to channel the ultrasonic energy into the bonding site without causing deformation of the wire. The specific force magnitude selected depends on the size and design of the bonding tool face, the size and hardness of the wire, and the sensitivity of the substrate. High power and a short bonding time are usually preferred to avoid metal fatigue and to prevent the initiation of internal cracks. Lower power nevertheless gives a good surface finish and a large pull strength [5].

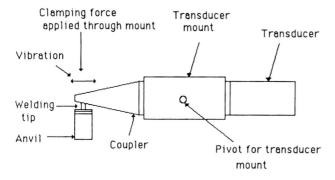

**Figure 5.5** Ultrasonic wedge bond.

## Combination of Thermocompression and Ultrasonic Bonding

The third bonding method is the combination of ultrasonic and thermocompression wire bonding. In the ultrasonic ball bonders, the ultrasonic heat is identical to that in the usual type of ultrasonic bonders, except a straight-wire capillary is used to feed the wire, as on the thermocompression bonder. Also included is the flame-off device necessary to form the ball on the gold wire. Whereas in straight ultrasonic gold wire bonding it is difficult to bond gold wire of less than 0.002 in. diameter, an ultrasonic ball bonder permits gold wire of 0.001 in. diameter to be used. The differences are in the capillary design and in the use of a heated stage. This is almost a complete combination of both thermocompression and ultrasonic methods, i.e., a heated stage, a capillary-type tool, and an ultrasonic transducer. The only device missing is the heated capillary, which becomes unnecessary with an ultrasonic transducer.

### 5.3.3 Wire Bond Failure Mechanisms

Wire bond failures can be divided into two categories. The first are the premature failures resulting from a poorly controlled or poorly designed manufacturing process. The second category consists of the failure modes that cause adequately made bonds to fail by contamination and/or wear-out types of environmental stresses during the operating life of the device.

### Failure Mechanisms Due to the Manufacturing Process

The bond strength depends on the materials and process variables associated with the substrate-metallization-wire composite structure. For example, an adequate gold bond requires a bonding load large enough to produce good interfacial conformity and a bond interface temperature high enough to effect contaminant dispersal. Compression in the bonding process is always used to increase the area of contact so as to produce a bond between area elements of fresh metal surfaces. The surface films are disrupted during the process and the bond occurs between patches of fresh metal. In the process of thermocompression bonding, English [8] found that lower than optimum bonding temperatures or lower tool loads resulted in inadequate bonds. If bonding was attempted at an extra-low temperature with a corresponding load increase, little or no bonding took place. Heating was not required for welding very clean surfaces. These observations are consistent with the view that heating is required for dispersing surface contamination. Alternatively, use of very low tool loads and high temperature also resulted in bond failure and/or low bond strength. The bonding process was found to be due to a shear displacement at the intended bond interface that disrupted the contaminant layers and contributed to the bond formation. English [8] noted that leads and metallized substrates were typically stored in air for days prior to interconnection. The surface therefore carried adsorbed gases, in particular water vapor. When the substrate and the lead frame were heated the bond strength increased with temperature and bake time. However, the postbond baking did not increase the bond strength significantly if the tool load was too low. It did, however, increase the strength of a bond made at a low bonding temperature. Figure 5.3b shows a poorly adhered ball bond.

Lang and Pinamaneni [9] identified the manufacturing parameters that affect wire bond strength. These include cleaning and copper plating of the lead frames, die attach cure conditions, atmosphere during bonding, surface finish of the lead frame, bonding time, bonding force, bonding pressure, and temperature. They found that the presence of an inert atmosphere was essential to prevent oxidation of the lead frame and that a lead frame with coarse surface finish gave greater bond strength than a smooth surface.

Weiner and Clatterbaugh [10] defined the machine parameters that could affect the shear strength of ball bonds. It was found that an increase in the ultrasonic power resulted in an increase in the shear strength of the bond. The substrate temperature was also found to affect the bond shear strength. The pedestal that supports the substrate during bonding is heated to enhance the formation of metallurgically sound bonds. The increase in the pedestal temperature increased the shear strength. Occasionally it was found necessary to leave the substrate on the heated pedestal for periods longer than can be considered normal. Even with extended residence times up to 3 hours, there was no significant change in the bond shear strength. The effect of contamination and cleaning techniques and burn-in on the ball shear strength was observed. An increase in the contaminant concentration resulted in a decrease in the shear strength. The cleaning procedure used to remove the contaminants was found to vary in effectiveness, as measured by the restoration of the shear strength, depending on both metallization and the contaminant type. Solvent cleaning was the least effective method for restoring the ball shear strength to uncontaminated levels. The ultraviolet-ozone used for the cleaning process was found to improve the bond shear strength most significantly. The shear strength of the bonds increased after burn-in and was highly dependent on both the type of contaminant and the substrate metallization.

Cratering is another failure mechanism which occurs predominantly as a result of improper control of bonding parameters. Excessive ultrasonic energy has been cited as a cause of cratering. Cratering in thermocompression bonds could also be caused by high bonding force, too great a tool-to-substrate impact velocity, or too small a ball which allows the hard bonding tool to contact the metallization. In studying the ultrasonic process, Winchell [12,39], found that even though metal flow is equal in all directions, stacking faults in silicon occur perpendicular to the direction of the ultrasonic bonding tool motion, thus verifying that ultrasonic energy was a major cause of the problem and capable of introducing defects into the single crystal silicon. Koyama [23] found that force and temperature could not cause silicon nodule cratering without ultrasonic energy. Contaminated bond pads are found to require more ultrasonic energy and higher temperature for making strong bonds. In general Kale [92] found that, for wedge bonds, too high or too low a static bonding force can result in cratering. Winchell [12,393] was the first to recognize the effect of bond pad thickness on the phenomenon of cratering. The bond pad was found to serve as a cushion protecting the underlying Si, $SiO_2$, polysilicon, GaAs from the stresses of the bonding process. The tendency to crater was found to be predominant in tin metallizations. The bond pad hardness has also been found to affect the phenomenon of cratering. A softer bond pad has been found to inhibit cratering by absorbing the energy from the bonding process. A harder bond pad, on the other hand, would readily transmit the energy from the bonding process to the substrate. Ravi [91] and Hirota [40] found that the best bonds were formed when the hardness of the wire and pad were reasonably matched.

Plastic-encapsulated components are covered in epoxy after the bonding process is complete. In an attempt to evaluate the effect of molding parameters on wire bonds, Koch et al. [13] conducted experiments on a thermolding process of epoxy encapsulation using a 28 chase mold. The molding parameters considered were transfer time, mold temperature, mold compound preheat temperature, and transfer pressure. Other factors considered included material flow characteristics and the kinetics of the molding compound. The experimental data showed that too short a transfer time as measured with the mold compound increases the number of bond failures. Further, it was evident that high material preheat temperatures and high transfer pressure increased the number of bond failures.

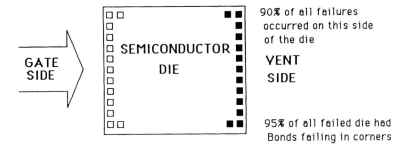

90% of all failures occurred on this side of the die

VENT SIDE

95% of all failed die had Bonds failing in corners

**Figure 5.6** Location of failed bond pads in relation to incoming material flow during molding. Most of the failures occurred on the vent side of the die [2].

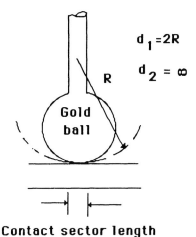

$d_1 = 2R$

$d_2 = \infty$

R

Gold ball

Contact sector length

**Figure 5.7** Contact pressure at the ball pad interface [1].

The results showed that temperature variations across the cavities increased the variation of bond failures from one cavity to the next. It was also found that these parameters were interdependent because viscosity and flow characteristics are dependent on heat transfer and hydrostatic pressure. Also, the analysis of the material inside the mold showed that more than 90% of the failures occurred on the opposite side of the die from the mold gate (Figure 5.6).

Ching and Schroen [1] developed a theoretical bond model based on the Hertz theory of contact pressure between a sphere and a flat surface (Figure 5.7). This model was used to study the stresses induced during the bonding process. The assumptions made were that both surfaces were elastic and that the intermetallic region possessed the same properties as the gold ball. It was further assumed that contact is made between a part of the sphere of the intermetallic and the flat silicon pad. The ultrasonic effects were also neglected. The area of contact is obtained as

$$A = \left[ \frac{3F((1 - v_1^2)/E_1) + ((1 - v_2^2)/E_2)}{8((1/D_1) + (1/D_2))} \right]^{1/3}$$

where

$$D_1 = 2R = B + \frac{C^2}{4B}$$

Principal stresses due to contact pressure are

$$\sigma_x = \sigma_y = -P_{max}\left[1 - Z/A\left(\tan^{-1}\left(\frac{1}{Z/A}\right)\right)(1 + U) - (1/2(Z^2/A + 1))\right]$$

where

$$Z = \frac{-P^{max}}{1 + Z/A} \quad \text{and} \quad P^{max} = \frac{-3F}{2\pi A^2}$$

$$\tau_{xy} = \tau_{xz} = \frac{\sigma_x - \sigma_y}{2} = \frac{\sigma_x - \sigma_z}{2}$$

and

$Z$ = distance below the plane of contact
$E$ = modulus of elasticity
$D$ = diameter of the spheres
$F$ = applied force
$B$ = depth of the intermetallic = 1 mil (approximately)
$C$ = sector length of the intermetallic
$v$ = Poisson's ratio

Subscripts 1 and 2 represent the two materials in contact

The conclusion drawn from this model was that to reduce bond pad cracking, a large contact area at touchdown with minimum force exerted on the bond pad should be achieved. The manufacturing variables include time to reach touchdown after ball formation and moisture content. It was shown that hardness of the gold ball at touchdown also contributed to stress exerted on the pad. The factors that contributed to the hardness of the gold ball were the wire impurity level and the temperature of the gold ball at touchdown, and the grain size was determined by the rate of cooling.

## Failure Mechanisms Due to Environmental Stresses and Other Conditions During Operating Life

One significant failure mechanism in operation is cracking of the bond pad. The bond failure in this mechanism is characterized by cracking of the underlying pad structure. Koch and Richling found that silicon nodules precipitate from the metallization in the pad and acted as points of high stress during the bonding of the wire to the pad regions. Silicon nodules with a diameter of approximately 1 µm, which grew by annealing after metal deposition, were distributed uniformly before bonding. After the bonding process was complete, these nodules decreased in the bonding region and damage was observed on the insulation.

Another failure mechanism is lifting of the bond due to shear stresses induced between the wire and the bond pad and between the bond pad and the substrate because of temperature and power cycling [14–16]. It was also observed that the gold-aluminum intermetallic that is formed during bonding continues its growth during baking and consumes all the aluminum that is left on the pad. The bond pads have oxides below the metallization that act as an insulating medium. This permits the multilevel oxide (MLO) to come into direct contact with the Au-Al intermetallic. As the devices are subjected to additional shear stress during temperature cycling, the adhesion between these two materials weakens, the ball is lifted, and the oxide is exposed.

A bond failure mechanism identified by Harman [17] in Al ultrasonic bonds was metallurgical cracks in the heel. Cracks were found to be a result of excessive flexing of the wire during loop formation, especially when the second bond was significantly lower than the first, either by operator motion of the micropositioner or by bonding machine vibration just before or during bonding tool lift-up from the first bond. The sharp metallurgical microcracks were found to propagate through the wire and cause failure during device operating life. Poonawala [11] identified the failures in cannon-launched devices due to long wire distances and die misalignment from the package cavity center. The inertial forces resulted in two failure mechanisms: plastic collapse of the wires creating a possibility of shorting to the cavity bottom and lateral collapse of the wires creating a possibility of shorting to the adjacent bond pads or adjacent wire bonds.

A wire bond subjected to temperature change experiences sheer stresses between the wire, the bond pad and the substrate as a result of differential thermal expansion [43]. Occasionally when gold and aluminum are bonded together, intermetallics form at the interface due to the interdiffusion of gold and aluminum. Gold-aluminum intermetallics are stronger than pure metals and their growth is enhanced by increased temperature and temperature cycling [43]. The difference in the hardness and the material properties including the coefficient of thermal expansion between the intermetallic and the surrounding metal, make the interface a potential site for failure.

Temperatures above 300°C accelerate the rate of intermetallic compound formation. Further, the time to decrease bond strength by 40% of the "as bonded strength" is found to decrease with increases in temperature.

Electrical leakage failure during functional test constitutes another failure mechanism of wire bonds. Bonds with no visible evidence of damage or mechanical weakness were found to have intermittent electrical leakage. Leakage failure became significant in devices with a multilayer oxide-free bond pad. Analysis of these leakage failures revealed that there were no cracks on pads using the same bonding conditions. The leakage problem is the result of poor insulation from the Si substrate due to the lack of a multilayer underneath the bond pad.

Cunningham [19] suggested that metallurgical (Kirkendall) voids were a cause of bond failure. These voids were formed by the different interdiffusion rates of gold and aluminum. Under various circumstances the voids may appear either in the gold or in the aluminum.

The aluminum wire bonded to a conventional gold metallization cavity in CER-DIPs (ceramic dual in-line packages) [20] has produced a well-known reliability hazard known as purple plague. This is a brittle gold-aluminum intermetallic that sometimes forms at an interface of a gold-aluminum thermocompression bond. This intermetallic appears purple in the crystalline form. Two types of plague-induced bond failures have been observed. In the first, the bond may be mechanically strong, but it can have a high electrical resistance or even be open-circuited. In this case, which typically occurs with gold wire bonded to thin aluminum metallization, voids form around the periphery, limiting the available conduction paths. In the second type of failure, the voids lie beneath the bond, causing the bond to fail due to mechanical weakness.

The Ag-Al system failure is very different from the Au-Al system failure, which is due to Kirkendall voiding of the diffusion front [20]. The high resistance in the Ag-Al

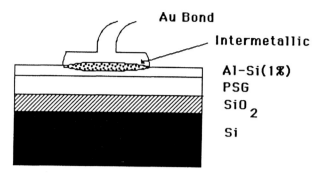

**Figure 5.8** Schematic cross-section of the bond pad structure.

bond is due to oxidation of the Ag-Al alloy, resulting in a thin, insulating oxide layer that completely envelops the alloyed zone.

According to Griffith's theory of brittle fracture, the fracture stress is directly related to the Young's modulus of the material. The phosphosilicated glass (PSG) layer beneath the bond pad has a lower Young's modulus than the thermal oxide layer because of the inherently lower density and high impurity concentration of phosphorus in the PSG oxide (Figure 5.8). The number of bond failures increases as a function of phosphorus concentration in the PSG film. The PSG layer fractures and the cracks propagate at a lower applied stress then in the thermally grown $SiO_2$ layer.

A failure mechanism in dry air was found to be selective oxidation of the Ag-Al alloy, and activation energies were measured for various atmospheres. Moisture was shown to decrease the activation energy. When the package absorbs water before soldering, soldering heat stress causes peeling off of the wire ball from the Si substrate or insulator. The quality of bond will affect the bond failure rate. The Ag-Al substrate bond system shows high resistance as a result of a thermally activated process typically used to assess the long-term reliability of IC devices, which causes the resistance of bond to change from negligible (0.1 $\Omega$) to 20 $\Omega$ or higher. Forrest [21] found that there is no discernible resistance change until a critical time is reached at which the resistance rises in a dramatic manner to values as high as 20 $\Omega$ or more.

Shukla and Deo [22] found that the failure mechanism in dry air is due to selective oxidation of the Ag-Al alloy. The expression for critical time was given by an Arrhenius relationship as

$$t_{cr} = t_0 \exp\left(\frac{\Delta H}{kT}\right)$$

where $t_0$ is a temperature-independent constant, $\Delta H$ is the activation energy, $k$ is the Boltzmann constant, and $T$ is the absolute temperature. The change in the resistivity of the Ag-Al binary system was found to be negligible until a particular critical time was reached, after which it increased to a very large value.

Forrest [21] noted that another failure mechanism was that of corroded wire bonds. Corrosion opened one end of the wire completely, and occasionally both ends of the wire, permitting the wire to move freely within the package volume and causing intermittent electrical short circuits. It was found that chlorine ions concentrated around wire bonds during the high-purity water rinse. Capillary action of the wire bond–water interface concentrated any dissolved chlorine at that point, causing the formation of $AlCl_3$ during elevated temperatures encountered during burn-in. Exposure of the conductor material to a chlorine environment caused a replacement chemical reaction to convert copper oxide to copper chloride at the substrate interface; the presence of the copper chloride caused the Al wire bonds to corrode or develop high-resistance intermetallics.

Moore [24] found that hybrid circuit metallization was very susceptible to aqueous corrosion. Contributing factors included the applied potential of the circuit power source to drive the corrosion reaction, the close proximity of the biased circuit conductors, ionic process residuals, microscopic and macroscopic galvanic couples, and the small mass of the conductors. It was stated that under these conditions any quantity of electrolyte providing ionic transport could present a significant reliability problem. The corrosion reaction, dissolution and plating, was found to proceed at a distance up the wire from the die surface. The effect was due to a thin layer of die coat that had wicked up the wire surface.

Another failure mechanism noted was electrical noise in the output of the circuit. The cause was determined to be the formation of intermetallics due to high chlorine concentrations around bonds in a ball-and-socket configuration. This type of bond exhibits high mechanical strength in conjunction with low conductivity due to formation of resistive compounds at the interface. When such a bond was subjected to a nondestrucive bond pull test, an apparent healing of noise occurred through a reduction of the bonding resistance by motion of the wire relative to the bonding surface.

Another failure mechanism was that of silver dendrite growth from the wire bond pads of an integrated circuit. An epoxy was used to attach the gold-backed die to the chip carrier die pad. After the epoxy cure the package was cleaned with oxygen plasma rinsed in deionized (DI) water. The wet package was then placed in an oven to dry. It was at the drying stage that silver dendrite growth was observed. The dendrites extended out over the glass passivation layer.

Harman [17] found that vibration forces that occur in the field are seldom severe enough to cause metallurgical fatigue or other bond damage. In general, large components of assembled systems failed before such forces were sufficient to damage the bonds. Schafft [5] calculated the resonant frequency as well as centrifuge-induced forces for gold and aluminum wire bonds of various geometries. The minimum excitation frequency that might induce resonance and thus damage gold wire bonds having typical geometries was found to be in the range of 3 to 5 kHz. For most aluminum wire bond geometries, the resonant frequencies required to damage the bonds were greater than 10 kHz.

Harman [17] stated that in the case of hermetic devices, even if the package does not contain any corrosive materials, metallurgical bond failure modes may result from the effect of high temperature or cyclic temperature changes. If the external temperature is greater than about 150°C for long periods of time, the wire bond partially anneals, producing a bond that is mechanically weaker in a bond pull test. Coucoulas [25], however, found that in the case of ultrasonic bonds, the work hardening and other strains in the thinned layer were partially annealed, resulting in a more reliable bond.

Wire bond failures due to temperature cycling were studied by Gaffney [26], Villella and Nowakowaski [27,28], Ravi and Philofsky [29], and Phillips [30] (see also Refs. 14–16). All of them investigated metallurgical flexure-fatigue failures in 0.001-in.-diameter aluminum wires with 1% silicon impurities. These failures resulted from repeated wire flexing due to the different coefficients of thermal expansion between the aluminum wire and the header as the device heated up and cooled down. The maximum flexure and therefore the failure were found to occur at the thinned bond heels. The heel of the chip bond experienced a greater temperature excursion and therefore was more prone to fail than the heel of the lead frame bond pad. Villella and Nowakowski [27] ran extensive statistical tests with cycled devices and determined that aluminum ultrasonic bonds were more reliable in this service than aluminum thermocompression bonds. Ravi and Philofsky [29] experimentally investigated the metallurgical flexure fatigue of a number of aluminum alloy wires and showed that aluminum wire with 0.1% magnesium was superior to the commonly used aluminum–1% silicon alloy. Phillips [30] calculated wire bond geometry effects and recommended that the loop height be at least 25% of the bond-to-bond spacing to minimize the bond flexure.

Another wire bond metallurgical failure mode was identified by Adams [31] (see also Refs. 14–16) for gold wire in plastic-encapsulated devices. A typical case of metal fatigue was encountered when the device was made to undergo thermal cycling. Adams calculated that for a $\Delta T$ of 100°C the axial stress in the wire due to different expansion coefficients of the wire and plastic would almost equal the breaking load of the wire, assuming that the plastic was bonded to the wire.

Mantese and Alcini [32] found that accelerated Al oxidation occurs as the temperature of the bond material is elevated, causing degradation of contact. Aluminum melts at 660°C and oxidizes readily, making it unsuitable for devices that experience high temperature. Other parameters found to affect bond quality during operation are the bonding time, ultrasonic power, tool length, tool wear, and type of substrate material.

A large fraction of microelectronic package failures is due to wire bond failures. Figure 5.9 shows the percentage of failures due to interconnections. Wire bond failures are a result of shear between the bond pad and the wire, flexure of the wire, excessive intermetallics, manufacturing parameters, galvanic corrosion, and chip-outs in Si/SiO$_2$. Plastic-encapsulated packages also fail as result of axial stresses in the wire resulting from the differential expansion of the epoxy encapsulation and the wire.

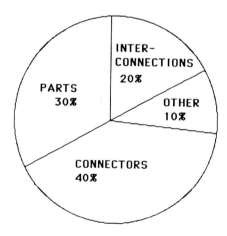

**Figure 5.9**   Failure and causes in electronic systems [AFFDL-TR-71-32].

## 5.4  TAPE-AUTOMATED BONDING

Another approach to fine-pitch interconnection of chip to lead frame is tape-automated bonding (TAB), shown in Figures 4.21–4.25 in Chapter 4. The interconnections are patterned on a polymer tape. The tape is positioned above the chip to be bonded so that the metal tracks on the polymer tape correspond to the bonding sites on the chip. A thermocompression process is used for bonding, and all joints are made in a single operation. Unlike the case of wire bonds, ultrasonic energy cannot be used for TAB. Machine characteristics such as bonding planarity, mechanical and thermal properties of the bonding tool, and temperature uniformity strongly affect the quality of thermocompression tape bonds. The bonding process can be divided into three steps: inner lead bonding, testing of chip to ensure functionality, and outer lead bonding. The bonding process uses a heated pressure head that forms a thermocompression bond to all the pads simultaneously. Bond pads are plated to raise the pad level above the surface of the chip, forming a bump. The tape-automated bonding process provides several advantages over the wire bonding process. These include a smaller bonding pad or chip, smaller bonding pitch on the chip, decreased quantity of gold used because of the smaller dimension of the interconnection, elimination of wire loop because of low-profile assembly, reduced vibration in bond geometry, increased production rate because of gang bonding, and stronger and uniform inner lead bond strength. Further, TAB bonds have better electrical performance than wire bonds, better high-frequency performance, reduced labor costs, higher I/O counts, higher weight, and decreased printed wiring board real estate. The chip can be attached in a face-up or face-down configuration.

The TAB process suffers from several disadvantages. These include requirements for a special tape and tool design for each chip design; bumped wire sites, which means additional wafer processing; larger capital investment; and special materials and equipment. Further, the beam lead reliability requirement is greater because of reduced thickness.

### 5.4.1  Types of TAB

There are four major types of TAB: area TAB, bumped-chip TAB, bumped-tape TAB, and multichip packaging. TAB is characterized by a bonding projection and bump between the chip and the lead. This bump forms the tape-automated bond and provides the required metallurgy for the bond. The bump also provides the physical standoff, preventing a short circuit between lead and chip. When the bump is placed on the chip, the method is known as bumped chip TAB. The bump may also be placed on the tape, in which case the method is known as bumped-tape TAB. In area TAB a multilayered tape provides interconnection between the grid of solder bumps of a VLSI chip and the bonding pads on a ceramic package. Multichip packaging consists of thick fiber interconnectors on a ceramic substrate that connect chips inside a multichip module. Multichip modules have increased I/O counts, increased density, and better performance.

### 5.4.2  Types of Tape

TAB tape has three major forms of construction: one-layer, two-layer, and three-layer tapes. The type of tape used depends on the application area.

One-layer tape is made on a metal foil with lead patterns photoetched on it. The metal is usually copper 0.07 mm thick. The tape can be bumped, or a bump can be placed on the chip. One-layer tape is usually well suited to low-cost assembly and high-speed inner lead bonding. Sprocket holes are etched simultaneously with circuits.

Two-layer tape is typically 1-ounce copper on a $0.5 \times 10^{-3}$ to $2 \times 10^{-3}$ in. thick plastic film. Polyimide is the material best suited for this process [33]. This type of tape may be manufactured by screening polyimide film on copper, casting a liquid polyimide film onto copper, or plating copper onto a polyimide film.

Sprockets and windows in the copper foil are etched simultaneously by chemical etching, while lead patterns are formed by a normal photoetching process. The type of tape allows design freedom along with testability for functionality of interconnection. The polyimide film also supports the delicate TAB leads.

Three-layer tape is usually 0.076 to 0.15 mm thick polyimide film coated with adhesive $0.5 \times 10^{-3}$ to $1 \times 10^{-3}$ in. thick. The polyimide film acts as a dielectric barrier and electrically isolates the leads to allow functional testing before the IC is finally connected to the package.

## 5.5 CHIP-TO-BOARD INTERCONNECTIONS

Chip-to-board interconnections are used to attach the integrated circuits (ICs) to PWBs. This category includes the dual in-line package (DIP), pin-grid array (PGA), plastic-leaded chip carrier (PLCC), leadless chip carrier (LCC), leaded ceramic chip carrier (LCCC), quad flatpacks (QFP), and tape-automated bonding. DIP uses through hole technology and is a widely used chip connector. It incorporates one row of pins arranged on 100-mil centers along each of the longer sides of the rectangular package. DIPs are available in sizes from 8 to 64 pins. The pins are short with rectangular cross sections. The longer cross-sectional dimension is aligned along the length of the DIP package. This package is limited to 64 pins and is a connector choice for less powerful applications.

The pin-grid array also uses through hole technology because of space requirements and large insertion forces required for component assembly. PGAs are substituted for DIPs when larger I/O counts are required or lower thermal resistance is necessary. The package is fabricated from ceramic with leads around 20 to 25 mils in diameter. This package is used for microprocessors and other more expensive applications.

The PLCC is used in multicontact sizes in less expensive IC applications. It is basically common to applications that do not generate much heat, because of its plastic construction. CER-DIPs are ceramic DIP for through hole mounting, and ceramic flatpacks and quad-packs are surface mounted. PLCCs are generally found in J-lead configurations.

Quad flatpacks are surface mount technology and are used for devices with higher frequencies and higher lead counts. Leads are mounted along the side of the package on 50-mil centers. The flatpack is nearly twice as efficient as the DIP in utilization of circuit board area.

TAB is another approach to chip-to-board interconnection. A thin layer of tape with connections for IC leads is applied to the chip and then bonded to the board. The assembly is covered by encapsulant to prevent contamination.

DIP switches are miniature switches that mount directly on a printed circuit board. The main use of DIP switches is to program electronic circuits (Figure 5.10). These switches have the advantage of small size and can be mounted directly on the printed circuit board. DIP switches are used to satisfy a programming function without consuming much space, eliminating messy jumper wires on board to change functions. Further, DIP

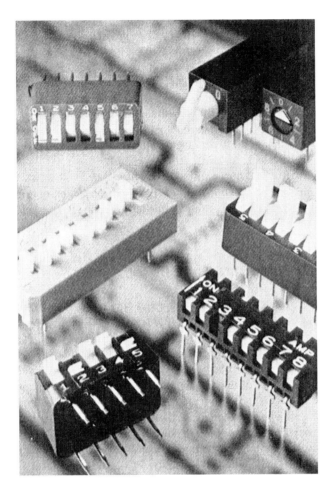

**Figure 5.10**   DIP switches. (Courtesy of AMP Inc.)

switches can be used to set printed circuit board (PCB) configurations, configure I/O circuits, and adjust products before shipment. Typical applications include computers, electronic printers, facsimile machines, processors, and vending machines. DIP switches in a computer are used to indicate the actual memory capacity and the number and type of drives installed. A DIP switch in a garage door is used to select a frequency on an individual opener. Security systems use DIP switches for a large number of coding combinations. DIP switches set prices of items. The advantage is that the user can change switch settings in the field as needed. A DIP switch is actually an assembly of several individual switches in a single housing. The number of switches ranges from 2 to 12, and 8 is most common. The switches have clearly labeled switch numbers and on/off positions to make identification easy. The switch is sealed to prevent any damage to it during wave soldering and flux cleaning. Figure 5.11 shows some commerical designs for DIP switches.

The ICs can be soldered to the board or mounted in sockets. Sockets have the advantages of ease of replacement and repair, lower service costs, ease of reprogramming

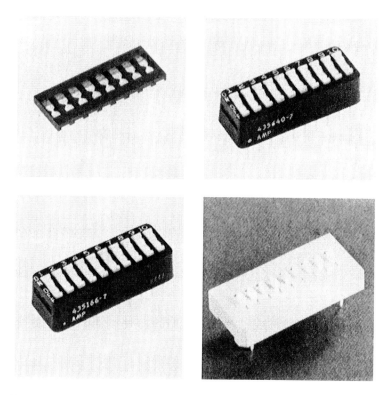

**Figure 5.11** Commercial designs for DIP switches. (Courtesy of AMP Inc.)

memory chips, and room for future changes to the circuitry. Sockets are not preferable in military applications and represent a severe thermal penalty for conductively cooled modules. Figure 5.12 shows the range of available densities for IC sockets. Sockets are designed to have a large target area to ease engagement and prevent damage to the IC's I/O pins during insertion. This is usually accomplished by using an oversized extreme (Figure 5.13). Twisted leads are a constant hazard to sockets. Stress relief features are

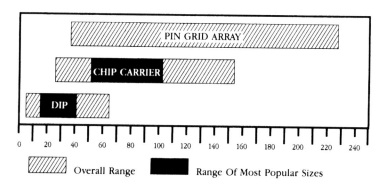

**Figure 5.12** Range of available pin densities for IC sockets. (Courtesy of AMP Inc.)

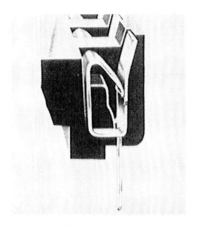

**Figure 5.13** Large target area on sockets eases engagement. (Courtesy of AMP Inc.)

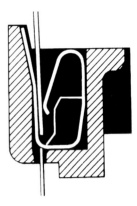

**Figure 5.14** Antiover-stress feature prevents damage to socket contacts from bent or twisted leads and angular or extraction of the packages. (Courtesy of AMP Inc.)

built into sockets to prevent damage (Figure 5.15). The main body of the sockets is usually raised off the PWB to facilitate cleaning after soldering. The bottom of the socket is kept closed to prevent the entry of dust and dirt into the contact area by capillary action (Figure 5.15). A fairly large contact force is maintained during insertions and extractions to remove the oxide layers and other contaminants on the surface of the contact by wiping action (Figure 5.16). Sockets are usually through hole or surface mount devices. Usually an

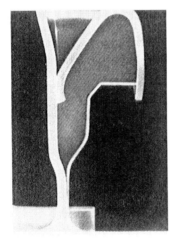

**Figure 5.15** Closed bottom design in sockets prevents flux contamination and solder wicking. (Courtesy of AMP Inc.)

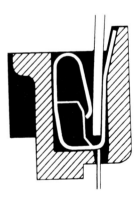

**Figure 5.16** Dual-wiping action of sockets remove oxide layers and contaminants from the surface of the connector contacts. (Courtesy of AMP Inc.)

**Figure 5.17** Polarization in sockets facilitates proper insertion and allows easy access probing. (Courtesy of AMP Inc.)

optional retention feature is placed on the socket's solder legs to hold it in place until it is soldered. Sockets are polarized; i.e., they have a distinguishing feature such as a chamfered edge, notch, or rounded corner that prevents incorrect insertion of an IC into the socket (Figure 5.17).

A potential drawback in DIPs is the large insertion and grid extraction forces. Large forces make ICs susceptible to leg damage. Low insertion force (LIF) or zero insertion force (ZIF) sockets are designed to ease insertion and extraction of components from sockets. LIF or ZIF sockets usually have mechanisms to hold two leaves of contact apart so that the DIP can be inserted with little or no force (Figure 5.18). Sockets are classified according to the component they are supposed to house. A usual classification is DIP sockets (Figure 5.19), burn-in sockets (Figure 5.20), PLCC sockets (Figure 5.21), LCC sockets (Figure 5.22), and PGA sockets (Figure 5.23). A miscellaneous category may include miniature spring sockets (Figure 5.24), single in-line memory module (SIMM) sockets (Figure 5.25), and ZIF or LIF sockets. The choice of socket depends on the application in addition to the component it is supposed to house. LIF and ZIF connectors are used for applications requiring many mating and unmating cycles, and burn-in sockets may be used in applications requiring high temperatures. Connections from chip to board can be made using solderless interconnects. These include buttons and uniaxial conductive elastomers. Buttons are usually made of copper-silver or beryllium-copper and are usually

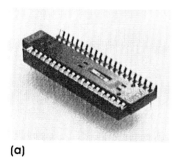
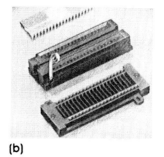

**(a)** **(b)**

**Figure 5.18** (a) LIF sockets are typically screwdriver actuated and can be used to advantage with expensive side brazed ceramic DIP's. (b) ZIF sockets with camming actions that can be gang-actuated are well suited to test and burn-in applications. (Courtesy of AMP Inc.)

**Figure 5.19** DIP sockets provide high quality at low cost. (Courtesy of AMP Inc.)

**Figure 5.20** Burn-in sockets require low insertion force contacts, high temperature plastics, and some form of IC extraction aid like the sword channel between contact rows on the sockets. (Courtesy of AMP Inc.)

multiple springs consisting of cantilevered beams, columns, and torsion springs. The random wire formation design provides a low inductance and a low resistance. Figure 5.26 shows the interconnection of a PGA using button technology. This interconnection technique can be extended to higher levels of packaging. Figure 5.27 shows board-to-board interconnections using button technology.

## 5.6 BOARD-TO-BOARD INTERCONNECTIONS

These connectors are used to attach two or more printed circuit boards. They include male strips, shrouded headers, female sockets for male pin strips and headers, card edge connectors, PCB connectors, and back plane connectors.

Male pin strips may be attached by compliant solder or press-fit technology. They are usually used where space on the printed circuit board or backplane is at a premium.

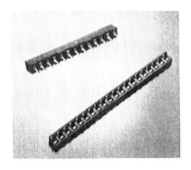

**Figure 5.21** PLCC sockets. (Courtesy of AMP Inc.)

**Figure 5.22** LCC sockets. (Courtesy of AMP Inc.)

**Figure 5.23** Pin grid array sockets. (Courtesy of AMP Inc.)

**Figure 5.24** Miniature spring sockets can be applied in any pattern as close as 0.085-inch between centers, providing pluggability for a wide range of component types. (Courtesy of AMP Inc.)

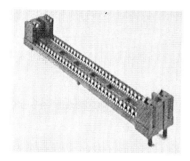

**Figure 5.25** SIMM sockets are designed for robotic applications and can increase packaging density by a factor of three over conventional DIP configurations. (Courtesy of AMP Inc.)

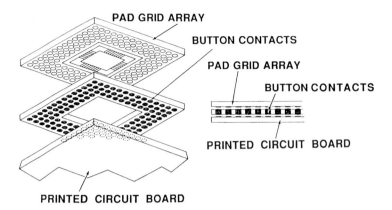

**Figure 5.26**  Chip-to-board interconnection using button technology. (Courtesy of CINCH.)

Card edge connectors are extensions of the edge of the printed circuit board and no pins are used. These are not durable, so they are not used in places where there is a frequent need to remove the connection. These connectors are common on personal computer mother boards or less expensive backplanes. They have a long rectangular housing and a slot into which the daughter board is inserted. Located within the housing are rows of contact springs that engage conductive pads located on the daughter board. On the back end of the housing the contact springs extend out to form a solder or solderless wrap. The pin count rarely exceeds 50 and installation of the mother board involves use of molded mounting loops and mechanical fasteners such as nuts and bolts.

Printed circuit board connectors may be permanently soldered on boards with the cable attached. They are often used in applications in which cost is a factor and no disconnection is required. PCB or PWB connectors are classified as either one-piece and

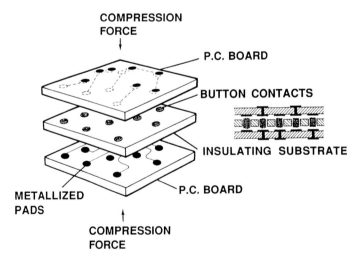

**Figure 5.27**  Board-to-board interconnection using button technology. (Courtesy of CINCH.)

**Figure 5.28**  One-piece PC connector. (Courtesy of AMP Inc.)

two-piece edge connectors. The one-piece edge connector is a receptacle containing female contacts designed to receive the edge of the printing wiring board on which the male contacts are printed (Figure 5.28). In the two-piece type, the male part is usually permanently mounted to the board (Figure 5.29).

One way of connecting board to board or wire to board is by using posts and receptacles. Posts are available in two varieties, machine applied and headers. Headers are single or double row, vertical or right angle, shrouded or unshrouded. Shrouded headers are common to high-pin-count applications in which protection of pins and guidance of the mating connector are important considerations, e.g., connection of a daughter board to a mother board in a personal computer. Figure 5.30 shows the various header configurations.

Machine posts are different because they stand alone on a board rather than in a header (Figure 5.31). The receptacle is the other half of the mating connection. Figure 5.32 shows various receptacle designs along with mating styles. The posts and receptacles are available in a variety of materials and platings. Headers and posts are usually made from a copper alloy, and machine posts are usually made of bronze. Receptacles are usually beryllium-copper or copper alloy. Figure 5.33 illustrates the use of posts and receptacles in wire-board interconnections.

Backplane connectors are pin-in-socket or card edge connectors, typically attached with compliant pin or press-fit technology. Backplane connectors are used in sophisticated PCBs with pin counts above 100.

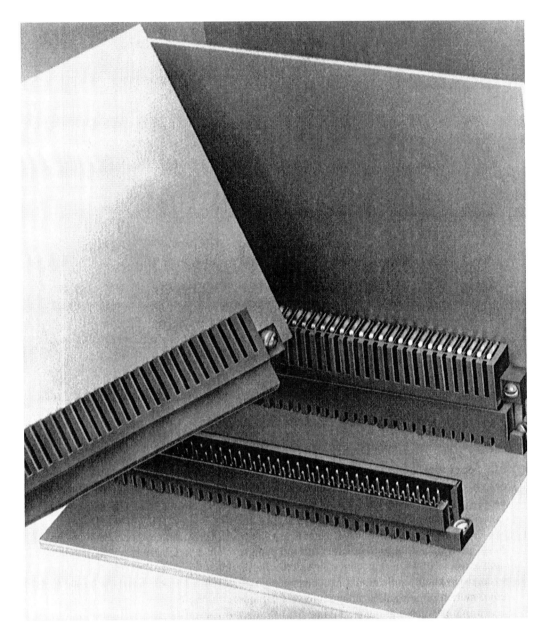

**Figure 5.29**  Two-piece PC connector. (Courtesy of AMP Inc.)

Board stacking connectors are male pin and female socket connectors used to connect boards directly without cable. Additional memory modules can be added to computers with this kind of connector.

SIMMs consist of clusters of active devices soldered to one or both sides of the PCB. These connections terminate on one edge, allowing vertical arrangement on the PCB. These sockets are designed for vertical or inclined mounting and are often ganged together,

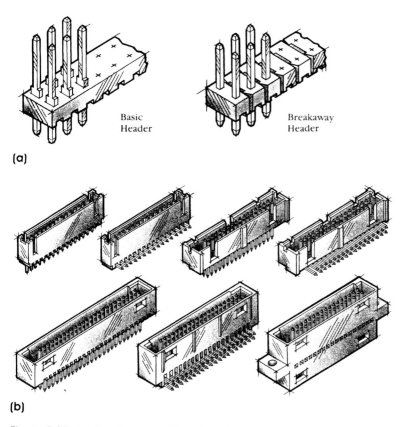

(a)

(b)

**Figure 5.30** Various header configurations for PCB's. (Courtesy of AMP Inc.)

minimizing the board real estate while maximizing memory capacity. They are characterized by low insertion and extraction forces. Installation of SIMMs generally involves
insertion and rotation into position. The rotation provides a desirable wiping action on
connects. Enormous stresses are imposed on the SIMM housing and locking latches. These
sockets are able to withstand the elevated temperatures encountered in solder reflow
processes.

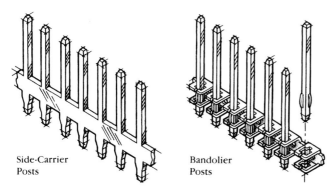

Side-Carrier
Posts

Bandolier
Posts

**Figure 5.31** Machine applied post offers fast application—up to
1400 posts per hour. (Courtesy of AMP Inc.)

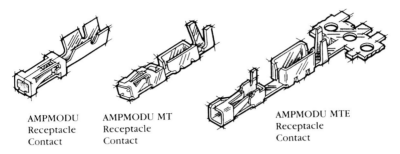

**Figure 5.32** Free-standing receptacles come in three mating styles. (Courtesy of AMP Inc.)

## 5.7 BOARD-TO-CABINET CONNECTORS

These are used to interconnect a printed circuit board and cabinet or two electronic hardware components. They include rack-and-panel, cylindrical, coaxial, and circular connectors.

Rack-and-panel connectors typically connect a panel or backplane to a drawer, rack, or module. Among all types of connectors, the rectangular shell allows the highest possible density per unit space. These connectors are common to applications in which reliability and space are critical, e.g., in telecommunications, modems, and printers. Figure 5.34 shows a common rack-and-panel connector.

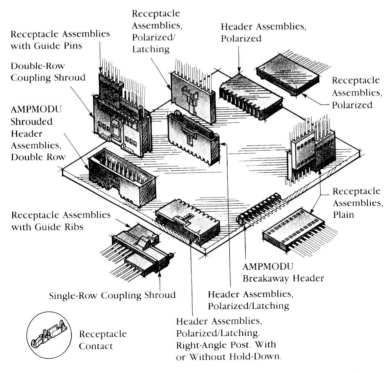

**Figure 5.33** Posts and receptacles for wire-to-board interconnections. (Courtesy of AMP Inc.)

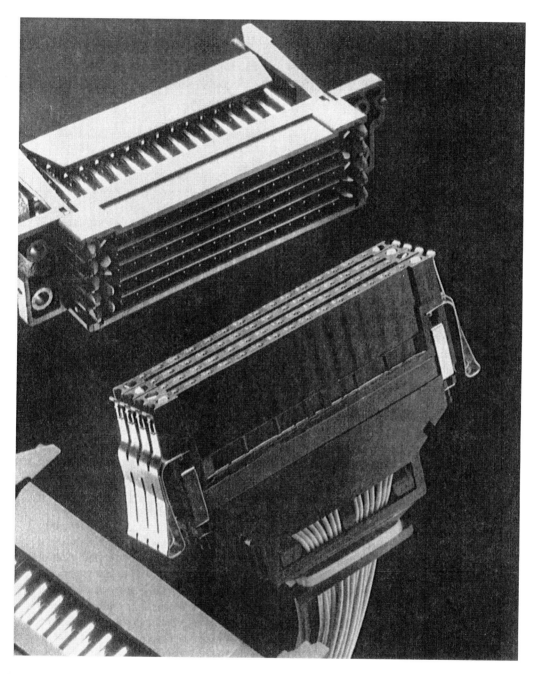

**Figure 5.34**  Rack and panel connector. (Courtesy of AMP Inc.)

Subminiature pin and socket connectors are characterized by their D-shaped mating face (Figure 5.35). These connectors are widely used in serial communications, telecommunications, and beat area networks. They are rugged, with the D face providing subminiature polarization during the mating process. Subminiature pin and socket connectors

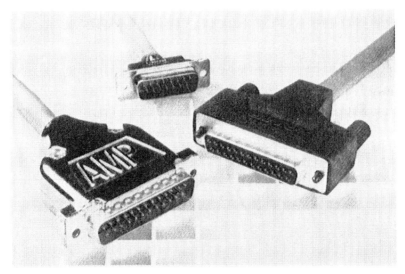

**Figure 5.35**  Distinguished by their D-shaped mating face, subminiature-D connectors are used for I/O applications. (Courtesy of AMP Inc.)

connect the electronics in the equipment to the outside world. Figure 5.36 shows typical subminiature D pin and socket connectors. I/O points at the back of the printed circuit are typically subminiature D connectors (Figure 5.37). The subminiature D connectors mounted on the cable are the cable mount connectors (Figure 5.38), and the board-mounted connectors form the other half of the mating connection (Figure 5.39). Connectors are available in straight and right-angle versions. The mating faces are sometimes charmfered for ease of engagement (Figure 5.40).

One of the earliest examples of cylindrical connectors is the MIL-C-5015 connector, a standard for airborne applications. Cylindrical connectors are still favored for military and aerospace applications because of their ruggedness and surviability in hostile environments, especially as the external connection between computer cabinets and related peripherals. Figure 5.41a and b show a typical cylindrical connector.

Coaxial connectors have a solid or stranded center conductor surrounded by a dielectric, which is typically covered with a conductive shield made of metal braid or tape and a layer of insulation. These connectors offer higher-frequency transmission and electromagnetic interference (EMI) protection. Figure 5.42 shows a typical coaxial connector. Coaxial cables are the standard transmission medium for high frequencies because of their high performance at high frequencies (impedance control) and shielding and EMI control even at lower frequencies. The basic cable consists of four parts: the inner conductor, the outer conductor, the dielectric separating the inner and outer conductors, and the jacket, which is the outer insulation (Figure 5.43). The characteristic impedance of the cable is a function of geometry and materials. The most efficient transfer of energy from source to load occurs when all parts have the same impedance. Mismatch in impedance produces inefficient energy transfer because of reflection of the energy back to the direction from which it came. Reflections occur in both directions, and such reflections and re-reflections produce a standing wave on the cable. A measure of the standing waves is the voltage standing wave ratio (VSWR), which is a function of the nonuniformity of the cable. Reflections are also a result of variations in cable dimensions and dielectric constant.

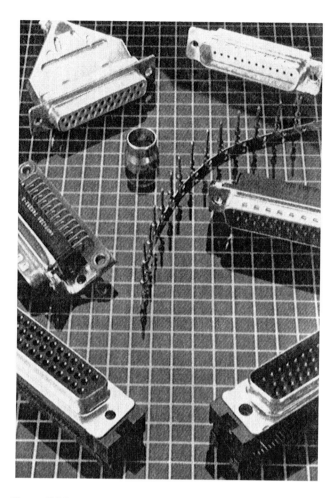

**Figure 5.36** A few typical subminiature-D connectors. (Courtesy of AMP Inc.)

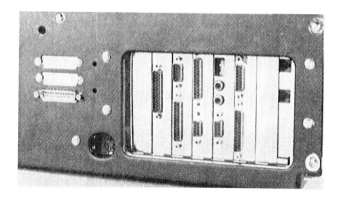

**Figure 5.37** I/O ports at the back of the PC are typically subminiature-D cable-mount connectors. (Courtesy of AMP Inc.)

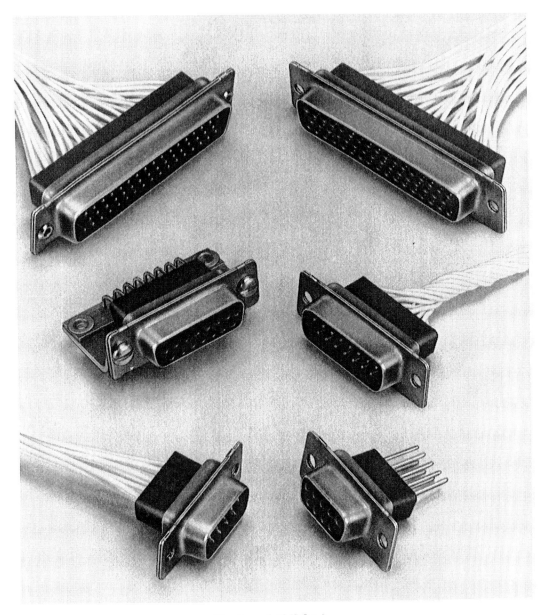

**Figure 5.38**   Cable-mount connectors. (Courtesy of AMP Inc.)

Figure 5.44 shows the various cable designs for the coaxial cable. The coupling method for coaxial cables determines the procedure for joining the two mating connector portions. The common types are bayonet, screw, and snap-on couplings. The bayonet or snap-on design allows fast connection with a simple twist or push/pull. Screw coupling offers a greater degree of safety against vibration. There are three main methods of terminating a coaxial cable: soldering and clamping, soldering, and crimping. Soldering and crimping is the most common procedure (Figure 5.45). It involves soldering of the

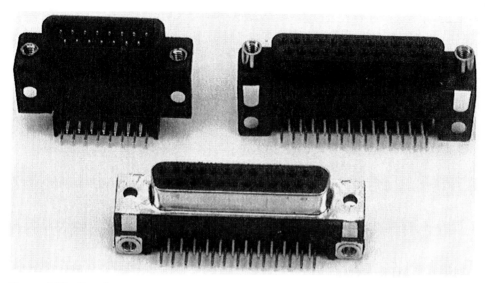

**Figure 5.39** Board-mount connectors. (Courtesy of AMP Inc.)

center conductor and crimping of the braid to the coupling. Materials used include brass, beryllium-copper, and stainless steel for base material. Platings are usually gold or silver, and the dielectric is Teflon or polypropylene. Figure 5.46a–d show some commercial coaxial connector designs. The metal braid and insulation provide low-impedance isolation of the outer shield through the connector interface to the mating assembly. Coaxial connectors are frequently found in audio applications and test equipment such as oscilloscopes.

In conventional connector design it takes approximately 100 grams of force to push one single pin into its respective socket. Hence, the maximum pin count is limited to about 50 if manual connector engagement is required. In practice, either some sort of

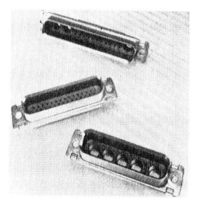

**Figure 5.40** Mating faces are chamfered for ease of engagement. (Courtesy of AMP Inc.)

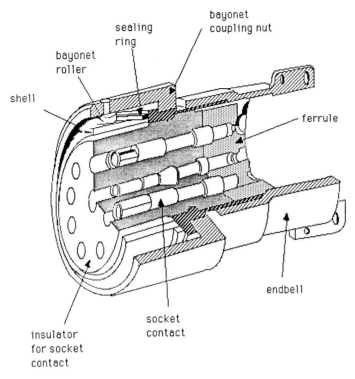

**Figure 5.41(a)** A schematic of a cylindrical connector.

mechanical aid such as a screw or lever or a special connector such as an LIF or ZIF will be used if the pin count exceeds 50.

The LIF connector is a cylindrically shaped socket with twisted strands of wire along the longitudinal axis [7,23]. Figure 5.47 is a schematic representation of a female part of an LIF connector. When the male pin enters the female socket, the wires stretch and conform to the shape of the pin. The insertion force is only about 43 grams. However, the mechanical complexity of LIF means high initial costs and extra weight.

The ZIF connector configuration eliminates insertion forces because there is no mechanical contact during mating. After insertion, a mechanism is activated to engage the contact pairs in a direction normal to the mating surface [34], as shown in Figure 5.48a and b. ZIF is ideal for connector pin counts of more than 300, where LIFs are not always feasible. However, because there is no sliding motion between the pin and the spring contact, there are no wiping forces during the mating process to clean the contact surfaces for good electrical contact. Lack of wiping forces is the chief drawback of ZIF connectors.

A combination of ZIF and large normal force is achieved using shape memory alloys. Shape memory alloys have the ability to change from a deformed shape to the original shape when triggered thermally. The deformation and recovery cycles can be repeated thousands of times without degradation if proper conditions are maintained. Shape memory contact components resemble standard contacts in appearance and are manufactured by similar processes. Shape memory alloys provide a zero insertion force along with high normal forces and outstanding performance in temperature cycling, shock, and vibration.

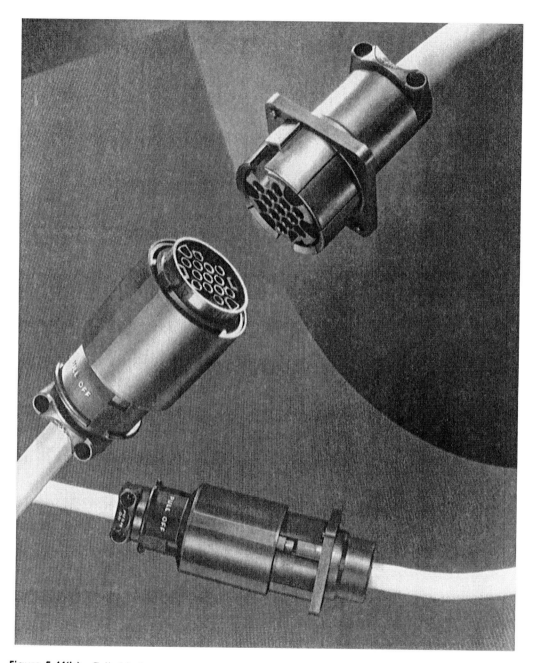

**Figure 5.41(b)** Cylindrical connector. (Courtesy of AMP Inc.)

Figure 5.49 shows a multicontact shape memory connector comprising three basic components: a shape memory element, a closing spring, and a flexible circuit (Kapton). The shape memory element uses a shape memory alloy that triggers above 125°C for applications requiring operation at − 55 to 125°C. The closing spring is made of stainless

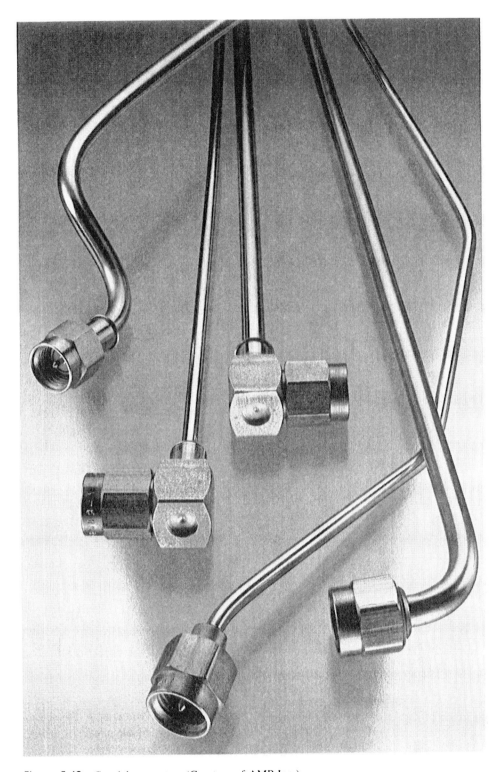

**Figure 5.42**   Coaxial connector. (Courtesy of AMP Inc.)

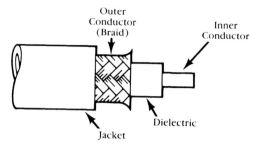

**Figure 5.43**   The basic coaxial cable. (Courtesy of AMP Inc.)

steel or beryllium copper. The circuit is AC or DC powered. When powered, the heater causes the memory alloy element to move to its original flat shape, facilitating the removal or insertion of the circuit board. When the heater is unpowered the connector closes, engaging the contacts with a high normal force.

## 5.8   CONNECTOR PROPERTIES

The most important properties to consider when selecting a connector, in addition to cost, weight, and size, are the insertion and extraction forces, wiping forces, electrical impedance and resistance, and reliability.

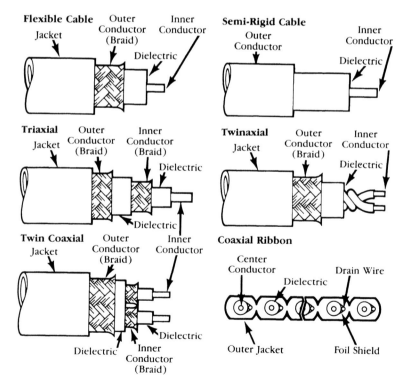

**Figure 5.44**   Various cable designs for coaxial cable. (Courtesy of AMP Inc.)

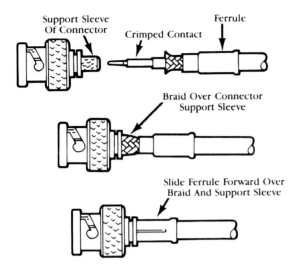

**Typical Assembly Procedure**
1. Slide ferrule on cable, then strip cable. Crimp center contact.
2. Insert contact into connector with braid over support sleeve.
3. Slide ferrule over braid and crimp ferrule.

**Figure 5.45** Soldering and crimping procedure for the co-axial cable. (Courtesy of AMP Inc.)

## 5.8.1 Insertion and Extraction Forces

Contact between connector contacts generally occurs in the plane of the mating surfaces. The insertion force, $F$, is given by

$$F = 2N \frac{\sin \alpha + \mu \cos \alpha}{\cos \alpha - \mu \sin \alpha} \tag{5.1}$$

where $\alpha$ is the angle of approach, $\mu$ is the coefficient of friction, and $N$ is the normal force at the contact point.

Once the contact reaches the flat parallel part of the pin, the angle of approach becomes zero and the insertion force reduces the simple frictional force. Figure 5.50 is a plot of insertion force versus contact engagement distance. The insertion force peaks where the flat and the inclined portion meet and then drops to simple frictional force as the spring member moves along the top of the connector pin.

For ease of engagement, the insertion force is designed large enough to break the insulating oxide film on the metal surfaces to make good electrical contact. The insertion force is determined by the lubrication of the connectors (so $\mu$ is kept small) and reduction of the contact angle $\alpha$. For a well-lubricated surface, the friction coefficient can be as low as 0.1, although it may run as high as 0.74 for an unlubricated surface.

Extraction forces are lower than insertion forces because the spring contact is moved along the flat portion of the pin to separate the two-part connector. Once the contact is on the tip incline, the remaining normal forces are sufficient to overcome contact friction and separate the mating surfaces. The extraction force, $F_e$, is

$$F_e = 2\alpha N \tag{5.2}$$

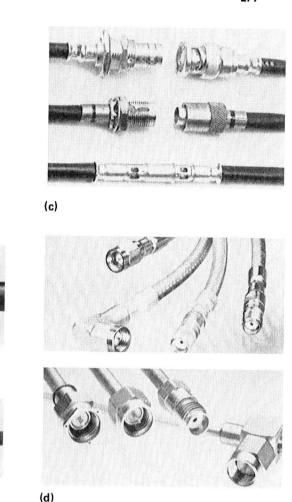

**(c)**

**(a)**

**(b)**                               **(d)**

**Figure 5.46(a–d)** Commercial coaxial connectors. (Courtesy of AMP Inc.)

## 5.8.2 Wiping Forces

A film of oxide, oil, and dirt may be present on the contact surfaces prior to mating. This film will decrease the electrical contact between the two mating surfaces and can prevent electrical transfer. When contacts slide against each other during contact mating, the surfaces can be cleaned by a sufficiently large wiping force between the contacts. Wiping action scrapes the surface debris, providing a clean surface for the contact. A connector is usually designed to provide the normal force for the wiping action. The contact surface area of a connector is usually based on cost, noise, size, and reliability. The magnitude of the wiping forces depends on the thickness and strength of the film, the surface plating materials, the degree of environmental cleanliness, and the presence (or absence) of lubricants. Experiments [20] show that wiping forces of less than 30 grams do not break through the common oxide films.

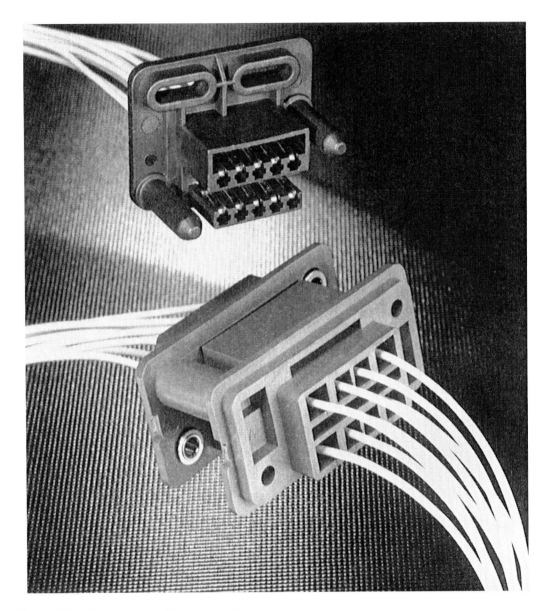

**Figure 5.47**   LIF connector. (Courtesy of AMP Inc.)

### 5.8.3   Electrical Impedance and Resistance

The impedance of a circuit is a measure of its opposition to current flow when an AC voltage is applied across it. Impedance matching is an important subject in connector design, since mismatched impedance can seriously degrade the integrity of the signal.

The resistance of a connector consists of two parts: bulk resistance and contact resistance. Bulk resistance is the resistance in the bulk of the material and constitutes a large part of the total resistance.

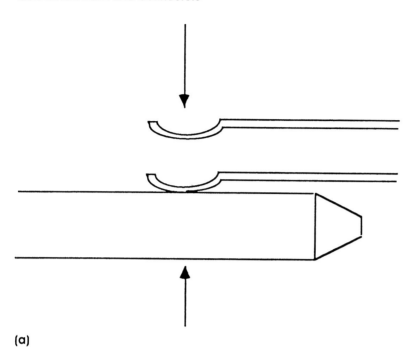

(a)

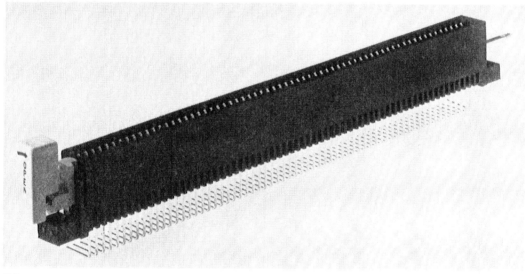

(b)

**Figure 5.48**   (a) ZIF contact engagement  and (b) ZIF connector. (Courtesy of AMP Inc.)

The contact resistance of a connector is composed of constriction resistance and film resistance, both of which impede the current flow. Constriction resistance is associated with the number and size of contact spots between two surfaces in contact. Film resistance is associated with the oxide film accumulated on the surfaces of two bodies in contact. Contact resistance is independent of the contact size and is dominated by constriction

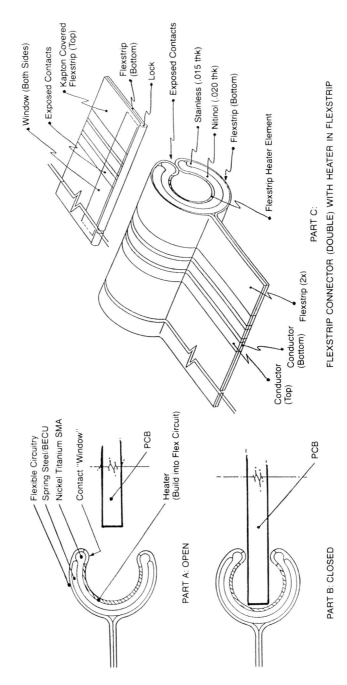

**Figure 5.49** Schematic diagram of a multiple-contact shape memory connector. (Courtesy of Beta Phase Inc.)

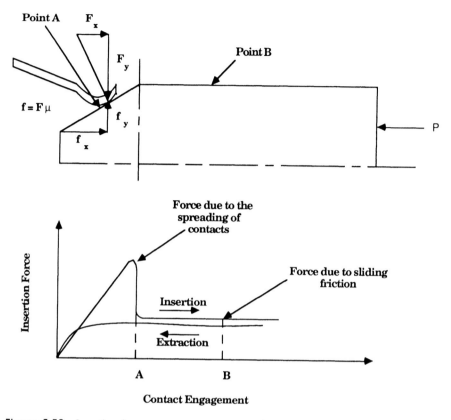

**Figure 5.50** Insertion force vs. contact engagement.

resistance of the tiny areas between the two connector halves. Figure 5.51 shows the variation of connector resistance with contact force.

Holm [45] presented a model for the total contact resistance $R_c$ as a function of the alloy and film resistivity, the number and area of the contact points, the hardness of the mating members, and the applied force:

$$R_c = R_{constr} + R_{film}$$

$$= \frac{\rho}{2} \sqrt{\pi H/nH} + \sigma H/N$$

(5.3)

where $\rho$ is the resistivity ($\Omega - $ m), $N$ is the load (N), $H$ is the hardness of the metal (N/ $m^2$), $n$ is the number of circular contact spots, and $\sigma$ is the film resistance ($\Omega^2$). The total resistance of a mated metal contact is generally a few milliohms. For practical purposes, 0.05 $\Omega$ may be used as the maximum allowable resistance value in most design cases [35].

## 5.8.4 Contact Sequencing

Contact sequencing implies the making of connections in a particular sequence within one physical connector. It is desirable in some applications to bring power to the connector

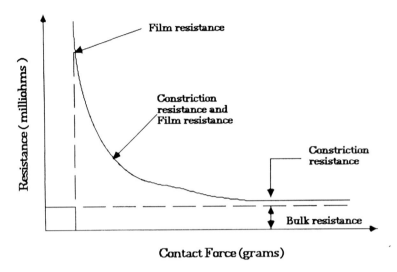

**Figure 5.51**  Contact resistance vs. contact forces.

in a particular order to prevent error or damage. For example, the ground connections are made before signal connections to protect against electrostatic discharge. Further, applications involving large currents encounter arcing while making or breaking the contact, resulting in degradation of the surface. For this reason the current flow is made or broken sequentially as the connector is inserted, so that current drain builds up gradually. The final contact point for each connector is kept as far removed from the initial contact point as possible. A large resistance, which decreases to zero at completion of the connector actuation, is provided in series to limit current and associated arcing.

### 5.8.5  Insulation Resistance

Insulation resistance is specified for each connector for operation in a high-voltage environment. This includes a contact-to-contact resistance and a contact-to-housing resistance. Further, a current rating is specified for contacts bringing power to a component. The maximum current rating is measured by the maximum allowable temperature rise when the connector is carrying the maximum specified current.

### 5.8.6  Noise

High-performance advanced electronic systems impose stringent requirements on the performance characteristics of electronic signal connectors in terms of noise. The noise generated in connectors may be identified as coupled noise, reflected noise, or radiated noise. Coupled noise is associated with an undriven connector when the surrounding connectors are carrying maximally adding signals with fast rise times. Reflected noise is associated with the internal inductance and the capacitance of the connector. Reflected noise occurs when the signal-carrying paths on either side of the connector behave as a transmission line, and the connector appears as a discontinuity because of its inductance and capacitance. Radiated noise is associated with the connector's electromagnetic compatibility.

## 5.8.7 Delay

There is a delay associated with the propagation of a signal through the connector. In low-performance systems, this delay is due to the connector capacitance. In high-performance systems, in which the connector impedance is matched to the transmission line, the delay is just a propagation delay through the connector.

## 5.8.8 Retention

A connector has a retention mechanism to prevent vibration, shock, and gravity from affecting performance. A relative contact motion caused by vibration or thermal expansion could cause arcing to occur, resulting in degradation of the contact resistance and heating of the contacts. The retention mechanism may be a friction device or a locking device, depending on the application.

## 5.8.9 Modularity

In applications in which a range of similar connections is made, it is desirable to make the connector assembly modular. This facilitates accommodation of a wide range of connections at a minimal expense. Further, it reduces excessive tolerance buildup.

## 5.8.10 Alignment

The mechanical interface between the two connectors must provide accurate registration to ensure correct contacts. The connector is so designed that the operator provides the initial coarse alignment and the connector system guides the leads on a more precise path, eventually connecting the mating parts together in precise alignment.

The contact area is frequently a compromise between cost, noise, and system size for minimum contact size. The contact area determines the accuracy with which the two mating parts can be brought together. Larger contact area increases reliability because it permits redundant contacts.

## 5.8.11 Process Compatibility

Connectors are usually joined to PWBs at the same time that other system components are joined. The connector materials therefore should be capable of withstanding the process chemicals and temperatures. For example, in the case of a board bearing vapor-phase-soldered surface mount components, connectors should have a material capable of standing such an environment.

## 5.8.12 Corrosion

Corrosion of electrical connectors causes a large number of electrical failures. These can be minimized by (1) installing connectors in a horizontal position, (2) placing a loop in the wire going into the connector so that water would not flow into the connector along the wire, and (3) using inhibitors on connector pins and receptacle interior mating areas. Sliding of two contact surfaces can result in prowing, galling, and burnishing wear. Prowing is a tear dislocation or shearing of part of the surface by a contact moving on the surface;

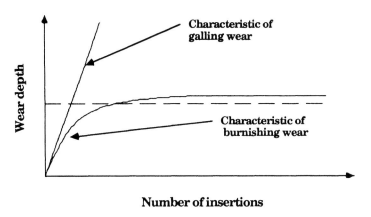

**Figure 5.52** Wear depth vs. number of insertions.

the tear adheres to the contact or is reattached to the surface and ridden by the contact again. Galling is an advanced stage of adhesive wear that tears up the surface. Burnishing wear is the compacting and smoothing of one surface by rubbing another surface against it. The objective of good connector design is to achieve burnishing wear.

Prowing and galling produce surface wear debris and work hardening of contact surfaces, thus changing the physical properties and topography of the surfaces. These types of wear result from an abnormally high contact force, poor lubrication, similar contact metallurgies of the mating surfaces, and rough and contaminated surfaces. Prowing and galling are undesirable for these reasons. Burnishing wear is enhanced by an optimum contact force, good lubricity at the contact interface, dissimilar contact metallurgies of the mating surfaces, and clean and smooth surfaces. Figure 5.52 shows the wear depth versus number of insertions for various types of wear.

## 5.9  FACTORS AFFECTING CONNECTOR RELIABILITY

The major factor which determines the overall reliability of a connector is time-dependent reliability—a measure of the failure rate of connectors relative to the length of time the connectors operate—is the predominant cause of connector failure. Connector service life is a function of the thickness of plating, ambient temperature, humidity, shock, and vibration. Product life is normally expressed in terms of the number of withdrawal/insertion cycles or in years of service life. Further, plug-dependent failures may be divided into two classes: those that recover with subsequent insertions and those that never recover, necessitating repair or replacement.

Connector reliability is a function of factors related to connector use, design, and manufacturing characteristics. These include circuit sensitivity, cleanliness, environment, twin contacts, and number of insertions of the contacts.

Tolerance of a circuit to high resistance decreases the probability of contact failure. The contacts are functional as long as their resistance is less than 1 $\Omega$. The causes of high resistance may include surface contaminants, such as solder flux, cover coat material, particles from plastic insulators, airborne particles, or excessive solid lubricants; corrosion films, such as oxides or sulfides formed at the contact finish or from substrate metals that are expressed or deposited at interface during connector wear; or mechanical damage that prevents sufficient contact force between contact members. The total resistance of a connector contact pair is typically in the range of 5 to 30 m$\Omega$. Defective connectors can often be made to work by unplugging and replugging them. This kind of reseating causes some contaminated contacts to decrease in resistance while causing others to increase, thus shifting the problem from one contact to another.

Dirt is another single cause of connector failure. Contamination during assembly operations may show up during equipment installation and testing and is a major contributor to infant mortality. Such contacts can usually be restored by cleaning.

The environment in which the connector operates determines the hazard rate of the connector. An office environment with air conditioning and a filtered air intake with controlled temperature and humidity would cause a lower hazard rate than an environment with no controls on temperature and humidity, airborne particulate matter, or any other environmental effects.

Twin contacts are often used to lower the probability of failure of contacts. In this case two separate contact areas are used instead of one, providing redundancy that lowers the probability of failure. Twin contacts are the most reliable design for critical applications.

Extended periods of exposure to the environment when left unmated cause the development of contamination or corrosion on the contact surface. Further, the probability of failure may be altered by the number of times the connector is plugged or unplugged.

Excessive temperature can cause a breakdown of insulation or a breakdown in the continuity of the conductors. Either of these two malfunctions can result in partial or total failure of the entire system. As the operating temperature increases, the electrical resistance of the conductor increases. The higher resistance causes the temperature of the conductor and its insulation to rise further. Complete breakdown occurs when the conductor melts or the insulation fails, leading to a short. Proper maintenance and common sense can help to prolong the connector service life. For instance, a connector is not a switch and should not be treated as such. Connectors are designed for continuous contact.

Connector metallurgy is probably the most important factor in reliable connectors. High-quality connectors usually have a noble metal such as gold or palladium plated over a diffusion barrier of nickel. The diffusion barrier works to prevent the connector base metal, usually a spring metal, from diffusing through the noble metal plating and contaminating the contact interface. Nickel diffusion barriers, usually 1.25 to 2.5 $\mu$m thick, are plated over the base metal. The contact interface metal must be thick enough to resist wear and corrosion during contact life. Gold or palladium on the order of 0.4 to 2.5 $\mu$m thick is normally used in high-quality contact systems. The exact thickness specified is determined by the particular application environment and the quality of plating of operation. Connectors designed for more corrosive environments must be sealed or have sufficient plating thickness to ensure a pore-free plating contact area. The porosity of the plating metal may require additional thickness at the contact interface. The thickness at the interface is also determined by the number of mate and unmate cycles. Even a small amount of relative motion at the contact interface can cause failures due to metal oxide formation.

## 5.10 CONTACT THEORY

Electrical contact theory was first developed in the early 1920s with Holm's research on the resistance to electrical current flow of two metallic bodies in contact [36]. When two metallic bodies are placed in contact with a low force normal to the contacting surfaces, contact is made at only a few protruding spots, known as asperities. As the force increases, the size and number of the asperities increase and the resistance decreases (Figure 5.51). The electrical resistance $R$ (ohms), of a material is

$$R = \frac{l}{\sigma A} \qquad (5.4)$$

where $l$ is the length of the conductor (cm), $\sigma$ is the electrical conductivity (ohm$^{-1}$ cm$^{-1}$), and $A$ is the cross-sectional area of the conductor (cm$^2$). Electrical resistance should be kept as low as possible. However, physical constraints such as the ease of mating will limit the allowable contact force. Normally the contact force per pin is about 1 to 2 and is seldom goes above 5 [37].

Metallic surfaces are often covered by a nonconducting layer such as an oxide film. When the thickness of the film is less than 2 to 3 nm [38], electrons can penetrate or tunnel through the nonconducting film. The tunnel effect has been studied extensively by Holm [36]. If the nonconducting layer on a metallic surface is discontinuous, current is able to flow through these discontinuities.

In a highly ideal case of a single circular contact spot with identical metals having a uniform film, the overall contact resistance is

$$R = R_c + R_F \qquad (5.5)$$
$$= \frac{\rho}{2r} + \frac{\sigma}{\pi r^2}$$

where $R_c$ is the constriction resistance (ohms), $R_f$ is the film resistance (ohms), $r$ is the radius of an asperity (cm) and $\rho$ is the bulk resistivity of the contact metal (ohm cm$^2$). For clean surfaces, the contact resistance is dominated by the constriction resistance $R_c = \rho/(2r)$, which is a function of temperature because both $\rho$ and $r$ depend on temperature. The resistivity $\rho$ decreases as the temperature is lowered, while the true contact area decreases because of thermal contraction. Matthiessen's empirical rule states that the total resistivity of a crystalline metallic specimen is the sum of the resistivity due to thermal agitation of the metal ions of the lattice and the resistivity due to the presence of imperfections in the crystal. The total resistivity, $\rho_{total}$ (ohm cm), is then given by

$$\rho_{total} = \rho_{th} + \rho_{imp} + \rho_{def} \qquad (5.6)$$
$$= \rho_{th} + \rho_{res}$$

where $\rho_{th}$ is the thermally induced resistivity, also known as the ideal resistivity; $\rho_{imp}$ is the resistivity due to imperfections in the crystal such as impurities; $\rho_{def}$ is the resistivity due to defects in the crystal; and $\rho_{res}$ is the residual resistivity.

Residual resistivity, $\rho_{res}$, arises from imperfections in the crystal and is independent of ambient temperature. $\rho_{th}$ is an intrinsic property of the material and cannot be altered. However, it is possible to lower the total resistivity by reducing the residual resistivity, $\rho_{res}$, through metallurgical processes such as grain growth and recrystallization to reduce imperfections of the crystal.

## REFERENCES

1.  Ching, T. B., and Schroen, W. H., "Bond Pad Structure Reliability," in *International Reliability Physics Symposium*, IEEE, 1988, pp. 64–70.
2.  Howell, J. R., and Slemmons, J. W., "Evaluation of Thermocompression Bonding Processes,"

presented to the 9th Welded Electric Packaging Association Symposium. Santa Monica, Calif. February 27, 1964; Autonetics Report No. T4-240/3110, March 1964, pp. 16, 19, 27, 28, 30.

3.  Hill, P., "Uniform Metal Evaporation," *Proc. Conference on Reliability of Semiconductor Devices and Integrated Circuits*, Vol. 2, Sect. 27, pp. 1–27, June 1964.

4.  Philofsky, E., "Design Limits When Using Gold-Aluminum Bonds," in *9th Annual Proc. Reliability Physics Symposium*, Las Vegas, IEEE, 1971, pp. 17, 18, 22.

5.  Schafft, H. A., "Testing and Fabrication of Wire Bonds Electrical Connections—A Comprehensive Survey," *Natl. Bur. Stand. (U.S.) Tech. Note 726*, pp. 80 and 106–109 (1972).

6.  Harper, C. A., ed., *Handbook of Materials and Processes for Electronics*, McGraw-Hill, New York, 1970.

7.  Harper, C. A., ed., *Handbook of Components for Electronics*, McGraw-Hill, New York, 1977.

8.  English, A. T., "Studies of Bonding Mechanisms and Failure Modes in Thermocompression Bonds of Gold Plated Leads to Ti-Au Metallized Substrates," in *International Reliability Physics Symposium*, IEEE, pp. 178–186.

9.  Lang, B., and Pinamaneni, S., "Thermosonic Gold Wire Bonding to Precious Metal Free Copper Leadframes," in *38th Electronic Component Conference*, IEEE, pp. 546–551, 1988.

10. Weiner, J. A., and Clatterbaugh, G. V., "Gold Ball Bond Shear Strength—Effects of Cleaning, Metallization and Bonding Parameters," IEEE, 1983.

11. Poonawala, M., "Evaluation of Gold Wire Bonds in IC's Used in Cannon Launched Environment," in *Electronic Component Conference*, IEEE, 1983, pp. 189–192.

12. Winchell, V. H., "An Evaluation of Silicon Damage Resulting from Ultrasonic Wire Bonding," 14th Annual Proceedings Reliability Physics Symposium, Las Vegas, Nevada, 1976, pp. 98–107.

13. Koch, T., Richling, W., Whitlock, J., and Hall, D., "A Bond Failure Mechanism," in *International Reliability Physics Symposium*, IEEE, 1986, pp. 55–60.

14. "Technical Report for Reliability Analysis/Assessment of Advanced Technologies," Rome Air Development Center, Griffin AFB, NY, 13441, (1989).

15. Lall, P., Barker, D., Dasgupta, A., Pecht, M., and Whelan, S., "Practical Approaches to Microelectronic Package Reliability Prediction Modelling," *ISHM Proc. 1989*, pp. 126–130, 1989.

16. Lall, P., Pecht, M., and Dasgupta, A., "A Failure Prediction Model for Wire Bonds," *ISHM Proc. 1989*, pp. 607–612, 1989.

17. Harman, G. G., "Metallurgical Failure Modes of Wire Bonds," *12th International Reliability Physics Symposium*, IEEE, 1974, pp. 131–141.

18. Panousis, N. T., and Bonham, H. B., "Bonding Degradation in Tantalum Nitride Chromium-Gold Metallization System," in *11th Annual Proceedings IEEE Reliability Physics Symposium*, IEEE, pp. 21–25, (1973).

19. Cunningham, J. A., "Expanded Contacts and Interconnections to Silicon Monolithic Integrated Circuits," *Solid State Electron.* 8:735–745 (1965).

20. ASME Handbook, *Metals Properties*, McGraw-Hill, New York, 1954.

21. Forrest, N. H., "Reliability Aspects of Minute Amounts of Chlorine on Wire Bonds Exposed to Pre-Seal Burn-in," Martin Marietta Aerospace Corp., Denver, pp. 549–551, (1982).

22. Shukla, R., and Deo, J. S., "Reliability Hazards of Silver-Aluminum Substrate Bonds in MOS Device, in *International Reliability Physics Symposium*, IEEE, 1982, pp. 122–127.

23. Koyama, H., Shiozaki, H., Okumura, I., Mizngashira, S., Higuchi, H., and Ajiki, T., "A New Bond Failure Wire Crater in Surface Mount Device, in *International Reliability Physics Symposium*, IEEE, 1988, pp. 59–63.

24. Moore, K. D., "Interconnection Failures in Circuit Assemblies," in *Electronic Component Conference*, IEEE, 1988, pp. 521–526.

25. Coucoulas, A., "Ultrasonic Welding of Aluminum Leads to Tantalum Thin Films," *Trans. Metall. Soc. AIME 236*:587–589 (1966).

26.  Gaffney, J., "Internal Lead Fatigue Through Thermal Expansion in Semiconductor Devices,"
     *IEEE Trans. Electron. Devices ED-15*: 617 (1968).
27.  Villela, F., and Nowakowski, M. F., "Investigation of Fatigue Problems in 1 Mil Diameter
     Thermocompression and Ultrasonic Bonding of Al wire," *NASA Tech. Memo. TM-X-64566*
     (1970).
28.  Nowakowski, M. F., and Villela, F., "Thermal Excusion Can Cause Bond Problems," in
     *9th Annual Proceedings IEEE Reliability Physics Symposium*, IEEE, 1971, pp. 172–177.
29.  Ravi, K. V., and Philofsky, E. M., "Reliability Improvement of Wire Bonds Subjected to
     Fatigue Stresses," in *10th Annual Proceedings IEEE Reliability Physics Symposium*, 1972,
     pp. 143–149.
30.  Phillips, W. E., ed., Microelectronic Ultrasonic Bonding," *Natl. Bur. Stand. (U.S.) Spec.
     Publ. 400-2*, pp. 80–86, 1974.
31.  Adams, C. N., "A Bonding Wire Failure Mode in Plastic Encapsulated Integrated Circuits,"
     in *11th Annual Proceedings Reliability Physics*, IEEE, 1973, pp. 41–44.
32.  Mantese, J. H., and Alcini, W. V., "Platinum Wire Wedge Bonding: A New IC and Mi-
     croprocessor Interconnect, *J. Electron. Mater. 17* (4):(1988).
33.  Smith, J. M., et al., "Hybrid Microcircuit Tape Chip Carrier Materials/Processing Trade-
     offs," *IEEE Trans. Parts Hybrid Packag. PHP-13* (3):257–268 (1977).
34.  Thompson, R. J., Cropper, D. R., and Whitaker, B., Boundability Problems Associated with
     the Ti-Pt-Au Metallization of Hybrid Microwave Thin Film Circuits, in *31st Electronic Com-
     ponent Conference*, IEEE, 1981, pp. 1–8.
35.  Lawrence, R., *Electrical/Electronic Interconnection Systems, a Guide to Connector Design
     and Techniques*, 4th ed., p. 134, Deutsch Company, Banning, Calif., 1979.
36.  Holm, R., *Electric Contacts, Theory and Applications*, 4th ed., Springer-Verlag, New York,
     1967.
37.  *Connectors and Interconnection Handbook*, Vol. 1, Electronic Connector Study Group, Phil-
     adelphia, 1978.
38.  Scroggie, M. G., and Johnstone, G. G., *Radio and Electronic Laboratory Handbook*, 9th
     ed., Newnes-Butterworth, London, 1980.
39.  Winchell, V. H. and Berg, H. M., "Enhancing Ultrasonic Bond Development," IEEE Trans-
     actions on Components, Hybrids and Manufacturing Technology, CHMT-1, 1978, pp. 211–
     219.
40.  Hirota, J., Machida, K., Okuda, T., Shimotomai, M. and Kawanaka, R., "The Development
     of Copper Wire Bonding for Plastic Molded Semiconductor Packages," 35th Electronic Com-
     ponent Conference Proceedings, Washington DC, 1985, pp. 116–121.
41.  Ravi, K. V. and White, R., Reliability Improvement in 1-mil Aluminum Wire Bonds for
     Semiconductors," Final Report, Motorola SPD, NASA Contract NAS8-26636, December 6,
     1971.
42.  Kale, V. S., "Control of Semiconductor Failures Caused by Cratering of Bond Pads," Pro-
     ceedings of the 1979 International Microelectronics Symposium, Los Angeles, California,
     1979, pp. 311–318.
43.  Philosky, E. M., and Ravi, K. V., "On Measuring the Mechanical Properties of Aluminum
     Metallization and Their Relationship to Reliability Problems," 11th Annual Proceedings of
     Reliability Physics Symposium, pp. 33–40, 1973.

## SUGGESTED READING

White, D. R. J., EMC Handbook, Vol. 3, *Electromagnetic Interference and Compatibility*, Inter-
    ference Control Technologies, Gainsville, Va. 1986.
Bilotta, A. J., *Connections in Electronic Assemblies*, p. 188, Marcel Dekker, New York, 1985.
Taylor, R. N., "High Density Connectors That Maintain a Low Profile," *15th Annual Connectors
    & Interconnection Technology Symposium Proc.*, 1982.
Lightner, L. S., "A Module Zero Insertion Force Cable Connector," *Proc. IEEE 25th Electronic
    Components Conference*, pp. 190–194, 1975.

MIL-C28840A(EC), Military Specification, "General Specification for Connectors, Electrical, Circular Threaded, High Density, High Shock Shipboard," Class D, Naval Electronic Systems Command, Department of Navy, Washington, D.C., 1981.

Mazda, F. F., *Electronic Instruments and Measurement Techniques*, pp. 237–238, Cambridge University Press, Cambridge, 1987.

Aujla, S., and Wiltshire, B., "Connector Insertion Force Characteristics," *Proc. 31st IEEE Holm Conference on Electrical Contacts*, IEEE, 1985, pp. 169–174.

Evans, C. J., "Connector Finishes: Tin in Place of Gold," *IEEE Trans. Components, Hybrids Manuf. Technol. 1 ChMT-3* (2):226–232 (1980).

Fung, Y. C., *Foundations of Solid Mechanics*, Prentice-Hall, Englewood Cliffs, N.J., 1965.

Zienkiewicz, O. C., *The Finite Element Method*, McGraw-Hill, New York, 1977.

Lee, J. D., Three Dimensional Finite Element Analysis of Damage Accumulation in Composite Laminate, *Comput. Struct. 15*(3):(1982).

Hinton, E., and Owen, D. R. J., *Finite Element Programming*, Academic Press, New York, 1977.

Solomon, H. D., "Influence of Hold Time and Fatigue Cycle Wave Shape on the Low Cycle Fatigue of 60/40 Solder," *38th Electronic Component Conference*, IEEE, 1988, pp. 7–12.

Philofsky, E. M., and Ravi, K. V., "On Measuring the Mechanical Properties of Aluminum Metallization and Their Relationship to Reliability Problems," in *11th Annual Proceedings of Reliability Physics*, IEEE, 1973, pp. 33–40.

## EXERCISES

1. What are the six levels of interconnection?
2. Justify the use of TAB in chip-to-lead-frame interconnections.
3. Outline the major connector properties.
4. What are the possible connector choices for
   (a) Chip-to-lead-frame interconnections
   (b) Chip-to-board interconnections
   (c) Board-to-board interconnections
   (d) Board-to-cabinet interconnections
5. List the various connector plating materials.
6. What are the properties for a good connector plating material?
7. What are the factors affecting connector reliability?
8. How does application influence the choice of plating material?
9. List the various connector materials.
10. What properties should a good contact material possess?
11. What is meant by polarization in IC sockets?
12. What is meant by mismatched impedance and how does it effect connector performance?
13. What parts is connector resistance compound of, and how does it vary with contact engagement?
14. Explain the contact theory for interconnections.
15. List the various types of wear observed in connectors. Which type of wear is desirable and why?
16. What is the importance of wiping forces in contact engagement?
17. What is meant by contact sequencing?
18. What are the various manufacturing processes for wire bonds?
19. List the various failure modes and mechanisms for wire bonds.
20. What are gold-aluminum intermetallics?
21. List the various bonds types.
22. List the factors affecting wire bond integrity.

# 6
# Layout

**Michael D. Osterman and Michael Pecht** *CALCE Center for Electronic Packaging, University of Maryland, College Park, Maryland*

The layout process is the link between the development of the circuit logic and the development of the conditions necessary for the system to be physically realized. Layout requires an understanding of the specific electronic technology being employed, the electrical constraints imposed on the circuitry, the physical properties of the workspace, and the tools available for manufacturing. The layout process is typically divided into partitioning and assignment, placement, and routing processes.

In the partitioning process, the logic diagram is partitioned into logic elements or groups of logic elements that are used to select electronic modules which perform the desired functions. In addition, the signal sets, which define specific component pin groupings, must be defined. The assignment process begins once the list of electronic modules has been established. Groups of modules are assigned to workspaces (i.e., components are assigned to printed wiring boards, boards are assigned to assemblies, etc.), which provide a physical support for the modules and a base on which the interconnection paths are laid out. Once the signal sets are determined, the modules must be located (placed) within the workspace.

After a placement configuration is determined, the final step in the layout process is routing. In this step, the paths required to interconnect the signal sets of the modules are laid out. With good placement, routing strategies have advanced to a point of achieving nearly complete solutions. Without good placement, routing algorithms often struggle and the effects on reliability, maintainability, and producibility can be dramatic. Although automated placement has been incorporated into layout systems, placement remains a major weakness in the total effectiveness of these systems.

This chapter begins by reviewing partitioning and assignment of logic elements. Several placement techniques for routability and reliability are then developed. Methods of combining placement for reliability and routability and placement for producibility are also reviewed. Finally, routing is addressed and some of the standard interconnection techniques employed in interconnecting components are discussed.

## 6.1 PARTITIONING AND ASSIGNMENT

Electronic systems are typically made up of a large number of devices that are arranged into subassemblies. Previously, discrete devices such as transistors, resistors, capacitors, and inductors soldered to a printed wiring board encompassed a subassembly. Today, the integrated circuits are themselves subassemblies of an electronic system and the multilayer board is a higher-level assembly. At both levels, the separation of devices into functional groups for optimum design is one of the many design activities that constitute layout. The activities involved in the separation of devices into particular groupings falls under partitioning and assignment.

Partitioning is the separation of electrical functions and their assignment to various building blocks to satisfy the design requirements. One objective of partitioning may be to make efficient use of available planar space. Another could include minimization of lead lengths or external connections. In addition, isolation of heat or electromagnetic field sources is a possible partitioning objective. Other objectives of partitioning include determining the module size of integrated circuits; finding the maximum number of repeatable modules; testability; reduction of internal connections; and the separation of system circuits into separate modules based on cost, maintainability, and reliability considerations.

Partitioning is a critical step in the design of electronic systems and directly affects such factors as cost, performance, maintainability, and reliability. Partitioning can facilitate design by defining subsections of the design that can be developed concurrently by different designers. A reduction in manufacturing costs can also be achieved with separate sections or modules. Partitioning operations can generally be considered to fall under one of two modes: vertical and horizontal.

Vertical partitioning focuses on separating the electronic design hierarchically into different levels of sophistication or complexity. For example, the lowest level of design generally includes the discrete devices and monolithic integrated circuit, which are the fundamental building blocks of the electronic system. The next level may be the printed wiring board or thin-film substrates, which consist of a large number of circuit elements. The printed wiring boards or thin-film packages are combined to form system units. Combinations of various units form subsystems or the total system. Vertical partitioning is concerned with the number of modular levels of a system and the modular size of each level. Consideration focuses on interfacing, connections, cost, maintainability, reliability, and performance. In addition, the throwaway level at which the cost to repair is greater than the cost to replace must be considered. As the cost to provide skilled technicians to locate and repair inoperative elements increases, the size of the economical throwaway module increases. Furthermore, the development of internal diagnostic elements versus the cost of the system and the criticality of the system must be considered.

Horizontal partitioning focuses at each module level and the division of circuits among the modules of that level. Determining the maximum number of repeatable circuit element groupings is one important activity in horizontal partitioning. Determining a standard set of gates and functions to be implemented by integrated circuits and selecting standard integrated circuits are other important activities. The related horizontal and vertical partitioning activity of establishing the size of the lowest-level module that will appear in the largest numbers strongly influences system cost and reliability.

## 6.2 PLACEMENT

Solving the placement problem involves positioning modules on a workspace with respect to some measure of success while observing constraints and guidelines. Good placement configurations meet performance requirements while providing sound thermal management, high reliability, minimization of board layers and signal paths, and ease and neatness of producibility at an acceptable cost.

Historically, placement techniques have been developed primarily on the basis of routability. Placement is also based on electrical, thermal, and mechanical guidelines. Electrical guidelines include minimizing the total wire length on the workspace, restricting the maximum length of any routing path, clustering functionally related modules to conform to speed and transmission line requirements, designing signal paths to meet required electrical characteristics (i.e., resistance, current capacity, and capacitances), keeping a strict tolerance on congestion, and keeping analog and digital functions shielded to prevent cross talk. Thermal guidelines include eliminating high thermal gradients on the workspace, placing components in such a way as to allow for adequate spreading of the heat being dissipated, ensuring that all component junction temperatures are kept well within their individual design specifications, and avoiding critical system failures based on the temperature-sensitive failure rates of the individual modules. Vibrational and thermally or mechanically induced stresses that lead to fatigue, cracking, and other related failures must also be avoided. Furthermore, with the demand for high reliability (performance over time), designers have been required to investigate techniques for improving placement based on the reliability characteristics of the board and components. The goal is to improve reliability by decreasing the total failure rate of the system. Overall, the placement configuration should reduce the life cycle cost of manufacturing and maintenance in the field while providing performance over time.

To achieve good placement, design engineers must be aware of the characteristics of individual modules; the logic design of the system; the electrical constraints; the thermal control being employed; physical constraints placed on the size, shape, and weight of the board; manufacturing processing; the environmental characteristics to which the electronic system will be exposed; and maintenance procedures. With the complexity of modern electronic designs that consist of hundreds of modules, the placement of modules involves difficult decision making that is often beyond the expertise of the design engineer. To incorporate all of the desired attributes of placement into a coherent mathematical model is extremely difficult. In addition, the mathematical models tend to be unaccommodating to standard problem-solving techniques. Thus, heuristic approaches are often employed. In addition, computer-aided design (CAD) techniques have been developed to help the design engineer.

Placement techniques are often categorized as continuous and noncontinuous. Continuous techniques treat the workspace as a continuous surface on which the modules are free to reside. Noncontinuous techniques partition the workspace into slots into which the modules are assigned. The noncontinuous approach is extremely useful for workspaces in which most of the modules to be placed are of similar shape and size, which simplifies the geometric problems encountered in placement.

All placement algorithms employed in the solution of the placement problem, whether continuous or noncontinuous, fall into two classes: constructive or iterative. Constructive procedures are typically applied to establish an initial placement configuration. Iterative techniques operate on the initial placement configuration, which is generated either randomly or by constructive procedures, and seek to improve that placement configuration.

Placement techniques operate on some set of measurable figures of merit. Figures of merit include routability, reliability, producibility, costs, and time to generate a placement.

Routability is a measure of the ease and neatness of specifying interconnection paths on a workspace. The total wirelength is an accepted estimate of the routability. Wirelength is dependent on the type of interconnection tree being employed. The interconnection tree is made up of vertices (connection points) that are connected by edges (wires). The most common interconnection tree for printed circuit boards (PCBs) or printed wiring boards (PWBs) is the minimum spanning tree, which is the interconnection network of minimum length (sum of the lengths of all edges) with vertices restricted to the pins of the individual signal set. The most common measure of distance used in layout is the Manhattan or rectilinear distance.

Reliability prediction estimates the life expectancy of the system or the performance of the system over time. It is affected by the quality of design and manufacture, thermal and mechanical stresses, and the environment. Reliability prediction equations for electronic components consider electrical, mechanical, and thermal factors that affect the life expectancy of the individual components. The failure rate is used to measure reliability by statistically measuring the number of failures over a given period of time at specific operating conditions. The total failure rate of a system is dependent on the logic configuration of the system. For a system that operates serially, the total failure rate is simply the sum of the failure rates of the individual modules of which the system is composed.

Producibility is a measure of how well the final layout adapts to the manufacturing and assembly technologies available for production. For example, array layouts with a majority of similar module types and shapes will reduce the cost of manufacturing significantly. For PWB design, minimizing the number of vias, minimizing the number of routing layers, and aligning components in rows on the board for automated placement help reduce the cost of manufacturing.

Associated with a given placement technique are the computer and computer operator cost to determine the placement, the cost to manufacture and assemble the product, and the cost to test. These life cycle costs should not be ignored, especially if the percentage gain in the other figures of merit is small.

## 6.3  PLACEMENT FOR ROUTABILITY

Placement for routability is generally based on minimizing a figure of merit or "cost" function that is related to the interconnection norm being employed, the congestion of interconnection tracks, and the total circuit area. Restrictions on component spacing, track width, and track spacing must also be satisfied.

The most common interconnection norm is the total wirelength. A common distance function used to measure wirelength is the Manhattan or rectilinear distance. For two connections located at positions $(x_1, y_1)$ and $(x_2, y_2)$, the Manhattan distance is

$$d_{12} = |x_1 - x_2| + |y_1 - y_2| \tag{6.1}$$

Another distance function that can be used is the Euclidean distance,

$$D_{12} = \sqrt{(x_1 - x_2)^2 + (y_1 - y_2)^2} \qquad (6.2)$$

For a planar workspace, such as a PWB, the out-of-plane wiring distance $z_1 - z_2$ is usually neglected, even if the PWB is multilayered. It is essential when positioning components to have an understanding of the interconnection rules. Nets used to interconnect pins of the signal sets are generally dependent on the technology being employed. For example, emitter-coupled logic (ECL) typically allows only two or three wires to be connected to any one pin. In other cases, it is required that interconnections initiate from a source or driver pin and connect to all other pins in the signal set.

In graph theory, the pin locations of a signal set represent nodes or vertices. A branch or edge can be formed between any two nodes. For $N$ nodes, a complete graph requires $N(N - 1)/2$ edges. The interconnection net is defined as a tree based on the nodes. A tree is a subset of the total graph interconnecting the nodes. Whereas a complete graph interconnects every node with every other node, a tree does not. A tree requires $N - 1$ edges. Trees employed in the interconnection schemes include the minimum spanning tree, the Steiner tree, and special trees.

The minimum spanning tree is defined as the tree of minimum length (sum of all branches) that employs only the $N$ pin locations of the signal set as nodes. An example of this tree is depicted in Figure 6.1(a). The minimum spanning tree distance is typically

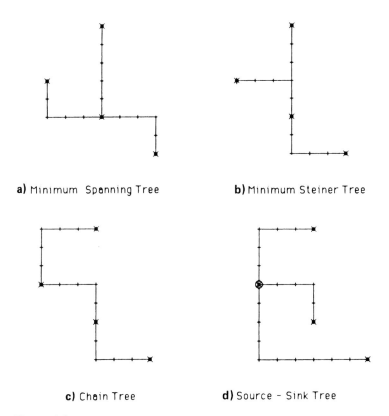

a) Minimum Spanning Tree   b) Minimum Steiner Tree

c) Chain Tree   d) Source – Sink Tree

**Figure 6.1** Connection tree types.

employed as a measure of a placement procedure's success. The Steiner tree interconnection graph employs the pin locations and, if necessary, auxiliary positions as nodes to minimize the total length of the tree. An example of the Steiner tree is shown in Figure 6.1(b).

Special-purpose trees typically satisfy topographical restrictions on the number of branches allowed to connect to any one pin or electrical requirements on the signal paths. For example, a chain tree is the minimum path that has at most two branches attached to any one node (Figure 6.1(c)). A source-sink tree (Figure 6.1(d)) requires that signal source pins connect all sink pins in the signal set with minimum length. The particular tree used in the routing of the board will determine the placement algorithm for routability.

Placement algorithms for routability generally require that the input information consist of a set of components $E = E_1, E_2, \ldots , E_N$, a set of interconnection sets $S = S_1, S_2, \ldots , S_m$ (i.e., signal sets of net lists), the physical dimensions of the board, and engineering restrictions such as the minimum signal distance between components. The component dimensions and number of pins are specified in the component list $E$. Each set of the interconnection set contains at least two components and their associated pin numbers that have to be connected. In addition, each element, $S_i$, of the interconnection set may carry an interconnection weight $\rho_i$, which designates the importance of the interconnection. From the interconnection set, $S$, a connectivity matrix, $C$, can be formed. The elements of the connectivity matrix, $C_{ij}$, specify the number of interconnections and the connection strength between components $i$ and $j$. The definition and particular use of the connectivity matrix can define a particular placement algorithm. Steinberg [1] and Hanan and co-workers [2–4] pose the noncontinuous placement problem of finding an arrangement $s$ such that the total wirelength

$$W(s) = \sum C_{ij} d_{s(i)s(j)} \tag{6.3}$$

is minimum.

In the noncontinuous case, where the board workspace is divided into slots, the problem reduces to the classical quadratic assignment problem. For $n$ components, there are $n!$ possible assignments that must be examined to determine the arrangement that yields the minimum value of $W$. Engineering restrictions may be employed to reduce the number of possible assignments. Various placement algorithms and heuristics have been developed to reduce the size of the problem.

The connectivity matrix is symmetric ($C_{ij} = C_{ji}$) and $C_{ii} = 0$. The connectivity matrix entries, $C_{ij}$, can be defined as the number of connections between components $i$ and $j$. If $\delta_k(E_i,E_j) = 1$ when $E_i$ and $E_j$ are both elements of $S_k$, and $\delta_k(E_i,E_j) = 0$ for any other circumstance, then

$$C_{ij} = \sum_{k=1}^{m} \delta_k(E_i,E_j) \tag{6.4}$$

where $m$ is the total number of interconnection sets.

In some cases, a weighted connectivity matrix may be desired. Under these circumstances, each element of the connectivity matrix is defined as

$$C_{ij} = \sum_{k=1}^{m} \rho_k \delta_k(E_i,E_j) \tag{6.5}$$

where $\rho_k$ is an interconnection weighing factor.

In some circumstances, the number of components in each interconnection set is considered important for placement considerations. Hanan and Kurtzberg [3] define connectivity matrix elements of the form

$$C_{ij} = \sum_{k=1}^{m} \frac{J_k + \lambda}{J_k} \rho_k \delta_k(E_i, E_j) \tag{6.6}$$

or

$$C_{ij} = \sum_{k=1}^{m} \frac{J_k + \lambda}{J_k} \delta_k(E_i, E_j) \tag{6.7}$$

where $\lambda$ is an integer greater than or equal to $-1$ and $J_k$ is the size of each interconnection set. Quinn and Breuer [5] defined the elements of the connectivity matrix to be

$$C_{ij} = \sum_{k=1}^{m} \frac{2}{J_k} \delta_k(E_i, E_j) \tag{6.8}$$

which is a fraction of the $J_k(J_k - 1)/2$ possible edges of the complete graph of $J_k$ nodes. The connectivity matrix may also simply be represented by 1's if any connection exists and 0's for no connection.

The connectivity matrices are sometimes employed to determine other measures of connectivity. For example, seed components for cluster placement procedures may be determined by calculating the overall connectivity of any one component. This is done by calculating the connectivity strength of each component determined by

$$C_i = \sum_{j=1}^{n} C_{ij} \tag{6.9}$$

where $n$ is the total number of components. The connectivity matrix used to determine $C_i$ depends on the placement routine being employed. In any case, components with the highest values of $C_i$ are generally used as seed elements in cluster development–type placement procedures.

## 6.3.1 Constructive Placement Methods

In constructive placement methods for routability, components from a set of unplaced components are selected sequentially based on their connectivity with respect to placed and unplaced components. The selected component(s) is then placed on the board. The choice of selection and positioning rules defines the specific placement method. Generally, constructive placement is of the noncontinuous type. The slot size is determined by the predominant or average size of the components to be placed plus additional space required for tolerance restrictions. Adjustments may be required for irregular and large components.

If the final placement configuration produced by constructive placement techniques is employed as the initial placement configuration used by iterative improvement schemes, consideration should be given to the time required to perform constructive placement. The following two constructive placement procedures developed by Hanan and Kurtzberg [3] provide satisfactory placement in a reasonable short amount of time.

## Pair-Linking Method

Hanan and Kurtzberg's pair-linking method [3] is based on the connectivity matrix determined by Eq. (6.7). Placement is initiated by selecting the pair of components that has the largest entry in the connectivity matrix, $C$. A subset of components may also be preassigned positions manually (i.e., edge connectors, test site connectors, irregularly shaped components) as a starting point.

If two or more entries in the connectivity matrix have the same value, they are said to be tied. In the event of ties during the initial selection of components, all tying component pairs are placed in a tie table and the entries along the rows (or columns) of the tying components in the connectivity matrix, excluding the tying entries, are examined. Again, selection is based on the largest entry found. If this again results in a tie, another tie table is established, the search continues in the manner denoted by the first tie, and all entries that brought about the ties are excluded. This process is a means of looking ahead to future selections.

Another tie resolution method sums the rows or columns of each tying pair. The pair with the larger summed value is selected for placement. The advantage is in the speed of calculation, since the matrix does not have to be searched and a record kept for each find. The drawback is that only first-level tie breaking is addressed. If another tie resulted from the tie resolution method, an arbitrary selection would occur between the pairs that have tied summed values.

After the initial pair of components is selected, the location of the components on the board must be determined. Generally, selected components are placed as near the center of the board as possible. The components of the selected pair are placed immediately next to each other, with the component belonging to the larger number of interconnection sets given the preferred position.

Once the initial components are placed, a repetitive selection and placement of unplaced components is performed. This is accomplished by examining the rows (or columns) in the connectivity matrix of the placed components and selecting the component from the unplaced set that has the largest entry.

In the event of ties, a search is conducted of the rows or columns corresponding to unplaced components that have tied, excluding the tying entries produced by components that have already been placed. All entries produced by placed components are ignored. In the event of another tie, a second tie table is set up, and only entries not associated with the previous tie table or any placed component are examined for the largest value.

An alternative approach to tie resolution requires that all columns or rows of the tying unplaced component are summed individually, excluding all entries resulting from placed components, and compared. The unplaced component with the larger summed value is selected. In case of another tie, an arbitrary selection is made between the unplaced components with tying sums.

After a selection is made, the selected unplaced component is positioned on the board. The selected component is placed as close as possible to the already placed component associated with it through the selection process. The growth in any direction of the placement configuration is restricted by the input boundary parameters $L$ and $W$ (length and width of the available area). If two or more positions are equally distant from the associated placed component, a weighted average of the locations of the previously placed components to which the selected component should be connected is calculated. The weighing factors for the placed component locations are given by the associated entries of the connectivity matrix. The approximate placement location $(\bar{x}_j,\bar{y}_j)$ for the selected component $j$ is given by

$$\bar{x}_j = \frac{\sum_{i\in A} C_{ij}x_i}{\sum_{i\in A} C_{ij}}$$

$$\bar{y}_j = \frac{\sum_{i\in A} C_{ij}y_i}{\sum_{i\in A} c_{ij}} \tag{6.10}$$

where $A$ is the set of placed components and the connectivity matrix is defined by Eq. (6.7). The selected component is placed in the open location as close to $(\bar{x}_j,\bar{y}_j)$ as possible. In the case of ties between open locations, the open location with the most surrounding open locations is selected. The processing time of this particular placement method is a direct function of the number of components and the total number of interconnection sets.

To illustrate the pair-linking method, consider 10 components joined by five signal sets. The signal sets are defined in Table 6.1. It should be noted that the signal set depicted in Table 6.1 is really incomplete. A true signal set would specify component and pin numbers for each signal set. For brevity and simplicity, the actual pin numbers will be neglected in all signal sets given in this chapter. Wirelength will be approximated by center-to-center distances between connected components. Letting $\lambda = -1$, the entry in the connectivity matrix for ((U2)(U7)) is calculated as an example in Figure 6.2. The complete connectivity matrix is given in Table 6.2. The components are to be placed into a 3 by 4 array of slots as shown in Figure 6.3. Tie resolution is performed by summing the rows or columns. The largest entry, 2.13, in the connectivity matrix corresponds to components (U2,U7), and these two components are placed in the geometric center of the array. The next component selected is U6, which has an entry value of 1.63. U6 is positioned adjacent to U7, which was the final selection to which it was paired. The next component selected is U5, which is paired with U2. All three open adjacent positions are equally distant from the desired placement position as calculated by Eq. (6.10). The selection then reduces to the position that has the highest number of adjacent open positions, which yields the position immediately left of component U2. The rest of the placement procedure follows along a similar pattern. The selection order after component U5 is U10, U3, U9, U4, U8, and U1. The final placement configuration is shown in Figure 6.4. The actual calculation of the total wirelength is carried out as an example in Figure 6.5. The total wirelength of the final placement configuration is 17 units. The optimum value is 15 units.

**Table 6.1**  Example Signal Set

| Signal set no. | Components | Number | Weight |
| --- | --- | --- | --- |
| 1 | U2, U3, U5, U6, U7, U10 | 6 | 1 |
| 2 | U2, U4, U6, U7, U9 | 5 | 1 |
| 3 | U1, U2, U5, U8 | 4 | 1 |
| 4 | U4, U9, U10 | 3 | 1 |
| 5 | U2, U7 | 2 | 1 |

$$C_{(U2)(U7)} = \sum_{k=1}^{5} \frac{J_k + \lambda}{J_k} \delta_k(U2,U7)$$

$$= \frac{6 + (-1)}{6} \delta_1(U2,U7) \nearrow^1 + \frac{5 + (-1)}{5} \delta_2(U2,U7) \nearrow^1$$

$$+ \frac{4 + (-1)}{4} \delta_3(U2,U7) \nearrow^0 + \frac{3 + (-1)}{3} \delta_4(U2,U7) \nearrow^0$$

$$\frac{2 + (-1)}{2} \delta_5(U2,U7) \nearrow^1$$

$$= \frac{5}{6} + \frac{4}{5} + \frac{1}{2}$$

$$= 2.13$$

**Figure 6.2** Sample calculation of connectivity matrix element between components U2 and U7.

## Cluster Development Method

In Hanan and Kurtzberg's cluster development technique [3], placement is based on a component connectivity with respect to all components that have already been placed on the board. The component with the largest value is selected for placement. The selected component is then placed as close as possible to the weighted center of gravity of the previously positioned components that belong to the same signal sets. The measure of connectivity is then recalculated over the reduced set of unplaced components.

**Table 6.2** Connectivity Matrix

|     | U1   | U2   | U3   | U4   | U5   | U6   | U7   | U8   | U9   | U10  |
|-----|------|------|------|------|------|------|------|------|------|------|
| U1  | 0.00 | 0.75 | 0.00 | 0.00 | 0.75 | 0.00 | 0.00 | 0.75 | 0.00 | 0.00 |
| U2  | 0.75 | 0.00 | 0.83 | 0.80 | 1.58 | 1.63 | 2.13 | 0.75 | 0.80 | 0.83 |
| U3  | 0.00 | 0.83 | 0.00 | 0.00 | 0.83 | 0.83 | 0.83 | 0.00 | 0.00 | 0.83 |
| U4  | 0.00 | 0.80 | 0.00 | 0.00 | 0.00 | 0.80 | 0.80 | 0.00 | 1.47 | 0.67 |
| U5  | 0.75 | 1.58 | 0.83 | 0.00 | 0.00 | 0.83 | 0.83 | 0.75 | 0.00 | 0.83 |
| U6  | 0.00 | 1.63 | 0.83 | 0.80 | 0.83 | 0.00 | 1.63 | 0.00 | 0.80 | 0.83 |
| U7  | 0.00 | 2.13 | 0.83 | 0.80 | 0.83 | 1.63 | 0.00 | 0.00 | 0.80 | 0.83 |
| U8  | 0.75 | 0.75 | 0.00 | 0.00 | 0.75 | 0.00 | 0.00 | 0.00 | 0.00 | 0.00 |
| U9  | 0.00 | 0.80 | 0.00 | 1.47 | 0.00 | 0.80 | 0.80 | 0.00 | 0.00 | 0.67 |
| U10 | 0.00 | 0.83 | 0.83 | 0.67 | 0.83 | 0.83 | 0.83 | 0.00 | 0.67 | 0.00 |

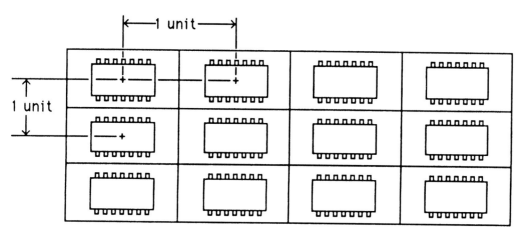

**Figure 6.3**  Component placement board.

The first placed, or seed, component is determined by calculating the overall connectivity of each component. This is done by summing the connectivity matrix along each row or column. The connectivity $C_i$ is defined by Eq. (6.9). The seed component is selected based on the maximum $C_i$. In the event of a tie, an arbitrary selection may be made. A set of components may also be prepositioned on the board manually, and the selection of unplaced components for placement may be carried out without the selection of a seed component.

Upon completion of the initial selection, a new function $F_i$, which is a measure of the expected number of lines of the interconnection nets between the unplaced component $i$ and the placed components, is calculated for each unplaced component. For every partially placed signal set $k$, there must be at least one link between the unplaced component and a placed component [3]. If $T_k$ is the maximum number of lines for signal set $k$, for a chain tree, then $T_k$ is

$$T_k = \begin{cases} \min[2p_k, 2(J_k - p_k)], & p_k \neq \dfrac{J_k}{2} \\ 2p_k - 1, & p_k = \dfrac{J_k}{2} \end{cases} \qquad (6.11)$$

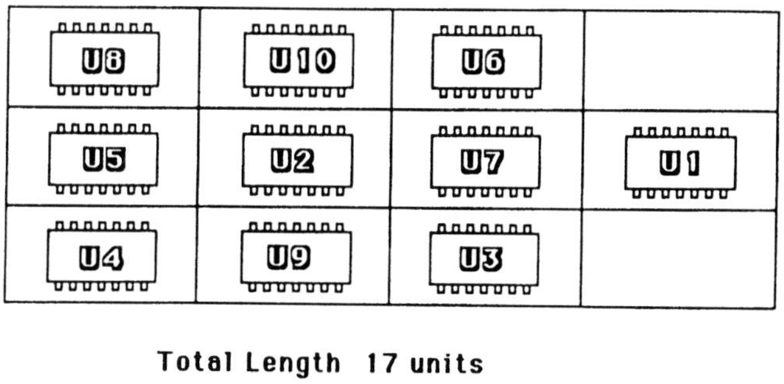

| U8 | U10 | U6 | |
| U5 | U2 | U7 | U1 |
| U4 | U9 | U3 | |

**Total Length  17 units**

**Figure 6.4**  Final placement configuration for pair-linking example.

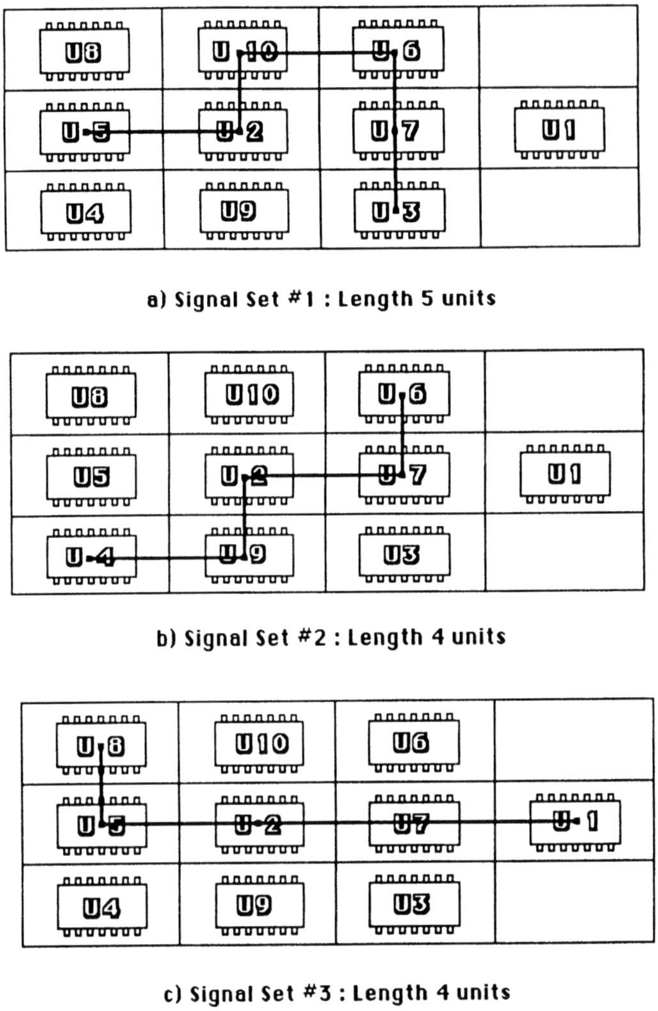

a) Signal Set #1 : Length 5 units

b) Signal Set #2 : Length 4 units

c) Signal Set #3 : Length 4 units

**Figure 6.5**   Calculation of total wirelength, part 1.

where $p_k$ is the number of placed components of the $k$th signal set and $J_k$ is the total number of components that make up the $k$th signal set. If the minimum spanning tree is employed, $T_k = J_k - 1$. Note that $T_k = 0$ if either no components of the $k$th signal set are placed or all components of the $k$th signal set are placed.

The function $F_i$ is defined by

$$F_i = \sum_{i \in D} \frac{T_k + 1}{T_k} \qquad (6.12)$$

where $D$ represents the set of partially placed signal sets $k$ of which component $i$ is a member. An iteration of the cluster development procedure involves the calculation of $F_i$ for each unplaced component. The unplaced component corresponding to the largest $F$ value is selected for placement. The selected component is placed on the board in an

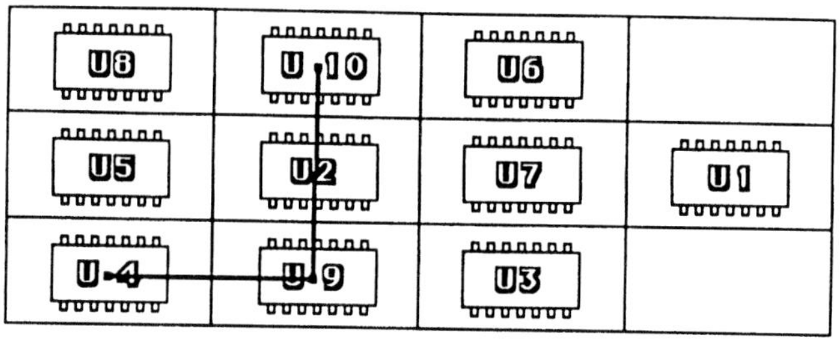

d) Signal Set #4 : Length 3 units

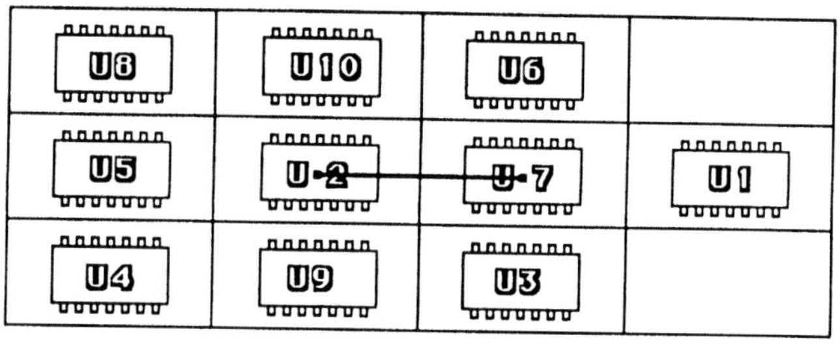

e) Signal Set #5 : Length 1 unit

| Signal Set # | Length (units) |
|---|---|
| 1 | 5 |
| 2 | 4 |
| 3 | 4 |
| 4 | 3 |
| 5 | 1 |
| Total | 17 |

Figure 6.5 (cont'd)

open location as close to the geometric center as possible. Hanan and Kurtzberg [3] suggests using the $C_i$ values as weighting factors for determining the target placement location $(x, y)$. However, the original definition of $(\bar{x}, \bar{y})$ may be used. The target location is calculated for each selected component, and the component is placed as close to the target position as possible. After a selected component is positioned on the board, the values of the $T_k$'s and $F_i$'s are recalculated based on the new partially placed signal sets. The process continues until all components are placed on the board. In the event of ties, an arbitrary selection is made between the components of tying $F$ values. The process time of the cluster development procedure is a direct function of the total number of components to be positioned.

To illustrate the cluster development procedure, the example problem used in the pair-linking method is employed. The connectivity for each component is determined by

**Table 6.3**  C List Connectivity Measure

| Component no. | Connectivity value |
|---------------|--------------------|
| U1            | 2.25               |
| U2            | 10.12              |
| U3            | 4.17               |
| U4            | 4.53               |
| U5            | 6.42               |
| U6            | 7.37               |
| U7            | 7.87               |
| U8            | 2.25               |
| U9            | 4.53               |
| U10           | 5.50               |

**Selected Seed U2**

$p_1 = 1$  $T_1 = \text{Min}(2, 2(6-1)) = 2$
$p_2 = 1;$ $T_2 = \text{Min}(2, 2(5-1)) = 2$
$p_3 = 1;$ $T_3 = \text{Min}(2, 2(4-1)) = 2$
$p_4 = 0;$ $T_4 = 0$
$p_5 = 1;$ $T_5 = 2-1 = 1$

**Calculation of F's**

$$F_i = \sum_{i \in D} \frac{T_k + 1}{T_k}$$

For Iteration #1 :  D = (1,2,3,5)

U1 : (3);  $F_{(U1)} = \dfrac{2+1}{2} = 1.5$        U2 : Placed ;  $F_{(U2)} = X$

U3 : (1);  $F_{(U3)} = \dfrac{2+1}{2} = 1.5$        U4 : (2);  $F_{(U4)} = \dfrac{2+1}{2} = 1.5$

U5 : (1,3);  $F_{(U5)} = \dfrac{2+1}{2} + \dfrac{2+1}{2}$        U6 : (1,2);  $F_{(U6)} = \dfrac{2+1}{2} + \dfrac{2+1}{2}$

$\qquad\qquad\qquad = 3.0$        $\qquad\qquad\qquad = 3.0$

U7 : (1,2,5);  $F_{(U7)} = \dfrac{2+1}{2} + \dfrac{2+1}{2}$        U8 : (3);  $F_{(U8)} = \dfrac{2+1}{2} = 1.5$

$\qquad\qquad + \dfrac{1+1}{1} = 5.0$

U9 : (2);  $F_{(U9)} = \dfrac{2+1}{2} = 1.5$        U10 : (1);  $F_{(U10)} = \dfrac{2+1}{2} = 1.5$

**Figure 6.6**  Sample calculation of $F$ value for first iteration of cluster placement example.

summing along either the row or column of the connectivity matrix given in Table 6.2. The resulting connectivity value for each component is given in Table 6.3. Since component U2 has the largest connectivity, it is selected as the seed and is placed as close to the geometric center as possible. The $F_i$ values for the first iteration are calculated in Figure 6.6. The $F_i$ values for each iteration are given in Table 6.4. The X entries correspond to placed components. Component U7 is then selected for having the largest $F$ value and placed adjacent to component U2 toward the open side. In the next iteration, component U5 is selected and placed directly below component U2. In the third iteration, component U6 is selected and placed directly below U7. In the fourth iteration, U1 and U8 tie. In the selection program, the last component to have the maximum $F$ value is selected. Thus, component U8 is selected and, by the weighted positioning method, placed directly above component U2. In the fifth iteration, component U1 achieves the maximum $F$ value and is placed immediately to the left of component U2. In the sixth iteration, all the remaining components tie in the value of $F$ and U10 is selected for placement. Component U10 is placed immediately to the right of U6. In the seventh iteration, components U4 and U9 tie. Component U9 is selected and placed immediately to the right of component U7. In the eight iteration, component U4 is selected and placed directly above component U7. The remaining component, U3, is finally selected and placed directly below component U1. The total wirelength based on a minimum spanning tree is found to be 16 units, which is one unit length less than in the pair-linking example. The final placement con-figuration is shown in Figure 6.7. Tie resolution involves placing the component in the target location that provides the largest number of surrounding open locations.

## 6.3.2  Iterative Placement Techniques

Iterative placement techniques operate on an initial placement configuration and reposition a subset of the components. If a trial repositioning significantly optimizes some figure of merit (cost), such as the total wirelength, then the change of positions is accepted. Typically, iterative placement techniques require significantly more computational time than con-structive placement techniques because of the extended iterative process. Some iterative placement techniques are explored in the following sections.

**Table 6.4**  $F$ Values per Iteration

| Component | Iteration Number | | | | | | | | |
|---|---|---|---|---|---|---|---|---|---|
| | 1 | 2 | 3 | 4 | 5 | 6 | 7 | 8 | 9 |
| U1 | 1.50 | 1.50 | 1.33 | 1.33 | 1.50 | X | X | X | X |
| U2 | X | X | X | X | X | X | X | X | X |
| U3 | 1.50 | 1.25 | 1.20 | 1.25 | 1.25 | 1.25 | 1.50 | 1.50 | 1.50 |
| U4 | 1.50 | 1.25 | 1.25 | 1.25 | 1.25 | 1.25 | 2.75 | 3.00 | X |
| U5 | 3.00 | 2.75 | X | X | X | X | X | X | X |
| U6 | 3.00 | 2.50 | 2.45 | X | X | X | X | X | X |
| U7 | 5.00 | X | X | X | X | X | X | X | X |
| U8 | 1.50 | 1.50 | 1.33 | 1.33 | X | X | X | X | X |
| U9 | 1.50 | 1.25 | 1.25 | 1.25 | 1.25 | 1.25 | 2.75 | X | X |
| U10 | 1.50 | 1.25 | 1.20 | 1.25 | 1.25 | 1.25 | X | X | X |

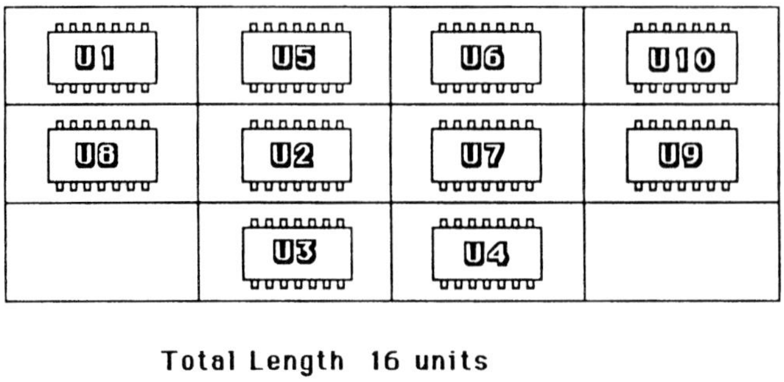

**Total Length   16 units**

**Figure 6.7**   Final placement for cluster development.

## Steinberg's Algorithm

Steinberg's placement algorithm [1] seeks to improve the placement by repositioning independent sets of components based on reducing the wirelength. An independent set is composed of components that do not have any common signal set. The placement algorithm operates on the largest independent sets first.

An iteration in Steinberg's algorithm involves removing all components of a selected independent set from the board. The removed components are then positioned in every open position arrangement on the board. At each test position, the wirelength of all nets involving the test component is calculated. For an independent set of $m$ components, there will generally be at least $m$ associated available test positions, resulting in $m!$ possible configurations. For boards with more open positions than components, dummy components are established which belong to every independent set. The cost of placing a dummy component in any available position in terms of wirelength is assumed to be much greater than that of placing any real component in an available position.

After each independent set is tested, an assignment cost matrix is established using the cost (wirelength) of placing each test component in each test position. The assignment problem can be solved by operating on the assignment cost matrix using standard solutions to the assignment problem [6–8]. The goal is to find the specific assignment of each set such that the total cost is minimized under the condition that no two components can share a given position.

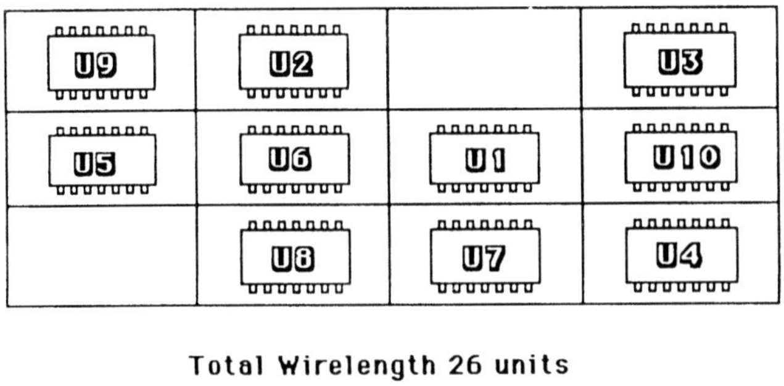

**Total Wirelength 26 units**

**Figure 6.8**   Initial placement configuration.

For large systems, the size of the independent sets may be restricted and subsets of the independent sets are formed. These subsets are treated as complete sets. This process tends to reduce computational time.

An example of this procedure is given for the placement problem defined in the pair-linking section. The initial placement configuration is depicted in Figure 6.8. The independent sets of components are listed in Table 6.5. From the placement configuration, there are two empty locations. Thus, two dummy components are added to each independent set.

For the first iteration, components U1, U3, and U4 are selected for repositioning. The components are removed from the board and tested in each available location. The wirelength cost is evaluated using the minimum spanning tree distance. In Figure 6.9a, the board is depicted with the independent components removed. In Figure 6.9b, the wirelength cost of placing U1 in the first open space is determined. Figure 6.9c depicts the wirelength cost calculation for U1 in the next open spot. Finally, Figure 6.10 depicts the wirelength calculation for U1 in the remaining open positions. The assignment matrix is shown in Figure 6.11a and the position assignment, determined by the Munkres [8] algorithm, is displayed in Figure 6.11b. The resulting placement configuration, shown in Figure 6.11c, has an overall reduction in wirelength of 5 units. The results of next five iterations are displayed in Figures 6.12 to 6.16. The final placement configuration depicted in Figure 6.17 has a total wirelength of 17 units, which is an overall reduction of 9 units from the initial placement configuration.

## Pairwise Interchange

The pairwise interchange procedure [3, 4] seeks to improve the initial placement configuration by iteratively interchanging pairs of components and measuring the improvement in terms of some figure of merit such as the total wirelength. At each iteration, a pair of components is selected and the components temporarily exchange positions. If the interchange results in an improvement (i.e., a reduction in total wirelength), the interchange is accepted and made permanent. For $m$ components, there are $m(m - 1)/2$ possible pairs that can be interchanged.

**Table 6.5** Independent Component Sets

| No. | Components |
| --- | --- |
| 1 | U1, U3, U4 |
| 2 | U1, U3, U9 |
| 3 | U3, U4, U8 |
| 4 | U3, U8, U9 |
| 5 | U1, U6 |
| 6 | U1, U7 |
| 7 | U1, U10 |
| 8 | U4, U5 |
| 9 | U5, U9 |
| 10 | U6, U8 |
| 11 | U7, U8 |
| 12 | U8, U10 |

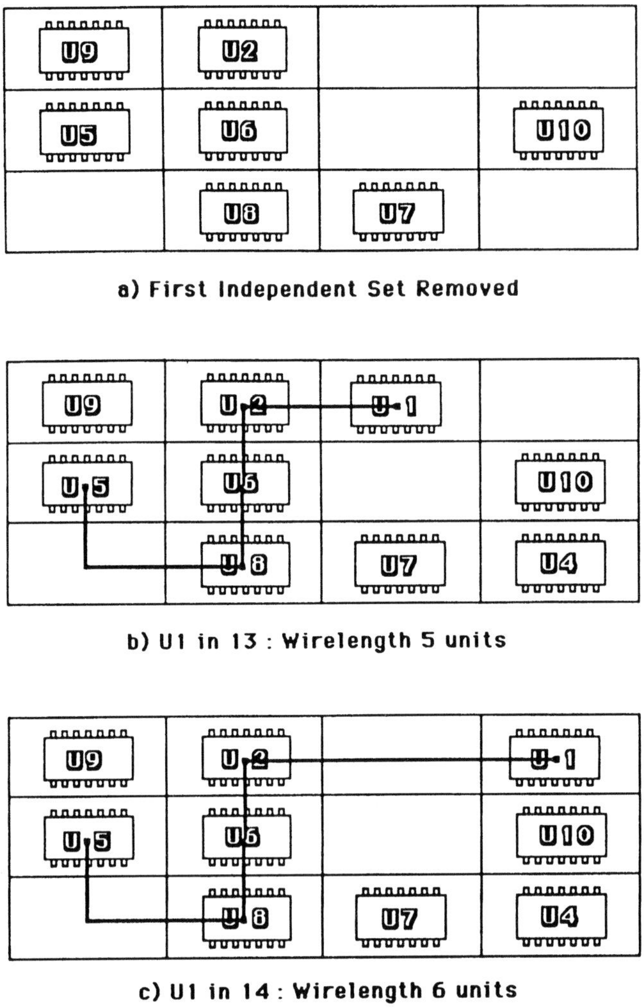

a) First Independent Set Removed

b) U1 in 13 : Wirelength 5 units

c) U1 in 14 : Wirelength 6 units

**Figure 6.9** Example of wirelength calculation for independent component U1 in Steinberg's algorithm.

The component that initiates the trial interchange process is known as the primary component. Secondary components are the components with which the primary component is interchanged. Each component may in turn be chosen as the primary component and interchanged with all others.

The selection of the primary component and secondary component can determine the particular interchange method. One method of sequencing the primary components is to measure each component's connectivity and select primary components in descending order of connectivity [3]. The connectivity measure for a component can be determined by summing the entries of the column or row of that individual component. The connectivity is defined by Eq. (6.9). The component with the largest $C_i$ value (most connected component) is selected as the primary component and is interchanged with all components that have equal or lower values. The order defines the sequence of selecting primary components and secondary components.

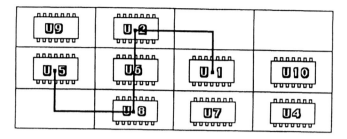

a) U1 in 23 : Wirelength 6 units

b) U1 in 31 : Wirelength 4 units

c) U1 in 34 : Wirelength 6 units

**Figure 6.10** Example of wirelength calculation for independent component U1 in Steinberg's algorithm.

The pairwise interchange technique can also be applied to restricted portions of a board. For example, it can be employed on rows or columns of components on a board that have thermal or reliability constraints. Restricting the set of secondary components reduces the computing time.

The following example illustrates the pairwise interchange procedure. Using the initial placement configuration depicted in Figure 6.18a, the $C$ list is generated using Eq. (6.9). The values of the $C$ list are given in Figure 6.4. From the list, component U2 is selected because it has the highest connectivity. Component U2 is interchanged with all other components. Component U7 is the first secondary component selected for interchange, and the interchange results in a reduction of total wirelength (depicted in Figure 6.18b). U2 is then interchanged with U6 and again a reduction in wirelength is achieved (depicted in Figure 6.18c). For all other interchanges with U2 as the primary component, no further

Components                           Positions

| | 13 | 14 | 23 | 31 | 34 |
|---|---|---|---|---|---|
| U1 | 5 | 6 | 6 | 4 | 6 |
| U3 | 7 | 7 | 5 | 7 | 6 |
| U4 | 9 | 10 | 8 | 12 | 10 |
| U11 | ∞ | ∞ | ∞ | ∞ | ∞ |
| U12 | ∞ | ∞ | ∞ | ∞ | ∞ |

a) Assignment Matrix

| Component | Position |
|---|---|
| U1 | 31 |
| U3 | 34 |
| U4 | 23 |

b) Position Selection

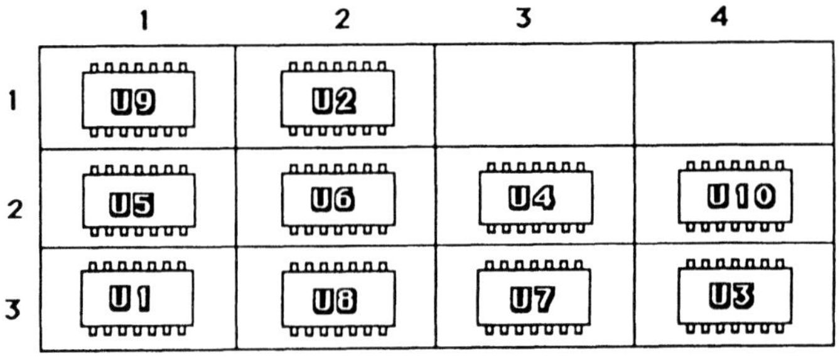

| | 1 | 2 | 3 | 4 |
|---|---|---|---|---|
| 1 | U9 | U2 | | |
| 2 | U5 | U6 | U4 | U10 |
| 3 | U1 | U8 | U7 | U3 |

c) Final Placement

**Figure 6.11**   Example of Steinberg's algorithm first independent set.

reduction in total wirelength is achieved. The final placement configuration resulting from U2 as the primary component is given in Figure 6.19. The total wirelength is now 21 units, which represents a reduction of 5 units. The second primary component selected is U7. The accepted interchanges occur between U7 and U8 and between U7 and U1. The final placement configuration at the end of the U7 interchange is depicted in Figure 6.20. The total reduction in wirelength is 2 units.

The next component selected as the primary component is U6. Accepted interchanges occur between U6 and U9 and between U6 and U8. The final placement configuration resulting from U6 as the primary component is depicted in Figure 6.21; the total wirelength reduction is 4 units. The rest of the components are selected, in turn, as primary components, but no further interchanges are accepted. The final total wirelength is 15 units, which means a total reduction of 11 unit lengths.

**Components** **Positions**

| | 11 | 13 | 14 | 31 | 34 |
|---|---|---|---|---|---|
| U1 | 4 | 5 | 6 | 4 | 6 |
| U3 | 7 | 7 | 7 | 7 | 6 |
| U9 | 8 | 6 | 7 | 9 | 6 |
| U11 | ∞ | ∞ | ∞ | ∞ | ∞ |
| U12 | ∞ | ∞ | ∞ | ∞ | ∞ |

**a) Assignment Matrix**

| Component | Position |
|---|---|
| U1 | 31 |
| U3 | 34 |
| U9 | 13 |

**b) Position Selection**

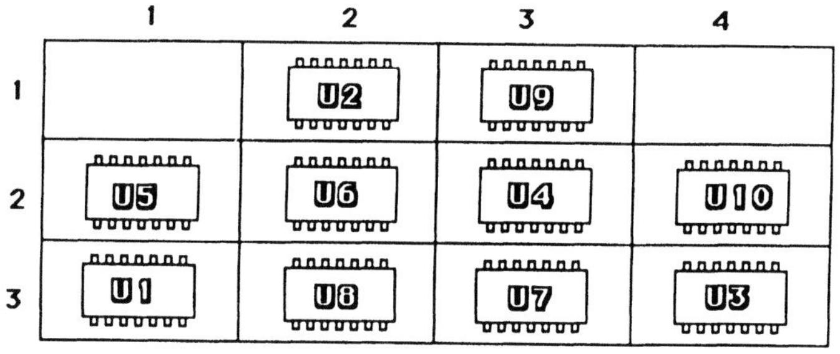

**c) Final Placement**

**Figure 6.12** Example of Steinberg's algorithm second independent set.

## 6.3.3 Force-Directed Placement Methods

Force-directed placement procedures are based on the principle that fictitious forces that are proportional to the distance between components can be utilized to minimize some figure of merit, such as wirelength. The goal is to find the relative position of each component on the board that results in a zero force.

### Force-Directed Placement

Force-directed equations are modeled after Hooke's law, which states that, if two particles are connected to each other by a spring, the attractive force between the two particles is equal to the spring constant times the distance between them. For placement, the components are considered to be particles and the spring constant can be a function of the

Components                          Positions

|      | 11 | 14 | 23 | 32 | 34 |
|------|----|----|----|----|----|
| U3   | 7  | 7  | 5  | 6  | 6  |
| U4   | 9  | 7  | 6  | 9  | 8  |
| U8   | 3  | 5  | 5  | 4  | 6  |
| U11  | ∞  | ∞  | ∞  | ∞  | ∞  |
| U12  | ∞  | ∞  | ∞  | ∞  | ∞  |

a) Assignment Matrix

| Component | Position |
|-----------|----------|
| U3        | 34       |
| U4        | 23       |
| U8        | 11       |

b) Position Selection

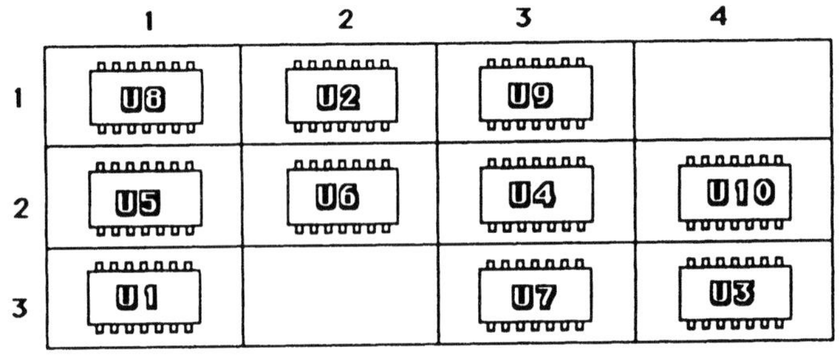

c) Final Placement

**Figure 6.13**   Example of Steinberg's algorithm third independent set.

number of signal sets (connectivity) the two components have in common, assuming the goal is to minimize the total wirelength.

Consider a component $i$ placed at position $(x_i, y_i)$ and another component $j$ at position $(x_j, y_j)$. The rectilinear distance between components is $\Delta x_{ij} = (x_j - x_i)$ and $\Delta y_{ij} = (y_j - y_i)$. The total distance between the two components is given in vector form as

$$\Delta \bar{S}_{ij} = \begin{Bmatrix} \Delta x_{ij} \\ \Delta y_{ij} \end{Bmatrix} \tag{6.13}$$

If $\bar{F}_{ij}$ is the force exerted on component $i$ by component $j$, and $K_{ij} = K_{ji}$ is proportional to the number of signals shared by components $i$ and $j$, then the force is expressed by

(Note: No summation on repeated subscripts.)

$$\bar{F}_{ij} = - K_{ij} \Delta \bar{S}_{ij} \tag{6.14}$$

Components Positions

| | 11 | 13 | 14 | 32 | 34 |
|---|---|---|---|---|---|
| U3 | 7 | 7 | 7 | 6 | 6 |
| U8 | 3 | 4 | 5 | 4 | 6 |
| U9 | 8 | 6 | 7 | 7 | 6 |
| U11 | ∞ | ∞ | ∞ | ∞ | ∞ |
| U12 | ∞ | ∞ | ∞ | ∞ | ∞ |

a) Assignment Matrix

| Component | Position |
|---|---|
| U3 | 34 |
| U8 | 11 |
| U9 | 13 |

b) Position Selection

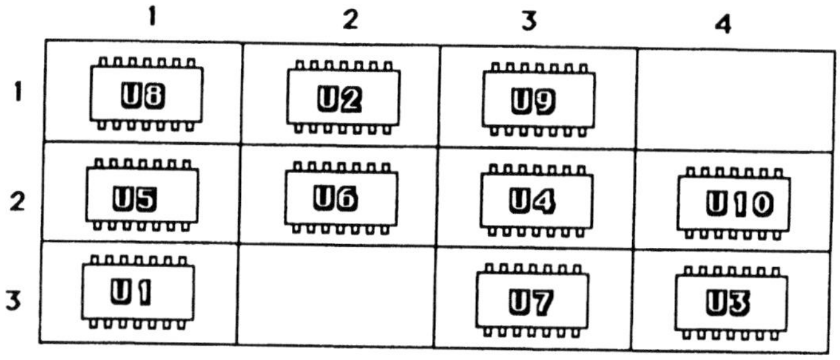

c) Final Placement

**Figure 6.14** Example of Steinberg's algorithm fourth independent set.

The magnitude of $\Delta\bar{S}$, denoted as $|\Delta\bar{S}|$, can be expressed in rectilinear or Euclidean terms. Quinn and Breuer [5] found that the rectilinear form has analytical as well as computational advantages. Noting that the expression of the force is a vector quantity, the coordinate components of the force expression in $x$ and $y$ directions are

$$Fx_{ij} = \frac{\bar{F}_{ij}\,\Delta x_{ij}}{\Delta\bar{S}_{ij}} = -K_{ij}\,\Delta x_{ij}$$

and

$$Fy_{ij} = K_{ij}\,\Delta y_{ij} \tag{6.15}$$

Components                          Positions

|      | 14 | 22 | 31 | 32 |
|------|----|----|----|----|
| U1   | 4  | 3  | 3  | 4  |
| U6   | 11 | 10 | 12 | 11 |
| U11  | ∞  | ∞  | ∞  | ∞  |
| U12  | ∞  | ∞  | ∞  | ∞  |

a) Assignment Matrix

| Component | Position |
|----------|----------|
| U1       | 31       |
| U6       | 22       |

b) Position Selection

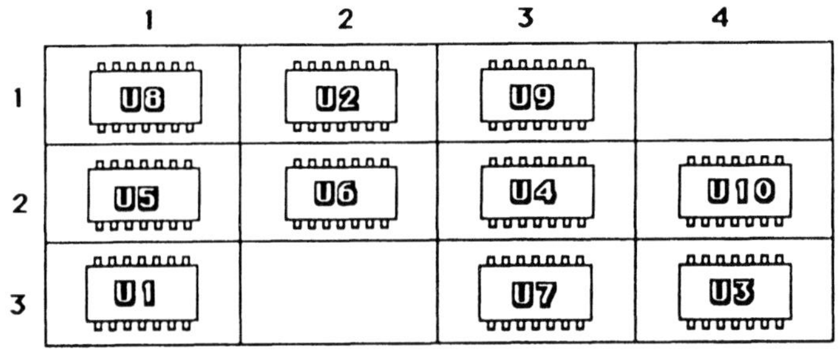

c) Final Placement

**Figure 6.15**  Example of Steinberg's algorithm fifth independent set.

For $M$ movable components, the total force on each component is a summation of forces over all components on the board. The expressions for the force on component $i$ in the $x$ and $y$ directions are

$$Fx_i = -\sum_{j=1}^{M} K_{ij} \Delta x_{ij}$$

and

$$Fy_i = -\sum_{j=1}^{M} K_{ij} \Delta y_{ij} \quad \text{for } i = 1, 2, \ldots M \tag{6.16}$$

Determining the solution of these sets of equations requires finding the locations of each component $i$ such that $Fx_i = Fy_i = 0$.

Components                          Positions

|      | 14 | 31 | 32 | 33 |
|------|----|----|----|----|
| U1   | 4  | 3  | 4  | 5  |
| U7   | 12 | 14 | 12 | 13 |
| U11  | ∞  | ∞  | ∞  | ∞  |
| U12  | ∞  | ∞  | ∞  | ∞  |

a) Assignment Matrix

| Component | Position |
|-----------|----------|
| U1        | 31       |
| U7        | 14       |

b) Position Selection

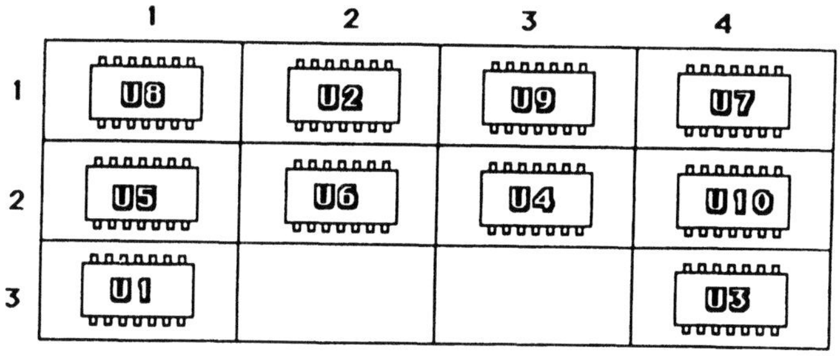

c) Final Placement

**Figure 6.16**  Example of Steinberg's algorithm sixth independent set.

Many iterative improvement algorithms have employed the force-directed equations in various forms. The force-directed equations can be used to determine target locations that produce a zero force on the components within a placement configuration. One form of the force-directed equations is

$$\bar{x}_i = \frac{\sum_j K_{ij} x_j}{\sum_j K_{ij}}$$

and

$$\bar{y}_i = \frac{\sum K_{ij} y_j}{\sum K_{ij}} \tag{6.17}$$

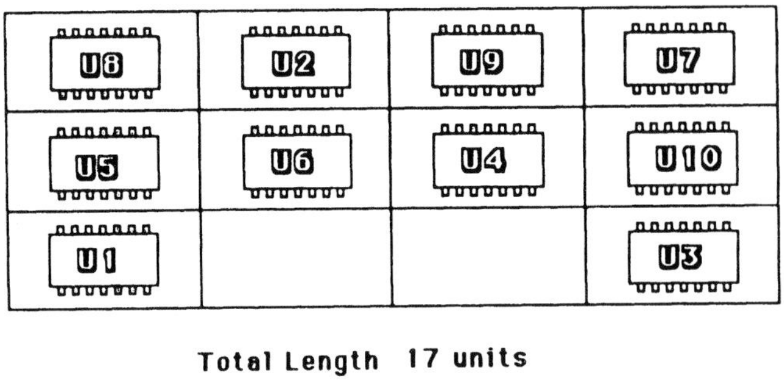
U8
U2
U9
U7
U5
U6
U4
U10
U1
U3
Total Length 17 units

**Figure 6.17** Example of Steinberg's algorithm final placement.

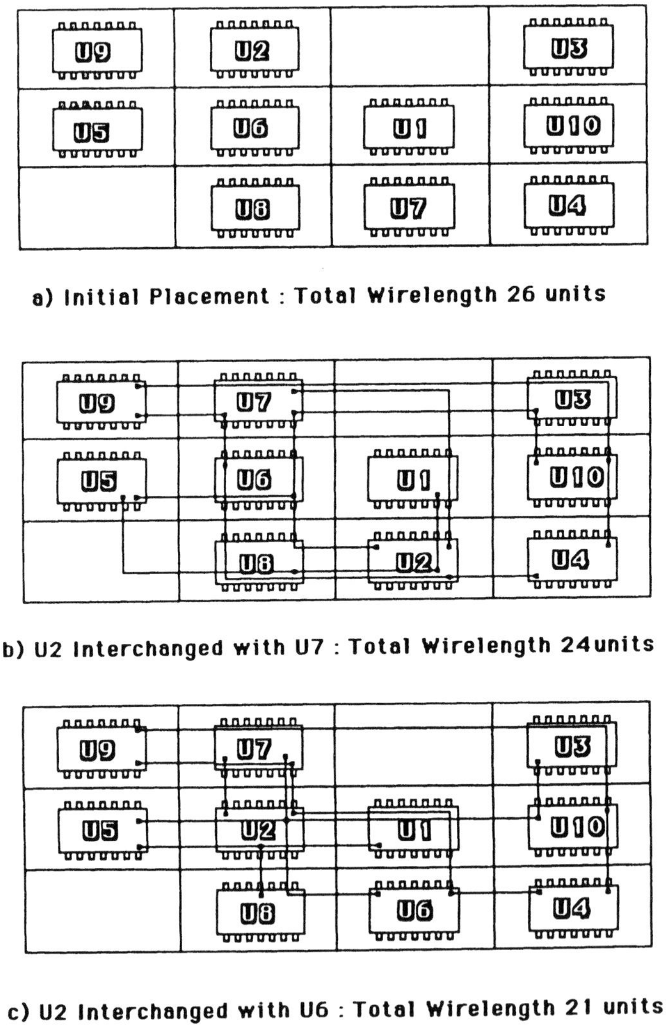
U9
U2
U3
U5
U6
U1
U10
U8
U7
U4
a) Initial Placement : Total Wirelength 26 units
U9
U7
U3
U5
U6
U1
U10
U8
U2
U4
b) U2 Interchanged with U7 : Total Wirelength 24units
U9
U7
U3
U5
U2
U1
U10
U8
U6
U4
c) U2 Interchanged with U6 : Total Wirelength 21 units

**Figure 6.18** Example of pairwise interchange with U2 as the primary component.

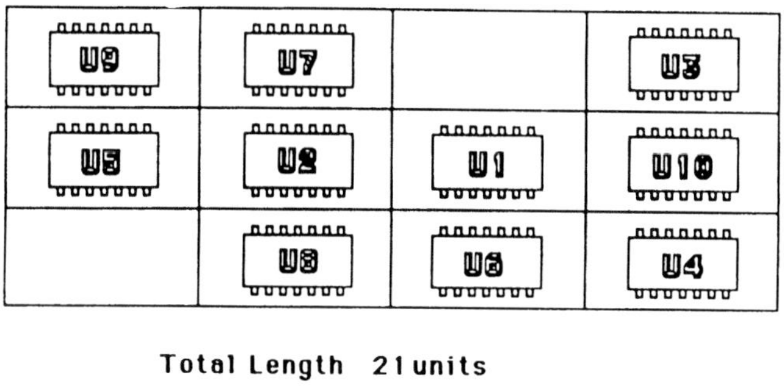

Total Length   21 units

**Figure 6.19**   Placement for pairwise interchange with U2 as the primary component.

Total Length   19 units

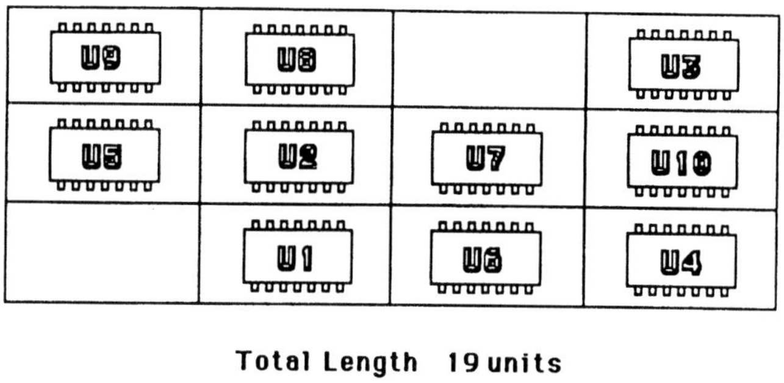

**Figure 6.20**   Placement for pairwise interchange with U7 as the primary component.

where $\bar{x}_i$ and $\bar{y}_i$ denote the target location and $x_j$ and $y_j$ are the locations of the components that share signal sets with component $i$. These equations are similar to the weighted average location described in the pair-linking procedure.

In order to prohibit the placement configuration from collapsing to a single point if all components are free to move in the plane, an initial unstretched spring length may be selected. This length is typically related to the component size. If components get too close, the connecting springs compress and push the components apart, thus restricting any overlap.

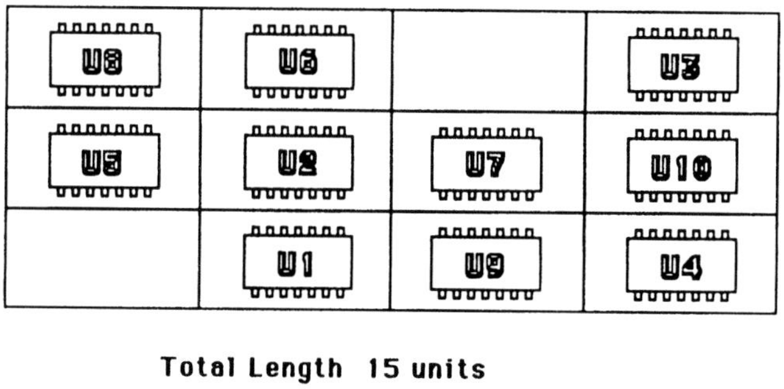

Total Length   15 units

**Figure 6.21**   Placement for pairwise interchange with U6 as the primary component.

Another method of reducing overlap is the inclusion of a repulsive term in the force equations. There are two basic forms of repulsion: one inversely proportional to the distance between unconnected components and the other a constant. Quinn and Breuer [5] determined that both forms yield about the same result. However, the constant repulsion term requires less computation time to employ.

The force equation employing a constant repulsive term is defined by

$$\bar{F}_{ij} = -K_{ij}\,\Delta\bar{S}_{ij} + \bar{R}_{ij} = -K_{ij}\begin{pmatrix}\Delta x_{ij}\\\Delta y_{ij}\end{pmatrix} + \frac{R_{ij}}{\Delta S_{ij}}\begin{pmatrix}\Delta x_{ij}\\\Delta y_{ij}\end{pmatrix}$$

where

$$R_{ij} = \begin{cases} 0 & \text{for } K_{ij} \neq 0 \\ 0 & \text{for } i = j \\ R & \text{for } K_{ij} = 0 \end{cases} \tag{6.18}$$

The repulsive force is in the direction of the line connecting components $i$ and $j$. Its magnitude, $R$, is

$$R = \left(\frac{1}{\mathrm{Cr}}\right)\left(\sum_{i=1}^{M}\sum_{j=1}^{M} K_{ij}\right)\left(\frac{1}{T}\right) \tag{6.19}$$

where $T$ equals the number of $K_{ij}$'s equal to zero and Cr is a controlling parameter. The optimum value of Cr has been found [5] to be in the range $1 < \mathrm{Cr} < 2$.

The force equation can be broken down into a component form,

$$Fx_{ij} = -K_{ij}\,\Delta x_{ij} + R_{ij}\frac{\Delta x_{ij}}{\Delta S_{ij}} \tag{6.20}$$

and

$$Fx_{ij} = -K_{ij}\,\Delta y_{ij} + R_{ij}\frac{\Delta y_{ij}}{\Delta S_{ij}}$$

The repulsive forces in the $Fx_{ij}$ and $Fy_{ij}$ equations are scaled by the cosine of the angle that the connecting line makes with the $x$ and $y$ axes, respectively. The presence of the $\Delta S_{ij}$ term couples the resulting force equations.

The total force acting on any component is determined by adding the force equations over all the modules in the board. The force acting on any component $i$, on a board consisting of $M$ movable components, is

$$Fx_i = \sum_{j=1}^{M} Fx_{ij} \quad \text{and} \quad Fy_i = \sum_{j=1}^{M} Fy_{ij} \tag{6.21}$$

The placement problem is formulated by setting the total forces acting on all of the components equal to zero. Components restricted to a fixed location and components restricted to move in only one direction can be handled by this procedure. However, Quinn and Breuer [5] noted that restricting the motions of components can adversely affect the results. The movable components tend to cluster around the fixed or restricted components. To compensate for this clustering effect, they suggest using a center-of-mass term that can be broken down into the $x$ and $y$ forces. These terms are denoted as FCMX and FCMY and are defined as

$$FCMX = \sum_{i=1}^{M} \sum_{j=M+1}^{N} -K_{ij} \, \Delta x_{ij}$$

$$FCMY = \sum_{i=1}^{M} \sum_{j=M+1}^{N} -K_{ij} \, \Delta y_{ij}$$

(6.22)

where $M$ is number of movable components and $N$ is the total number of components on the board. An alternative form for FCMX and FCMY is

$$FCMX = \sum_{i=1}^{M} Fx_i$$

$$FCMY = \sum_{i=1}^{M} Fy_i$$

(6.23)

To hold the center of the movable mass fixed, a portion of the FCMX and FCMY is subtracted from the movable-component force equations (6.21). For $M$ movable components, FCMX/$M$ is subtracted from the $Fx_i$ component force equations and FCMY/$M$ is subtracted from each $Fy_i$ component force equation. This results in the following sets of equations [5]:

$$Fx_i = \left[ \sum_{j=1}^{N} -K_{ij} \, \Delta x_{ij} + R_{ij} \frac{\Delta x_{ij}}{\Delta S_{ij}} \right] - \frac{FCMX}{M}$$

$$Fy_i = \left[ \sum_{j=1}^{N} -K_{ij} \, \Delta y_{ij} + R_{ij} \frac{\Delta y_{ij}}{\Delta S_{ij}} \right] - \frac{FCMY}{M}$$

(6.24)

for all $i = 1, 2, \ldots, M$.

## A Force-Directed Placement Procedure

Quinn and Breuer [5] developed a force-directed placement procedure consisting of two phases. In phase one, the relative positions of the components are determined by setting the coupled set of nonlinear equations given by Eq. (6.24) to zero and iteratively solving the system for the $x_i$ and $y_i$ values of the movable components. A modified Newton-Raphson method can be used to solve the simultaneous equations. The values of $x_i$ and $y_i$ obtained through the solution of the set of force-directed equations assume the board is treated as a continuous plane. The dimensions of the individual components are not considered by the force-directed equations. In phase two, components are placed into slots (rows and columns) to ensure minimum displacement from the force-directed assignment.

As an example, a randomly configured board with 25 movable components is considered. The net list is given in Table 6.6 and the placement results are depicted in Figure 6.22.

The repositioning process operates iteratively on pairs of components that overlap in the continuous case. If there are several overlaps on a component, each is treated separately. Overlaps are treated by component pairs. The component closest to the mass center of the board of the pair to be resolved remains fixed. Figure 6.23a shows the overlap resolution process for two overlapping components. Movement per iteration is restricted as shown in Figure 6.23b to ensure that larger components do not cover the smaller ones. The equations that determine the movement per iteration [5] are

$$\Delta X_{C2} = \sigma_1 \min(\text{overlap}, W) \qquad\qquad (6.25)$$

$$\Delta Y_{C2} \; \sigma_2 \min(\text{overlap}, H)$$

where

$$\sigma_1 = \cos\theta = \frac{X_{C2} - CMX}{S_2}$$

$$\sigma_2 = \sin\theta = \frac{Y_{C2} - CMY}{S_2}$$

and $W$ is the maximum movement in the $x$ direction and $H$ the maximum movement in the $y$ direction. Components are adjusted until all overlapping has been resolved. Other methods for resolving the overlap of components are discussed by Wippler et al. [10].

## Force-Directed Relaxation

The force-directed relaxation technique is an iterative placement improvement technique that employs the force equations (6.17) to determine a target point that produces a zero force for a selected component. The selected component is then positioned in the closest available target location that is not occupied by a locked component. When target locations tie, the location that produces the minimum cost based on interconnection is selected. Once a selected component is positioned, the location the component occupies is locked. By this process, the order of selection is based on the $C_i$ values associated with the individual components.

**Table 6.6**    Interconnection List, 25-Component Example

| Signal set | | Signal set | |
|---|---|---|---|
| No. | Component | No. | Component |
| 1 | U1, U2 | 21 | U13, U14 |
| 2 | U2, U3 | 22 | U14, U15 |
| 3 | U3, U4 | 23 | U11, U16 |
| 4 | U4, U5 | 24 | U12, U17 |
| 5 | U1, U6 | 25 | U13, U18 |
| 6 | U2, U7 | 26 | U14, U19 |
| 7 | U3, U8 | 27 | U15, U20 |
| 8 | U4, U9 | 28 | U16, U17 |
| 9 | U5, U10 | 29 | U17, U18 |
| 10 | U6, U7 | 30 | U18, U19 |
| 11 | U7, U8 | 31 | U19, U20 |
| 12 | U8, U9 | 32 | U16, U21 |
| 13 | U9, U10 | 33 | U17, U22 |
| 14 | U6, U11 | 34 | U18, U23 |
| 15 | U7, U12 | 35 | U19, U24 |
| 16 | U8, U13 | 36 | U20, U25 |
| 17 | U9, U14 | 37 | U21, U22 |
| 18 | U10, U15 | 38 | U22, U23 |
| 19 | U11, U12 | 39 | U23, U24 |
| 20 | U12, U13 | 40 | U24, U25 |

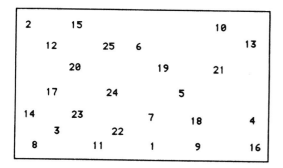

a) **Initial Placement Configuration**

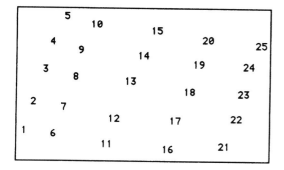

b) **Force-Directed Placement Results**

| 5 | 10 | 15 | 20 | 25 |
|---|----|----|----|----|
| 4 | 9  | 14 | 19 | 24 |
| 3 | 8  | 13 | 18 | 23 |
| 2 | 7  | 12 | 17 | 22 |
| 1 | 6  | 11 | 16 | 21 |

c) **Final Placement Configuration**

**Figure 6.22** Force-directed placement for 25-component example.

Initially, all positions on the board are unlocked and a displaced component is selected for relaxation. If a target point is not occupied by a component, the selected component is positioned in the empty slot and the component with the next largest $C_i$ value is selected. The cycle terminates when all components have been selected. Figure 6.24 illustrates part of this process with the original placement configuration shown in Table 6.3. Since component U2 has the largest $C_i$ value (Table 6.3), it is selected first. The calculation of

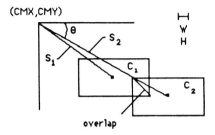

## a) Overlap Terminology

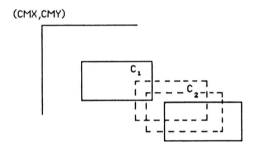

## b) Overlap Resolution

**Figure 6.23**   Resolution of component overlap.

its target location is given in Figure 6.25. From this calculation the target location is found to contain U1, which is replaced by U2. The resulting interconnection of the new configuration is given in Figure 6.24. Component U1 is then selected for relaxation and the process repeated.

### Force-Directed Pairwise Relaxation

The force-directed pairwise relaxation technique [3, 4] is a pairwise interchange procedure that employs the concept of the force equations and the zero-force target point. Target points are calculated for each component on the board. If component A has a target location near component B and B has a target location near A, then the components are interchanged. As in the pairwise interchange technique, the trial interchange is accepted only if a reduction in the interconnection cost is achieved. The zero-force locations are then recalculated and another cycle is initiated. The entire procedure terminates when interchanges can no longer be found.

### 6.3.4   Min-Cut Techniques

Although minimizing the total wirelength is one standard for placement, it does not necessarily provide the optimal placement configuration in terms of routing. Another aspect of routability is the wire density. High wire density makes routing difficult and

$$\bar{x} = \frac{\sum_{j=1}^{10} C_{2j} x_j}{\sum_{j=1}^{10} C_{2j}} \qquad \bar{y} = \frac{\sum_{j=1}^{10} C_{2j} y_j}{\sum_{j=1}^{10} C_{2j}}$$

$$\sum_{j=1}^{10} C_{2j} x_j = 0.75 \times 4 + 0.83 \times 4 + 0.80 \times 4 + 1.58 \times 1 + 1.63 \times 2$$
$$\qquad 2.13 \times 3 + 0.75 \times 2 + 0.80 \times 1 + 0.83 \times 4$$
$$= 25.62$$

$$\sum_{j=1}^{10} C_{2j} y_j = 0.75 \times 2 + 0.83 \times 1 + 0.80 \times 3 + 1.58 \times 3 + 1.63 \times 2$$
$$\qquad 2.13 \times 3 + 0.75 \times 3 + 0.80 \times 1 + 0.83 \times 2$$
$$= 22.25$$

$$\sum_{j=1}^{10} C_{2j} = 10.1$$

$$\bar{x} = \frac{25.62}{10.1} = 2.536 \sim 3 \qquad \bar{y} = \frac{22.25}{10.1} = 2.20 \sim 2$$

**Figure 6.24** Placement with force-directed relaxation.

can be detrimental in terms of electrical cross talk, impedances, etc. To attack the problem of wire density, a class of placement techniques known as min-cut have been developed. Min-cut placement techniques are utilized to distribute the wires uniformly.

Min-cut placement is an iterative improvement placement technique that partitions the board with imaginary lines and then seeks to minimize the number of wires crossing the lines by repositioning the components. For example, consider the board and the vertical line $c'$ shown in Figure 6.26. Line $c'$ is called a cut line. The components on the board fall into two sets, $A$ and $B$, to the left and the right of line $c'$, respectively. The number of wires that cross $c'$ is denoted as $\sigma(c')$. The objective of min-cut is to minimize $\sigma(c')$ by appropriately interchanging components in set $A$ and set $B$. The resulting sets are denoted as $A'$ and $B'$. Once $\sigma(c')$ is minimized, horizontal cut line $c''$ may be drawn as shown in the lower part of Figure 6.26. The board is now divided into four sets, $A_1$, $A_2$, $B_1$, and $B_2$. Letting $\sigma(c'')$ equal the number of wires crossing line $c''$, the objective is to minimize $\sigma(c'')$. In this case, we must operate on two groups of two sets. The components in set $A'$ are divided into subsets $A_1$ and $A_2$ and the components of set $B'$ are divided into subsets $B_1$ and $B_2$. To minimize $\sigma(c'')$, the components in $A_1$ and $A_2$ and the components in $B_1$ and $B_2$ are redistributed. Because $\sigma(c')$ has been minimized, interchange between components in the $A$ subsets and the $B$ subsets is not allowed. As a result, the minimization routine may operate with the subset in $A'$ and $B'$ concurrently. Further subdivision follows a similar pattern.

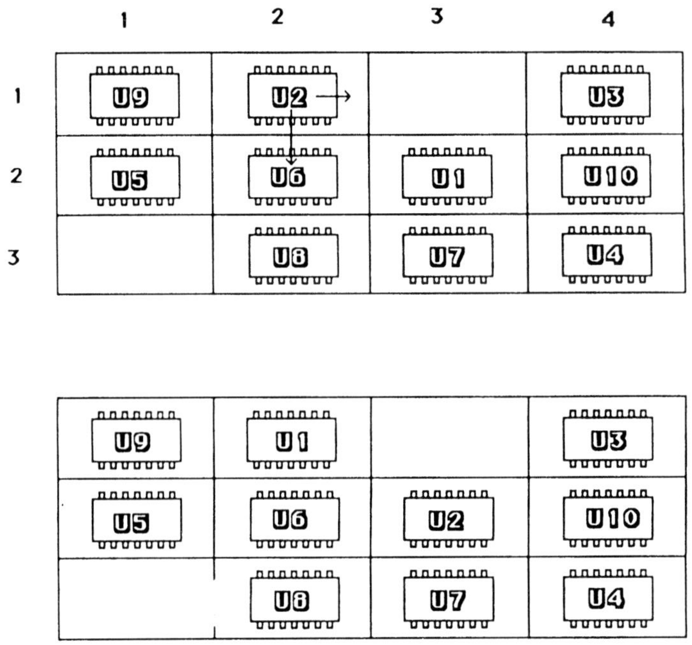

**Figure 6.25**   Calculation of target location for component U2.

The number of cut lines employed and their positions and orientations define the particular min-cut procedure. If the board is naturally placed in rows and columns and the cut lines divide all rows and columns, then the min-cut procedure can be shown to minimize the total wirelength. However, min-cut techniques generally stop well before the division becomes so extreme.

Kernighan and Lin [11] examined the problem of minimizing $\sigma(c)$. They developed an algorithm for partitioning graphs to minimize the branches by dividing the graph into two subsets and redistributing the nodes between the two subsets. This procedure is now examined in terms of component placement.

Let a board be divided by a cut line $c$ into two sets of components $A$ and $B$. Let $C_{ij}$ be the connectivity matrix defined by Eq. (6.4). For each component $\alpha$ within region $A$, or $\alpha \in A$, the external cost $E_\alpha$ in terms of the number of wires crossing due to component $\alpha$ is defined by

$$E_\alpha = \sum_{y \in B} C_{\alpha y} \tag{6.26}$$

and the internal cost $I_\alpha$ is defined by

$$I_\alpha = \sum_{x \in A} C_{\alpha x} \tag{6.27}$$

The difference, $D_\alpha$, between the number of external and internal connections for each component $\alpha$ is defined by $D_\alpha = E_\alpha - I_\alpha$. The gain resulting from interchanging two components, $\alpha_i$ from region $A$ and $\rho_i$ from region $B$, is

$$g_i = D_{\alpha_i} + D_{\rho_i} - 2C_{\alpha_i \rho_i} \tag{6.28}$$

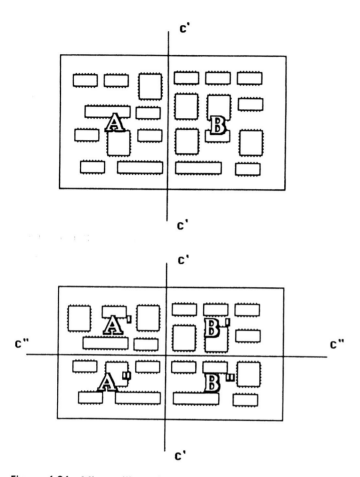

**Figure 6.26**  Min-cut illustration.

The first step in the procedure to minimize the number of wires crossing between set $A$ and set $B$ is to calculate $D_i$ for all components in both sets. Next, a component from $\alpha_i \in A$ and a component from $\rho_i \in B$ are selected such that the net gain $g_i$ is maximum. Then let $\alpha_1' = \alpha_i$, $\rho_1' = \rho_i$, and remove $\alpha_1'$ and $\rho_1'$ from the sets

$$A' = A - \{\alpha_1'\} \quad \text{and} \quad B' = B - \{\rho_1'\}$$

Recalculating $D_i$ for all components in $A'$ and $B'$ gives

$$D_x' = D_x + 2C_{x\alpha_1} - 2C_{x\rho_1}, \qquad x \in A' \tag{6.29}$$

$$D_y' = D_y + 2C_{y\rho_1} - 2C_{y\alpha_1}, \qquad y \in B' \tag{6.30}$$

After the $D_i$'s have been recalculated, another pair of components $\alpha_i \in A'$ and $\rho_i \in B'$ is selected such that net gain $g_2$ is maximum. We then set $\alpha_2' = \alpha_i$, $\rho_2' = \rho_i$, and remove $\alpha_2'$ and $\rho_2'$ from further consideration by removing them from sets $A'$ and $B'$, respectively. The process continues to calculate $g_i$ until all components in one or both sets have been removed from consideration.

Once the identification process is exhausted, interchanging $\alpha_i'$ and $\rho_i'$ decreases the cost by $g_i$. Because the gains have been determined in descending order, the objective is to determine the number $k$ such that

$$G = \sum_{i=1}^{k} g_i \qquad (6.31)$$

is maximized.

Once the value of $k$ is determined, components $\{\alpha_1', \alpha_2', \ldots, \alpha_k'\}$ are interchanged with components $\{\rho_1', \rho_2', \ldots, \rho_k'\}$. The net reduction in wires crossing the cut line $c$ is exactly $G$. The one problem that remains is how to determine which pair of components yield the maximum gain at any point in the exhaustive search. One method suggested by Kernighan and Lin [11] is to order the $D_\alpha$'s and the $D_\rho$'s in descending order. Thus, only a few component pairs need to be considered, starting with the components that produce the maximum values of $D_\alpha$ and $D_\rho$. The gains are calculated from the components in set $A$ and set $B$ by appending components to the sets in descending order of the $D$'s. Once the gain is found to be less than a previously considered gain, the search terminates and the previous component pair is selected for producing the maximum gain.

When considering the component placement problem, some components may be located on the board at specific locations for various engineering reasons. Min-cut techniques operate only on the components that are free to reside anywhere on the board. Thus, when performing a min-cut procedure on a board, part of the preprocessing operations is to specify the set of movable components. Breuer [12] discusses these restriction and other considerations for employing the Kernighan and Lin procedure.

The min-cut placement techniques depend on the definition of the sequence of cut lines used to segment the board. Breuer [12] defined two types of cut lines: the slice line, which segments a fixed number of components, and the bisection line, which attempts to divide components evenly on the board. Cut lines may be horizontal and vertical in the plane of the board. Schmidt and Druffel [13] employed a sequence of vertical and horizontal bisecting cut lines. Breuer [12] defined a similar cut line sequence and proposed a continuous quadratic placement procedure until all components are assigned to rows and columns. Breuer also defined a slice/bisection placement procedure that uses a set of slice cuts oriented in one direction to divide the board into sets of $K$ and $(N - iK)$ components until all components are assigned to a row. The components are then assigned to columns by bisections that are orthogonal to the slice lines.

The min-cut placement procedures can be employed following the use of traditional placement techniques. Wippler et al. [10] discuss the use of force-directed placement and min-cut in very large scale integration (VLSI) layout.

### 6.3.5 Comparison of Routability Placement Techniques

Based on an experimental study of placement techniques, Hanan et al. [4] show the application of several placement algorithms to six separate placement problems. These problems include three levels of computer logic packaging: logic gates on a chip, integrated circuit packs on a card, and cards on a board. The problem size ranged from 60 to 1300 components. The force-directed pairwise relaxation technique operating on the associative quadratic assignment problem produced the minimum or near-minimum interconnection cost on all six problems. The force-directed relaxation placement was also the best algorithm in terms of computing cost. In a study by Quinn and Breuer [5], their placement procedure was found to be comparable in computing cost to force-directed relaxation placement and typically produced better interconnection results.

It has been debated whether it is better to use several random starts or a constructive initial placement start in iterative placement improvement techniques. In Ref. 4, the use of constructive initial placement for large problems results in computing improvements of a factor of 10 and sometimes as much as 20 over iterative techniques. Breuer [5] suggests that for problems with over 150 components a constructive initial placement procedure should be used, followed by an iterative placement improvement procedure. An overview of the component placement methodologies can be found in Hanan and Kurtzberg [3] and in Preas and Karger [45].

Comparing the quadrature, slice-bisection, and bisection procedures, Breuer determined that quadrature yielded the best results. Corrigan [14], using a CAD layout package called PLACE developed within the Hughes CAD system, found that the slice-bisection procedure yielded the best results.

## 6.4  PLACEMENT FOR RELIABILITY

In the placement of electronic components on a PWB, reliability prediction and analysis have typically been treated as postprocesses. However, reliability is a critical part of the design procedure and is directly affected by the layout process. The arrangement of components on the board for improved reliability has been the topic of several articles [15–21]. In particular, Pecht et al. [21] examined placement for reliability on convectively cooled boards and developed routines for determining near-optimum placement configurations. Dancer and Pecht [15] examined and compared several optimization schemes for convectively cooled PWBs for computational accuracy and speed. Mayer and co-workers [16] examined optimization of reliability and life cycle cost based on the thermal design in avionics. In addition, MIL-HDBK-217E [22] and the *Reliability Design Handbook* [23] stress the importance of thermal considerations in the design of all electronic systems.

The reliability of an electronic card assembly (ECA) is inherently based on the failure mechanisms of the individual parts of the ECA, although interaction effects related to fan-out, noise margins, and software can arise. For components, electrical overload, temperature-dependent mechanical failures, and shock or vibration stresses have been identified as some of the primary causes of failure. From a mechanical standpoint, the thermal, shock, and vibration-dependent failures may be addressed in the layout of components on printed wiring boards. Hazard rates of components in a probabilistic model are generally monotonically increasing functions of temperature or a specified temperature difference. In addition, the number of thermal cycles to failure for the fatigue of wire bonds in components and solder joints is a monotonically decreasing function of a cyclic temperature difference.

Pecht et al. [24] have compiled some of the common failure mechanisms of microelectronic components that are dependent on temperature. The strong dependence on temperature has generally been modeled by an Arrhenius relation [22], which fits the philosophy that microelectronic failures are governed by chemical and diffusion processes. However, studies of failure mechanisms in microelectronics suggest that other functional relationships more accurately model the failures [25].

Temperature-induced failure can result from thermal-mechanical stresses, corrosion, and electrical stresses. Thermal-mechanical failures generally involve stresses produced by thermal expansions that cause tensile, compression, bending fractures and fatigue. Corrosion is a diffusion and chemical process involving moisture ingress and galvanic and ionic corrosion. Corrosion may also be accelerated by attacking cracks formed under stress and fatigue. In terms of electrical failures, temperature is involved in electrical overstress, dielectric breakdown, and electromigration. Depending on the specific operating conditions, any of these mechanisms or a combination may be directly involved in component failure. For example, systems with a high frequency of power cycles may fail due to thermally induced fatigue, while systems with a low frequency of power cycles may fail due to corrosion and/or electrical overstress.

Assuming that the reliability, $R$, of an ECA is given by the product of the individual part (i.e., component, interconnects, PWB) reliabilities, it can be defined as

$$R = \prod_i R_i \tag{6.32}$$

An individual component reliability, $R_i$, is defined by

$$R_i(t) = \exp - \int_0^\tau H_i(T, \Delta T, \tau)\, d\tau \tag{6.33}$$

where $H_i(T,\Delta t,\tau)$ is the component hazard rate based on the failure mechanisms, $T$ is the component operating temperature, $\Delta T$ is a specified component temperature difference, and $\tau$ is the time variable. The time dependence of the hazard rates of the individual components may be independent of the individual components and the hazard rate may be written as

$$H_i(T,\Delta T,t) = h_i'(T,\Delta T) \cdot f_i(t) \tag{6.34}$$

where $f_i(t)$ is the time relation and $h_i'(T,\Delta T)$ is the temperature-dependent part. Under these assumptions, the reliability of the ECA can be written as

$$R(t) = \exp - \sum_i h_i'(T,\Delta T) \int_0^t f(\tau)\, d\tau \tag{6.35}$$

Therefore, one approach for maximizing the reliability of the ECA is to minimize the total hazard rate,

$$h_T = \sum_{i=1}^N h_i'(T,\Delta T)(F_i(t) - F_i(0)) \quad \text{or} \quad h_T = \sum_{i=1}^N h_i(T,\Delta T) \tag{6.36}$$

where $N$ is the number of components on the system and $h_i(T,\Delta T)$ is the hazard rate evaluated at the specified mission time. If the hazard rate is constant over a particular operating period, it is sometimes referred to as the failure rate. In general, the failure rate prediction equation for an electronic component or device assumes the Arrhenius model in the form

$$\lambda_i(T_{jc_i}) = D_i + B_i \exp\left[ -A_i\left( \frac{1}{T_{jc_i} + 273} \right) \right] \tag{6.37}$$

where $T_{jc_i}$ is the junction temperature of the component and $A_i$, $B_i$, and $D_i$ are constants determined by the package type and electrical and thermal characteristics of component *i*. The components are assumed to be serially interconnected so that the total failure rate of the system is

$$\lambda_T = \sum_{i=1}^{N} \lambda_i(T_{jc_i}) \tag{6.38}$$

The objective is to improve the total reliability by minimizing the total failure rate. For the most part, the concepts employed in placement for reliability have been based on innovative interpretations of the scheduling problem addressed by Rothkopf [26], Rothkopf and Smith [27], and McNaughton [28]. In the scheduling problem, if the cost function is a linear or exponential function of some variable *t*, then a priority ordering scheme exists that guarantees an optimum solution. Since the hazard rate prediction equations are monotonically increasing with respect to component temperature, the problem is to modify the sequencing scheme to include the heat transfer equations of the particular cooling mode being applied to the system. In the following sections, reliability placement is examined for both the convective and conductive cooling modes. Initially, the problem of placement will be reduced to one dimension and the techniques for placement will be developed for a single sequence of devices or row of components of a PWB. Placement procedures for an entire PWB will be discussed based on the implications of the single-row placement procedures.

## 6.4.1  Convection Cooling Model

For the convection cooling model, components are assumed to be placed on the top surface of the PWB and cooled by a working fluid passing over the components or through a fin structure attached to the bottom surface of the board. All operating and thermal conditions are assumed to be at steady state. In the general case, for *N* components and *N* locations, *N*! possible arrangements must be examined to determine the arrangement that yields the minimum total hazard rate. In the following paragraphs, the placement of components on a single row is considered in terms of minimizing the total failure rate of the system by examining the effect of switching two adjacent components on a row placed to yield the minimum total failure rate.

To calculate the total hazard rate of the components in a system, the individual component hazard rates must be calculated. To calculate the component hazard rates, the component temperatures must be known. The temperature of an individual component depends on the component's thermal resistance from the point of interest on the component to the working fluid, the amount of heat dissipated by the component, and the temperature of the fluid associated with the component. The thermal resistance between the point of interest and the fluid at steady state is assumed to be constant, and the junction temperature is primarily a function of the fluid temperature under the component.

For steady-state convection cooling, the thermal analysis necessary to calculate the fluid temperature is typically performed on a row-by-row basis. The fluid temperature at any location on the PWB depends on the heat dissipated by the upstream components because the fluid temperature increases as it passes each consecutive component. The fluid temperature at position *i* can be approximated by

$$T_{f_i} = T_{in} + \sum_{j=0}^{i} \frac{q_i}{\dot{m}c_p} \tag{6.39}$$

where $T_{in}$ is the inlet fluid temperature, $q_j$ is the heat dissipated at position $j$, $\dot{m}$ is the mass flow rate of the fluid, and $c_p$ is the specific heat of the fluid. The summation includes all the heat dissipated by components placed on the row before and including component $i$. The temperature for component $i$ at position $k$ along the board, $T_k^i$, is defined as

$$T_k^i = T_{f_k} + q_i(Rxx_i) \tag{6.40}$$

where $Rxx$ is the resistance between the point of interest on the component and the cooling fluid.

Expanding the hazard rate prediction equation (6.34) of an individual component $i$ around an estimated operating temperature $T_o^i$ using the Taylor series yields

$$h_i(T_i^i) = h_i(T_o^i) + \sum_{k=1}^{\infty} \frac{dh_i^k(T_o^i)}{dT^k} \frac{T_i^i - T_o^i}{k!} \tag{6.41}$$

where $T_i^i$ is the new component temperature. If the minimum total hazard rate, $h_T$, is generated by placing the components on the row in ascending order (i.e., 1, 2, 3, . . . , $N$), then the result of switching any two adjacent components $i$ and $i + 1$ on a row is

$$h_T' - h_T = h_i(T_{i+1}^i) - h(T_i^i) + h_{i+1}(T_i^{i+1}) - h_{i+1}(T_{i+1}^{i+1}) \tag{6.42}$$

Using the Taylor series expansion expressed in Eq. (6.41), $h_i(T_{i+1}^i)$ and $h_{i+1}(T_i^{i+1})$ are expanded around $T_i^i$ and $T_i^{i+1}$, respectively. Replacing the original component hazard rates with the expansion in Eq. (6.42) and multiplying through by $\dot{m}c_p/q_i q_{i+1}$ yields

$$\frac{(h_T' - h_T)\dot{m}c_p}{q_i q_{i+1}} = \sum_{k=1}^{\infty} \frac{dh_i^k(T_i^i)}{dT^k} \left(\frac{1}{k!q_i}\right)\left(\frac{q_{i+1}}{\dot{m}c_p}\right)^{k-1}$$
$$- \sum_{k=1}^{\infty} \frac{dh_{i+1}^k(T_{i+1}^i)}{dT^k} \left(\frac{1}{k!q_{i+1}}\right)\left(\frac{q_i}{\dot{m}c_p}\right)^{k-1} \tag{6.43}$$

Under the assumption that the original placement configuration is optimal, the left-hand side of Eq. (6.43) must be greater than or equal to zero. This implies that the inequality

$$\sum_{k=1}^{\infty} \frac{dh_i^k(T_i^i)}{dT^k} \left(\frac{1}{k!q_i}\right)\left(\frac{q_{i+1}}{\dot{m}c_p}\right)^{k-1} \geq \sum_{k=1}^{\infty} \frac{dh_{i+1}^k(T_i^{i+1})}{dT^k} \left(\frac{1}{k!q_{i+1}}\right)\left(\frac{q_i}{\dot{m}c_p}\right)^{k-1} \tag{6.44}$$

must be valid as $i$ is taken from 1 to $N - 1$. If this inequality is not true, the left-hand side of Eq. (6.44) is negative and $h_T$ cannot be the minimum hazard rate. However, if the inequality is true for all the components placed along the row, the minimum total hazard rate is guaranteed under the assumptions made in this derivation.

From the inequality defined in Eq. (6.44), a priority metric of any order $n$ may be written as

$$PR_i^n = \sum_{k=1}^{n} \frac{dh_i^k(T_i)}{dT^k} \left(\frac{1}{k!q_i}\right)\left(\frac{q_{i+1}}{\dot{m}c_p}\right)^{k-1} \tag{6.45}$$

where $q_{i+1}$ is the heat dissipated from the component immediately downstream of component $i$. If the components are placed so that the individual priority metric of an upstream component is always greater than that of the downstream component evaluated at this upstream position, then the placement configuration is guaranteed to yield the minimum total hazard rate.

The first-order priority metric ($n = 1$) is

$$PR_i = \frac{dh_i(T_i)}{dT}\left\{\frac{1}{q_i}\right\} \qquad (6.46)$$

This priority metric is the simplest to use because it is independent of the other components. All higher-order priority metrics defined in Eq. (6.45) are dependent on the heat dissipation rate of the downstream component adjacent to component $i$.

In the following paragraphs, two procedures for placement along a row of $N$ components are discussed. The first is based on the priority metric defined by Eq. (6.46). The second placement procedure suggests a way to utilize the general priority metric defined by Eq. (6.45). A placement strategy of an entire PWB for reliability based on minimizing the sum of the individual component hazard rates is also discussed.

The simplest placement procedure for minimizing the total hazard rate on a row consists of placing components according to the priority metric defined by Eq. (6.46), using the approximate derivative of component hazard rate with respect to temperature,

$$\frac{dh_i(T^i)}{dT} = \frac{h_i(T_{out}) - h_i(T_{in})}{T_{out} - T_{in}} \qquad (6.47)$$

where

$$T^i_{out} = T_{out} + q_i(Rxx_i) \qquad (6.48)$$

and

$$T^i_{in} = T_{in} + q_i(Rxx_i) \qquad (6.49)$$

Here $T_{out}$ is the outlet fluid temperature and $T_{in}$ is the inlet fluid temperature. Components are placed starting at the location closest to the inlet in descending order of their priority number toward the outlet end of the board.

This procedure can be improved by progressively updating the fluid temperature due to the placement of each component. The priority metric defined in Eq. (6.46) is employed and a set of priority numbers is determined for all unplaced components based on the increased temperature. For convective cooling, the increased fluid temperature is given by

$$T_{f_i} = T_{in} + \frac{q_i}{\dot{m}c_p} + \sum_{j \in A} \frac{q_j}{\dot{m}c_p} \qquad (6.50)$$

where $A$ is set of placed components. This result is introduced into Eq. (6.40) to determine the individual component temperatures. The component with the highest $PR_j$ is selected and placed in the open location closest to the inlet. In case of a tie, an arbitrary selection may be made. This process is repeated until all components are placed.

To obtain a more accurate ordering scheme, the higher-order terms must be used. However, the sequence of placement is dependent on the placement itself because the higher-order terms have the heat dissipated by a trailing component in the priority number. The method is outlined below.

Initially, all components are assumed to be unplaced and the temperature for each component is determined as if the component were placed in the leading edge position. The fluid temperature is determined using Eq. (6.50) and the priority number for each component is determined using Eq. (6.46). The component with the highest $PR_i$ is selected for placement in the leading edge position. Once the initial component is selected, the placement process becomes a repetitive procedure using a priority metric of any order $n$ based on Eq. (6.45). The objective of this particular procedure is to check the placement as the components' positions are being assigned using Eq. (6.43). In this procedure, the selected component is held in a temporary status until the assignment is approved. The approval depends on satisfying the inequality given by Eq. (6.44). This is accomplished by defining a value $\mathrm{DPR}_{su}^{n}$ as

$$\mathrm{DPU}_{su}^{n} = \mathrm{PR}_s^n - \mathrm{PR}_u^n = \sum_{k=1}^{n} \frac{dh_s^k(T^s)}{dT^k}\left\{\frac{1}{q_s k!}\left(\frac{q_u}{\dot{m}c_p}\right)^{k-1}\right\} - \sum_{k=1}^{n} \frac{dh_u^k(T^u)}{dT^k}\left\{\frac{1}{q_u k!}\left(\frac{q_s}{\dot{m}c_p}\right)^{k-1}\right\} \tag{6.51}$$

where s represents the selected component and u represents any unplaced component. The $\mathrm{DPU}_{su}^{n}$ is calculated for all unplaced components. If all the DPU values are positive, the selected component is approved for placement. However, if $\mathrm{DPU}_{su}^{n}$ is negative, the unplaced component represented in the occurring negative DPU becomes the selected component and the DPU values are recalculated for all unplaced components, including the formerly selected component. Once a component is approved for placement, the next component selected is the unplaced component that represents the smallest DPU. The fluid temperature is augmented by the approved components upstream. The process repeats with positions immediately adjacent to the approved component as the new placement position. The process continues until all components are placed in the row.

Both placement procedures have been tested with several example cases. Of the two procedures, the second provides the better results and determines the optimum placement in all cases. The first method, however, is much easier to apply. In addition, a check can be made for the placement in the first procedure by applying the inequality in (6.44) down the row of components. If the inequality is held, the placement configuration is indeed optimum. When the procedures fail, the total hazard rate is within 5% of the minimum for the test examples that have been examined.

Placement for an entire board requires placing components in positions relative to the inlet and outlet ends of the board. Near-optimum solutions can be determined for an entire convectively cooled board by distributing components into rows and using one of the sequencing procedures to place components for each row.

Target locations are set by columns that are defined to be perpendicular to the fluid flow. Rows are parallel to the fluid flow. The intersections of the rows and columns create slots, which often serve as guidelines for component placement. In some cases, a component may take up more than one slot and adjustments in component placement will be required. The initial target locations are adjacent to the inlet edge of the board. The target location selection is based on the proximity to the inlet edge and the current heat loads of the individual rows. The placement procedure is described in the following paragraphs.

Initially, the board is divided into rows and columns based on the predominant widths and lengths of the components to be placed on the board. In addition, it is assumed that the components can be placed in any open slot in any row or column provided that component overlap does not occur. The slots are initially all open (i.e., all components are initially unplaced) and the set $A$ of placed components is initially empty. Target placement positions are selected based on their proximity to the inlet location and current heat load of the rows. The first selected target position is arbitrarily selected from the slots in the first column that are adjacent to the inlet edge of the board. Before the placement process is initiated, the thermal resistance $Rxx$ is evaluated for each component using a typical resistance networking technique.

Once the resistances are determined, the temperatures of the individual rows are determined from Eq. (6.50) with the modification that only components placed in the row of interest are used in determining the fluid temperature. With the fluid temperature, the approximate temperatures for all unplaced components at the target location are determined using Eq. (6.40). Then the priority number of every unplaced component is calculatd using Eq. (6.46). The component with the largest priority number is selected and placed in the target slot. Once the component is selected and placed, it is removed from the unplaced set of components.

The temperatures and priority numbers are determined for each unplaced component in each open slot in the first column until all open slots are filled. If there are $M$ open slots in the first column, then the first $M$ components with the highest priority numbers can be selected for placement. The placement of the selected components in the available slots is arbitrary. However, a two-dimensional thermal analysis along the column could be employed to improve the placement results.

Once all slots adjacent to the inlet edges are filled, the heat loads in each row may be balanced to ensure relatively uniform heating across the board. Thus, target locations in a column are selected based on the heat loads in the individual rows. The open slot in the row with the lowest heat load is selected as the target location. If a row becomes filled, the row with the next lowest heat load is selected. The heat load in any row is determined by summing the heat dissipation rates of the placed components within the row. The heat loads of the rows are recalculated after each placement.

Once the target location is selected, the fluid temperature is calculated for every unplaced component using the modified version of Eq. (6.50). At this point, either row placement procedure described previously may be employed to select the next component for placement. When using the $n$-order procedure, the placed component on the row upstream of the target location is used in calculating $DPR''_{su}$. Placement continues until all components are placed on the board.

## 6.4.2  Conduction Cooling Model

In this section, a scheme is developed for component placement for reliability in terms of the thermal response of conductively cooled PWBs operating at steady-state conditions. The development is similar to that of the convectively cooled model, but the heat transfer equations for conduction make the development of the resulting placement procedure more difficult and less accurate. Initially, the problem of placement on a single row is examined and the implications of the placement methodology are applied to component placement for the entire board.

We assume that components are placed on the PWB in relatively thermally independent rows. This assumption is good if heat rails are employed on the individual rows or if the rows are thermally matched. Heat is transferred from the components, through the rails, to constant-temperature ($T_s$) heat sinks located at each end of the row.

As was done in the convection case, the components are initially assumed to be placed in a row on a PWB to yield the minimum total hazard rate $h_T$. If the components are optimally placed, then switching any two components will result in an increased total hazard rate $h'_T$. Assuming components $i$ and $i + 1$ are switched, the difference in the total hazard rate is

$$h'_T - h_T = \sum_{k=1}^{N} [h_k(T'_k) - h_k(T_k)] \tag{6.52}$$

where $T'_k$ is the temperature of component $k$ which occurs as a result of interchanging $i$ and $i + 1$, and $T_k$ is the temperature of component $k$ in the original placement configuration. Since $h_T$ is assumed to be the minimum total hazard rate, the right-hand side of Eq. (6.52) must be positive. Therefore, the objective of placement for reliability is to ensure that the result of interchanging any two components on the PWB results in a positive value for Eq. (6.52).

Unlike the convective case, the placement of a component on a conductively cooled PWB affects the temperature of all the other placed components. For a single row, the temperature at any point under steady-state conditions can be determined by adding up the effects of all participating factors. To determine the thermal effect of a component $p$ at any position $x_p$, we consider the component to be a point source dissipating heat at a rate of $q_p$ and make the following first-order approximations:

$$q_{pL} = \frac{T_p - T_{ls}}{x_p/kA} \tag{6.53}$$

and

$$q_{pR} = \frac{T_p - T_{rs}}{(L - x_p)/kA} \tag{6.54}$$

where $q_{pL}$ is the rate at which heat is transferred to the left-hand sink, $q_{pR}$ is the rate at which heat is transferred to the right-hand sink, $T_p$ is the board temperature generated at position $x_p$, $k$ is the thermal conductivity of the board, $L$ is the length of the board, $A$ is the cross-sectional area of the row, and $T_{ls}$ and $T_{rs}$ are the left and right sink temperatures, respectively. From conservation of energy principles, the rate of heat $q_p$ dissipated by a component $p$ must be equal to the rate at which heat is transferred to the sinks. Therefore,

$$q_p = q_{pL} + q_{pR} \tag{6.55}$$

Substituting Eq. (6.53) and (6.54) into Eq. (6.55) and rearranging the terms yields

$$T_p = \frac{q_p(x_pL - x_p^2)}{kAL} + T_{ls}\left(1 - \frac{x_p}{L}\right) + T_{rs}\left(\frac{x_p}{L}\right) \tag{6.56a}$$

Without loss of generality, the two heat sinks are assumed to be at 0°C. The temperature contribution of component $p$ at position $x_p$ can then be written as

$$T_p = \frac{q_c(x_pL - x_p^2)}{kAL} \tag{6.56b}$$

The contribution of component $p$ at any location $x_k$ is given as

$$T_c(x_k) = \begin{cases} \dfrac{T_p x_k}{x_p} & \text{for } x_k \leq x_p \\[2ex] \dfrac{T_p(L - x_k)}{L - x_p} & \text{for } x_k \geq x_p \end{cases}$$

or

$$T_c(x_k) = T_p \min\left(\frac{x_k}{x_p}, \frac{L - x_k}{L - x_p}\right) \qquad (6.57)$$

The board temperature $T(x)$ at any position $x$ on the row can be determined from

$$T(x) = T_s + \sum_{j=1}^{N} T_c(x) \qquad (6.58)$$

where $T_s$ is the heat sink temperature.

The temperature $T_k$ of component $k$ is equal to the board temperature under the component plus the temperature increase between the board and the component,

$$T_k(x) = T(x) + q_k(Rxx_k) \qquad (6.59)$$

where $Rxx_k$ is a constant that specifies the thermal resistance between the point of interest on the component and the board for component $k$. Thus, the temperature of any component $k$ is a function of its position on the board, and the only change in the temperature of a component resulting from moving the components is picked up from the board temperature.

The hazard rate prediction equation can be approximated by a Taylor series expansion around an operating temperature $T_o$ such that

$$h_i(T) = h_i(T_o) + \sum_{m=1}^{\infty} \frac{dh_i^m(T_o)}{dT^m} \frac{(T - T_o)^m}{m!} \qquad (6.60)$$

Using the Taylor series expansion to express the hazard rates of all the components resulting from switching components $i$ and $i + 1$ to positions $i + 1$ and $i$, respectively, and substituting the results into Eq. (6.52) yields

$$
\begin{aligned}
h_T' - h_T = \sum_{m=1}^{\infty} \Bigg\{ & \frac{dh_i^m(T_i)}{dT^m} \frac{R^m}{m!} \left[ q_i(L - 2x_i - \delta x) \right. \\
& \left. + \sum_{j=i+2}^{N} q_j(L - x_j) - \sum_{j=1}^{i-1} q_j x_j \right]^m \\
& - \frac{dh_{i+1}^m(T_{i+1}')}{dT^m} \frac{R^m}{m!} \left[ q_{i+1}(L - 2x_i - \delta x) \right. \\
& \left. + \sum_{j=i+2}^{N} q_j(L - x_j) - \sum_{j=1}^{i=1} q_j x_j \right]^m \\
& + \sum_{k=1}^{i-1} \left[ \frac{dh_k^m(T_k)}{dT^m} \frac{(-R)^m x_k^m}{m!} (q_i - q_{i+1})^m \right] \\
& + \sum_{k=i+2}^{N} \left[ \frac{dh_k^m(T_k)}{dT^m} \frac{R^m(L - x_k)^m}{m!} (q_i - q_{i+1})^m \right] \Bigg\}
\end{aligned} \qquad (6.61)
$$

where $R = \delta x/kAL$ and $\delta x$ is the distance between $x_{i+1}$ and $x_i$.

Since the original placement configuration is assumed to generate the minimum total hazard rate, the right-hand side of Eq. (6.61) must be greater than or equal to zero. If this is not true, then the original placement of components cannot be optimal with respect to the total hazard rate. Equation (6.61) can thus be used to prove that a particular placement arrangement is optimal with respect to the total hazard rate. However, Eq. (6.61) does not explicitly predict the optimal placement configuration.

In examining the terms of Eq. (6.61), we note that under normal operating conditions, such as those given in MIL-HDBK-217E [22], the derivatives of the hazard rates with respect to the junction temperature are always positive. The signs of the terms being multiplied by the derivatives are dependent on the positions of $x_i$ and $x_{i+1}$, the magnitudes of $q_i$ and $q_{i+1}$, and the heat dissipation rates and positions of all other components. For example, the quantity

$$(L - 2x_i - \delta x) \tag{6.62}$$

has a sign change when $x_i$ passes through the point $x_p$, defined by

$$x_p = \frac{L - \delta x}{2} \tag{6.63}$$

For

$$x_p < \frac{L - \delta x}{2} \tag{6.64}$$

the value of the quantity is positive. Either end of the row can be assumed to be the starting point. If the placement is symmetric on both sides of the board in terms of position and heat dissipation rates, the value of the summations defined by

$$\sum_{j=i+2}^{N} q_j(L - x_j) - \sum_{j=1}^{i-1} q_j x_j \tag{6.65}$$

goes to zero as $x_i$ goes to $L/2$. If the placement is nearly symmetric, then these summations generally result in a positive quantity for $x_i$ between zero and $L/2$ and a negative quantity for $x_i$ between $L/2$ and $L$. However, if the position reference is reversed by setting zero equal to $L$ and $L$ equal to zero and the positions of the components are redefined, then both sign changing quantities defined by (6.62) and (6.65) can be maintained as positive values for any $x_p$ between zero and $(L - \delta x)/2$. Finally, a sign change in Eq. (6.61) can also be produced by the difference in the heat dissipation rates of components $i$ and $i + 1$ in the hazard rate derivative terms of all components excluding $i$ and $i + 1$ in the right-hand side of Eq. (6.61).

The key for positioning the components for reliability based on the thermal response of the board is that for interchanging any two adjacent components, Eq. (6.61) must be greater than or equal to zero. If we assume that the second- and higher-order terms are much smaller than the first-order terms, Eq. (6.61) can be written as

$$
\begin{aligned}
\frac{(h'_T - h_T)kAL}{\delta x} \cong{} & \frac{dh_i(T_i)}{dT}\left[ q_i(L - 2x_i - \delta x) \right. \\
& \left. + \sum_{j=i+2}^{N} q_j(L - x_j) - \sum_{j=1}^{i-1} q_j x_j \right] \\
& - \frac{dh_{i+1}(T_{i+1})}{dT}\left[ q_{i+1}(L - 2x_i - \delta x) \right. \\
& \left. + \sum_{j=i+2}^{N} q_j(L - x_j) - \sum_{j=1}^{i-1} q_j x_j \right] \\
& - \sum_{k=1}^{i-1} \frac{dh_k(T_k)}{dT} x_k(q_i - q_{i+1}) \\
& + \sum_{k=i+2}^{N} \frac{dh_k(T_k)}{dT} (L - x_k)(q_i - q_{i+1})
\end{aligned}
\tag{6.66}
$$

From Eq. (6.61), we notice that quantities being multiplied by the failure rate derivatives of both components $i$ and $i + 1$ have the same form. However, the signs are opposite. Thus, for the placement to be optimal in terms of reliability with respect to the thermal response of the PWB, the following inequality must be valid:

$$
\begin{aligned}
&\frac{dh_i(T_i)}{dT}\left[ q_i(L - 2x_i - \delta x) + \sum_{j=i+2}^{N} q_j(L - x_j) - \sum_{j=1}^{N} q_j x_j \right] \\
&+ \sum_{k=i+2}^{N} \frac{d\lambda_k(T_{jck})}{dt}(L - x_k)q_i - \sum_{k=1}^{i-1} \frac{dh_k(T_k)}{dT} x_k q_i \\
&\geq \frac{dh_{i+1}(T_{jci+1})}{dT}\left[ q_{i+1}(L - 2x_i - \delta x) + \sum_{j=i+2}^{N} qj(L - x_j) \right. \\
&\left. - \sum_{j=1}^{N} q_j x_j \right] + \sum_{k=i+2}^{N} \frac{dh_k(T_k)}{dT}(L - x_k)q_{i+1} - \sum_{k=1}^{i-1} \frac{dh_k(T_k)}{dT} x_k q_{i+1}
\end{aligned}
\tag{6.67}
$$

The application of this inequality in a placement scheme presents a number of difficulties that may reduce the accuracy of placement. Although the inequality defined by Eq. (6.67) verifies that a certain placement configuration is optimal, it does not explicitly predict the optimum solution. With regard to any constructive placement scheme, the positions of components are unknown until they are assigned a position on the board. Thus, all quantities that are determined by the positions of the components on the board are unknown. However, a placement scheme can be developed using approximations for the effect of the unknown placement configuration on the temperatures of the components and the unknown positions of the components.

The placement scheme is based on a priority metric defined by

$$
\begin{aligned}
\text{PR1}_m &= \frac{dh_m(T_m)}{dT}\left[ q_m(L - 2x_m - \delta x) + \sum_{k=i+2}^{N} q_k(L - x_k) - \sum_{k=1}^{N} q_k x_k \right] \\
&+ \sum_{k=i+2}^{N} \frac{dh_k(T_k)}{dT}(L - x_k)q_m - \sum_{k=1}^{i-1} \frac{dh_k(T_k)}{dT} x_k q_m
\end{aligned}
\tag{6.68}
$$

This priority metric is still difficult to apply to an actual placement scheme. To simplify the priority metric, we note that derivatives of the components on each side of component $m$ are being multiplied by the heat dissipation rate of component $m$. In addition, the quantity resulting from the other component derivatives reflects the effect of the heat dissipation rate of component $m$. The effect of the heat dissipation rate of component $m$ is also reflected in the first quantity in the priority number. Therefore, it may be appropriate to neglect the effect of the derivatives of the other components. In this case, the priority metric may be simplified to

$$
\text{PR2}_m = \frac{dh_m(T_m)}{dT}\left[ q_m(L - 2x_m - \delta x) + \sum_{k=i+2}^{N} q_k(L - x_k) - \sum_{k=1}^{N} q_k x_k \right]
\tag{6.69}
$$

However, neglecting the other component derivative terms may produce inaccurate assignments in the placement scheme. In addition, the unknown placement configuration must still be approximated for a placement scheme to be employed. Again, it is possible to simplify the equation by neglecting the unknown placement of all components not under consideration. If we neglect the effect of the heat dissipation rates, the positions, and the derivatives of the components not under consideration. Eq. (6.61) simplifies to

$$\frac{(h_T' - h_T)kAL}{\delta x(L - 2x_i - \delta x)} \cong \frac{dh_i(T_i)}{dT} q_i - \frac{dh_{i+1}(T_{i+1})}{dT} q_{i+1} \tag{6.70}$$

Equation (6.70) allows us to develop a placement procedure that is dependent only on the component under consideration. If $h_T$ is the minimum total hazard rate, the left-hand side of Eq. (6.70) must be positive for $x_i$ less than $(L - \delta x)/2$ referenced from either heat sink. Thus, a placement scheme can be developed using the priority metric $\mathrm{PR}_m$ defined by

$$\mathrm{PR3}_m = \frac{dh_m(T_m)}{dT} q_m \tag{6.71}$$

Here, only the approximate temperature of component $m$ is needed to calculate the priority number. The ordering of components is dependent only on the individual component derivative of the hazard rate with respect to temperature and the heat dissipation rate of the component.

Limitations on the use of the priority numbers defined by Eqs. (6.68), (6.69), and (6.71) arise because only the first-order terms were included in their development. Therefore, any method based on the priority metrics given above does not guarantee an optimum placement solution. Furthermore, the ability to predict the temperatures of the components accurately without knowing the actual placement reduces the accuracy of any placement scheme.

In this section, two constructive placement procedures for minimizing the total hazard rate on a row are introduced. In each, a target location is determined, a component is selected from the set of the unplaced components and placed in the target location, and a new target location is determined. Once a component is placed, it is removed from any further considerations. Initially, all components are assumed to be in the unplaced component set. Selection is based on the priority metrics developed earlier.

In this placement procedure, the priority metric $\mathrm{PR3}_m$ given by Eq. (6.71) defines the selection criterion. The first target location is the open position closest to either of the two heat sinks. The board temperature at the target location is approximated by assuming that the heat dissipated by all of the components to be placed, excluding the component under consideration, is added together, and the result is treated as a single source located at the center of the board. Thus, the board temperature at the target location, $x_1$, is

$$T_{b_i}(x_1) = \frac{(q\mathrm{sum}_T - q_i)x_1}{2kA} + \frac{q_i(x_1 L - x_1^2)}{kAL} + T_s \tag{6.72}$$

where $q\mathrm{sum}_T$ is the sum of the heat dissipation rates of all components and $q_i$ is the heat dissipation rate of the component under consideration. The temperature for component $i$ is determined by

$$T_i(x_1) = T_{b_i}(x_1) + q_i(Rxx_i) \tag{6.73}$$

The priority number, $\mathrm{PR3}_m$, of Eq. (6.71) is evaluated for all components in the unplaced set, and the component with the maximum priority number is selected and placed in the target location.

The second target location is the open position closest to the other heat sink. The board temperature is determined by an equation similar to Eq. (6.72) that includes the thermal contribution resulting from the placed component. The junction temperature for any component $i$ placed in the target location $x_N$ is approximated by

$$T_i(x_N) = \frac{(qsum_T - q_i)(L - x_N)}{2kA} + \frac{q_i(x_N L - x_N)^2}{kAL}$$
$$+ \frac{q_1 x_1 (L - x_N)}{kAL} + T_s + q_i(Rxx_i) \tag{6.74}$$

where $qsum_T$ is the sum of the heat dissipation rates of all unplaced components, $q_i$ is the heat dissipation rate of the component under consideration, $x_N$ is the position closest to the heat sink, and $q_1$ and $x_1$ correspond to the heat dissipation and position of the first component selected.

Again, the priority number for each unplaced component is evaluated and the component with the maximum priority number is selected. The sum of the heat dissipations of placed components multiplied by their positions relative to the closest sink on both sides of the board should be tabulated by

$$QXsum_1 = \sum_\alpha q_i x_i \tag{6.75}$$

where $\alpha$ is the set of all placed components on the range $0 < x < L/2$ and

$$QXsum_2 = \sum_\beta q_i(L - x_i) \tag{6.76}$$

where $\beta$ is the set of all placed components on the range $L/2 < x < L$. By comparing values of $QXsum_1$ and $QXsum_2$, the next target location is selected on the side with the smaller $QXsum$. Target locations on the left side are order $\{x_2, x_3, x_4, \ldots\}$ and target locations on the right side of the board are order $\{x_{N-1}, x_{N-2}, x_{N-3}, \ldots\}$. For components in the unplaced set, the components' temperatures are approximated by

$$T_i(x_t) = T_s + \frac{(qsum_T - q_i)x_t}{2kA} + \frac{q_i(x_t L - x_t)^2}{kAL}$$
$$+ QXsum_1 \frac{L - x_t}{kAL} + QXsum_2 \frac{x_t}{kAL} \tag{6.77}$$
$$+ q_i(Rxx_i)$$

where $x_t$ is the new target location, $qsum_T$ is again the sum of the heat dissipation rates of all unplaced components, and $i$ is the component under consideration.

The placement procedure selects the target location on the side with the lower $QXsum$. Once the target location is determined, the board temperature is approximated at the target location for each unplaced component. Then the priority number is evaluated for each unplaced component. The unplaced component with the highest priority number at the target location is selected and placed in the target location. The procedure continues until all components are placed.

When $x$ is at $(L - \delta x)/2$ as measured from either heat sink or inside the region around $L/2$, ordering takes place by selecting components with the lowest priority numbers. At the zero point, only two components are now in question and a thermal check of the entire row will reveal the actual positioning.

A similar placement procedure can be carried out using the priority metric defined by Eq. (6.69). In this case, the position of the target location along with the heat dissipation rates and positions of the placed components must be considered in evaluating the priority metric. For the selection of the first two components, which are placed at $x_1$ and $x_N$, the procedure follows the same method as in the previous procedure. The selection of target locations is also the same. The difference lies in the evaluation of the priority metric after the first two components have been selected.

For all placed components, the priority number is thus evaluated by

$$PR2_m = \frac{dh_m(T_m)}{dT}\left[ q_m(L - 2x_s - \delta x) + \frac{(qsum_T - q_m)L}{2} + QXtot \right] \qquad (6.78)$$

where $QXtot$ is defined by

$$QXtot = \begin{cases} QXsum_2 - QXsum_1 & \text{for } x_t \leq \dfrac{L}{2} \\[2ex] QXsum_1 - QXsum_2 & \text{for } x_t \geq \dfrac{L}{2} \end{cases}$$

$x_s$ is defined by

$$x_s = \begin{cases} x_t & \text{for } x_t \leq \dfrac{L}{2} \\[2ex] L - x_t & \text{for } x_t \geq \dfrac{L}{2} \end{cases}$$

$x_t$ is the target location, and $QXsum_1$ and $QXsum_2$ were defined in the first procedure. Again, the selection is based on the maximum value of the priority number. At the zero points, it is necessary to keep the $QXsum$ values of both sides approximately equal.

In placement for reliability for a PWB, the board is divided up into slots that form rows and columns that can accommodate individual components. Components are placed on the board by using placement procedures similar to those previously defined.

Initially, all components are assigned to the unplaced set. The board temperature at the open slots adjacent to one of the heat sinks is calculated following Eq. (6.72). The heat dissipated by all the components is considered as a single source at the center of the board. The priority metric defined by Eq. (6.71) is employed to evaluate a priority number for all of the components. Components are then symmetrically assigned to the open slots adjacent to the heat sinks based on their priority number and removed from the unplaced set. The selection is based on the highest priority number in the unplaced set. Target locations are selected from the columns of open locations adjacent to the heat sinks. The procedure is to fill the edge locations and equally distribute the components with respect to the sum of the heat dissipation rates along each individual row. The first two components selected are placed in opposite corners on the board. The procedure continues until the two columns adjacent to the heat sinks are filled. During this process, the $QXsum_1$'s and $QXsum_2$'s are calculated as in procedure 1 for each individual row.

The selection of the next target location is based on $QXsum_t$, the sum of the $QXsum_1$ and $QXsum_2$ for each individual row. The row with the lowest $QXsum_t$ is selected. Column selection is based on $QXsum_1$ and $QXsum_2$. The open location nearest to the placed component on the side with the lowest $QXsum$ is selected. The board temperatures at this open slot are calculated following Eq. (6.72). For components on the selected row and unplaced components, Eq. (6.72) is not modified. The difference occurs for placed components not on the selected row. In this case, the temperature contributions are scaled by the distance from the placed components to the selected row. The placed components closest to the row have a greater effect than those farther away from the selected row.

Once the board temperature at the selected position is approximated, the priority metric defined by Eq. (6.71) is employed to evaluate the priority numbers of the unplaced components in the target location. The component with the highest priority number is selected and assigned to the target location. The selected component is removed from the unplaced set, the $QX$sum's are recalculated, and a new target location is selected. The selection of a new target location and the evaluation of board temperature at the new target location are repeated as previously described. Selections are always based on the component in the unplaced set with the highest priority number. The process continues until all components are placed on the board. Again, it should be pointed out that this method does not guarantee the absolute minimum total hazard rate.

### 6.4.3 Summary of Placement for Reliability

The developed placement procedures do not guarantee the minimum total hazard rate placement. However, the placement configurations yield total hazard rates that are usually within a few percent of the actual minimum total hazard rate. The failure to predict the optimum configuration stems from neglect of the higher-order terms and from the estimation of the component temperatures in the placement procedures.

Although the theoretical derivation was restricted to single rows, components on an entire board using the given PWB placement procedure can be placed to yield near-minimum total hazard rate. Overall, the methods described in this section provide insight into the problem of minimizing the total hazard rate of components on convectively and conductively cooled PWBs.

## 6.5 PLACEMENT FOR PRODUCIBILITY

Placement for producibility is dependent on the manufacturing and assembly technologies being employed. The producibility of a board is also dependent on placement for routability, in terms of the cost of routing the board. The majority of producibility considerations involve geometric constraints in terms of component spacing clearance on the board and tolerances for the top and bottom surfaces of the board dependent on the board assignment in a larger electronic system. Placement for producibility may be a postprocess after the placement for routability and reliability or may be added to the other placement processes in terms of defining placement constraints. In any case, the final placement configuration affects producibility in terms of initial board design cost, production time costs, and the additional cost for special construction requirements.

In most cases, placement for producibility procedures are heuristic, with the goal of providing a placement configuration that is easy to construct and inspect. For example, the board area should be conserved. Board area is at a premium. Area is needed for routing corrections and in case additional components are added. Components that are similar in shape and size should be grouped on a board. Figure 6.27 show area conservation for a circuit board.

When considering placement for producibility, a grid pattern is often superimposed on the carrier. For printed wiring boards, the grid pattern normally has a 50-mil (0.050-in.) or 100-mil (0.100-in.) spacing as shown in Figure 6.28. However, grid spacing for some boards has reached 25 mils and the current trend is to move to even smaller grid sizes. Final placement configurations should use the grid pattern to provide standard spacing

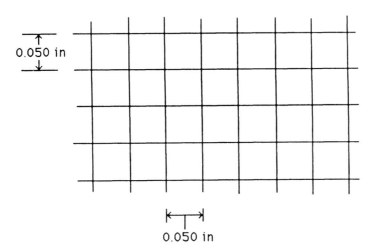

0.050 in

0.050 in

**Figure 6.27**   Area conservation example.

between components. The grid pattern naturally allows for placement patterns to fall into row-column orderings. The grid pattern may also be employed in the routing stage with interconnection paths laid out on the grid lines.

Boards have two principal directions defined by the superimposed grid pattern. Rectangular integrated circuit components should never be placed askew of the *principal* directions. Not only does this generally waste board area, it also makes automated placement difficult and can have adverse effects in terms of thermal and mechanical stress (see Figure 6.27).

In general, all integrated circuit components are oriented so that their denoted first pin is located in the same position with respect to the component's center. This provides easier placement of the physical component by robotic placement machines and easier identification of component placement for human interaction. Exceptions in the pin location rule are made in some cases to improve routability. In addition, board construction in terms of drilling holes for component pin locations is less complicated because all dual in-line packages (DIPs) will have the through holes in a preferred spacing arrangement. Finally, components should be evenly distributed over the surface of the board to provide uniform stiffness and equal weight distribution.

## 6.6   PLACEMENT TRADE-OFFS

### 6.6.1   An Example

Pecht et al. [21], investigated component placement trade-offs between routing and reliability. A test PCB, which is typical of PCBs utilized for ultra-high-reliability radar applications, was obtained from Westinghouse Electronic Systems Group. The PCB consisted of 59 active digital components, 7 spares, and 32 passive (resistive and capacitive) components. There were 1229 pin connections in the net list, including power and ground connections.

A common routing algorithm was then used to route the board with the assumption of a six-signal-layer PCB with power and ground planes. A 12.5-mil grid spacing was

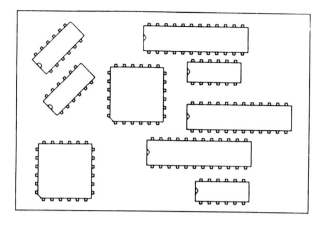

**a) Poor Layout Configuration**

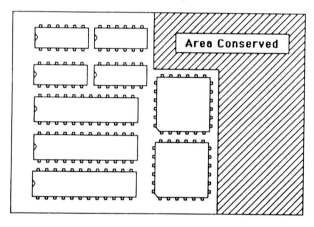

**b)  Improved Layout Configuration**

**Figure 6.28**  Typical grid spacing.

employed. Restricted plated through holes (no vias allowed under components) were imposed to meet military standards. There were 126 off-grid elements. All track widths were fixed. Standard fine-line technology was used with 8-mil lines, 8-mil spaces, and 40-mil feedthroughs. Three differential transmitters were constrained for each case to the bottom of the PCB in order to satisfy electrical requirements.

The PCB was thermally analyzed assuming conduction cooling with constant external heat sink temperatures of 21.8°C along the sides of the PCB. The top and bottom of the PCB were assumed insulated. A finite-difference algorithm [29] was employed to solve the thermal resistive network. Both component case and junction temperatures were calculated for output. The failure rate was calculated using the analysis equations for individual component failure rates specified in MIL-HDBK-217E, assuming the components operate in a serial manner.

A parts stress analysis for a fixed case (environmental) temperature of $T_c = 35°C$ gave a failure rate of 166 fr/mh (failures per million hours) for the PCB. This information aids in the sensitivity analysis for optimal placement based on thermal reliability analysis. The actual component failure rate follows from the junction temperature through the case temperature relative to component location within the PCB, the cooling mechanism and associated resistive network, and the heat (power) dissipated by the component.

Four cases were tested using the same board outline and components but with different component arrangements. Comparison guidelines included (1) the total routed wirelength as calculated at routing completion, (2) the number of vias required to route the PCB, and (3) the board failure rate.

The case I PCB was laid out according to traditional routing and thermal guidelines. This case was included to give an estimate of measures that might be typically generated by industry.

Components were placed using the autoplacer in case II. This was included to provide an estimate of the lower bound on the routing measures of total wirelength and number of vias.

Components were placed in case III to generate a lower bound on the total PCB thermal failure rate. A thermal analysis program was needed to calculate the component junction temperatures for input into a reliability analysis routine. Component locations were then determined from a thermal reliability sensitivity analysis.

Test case IV was conducted to provide information on a PCB designed for routing and thermal reliability partial optimization. This case required an initial thermal reliability sensitivity study to determine components that have (1) a high junction temperature sensitivity of the failure rate, (2) a large base failure rate, and (3) a heat dissipation factor that strongly affects the board temperature profile.

The results of this study are summarized in Figures 6.29 to 6.31, which show the various distributions of components for the test cases. The failure rate improved when components were placed according to their failure rate sensitivity as a function of tem-

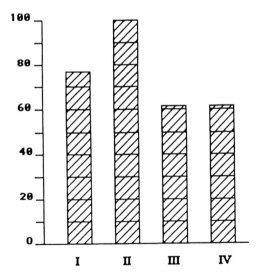

**Figure 6.29**  Failure rate comparison.

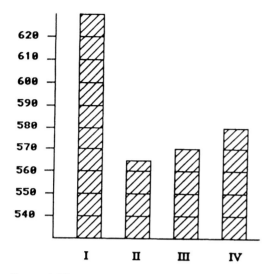

**Figure 6.30** Number of vias.

perature. In fact, the PCB placed for reliability (case III) showed a 36% improvement over the PCB placed for routability (case II) and a 19% improvement over the industrially designed PCB (case I).

Six sensitive components contributed about 60 to 80% of the total PCB failure rate, depending on placement. The remaining components had failure rate versus temperature curves that were nearly coincident around the operating temperature of the PCB. Once the six components were placed, locations of the remaining components had a small effect on the total failure rate. In fact, only a 0.3% variation was observed between the reliability constrained PCB (case IV) and the PCB placed for thermal reliability optimization (case

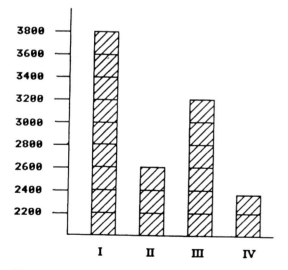

**Figure 6.31** Total printed wirelength.

III). Furthermore, the routing results for the constrained PCB were impressive. When the PCB was autoplaced with respect to routability (case II), the total wirelength was 2,658 and the number of vias was 562. This represents an improvement of 19% in wirelength and a savings of 7 vias over the optimum reliability PCB (case III), an improvement of 31% in wirelength and a savings of 58 vias over the industrial case (case I), and a savings of 21 vias but a 12% increase in wirelength over the reliability constrained case (case IV). A complete trade-off study would require a determination of the manufacturing cost per via and track per unit length layout. However, it is believed that the gain in reliability observed in test case IV, especially when life cycle costs are included, outweighs the small up-front increase in the number of vias.

The investigation exemplifies how the arrangement of components on a PCB affects the routability measured by the total wirelength and number of vias and the reliability expressed in terms of the failure rate. Although a PCB placed to minimize the failure rate could be unroutable, and a PCB placed for routability could have the worse-case failure rate, this study suggests that appropriate design trade-offs through a sensitivity analysis may locate a compromise to satisfy design requirements.

## 6.6.2  Coupled Placement for Reliability and Routability

The reliability placement procedures can be used to achieve placement configurations that have near-minimum total failure rates for conduction and convection cooling models on an entire board. Using a force-directed placement methodology, a global placement procedure for both routability and reliability is developed. To apply the force-directed placement procedure to the problem of placement for reliability, the reliability placement procedures discussed previously are used to produce a placement pattern on the board based on the number of rows or columns desired on the board and the dimensional constraints of the board and the components. A fictitious connectivity matrix is then generated to yield the same placement configuration under force-directed placement processing. The fictitious connectivity matrix is called the position-adjacent matrix, denoted by $K'_{ij}$. The actual connectivity matrix, $K_{ij}$, is developed by examining the interconnection nets or signal sets.

To generate the position-adjacent matrix $K'_{ij}$, a placement configuration is established based on minimizing the total failure rate. Then the position-adjacent matrix is generated from the relative positions of the components on the board and between themselves. In effect, the procedure is a method of working backward in the placement procedure.

Initially, the entire position-adjacent matrix is set to zero and each component is examined for its spatial relationship with all other components. The value of an entry, $\xi$, in the position-adjacent matrix is determined by the average of the nonzero entries of the connectivity matrix. For example, if component $i$ is adjacent to component $j$ in the $x$ or $y$ direction, then the entry at $K'_{ij}$ is set to $\xi$. If component $i$ is not adjacent to component $j$, then the entry at $K'_{ij}$ is set to zero. In addition, the $K'_{ii}$ entry for components that are on the working fluid inlet edge in the convection case or on one of the heat sink edges in the conduction case is $\xi$.

Once the position-adjacent matrix is completed, the repulsive matrix, $R'_{ij}$, for the position-adjacent matrix is established. The repulsive matrix is calculated exactly as described by Quinn and Breuer [5]. The difference between the position-adjacent matrix and the connectivity matrix is that an external force acts on selected edge components in the direction parallel to the primary heat flow. This external force acts on the components that have $K'_{ii}$ equal to $\xi$ and acts to pull these components to the positions on the board adjacent to the inlet end for convection and the selected heat sink for conduction. The external force is calculated from

$$Fx_i = -K'_{ii} \Delta x_{is} = K'_{ii}(x_i - x_s) \tag{6.79}$$

where $x_s$ is the position next to the heat sink or inlet edge for the conduction or convection case, respectively. In the placement procedure, $\Delta x_{ii}$ is set equal to $\Delta x_{is}$. The external force is employed to orient the placement configuration properly on the board regardless of the initial component assignments.

Once the position-adjacent matrix and the repulsive matrix for the reliability placement configuration are established, the force-directed placement procedure is initiated using the connectivity matrix and the position-adjacent matrix to calculate the forces on each component. Placement is controlled by a weighting factor $\omega$ that is restricted to the range $0 \le \omega \le 1$. The force equations are given by

$$Fx_i = \left[ \sum_{j=1}^{N} -(\omega K_{ij} + (1 - \omega)K'_{ij}) \Delta x_{ij} + (\omega R_{ij} + (1 - \omega)R'_{ij}) \frac{\Delta x_{ij}}{\Delta S_{ij}} \right] \tag{6.80}$$

and

$$Fy_i = \left[ \sum_{j=1}^{N} -(\omega K_{ij} + (1 - \omega)K'_{ij}) \Delta y_{ij} + (\omega R_{ij} + (1 - \omega)R'_{ij}) \frac{\Delta y_{ij}}{\Delta S_{ij}} \right] \tag{6.81}$$

for $i = 1, 2, \ldots, M$. To ensure that the center of mass of the system remains fixed, the center-of-mass forces are calculated from

$$FCMX = \sum_{i=1}^{N} Fx_i \quad \text{and} \quad FCMY = \sum_{i=1}^{N} Fy_i \tag{6.82}$$

Once the center-of-mass forces are calculated, a portion of the center-of-mass forces is subtracted from the force equations of the movable components. All that remains is to solve for the positions that make the forces on the movable components equal to zero. The procedure operates on

$$Fx_i = \left[ \sum_{j=1}^{N} -(\omega K_{ij} + (1 - \omega)K'_{ij}) \Delta x_{ij} + (\omega R_{ij} + (1 - \omega)R'_{ij}) \frac{\Delta x_{ij}}{\Delta S_{ij}} \right] \\ - \frac{FCMX}{M} = 0 \tag{6.83}$$

and

$$Fy_i = \left[ \sum_{j=1}^{N} -(\omega K_{ij} + (1 - \omega)K'_{ij}) \Delta y_{ij} + (\omega R_{ij} + (1 - \omega)R'_{ij}) \frac{\Delta y_{ij}}{\Delta S_{ij}} \right] \\ - \frac{FCMY}{M} = 0 \tag{6.84}$$

for $i = 1, 2, \ldots, M$. This set of coupled equations can be solved by a modified Newton-Raphson method for nonlinear equations [5]. Thus, the coupled placements compete with each other for dominance in the placement configuration. The weighting factor allows the designer to bias the final placement configuration outcome for routability (i.e., $\omega = 1$) or for reliability (i.e., $\omega = 0$).

The combined force-directed placement procedure has two process stages. In the first stage, components that have specific position requirements based on electrical, mechanical, or thermal factors are located. The remaining group of components are movable. The movable components can be placed using a constructive placement for routability routine as described by Hanan and Kurtzberg [3] or can be randomly placed. This initial placement primarily affects the computational time required to find the equilibrium component arrangement defined by Eqs. (6.83) and (6.84).

After the combined force-directed placement procedure is executed, overlapping components and their violations of placement constraints must be resolved. The objective of this second stage is to resolve the spatial interrelationships of the components by minimizing the total displacement of all the components on the board. Interaction with the layout designer is desirable to resolve other placement deficiencies.

As an example, the placement procedure described above will be applied to two placement cases. To test the placement for reliability, the components were randomly assigned data from MIL-HDBK-217E [22]. The values of the pi factors used in the calculation of the component failure rates are $\pi_Q = 0.25$, $\pi_L = 1.0$, $\pi_E = 1.0$, and $\pi_v = 1.0$.

The combined force-directed placement procedure is applied for both convection and conduction cooling models using weighting factors, $\omega$, of 0, 0.25, 0.5, 0.75, and 1.0. The final placement arrangements are set in array form on the board from the rough placement configurations resulting from the combinational force-directed placement procedure.

Once the components are positioned on the board, the placement is analyzed in terms of the approximate total wirelength, TWL, and the total failure rate, $\lambda_T$. To approximate the total weighted wirelength, the center-to-center Manhatten distance is measured between each pair of interconnected components and summed over all interconnections. This procedure yields a measure of the complete graph needed to connect all components in the various signal sets. In the convection case, the working fluid is air with an inlet fluid temperature of 20°C and a desired average outlet temperature of 40°C. Although the rows are not physically independent, this treatment yields results with only marginal errors for

**Table 6.7** Reliability Placement for Convection-Cooled 10-Component Example

| Component name | Row | Column |
|---|---|---|
| U1 | 1 | 4 |
| U2 | 2 | 3 |
| U3 | 1 | 2 |
| U4 | 3 | 2 |
| U5 | 2 | 2 |
| U6 | 3 | 1 |
| U7 | 2 | 1 |
| U8 | 1 | 3 |
| U9 | 1 | 1 |
| U10 | 3 | 3 |

high flow rates. The conduction cooling analysis is performed using a finite-difference model on the board. The boards have sinks at opposite ends along their lengths. The sink temperature is set at 30°C. The sides of the boards are assumed to be insulated. From the thermal analysis, the component junction temperatures are calculated and the total failure rate of the boards assuming a serial connected system are calculated for each placement arrangement.

Before examining the two examples, the development of the position-adjacent matrix along with the accompanying repulsive matrix will be discussed for the 10-component example. The 10 components are to be placed on a convectively cooled board that is cooled from left to right. In the reliability placement procedure, the placement configuration is set to have three rows parallel to the flow direction. The convection placement procedure results are given in Table 6.7. From these results, the position-adjacent matrix can be established. Since component U1 is adjacent to component U8, a 1 is placed in the position-adjacent matrix in row 1 column 8 and row 8 column 1. Since component U7 is in column 1 and row 2 on the board as a result of the placement procedure, it is adjacent to components U6 and U9, which are also placed in column 1 on the board, and component U5, which is to be positioned in row 2 and column 2 on the board. Since component U7 is in column 1 on the board, which is on the inlet edge, an entry is made in the position-adjacent matrix at column 7 and row 7. The entire position-adjacent matrix is established and presented in Table 6.8a. The corresponding repulsive matrix is given in Table 6.8b. The entries in the position-adjacent matrix are given as 1 because for this particular case only the reliability placement is tested. If a combinational placement procedure is desired, the two matrices presented in Table 6.8 are multiplied by the average connectivity matrix entry.

Using the matrices in Table 6.8, the 10-component problem is placed using the force-directed placement method. The initial random placement configuration is depicted in Figure 6.32a. The force-directed placement procedure is applied to the random placement configuration and allowed to operate until the convergence criterion is calculated to be less than one. The rough placement configuration is depicted in Figure 6.32b. The resulting final placement is depicted in Figure 6.32c. Notice that the results given in Table 6.7 are now depicted graphically in Figure 6.32c.

To combine placement for routability, the connectivity matrix must also be established. For the 10-component problem the net list is given in Table 6.9. From the net list, the connectivity matrix is developed and depicted in Table 6.10. The component data for the 10-component example are depicted in Table 6.11. The results of placement for the convection and conduction cases are shown in Table 6.12. The convection and conduction results are shown graphically in Figures 6.33 and 6.34, respectively.

A 25-component problem is illustrated in Tables 6.13 to 6.15 and Figures 6.35 and 6.36. In Table 6.13 the component data are presented. Table 6.14 shows the interconnection list for the 25-component example, and Table 6.15 contains the placement results for the convection and conduction cooling models. Figures 6.35 and 6.36 depict graphically the results of increasing the weighting factor for the convection and conduction cooling models, respectively. In the graphs, the total wirelength, TWL, and the total failure rate, $\lambda_T$, are defined as percentages off the minimum value calculated for each example. For example, if $\lambda_{Tmin}$ is the minimum calculated failure rate and for a particular weighting factor a failure rate of $\lambda_{Tcal}$ is calculated, then the percent off is given by

$$\lambda_{Tpercent} = \frac{\lambda_{Tcal} - \lambda_{Tmin}}{\lambda_{Tmin}} \times 100 \qquad (6.85)$$

**Table 6.8**  Reliability Connectivity Matrix for 10-Component Example

(a) Position-Adjacent Matrix

|      | U1 | U2 | U3 | U4 | U5 | U6 | U7 | U8 | U9 | U10 |
|------|----|----|----|----|----|----|----|----|----|-----|
| U1   | 0  | 0  | 0  | 0  | 0  | 0  | 0  | 1  | 0  | 0   |
| U2   | 0  | 0  | 0  | 0  | 1  | 0  | 0  | 1  | 0  | 1   |
| U3   | 0  | 0  | 0  | 0  | 1  | 0  | 0  | 1  | 1  | 0   |
| U4   | 0  | 0  | 0  | 0  | 1  | 1  | 0  | 0  | 0  | 1   |
| U5   | 0  | 1  | 1  | 1  | 0  | 0  | 1  | 0  | 0  | 0   |
| U6   | 0  | 0  | 0  | 1  | 0  | 1  | 1  | 0  | 0  | 0   |
| U7   | 0  | 0  | 0  | 0  | 1  | 1  | 1  | 0  | 1  | 0   |
| U8   | 1  | 1  | 1  | 0  | 0  | 0  | 0  | 0  | 0  | 0   |
| U9   | 0  | 0  | 1  | 0  | 0  | 0  | 1  | 0  | 1  | 0   |
| U10  | 0  | 1  | 0  | 1  | 0  | 0  | 0  | 0  | 0  | 0   |

(b) Position-Adjacent Repulsive Matrix

|      | U1    | U2    | U3    | U4    | U5    | U6    | U7    | U8    | U9    | U10   |
|------|-------|-------|-------|-------|-------|-------|-------|-------|-------|-------|
| U1   | 0.000 | 0.286 | 0.286 | 0.286 | 0.286 | 0.286 | 0.286 | 0.000 | 0.286 | 0.286 |
| U2   | 0.286 | 0.000 | 0.286 | 0.286 | 0.000 | 0.286 | 0.286 | 0.000 | 0.286 | 0.000 |
| U3   | 0.286 | 0.286 | 0.000 | 0.286 | 0.000 | 0.286 | 0.286 | 0.000 | 0.000 | 0.286 |
| U4   | 0.286 | 0.286 | 0.286 | 0.000 | 0.000 | 0.000 | 0.286 | 0.286 | 0.286 | 0.000 |
| U5   | 0.286 | 0.000 | 0.000 | 0.000 | 0.000 | 0.286 | 0.000 | 0.286 | 0.286 | 0.286 |
| U6   | 0.286 | 0.286 | 0.286 | 0.000 | 0.286 | 0.000 | 0.000 | 0.286 | 0.286 | 0.286 |
| U7   | 0.286 | 0.286 | 0.286 | 0.286 | 0.000 | 0.000 | 0.000 | 0.286 | 0.000 | 0.286 |
| U8   | 0.000 | 0.000 | 0.000 | 0.286 | 0.286 | 0.286 | 0.286 | 0.000 | 0.286 | 0.286 |
| U9   | 0.286 | 0.286 | 0.000 | 0.286 | 0.286 | 0.286 | 0.000 | 0.286 | 0.000 | 0.286 |
| U10  | 0.286 | 0.000 | 0.286 | 0.000 | 0.286 | 0.286 | 0.286 | 0.286 | 0.286 | 0.000 |

The same procedure is employed to calculate the percent off the total wirelength. From Figures 6.33 to 6.36, the total wirelength experiences a smooth transition of TWL while the weighting factor is increased.

The combined force-directed placement procedure allows the layout designer to place components for either reliability or routability, depending on the selected weighting factor.

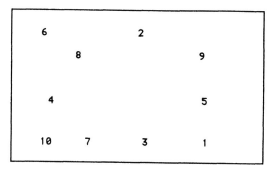

## a) Initial Placement

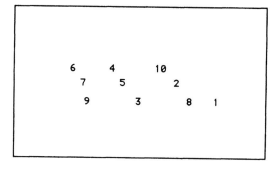

## b) Force-Directed Rough Placement

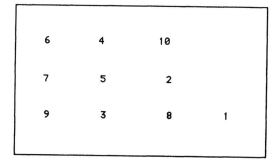

## c) Final Placement Configuration

**Figure 6.32**  Placement for reliability in 10-component example.

In addition, the combined placement procedure can be used as a tool to place components and then determine the trade-offs that can be made in placing for reliability and routability. As the weighting factor is increased to 1 (toward placement for routability), the total wirelength decreases as expected. The total failure rate generally increases with increasing value of the weighting factor. The rate of increase in the total failure rate with increasing weighting factor is generally more erratic.

**Table 6.9**      Interconnection
List for 10-Component Example

| Signal Set | |
|---|---|
| No. | Component |
| 1 | U1, U10 |
| 2 | U1, U6 |
| 3 | U1, U8 |
| 4 | U2, U6 |
| 5 | U2, U7 |
| 6 | U3, U8 |
| 7 | U3, U9 |
| 8 | U3, U10 |
| 9 | U4, U5 |
| 10 | U4, U9 |
| 11 | U4, U10 |
| 12 | U5, U7 |
| 13 | U5, U10 |
| 14 | U6, U7 |
| 15 | U6, U8 |
| 16 | U6, U10 |
| 17 | U6, U10 |
| 18 | U6, U10 |
| 19 | U9, U10 |

When applying the force-directed placement procedure to highly connected systems, components tend to cluster and the resolution of components into positions on the board becomes difficult. In some cases, the placement procedure must be performed in several stages. Initially, all components are allowed to move freely on the board until the system converges. Then components on the outer edges of the clusters are assigned board slots relative to their positions with respect to the other components on the board. Once components are assigned border locations, these components are locked in place and the force-directed placement algorithm is repeated. If overlap is still a problem, more components on the outer edges of the clusters are assigned positions based on their positions relative to the other components on the board. These components are then locked and the force-directed placement algorithm is repeated. The procedures described above are continued until all components' positions are resolved.

The methodology for placement of electronic components presented here is based on minimizing the total failure rate of the system and on minimizing the total weighted wirelength. The methodology of placement for reliability differs from the existing heuristic methods of placement for reliability in that it addresses the physics of the problem rather than methodically guessing at better placement configurations. Thus, significant improvement in computational speed as well as an insightful understanding of the problem is gained. The selection of the force-directed placement procedure stems from the procedure's capability to combine placement for reliability and routability. For an actual placement procedure to be effective, consideration of other electrical and mechanical constraints must be included in the hierarchy of the placement system.

**Table 6.10** Connectivity Matrix for 10-Component Example

(a) Connectivity Matrix

|     | U1 | U2 | U3 | U4 | U5 | U6 | U7 | U8 | U9 | U10 |
|-----|----|----|----|----|----|----|----|----|----|-----|
| U1  | 0  | 1  | 0  | 0  | 0  | 1  | 0  | 1  | 0  | 0   |
| U2  | 1  | 0  | 0  | 0  | 0  | 1  | 1  | 0  | 0  | 0   |
| U3  | 0  | 0  | 0  | 0  | 0  | 0  | 0  | 1  | 1  | 1   |
| U4  | 0  | 0  | 0  | 0  | 1  | 0  | 0  | 0  | 1  | 1   |
| U5  | 0  | 0  | 0  | 1  | 0  | 0  | 1  | 0  | 0  | 1   |
| U6  | 1  | 1  | 0  | 0  | 0  | 0  | 1  | 1  | 0  | 3   |
| U7  | 0  | 1  | 0  | 0  | 1  | 1  | 0  | 0  | 0  | 0   |
| U8  | 1  | 0  | 1  | 0  | 0  | 1  | 0  | 0  | 0  | 0   |
| U9  | 0  | 0  | 1  | 1  | 0  | 0  | 0  | 0  | 0  | 1   |
| U10 | 0  | 0  | 1  | 1  | 1  | 3  | 0  | 0  | 1  | 0   |

(b) Repulsive Matrix

|     | U1 | U2 | U3 | U4 | U5 | U6 | U7 | U8 | U9 | U10 |
|-----|-----|-----|-----|-----|-----|-----|-----|-----|-----|-----|
| U1  | 0     | 0     | 0.731 | 0.731 | 0.731 | 0     | 0.731 | 0     | 0.731 | 0.731 |
| U2  | 0     | 0     | 0.731 | 0.731 | 0.731 | 0     | 0     | 0.731 | 0.731 | 0.731 |
| U3  | 0.731 | 0.731 | 0     | 0.731 | 0.731 | 0.731 | 0.731 | 0     | 0     | 0     |
| U4  | 0.731 | 0.731 | 0.731 | 0     | 0     | 0.731 | 0.731 | 0.731 | 0     | 0     |
| U5  | 0.731 | 0.731 | 0.731 | 0     | 0     | 0.731 | 0     | 0.731 | 0.731 | 0     |
| U6  | 0     | 0     | 0.731 | 0.731 | 0.731 | 0     | 0     | 0     | 0.731 | 0     |
| U7  | 0.731 | 0     | 0.731 | 0.731 | 0     | 0     | 0     | 0.731 | 0.731 | 0.731 |
| U8  | 0     | 0.731 | 0     | 0.731 | 0.731 | 0     | 0.731 | 0     | 0.731 | 0.731 |
| U9  | 0.731 | 0.731 | 0     | 0     | 0.731 | 0.731 | 0.731 | 0.731 | 0     | 0     |
| U10 | 0.731 | 0.731 | 0     | 0     | 0     | 0     | 0.731 | 0.731 | 0     | 0     |

## 6.7 ROUTING

Interconnection problems in electronics are analogous to traveling salesman or transport problems. The problem of fetching a pail of water by the shortest path between two points near a bank of a stream was solved by the ancient Greeks. The problem of interconnecting

**Table 6.11**  Component Data for 10-Component Example

| Component name | Type | Heat dissipation (W) | $R_{jc}$ (°C/W) | A1 | C1 | C2 | Complex | No. of pins |
|---|---|---|---|---|---|---|---|---|
| U1 | 00108C | 0.24 | 50 | 4635 | 0.01 | 0.0048 | 6G | 14 |
| U2 | 00404C | 0.18 | 50 | 4635 | 0.01 | 0.0048 | 3G | 14 |
| U3 | 02601C | 0.04 | 50 | 5214 | 0.01 | 0.0048 | 4G | 14 |
| U4 | 02802C | 0.12 | 50 | 5214 | 0.01 | 0.0048 | 3G | 14 |
| U5 | 65001C | 0.30 | 50 | 6373 | 0.01 | 0.0048 | 4G | 14 |
| U6 | 40301J | 1.00 | 40 | 5794 | 0.16 | 0.0090 | 16384B | 24 |
| U7 | 23105E | 0.94 | 50 | 5214 | 0.025 | 0.0056 | 1024B | 16 |
| U8 | 08101C | 0.48 | 50 | 5214 | 0.01 | 0.0048 | 2G | 14 |
| U9 | 10104C | 0.40 | 50 | 7532 | 0.01 | 0.0048 | 29T | 14 |
| U10 | 07904E | 0.34 | 50 | 5214 | 0.01 | 0.0056 | 15G | 16 |

three villages by a road network of minimum total length was solved by Jacob Steiner (1796–1863).

A powerful and commonly used tool in routing of printed wiring boards is the heuristic Lee maze-running algorithm. Super Apricot, developed by Hosking et al. [30], is a routing algorithm valid for two-layer boards. The method calculates the crossing points for all connections and plans rough routes; designer intervention is required. A topography-based nonminimum routing algorithm developed by Doreau and Abel [31] has the capability of

**Table 6.12**  Placement Results for 10-Component Example

(a) Convection Example

| $\omega$ | Total wirelength (units) | $\lambda_T$ (fr/mhr) |
|---|---|---|
| 0.00 | 86 | 0.22440 |
| 0.25 | 74 | 0.22550 |
| 0.50 | 62 | 0.24100 |
| 0.75 | 54 | 0.35699 |
| 1.00 | 50 | 0.35720 |

(b) Conduction Example

| $\omega$ | Total wirelength (units) | $\lambda_T$ (fr/mhr) |
|---|---|---|
| 0.00 | 110 | 0.08290 |
| 0.25 | 80 | 0.08307 |
| 0.50 | 62 | 0.08350 |
| 0.75 | 54 | 0.08340 |
| 1.00 | 46 | 0.08371 |

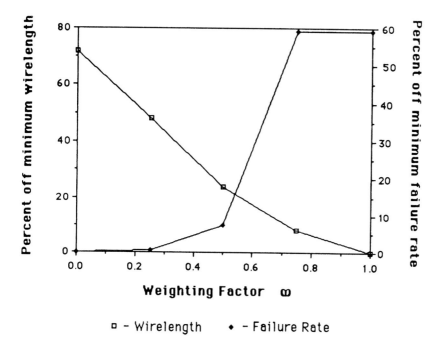

**Figure 6.33**   Placement results for convection-cooled 10-component example.

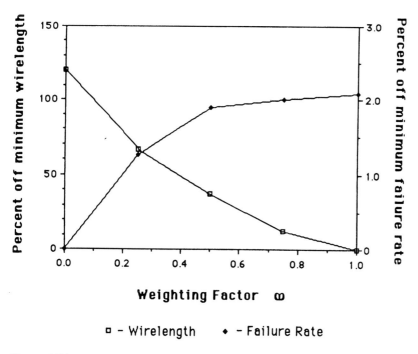

**Figure 6.34**   Placement results for conduction-cooled 10-component example.

**Table 6.13**   Component Data for 25-Component Example

| Component name | Type | Heat dissipation (W) | $R_{jc}$ (°C/W) | A1 | C1 | C2 | Complex | No. of pins |
|---|---|---|---|---|---|---|---|---|
| U1 | 02803F | 0.27 | 68 | 5214 | 0.01 | 0.0047 | 72G | 16 |
| U2 | 31306D | 0.08 | 70 | 5794 | 0.01 | 0.0037 | 4G | 14 |
| U3 | 01005F | 0.34 | 68 | 4635 | 0.01 | 0.0047 | 18G | 16 |
| U4 | 20202F | 0.72 | 68 | 4635 | 0.035 | 0.0047 | 1024B | 16 |
| U5 | 30102D | 0.05 | 70 | 5794 | 0.01 | 0.0037 | 12G | 14 |
| U6 | 04602F | 0.14 | 68 | 5214 | 0.01 | 0.0047 | 17G | 16 |
| U7 | 31201F | 0.21 | 68 | 5794 | 0.01 | 0.0047 | 42G | 16 |
| U8 | 00602F | 0.55 | 68 | 4635 | 0.01 | 0.0047 | 36G | 16 |
| U9 | 20402F | 0.79 | 68 | 5214 | 0.035 | 0.0047 | 2048B | 16 |
| U10 | 08201F | 0.60 | 68 | 5214 | 0.01 | 0.0047 | 31G | 16 |
| U11 | 04101D | 0.01 | 70 | 5214 | 0.01 | 0.0037 | 6G | 14 |
| U12 | 32801D | 0.30 | 70 | 5794 | 0.01 | 0.0037 | 10G | 14 |
| U13 | 00502D | 0.10 | 70 | 4635 | 0.01 | 0.0037 | 6G | 14 |
| U14 | 23004F | 0.41 | 68 | 4635 | 0.025 | 0.0047 | 256B | 16 |
| U15 | 00107D | 0.16 | 70 | 4635 | 0.01 | 0.0037 | 4G | 14 |
| U16 | 02907F | 0.07 | 68 | 5214 | 0.01 | 0.0047 | 18G | 16 |
| U17 | 44201F | 0.83 | 68 | 5794 | 0.01 | 0.0047 | 24G | 16 |
| U18 | 00101D | 0.04 | 70 | 4635 | 0.01 | 0.0037 | 1G | 14 |
| U19 | 23103F | 0.47 | 68 | 5794 | 0.025 | 0.0047 | 1024B | 16 |
| U20 | 00206D | 0.11 | 70 | 4635 | 0.01 | 0.0037 | 11G | 14 |
| U21 | 08101D | 0.48 | 70 | 5214 | 0.01 | 0.0037 | 2G | 14 |
| U22 | 30106F | 0.15 | 68 | 5794 | 0.01 | 0.0047 | 36G | 16 |
| U23 | 47401F | 0.50 | 68 | 6373 | 0.01 | 0.0047 | 27G | 16 |
| U24 | 10405F | 0.40 | 68 | 7532 | 0.01 | 0.0047 | 8T | 16 |
| U25 | 23105F | 0.94 | 68 | 5214 | 0.024 | 0.0047 | 1024B | 16 |

routing large complex boards. An algorithm that joins points in a point cluster has been proposed by Rosa and Lucio [32]. Hightower [33] developed a channel router that is faster than Lee's but does not always yield complete routing for a board. A program for efficient routing between fixed vias has been developed by Foster [34]. A combinational routing system based on line search and maze-running algorithms has been developed at the Sharp Corporation in Japan by Nishioka et al. [35]. This system is used predominantly on two-layer boards that are not excessively large. A method for finding the best routing path, squeezing and shoving, and ripping up previously made paths, followed by a cleanup routine that results in a 99% success rate, has been discussed by Bollinger [36]. A router based on a line search and wave front approach, which routes along a wave front initiated at the edge connector, has been developed by Patterson and Phillips [37]. Finally, a global router that routes all networks simultaneously has been developed by Soukup [38].

A typical routing layout for a multilayer board is shown in Figure 6.37. Interconnection paths may be laid out on more than one layer. To accomplish multilayer routing, feedthrough pins or vias are used. For DIP components, pins generally are assumed to extend through all layers of the board. In certain cases, a solid connection path, which is called a via, is made through the board. The via and the pin through hole are shown in Figure 6.38. The number of vias is generally limited by manufacturing considerations and struc-

**Table 6.14** Interconnection List for 25-Component Example

| Signal Set | | Signal Set | |
|---|---|---|---|
| No. | Component | No. | Component |
| 1 | U1, U2 | 21 | U13, U14 |
| 2 | U2, U3 | 22 | U14, U15 |
| 3 | U3, U4 | 23 | U11, U16 |
| 4 | U4, U5 | 24 | U12, U17 |
| 5 | U1, U6 | 25 | U13, U18 |
| 6 | U2, U7 | 26 | U14, U19 |
| 7 | U3, U8 | 27 | U15, U20 |
| 8 | U4, U9 | 28 | U16, U17 |
| 9 | U5, U10 | 29 | U17, U18 |
| 10 | U6, U7 | 30 | U18, U19 |
| 11 | U7, U8 | 31 | U19, U20 |
| 12 | U8, U9 | 32 | U16, U21 |
| 13 | U9, U10 | 33 | U17, U22 |
| 14 | U6, U11 | 34 | U18, U23 |
| 15 | U7, U12 | 35 | U19, U24 |
| 16 | U8, U13 | 36 | U20, U25 |
| 17 | U9, U14 | 37 | U21, U22 |
| 18 | U10, U15 | 38 | U22, U23 |
| 19 | U11, U12 | 39 | U23, U24 |
| 20 | U12, U13 | 40 | U24, U25 |

**Table 6.15** Placement Results for 25-Component Example

(a) Convection Example

| $\omega$ | Total wirelength (units) | $\lambda_T$ (fr/mhr) |
|---|---|---|
| 0.00 | 268 | 0.18947 |
| 0.25 | 208 | 0.19659 |
| 0.50 | 160 | 0.22480 |
| 0.75 | 142 | 0.23100 |
| 1.00 | 80 | 0.24500 |

(b) Conduction Example

| $\omega$ | Total wirelength (units) | $\lambda_T$ (fr/mhr) |
|---|---|---|
| 0.00 | 276 | 0.09450 |
| 0.25 | 238 | 0.09456 |
| 0.50 | 150 | 0.09477 |
| 0.75 | 106 | 0.09474 |
| 1.00 | 80 | 0.09482 |

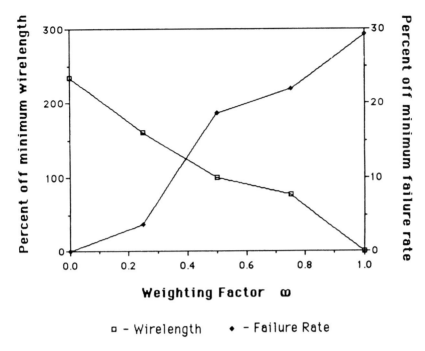

**Figure 6.35**   Placement results for convection-cooled 25-component example.

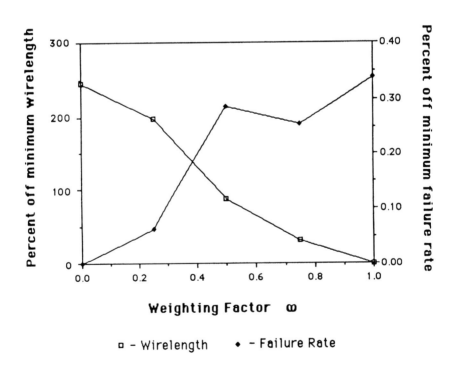

**Figure 6.36**   Placement results for conduction-cooled 25-component example.

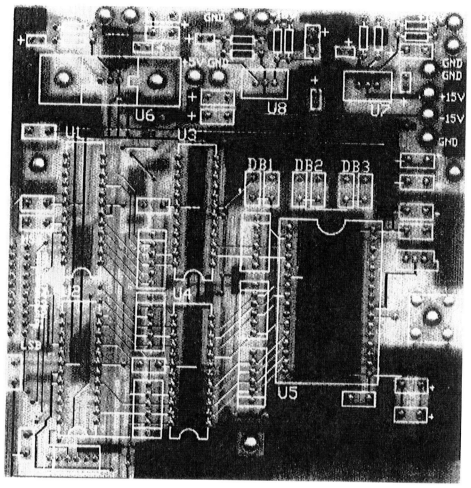

**Figure 6.37** Multilayer printed wiring board.

tural integrity. For this reason, routing procedures often aim at maximizing the number of connection paths that can be laid out on a single layer.

To keep a minimum distance between interconnection paths, the interconnection paths are generally laid out on a grid structure imposed on the carrier. For organic printed wiring boards, the grid lines are normally spaced at 50 mils (0.050 in.) with pins located on

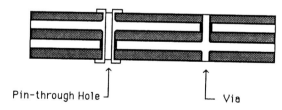

Pin-through Hole ⌐       ⌐ Via

**Figure 6.38** Multilayer connection structures.

100-mil centers. The use of a grid generally prohibits curved or diagonal lines, which are often seen in manual layouts. In addition, the grid naturally layers interconnection paths and distances are measured using Manhattan distance.

Interconnecting terminals with the shortest possible interconnection paths is essential in terms of producibility and efficient operation of large-scale integration and communication networks. Efficient design of the interconnection paths becomes increasingly difficult as the component density and critical circuit speed increase. Routing is a process of determining the interconnection paths needed for component networking. The basic problem is to lay out the necessary connection paths on the carrier after the components have been placed. The objective is to achieve the interconnections specified by the signal sets under all imposed constraints. Typical constraints include the width of the connector, the number of connectors that can impinge on a pin or via, the number of pins or vias that a connector can connect, the fact that connection paths cannot cross minimization of the number of vias or feedthroughs, the number of layers on which interconnection paths can be laid, and separation of connection paths by a specified distance. Routing algorithms are iterative procedures that generally include four basic operations: path determination, layering, path prioritizing, and path definition. Input parameters include constraint information, signal sets, component positions, and pin locations. The end product of an algorithm is a concise definition of all interconnection paths on the board.

Before discussing the tools used in routing, some of the common terminology is reviewed. Pins may be called nodes or vertices of a graph. Graphical structures that interconnect more than one pin are commonly referred to as subtrees. A net is a partial or complete interconnection of all pins in a signal set. A connection between two nodes is a branch. A channel is a line along one of the principal directions of the board. Finally, an interconnection tree is a special type of graph that connects a group of nodes.

### 6.7.1  Path Determination

Path determination and distance approximation are dependent on the interconnection tree that is being employed. Some of the types of interconnection trees have been defined in Section 6.3. For a simple signal set composed of only two pins, a wire (or track) is simply defined as passing between the two pins. However, if the signal set is composed of more than two pins, a net must be defined or approximated for all of the pins of the signal set. Procedures have been developed to approximate the various interconnection trees and can define the wire list for an individual signal set. In routing, path determination may be iteratively calculated as the router operates. Thus, a complete list of wires is not determined until the routing algorithm is completed.

The minimum tree algorithm closely approximates the minimum spanning tree. This method is governed by examining the distances between pairs of nodes and defining branches composed of at least two nodes. Initially, the distances between all possible pairs of nodes are compiled. The node pairs are then sorted in ascending order of distance and are considered for branches from the sorted list. Under these rules, the node pair with the shortest distance is defined as the first subtree. The remaining node pairs are then considered in the sorted order. Four rules govern the consideration of the node pairs in forming the interconnection tree [39]. (1) If neither of the nodes appears in the defined subtrees, a new subtree containing the node pair under consideration is defined. (2) If one of the nodes appears in a previously defined subtree, the node of the node pair that is not contained in the defined subtree is added to the subtree by the branch defined by the node pair under consideration. (3) If the nodes are contained in separate subtrees, the subtrees are joined by the branch defined by the node pair under consideration to form a larger single subtree. (4) If both nodes are present in a single subtree, the next node pair is considered. When all nodes of the signal set are defined in a single tree, the minimum tree has been defined. For $n$ nodes, $(n - 1)$ branches are formed in the complete tree. To provide a better understanding of the minimum tree algorithm, an example is shown in Figure 6.39.

The Steiner tree is constructed from a signal set of $n$ pins and the addition of extra connection points. The Steiner tree has been addressed by Hanan [2]. Unfortunately, an exact solution does not exist for the general Euclidean or Manhattan case. In addition, the extra $k$ points must be added to the wire list and their locations must be added to the data base for routing.

To approximate the Steiner tree, a connection network can be constructed using a few simple rules. Prim [40] suggests two simple principles that can be modified to determine the connection network for a signal set: (1) isolated nodes or unconnected nodes can be connected to the nearest node, and (2) isolated branches or a connection between two or more nodes can be connected to the node nearest to the branch. The idea in the modifications is to let connections be made to points other than the pins. As in the subtree algorithm, the lengths involved in connecting each pin to all other pins of a signal set are determined. The branch that has the short distance is accepted and recorded. A set of lines that includes a section or all of the recorded branches is defined on the board. Measures for all unconnected pins are defined by the shortest distance to the defined lines. The pin that has the shortest measured distance is selected and a branch is constructed to connect the pin to the existing branch. If a new line can be defined by the construction from a branch, then the line is added to the set of existing lines. When the set of lines has been updated, the distances are recalculated for all unconnected pins. The process continues until all pins are connected to the tree. Connections between nodes and existing branches are favored over connections to simply extended lines. In some cases, the initial selection of a branch may be weighted to improve the overall success of the process. For a signal set, the pin located closest to the geometric center of the signal set may be selected as having to be contained in the initial branch. An example of this procedure is shown in Figure 6.40.

Chain trees are approximated by the traveling salesman problem [41]. The traveling salesman problem is presented classically as the problem of determining the shortest tour by which a salesman starting at one city can visiting $N$ cities and return to the starting city. There are at most $(N - 1)!/2$ possible tours. For large values of $N$, it is impractical to consider all possibilities. To date, there are no acceptable exact computational solution methods for this problem. Certain branch-and-bound techniques appear to be efficient for some specific problems. However, the computation time involved is unpredictable and increases very rapidly with $N$. Numerous authors have used different techniques to obtain near-optimal solutions by a series of approximations and have been able to obtain optimal solutions for some specific problems. Results, which depend on inspectional work, are usually heuristic in nature and also highly problem dependent, which gives rise to programming difficulties. One method of attacking the problem in terms of routing has been suggested by Aker [42]. In this method, a starting pin is selected and the program selects the closest pin to the starting pin to form a path. The most recently selected pin is then used to select a new pin, which is not in the chain, based on the shortest distance. Although this method does not ensure that the shortest chain is obtained, it is quite fast and allows numerous starting pin selections. An example of this process is shown in Figure 6.41.

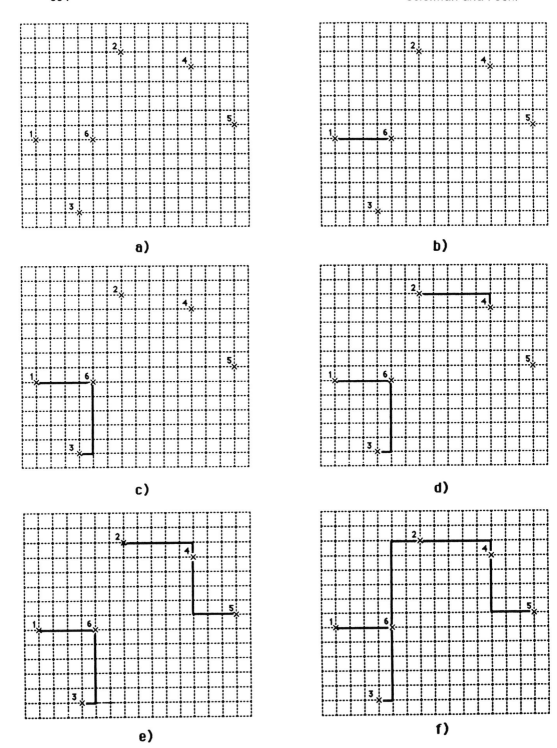

**Figure 6.39**   Minimum spanning tree method.

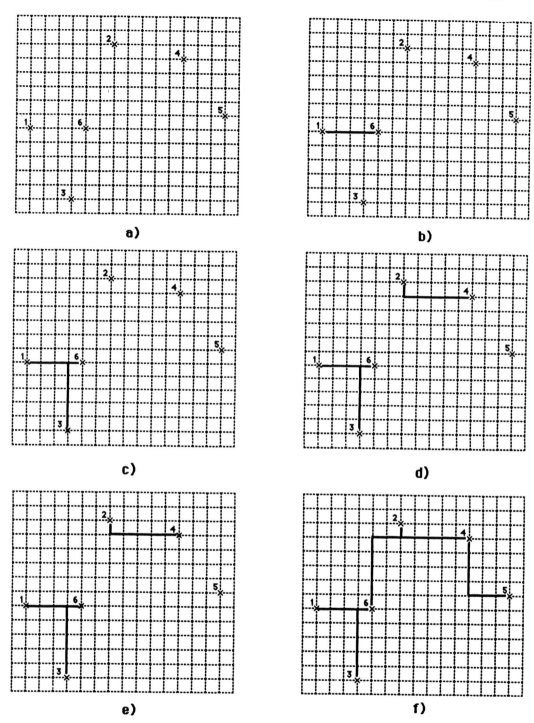

**Figure 6.40** Approximation of Steiner tree.

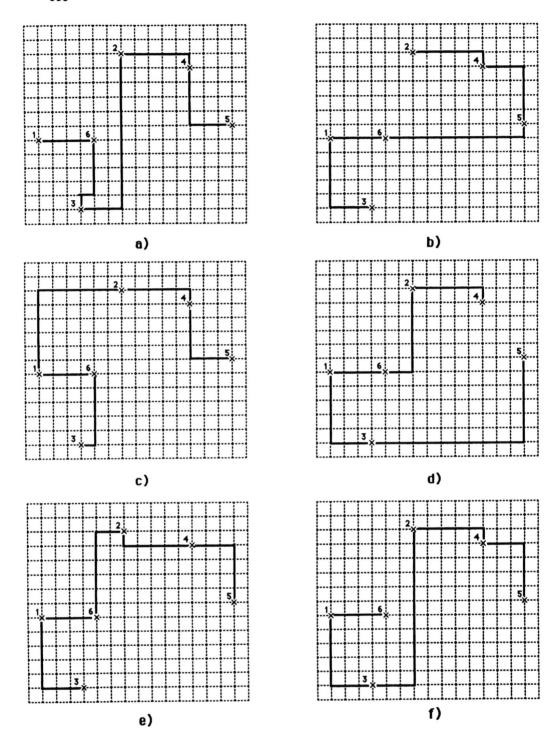

**Figure 6.41** Chaining tree (traveling salesman problem).

Another path determination problem involves the edge connector. Boards normally contain a number of pin groups that are located along the edges of the board. The groups represent edge connectors that allow data and signal transfer between the board and the rest of the system to which the board is connected. When an element pin must be linked to a connector pin, generally any connector pin may be selected. The problem that arises is how to make the most efficient use of the pin-to-connector assignments.

The goal of all routing is to make the process as simple as possible and to avoid crossovers when connecting the pins to the connectors. Aker [42] describes a process for connecting pins to edge connectors by the rotation of a vector extending from a selected connection pin. The connection pins are selected starting at the leftmost connection pin and moving to the right. The vector is rotated counterclockwise until it comes in contact with an element pin that must be connected. The path along the vector between the connector pin and the element pin is assigned. The base point of the vector is then moved to the next connector pin to the right of the previous connector pin, and the vector begins to rotate again until another element pin that must be connected to the connector is found. Again, the path is accepted and the base of the vector is moved to the next connector pin. The process continues until paths are accepted for all element pins that must be connected to a connector pin.

A problem that may occur in cases in which there are more connector pins than element pins is a skewed assignment of paths from element to connector pins. To alleviate this problem, Aker suggests using a slack ratio defined as the number of extra unassigned connectors over the number of remaining element pins that must be connected. If the slack ratio is greater than or equal to one and if the selected unassigned element pin is closer to another connector, then the current connector pin is skipped.

## 6.7.2 Layering

Layering of the routing paths involves determining the layer on which an entire or a partial interconnection path must be laid out. If the complete set of interconnection wires has been determined, layering involves grouping sets of wires onto different layers of the board. Grouping is determined by attempting to eliminate overlap of wires on the board. One natural grouping of wires is by principal directions defined by the superimposed grid. One layer will contain the wires that are predominantly aligned in one principal direction, and another layer will be composed of wires that predominantly run in the other principal direction. Problem wires that cannot fit on either of the first two layers without overlap can be assigned to available additional layers.

In iterative routing, the assignment of layers may be determined as the router proceeds to route the board. For example, if the shortest path between two nodes is obstructed, the router may decide to attempt a routing path on another layer. The time and method for making the routing decision could be determined by the layout engineer or the program. Figure 6.42 shows the cost of routing per layer in man-hours, which increases with additional board layers.

## 6.7.3 Path Prioritizing

The order in which the connection paths are constructed is thought to play a significant role in the overall success of the routing procedure. The determination of the order is generally based on the length of the connection path. Shorter paths are usually considered

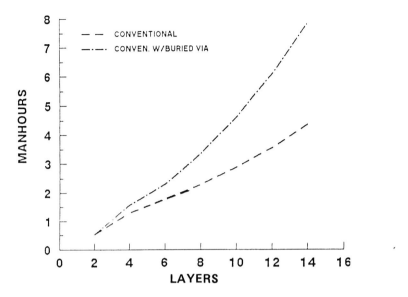

**Figure 6.42**   Trend in man-hours per board layer, cost computation.

first, followed by increasingly longer paths. To appreciate the importance of path prioritizing, consider the example shown in Figure 6.43. In this case, two pairs of pins must
be connected. If $A$ and $A'$ are connected using a horizontal line followed by a vertical
line, then the minimum connection of $B$ to $B'$ will obstructed. However, if $B$ and $B'$ are
connected first, then $A$ and $A'$ can still be connected by the minimum length.

Aker [42] suggests a simple rule for prioritizing the connections. The priority number
of any connection is equal to the number of pins enclosed by the rectangle defined by
the connection. Connection paths are laid out in ascending order of their priority numbers.
To demonstrate this procedure, consider the connection of four pin pairs $(A,B,C,D)$ in
Figure 6.44. When the four rectangles are drawn, the priority numbers of the four pin
pairs are 1, 0, 5, and 2, respectively. Therefore, the prioritized layout order is $B$, $A$, $D$,
$C$.

From the examples considered, it can be seen that the order of constructing interconnection paths is important in terms of successfully routing for minimal distances.
However, Abel [43] has presented statistical evidence that the performance of the router,
in terms of defining the minimum (or ideal) lengths of connections completely, is in fact
independent of the order in which the connections are attempted. For iterative routing
procedures, a complete list of connections may not be available for the entire board. In
addition, nets may be formed using points that are not defined until a connection is made.

## 6.7.4  Track Layout

Lee's maze-running algorithm [44] is one of the most important tools in routing. Lee's
algorithm was initially implemented on a single-layer router and later extended to multilayer
routing. The algorithm is conceptually easy to grasp and is best presented by means of
an example. Consider Figure 6.45, where each individual cell represents a crossing point
of the grid structure imposed on the board. The dark cells represent pins or recorded

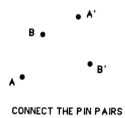

CONNECT THE PIN PAIRS

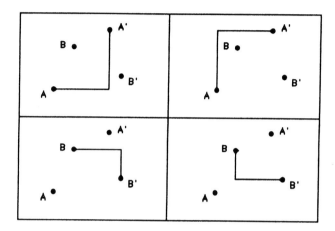

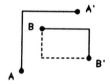

**Figure 6.43** Prioritizing connections.

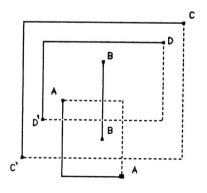

Priority Numbers
and Rank

| Pin | Priority No. | Rank |
|-----|-------------|------|
| A | 1 | 2 |
| B | 0 | 1 |
| C | 5 | 4 |
| D | 2 | 3 |

**Figure 6.44** Aker's priority scheme.

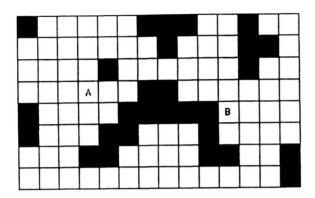

**a) Track Layout Problem**

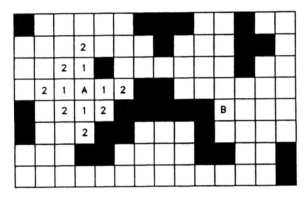

**b) Lee's Algorithm Initiation**

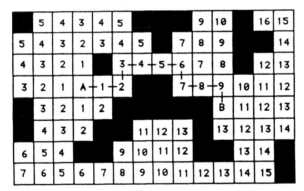

**c) Completed Lee's Algorithm**

**Figure 6.45**  Lee's algorithm example.

paths. Empty cells represent free routing area. In this example, the objective is to interconnect node *A* with node *B* by the shortest possible path. The shortest path is the path
that includes the smallest number of empty cells. For future reference, cell will be considered adjacent if they share common edges.

In Lee's algorithm, one of the two nodes that must be connected is selected as the starting node. In the example, we assume that node $A$ is selected. Once a node is selected, a 1 is written in all empty cells adjacent to the starting node. After all available adjacent cells are labeled, a 2 is written in all available empty cells adjacent to cells containing 1. Then a 3 is written in all available empty cells adjacent to cells containing number 2. The process continues in this manner until a number is entered into the cell containing node $B$ or until no empty cells remain. In the first case, the algorithm has defined the path and the optimum length. In the example, the order of the optimum length is 9 (Figure 6.45c). To determine the path, start at node $B$ and move to successively lower numbers until $A$ is reached. It should be noted that there may be choices of lower-numbered cells to select. In this case, the program is normally given a preferred direction in terms of selecting the next lower-numbered cell.

Various modifications of Lee's algorithm have been proposed and are employed. One modification is the application to multilayer boards. This is accomplished simply by redefining the two-dimensional cells as three-dimensional cubes. The procedure of cell numbering is the same. The board layers are represented by layers of cells. Restrictions in numbering can be made so that only certain cells allow transferring between layers.

One of the primary criticisms of Lee's algorithm is the excessive amount of computer memory required for its application. For every cell being processed, there must be a corresponding memory location within the computer. Other criticisms are directed at the time required to determine the interconnection path and the fact that the algorithm may fail to find a path.

To reduce the memory requirements and speed up the path definition, the two nodes that must be connected may be framed. By framing, only a specified portion of the board containing both nodes is considered for applications of Lee's algorithm. An artificial boundary is set up in this procedure. Normal frame sizes are 10 to 20% greater than the rectangle containing the two nodes. If Lee's algorithm fails to find a path, the frame may be enlarged or removed and the process can continue. Another strategy for speeding up Lee's algorithm involves a wise choice in the selection of the starting node. Nodes closest to the boundary of the board require less time than those closer to the center of the board. A third method of improving the speed is to allow a start at both nodes. This method is called a double fan-out. In this process, the labeling of empty cells continues until a contact is made between the two label processes. This requires a more sophisticated program but it does reduce the time required.

Another modification, suggested by Aker [42], is to redefine the labeling sequence so that only four states for any specific memory location are required. The idea stems from the observation that for any labeled cell all adjacent cells are either one unit greater or one unit lower than the chosen cell. From this observation, Aker noted that it is sufficient when considering an individual cell that the labels of the precedings cells be different from the labels of succeeding cells. The simplest sequence that Aker suggests is 1 - 1 - 2 - 2 - 1 - 1 - 2 - 2 - 1 - · · · . An example of this approach is shown in Figure 6.46.

## 6.7.5  Routing Schemes

When the components have been placed on the carrier, the signal sets must be processed to establish the interconnections between pins. The signal sets may be completely processed initially or iteratively processed and updated during routing. In complete processing, the entire wire list is determined initially. From the wire list, the layers on which the wires

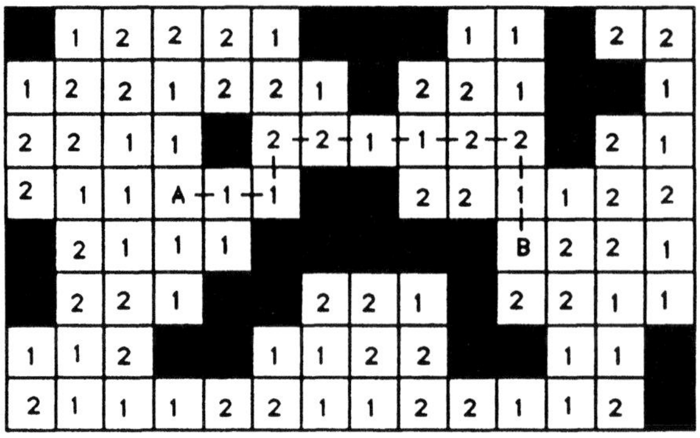

**Figure 6.46** Aker's method (sequence {1-1-2-2-1-1 · · ·})

reside are determined. Once the wires are assigned to layers, the wires on each layer are prioritized to determine the order in which they will be routed. Finally, the wires are routed. Complete processing generally results in a longer total wirelength and is not usually employed.

In the iterative case, the routing process combines signal set processing and wire determination directly with ordering, layering, and path definition. In this process, the path determination stage examines each signal set and approximates branch distances for all unconnected pins. The branches of each signal set are prioritized, with the shortest branches getting the highest priority. The branch in each signal set with the highest priority is routed and recorded. Once a branch is recorded in a signal set, the priority values of the remaining unconnected nodes of the signal set are reevaluated based on the distance between the unconnected nodes and related accepted branches. The path definition stage than attempts to make a connection between the unconnected node with the highest priority value and the channels defined by the accepted branches of the signal set. Priority values are highest for the shortest paths. Layering is allowed in the path definition stage. The iterative process continues to route and update priority values for unconnected nodes until all signal set nets are completed.

## 6.8 THE FUTURE

There are many alternatives to the algorithms defined here, and the study and formulation of new placement algorithms are continuing. Placement algorithms for two-sided boards still need to be developed and algorithms improved. Furthermore, more studies of incorporating other placement goals, such as minimizing congestion and preserving heat dissipation levels, into placement algorithms are required.

## EXERCISES

6.1. Determine the connectivity matrix for the set of nine components A, B, C, D, E, F, G, H, I. Use Eq. (6.6) and the signal set provided below. Assume $\lambda = -1$.

**Table 6.16**

| Components | Weighting factor |
|---|---|
| B,D,E,F,H | 1 |
| A,B,C,E | 1 |
| A,D,G,E | 1 |
| C,E,F,I | 1 |
| E,G,H,I | 1 |
| A,B,D | 1 |
| B,C,F | 1 |
| D,G,H | 1 |
| I,H,F | 1 |

6.2. Find the positions of the components using the pairwise linking method for the components and signal sets given in Exercise 6.1. Use Figure 6.47 as a reference for placement.

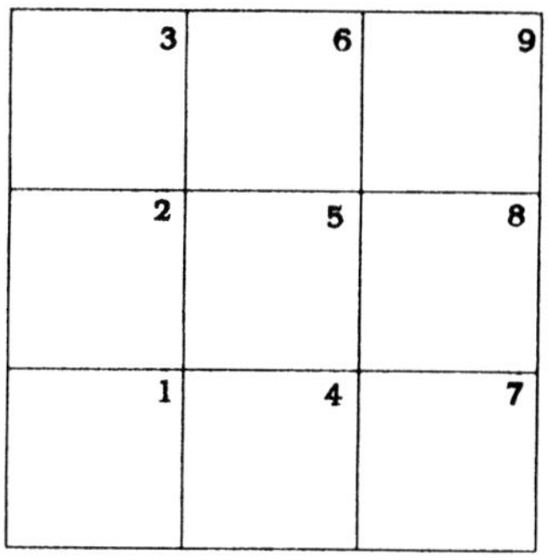

**Figure 6.47**

6.3. Determine the total wirelength based on center-to-center distances. Assume that the distance between the centers of two adjacent blocks is 1 unit.

6.4. Using the initially placed layout in Figure 6.48, apply the pairwise interchange method to determine the final placement. The signal set is given in Table 6.17.

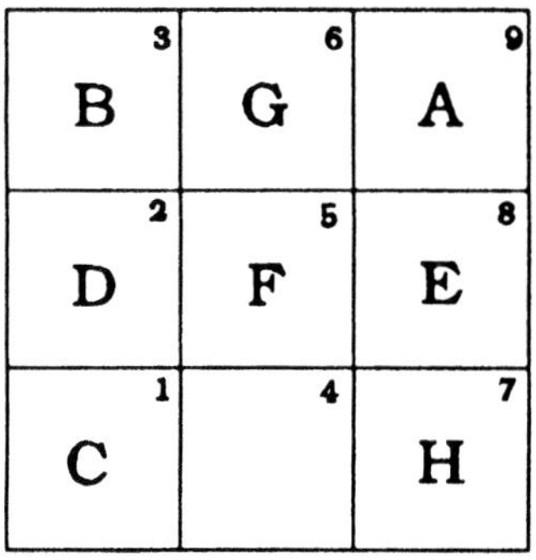

**Figure 6.48**

**Table 6.17**

| Components | Weighting factor |
|---|---|
| B,D,E,F,H | 1 |
| D,E,G,H | 1 |
| A,B,C | 1 |
| A,F,G | 1 |

6.5. Determine the independent set grouping for the eight components given in Exercise 6.4. Find the placement configuration after performing Steinberg's placement method on the largest independent set.

6.6. Calculate the force vector on component E from Exercise 6.4, using Eq. (6.9), to calculate the connectivity matrix.

6.7. Determine the optimal arrangement of a row for the three-component system given in Table 6.18.

**Table 6.18**

| Component name | $q$ (W) | $Rxx$ (°C/W) | $A$ (No. of failures/$1000^3$ hr) |
|---|---|---|---|
| A | 1.0 | 15 | 0.0020 |
| B | 1.5 | 10 | 0.0015 |
| C | 2.0 | 10 | 0.0030 |

Assume the hazard rate of each component is given by

$$h_i(T) = A_i(T)^2$$

The row is cooled convectively with air at an inlet temperature of 20°C, a mass flow rate of 0.03 kg/min, and $c_p = 1$ kJ/kg °C. Use the reliability placement procedure and an exhaustive search.

6.8. Find the reliability placement for a row of five components under convection cooling. Use the components provided in Table 6.19. Assume the inlet temperature is 30°C, the mass flow rate is 0.3 kg/min, and $c_p$ is 1000 J/kg °C.

**Table 6.19**

| Component | $A$ (kelvin) | $B$ (fr/mhr) | $D$ (fr/mhr) | $Rxx$ (°C/W) | $q$ (W) |
|---|---|---|---|---|---|
| 1 | 1000 | 1.00 | 0.1 | 30 | 1.50 |
| 2 | 2000 | 1.50 | 0.5 | 20 | 0.50 |
| 3 | 1000 | 2.00 | 0.1 | 40 | 1.00 |
| 4 | 1500 | 1.00 | 0.5 | 50 | 1.25 |
| 5 | 3000 | 1.25 | 0.1 | 20 | 2.00 |

6.8. Use the following equation to determine the component hazard rates:

$$h_i(T_i) = D_i + B_i e^{-A_i/T_i}$$

6.9. Use Aker's Prioritizing method to determine the order in which the pairs of terminal points should be routed in Figure 6.49.

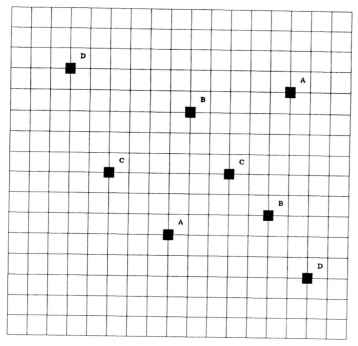

**Figure 6.49**

6.10. Determine the minimum spanning tree which connects the five nodes depicted in Figure 6.50.

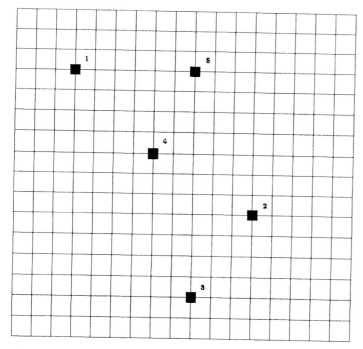

**Figure 6.50**

6.11. Determine the shortest path between the two connection points A by manually performing Lee's Algorithm on Figure 6.51.

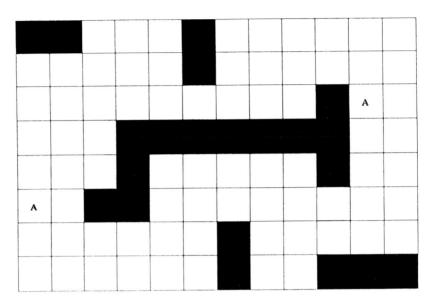

**Figure 6.51**

## REFERENCES

1.  Steinberg, L., "The Backboard Wiring Problem: A Placement Algorithm," *SIAM Rev.* 3(1):37–50 (1961).
2.  Hanan, M., "On Steiner's Problem with Rectilinear Distance," *J. Soc. Ind. Appl. Math.* 14:225–265 (1966).
3.  Hanan, M. and Kurtzberg, J., "Placement Techniques," in *Design Automation of Digital Systems: Theory and Techniques*, Breuer, M. A. (ed.), Prentice-Hall, Englewood Cliffs, N.J., 1972, chapter 5.
4.  Hanan, M., Wolff, P. and Agule, B., "A Study of Placement Techniques for Computer Logic Graphs," *Proc. 13th Design Automation Conference*, San Francisco, June 1976, pp. 214–224.
5.  Quinn, N. R. and Breuer, M., "A Force Directed Component Placement Procedure for Printed Circuit Boards," *IEEE Trans. Circuits Syst.* CAS-26(6):377–388 (1979).
6.  Kuhn, H. W., "The Hungarian Method for the Assignment Problem," *Naval Res. Logist. Q.* 11:83–97 (March–June 1955).
7.  Kurtzberg, J. M., "On Approximation Methods for the Assignment Problem," *J. ACM* 9(4):419–439 (1982).
8.  Munkres, J., "Algorithms for the Assignment and Transportation Problems," *J. Soc. Ind. Appl. Math.* 5:32–38 (1957).
9.  Breuer, M. A., "General Survey of Design Automation of Digital Computers," *Proc. IEEE* 54(12):1708–1721 (1966).
10. Wippler, G., Wiesal, M. and Mylnski, D., "A Combined Force and Cut Algorithm for Hierarchical VLSI Layout," *Proc. 19th Automation Conference*, June 1982, pp. 671–677.
11. Kernighan, B. W. and Lin, S., "An Efficient Heuristic Procedure for Partitioning Graphs," *Bell Syst. Tech. J.* 49:291–307 (1970).

12. Breuer, M. A., "A Class of Min-Cut Placement Algorithms," *Proc. 14th Design Automation Conference*, New Orleans, June 1977, pp. 284–290.

13. Schmidt, D. and Druffel, L., "An Iterative Algorithm for Placement of Integrated Circuits," *Proc. 12th Design Automation Workshop*, June 1975, pp. 361–368.

14. Corrigan, L. I., "A Placement Capability Based on Partitioning." *Proc. 16th Design Automation Workshop*, June 1979, pp. 406–413.

15. Dancer, D., and Pecht, M., "Component Placement Optimization for Convictively Cooled Electronics," *IEEE Trans. Reliab.*, vol. 38, no. 2, June 1989, pp. 199–205.

16. Zara, E., Midolo, L. and Mayer, A., "Advanced Environmental Control for Avionic Systems," *NAECON '77 Record*, 1977, pp. 60–67.

17. Osterman, M. D. and Pecht, M., "Placement for Reliability, Based on Physics of Failure Concepts," *NAECON 90*, May 1990, pp. 1021–1027.

18. Osterman, M. and Pecht, M., "Placement for Reliability and Routability of Convectively Cooled PWBs," *IEEE Transactions on Computer-Aided Design of Integrated Circuits and Systems*, vol. 9, no. 7, July 1990, pp. 734–744.

19. Osterman, M. D. and Pecht, M., "Component Placement for Reliability on Conductively Cooled Printed Wiring Boards," *Trans. ASME J. Electron. Packag. 111*:149–156 (1989).

20. Pecht, M., "Computer Aided Design for PCB Reliability," *Proc. 33rd Annual Institute of Environmental Sciences Conference*, May 2–5, 1987.

21. Pecht, M., Palmer, M. and Naft, J., "Thermal Reliability Management in PCB Design," *Proc. 1987 Annual Reliability and Maintainability Symposium*, January 27–29, 1987.

22. "Reliability Prediction of Electronic Equipment," *Military Handbook MIL-HDBK-217E*, Rome Air Development Center, New York, February 1987.

23. Research Analysis Center, *Reliability Design Handbook*, Cat. No. RDH-376, IIT Research Institute, Chicago, March 1976.

24. Pecht, M., Lall, P. and Whelan, S., "Temperature Dependency of Microelectronic Device Failures," *Quality and Reliability Engineering International*, vol. 6, 1990, pp. 275–284.

25. "Reliability Assessment of Advanced Technologies," RADC study conducted by Westinghouse and the University of Maryland.

26. Rothkopf, M. H., "Scheduling Independent Tasks on Parallel Processors," *Manage. Sci. 12*(8):437–447 (1966).

27. Rothkopf, M. H. and Smith, S.A., "There Are No Undiscovered Priority Index Sequencing Rules for Minimizing Total Delay Costs," *Oper. Res. 32*(2):451–456 (1984).

28. McNaughton, R., "Scheduling with Deadlines and Loss Functions," *Manage. Sci. 6*:1–12 (1959).

29. Pecht, M., RAMCAD: "Reliability and Maintainability Computer Aided Design," University of Maryland, 1984.

30. Hosking, G., et al., "An Automatic Routing Program for Very Complex Double Sided PC Boards," Third International Conference and Exhibition on Computers in Engineering and Building Design, Brighton, England, March 14–16, 1978.

31. Doreau, M. T. and Abel, L. C., "A Topologically Based No Minimum Distance Routing Algorithm," *Proc. 14th Design Automation Conference*, New Orleans, June 1977, pp. 92–99.

32. Rosa, R. C. and Lucio, T. P., "Programs for Two-Sided Printed Circuit Design," *Comput. Aided Des. 11*:297–303 (1979).

33. Hightower, D. W., "A Solution to Line-Routing Problems on the Continuous Plane," *Proc. 6th Design Automation Workshop*, June 1968, pp. 1–24.

34. Foster, J. C., "A Router for Multilayer Printed Wiring Backplanes," *Proc. 10th Design Automation Conference*, June 1973, pp. 44–45.

35. Nishioka, I., et al., "A Minicomputerized Automatic layout System for Two-Layer PWC," *Proc. 14th Design Conference*, June 1977, pp. 1–11.

36. Bollinger, H., "Designing Circuit Boards the Easy Way," *Mach. Des. 50*:110–115 (1978).

37. Patterson, G. L. and Phillips, B. H., "Computer Aided Design of PWC," GEC *J. Sci. Technol.*, pp. 107–110 (1977).

38. Soukup, J., "Global Router," *Proc. 16th Design Automation Conference*, June 1979, pp. 481–484.

39. Brinsfield, J. G. and Tarrant, S. R., "Computer Aids for Multilayer Printed Wiring Board Design," *Proc. 12th Design Automation Conference*, June 1975, pp. 269–305.

40. Prim, R. C., "Shortest Connection Networks and Some Generalizations," *Bell Syst. Tech. J. 36*:1389–1401 (1957).

41. Flood, M. M., "The Travelling Salesman Problem," *J. Oper. Res. 4*:61–75 (1956).

42. Aker, S., "A Modification of Lee's Path Connection Algorithm," *IEEE Trans. Electron. Comput. EC-16*(1): (1967), pp. 97–98.

43. Abel, L., "On the Ordering of Connections for Automatic Wire Routing," *IEEE Trans. Comput.* C-21(11), 1972, pp. 1227–1233.

44. Lee, C., "An Algorithm for Path Connections and Its Applications," *IRE Trans. Electron. Comput.*, EC-10(3), 1961, pp. 346–355.

45. Preas, B., and Karger, P., "Automatic placement: A review of current techniques," *IEEE 23rd Design Automation Conference*, 1986, pp. 622–629.

## SUGGESTED READING

Ablasser, I. and Jager, U., "Circuit Recognition and Verification Based on Layout Information," *Proc. 18th Design Automation Conference*, June 1981, pp. 334–336.

Adachi, T., Kitazawa, H., Nagatani, M. and Sudo, T., "Hierarchical Top Down Layout Design Method for VLSI Chip," *Proc. 19th Design Automation Conference*, June 1982, pp. 785–791.

Aho, A., Garey, M. and Hwang, F., "Rectilinear Steiner Trees: Efficient Special Case Algorithms," *Networks* 7:37–58, 1977.

Akers, S., Geyer, J. and Roberts, D., "IC Mask Layout with a Single Conduct Layer," *Proc. 7th Annual Design Automation Workshop*, San Francisco, 1970, pp. 7–16.

Alexander, D., "A Technology Independent Design Rule Checker," *Proc. 3rd USA–Japan Computer Conference*, 1978, p. 412.

Aranoff, S. and Abulaffio, Y., "Routing of Printed Circuit Boards," *18th Design Automation Conference*, Paper 9.1, June 1981.

Baird, H. S., "Fast Algorithm for LSI Artwork Analysis," *Proc. 14th Design Automation Conference*, June 1977, pp. 303–311.

Barney, C., "Plaudits and Pans Greet Engineering Work Station," Special Report, *Electronics*, July 29, 1985, pp. 51–52.

Berge, C., *The Theory of Graphs and Its Applications*, Wiley, New York, 1962.

Bowlby, R., "The DIP May Take Its Final Bows," *IEEE Spectrum*, June 1985, pp. 31–42.

Brayton, R. K. and McMullen, C., "The Decomposition and Factorization of Boolean Expressions," *IEEE, International Symposium on Circuits and Systems*, Rome, May 10–12, 1982, pp. 49–54.

Breuer, M. A., "The Formulation of Some Allocation and Connection Problems as Integer Programs," *Naval Res. Logist. Q. 13*:83–95 (1966).

Breuer, M. A., ed., *Design Automation of Digital Systems: Theory and Techniques*, Prentice-Hall, Englewood Cliffs, N.J., 1972.

Breuer, M. A., "Min-Cut Placement," *J. Des. Autom. Fault Tol. Comp. 1*(4):343–362 (1977).

Brinkmann, K. D. and Mlynski, D. A., "Computer Aided Chip Minimization for IC-Layout," *Proc. IEEE International Symposium on Circuits and Systems*, 1976, pp. 650–653.

Brooks, R. L., Smith, C., Stone, A. B. and Tutte, W. T., "The Dissection of Rectangles into Squares," *Duke Math. J. 7*:312–340 (1940).

Burstein, M. and Pelavin, R., "Heirarchical Wire Routing," *IEEE Trans. Computer-Aided Des. Integrated Circuits Syst. CAD-2*(4):223–234 (1983).

Chandrasekhar, M. and Breuer, M., "Optimum Placement of Tow Rectangular Blocks," *Proc. 19th Design Automation Conference*, 1982, pp. 879–886.

Chang, C. S., 'LSI Layout Checking Using Bipolar Device Recognition Technique," *Proc. 16th Design Automation Conference*, June 1980, pp. 95–101.

Chang, S. "The Generation of Minimal Trees with a Steiner Topology," *Journal of the Association for Computing Machinery*, *19*(4):699–711, October 1972.

Chen, K., Ferer, M., Khokhani, K. Nan, K. and Schmidt, S., "The Chip Layout Problem: An Automatic Wiring Procedure," *Proc. 14th Design Automation Conference*, June 1977, pp. 298–302.

Chawla, B. R., Gummel, H. K. and Kozak, P., "MOTIS—an MDS Timing Simulator," *IEEE Trans. Circuits Syst.* *CAS-22*(12):901–910 (1975).

Chiba, T., Okuda, N., Kambe, T., Nishioka, I., Inufushi, T. and Kimura, S., "SHARPS: A Hierarchical Layout System for VLSI," *Proc. 18th Design Automation Conference*, June 1981, pp. 820–827.

Christensen, C. and Pinson, E. N., "Multi-Function Graphics for a Large Computer System," *Proc. FJCC.* 31: 697–711 (1967).

Chyan, D. and Breuer, M., "A Placement Algorithm for Array Processors," *Proc. 20th Design Automation Conference*, 1983, pp. 182–188.

Ciampi, P. "A System for Solution of the Placement Problem," *Proc. 12th Design Automation Conference*, 1975, pp. 406–413.

Cocker, N., "Computer Aided Placement and Routing of High Density Interconnection Systems," Technical Report, International Computer Limited, Microsystems Division, Research and Development Organization, England August 1972.

Cole, B. C., "A Chip Business That Is Still Growing," Special Report, *Electronics*, July 22, 1985, pp. 40–45.

Collett, R., "Board Layout Systems Grapple with Advanced Technology," *Digital Des.* January 1986, pp. 55–58.

Cote, L. and Patel, A., "The Interchange Algorithms for Circuit Placement Problems," *Proc. 17th Design Automation Conference*, 1980, pp. 528–534.

Cox, G. and Carroll, B., "The Standard Transistor Array (Star), Part II: Automatic Cell Placement Techniques," *Proc. 17th Design Automation Conference*, 1980, pp. 451–457.

Dantzig, G., et al., "Solution of a Large Scale Travelling Salesman Problem," *J. Oper.* Res. 2: 394–410 (1954).

Davis, R. M., "The DOD Initiative in Integrated Circuits," IEEE, *Computer*, July 1979, pp. 74–79.

Deutsch, D., "A Dogleg Channel Router," *Proc. 13th Design Automation Conference*, 1976, pp. 425–433.

Donath, W. E., "Statistical Properties of the Placement of a Graph," *SIAM J. Appl. Math.* *16*(2): 376–387 (1968).

Druffel, L. E., Schmidt, D. C. and Wagner, R. A., "A Simple, Efficient Design Automation Processor," *Proc. 11th Design Automation Workshop*, June 1974, pp. 127–136.

Eisenberg, H., "CADMON: Improving the CAD System Human Interface," *Proc. 15th Design Automation Conference*, June 1978, pp. 353–358.

Fiduccia, C. M. and Mattheyses, R. M., "A Linear-Time Heuristic for Improving Network Partitions," *Proc. 19th Design Automation Conference*, June 1982, pp. 175–181.

Fisk, C. J., Caskey, D. L. and West, L. L., "ACELL: Automated Circuit Card Etching Layout," *Proc. IEEE* 55 (11):1971–1982 (1967).

Fletcher, R. and Reeves, C. M., "Function Minimization by Conjugate Gradients," *Comput. J.* 7:149–154 (1964).

Foley, E., "Designing MPU Boards for Testability," *Electron Test* 2:48–54 (1979).

Foley, E., "The Effects of the Microelectronics Revolution on Systems and Board Test," IEEE, *Computer*, October 1979, pp. 32–38.

Ford, L. and Fulkerson, D., *Flows in Networks*, Princeton University Press, Princeton, N.J., 1962.

Foster, J. C., "Prerouting Analysis Programs," Proc. *12th Design Automation Conference*, June 1975, pp. 306–310.

Foster, J. C., "A Unified CAD System for Electronic Design," *Proc. 21st Design Automation Conference*, June, 1984, pp. 365–373.

Funkunanga, K., Yamada, H., Stone, H. and Kasai, T., "Placement of Circuit Modules Using a Graph Space Approach," *Proc. 20th Design Automation Conference*, 1983, pp. 465–471.

Gamblin, R. L., Jacobs, M. Q. and Tunis, C. J., "Automatic Packaging of Miniaturized Circuits," in *Advances in Electronic Circuit Packaging*, Vol. 2, Walker, G. A. (ed.), Plenum, New York, 1962, pp. 219–232.

Garey, M. R. and Johnson, D. S., *Computers and Intractability*, Freeman, San Francisco, 1979.

Garside, R. G. and Nicholson, T. A., "Permutation Procedure for the Backbord Wiring Problem," *Proc. IEEE*, January 1968, pp. 27–30.

Gheewala, T. and MacMillan, D., "High-Speed GaAs Logic Systems Require Special Packaging," in *Everything Designers Need*, GigaBit Logic Inc., May 17, 1984.

Gilmore, P. C., "Optimal and Suboptimal Algorithms for the Quadratic Assignment Problem," *J. Soc. Ind. Appl. Math. 10*(2):305–313 (1962).

Glaser, R. H., "A Quasi-Simplex Method for Designing Suboptimal Packages for Electronic Building Blocks," *Proc. 1959 Computer Applications Symposium*, Armour Research Foundation, Illinois Institute of Technology, 1959, pp. 100–111.

Gross, A. G., Raamot, J. and Watkins, S. B., "Computer System for Pattern Generator Control," *Bell Syst. Tech. J. 49*(9):2011–2029 (1970).

Hanan, M., "Layout, Interconnection and Placement," *Networks*, 5:85–88 (1975).

Hashimoto, A. and Stevens, J., "Wiring Routing by Optimizing Channel Assignment within Large Apertures," *Proc. 18th Design Automation Conference*, 1971, pp. 155–169.

Hauser, J. E., "Hermetic Chip Carrier/Epoxy-Glass Multilayer Board Thermal Cycle Test Evaluation Study," General Electronic Ordnance Systems, May 9, 1980.

Heiss, S., "A Path Connection Algorithm for Multi-layer Boards," *Proc. 5th Design Automation Workshop*, 1968, pp. 6–14.

Held, M. and Karp, R., "A Dynamic Programming Approach to Sequencing Problems." *J. Soc. Ind. Appl. Math. 10*:196–210 (1962).

Heller, W. R., Sorkin, G. and Maling, K., "The Planar Package Planner for System Designer," *Proc. 19th Design Automation Conference*, June 1982, pp. 253–260.

Heller, L. G., Griffin, W. R., Davis, J. W. and Thoma, N. G., "Cascode Voltage Switch Logic—a Differential CMOS Logic Family," *Proc. IEEE International Solid-State Circuits Conference*, San Francisco, February 1984, pp. 16–17.

Hightower, D. W., "The Interconnection Problem—A Tutorial," *Computer*, 7(4):18–32 (1974).

Hightower, D. and Boyd, R. "A Generalized Channel Router," *Proc. 17th Design Automation Conference*, 1980, pp. 12–21.

Hillier, F. S. and Connors, M. M., "Quadratic Assignment Problem Algorithms and the Location of Indivisible Facilities," *Manage. Sci. 13*(1):42–57 (1966).

Horng, C. S. and Lie, M., "An Automatic/Interative Layout Planning System for Arbitrarily-Sized Rectangular Building Blocks," *Proc. 18th Design Automation Conference*, June 1981, pp. 293–300.

Hotchkiss, J., "The Roles of in-Circuit and Functional Board Test in the Manufacturing Process," *Electron. Packag. Prod. 19*(1):47–66 (1979).

Hung, M. and Rom, W., "Solving the Assignment Problem by Relaxation," *Operation Research*, vol. 28, no. 4, July–August 1980, pp. 969–982.

Hwang, F. K., "On Steiner Minimal Trees with Rectilinear Distance," *SIAM J. Appl. Math. 30*: 104–114 (1976).

Hwang, F. K., "An O(n log n) Algorithm for Suboptimal Rectilinear Steiner Trees," *IEEE Trans. on Circuit and Systems CAS-26* (1):75–77 (1979).

Hwang, F. K., "The Rectilinear Steiner Problem," *Design Automation and Fault Tolerant Computing*, 2(4):303–310 (1978).

Ji-Guang, X. and Kozawa, T., "An Algorithm for Searching Shortest Path by Propagating Wavefronts in Four Quadrants," *Proc. 18th Design Automation Conference*, June 1981, p. 29.

Kazuhiro, U., Kitazawa, H. and Harada, I., "CHAMP: Chip Floor Plan for Hierarchical VLSI Layout Design," *IEEE Trans. Computer-Aided Des.* 4(1): pp. 12–22 (1985).

Keister, F. Z., "Chip Carrier Study," Internal Hughes Aircraft Report to R. Y. Scapple, January 19, 1979.

Kernighan, B. W., "Some Graph Partitioning Problems Related to Program Segmentation," Ph.D. Thesis, Princeton University, January 1969, pp. 74–126.

Kitazawa, H. and Ueda, K., "Chip Area Estimation Method for Chip Floor Plan," *Electron. Lett.* 20(3):137–139 (1984).

Kodres, U. R., "Geometrical Positioning of Circuit Elements in a Computer," Conference Paper No. CP 59–1172, AIEE Fall General Meeting, October 1959.

Kodres, U. R., "Formulation and Solution of Circuit Card Design Problems Through Use of Graph Methods," in *Advances in Electronics Circuit Packaging*, Vol. 2, Walker, G. A. (ed.), Plenum, New York, 1962.

Koren, N., "Pin Assignment in Automated Printed Circuit Board Design," *Proc. 9th Design Automation Workshop*, 1972, pp. 72–79.

Korn, R. K., "An Efficient Variable-Cost Maze Router," *Proc. 19th Design Automation Conference*, June 1982, pp. 425–431.

Kozama, R. J., "IBM's Entry into CAE Worries Rivals," Special Report, *Electronics*, July 29, 1985, pp. 53–54.

Krauss, A. and Bar-Cohen, A., *Thermal Analysis and Control of Electronic Equipment*, McGraw-Hill, New York, 1985.

Kruskal, J. B., "On the Shortest Spanning Subtree of a Graph and the Traveling Salesman Problem," *Proc. Am. Math. Soc.* 7:48–50 (1956).

Kruskal, J. B., "Multi-Dimensional Scaling by Optimization Goodness of Fit to a Non-Metric Hypothesis," *Psychometrika* 29(1):1–27 (1964) and 29(2):115–129 (1964).

Kulkarni, K. G. and Prabhakar, A., "Automated Routing of Printed Circuit Boards," *J. Inst. Elec. Tele. Eng.* 24:158–163 (1978).

Kurtzberg, J. M., "Algorithms for Backplane Formation," in *Microelectronics in Large Systems*, Washington, D.C., Spartan Books, 1965, pp. 51–76.

Kurtzberg, J. M. and Estes, B., "Initial Card Placement Algoriths: An Evaluation," Burroughs Report TR 61–44, August 21, 1961.

Kurtzberg, J. M. and Seward, J., "Program for Star Cluster Wiring of Backplane," Burroughs Internal Report, January 1964.

Landeau, I. Ya., *Use of Computer for the Design of Computers*, Energiya, Moscow, 1974.

LaPaugh, A., "A Polynomial Time Algorithm for Optimal Routing around a Rectangle," *Proc. 21st Annual Symposium on Foundations of Computer Science*, 1980, pp. 282–293.

Lauther, U., "A Min-Cut Placement Algorithm for General Cell Assembliers Based on a Graph Representation," *Proc. 16th Design Automation Workshop*, June 1979.

Lawler, E. L., "The Quadratic Assignment Problem," *Manage. Sci.* 9:586–599 (1963).

Lawler, E. L. and Wood, D. E., "Branch and Bound Methods: A Survey," *J. Oper. Res. 14*: 699–719 (1966).

Leblond, A., "CAF: A Computer-assisted Floorplanning Tool," *Proc. 20th Design Automation Conference*, June 1983, pp. 747–753.

Leiserson, C., "Area-efficient Layouts (for VLSI)," *Proc. 21st Annual Symposium on Foundations of Computer Science*, 1980, pp. 270–281.

Lin, S., "Computer Solution of the Traveling Salesman Problem," *Bell Syst. Tech. J.* 44:2245–2269 (1965).

Little, J. D. C., Murty, K. G., Sweener, D. W. and Karel, C., "An Algorithm for the Traveling Salesman Problem," *J. Oper. Res. 11*:972–989 (1963).

Loberman, H. and Weinberger, A., "Formal Procedures for Connecting Terminals with a Minimum Total Wire Length," *J. ACM 4*:428–437 (1957).

Magnuson, W., "A Comparison of Constructive Placement Algorithms," *IEEE Region 6 Conference Record*, 1977, pp. 28–32.

Maling, K., Mueller, S. H. and Heller, W. R., "On Finding Most Optimal Rectangular Package Planes," *Proc. 19th Design Automation Conference*, June 1982, pp. 663–670.

Mamelak, J. S., "The Placement of Computer Logic Modules," *J. ACM 13*:615–629 (1966).

Manuel, T., "The Pell-Mell Rush into Expert Systems Forces Integration Issue," Special Report, *Electronics*, July 1, 1985, pp. 54–59.

Mattison, R. L., "A High Quality, Low-Cost Router for MOS/LSI," *Proc. 9th Design Automation Workshop*, Dallas, 1972, pp. 94–103.

McGrath, E. and Whitney, T., "Design Integrity and Immunity Checking," *Proc. 17th Design Automation Conference*, 1980, pp. 263–268.

Miron, G. J. and Tarrant, S. R., "The Automatic Printed Wiring Routing System of BACKIS," *Proc. 12th Design Automation Conference*, June 1975, pp. 311–316.

Moore, F. E., "Shortest Path Through a Maze," *Ann. Harvard Computation Lab. 30*(11):285–292.

Morozov, K. K., *Computer-Aided Design of Printed Circuit Boards*, Sovetskoye Radio Press, Moscow, 1978.

Nagatani, M., Miyashita, H., Okamoto, H., Tansho, K. and Sugiyama, Y., "An Automated Layout System for Functional Blocks: PLASMA," Monograph of Technical Group on Design Technology of Electronics Equipment, Information Processing Society, Japan, 1980, pp. 41–110.

Newman, W. and Sprull, R., *Fundamentals of Interactive Computer Graphics* (transl.), Mir Press, Moscow, 1976.

Nugent, C. E., Vollman, T. E. and Ruml, J., "An Experimental Comparison of Techniques for the Assignment of Facilities to Locations, *J. Oper. Res. 16*:150–173 (1968).

Odawara G., Iijima, K. and Wakabayashi, K., "Knowledge Based Placement Technique for Printed Wiring Board," *Proc. 19th Design Automation Conference*, June 1985, pp. 616–622.

Ohtsuki, T. and Sato, M., "Gridless Routers for Two-Layer Interconnection," *Proc. IEEE International Conference on Computer Aided Design*, Nov. 1984, pp. 76–78.

Ore, O., *Theory of Graphs*, American Mathematical Society, Providence, R.I., 1962.

Otten, R. H. J. M., "Automatic Floorplan Design," *Proc. 19th Design Automation Conference*, June 1982, pp. 261–267.

Palczewski, M., "Performance of Algorithms for Initial Placement," *Proc. 21st Design Automation Conference*, 1984, pp. 399–404.

Perl, Y. and Shiloach, Y., "Finding Two Disjoint Paths between Two Pairs of Vertices in a Graph," *Journal of the Association for Computer Machinery 25*:1–9 (1978).

Persky, G., Deutsch, D. N. and Schweikert, D. G., "LTX-A Minicomputer Based System for Automated LSI Layout," *J. Des. Autom. Fault Tol. Comp. 1*(3):217–256 (1977).

Persky, G., "PRO—An Automatic String Placement Program for Polycell Layout," *Proc. 13th Design Automation Conference*, 1976, pp. 417–424.

Pinter, R. "On Routing Two-Point Nets Across A Channel," *Proc. 19th Design Automation Conference*, 1982, pp. 894–902.

Pomentale, T., "An Algorithm for Minimizing Backboard Wiring Functions," *Commun. ACM 8*(11):699–703 (1965).

Pomentale, T., "The Minimization of Backboard Wiring Functions," *SIAM Rev. 9*(3):564–568 (1976).

Preas, B. T. and Gwyn, C. W., "Methods for Hierarchical Automatic Layout of Custom LSI Circuit Masks," *Proc. 15th Design Automation Conference*, Las Vegas, June 1978, pp. 206–212.

Preston, G. W., "High-Speed Integration Circuits for Military Applications," IDA Paper P-1423, November 1979.

Prince, J. L., "VLSI Device Fundamentals," in *VLSI Fundamentals and Applications*, Barbe, D. F. (ed.), Springer-Verlag, New York, 1980.

Quinn, N. R., Jr., "The Placement Problem as Viewed from the Physics of Classical Mechanics," *Proc. 12th Design Automation Conference*, June 23–25, 1975, pp. 173–178.

Radar, J. A., "Dynamic Allocation of Arrays in FORTRAN," *Eleventh Annual Asilomar Conference on Circuits, System, and Components*, 1977, pp. 295–298.

Reiter, S. and Sherman, G. "Discrete Optimization," *J. Soc. Ind. Appl. Math. 13*:864–889 (1965).

Rhea, J., "VHSIC—Advanced Components for a New Generation of Weapon Systems," *Intel. Def. Rev. 6*:887–889 (1980).

Richardson, R. E., "Modeling of Low-Level Rectification RFI in Bipolar Circuitry," IEEE Trans. Electromagnetic Compatibility *EMC-21*:307–311 (1979).

Rivert, R. and Fiduccia, C. M., "A Greedy Channel Router," *Proc. 19th Design Automation Conference*, Las Vegas, June 1982, pp. 418–424.

Rosenberg, R., "New Tools Make CAE Systems Affordable," Special Report, *Electronics*, July 29, 1985, pp. 46–50.

Rosenthal, C. W., "Increasing Capabilities in Interactive Computer Graphics Terminals," *Proc. 9th Design Automation Workshop*, June 1972, pp. 317–325.

Rosenthal, C. W., "Physical Design and Manufacturing Information Aspects," *Proc. 21st Design Automation Conference*, 1984, pp. 374–383.

Rubin, F., "The Lee Path Connection Algorithm," *IEEE Trans. On Comput., C-23*:907–914 (1974).

Sahni, S. and Bhatt, A., "The Complexity of Design Automation Problems," *Proc. 17th Design Automation Conference*, 1980, pp. 402–411.

Sato, K., Nagai, T., Shinoyama, H. and Yahara, T., "MIRAGE—a Simple Model Routing Program for the Hierarchical Layout Design of IC Masks," *Proc. 16th Design Automation Conference*, June 1979, pp. 297–304.

Sato, K., Nagai, T., Tachibana, M., Shimoyama, H., Ozaki, M. and Yahara, T., "MILD—a Cell Based Layout System for MOS LSI," *Proc. 18th Design Automation Conference*, June 1981, pp. 828–836.

Schuler, D. M. and Ulrich, E. G., "Clustering and Linear Placement," *Proc. 9th* Design Automation Workshop, Dallas, 1972, pp. 50–56.

Schweikert, D., "A 2-Dimensional Placement Algorithm for The Layout of Electrical Circuits," *Proc. 13th Design Automation Conference*, 1976, pp. 408–416.

Schweikert, D. G. and Kernighan, B. W., "A Proper Model for the Partitioning of Electrical Circuits," *Proc. 9th Design Automation Workshop*, Dallas, 1972, pp. 57–62.

Sha, L. and Dutton, R., "An Analytical Algorithm for Placement of Arbitrarily Sized Rectangular Blocks," *Proc. 22nd Design Automation Conference*, 1985, pp. 602–608.

Shiraishi, H. and Hirose, F., "Efficient Placement and Routing for Masterslice LSI," *Proc. 17th Design Automation Conference*, Minneapolis, June 1980, pp. 458–464.

Shteyn, M. Ye. and Shteyn, B. Ye., *The Methods of Machine Design of Digital Equipment*, Sovetskoye Radio Press, Moscow, 1973.

Shupe, C. F., "Automatic Component Placement in the NOMAD System," *Proc. 12th Design Automation Conference*, June 1975, pp. 162–168.

Shupe, C. F., "Automatic Component Placement in an Interactive Minicomputer Environment," *Proc. 18th Design Automation Conference*, June 1981, pp. 145–152.

Sloan, J., *Design and Packaging of Electronic Equipment*, Van Nostrand Reinhold, New York, 1985.

Soukup, J., "Circuit Layout," *Proc. IEEE: 69, (1)*:1281–1304 (Oct. 1981).

Soukup, J., "Fast Maze Router," *Proc. 15th Design Automation Conference*, 1978, pp. 100–102.

Steiglitz, K. and Weiner, P., "Some Improved Algorithms for Computer Solution of the Traveling Salesman Problem," *Proc. 6th Annual Allerton Conference*, October 1968, pp. 814–821.

Stevens, J. M., "Fast Heuristic Techniques for Placing and Wiring Printed Circuit Boards," CAC Document No. 58, Center for Advanced Computation, University of Illinois, Urbana (1972).

Sudo, T., Ohtsuki, T. and Goto, S., "CAD System for VLSI in Japan," *Proc. IEEE 71*:129–143 (1983).

Suen, L., "A Statistical Model for Net Length Estimation," *18th Design Automation Conference*, Paper 40.3, June 1981, pp. 769–773.

Supowit, K., "A Minimum-Impact Routing Algorithm," *Proc. 19th Design Automation Conference*, 1982, pp. 104–111.

Szymanski, T., "Dogleg Channel Routing is NP-Complete," *IEEE Trans. on Computer Aided Design*, CAD-4:31–40 (Jan. 1985).

Thompa, M., "An Optimal Solution to the Wire-length Problem," *Proc. 12th Annual ACM Symposium on Theory of Computing*, 1980, pp. 161–176.

Ueda, K., "Placement Algorithm for Logic Modules," *Electron. Lett.* *10*(10):206–208 (1974).

Ueda, K. and Kitazawa, H., "Algorithm for VLSI Chip Floor PLAN," *Electron. Lett.* *19*(3):77–78 (1983).

Ueda, K., Kitazawa, H., and Harada, I., "CHAMP: Chip Floor Plan For Hierarchical VLSI Layout Design," *IEEE Trans. on Computer Aided Design of Integrated Circuits and Systems*, (4): (1):12–22 (Jan. 1985).

Vajda, S., *Mathematical Programming*, Addison-Wesley, Reading, Mass., 1961.

Vecchi, M. and Kirkpatrick, S., "Global Wiring By Simulated Annealing," *IEEE Trans. on Computer Aided Design*, CAD-2:215–222 (1983).

Weinberger, A., "Large-scale Integration of MOS Complex Logic: A Layout Method," *IEEE Journal of Solid State Circuits*, SC-2:182–190 (Dec. 1967).

Welt, M. J., "NOMAD: A Printed Wiring Board Layout System," *Proc. 12th Design Automation Conference*, June 1975, pp. 152–161.

Wiseman, D., "To Simulate or Not to Simulate?" *Electron. Test* 2(3):45–64 (1979).

Yudin, D. B., Goryashko, A. P. and Nemirovskiy, A. S., *Mathematical Methods for Optimization of Devices and Algorithms Used in Automatic Control Systems*, Sovetskoye Radio Press, Moscow, 1982.

Zibert, K. "Ein Beitrag zum rechnergestutzten topologisschen Entwurf von Hybride-Schaltungen," Ph.D. Thesis, Technical University, Munich, 1974.

Zibert, K. and Seal, R., "On Computer Aided Hybrid Circuit Layout," *Proc. IEEE International Symposium on Circuits and Systems*, San Francisco, 1974, pp. 314–318.

# 7

# Thermal Design Analysis

**Dennis K. Karr**    *Airpax, Frederick, Maryland*, and *Cetar, Ltd., Columbia, Maryland*

**Milton Palmer, III**    *Ramsearch, College Park, Maryland*

**David Dancer**    *Cetar, Ltd., Columbia, Maryland*

## 7.1  INTRODUCTION

### 7.1.1  Aims of Thermal Analysis and Control

Thermal considerations in the design of electronic packages are becoming increasingly important. The desire for greater speed and power in electronic devices has given rise to a continued increase in circuit densities and power dissipation in modern electronic packages. Figure 7.1 graphically shows the trend of increasing complexity in electronic circuitry since 1950, from small-scale integration (SSI) levels to very large scale integration (VLSI) levels and beyond. These increasing power levels have initiated a revolution in the thermal control technologies used to solve the problems involved in cooling these assemblies. A case in point is digital computer packaging. Thermal control has evolved from an array of industrial fans used in 1946 to cool the ENIAC (using 18,000 vacuum tubes), to complex conduction modules such as the NEC LCM and the IBM thermal conduction module (TCM) (shown in Figure 7.2), to direct immersion cooling of circuit board–mounted components in dielectric fluorocarbon in the Cray II supercomputer (shown in Figure 7.3). Recently, even more esoteric technologies have found application in electronic cooling, such as the use of the thermoelectric effect and heat pipes to cool individual components.

Two main fields of thermal science are applied to the design of electronic packages: thermal control and thermal analysis. Thermal control is the selection of heat transfer technology and equipment to meet specific thermal requirements in the design of the electronic package being considered. The goals are typically removing the heat dissipated by the electrical and electronic components, constraining the maximum electronic component temperatures below specified values, and minimizing thermal gradients between components to influence thermal sensitivity. Dissipation of electrical energy in electronic

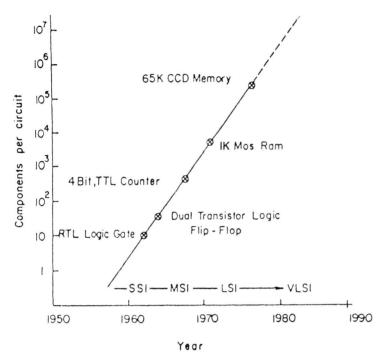

**Figure 7.1**   Circuit density trend. (From Ref. 1, © McGraw-Hill, Inc.)

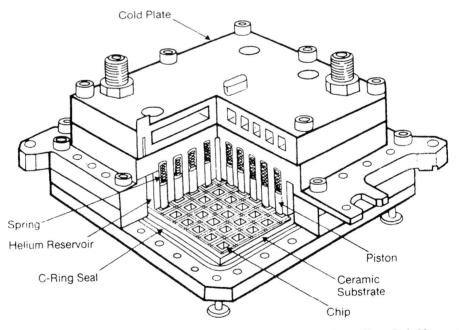

**Figure 7.2**   IBM cold plate–cooled thermal conduction module (TCM). (From Ref. 83, courtesy of ISHM.)

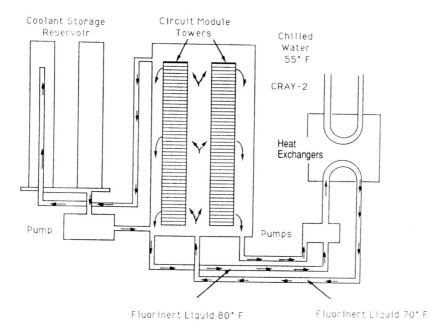

Coolant Storage
Reservoir

Circuit Module
Towers

Chilled
Water
55° F

CRAY-2

Heat
Exchangers

Pump

Pumps

Fluorinert Liquid 80° F

Fluorinert Liquid 70° F

**Figure 7.3**   Flow of coolant in Cray-2 supercomputer.

components results in an equivalent amount of heat. In order to control the temperature in the vicinity of the component, the heat must ultimately be transferred from the component to a heat sink. This heat transfer ranges from about 0.2 W/cm$^2$ for a typical printed wiring board to about 20 W/cm$^2$ for current VLSI, with future estimates of heat transfer on the order of 100 W/cm$^2$. Additional factors that often must be considered in the thermal control design include noise levels that limit fan selection and allowable temperature of coolant exhaust, as well as cost, weight, location, accessibility, and power consumption of the thermal control mechanisms themselves.

There are four reasons why it is desirable to know and control the temperatures of the components of an electronic system. First, the materials selected for the system should be compatible with the temperatures to be experienced in service. Melting or vaporization of materials is virtually never acceptable. Second, for some electronic components, such as transistors, a high junction temperature causes an increase in current, which in turn causes an increase in power dissipation, which in turn causes an increase in junction temperature, and so on. This can lead to what Kraus and Bar-Cohen [1] describe as a "thermal runaway" and catastrophic failure of the component.

Third, thermal control is necessary to ensure stable and predictable performance. High operating temperatures can result in loss of performance in two ways: loss of noise margin and reduced operation speed. For example, for a transistor connected in a common-emitter configuration, a prescribed value of collector-to-emitter voltage $V_{ce}$ is required for logic operation. If the collector current $I_c$ increases because of high junction temperature, the desired value of $V_{ce}$ may not be obtainable. Thus, if a particular device runs hotter than others in the same circuit, the difference in temperature can cause the logic levels to vary, resulting in a loss in noise margin. A maximum difference in temperature throughout a circuit of 35°C is commonly recommended. The speed of virtually all logic types goes down as the operating temperature increases [2]. Thermal control allows better estimation of power and current requirements, resulting in better system design.

Finally, much has been written about the inverse relationship between junction or component temperature and reliability. Operating temperatures and the failure rates of most electronic components are related. This may be a direct relationship or may be a dependence on a threshold value, a change in temperature, or a rate of change of temperature [3]. A commonly used rule of thumb, which is not recommended by these authors, is that a device's failure rate is halved by reducing its operating temperature by 10°C.

"Catastrophic thermal failures" may be caused by melting or vaporization of the components, thermal fracture of the mechanical support, mechanical separation of the leads, or plastic flow of the printed circuits. An upper temperature bound is generally specified to prevent failures such as these. For example, the "critical component temperature" (CCT) is defined by the United States Navy as 105°C for semiconductor devices and 20°C less than the "manufacturer's rating of outside surface" for other devices (reference: MIL-M-28787). The Commercial Airlines ARINC Specification 404 defines the maximum case temperature for a variety of component types (see Table 7.1).

Prolonged use at elevated operating temperatures can lead to failures in many electronic components due to creep in the bonding materials, parasitic chemical reactions in switches and connectors, and diffusion in solid-state devices [1]. Electrical failures may be caused by electrical overstress, dielectric breakdown, and electromigration [6]. The expression used commonly for the failure rate $\lambda_i$ of an electronic component is the Arrhenius relation, expressed as

$$\lambda_i(kT_j) = D_i + B_i e^{-E_i(1/kT_j)} \tag{7.1}$$

where $kT_j$ is the junction temperature, $E_i$ is the activation energy for the specific failure

**Table 7.1** Commercial Airlines ARINC Specification 404

| Part type | Maximum case temperature[a] |
|---|---|
| Capacitors | |
|    Tantalum—solid, foil, and sintered anode | $0.67 \times T_M = T_C$ |
|    Aluminum electrolytic, paper, ceramic, glass, mica | $0.72 \times T_M = T_C$ |
| Resistors | |
|    All | $0.68 \times T_M = T_C$ or 120°C, whichever is lower |
| Semiconductors, integrated circuits, diodes | $0.6 \times T_{MJ} - (\theta_{JC} \times P) = T_C$ |
| Relays and switches | $0.75 \times T_M = T_C$ |
| Transformer, coils and chokes | Min. of 35°C below manufacturer's hot spot temperature |

[a] $T_M$ = component manufacturer's maximum permissible body temperature at zero power (°C); $T_{MJ}$ = component manufacturer's maximum permissible semiconductor junction temperature at zero power (°C); $\theta_{JC}$ = component manufacturer's thermal resistance from junction to case (°C/W); $P$ = normal power dissipation for the specific circuit application (W); $T_C$ = external surface or case temperature of component (°C).
Source: Ref. 207. Copyright © Computer Science Press. Reprinted with permission.

mechanism under study, and $B_i$ and $D_i$ are constants based on package type, environment, and electrical characteristics. This has been developed as a monotonically increasing relationship between the failure rate and the junction temperature [3]. Many of the failure mechanisms related to overtemperature conditions have been modeled as Arrhenius relations based on the philosophy that these mechanisms are governed by chemical and diffusion processes [6]. Table 7.2 shows the typical activation energies that are associated with many of these types of failure mechanisms.

However, there are several electronic component failure mechanisms that are functions of temperature change, the rate of temperature change, and spatial temperature gradients for which the Arrhenius relation is inappropriate. These are primarily mechanical failure mechanisms and arise from differential thermal expansion between bonded materials, large time-dependent temperature changes, and large spatial temperature gradients, all of which can cause failures due to tension, compression, bending, fatigue, and fracture [6, 7]. In general, the nonelectrical failure mechanisms can be classified into package-related failures, die failures, and failures due to interconnects. Package failures may occur at the package seal, lid, body, lead frame, and external leads and in the encapsulant. Die failures may occur at the die, the die attach, or the substrate attach. The interconnect failures may occur at the wire and the wire bond and in the conductor paths in the die and the substrate [8].

Models for many of these failure mechanisms have been developed only recently [7, 8]. Many of these models, including those for bond wire flexure fatigue, shear fatigue at bond pad–substrate interface, shear fatigue at bond pad–wire interface, die brittle cracking, die attach fatigue, and substrate attach fatigue, show a dependence on $\Delta T$, or the temperature difference encountered. As a result, recommended values for the maximum operating $\Delta T$ have been established (see Table 7.3) [8].

In summary, then, many failure mechanisms in electronic components are temperature dependent. The temperature dependence of various failure mechanisms corresponding to different failure sites is shown in Table 7.4.

For these reasons, effective thermal management of electronic components and printed circuit board (PCB) assemblies is critical to ensure predictable and reliable performance. Thermal evaluation of printed circuit board hardware frequently involves the use of either thermocouples or infrared imaging equipment. However, the identification of excessive temperatures, temperature gradients, or temperature fluctuations at the hardware stage typically results in a need for costly and time-consuming redesign and/or a duplication of the layout process. It is far more desirable to be able to predict board and component temperatures prior to fabrication.

Therefore, thermal analysis is becoming an important part of the PCB design process. Computer-aided thermal analysis tools are increasingly being implemented in computer-aided design (CAD) system development environments. These tools are typically being integrated in the design system so that the PCB mechanical design—that is, the component layout and board geometry data—can flow into the thermal analysis routines and, through successive iterations, the board designer can optimize component placement and board design to manage the thermal conditions effectively [9, 10].

The aim of thermal analysis is to predict the temperatures and heat fluxes that will occur in an electronic assembly during field operation. The results of thermal analysis are needed to determine the reliability of the printed wiring board (PWB) or system, which is based on both the electronic reliability of the individual components in the predicted thermal environment and the mechanical reliability of the component connections and mechanical package. The results of the thermal analysis are also used to determine the

**Table 7.2** Time-Dependent Failure Mechanisms in Semiconductor Devices

| Device association | Process | Relevant factors[a] | Accelerating factors[a] | Typical activation energy (eV) | Model | Reference |
|---|---|---|---|---|---|---|
| Silicon oxide | Surface charges | Mobile ions | $T, V$ | 1.0 | Fitch et al. | 1A |
| Silicon–silicon oxide interface | Inversion, accumulation | $E/V, T$ | | | Peck | 2 |
| | Oxide pinholes | $E/V, T$ | $E, T$ | 0.7–1.0 (Bipolar) | 1984 WRS | 1B |
| | | | | 1.0 (Bipolar) | Hokari et al. | 5 |
| | Dielectric breakdown (TDDB) | $E/V, T$ | $E, T$ | 0.3–0.4 (MOS) | Domangue et al. | 3 |
| | | | | 0.3 (MOS) | Crook, D. L. | 4 |
| | Charge loss | $E, T$ | $E, T$ | 0.8 (MOS) EPROM | Gear, G. | 11 |
| Metallization | Electromigration | $T, J$ | $J, T$ | 1.0 | Nanda, et al. | 6 |
| | | Grain size | | Large grain Al (glassivated) 0.5 Small grain Al | Black, J. R. | 7 |
| | | Doping | | 0.7 Cu-Al/Cu-Si-Al (sputtered) | Black, J. R. | 12 |
| | Corrosion Chemical Galvanic Electrolytic | Contamination | $H, E/V, T$ | 0.6–0.7 (for electrolysis) E/V may have thresholds | Lycoudes, N. E. | 8 |
| Bond and other mechanical interfaces | Intermetallic growth | $T$, impurities Bond strength | $T$ | 1.0 (Au/Al) | Fitch, W. T. | 9 |
| Various wafer fab, assembly, and silicon defects | Metal scratches Mask defects, etc. Silicon defects | $T, V$ | $T, V$ | 0.5–0.7 eV | Howes, et al. | 13 |
| | | | | 0.5 eV | MMPD | 13 |

<sup></sup>$^a$ $V$ = voltage; $E$ = electric field; $T$ = temperature; $J$ = current density; $H$ = humidity

$^b$ Reference:

1A. 1.0-eV activation for leakage type failures.
Fitch, W. T., Greer, P., and Lycoudes, N., "Data to Support 0.001%/1000 Hours for Plastic I/C's." Case study on linear product shows 0.914-eV activation energy which is within experimental error of 0.9- to 1.3-eV activation energies for reversible leakage (inversion) failures reported in the literature.

1B. 0.7 to 1.0 eV for oxide defect failures for bipolar structures. This is under investigation subsequent to information obtained from 1984 Wafer Reliability Symposium, especially for bipolar capacitors with silicon nitride as dielectric.

2. 1.0-eV activation for leakage-type failures.
Peck, D. S., "New Concerns About Integrated Circuit Reliability," 1978 Reliability Physics Symposium.

3. 0.36 eV for dielectric breakdown for MOS gate structures.
Domangue, E., Rivera, R., and Shepard, C., "Reliability Prediction Using Large MOS Capacitors," 1984 Reliability Physics Symposium.

4. 0.3 eV for dielectric breakdown.
Crook, D. L., "Method of Determining Reliability Screens for Time Dependent Dielectric Breakdown," 1979 Reliability Physics Symposium.

5. 1.0 eV for dielectric breakdown.
Hokari, Y., et al., IEDM Technical Digest, 1982.

6. 1.0 eV for large grain Al-Si (compared to line width).
Nanda, Vangurd, Gj-P; Black, J. R.; "Electromigration of Al-Si Alloy Films," 1978 Reliability Physics Symposium.

7. 0.5 eV Al, 0.7 eV Cu-Al small grain (compared to line width).
Black, J. R., "Current Limitation of Thin Film Conductor," 1982 Reliability Physics Symposium.

8. 0.65 eV for corrosion mechanism.
Lycoudes, N. E., "The Reliability of Plastic Microcircuits in Moist Environments," 1978 Solid State Technology.

9. 1.0 eV for open wires or high-resistance bonds at the pad bond due to Au-Al intermetallics.
Fitch, W. T., "Operating Life vs Junction Temperatures for Plastic Encapsulated I/C (1.5 Mil Au Wire)," unpublished report.

10. 0.7 eV for assembly-related defects.
Howes, M. G., and Morgan, D. V., *Reliability and Degradation, Semiconductor Devices and Circuits*, Wiley, New York, 1981.

11. Gear, G., "FAMOS PROM Reliability Studies," 1976 Reliability Physics Symposium.

12. Black, J. R., unpublished report.

13. Motorola Memory Products Division, unpublished report.

Source: Ref. 200, courtesy of Motorola Semiconductor Product Sector.

**Table 7.3**  Recommended Values for Component Operating $\Delta T$

| Usage environment classification | | $\Delta T^a$ |
|---|---|---|
| MIL-HDBK-217E | Proposed | (°C) |
| $A_{IA}$ $A_{IB}$ $A_{IC}$ $A_{IF}$ $A_{IT}$ $A_{RW}$ | $A_I$ | 30 |
| $A_{UA}$ $A_{UB}$ $A_{UC}$ $A_{UF}$ $A_{UT}$ | $A_U$ | 55 |
| $C_L$ | $C_L$ | b |
| $G_B$ $G_{MS}$ | $G_B$ | 30 |
| $G_F$ | $G_F$ | 55 |
| $G_M$ $M_P$ | $G_M$ | c |
| $M_{FA}$ $M_{FF}$ $M_L$ | $M_F$ | b |
| $N_H$ $N_S$ $N_{SB}$ | $N_I$ | 50 |
| $N_U$ | $N_U$ | 55 |
| $U_{SL}$ | $N_{UL}$ | b |
| $N_{UU}$ | $N_{UU}$ | 35 |
| $S_F$ | $S_F$ | 35 |

[a] $\Delta T$ values are for use when thermal analysis or test data are not available.
[b] Application environments referring to this note are of short duration and have negligible effects on the package (nonelectrical) related failure mechanisms, for which the prelaunch storage conditions will have the dominant effect. Use $\Delta T = 5°C$ for storage under controlled storage conditions and $\Delta T = 20°C$ for uncontrolled storage conditions.
[c] Use $G_B$ application environment for equipment mounted in temperature-controlled compartments and $G_F$ for uncontrolled compartments.
Source: Ref. 8.

suitability of competing thermal control designs. A complete thermal analysis of an electronic package involves analysis of individual locations of energy dissipation (such as emitter-base junctions of transistors, diodes, or cores of resistances), carriers designed for mounting individual components (such as printed wiring boards) and heat exchangers selected for thermal control of component assemblies and the entire package. Thus, a complete thermal analysis requires consideration of each of the primary levels of packaging, as illustrated by Figure 7.4.

Thermal analysis involves the understanding of the fundamental heat transfer mechanisms used in the thermal control of electronic packages. Heat may be transferred from a body by three modes: conduction, convection, and/or radiation, as shown in Figure 7.5. Conduction is heat transfer that occurs when there is a temperature gradient in a body. Convection is heat transfer that occurs due to the transport of thermal energy by the bulk motion of a fluid. If the fluid motion is due to buoyancy forces, the convection is termed natural or free convection. Convective motion that is caused by external sources, such as pumps or fans, is called forced convection. Another form of free or forced convection occurs when the fluid undergoes a phase change, normally boiling. This method of heat removal is not very common in general electronics applications because of the large volume typically associated with the equipment (pumps, condensers, etc.), as was shown in Figure 7.3. However, phase-change convection is useful in environments with extremely

**Table 7.4**  Summary of the Failure Sites and Mechanisms Along with the Related Temperature Dependence

| Failure site | Failure mechanism | Form of temperature dependence |
|---|---|---|
| Wire | Flexure fatigue | Magnitude of temperature change, temperature, temperature gradient |
| Wire bond | Shear fatigue | Magnitude of temperature change, temperature, temperature gradient |
| Die | Fracture | Magnitude of temperature change, temperature, temperature gradient |
| Die adhesive | Fatigue | Magnitude of temperature change, temperature |
| Encapsulant | Reversion | Temperature |
| Package | Moisture ingress, leading to corrosion | Temperature, magnitude of temperature change |
| Die metallization | Electrolyte formation | Temperature |
|  | Corrosion | Temperature |
| Die passivation | Thermal mismatch fatigue | Magnitude of temperature change |
| Die | Hot spot formation | Temperature |
| Device oxide | Electrostatic discharge | Temperature |
| Device | Dielectric breakdown | Temperature |
|  | Ionic contamination | Temperature |
|  | Surface charge spreading | Temperature |
| Device substrate-oxide interface | Hot electrons | Temperature |
| Device metallization tracks | Electromigration | Temperature, temperature gradient |
|  | Metallization migration | Temperature |

Source: Ref. 7, © John Wiley & Sons, Inc. Reprinted by Permission.

high heat flux, such as mainframe computers and supercomputers, and in certain microwave electronic applications. Figure 7.6 shows the relative effectiveness of natural convection, forced convection, and boiling for the cooling of electronics under different conditions.

Radiation is heat that is transferred between two bodies by the exchange of electromagnetic radiation. Radiation takes place between any two bodies that have different temperatures and is the only form of heat transfer that occurs in a vacuum.

This chapter discusses all three modes of heat transfer—radiation, convection, and conduction. All three occur in the printed circuit board environment. As a broad rule of thumb, under common forced convection conditions, the relative proportions of these heat transfer modes are 10:60:30, respectively. In a natural convection environment, the

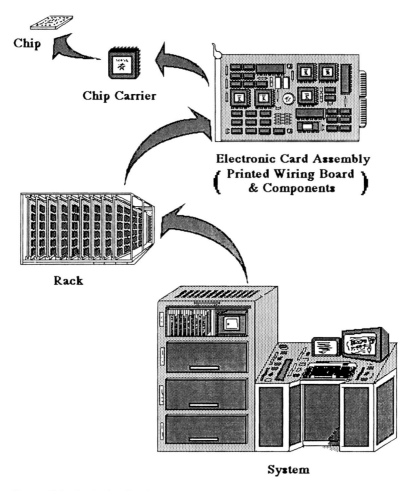

**Figure 7.4**  Packaging levels.

relative proportions become 20:40:40 [11]. The reader may note that more space is devoted to conduction than to the other heat transfer modes. This does not imply that these other modes of heat transfer are not important but merely reflects the fact that the state of the art for predicting conduction in electronic systems is much more advanced than the state of the art for predicting heat transfer by other modes. For example, to predict convection effects, an analyst must primarily extrapolate empirical data obtained with configurations that may be considerably different from the configuration of interest to the analyst.

In this chapter, therefore, the general laws and relationships associated with heat transfer in microelectronic applications will be established. First, the concept of the *thermal resistance* of a microelectronic device will be explored.

## 7.1.2  Thermal Resistance—Definitions

The industry standard description of the thermal performance of an integrated circuit device is the thermal resistance, referred to as $R_{ja}$, or $\theta_{ja}$. $R_{ja}$ is often broken out into two

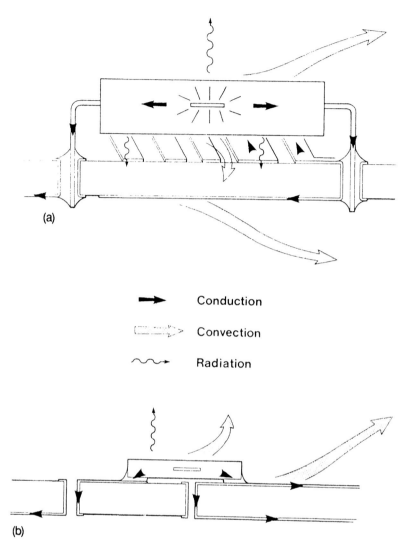

**Figure 7.5** (a) Heat dissipation from DIP mounting structure. (b) Thermal dissipation from components on PCB. (From Ref. 83, courtesy of ISHM.)

separate components, an internal resistance $R_{jc}$, or $\theta_{jc}$, and an external resistance $R_{ca}$, or $\theta_{ca}$. The largely conductive resistance $R_{jc}$ represents the thermal path from the active chip, through the chip support and chip bonding materials, to the case of the integrated circuit package. The primarily convective resistance $R_{ca}$ characterizes the thermal path from the case to the external cooling fluid, either directly or through a heat sink [12]. The semi-conductor junction temperature, $T_j$, is related to these values by

$$T_j = T_a + P_d(R_{jc} + R_{ca}) \tag{7.2}$$
$$= T_a + P_d(R_{ja}) \tag{7.3}$$

where $T_j$ is the junction temperature (°C), $T_a$ the local ambient temperature (°C), $P_d$ the package power dissipation (W), $R_{jc}$ the junction-to-case thermal resistance (°C/W), $R_{ca}$

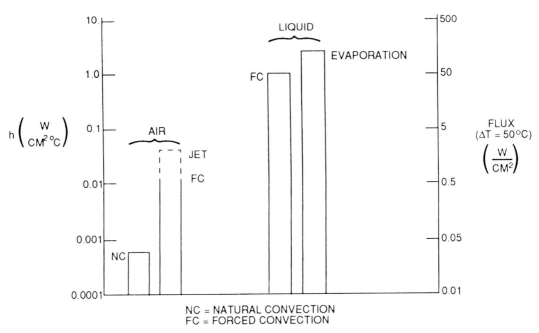

**Figure 7.6** Comparison of heat transfer coefficients and heat flux for various modes of cooling. (From Ref. 83, courtesy of ISHM.)

the case-to-ambient thermal resistance (°C/W), and $R_{ja}$ the junction-to-ambient thermal resistance (°C/W) [13].

The thermal resistance $R_{ja}$ can be defined [11] as

$$R_{ja} = \frac{T_j - T_f}{Q_c} \tag{7.4}$$

where $T_j$ is the junction temperature (°C), $T_f$ the coolant temperature (°C), and $Q_c$ the chip heat dissipation (W). Similarly, $R_{jc}$ can be defined as

$$R_{jc} = \frac{T_j - T_c}{Q_c} \tag{7.5}$$

where $T_c$ is the case temperature (°C).

Manufacturers of components typically list on their data sheets single typical values for $R_{ja}$, representing stand-alone components characterized in still air [2, 14]. Similarly, typical values are given for $R_{jc}$, measured by standard methods described in MIL-STD-883C, Method 1012, and SEMLI Standard Document 1321. These standard test methods allow the choice of either a fluid bath or a heat sink environment and assume that the resulting values of $R_{jc}$ are independent of these environmental changes [15].

However, studies have shown that the local circuit board-to-ambient temperature rise, $T_{ba}$, at the component has a first-order effect on $R_{ja}$. In turn, $T_{ba}$ is affected by board size, thermal conductivity, component placement, and total power dissipation. It has been shown that under constant cooling conditions, $R_{ja}$ can vary by a factor of 4 or more due to changes in the ratio of $T_{ba}$ to the package power dissipation $P_d$ [16].

Thus, $R_{ja}$ is useful only as a first approximation of the thermal performance under actual end-use conditions. Circuit designers who are unaware of these constraints and attempt to use the published $R_{ja}$ values to predict the thermal performance under all conditions may be faced with redesign efforts after the identification of thermal problems.

Similarly, tests have shown that $R_{jc}$ is not a constant. Instead, the internal thermal path, and thus $R_{jc}$, is dependent on the application environment, specifically the package attachment, the thermal characteristics of the second-level packaging technique, and the external convective heat transfer coefficient [12]. Data collected show that $R_{jc}$ varies as much as 40% just because of the variations allowable in the standard measurement specifications [15].

Thus, strictly speaking, $R_{jc}$ as presently defined is valid only when the package case is isothermal. When temperature variations are encountered on the external surfaces of the package, as is common, errors may be encountered when the reported values of $R_{jc}$ are used to predict chip temperatures [12].

### 7.1.3   New Approaches—Thermal Resistance Models

Improved models have been proposed. A more detailed model for the determination of $R_{ja}$ has been developed by Andrews [16]. In this model, as illustrated by Figure 7.7, $R_{ja}$ is broken into component parts and is defined as

$$R_{ja} = \left( R_{jh} = \frac{R_{ha}R_{hb}}{R_{ha} + R_{hb}} + \frac{R_{ha}}{R_{ha} + R_{hb}} \right) \frac{T_{ba}}{P_d} \tag{7.6}$$

where $R_{jh}$ is the junction-to-header thermal resistance (°C/W), $R_{ha}$ the header-to-ambient thermal resistance (°C/W), $R_{hb}$ the header-to-board thermal resistance (°C/W), and $T_{ba}$ the mounting surface temperature rise above ambient temperature (°C).

Using this modeling concept, Andrews [16] presents eight variables that should be listed on a data sheet for use in determining $R_{ja}$:

$$R_{ja} \quad R_{jh} \quad P_d \quad T_{ba} \quad S_p\,h \quad S_s\,R_{sao}$$
$$(°C/W)\ (°C/W)\ (W)\ (°C)\ (W/cm\,°C)\ (°C/W)$$

Here $h$ is the combined convection and radiation heat transfer coefficient (W/m$^2$°C), $S_p$ is the sensitivity of the junction-to-ambient temperature rise to changes in $T_{ba}$ with $P_d$ and $h$ constant,

$$S_p = \frac{T_{ja2} - T_{ja1}}{T_{ba2} - T_{ba1}} \tag{7.7}$$

$S_s$ is the sensitivity of socket-to-ambient temperature rise to changes in $T_{ba}$ with $P_d$ and $h$ constant (when socket is used), and $R_{sao}$ is the socket-to-ambient thermal resistance with $T_{ba} = 0$,

$$R_{sao} = \frac{T_{sa}}{Q_s} \qquad (T_{ba} = 0),\ °C/W \tag{7.8}$$

In this model, the heat generated by the chip, $P_d$, flows from the junction to the header, where it divides into heat flows $Q_a$ and $Q_b$. $Q_a$ is the package-to-air heat flow, dependent on the net conduction, convection, and radiation. $Q_b$ is the package-to-board

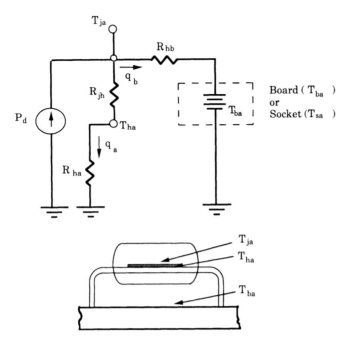

$$R_{ja} = R_{jh} + \frac{R_{ha}R_{hb}}{R_{ha} + R_{hb}} + \frac{R_{ha}}{R_{ha} + R_{hb}} * \frac{T_{ba}}{P_d}$$

**Figure 7.7**   Package thermal resistance model. (From Ref. 16, © 1988, IEEE.)

heat flow, which is dependent on the board temperature and the heat resistance path from the package to the board [2].

Bar-Cohen et al. [17] have likewise offered a modified model for the determination of $R_{jc}$. In a package used in actual operating conditions, heat may be nonuniformly generated, and two- and three-dimensional conduction effects are common in single and multichip packages. Thus, there is a multiplicity of internal paths and heat is distributed according to the boundary conditions imposed on the surfaces of the package. Therefore, an expression for chip temperature should be a function of elapsed time, thermal capacitance of each layer, power dissipation, and the temperature of each exposed surface, for every package type of interest. With this approach, a complex analytical relation would be established that would fully characterize the thermal behavior of the chip package, including both transient and steady-state responses, and would be indistinguishable from the finite-element model. In lieu of this fully developed approach, the chip package thermal resistance could be reasonably approximated by the use of the "lumped" thermal capacitance of each distinct material and division of the package surfaces into top, bottom, side, and lead areas.

To begin this approach, a model has been proposed that is limited to thermal steady state and includes heat transfer from the chip to just three distinct surfaces on the package: the top, bottom, and side. These conditions are approximated by a network of three thermal resistances connected in parallel, or

$$Q = \frac{T_\mathrm{j} - T_1}{R_1} + \frac{T_\mathrm{j} - T_2}{R_2} + \frac{T_\mathrm{j} - T_3}{R_3} \tag{7.9}$$

where $Q$ is heat flow (W); $T_\mathrm{j}$ is junction temperature (°C); $T_1$, $T_2$, and $T_3$ are top, bottom, and side surface temperature, respectively (°C); and $R_1$, $R_2$, and $R_3$ are top, bottom, and side thermal resistance respectively (°C/W). This equation can be algebraically manipulated to yield

$$T_\mathrm{j} = \frac{R_2 R_3 T_1}{R_\mathrm{s}} + \frac{R_3 R_1 T_2}{R_\mathrm{s}} + \frac{R_1 R_2 T_3}{R_\mathrm{s}} + \frac{R_1 R_2 R_3 Q}{R_\mathrm{s}} \tag{7.10}$$

where

$$R_\mathrm{s} = R_1 R_2 + R_2 R_3 + R_1 R_3 \tag{7.11}$$

Generalizing for $n$ elements along the surface, this becomes

$$T_\mathrm{j} = \sum_{k=1}^{n} A_k T_k + Q A_{n+1} \tag{7.12}$$

where $A_k$ is determined by the internal resistances of the chip package. For example, the coefficient of $T_1$ in the case of $n = 3$ would be

$$A_1 = \frac{R_2 R_3}{R_1 R_2 + R_2/R_3 + R_1 R_3} \tag{7.13}$$

These equations thus incorporate the contribution of each distinct package surface to heat removal from the chip. A modified junction-to-case thermal resistance can be written as

$$R_\mathrm{jc} = \frac{T_\mathrm{j} - \sum_{k=1}^{n} A_k T_k}{Q} \tag{7.14}$$

This model, by incorporating the thermal contribution of each distinct package surface to the junction temperature, is proposed by Bar-Cohen et al. [17] as a critical first step in the establishment of an improved, modified $R_\mathrm{jc}$.

In summary, then, current research efforts are addressing the relationships between the multiplicity of internal heat transfer paths, the effects of external surface temperatures and convective heat transport, and the effects of neighboring components on thermal impedance. Other work not detailed herein is directed toward improving the model for thermal capacitance to reflect the time-dependent heat accumulation that occurs during power switching [18]. It should be emphasized again that the current published data sheet values for thermal impedance should be used as first approximations only and that much care should be given to assumptions that are made when using these data.

## 7.2 CONDUCTION

### 7.2.1 Fourier's Law

When a temperature gradient exists in a medium, heat transfer takes place such that at the macroscopic level heat is transferred from the part of the medium at the higher temperature to the part at the lower temperature. The medium can be a solid or a fluid. This transfer of heat can also take place from one body at a higher temperature to another body that is in physical contact and at a lower temperature. At the microscopic level, conduction can be defined as the transfer of energy from the more energetic to the less energetic particles of a substance due to particle interaction. The energy level of the particles is related to the random translational motion, the internal rotational motion, and the vibrational motion. Higher energies are associated with higher temperatures. Therefore, in conduction, energy transfer takes place in the direction of decreasing temperature.

For steady-state one-dimensional conduction without any internal heat generation, the governing empirical law of conduction is

$$q = -kA \frac{dT}{dx} \tag{7.15}$$

where $q$ is the heat transfer rate, or power (W); $k$ is a property associated with a particular material called its thermal conductivity (W/m °C) and is positive for all known materials; $A$ is the cross-sectional area through which heat is flowing (m$^2$); $T$ is the temperature in the body (°C); and $x$ (m) measures the direction normal to the area $A$. This relationship is known as Fourier's law. It states that the rate of heat flow through a material is proportional to the area of material normal to the heat flow and the temperature gradient along the heat flow path.

The thermal conductivity is a very important material property in thermal analysis. A vacuum, in which $k = 0$, and the material diamond, for which $k = 2300$ W/m °C, represent the limits on the conductivity. Generally, the thermal conductivities of metals are higher than those of nonmetals, and crystalline materials have higher values of conductivity than amorphous materials. Tables 7.5 and 7.6 list the thermal conductivities for a number of materials relevant to electronics packaging.

For the more general case of transient three-dimensional conduction with internal heat generation (e.g., from the dissipation of electrical energy), heat conduction is described by

$$\rho C_p \frac{\partial T}{\partial t} = \nabla \cdot \mathbf{k} \, \nabla T + q''' \tag{7.16}$$

The operator

$$\nabla$$

**Table 7.5**  Thermal Conductiveness

| Material | Thermal conductivity | (W/m °C) |
|---|---|---|
| Air | 0.0257 | at 20°C |
| Alumina (90%) | 16.7 | at 20°C |
| Beryllia (99.5) | 168 | at 20°C |
| Copper | 386 | |
| Epoxy-glass | 0.294 | |
| Gold | 314 | |
| Kovar | 16.7 | |
| Silicon | 125 | |
| Solder (60/40) | 49.3 | |

Source: Ref. 83, courtesy of ISHM.

**Table 7.6** Equivalent Thermal Conductiveness

| Material | Equivalent thermal conductivity (W/m °C) |
|---|---|
| Epoxy-glass (no copper) | 0.294 |
| Epoxy-glass (1 oz. copper) | 9.11 |
| Epoxy-glass (2 oz. copper) | 17.71 |
| Epoxy-glass (4 oz. copper) | 35.13 |
| Finstrate (1 mm copper) | 249.1 |

Source: Ref. 83, courtesy of ISHM.

is defined as

$$\nabla = \mathbf{i}\frac{\partial}{\partial x} + \mathbf{j}\frac{\partial}{\partial y} + \mathbf{k}\frac{\partial}{\partial z}$$

in the Cartesian coordinate system, where $\mathbf{i}$, $\mathbf{j}$, and $\mathbf{k}$ are unit vectors. In Eq. (7.16), $\rho$ is the density of the material (kg/m$^3$), $C_p$ is the material specific heat (J/kg °C), $T$ is the temperature (°C), and k is the thermal conductivity (W/m °C). Special cases of Eq. (7.16) are applicable to several electronics heat transfer problems. If the thermal conductivity and the specific heat of the material are constant, then Eq. (7.16) reduces to:

$$\frac{\partial T}{\partial t} = \alpha \nabla^2 T + \frac{q'''}{\rho C_p} \tag{7.17}$$

where $\alpha = k/\rho C_p$ is defined as the thermal diffusivity and the quantity $q'''/\rho C_p$ represents the rate at which the temperature of the material would rise due to dissipation if thermal conduction was absent. If the equipment has reached steady operating conditions, Eq. (7.17) reduces to

$$\nabla^2 T + \frac{q'''}{k} = 0 \tag{7.18}$$

Equation (7.18) is called the Poisson equation. If the material under consideration is not dissipative, Eq. (7.17) reduces to

$$\frac{\partial T}{\partial t} = \alpha \nabla^2 T \tag{7.19}$$

Equation (7.19) is called Fourier's equation. Finally, if the material is nondissipative and in steady state, Eq. (7.17) reduces to

$$\nabla^2 T = 0 \tag{7.20}$$

Equation (7.20) is called the Laplace equation.

## 7.2.2  Boundary Conditions

Two different kinds of boundary conditions may exist in electronic heat transfer:

1.  Specified temperature: Temperature on the boundary surface may be specified as a function of time and position as follows:

$$T = T(s,t) \tag{7.21}$$

where $s$ denotes coordinates along the surface and $t$ is time.

2. Specified heat flux: The derivative of temperature in a direction normal to the boundary surface may be specified as follows:

$$\frac{\partial T}{\partial n} = f(s,t,T) \tag{7.22}$$

where $n$ is the direction of the local normal to the surface $s$. If the surface is insulated, Eq. (7.22) reduces to

$$\frac{\partial T}{\partial n} = 0 \tag{7.23}$$

If the surface is heated or cooled by convection, it is often assumed that Eq. (7.22) reduces to

$$\frac{\partial T}{\partial n} = -h(T_w - T_\infty)/k \tag{7.24}$$

where $T_w$ is the temperature of the surface, $T_\infty$ is the temperature of the surrounding fluid, and $h$ is the convection heat transfer coefficient. It should be noted that $h$ will usually be a function of $(T_w - T_\infty)$ and that $\partial T/\partial n$ is generally a nonlinear function of $(T_w - T_\infty)$. Carslaw and Jaeger [25] discuss various other boundary conditions that are often encountered in heat conduction.

### 7.2.3 Nondimensional Numbers in Conductive Heat Transfer

Consider one-dimensional time-dependent conduction in a semi-infinite slab as shown in Figure 7.8. Suppose that the thermal conductivity and specific heat of the material are constant, one face of the slab ($x = 0$) is thermally insulated, the other face ($x = L$) loses heat by convection to a surrounding fluid with temperature $T_\infty$, and the initial temperature of the material is $T_0$. Equation (7.17) becomes

$$\frac{\partial T}{\partial t} = \alpha \frac{\partial^2 T}{\partial x^2} + \frac{q'''}{\rho C_p} \tag{7.25}$$

The boundary and initial conditions are

$$T = T_0 \qquad \text{for } t = 0; \text{ all } x$$

$$\frac{\partial T}{\partial x} = 0 \qquad \text{for } t > 0, x = 0 \tag{7.26}$$

$$\frac{\partial T}{\partial x} = \frac{-h(T - T_\infty)}{k} \qquad \text{for } t > 0, x = L$$

These boundary conditions mean that the $x = 0$ face of this component (which could represent a long resistor) is thermally insulated, while the face at $x = L$ is cooled by forced convection. In order to understand the characteristics of the problem, the equation and the boundary conditions are normalized by characteristic dimensions as follows:

$$x^* = \frac{x}{L}, \qquad T^* = \frac{T - T_\infty}{T_0 - T_\infty}, \qquad t^* = \frac{\alpha t}{L^2} \tag{7.27}$$

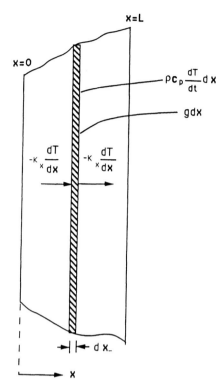

**Figure 7.8** One-dimensional transient conduction in a semi-infinite slab.

If Eqs. (7.27) are substituted in Eq. (7.25), the following nondimensional equation is obtained:

$$\frac{\partial^2 T^*}{\partial x^{*2}} + \text{Go} = \frac{\partial T^*}{\partial t^*} \tag{7.28}$$

where $t^* = \alpha t / L^2$ is called the Fourier number and $\text{Go} = q'''(L^2/k(T_0 - T_\infty))$ is called the generation number. The initial condition and boundary conditions of Eq. (7.28) reduce to

$$
\begin{aligned}
T^* &= 0 && \text{for } t^* = 0, \text{ all } x^* \\[6pt]
\frac{\partial T^*}{\partial x^*} &= 0 && \text{for } t^* > 0, x^* = 0 \\[6pt]
\frac{\partial T^*}{\partial x^*} &= -\text{Bi}\, T^* && \text{for } t^* > 0, x^* = 1
\end{aligned}
\tag{7.29}
$$

where $\text{Bi} = hL/k$ is the Biot number.

Depending on the values of the generation number, the Fourier number, and the Biot number, heat transfer problems can often be simplified. If the Fourier number is large and if the problem has a steady-state solution, then the temperatures would be approaching steady-state values and the time derivatives in Eqs. (7.16), (7.17), (7.19), (7.25), and (7.28) may be neglected. If the generation number Go is small, then the source term can be neglected in Eqs. (7.16), (7.17), (7.18), (7.25), and (7.28). Notice that the generation term may be negligible because of low dissipation or high thermal conductivity. The next section discusses the special case of small Biot numbers. Notice that the Biot number may be small if the object is small, the thermal conductivity is high, or the heat transfer coefficient is low.

Although the concept of nondimensional numbers is useful, it should be noted that there are widespread variations on the specific definitions of the nondimensional numbers. For example, for the semi-infinite slab discussed above, the Biot number is sometimes defined as $hL/k$ and sometimes as $h(L/2)/k$.

### 7.2.4 Small Biot Number Approximation

When the Biot number (Bi) is small (a common rule of thumb for small is Bi < 0.1), the thermal analysis may often be greatly simplified. For the example discussed in the previous section, Eqs. (7.2.11) and (7.2.12) reduce to the lumped parameter conditions as follows:

$$\rho C_p L \frac{dT}{dt} = -h(T - T_\infty) \tag{7.30}$$

with the initial condition $T = T_0$ at $t = 0$. The solution to Eq. (7.30) is

$$\frac{T - T_\infty}{T_0 - T_\infty} = e^{-\text{Bi}\, t^*} \tag{7.31}$$

It can be shown [1,26–28] that Eq. (7.31) holds for any homogeneous body if Bi is defined as $h(V/A)/k$ and $t^*$ is defined as $\alpha t/(V/A)^2$, where $V$ is the volume of the body and $A$ is the surface area of the body.

For small electronic components the characteristic heat-up and cool-down times can be estimated easily using Eq. (7.31). Section 7.9 contains one example problem.

### 7.2.5 Steady-State Analysis

If the Fourier number is large and if the heat generation rate is not a function of temperature, Eq. (7.18) describes the temperature profiles in a material. If the generation number is small, as is the case for points other than the active junctions in semiconductor devices or resistors, Eq. (7.18) further simplifies to Eq. (7.20). Even Eq. (7.20) is not easy to solve in three dimensions. Therefore, one- and two-dimensional approximations of Eq. (7.20) are often used in many design applications in electronics.

### 7.2.6 One-Dimensional Analysis

One-dimensional analysis of heat conduction is applicable in cases in which the temperature gradients in a particular direction are much greater than the temperature gradients in the other directions. One-dimensional analysis is also appropriate for cases in which the resistance to heat transfer in a particular direction is much lower than the resistance to heat transfer in the other directions. Two simple shapes will be considered.

First, a thin slab of material without heat generation as shown in Figure 7.9 is considered. The thickness $L$ of the slab is much less than the length and the width, satisfying the conditions of one-dimensional heat transfer.

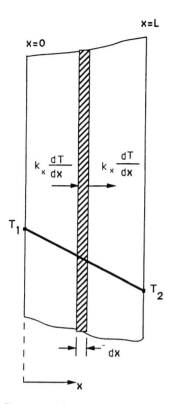

x=L

x=0

$k_x \dfrac{dT}{dx}$   $k_x \dfrac{dT}{dx}$

$T_1$

$T_2$

dx

x

**Figure 7.9** One-dimensional steady-state conduction in a semi-infinite slab without heat generation.

The equation governing the transfer of energy through the slab is

$$\frac{\partial}{\partial x}\left(k_x \frac{\partial T}{\partial x}\right) = 0 \tag{7.32}$$

where the subscript $x$ on the thermal conductivity indicates that the thermal conductivity may be different in the $y$ and $z$ directions. The boundary conditions for this problem are the specified wall temperatures as follows:

$$\begin{aligned} T = T_1, &\quad x = 0 \\ T = T_2, &\quad x = L \end{aligned} \tag{7.33}$$

Integrating equation (7.32) twice yields

$$k_x T = C_1 x + C_2 \tag{7.34}$$

The constants $C_1$ and $C_2$ are determined from the boundary conditions as

$$C_2 = k_x T_1, \qquad C_1 = \frac{k_x(T_2 - T_1)}{L} \tag{7.35}$$

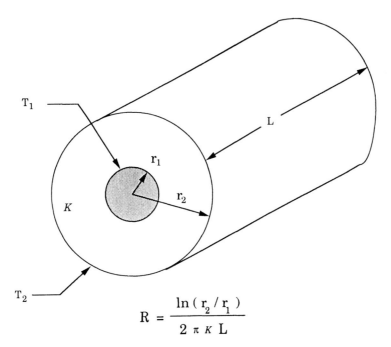

$$R = \frac{\ln(r_2 / r_1)}{2 \pi K L}$$

**Figure 7.10**   Radial conduction.

Substituting (7.35) into (7.34) and rearranging yields

$$T = T_1 + \frac{X}{L}(T_2 - T_1) \tag{7.36}$$

Equation (7.36) confirms the linear distribution shown in Figure 7.9. Using Eq. (7.15), the heat flux through the slab is

$$q = \frac{kA(T_1 - T_2)}{L} \tag{7.37}$$

Next, a hollow cylinder without heat generation as shown in Figure 7.10 is considered. Since cylindrical coordinates are needed to describe this geometry, referring to Eq. (7.15), $A$ can be found to be

$$A = 2\pi r L \tag{7.38}$$

Thus, from Eq. (7.15),

$$q = -2\pi k L r \frac{dT}{dr} \tag{7.39}$$

Equation (7.39) can be rearranged to yield

$$\frac{-g}{2\pi kL}\frac{dr}{r} = dT \tag{7.40}$$

With integration from $r_1$ and $r_2$, where the temperatures are $T_1$ and $T_2$, respectively, the result becomes

$$q = \frac{2\pi kL}{\ln (r_2/r_1)} (T_1 - T_2) \tag{7.41}$$

For some multidimensional heat transfer problems, it can be shown [25,26,28] that the solution is simply the product of the solutions of one-dimensional problems. For many shapes the solutions can be generalized as

$$q = kS(T_2 - T_1) \tag{7.42}$$

where $S$ is defined as the conduction shape factor, with dimensions of length. Figure 7.11 illustrates the general case, and Figure 7.12 shows several multidimensional shapes and the corresponding shape factor for each.

In general, the solution of transient one-dimensional problems is not as straightforward as the solution of steady-state one-dimensional problems. However, the literature [1,25,26,28–50] contains many graphs of the solutions of these transient problems. Of particular use are the 120 charts of Schneider [49], covering the temperature response under a variety of body shapes and boundary conditions and for a wide range of Fourier and Biot numbers.

## 7.2.7 Electrical Analogs

For one-dimensional, steady-state conduction in a plane wall with no heat generation and constant thermal conductivity, the temperature varies linearly with $x$. Thus, an analogy can be drawn between a thermal circuit and an electrical circuit.

Equation (7.37) describes the steady-state heat transfer through a slab in terms of the temperature difference at two different points and is an analog of Ohms law. The tem-

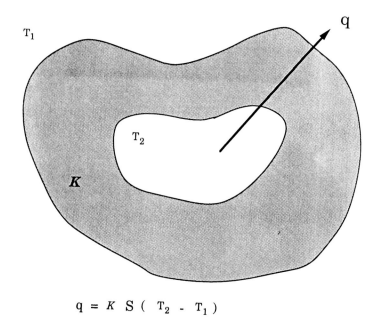

$$q = K \ S \ ( \ T_2 - T_1 )$$

**Figure 7.11** Multidimensional conduction. $q = ks(T_2 - T_1)$, where $S$ is the conduction shape factor, with dimensions of length.

PHYSICAL SYSTEM                    SCHEMATIC   SHAPE FACTOR  RESTRICTIONS

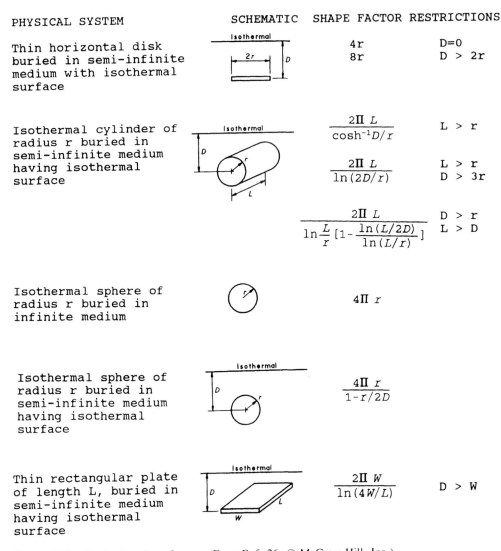

Thin horizontal disk
buried in semi-infinite
medium with isothermal
surface

$$4r \qquad D=0$$
$$8r \qquad D > 2r$$

Isothermal cylinder of
radius r buried in
semi-infinite medium
having isothermal
surface

$$\frac{2\Pi L}{\cosh^{-1}D/r} \qquad L > r$$

$$\frac{2\Pi L}{\ln(2D/r)} \qquad \begin{array}{l} L > r \\ D > 3r \end{array}$$

$$\frac{2\Pi L}{\ln\dfrac{L}{r}[1-\dfrac{\ln(L/2D)}{\ln(L/r)}]} \qquad \begin{array}{l} D > r \\ L > D \end{array}$$

Isothermal sphere of
radius r buried in
infinite medium

$$4\Pi\, r$$

Isothermal sphere of
radius r buried in
semi-infinite medium
having isothermal
surface

$$\frac{4\Pi\, r}{1-r/2D}$$

Thin rectangular plate
of length L, buried in
semi-infinite medium
having isothermal
surface

$$\frac{2\Pi W}{\ln(4W/L)} \qquad D > W$$

**Figure 7.12**  Conduction shape factors. (From Ref. 26, © McGraw-Hill, Inc.)

perature difference, $(T_2 - T_1)$, is analogous to an electrical potential difference; the heat flux, $q$, is analogous to the electrical current; and the thermal resistance, $L/k_x$, is analogous to the electrical resistance. Equation (7.37) can be rewritten as

$$q = \frac{\Delta T}{R} \tag{7.43}$$

where $R$ is the thermal resistance.

If the heat transfer process is steady state and approximately one-dimensional, then the electrical analog technique can be extended to assemblies consisting of one or more composite materials. Equation (7.43) can be generalized to

$$q = \frac{\Delta T}{\Sigma R} \tag{7.44}$$

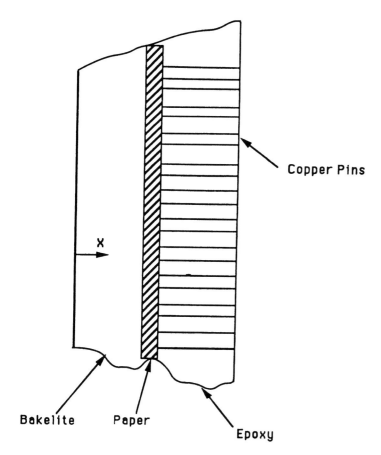

**Figure 7.13**   Conduction through a composite slab.

where $\Sigma R$ is the sum of the thermal resistances of the layers. Consider the example shown in Figure 7.13. It consists of a slab of bakelite in contact with a slab of paper, which is in contact with an epoxy board containing several copper pins. Since the conduction model is one-dimensional, the thermal resistances of the bakelite and paper slabs can be written as

$$R_1 = \frac{L_1}{k_1}, \qquad R_2 = \frac{L_2}{k_2} \tag{7.45}$$

For heat transfer through the composite material involving copper pins and epoxy board, the thermal resistance is modeled as a parallel combination as follows:

$$\frac{1}{R_3} = \frac{k_{3a}(1 - A_c)}{L_3} + \frac{k_{3b}(A_c)}{L_3} \tag{7.46}$$

where $k_{3a}$ and $k_{3b}$ are thermal conductivities of epoxy and copper, respectively, and $A_c$ is the fraction of the cross-sectional area occupied by copper.

In Figures 7.14 to 7.16 the electrical analogs of several problems are shown schematically. Figure 7.14 shows the electrical analog for heat conduction through several

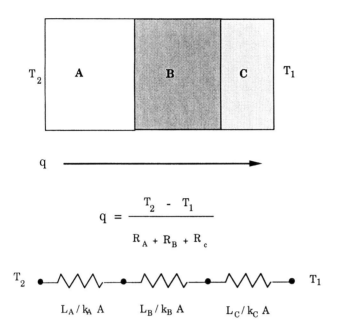

$$q = \frac{T_2 - T_1}{R_A + R_B + R_c}$$

$T_2$ ——/\\/\\/——•——/\\/\\/——•——/\\/\\/——• $T_1$

$\quad\quad L_A / k_A\ A \quad\quad\quad L_B / k_B\ A \quad\quad\quad L_C / k_C\ A$

**Figure 7.14**   Electrical analog of heat conduction through
several layers in series.

composite layers in series. Figure 7.15 shows the electrical analog for heat conduction
through several composite layers in series and in parallel. Figure 7.16 shows the electrical
analog for heat conduction in one layer with a convective resistance at both ends of the
layer.

   This technique is directly applicable to the thermal analysis of electronic packages
and is particularly useful for the estimation of junction temperatures.

   Figure 7.17 shows a typical plastic dual in-line package (DIP) construction. Figure
7.18 shows the four principal paths of heat flow by conduction in this package, specifically
(A) from the semiconductor junction through the plastic, (B) from the junction down the
metal tie bars and into the plastic, (C) from the junction down the leads and into the
plastic, and (D) from the function down the leads and into a socket or board. Figure 7.19
shows a typical air-cooled module and the equivalent thermal resistance network. Note
that this network shows both series and parallel thermal resistance paths, as is typically
found in electronics cooling applications.

   In transient one-dimensional heat conduction, there are also electrical analogies. The
quantity $\rho V C_p$ is called the thermal capacitance and is analogous to electrical capacitance.
The interested reader is referred to Kraus and Bar-Cohen [1] for more details of transient
thermal-electrical analogies.

## 7.2.8   Fins and Extended Surfaces

Often, in the cooling of discrete components, use is made of extended surfaces, or fins.
Fins give the heat a conduction path from the surface that is much more effective than
convection alone from the surface. Fins are normally analyzed by making use of a quantity
called a fin efficiency. Consider a fin surface such as shown in Figure 7.20. The air is

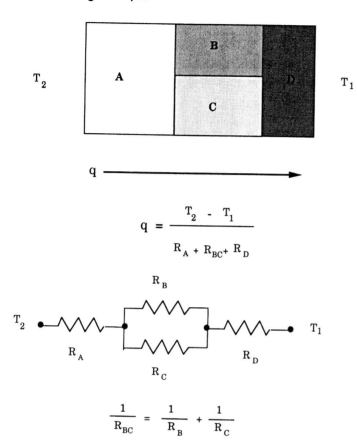

**Figure 7.15** Electrical analog of heat conduction through several layers in series and in parallel.

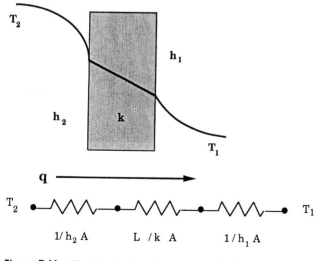

**Figure 7.16** Electrical analog of heat conduction through one layer with convective cooling at one end and convective heating at the other end.

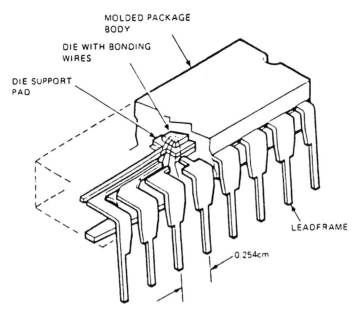

**Figure 7.17** Construction of a typical plastic package. (From Ref. 83, courtesy of ISHM.)

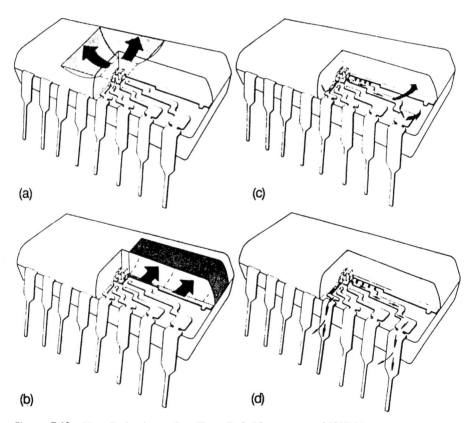

**Figure 7.18** Heat dissipation paths. (From Ref. 83, courtesy of ISHM.)

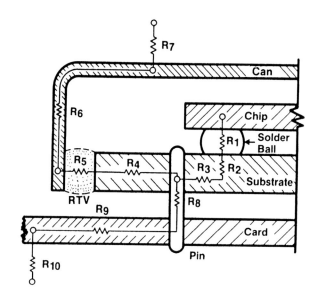

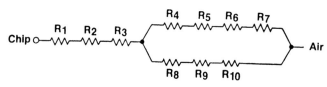

**Figure 7.19**   Thermal resistance network. (From Ref. 83, courtesy of ISHM.)

at temperature $T_0$, the heat into this fin from the components is $q$, and energy is lost by convection along the sides of the fin and from the end. An equation along the sides of the fin and from the end. An equation for the temperature will be determined. First, it is assumed that the fin is thin, and so the temperature is uniform at any vertical cut through the fin. Also, heat is assumed to be dissipated evenly across the base of the fin, so that

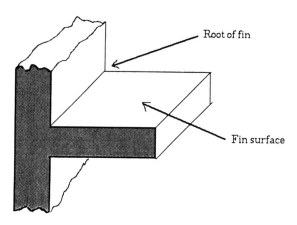

**Figure 7.20**   Schematic of a fin.

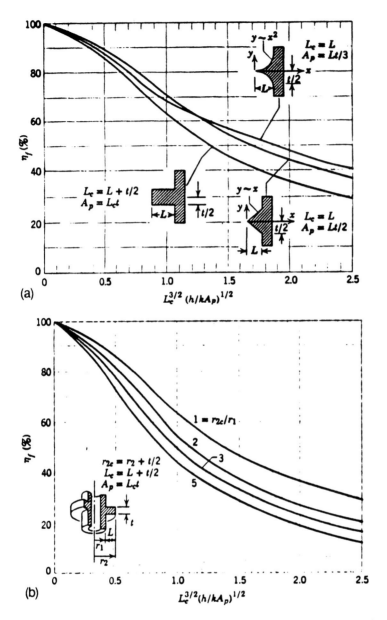

**Figure 7.21** Efficiencies of (a) straight fins (rectangular, triangular, and parabolic profiles) and (b) annular fins of rectangular profiles. (From Ref. 24, © John Wiley & Sons, Inc., reprinted by permission.)

the temperature is uniform throughout the depth of the fin. Hence, the temperature depends only on the location along the length of the fin, $x$, and the governing equation for conduction is $q = -kA \, dT/dx$ and for convection is $q = hA(T - T_0)$. From the principle of conservation of energy, the energy into the volume is equal to the energy out of the volume. Thus, for the volume described by $V = wt \, dt$,

$$-kWt\frac{dT}{dx} = -kWt\left(\frac{dT}{dx} + \frac{d^2T}{dx^2}\,dx\right) + hW\,dx\,(T - T_0)$$

or (7.47)

$$kwt\frac{d^2T}{dx^2} = hw(T - T_0)$$

This is the governing equation for the temperature of the fin. If the fin is of uniform thickness ($t$ is constant), the solution of Eq. (7.47) is

$$T - T_0 = C_1 \cosh\left(\frac{hx}{kt}\right) + C_2 \sinh\left(\frac{hx}{kt}\right)$$ (7.48)

The boundary conditions are

$$\frac{d(T - T_0)}{dx} = \frac{h}{k}(T - T_0) \quad \text{at } x = L$$ (7.49)

$$\frac{d(T - T_0)}{dx} = \frac{q}{kWt} \quad \text{at } x = 0$$ (7.50)

These two boundary conditions are used to determine $C_1$ and $C_2$ in Eq. (7.48) to obtain an equation for the temperature distribution of the fin. The base temperature is found by setting $x = 0$ in this equation.

An ideal fin is one whose entire convecting surface is at the root or base temperature, but all real fins operate at some lower effective surface temperature. The temperature along the fin will be less than that of the root, because the conduction of heat along the fin requires a temperature gradient. The fin efficiency, $\eta$, is defined as the ratio of the actual heat transferred by the fin to the heat that would be transferred by the fin if the entire surface was at the root or base temperature. Charts of fin efficiencies are available in the literature [26, 28, 51] for many fin geometries. See Figure 7.21 for examples of fin efficiency charts.

For use in electronics, fins are usually in the form of an extrusion, as shown in Figure 7.22, or a plate, as shown in Figure 7.23. Fins are used as an aid in convection-cooled boards. Extrusion fins are generally used as an aid to natural convection–cooled PCBs, and plate fins are generally used as an aid in forced convection–cooled PCBs. Since from Eq. (7.15) a larger fin area implies improved heat transfer from the fins, better heat transfer from the electronic components results. However, it can be shown [26, 51] that if the fin is used with high-velocity fluids or boiling liquids, it may actually result in a reduction in heat transfer.

When analyzing plate fins such as that shown in Figure 7.23, a characteristic convection heat transfer coefficient, $h$, is determined on the basis of the geometry of the fins. Using this coefficient, the heat transfer rate from the board is calculated. For plain fins, relations for duct flow can be used. Often, interrupted plate fins are used so that the flow is turbulent to take advantage of the better heat transfer characteristics of turbulent flow. For these types of fins, special correlations must be used. The root or base of the fin is the location where the fin is attached to the surface that is to be cooled.

For additional information on the applications of extended surfaces in the thermal design analysis of electronic packages, the reader is referred to Kern and Kraus [51], Holman [26], Kraus and Bar-Cohen [1], and Seely and Chu [27].

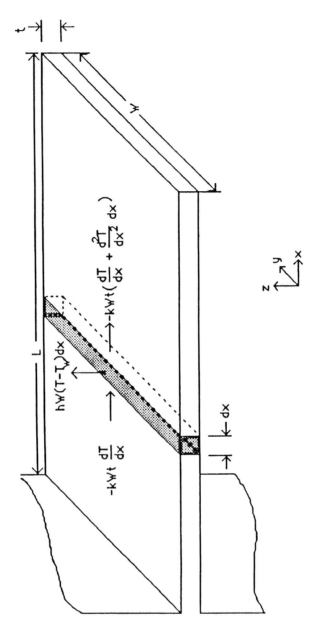

**Figure 7.22** Extrusion fin.

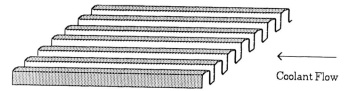

**Figure 7.23**   Plate fin.

## 7.2.9   Contact Resistance

When using conduction cooling, another consideration that must be addressed is thermal contact resistance. When two surfaces are brought into contact and heat is allowed to flow through the contact surface, an additional resistance to heat transfer exists. Figure 7.24 illustrates this condition. This resistance is due to the fact that the two surface areas are not completely in contact because of surface roughness. This additional thermal resistance is called contact resistance. An exaggerated view of this is shown in Figure 7.25. Although much empirical data and many qualitative and quantitative models of contact resistance exist [52–81], no single set of equations exists to allow the reliable computation of contact resistance. Typically, values for contact resistance may be estimated from the literature or are supplied by the manufacturers of certain devices. Reviews of the thermal contact resistance can be found in Refs. 82–85.

The contact resistance is dependent on the metrology of the surfaces—that is, whether they are smooth or rough, flat or wavy—and on whether deformation takes place in the surfaces when they mate [1]. To illustrate the working correlations that have been established for different general categories of surfaces, it has been shown by Cooper et al. [65] that for rough, nominally flat surfaces (in a vacuum)

$$h_c = \frac{1.45 k_s m (P/H)^{0.985}}{\sigma} \tag{7.51}$$

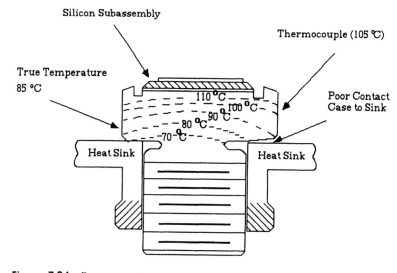

**Figure 7.24**   Poor contact.

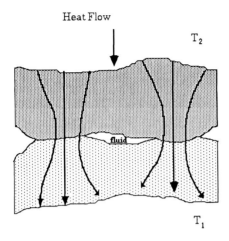

Heat Flow

$T_2$

$T_1$

$$q = (T_2 - T_1)/R_{contact}$$

$$R_{contact} = 1/h_c A_{apparent}$$

$h_c$ : Contact Coefficient

**Figure 7.25**   Contact resistance.

where $h_c$ is the contact coefficient ($W/m^2$ °C), $k_s$ is the harmonic mean solid conductivity (W/m °C), $P$ is the apparent contact pressure (kg/$m^2$), $H$ is the hardness of the softer solid (kg/$m^2$), $\sigma$ is the rms surface roughness (m), $m$ is the rms asperity slope (dimensionless), and $3.5 \times 10^{-4} < P/H < 10^{-2}$, $13.8 < k_s < 133.2$ (W/m °C), $1~\mu m < \sigma < 85~\mu m$, and $0.08 < m < 0.16$.

Contact resistance is a particulary important parameter when wedge locks or edge connectors are part of an electronic package. These devices are extensively used in PCB designs to conduct heat from the edges of a board to the cold plate. An edge connector is simply a plug that will accept the board edge, whereas a wedge lock is clamped down on the edge of the board by the application of pressure, usually using a mechanical mechanism such as a screw. The manufacturers of these devices supply the thermal resistance as a function of length of the connector. Wedge locks usually have a lower thermal resistance than edge connectors because the contact with the board surface is under mechanical pressure. When surfaces are brought into contact and there is concern for the thermal contact resistance, a thermal grease may be applied to the mating surfaces to reduce the contact resistance. The grease fills the gaps between the surfaces and, compared to air, is a low thermal resistance material. Figure 7.26 shows how thermal grease can be used to improve the thermal path between two materials that have rough surfaces.

## 7.2.10   Constriction Resistance

The one-dimensional heat transfer assumption discussed earlier is seldom valid in electronic components. Typically, the dissipation of energy occurs at small emitter-base junctions. The dissipated heat must then "spread" by two- and three-dimensional conduction to larger cross sections before being transferred to the cooling fluid. The construction resistance concept is shown schematically in Figure 7.27.

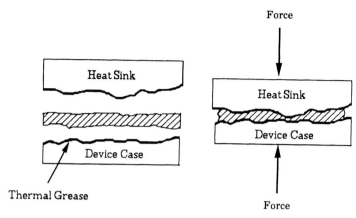

**Figure 7.26** Thermal interfaces between a semiconductor device, and insulator, and the sink before and after engagement.

Analytical methods of solving multidimensional problems generally rely on separation of variables and solution of the resulting ordinary differential equations with standard techniques. Analytical solutions are generally not available for practical thermal design of electronics. Therefore, multidimensional effects are often treated with approximations that involve one-dimensional heat transfer calculations with additional thermal resistances

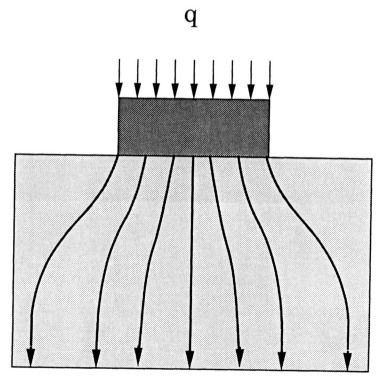

**Figure 7.27** Constriction resistance. This resistance is due to spreading of heat flow from a discrete source on the surface of a conducting medium.

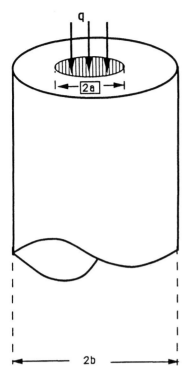

**Figure 7.28** Component junction
mounted on a chip.

at locations of significant multidimensional transfer. The additional resistances are called
constriction resistances [1] or spreading resistances [27].

Consider a component junction mounted on a chip shown in Figure 7.28. The diameter
of the junction, $2a$, is much smaller than the size of the chip, $2b$. The distance between
neighboring junctions is also large. In this case the constriction resistance is given by

$$R_{cs} = \frac{1}{\sqrt{\pi}\,ak} \tag{7.52}$$

If the size of the chip, $2b$, is not small compared to the junction diameter, a correction
factor for this effect is needed as follows:

$$R_{cs} = \frac{1}{\sqrt{\pi}\,2ak}\left(1 - \frac{a}{b}\right)^{1.5} \tag{7.53}$$

The chip itself may not be circular in cross section and modifications to Eq. (7.53)
to accommodate a circular junction on a rectangular chip may be needed. Kadambi and
Abuaf [86] discuss these modifications as well as empirical relations for complex situations.

Alternative geometries may be as shown in Figure 7.29; in this case, the constriction
resistance is given by

$$R_c = \frac{\ln[1/\sin(\pi a/2b)]}{\pi L k} \tag{7.54}$$

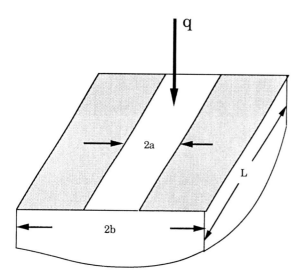

**Figure 7.29**  Constriction resistance.

For the general case of a heat source mounted on a base material as shown in Figure 7.30, a generic formula for the rough estimation of the constriction resistance is

$$R_c = 0.467 \frac{(1 - \epsilon)^{1.35}}{k\sqrt{A_j}}$$  (7.55)

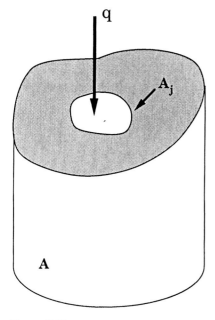

**Figure 7.30**  Generic formulas: constriction resistance.

where

$$\epsilon = \sqrt{\frac{A_j}{A}} \qquad\qquad (7.56)$$

### 7.2.11 Numerical Methods

The most powerful approximate methods for solving multidimensional heat conduction problems are based on either finite-element or finite-difference techniques. Detailed descriptions of finite-element methods applied to heat conduction are given in Refs. 87–94. The use of finite-difference methods to solve heat conduction problems is discussed in Refs. 25, 26, 28, 92–99, 101.

Both finite-difference and finite-element methods are used to solve partial differential equations in numerical thermal analysis. The primary differences between these methods with regard to their relative capabilities for thermal modeling are related to geometry, computation time, and memory storage requirements.

In the finite-difference method, the domain $D$ of the problem is covered by a mesh, usually square or rectangular in shape, as illustrated by Figure 7.31. Boundary conditions are imposed at every point of the boundary surrounding the domain $D$. The boundary conditions imposed are of three kinds:

1. Boundary value problem of the first kind: The solution $U$ is to satisfy the partial differential equation in $D$ and take on assigned values

$$U = Ub(x,y) \qquad\qquad (7.57)$$

   on the boundary $\partial D$ of the region.
2. Boundary value problem of the second kind: The solution $U$ is to possess assigned normal derivatives

$$\frac{\partial U}{\partial n} = Ub' \qquad\qquad (7.58)$$

   on the boundary $\partial D$.
3. Boundary value problem of the third kind: Normal derivatives of the solution $U$ take the mixed form

$$\frac{\partial U}{\partial n} = G + H*U \qquad\qquad (7.59)$$

   on the boundary $\partial D$.

Values of the numerical solution of the equation under consideration are then assigned to the intersections or nodes of the mesh. A finite-difference approximation to the differential equation is calculated in each node that is not on a boundary of the first kind. A series of algebraic equations, which are called finite-difference equations, results, for example,

$$\frac{\partial \rho}{\partial x} = \frac{\rho_{i+1} - \rho_{i-1}}{x_{i+1} - x_{i-1}} \qquad\qquad (7.60)$$

The finite-difference equations that result are then solved to produce the numerical solution. This generally requires the solution of a large number of linear algebraic equations

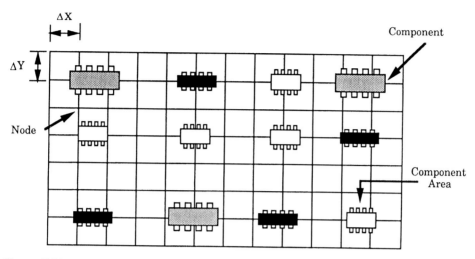

**Figure 7.31** Schematic of components and nodes.

by computer algorithms, the mathematics of which are referred to as numerical linear algebra.

The finite-element method, on the other hand, requires the division of the domain $D$ into finite subdomains $D_e$, or elements, with nodes located on interelement lines and/or inside the elements. Discrete values $u_n$ are calculated for the solution at each node. In each element $D_e$, then the solution is approximated, generally by a polynomial function, by interpolating between the values of $u_n$ that correspond to the nodes that belong to the element $D_e$. The final solution is expressed by a weighted sum

$$U(x) \cong \sum_{n=1}^{N} \Phi_n(x)u_n \qquad (7.61)$$

where the $\Phi_n(x)$ are known as basis functions. In this method, the node points may be irregularly distributed and the elements may be irregularly shaped when defined [102,103]. Thus, the finite-element method can be used to model more accurately real-world geometry that cannot be approximated using a rectangular mesh, and this is the advantage of this method over the finite-difference method. However, because of the complicated interpolation calculations that are required at each node for the finite-element method, longer computational time at each node and greater storage capacity are required. The time of computation for the finite-difference method can be as little as one-tenth of the time required for the finite-element method [2, 11].

In summary, an advantage of the finite-element method in computer-aided thermal modeling is the capability of modeling nonrectangular geometries. A disadvantage is slower computational speed and greater storage requirements.

In this section the finite-difference technique will be illustrated by considering conduction in printed circuit boards as examples. Two techniques for cooling PCBs by conduction are discussed. The first technique involves conduction across the plane of the PCB either to a heat sink or to cold-plate heat exchangers on the sides. The second method is by conduction rails placed under the components on the board. This provides paths for heat transfer.

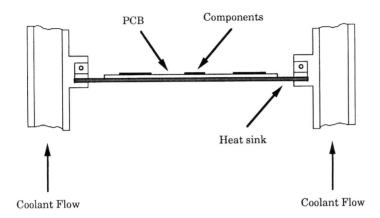

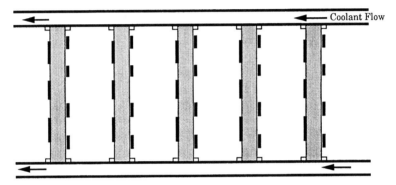

**Figure 7.32** (a) PCB cooled by conduction to cool plates. (b) Cold plate–cooled electronic enclosure.

*Example 1.* The simplest way to cool a PCB by conduction is to attach a heat sink to the entire area of the PCB and attach cold plates to this heat sink, as illustrated in Figure 7.32. The cold plates are usually heat exchangers through which a fluid such as water flows. Alternatively, instead of a heat sink the PCB may have layers of conductive material within it, so that the conductive layer is actually a particular layer of the PCB itself. The method of analysis for these two examples is the same. First, the temperature distribution of the heat sink is determined. Then the temperatures of the components' cases are determined based on the materials between the heat sink and the components.

The governing equation for this case is Eq. (7.18). Since the heat sink is generally larger in the planar directions than in the thickness direction, Eq. (7.18) can be written

$$0 = \frac{\partial^2 T}{\partial x^2} + \frac{\partial^2 T}{\partial y^2} + \frac{q'''}{k} \tag{7.62}$$

Discretizing (7.62) using a second-order finite-difference method and rearranging yields

$$\frac{(T_{i+1,j} + 2T_{i,j} + T_{i-1,j})}{\Delta x^2}$$

$$+ \frac{(T_{i,j+1} + 2T_{i,j} + T_{i,j-1})}{\Delta y^2} = -\frac{q'''}{k \, \Delta x \, \Delta y \, \Delta z} \quad (7.63)$$

where $q'''$ is the heat generation rate in volume $\Delta x \, \Delta y \, \Delta z$. Figure 7.33 illustrates the concept of a finite control volume. The heat sink substrate is divided into nodes of size $\Delta x$ by $\Delta y$. The term $\Delta z$ is the thickness of the heat sink. Equation (7.63) can be recast in terms of an electrical analogy with resistances given by $R = L/(kA) = \Delta x/(k \, \Delta y \, \Delta z)$ or $\Delta y/(k \, \Delta x \, \Delta z)$.

There is an equation such as (7.63) for each node that is not on a boundary. At the cold-plate boundary the equation becomes

$$T_{i,j} = T_{\text{cold plate}} \quad (7.64)$$

Similar equations are obtained for insulated boundaries at other edges. These equations, one for each node, form a system of equations that can be solved using a numerical method.

*Example 2.* A second way to cool PCBs conductively is by placing conductive rails under the components, as shown in Figure 7.34. The temperature distribution of the rails is determined using a resistance network. Conduction laterally in the $y$ direction to the epoxy board is neglected. This is possible because the conductivity of epoxy is on the order of 0.1 W/m °C, whereas that of the metallic rails, which are often made of copper, is on the order of 350 W/m °C.

$$\nabla^2 T + q/k = 0 \qquad \text{Finite Difference Method}$$

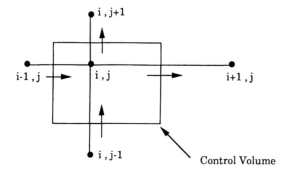

Energy conducted into control volume
+
energy produced

=

Energy conducted out of
control volume

**Figure 7.33** Numerical methods for steady conduction.

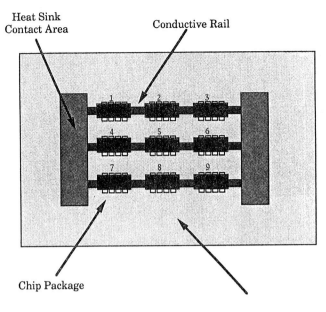

Heat Sink
Contact Area

Conductive Rail

Chip Package

Epoxy Circuit Board

**Figure 7.34** Use of conductive rails for cooling. Assume all heat is uniformly applied to the surface of the rails. The rails act as if they have uniform internal heat generation.

The conductivity of the copper is three orders of magnitude larger than that of the epoxy, so it is reasonable to assume that virtually all of the heat dissipated from the components is transferred through the copper.

As described above, the resistances in the network are of the form $R = L/kA$. Here $k$ is the conductivity of the metal conductor, $A$ is the thickness of the rail times the width of the rail, and $L$ is the distance between the centers of the components. Because all the resistances are known and the temperature of the heat sink is known, the network can be solved. First, the temperature of the rail under component 1 is found using the temperature of the heat sink, the power from component 1, and the rail resistance as follows:

$$T_{c1} = T_{\text{heat sink}} + q_{c1}\left(\frac{L}{kA}\right) \tag{7.65}$$

where $q_{c1}$ is the power dissipated from component 1. The temperature of the rail underneath components 3, 4, 6, 7, and 9 is then found in a similar manner.

The temperature of the rail under component 2 is next found by observing that part of the power from each of the components 2, 5, and 8 goes to the left and part goes to the right according to the following equation:

$$q_{c2} = q_{\text{left}} + q_{\text{right}} = (T_{c2} - T_{c1})\left(\frac{kA}{L}\right)_{\text{left}}$$
$$+ (T_{c2} - T_{c3})\left(\frac{kA}{L}\right)_{\text{right}} \tag{7.66}$$

Equation (7.66) can be rearranged to solve for $T_{c2}$. The temperature of the rail underneath components 5 and 8 is found in a similar manner. This procedure can be modified for any number of rails or components.

After the heat sink temperature distribution is calculated, the component case temperatures must be determined. The governing equation is

$$T_{case} = T_{sink} + qR_{board} \tag{7.67}$$

$R_{board}$ is the resistance provided by the materials between the heat sink and the bottom of the case. For a conduction-cooled PCB, $R_{board} = \Sigma_i \, L_i/(k_i A_i)$ where there is a term $i$ for each board layer. If a particular layer of the board is modeled as the heat sink, this summation includes only the layers between the sink and the bottom of the case. If the board has plated through holes, the layers of the board and the plated through holes are modeled as parallel resistances. The resistance of the plated through hole is

$$R_{pth} = \frac{H_{pth}}{k_{plating}\pi(r_o^2 - r_i^2)} \tag{7.68}$$

where $H_{pth}$ is the height of the hole through the board and $r_o$ and $r_i$ are the outer and inner radii of the plating. The effective resistance of the materials between the heat sink and the bottom of the case is then

$$R_{board} = \frac{1}{N/R_{pth} + 1/\Sigma_i \, L_i/(k_i A_i)} \tag{7.69}$$

where $N$ is the number of plated through holes under the component. The component's junction temperature is then found from

$$T_{junction} = T_{case} + qR_{jc} \tag{7.70}$$

where $R_{jc}$ is the junction-to-case resistance to heat transfer and is generally given by the manufacturer of the component. (See Sections 7.1.2, 7.1.3).

## 7.3 CONVECTION

When a fluid, i.e., either a gas (air) or liquid (water), passes over a solid surface that is at a higher temperature, energy is transferred to the fluid from the surface by convection heat transfer. First, energy is transferred to the fluid from the surface by random molecular motion (diffusion). Then this energy is carried or convected downstream by bulk or macroscopic motion of the fluid and is further diffused throughout the fluid by conduction. Convection, then, is a cumulative process consisting of heat transport by conduction and by bulk motion of the fluid.

As a result of the fluid-surface iteraction, a region exists where the velocity is zero at the surface to a finite value $u_\infty$ associated with the flow, as illustrated in Figure 7.35. This region is known as the hydrodynamic boundary layer or velocity boundary layer. If the surface and fluid temperatures are different, there will also be a region in the fluid in which the temperature varies from $T_s$ at the surface ($y = 0$) to $T_\infty$ in the outer flow region. This is the thermal boundary layer. In general, the thicknesses of the hydrodynamic and thermal boundary layers will be different. Conductive diffusion effects dominate near the

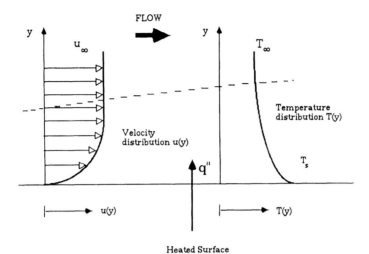

**Figure 7.35**  Boundary layer development in convection heat transfer.
(From Ref. 24, © John Wiley & Sons, Inc., reprinted by permission.)

surface, where the velocity is low. At the surface, heat transfer is by this mechanism only.

Convective heat transfer is classified by the nature of the flow. If the fluid flow is caused by an external means, such as by a pump, a fan, or a blower, the process is known as *forced convection*. In contrast, if the fluid flow is induced by buoyancy effects caused by density differences, the process is called *natural* or *free convection*. A combination of both is called *mixed convection*.

Furthermore, when the fluid moves in smooth layers, sheared slowly by the action of the fluid viscosity, it is referred to as *laminar flow*. When the flow is characterized by eddies throughout the flow region, it is referred to as *turbulent flow*. Typically, a *transition*

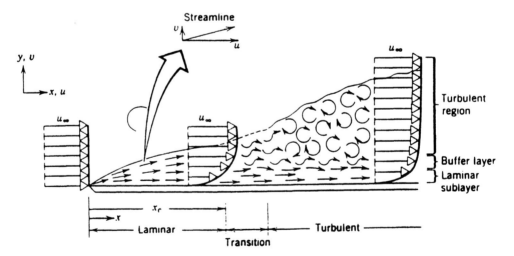

**Figure 7.36**  Velocity boundary layer development on a flat plate. (From Ref. 24, © John Wiley & Sons, Inc., reprinted by permission.)

*region* exists between laminar and turbulent flow regions. Figure 7.36 illustrates these flow characteristics, and Figure 7.37 shows how they may be observed in a typical electronics cooling application. In order to compute heat transfer rates, it is necessary to know whether the flow is of natural, forced, or mixed convection and whether it is laminar or turbulent [12, 23, 24, 100].

In general, the heat transferred by the convection heat transfer mode is internal thermal energy. However, there are processes in which there is latent heat exchange, associated with a phase change between the liquid and vapor states of the fluid. Boiling and condensation are convection heat transfer processes of the latent heat exchange type.

The appropriate rate equation for convection heat transfer is

$$Q = hA(T_s - T_\infty) \tag{7.71}$$

where $Q$ is the heat flow rate (W), $T_s$ the surface temperature (°C), $T_\infty$ the fluid temperature (°C), $h$ the convection heat transfer coefficient (also called the film conductance or film coefficient) (W/m$^2$ K), and $A$ the heat transfer surface area (m$^2$).

This equation is known as Newton's law of cooling. The proportionality constant $h$ encompasses all of the effects that influence the convection mode, including the surface geometry, the nature of the fluid motion, and a number of the fluid transport and thermodynamic properties. Typically, convection problems reduce to studies of how $h$ is to be determined [23, 24, 100]. Table 7.7 and Figure 7.38 show the relative magnitude and approximate values of heat transfer coefficients for different coolants and different modes of convection cooling.

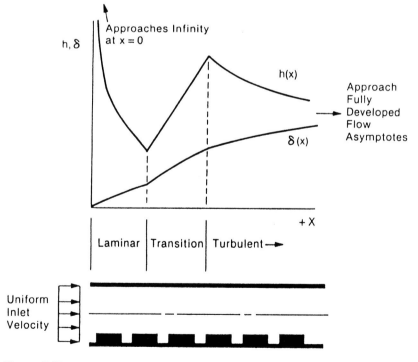

**Figure 7.37**  Boundary layer and heat transfer coefficient development in channel flow. (From Ref. 83, courtesy of ISHM.)

**Table 7.7**  Approximate Values of Convection Heat Transfer Coefficients

| Mode | $h$ | |
|---|---|---|
| | W/m² °C | BTU/hr ft² °F |
| Free convection, $\Delta T = 30°C$ | | |
| Vertical plate 0.3 m (1 ft) high in air | 4.5 | 0.79 |
| Horizontal cylinder, 5-cm diameter, in air | 6.5 | 1.14 |
| Horizontal cylinder, 2-cm diameter, in water | 890 | 157 |
| Forced convection | | |
| Air flow at 2 m/sec over 0.2-m square plate | 12 | 2.1 |
| Air flow at 35 m/sec over 0.75-m square plate | 75 | 13.2 |
| Air at 2 atm flowing in 2.5-cm-diameter tube at 10 m/sec | 65 | 11.4 |
| Water at 0.5 kg/sec flowing in 2.5-cm-diameter tube | 3,500 | 616 |
| Air flow *across* 5-cm-diameter cylinder with velocity of 50 m/sec | 180 | 32 |
| Boiling water | | |
| In a pool or container | 2,500–35,000 | 440–6,200 |
| Flowing in a tube | 5,000–100,000 | 880–17,600 |
| Condensation of water vapor, 1 atm | | |
| Vertical surfaces | 4,000–11,300 | 700–2,000 |
| Outside horizontal tubes | 9,500–25,000 | 1,700–4,400 |

Source: Ref. 26, © McGraw-Hill, Inc.

Passive cooling of electronic components by natural convection is considered the least expensive, quietest, and most reliable form of cooling and is used widely in telephone switching units, avionics packages, computers, and other electronic systems. The prediction of temperatures on a natural convection–cooled printed wiring board involves a problem of simultaneous heat transfer with interactions between the modes of convection, conduction, and radiation. This problem is made more difficult when board spacing is small with regard to the component package sizes, because of the need to predict the flow pattern in the narrow channel. These problems are among the most complex of physical problems because of the number of factors that affect the heat transfer, as shown in Figure 7.39.

The challenge of modeling the convection cooling environment will not have been met until a designer can predict the temperature of any element in an arbitrary three-dimensional array with a specified heat release on each element by defining only the layout geometry of the array, the board spacing, the power distribution, and the cooling flow, if forced convection. Computational fluid mechanics analyses, coupled with conduction and radiation analysis programs that acknowledge the detailed geometry and material characteristics, will probably provide the ultimate solutions. However, these programs are not available now. Most convection analysis programs now require the input of the

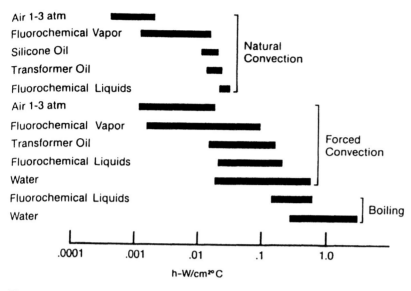

**Figure 7.38** Magnitude of heat transfer coefficients for various coolants and modes of convection. (From Ref. 83, courtesy of ISHM.)

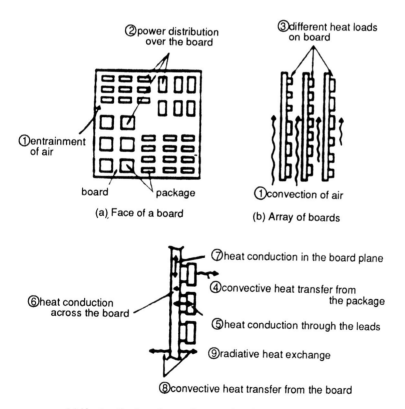

(c) Heat paths from the package to the air

**Figure 7.39** Factors that affect heat transfer. (From Ref. 100, © Hemisphere Publishing Corp., reprinted by permission.)

heat transfer coefficient at each point on the surface being analyzed. Because the heat transfer coefficient at a point on the surface is very situation specific and depends on the heat flux, the surface temperature, and a reference fluid temperature, all of which are affected by the distribution of temperature in the fluid above the point, the local fluid mechanics, and the upstream distribution of heat flux, much heat transfer research is spent trying to describe the heat transfer behavior of different shapes in benchmark situations. These benchmark situations typically involve either uniform or stepped wall temperatures (or heat flux) on smooth external surfaces or inside smooth-walled channels, with fully developed laminar or turbulent flow. Component packages are treated as heat sources. If the heat sources are assumed to be indistinguishable from each other over the surface of the board, the board can be modeled as either an isothermal or isoflux plate, depending on the thermal conductivity of the board. When the thermal conductivity across the board is assumed to be zero, the board is modeled as an asymmetric isoflux channel consisting of one isoflux wall and one insulated wall. An isothermal model is used when the board is modeled as having infinite thermal conductivity in its plane, and a symmetric or asymmetric isothermal channel model is used when the board is modeled as having either infinite or zero thermal conductivity across the board, in the direction normal to its plane. It has been shown that typical printed wiring board applications lie within the range established by these extremes [100,142].

Generally, natural convection cooling in real system applications with heat-dissipating components is influenced by the discrete nature of the heat sources arranged in an irregular array, the interaction between heat sources by substrate conduction and radiation, thermal wake-plume interactions, and the highly irregular, three-dimensional surface roughness of the component array and the walls.

With regard to forced convection, thermal analysis is typically done on the basis of correlations with benchmark situations that describe the heat transfer performance of a certain flow geometry, usually expressed using a Reynolds number. However, because the geometries encountered in electronics cooling are so diverse, a correlation for the actual geometry rarely exists, so choices are made from those available. The analytical solutions presented in the literature assume disturbance-free flow and most assume laminar or turbulent flow. Heat transfer in transitional or disturbed states is not well understood.

Most practical flows in forced convection, however, are highly disturbed. There is no widely accepted model for disturbed flow. A boundary layer under a disturbed stream is treated as turbulent, regardless of the Reynolds number, and the heat transfer rate that results is assumed to be underestimated.

The most common geometries found in electronics cooling applications are rectangles, i.e., DIPs and flatpacks, or short cylinders, i.e., resistors and capacitors, mounted either parallel to or perpendicular to the "wall" (the printed wiring board). These may be in regular or irregular patterns. The sizes of the elements may be uniform or nonuniform. The heat transfer behavior of isolated components, determined experimentally, is usually used to model the behavior of the components, even though the actual value of $h$ in the real array location will be different from that in the isolated element experimentation [41,100,142].

### 7.3.1 Natural Convection

Natural convection (also known as free convection) is the mode of cooling for most consumer products, as well as in many other electronic systems. Natural convection has many advantages—it costs nothing, it weighs nothing, it is quiet, and it requires no effort to implement or maintain. However, natural convection can dissipate only up to about 0.2 W/cm² using air. Some military and computer electronics dissipate a much higher heat flux than this, so that natural convection may have to be augmented for these more complex systems.

Even though natural convection is the simplest method of thermal control to employ, there is no simple way to determine its effect. There are no closed-form solutions for the equations for buoyancy-driven flow over a plate with individual heat sources protruding into the flow. We must therefore rely either on closed-form solutions for simple geometries and boundary conditions that may approximate the actual conditions in an electronic package or on empirically determined heat transfer effects.

It is convenient to divide natural convection into two major groups—external flows and internal flows. An external natural convective flow is a natural convective flow in which either the fluid is not bounded or the walls bounding the fluid are so distant from the object being heated or cooled that the fluid may be considered to be infinite in extent. An internal natural convective flow is a natural convective flow in which the walls bounding a fluid are so near the object being cooled or heated that wall effects must be considered.

Exact solutions of the continuity, momentum, and energy equations for buoyancy-driven boundary layer flow show that the relevant relationships for natural convection heat transfer may be expressed functionally as

$$\text{Nu}_x = f(\text{Gr}_x, \text{Pr}) \tag{7.72}$$

where $\text{Nu}_x$ is the local Nusselt number, $\text{Gr}_x$ is the local Grashof number, and Pr is the local Prandtl number. These quantities are dimensionless variables that represent the ratio of convective to conductive heat transfer, the ratio of buoyancy forces to viscous forces, and the ratio of viscous diffusion to thermal diffusion, respectively. The local Prandtl number is a property of the fluid and is defined as

$$\text{Pr} = \frac{\nu}{\alpha} \tag{7.73}$$

where $\nu$ is the kinematic viscosity of the fluid (m²/s) and $\alpha$ is the thermal diffusivity of the fluid (m²/s). Air at atmospheric conditions has a Prandtl number of approximately 0.7. The local Grashof number is defined as

$$\text{Gr}_x = \frac{g\beta\,\Delta T\,x^3}{\nu^2} \tag{7.74}$$

where $g$ is the local acceleration due to gravity (m/s²); $\beta$ is the volume coefficient of expansion of the fluid (1/K), which equals the inverse of the absolute temperature for an ideal gas; $x$ is a characteristic length (m); $\Delta T$ is a characteristic temperature difference (°C); and $\nu$ is the kinematic viscosity of the fluid (m²/s). The local Nusselt number is defined as

$$\text{Nu}_x = \frac{h_x X}{k} \tag{7.75}$$

where $h_x$ is the local heat transfer coefficient, $x$ is the characteristic length, and $k$ is the thermal conductivity of the fluid. It should be noted that the formula for the local Nusselt number is identical to that for the Biot number introduced in Section 7.2.4, except that

$k$ for the local Nusselt number is the thermal conductivity of the fluid and $k$ for the Biot number is the thermal conductivity of the solid being heated or cooled.

Frequently in free convection it is convenient to deal with a Nusselt number based on the average heat transfer coefficient. The average heat transfer coefficient over some length, $L$, is defined as

$$h = \frac{1}{L} \int_0^L h_x dx \tag{7.76}$$

and the Nusselt number based on the average heat transfer coefficient is defined as

$$\overline{Nu_L} = \frac{hL}{k} \tag{7.77}$$

The Rayleigh number Ra is the product of the Grashof and Prandtl numbers.

Unless the Prandtl number is small (e.g., $< 0.4$) or the Rayleigh number is small (e.g., $< 10^4$), natural convection heat transfer can usually be adequately described by

$$Nu = C(Gr\ Pr)^n = C\ Ra^n \tag{7.78}$$

where $C$ and $n$ are constants that depend on the geometry, the condition of the surface (i.e., isothermal or constant heat flux), and the flow regime (i.e., laminar or turbulent). As will be discussed in the next two sections, the characteristic length used for the Nusselt, Grashof, and Rayleigh numbers in Eq. (7.78) depends on the specific geometry being considered. Equation (7.78) implies that if the physical properties are constant, then the average Nusselt number between $x = 0$ and $x = L$ is

$$Nu = \frac{Nu_{x=L}}{3n} \tag{7.79}$$

For the special case of $n = 0.33$, Eq. (7.79) implies that the Nusselt number is independent of $x$. Many investigators have found that for turbulent flow $n \approx 0.33$.

For problems in which the local wall heat flux, $g_x$, is specified, it is convenient to use the modified local Grashof number and the modified local Rayleigh number. The modified local Grashof number is defined as

$$Gr_x^* = Gr_x\ Nu_x = \frac{g\beta q_x x^4}{k\nu^2} \tag{7.80}$$

and the modified local Rayleigh number is defined as

$$Ra_x^* = Ra_x\ Nu_x = Gr_x^*\ Pr_x \tag{7.81}$$

Substituting (7.81) into (7.78) yields

$$Nu_x = C'\ Ra_x^{*n'} \tag{7.82}$$

where $C' = C1/(1 + n)$ and $n' = n/(1 + n)$

Transition in a free convection boundary layer is related to the magnitude of the buoyancy and viscous forces in the fluid. For vertical plates in air, transition from laminar to turbulent flow begins when $Gr \approx 10^7$. The turbulence becomes fully developed at $Gr \approx 10^9$. For fluids other than air, it is not well established whether the transition to turbulence is a function of Grashof number only, Rayleigh number only, or both Grashof and Prandtl numbers. The transition from laminar flow to turbulent flow is discussed by Eckert and Soehngen [107], Gebhart [108, 109], and Mollendorf and Gebhart [110].

## 7.3.2 Empirical Correlations: External Flows

Most research in natural convection has been done for isolated objects being cooled or heated in a large fluid body. In this section, some representative results are presented.

For vertical plates and cylinders in which the length (height) is much greater than the width (diameter), McAdams [111] developed Eq. (7.83) for laminar natural convection and Warner and Arpaci [112] and Tsuji and Nagano [113] developed Eqs. (7.84) and (7.85), respectively, for turbulent natural convection.

$$Nu = 0.59Ra^{0.25}, \qquad 10^4 < Ra < 10^9 \tag{8.83}$$

$$Nu = 0.10Ra^{0.33}, \qquad 10^9 < Ra < 10^{13} \tag{7.84}$$

$$Nu = 0.11Ra^{0.33}, \qquad 10^9 < Ra < 10^{13} \tag{7.85}$$

In Eqs. (7.83), (7.84), and (7.85), the characteristic length is the length of the plate.

For horizontal plates, the coefficients in Eq. (7.78) depend on whether the plate is warmer or cooler than the surrounding fluid and on whether it is facing up or down. Based on the work of McAdams [111], improved accuracy may be obtained by altering the form of the characteristic length on which the correlations are based. If the characteristic length is defined as $L =$ plate area/plate perimeter, then the following correlations are recommended by Holman [26].

Heated plate facing up or cooled plate facing down:

$$Nu = 0.54Ra^{0.25}, \qquad 10^4 < Ra < 10^7 \tag{7.86}$$

$$Nu = 0.15Ra^{0.33}, \qquad 10^7 < Ra < 10^{11} \tag{7.87}$$

Heated plate facing down or cooled plate facing up:

$$Nu = 0.27Ra^{0.27}, \qquad 10^5 < Ra < 10^{10} \tag{7.88}$$

For horizontal cylinders, the characteristic length is the cylinder diameter. McAdams [102] recommends that

$$Nu = 0.4, \qquad 10^{-5} < Ra \tag{7.89}$$

$$Nu = 0.53Ra^{0.25}, \qquad 10^4 < Ra < 10^9 \tag{7.90}$$

$$Nu = 0.13Ra^{0.33}, \qquad 10^9 < Ra < 10^{12} \tag{7.91}$$

and Morgan [114] recommends that:

$$Nu = 0.68Ra^{0.058}, \qquad 10^{-10} < Ra < 10^{-2} \tag{7.92}$$

$$Nu = 1.02Ra^{0.15}, \qquad 10^{-2} < Ra < 10^2 \tag{7.93}$$

$$Nu = 0.85Ra^{0.19}, \qquad 10^2 < Ra < 10^4 \tag{7.94}$$

$$Nu = 0.48Ra^{0.25}, \qquad 10^4 < Ra < 10^7 \tag{7.95}$$

$$Nu = 0.12Ra^{0.33}, \qquad 10^7 < Ra < 10^{12} \tag{7.96}$$

It should be noted that for $10^{-5} < \text{Ra} < 10^4$, McAdams' [111] correlation is not easily described by a closed-form solution.

Based on the experiments of Yuge [115] with air and Amato and Tien [116] with water, Holman [26] recommends the following equations for natural convection from spheres, where the characteristic length is the sphere diameter:

$$\text{Nu} = 2 + 0.43\text{Ra}^{0.25}, \qquad 0.7 < \text{Ra} < 0.7 \times 10^5 \tag{7.97}$$

$$\text{Nu} = 2 + 0.50\text{Ra}^{0.25}, \qquad 3.0 \times 10^5 < \text{Ra} < 8.0 \times 10^8 \tag{7.98}$$

King [117] studied the heat transfer from plates, cylinders, blocks, spheres, and other geometries and gave a general correlation (''King's rule'') to predict the Nusselt number for laminar natural convection from an arbitrary shape. The relation can be employed when specific data for the body of the given shape are not available. The expression is

$$\text{Nu} = 0.6\text{Ra}^{0.25}, \qquad 10^4 < \text{Ra} < 10^9 \tag{7.99}$$

The characteristic dimension $L$ of the body is given by $1/L = 1/L_h + 1/L_v$, where $L_h$ and $L_v$ are the horizontal and vertical dimensions. Therefore, for an infinitely wide vertical plate, it becomes the length.

King's rule generally agrees well with experimental data. However, Sparrow and Ansari [118] present data refuting the rule. Lienhard [119] has proposed an alternative to King's rule that correlates the data using a characteristic length based on the distance a fluid particle travels in the boundary layer. Lienhard's technique appears to be more accurate than King's rule but is more difficult to apply.

For inclined surfaces, the interested reader is referred to Holman [26].

### 7.3.3 Empirical Correlations—Enclosures

Acceptable correlations for free convection heat transfer coefficients along the surface of isolated components have already been discussed, but these relations are generally unsuitable for determining the detailed thermal behavior of component arrays forming vertical or horizontal air channels. Packaging constraints and partitioning considerations, as well as system operating modes, lead to a wide variety of complex heat dissipation profiles along the channel walls. A symmetric isothermal or isoflux boundary condition, or use of an isothermal/isoflux boundary together with an insulated boundary condition along with the adjoining channel, can yield an acceptable model of such configurations.

In this section, analytical relations are presented for determining the natural convection Nusselt number or nondimensional heat transfer coefficient, maximum surface temperature, channel width, or spacing between surfaces forming a two-dimensional channel appropriate to various thermal constraints for symmetric and asymmetric, isothermal, and isoflux boundary conditions. Furthermore, it is demonstrated that the maximum component surface temperature attained on a natural convection, air-cooled PCB can be bounded from above and below by analytical relations related to the uniform heat flux, or isoflux, condition.

Bar-Cohen and Rohsenow [120, 121] have observed that in natural convection heat transfer in two-dimensional wide or short channels, heat transfer rates approach those associated with flow along isolated plates in infinite media. Alternatively, for narrow and/ or long channels, the boundary layers merge relatively close to the entrance and fully developed flow prevails along much of the channel. Based on these observations, Bar-Cohen and Rohsenow [120, 121] developed composite relations for the nondimensional heat transfer coefficient along the channel walls, under various thermal boundary conditions, by combining available relations for the fully developed flow and isolated plate asymptotes, respectively.

Bar-Cohen and Rohsenow [120, 121] derived that the channel Nusselt number for fully developed laminar natural convection for the symmetrical isothermal case (i.e., vertical channels formed by two isothermal plates) is

$$Nu = \left[ \frac{576}{(Ra'')^2} + \frac{2.873}{Ra''} \right]^{-1/2} \tag{7.100}$$

Here $Ra''$ is the modified channel Rayleigh number, equal to $Ra(b/L)^5$; $b$ is the effective spacing between the PCBs; and $L$ is the height of the plates. Similarly, they derived Eq. (7.101), (7.102), and (7.103) to predict the channel Nusselt number for fully developed laminar natural convection for the asymmetric isothermal case (i.e., vertical channels formed by an isothermal plate and an insulated plate), the symmetric isoflux case (i.e., vertical channels formed by two plates with uniform heat flux), and the asymmetric isoflux case (i.e., vertical channels formed by a plate with a uniform heat flux and an insulated plate), respectively:

$$Nu = \left[ \frac{144}{(Ra'')^2} + \frac{2.873}{Ra''} \right]^{-1/2} \tag{7.101}$$

$$Nu = \left[ \frac{48}{Ra''} + \frac{2.51}{(Ra'')^{0.4}} \right]^{-1/2} \tag{7.102}$$

$$Nu = \left[ \frac{24}{Ra''} + \frac{2.51}{(Ra'')^{0.4}} \right]^{-1/2} \tag{7.103}$$

Bar-Cohen and Rohsenow [120, 121] used equations (7.100) to (7.103) to derive the spacing between heat-dissipating vertical PCBs/cards in two-dimensional flow that maximizes the total rate of heat dissipation (i.e., the spacing that results in a maximum value of the product of total surface area and total heat transfer coefficient). They found that the optimum spacings for thin cards are

$$b_{opt} = 2.714 P^{-1/4} \quad \text{(symmetric isothermal case)} \tag{7.104}$$

$$b_{opt} = 2.154 P^{-1/4} \tag{7.105}$$

$$b_{opt} = 1.472 P^{-1/5} \quad \text{(symmetric isoflux case)} \tag{7.106}$$

$$b_{opt} = 1.169 P^{-1/5} \quad \text{(asymmetric isoflux case)} \tag{7.107}$$

where $P$, the system parameter, is defined as

$$P \equiv \frac{C_p \rho^2 g \beta \, \Delta T}{\mu k L} \tag{7.108}$$

$L$ is the length of the PCB/card, $b_{opt}$ is the optimum spacing divided by the card length, and the physical properties are evaluated at the average air temperature.

Bar-Cohen [122] concluded that the natural convection heat transfer in vertical channels formed by actual PCBs is generally characterized by relatively low air flow rate, laminar flow, and low average heat transfer coefficients. He further concluded that, as a result of these characteristics, the temperature rise along an isoflux surface is dominated by the temperature rise of the flowing air, which is, in turn, primarily determined by the integral of the surface heat flux from the inlet to the point of interest and is only moderately sensitive to the specific heat flux profile. In addition, because of the relatively low heat transfer coefficients associated with this flow, in-plane conduction and through-plane conduction can be expected to smooth the spatial distribution of heat flux and lead to a channel surface that approaches the symmetric isoflux boundary condition.

Based on the above observations, the expressions for the natural convection Nusselt number for vertical isoflux, parallel-plate channels are used by Bar-Cohen [122] to examine the thermal performance of vertical channels with imposed thermal boundary conditions that depart from uniform heat flux. Alternatively, it is the expectation of laminar flow in the channel and the relative insensitivity of laminar flow to surface roughness that make it possible to apply the same smooth-wall relations to the nonsmooth channel of components carrying PCBs.

The bounding relations for maximum surface temperature developed by Bar-Cohen [122] are based on a two-dimensional analysis of buoyant flow in smooth plate channels with uniform heat flux boundary conditions. The maximum surface temperature in an ideal isoflux channel is attained at the top of the channel, where the local air temperature is highest. Although the surface temperature is indeed determined primarily by the total heat addition to the air circulating in the channel, the secondary influence of the local heat flux can be expected to result in a lower surface temperature for the symmetric, than for the asymmetric, isoflux configuration. The assumption that all the dissipated heat is distributed uniformly on both sides of each PCB offers a lower bound on the expected maximum component surface temperature. The symmetric isoflux Nusselt number on the temperature $T_L$ at the top of the channel $(x = L)$ can be given by

$$\text{Nu} = \frac{(b/k)q''}{(T_L - T_0)} \tag{7.109}$$

where $k$ is the thermal conductivity of the air and $q''$ is the imposed heat flux.

Bar-Cohen [122] substituted Eq. (7.109) into Eqs. (7.102) and (7.103) to yield the following bounding relations:

$$(T_L - T_0) = \left(\frac{q''b}{k}\right)\left[\frac{48}{\text{Ra}''} + \frac{2.51}{(\text{Ra}'')^{0.4}}\right]^{1/2} \tag{7.110}$$

$$(T_L - T_0) = \left(\frac{q''b}{k}\right)\left]\frac{24}{\text{Ra}''} + \frac{2.51}{(\text{Ra}'')^{0.4}}\right]^{1/2} \tag{7.111}$$

Although Eqs. (7.110) and (7.111) are applicable to all channel dimensions, for extremely short channels or very widely spaced PCBs (including a single PCB) it may be more convenient to calculate the desired temperature differences directly from the expression appropriate to the isoflux isolated plate limit, which is given as

$$\Delta T = 1.59\left(\frac{q''L}{K}\right)(\text{Ra}_L^*)^{-1/5} \tag{7.112}$$

where $L$, the length of the PCB, is the characteristic length of $\text{Ra}_L^*$, the modified Rayleigh number.

Bar-Cohen [122] has compared his analytical bounding relations with the experimental data and found that the bounding relations are generally consistent with the experimental data, except at narrow PCB spacings.

Aung et al. [123] have proposed Eq. (7.113) for uniformly heated printed wiring boards in fully developed laminar natural convection channel flow and Eq. (7.114) for uniformly heated printed wiring boards in natural convection flow over an isolated vertical plate:

$$Nu = 0.144(Ra'')^{0.5} \qquad\qquad (7.113)$$

$$Nu = 0.524(Ra'')^{0.2} \qquad\qquad (7.114)$$

Equation (7.113) is applicable for relatively narrow or long channels where the boundary layers from adjacent circuit boards merge, and Eq. (7.114) is applicable for relatively wide or short channels where the boundary layers on adjacent circuit boards act independently of each other for the entire channel length.

Wirtz and Sultzman [124] have proposed Eq. (7.115) for uniformly heated printed wiring boards in laminar natural convection channel flow:

$$Nu = \frac{0.144(Ra'')^{0.5}}{b} \qquad\qquad (7.115)$$

where $b = [1 + 0.0156Ra^{0.9}]^{0.33}$. In the limiting case for fully developed laminar flow (i.e., as Ra'' approaches 0), Eq. (7.115) approaches Eq. (7.113). In the limiting case for the isolated vertical plate (i.e., large Ra''), Eq. (7.115) approaches Eq. (7.116):

$$Nu = 0.577(Ra'')^{0.2} \qquad\qquad (7.116)$$

As Ra'' approaches 0, Eqs. (7.102) and (7.103) discussed above reduce to

$$Nu = 0.63(Ra'')^{0.2} \qquad\qquad (7.117)$$

Johnson [125] compared Eqs. (7.113) through (7.117) with experimental data from uniformly heated models of printed wiring boards vertically mounted in a frame. The experimental models varied from uniformly heated quasi-smooth plates to actual circuit board arrays. Johnson found that Eqs. (7.115) and (7.116) are in good agreement with the experimental data for Ra < 300. For 300 < Ra'' < 1000, Johnson found that equations (7.115) and (7.116) slightly overpredict the heat transfer.

### 7.3.4  Effects of Miniaturization of Electronic Devices

The heat dissipation area of typical microchips is very small in two dimensions (e.g., 5 mm by 5 mm), so those components are usually modeled as uniform heat flux devices. The natural convection correlations discussed in the preceding sections are based on relatively large heating surfaces and may not give an accurate prediction of the heat transfer characteristics of miniature electronic devices.

Park and Bergles [126] experimentally investigated heat transfer characteristics of miniature electronic devices and observed that the experimental heat transfer coefficients were larger than the predicted heat transfer coefficients. They attributed this to three-dimensional boundary layer effects near the miniature electronic devices. They correlated their local heat transfer coefficients by

$$Nu_x = a\,Ra_x^{*b} \qquad\qquad (7.118)$$

where $a$ and $b$ are empirical constants that depend on the fluid being used. For refrigerant R-113,

$$a = 0.906(1 + c)^{0.2745} \tag{7.119}$$

$$b = 0.184(1 + d)^{-0.0362} \tag{7.120}$$

where $c = 0.00111/(W/W_x)^{3.965}$, $d = 2.64 \times 10^{-5}/(W/W_x)^{9.248}$, $W$ is the width of the component, and $W_x$ is 70 mm.

### 7.3.5  Numerical Methods

In principle, numerical methods can be used to solve problems involving natural convection in electronic systems. As was discussed in Section 7.2.11, the partial derivatives in the governing equations can be replaced with finite differences or finite elements and the resulting systems of algebraic equations can be solved. In practice, numerical methods are seldom practicable for natural convection in complex systems, such as typical electronic packages. The interested reader is referred to Shih [93], Roache [127], Patankar [128], Anderson et al. [129], de Vahl Davis and Jones [130], Upson et al. [131], Quon [132–134], and Mallison and de Vahl Davis [135].

### 7.4  FORCED CONVECTION

When the power dissipation of an electronic system is too large for the system to be properly cooled by natural convection, forced convection is usually the next option considered. Figures 7.40 and 7.41 show two forced convection cooling schemes for electronics cooling applications.

As was the case for the corresponding natural convection problem, there are no closed-form solutions for the equations of forced convection flow over a plate with individual heat sources protruding into the flow. We must therefore rely on either closed-form solutions for simple geometries and boundary conditions that may approximate the actual conditions in an electronic package or on empirically determined heat transfer effects.

As with natural convection, it is convenient to divide forced convection into two major groups—external flows and internal flows.

Exact solutions of the continuity, momentum, and energy equations for forced convection boundary layer flow show that the relevant relationships for forced convection heat transfer may be expressed functionally as

$$Nu_x = f(Re_x, Pr) \tag{7.121}$$

where $Nu_x$ is the local Nusselt number, $Re_x$ is the local Reynolds number, and $Pr$ is the local Prandtl number. As discussed previously, these quantities are dimensionless variables that represent the ratio of convective to conductive heat transfer, the ratio of inertia forces to viscous forces, and the ratio of viscous diffusion to thermal diffusion, respectively. The local Prandtl number and the local Nusselt number were previously defined in equations (7.73) and (7.75), respectively. The local Reynolds number is defined as

$$Re_x = \frac{u_x x}{\nu} \tag{7.122}$$

where $u_x$ is a characteristic fluid speed (m/s), $x$ is a characteristic length (m), and $\nu$ is the kinematic viscosity of the fluid (m²/s).

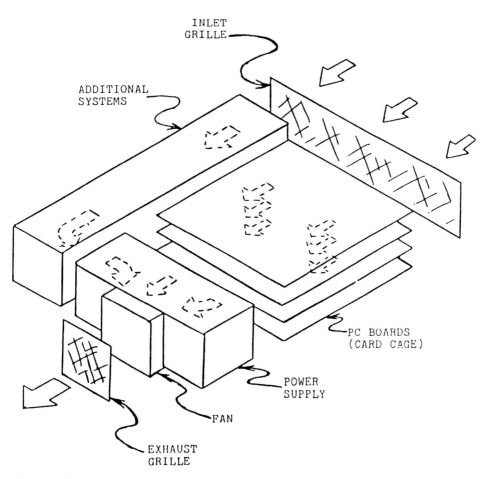

**Figure 7.40**  Air flow distribution in a small system. (From Ref. 83, courtesy of ISHM.)

Another nondimensional number often used in forced convection is the local Stanton number, which is defined as

$$St_x = \frac{Nu_x}{Re_x\,Pr} \tag{7.123}$$

As in free convection, it is convenient in forced convection to work with an average heat transfer coefficient, defined by Eq. (7.76), and a Nusselt number based on the average heat transfer coefficient, defined by Eq. (7.77).

For flow along a plate, transition to turbulence in a forced convection boundary layer begins when $Re \approx 5 \times 10^5$ and becomes fully developed at $Re \approx 10^6$, where the characteristic length is taken as the length along the plate. For flow in a pipe or duct, transition to turbulence begins when $Re \approx 2 \times 10^3$ and becomes fully developed when $Re \approx 1 \times 10^4$, where the characteristic length is the hydraulic diameter of the pipe or duct. The hydraulic diameter is four times the cross-sectional area divided by the wetted perimeter.

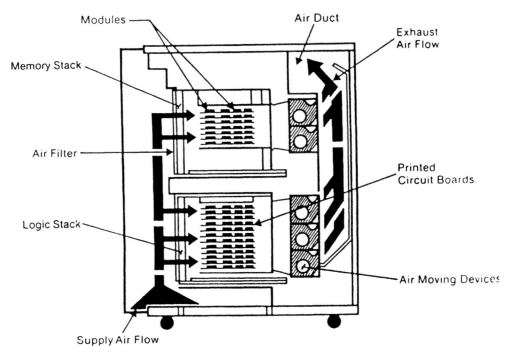

**Figure 7.41**   Forced convection design example. (From Ref. 83, courtesy of ISHM.)

Forced convection heat transfer is often correlated by

$$Nu = C\,Re^{n}\,Pr^{m} \tag{7.124}$$

where $C$, $m$, and $n$ are constants that depend on the geometry, the condition of the surface (i.e., isothermal or constant heat flux), and the flow regime (i.e., laminar or turbulent). As will be discussed in the next two sections, the characteristic length used for the Nusselt and Reynolds numbers in Eq. (7.124) depends on the specific geometry being considered.

### 7.4.1   Empirical Correlations—External Flows

For laminar forced convection flow over an isothermal flat plate:

$$Nu = 0.66\,Re^{0.25}\,Pr^{0.33} \tag{7.125}$$

and for turbulent forced convection flow over an isothermal flat plate, empirical data are well correlated by

$$Nu = 0.036\,Re^{0.8}\,Pr^{0.33} \tag{7.126}$$

Equations (7.125) and (7.126) are also applicable to moderately curved plates.

Holman [26] and Kraus and Bar-Cohen [1] present extensive compilations of correlations of forced convection flow over spheres and circular and noncircular cylinders. Robinson et al. [136, 137] present several correlations for flow of air over electron tubes and prismatic electronic components.

### 7.4.2 Empirical Correlations—Internal Flows

For fully developed laminar forced convection flow in circular tubes, the Nusselt number is constant and is 3.66 for uniform wall temperature and 4.36 for uniform wall heat flux. For fully developed turbulent forced convection flow in smooth circular tubes, Dittus and Boelter [140] recommend

$$Nu = 0.036 \, Re^{0.8} \, Pr^{n} \tag{7.127}$$

where $n = 0.4$ if the tube is being heated and $n = 0.3$ if the tube is being cooled.

Additional correlations are tabulated by Holman [26] and Kraus and Bar-Cohen [1] to account for pipe roughness, entrance effects, noncircular geometry, and viscosity temperature dependence.

### 7.4.3 Superposition Approach

The technique of using correlations with experimental lone-element results does not apply when the heat release is not uniformly distributed over an array of components, as is the case in many practical systems. Heat transfer is influenced not only by the fluid mechanics, set by the geometry, but also by the temperature distribution downsteam of each heated element.

As reported by Moffat et al. [141], the thermal wake of a heated element in an array has an effect on the temperature of its downstream neighbors, felt mainly in the column containing the heated element and in the first four or five rows downstream of the heated element. For a distance beyond five rows downstream, the effect is simply to increase the mean air stream temperature. A thermal wake function $\theta_{ij}$ is used to describe the temperature rise experienced as a function of position downwind of a heated component.

Arvizu and Moffat [142] earlier proposed a superposition approach to the prediction of component temperatures, based on a summing of the temperature rise due to self-heating and the temperature rise due to contributions from heated upstream neighbors. This work used a pressure coefficient, $C_p$, to relate the effects of a heated component on its downstream neighbors. First, the pressure coefficient was calculated for each row in the array. The array velocity for each row was then calculated, and the Reynolds number based on the array velocity was established. The Nusselt number was then determined. Next, the self-heating temperature rise was established based on the specified rating for each powered element, and the contributions from each element to their downstream neighbors were calculated.

Anderson and Moffat [143, 144] continued the superposition approach by proposing that the temperature rise above the channel inlet can be broken into two terms:

$$T_e - T_{in} = (T_e - T_{ad}) + (T_{ad} - T_{in}) \tag{7.128}$$

where $T_e$ is the element temperature (°C), $T_{ad}$ the adiabatic temperature (°C), $(T_e - T_{ad})$ the self-heating temperature rise at the element due to internal heating (°C), and $T_{ad} - T_{in})$ the adiabatic temperature rise due to the thermal wake effect of upstream heated elements (°C).

$$(T_e - T_{ad}) = \frac{q_{conv}}{h_{ad}} \tag{7.129}$$

where $q_{conv}$ is the convective heat transfer rate per unit area (W/m$^2$) and $h_{ad}$ the adiabatic heat transfer coefficient (W/m$^2$ °C).

A superposition kernel function $G$ is then described such that, for the adiabatic temperature rise on the $n$th row in a one-dimensional problem, where the heat flux varies only on a row-by-row basis,

$$(T_{ad} - T_{in})_n = \sum_{i=1}^{n-1} \frac{q_i G(n - i)}{mC_p}$$ (7.130)

where

$$G(n - i) = \frac{T_{ad,n-1} - T_{in}}{T_{mean,n-i} - T_{in}}$$ (7.131)

$$T_{mean} = T_{in} + \frac{q_{conv}}{mC_p}$$

and $q_i$ is the convective heat transfer on the $i$th row (W), $q_{conv}$ the convective heat transfer rate (W), $m$ the mass flow rate in the channel (kg/sec), and $C_p$ the specific heat of air (kJ/kg °C). The element temperature rise is then described as

$$(T_e - T_{in})_n = \frac{q_n}{h_{ad}} = \sum_{i=1}^{n-i} \frac{q_i}{mC_p} G(n - i)$$ (7.132)

For a two-dimensional array, the two-dimensional superposition kernel function g is defined as

$$(T_{ad} - T_{in})_{n,m} = \sum_{j=1}^{M} \sum_{i=1}^{n-i} \frac{q_{i,j}}{mC_p} g(n - i, m - j)$$ (7.133)

where $n$ is the row number, $m$ the column number, $M$ the total number of columns in the array, and

$$g(n - i, m - j) = \frac{T_{ad,(n-i,m-j)} - T_{in}}{T_{mean,(n-i,m-j)} - T_{in}}$$ (7.134)

Relating the superposition kernel function to the thermal wake function described earlier, for an element one row downstream and in the same column as the heated element

$$g(n - i, 0) = \theta_{n-i} \frac{mC_p}{h_{ad,i}A} + 1$$ (7.135)

$$g(n - i, 0) = \frac{\theta_{1,0}}{N} \frac{mC_p}{h_{ad,i}A_i} + 1$$ (7.136)

where $N = n - i$ is the number of rows behind the heated element, and $\theta_{1,0}$ is the wake on the element one row downstream and in the same row as the heated element [144].

Thus, a superposition method has been presented for calculating the temperature rise of a component in a regular array with arbitrary heating, in forced convection, as a sum of two terms: the self-heating rise due to internal heating and the adiabatic temperature rise due to the thermal wake effect of the upstream heated element. The method requires the input of empirical data, specific to the geometry under consideration, and currently applies only to regular in-line arrays [143, 144].

### 7.4.4   Barriers, Missing Modules, or Odd-Sized Modules

A commonly encountered heat transfer problem in electronic equipment is the cooling of an array of heat-generating, blocklike modules in the presence of either barriers, missing modules, or odd-sized modules. A schematic diagram of one such array is shown in Figure 7.42.

Sparrow et al [148, 149] reported experiments to determine the effect on the heat transfer coefficients for a uniform array of modules in the presence of odd-sized or missing modules. In Ref. 148 an array of modules of uniform height was investigated experimentally, and in Ref. 149 account was taken of the occasional presence of modules whose height differed from that of most of the other modules. They used a naphthalene sublimation technique to measure the mass transfer and then deduced the heat transfer by invoking the Reynolds analogy between heat transfer and mass transfer. A wide variety of arrangements of odd-sized modules, missing modules, or barriers was investigated.

Sparrow et al. [148] observed that, for a fully populated array with square modules and without any missing modules, odd-sized modules, or barriers, fully developed heat transfer coefficients were encountered for the fifth and all subsequent rows. These fully developed coefficients could be correlated by

$$Nu = 0.935 Re^{0.72} \tag{7.137}$$

where the characteristic length of the Reynolds number is the height of the modules and the characteristic length of the Nusselt number is the length of the square module. Sparrow et al. [148] also found that if there was a missing module in an array the heat transfer coefficients at other modules located near the site of the missing module were greater than if there was no missing module. The greatest enhancements occurred at lower Reynolds numbers and for modules just downstream of the site of the missing module. Implantation of a barrier in an array of modules was also found to be an effective means of enhancing

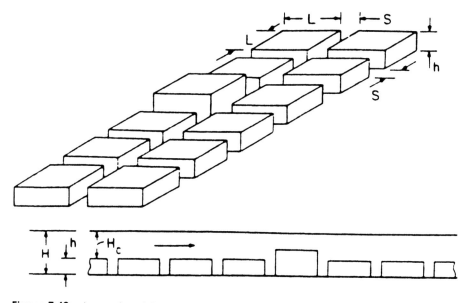

**Figure 7.42**   Array of modules with one odd-sized module.

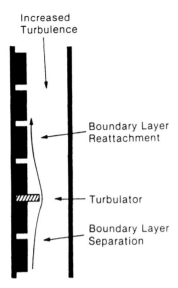

Increased
Turbulence

Boundary Layer
Reattachment

Turbulator

Boundary Layer
Separation

**Figure 7.43**  Use of a turbulator
to thermally enhance downstream
components. (From Ref. 83,
courtesy of ISHM.)

the heat transfer, as shown in Figures 7.43 and 7.44. However, the implantation of barriers
was also found to increase the pressure drop significantly. Sparrow et al. [149] found that
the presence of a module with a different height in an otherwise uniform array of modules
enhanced heat transfer compared to an array of modules with uniform height.

### 7.4.5  Fan Design

In forced convection cooling, the two most common air-moving devices used are fans
and blowers. Fans move air along their axis of rotation, whereas blowers usually move
air in a direction normal to their axis of rotation, either radially or tangentially. Fans are
usually used in low air flow applications, such as cooling individual products, and blowers
are used for high air flow applications, such as cooling large electronic systems. Blowers
are usually more efficient and quieter than fans, but blowers usually need larger mounting
spaces than fans.

Strassberg [150] has proposed the following rules of thumb to determine whether a
fan or blower is necessary:

1.  If a PCB generates less than 0.5 W/in.$^2$ (0.08 W/cm$^2$), natural convection is the
    best cooling method.
2.  If a PCB generates between 0.5 and 1.0 W/in.$^2$ (0.08 and 0.16 W/cm$^2$), natural
    convection might be adequate but a fan or blower should be considered.
3.  If the PCB generates between 1.0 and 2.5 W/in.$^2$ (0.16 and 0.39 W/cm$^2$), a fan
    or blower blowing air parallel to the plane of the PCB should provide sufficient
    cooling.
4.  If the PCB generates between 2.5 and 5.0 W/in.$^2$ (0.39 and 0.78 W/cm$^2$), a fan
    or blower blowing air normal to the plane of the board, directly at the hot com-
    ponents, should be considered.

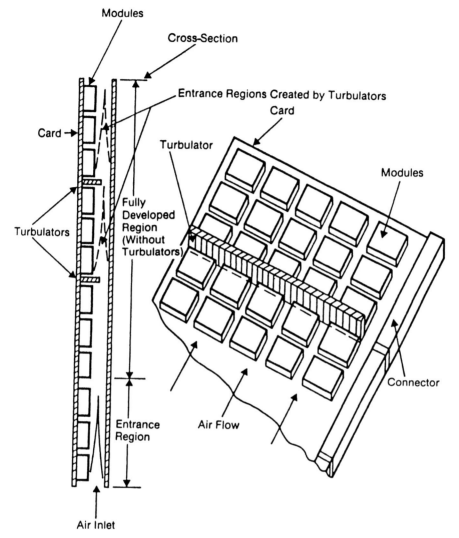

**Figure 7.44**  Turbulator for boundary layer control on an air-cooled card. (From Ref. 83, courtesy of ISHM.)

5. If the PCB generates more than 5.0 W/in.$^2$ (0.78 W/cm$^2$), auxiliary cooling methods, such as heat exchangers or heat pipes, will probably be required.

Detailed guidelines for selecting the size and type of fan or blower are presented by Welsh [151], Madison [152], Strassberg [150], Markstein [153], Kraus and Bar-Cohen [1], Ellison [154], Vogel [155], Weghorn [156], and Steinberg [157], as well as by the various fan and blower manufacturers.

In addition to selecting the size and type of fan or blower, one must decide whether to blow cooling air through the enclosure or to draw cooling air through the enclosure. When a fan or blower is used to blow air through filtered inlets into the enclosure, as shown in Figure 7.45, the entire enclosure can be pressurized with clean air. This pressurization helps to prevent dirt or dust from entering the enclosure. The pressurization

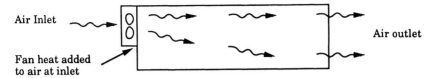

**Figure 7.45**  Fan used to blow air through enclosure.

also causes the air to be denser, providing a better cooling medium. However, the heat dissipation of the fan or blower adds to the heat load of the enclosure.

When a fan or blower is used to draw cooling air through the enclosure, as shown in Figure 7.46, the enclosure's interior will be at a negative gauge pressure, pulling air into the enclosure through small gaps and cracks as well as through the filtered inlets. Drawing air provides a greater flow of cooling air than does blowing air. However, drawing air into the enclosure results in more dust and dirt entering the enclosure than does blowing air.

### 7.4.6  Numerical Methods

The comments made in Section 7.3.5 about numerical methods for natural convection are generally applicable to numerical methods for forced convection. In principle, numerical methods can be used to solve problems involving forced convection in electronic systems. However, in practice, numerical methods are seldom practicable for forced convection in complex systems, such as typical electronic packages. The interested reader is referred to Shih [93], Roache [127], Patankar [128], and Anderson et al. [129]

### 7.5  MIXED CONVECTION

The previous sections assumed that natural convection and forced convection were the only modes of convective heat transfer. However, in many electronic cooling applications, both natural convection and forced convection are present. In some applications, one has aiding flow in which the forced and free convection currents are in the same direction. In other applications, one has opposing flow in which the currents are in the opposite direction.

A rule of thumb is that if $Gr/Re^2 > 10$, forced convection can be neglected, and that if $Gr/Re^2 < 0.1$, natural convection can be neglected. However, in many complex problems it may be difficult to define representative values of Gr and Re. Furthermore, it is not unusual for natural convection to predominate in some regions of an electronic system, forced convection to predominate in other regions, and both natural and forced convection to be important in the remaining regions.

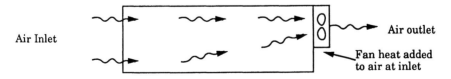

**Figure 7.46**  Fan used to draw air through enclosure.

The interested reader is referred to Metais and Eckert [158], Brown and Gauvin [159], Fand and Keswani [160], and Depew et al. [161].

## 7.6 BOILING AND CONDENSATION HEAT TRANSFER

Previous sections of this chapter have assumed that the heat transfer media were either homogeneous or single-phase materials. Much higher heat transfer rates may usually be attained if a fluid undergoes a change of phase.

Boiling may occur when a surface and a liquid come in contact and the surface is maintained at a temperature above the saturation temperature of the liquid. Figure 7.47 shows the four major regimes of boiling, based on experimentation done with an electrically heated platinum wire submerged in water [26]. Figure 7.47 shows the heat flux plotted against the temperature excess, $T_w - T_{sat}$, or surface superheat [1], which is the difference between the wire temperature and the saturation temperature of the fluid. In region 1, free convection is the primary heat transport vehicle because of the temperature difference between the hot surface and the fluid. At the boundary between regions 1 and 2, bubbles begin to form at nucleation sites on the surface because of the superheat. Through region 2, the available superheat results in the activation of many more bubbles at nucleation sites, and the fluid circulation caused by the departure of the bubbles leads to a steep rise in heat flux. However, at the boundary between regions 2 and 3, the bubble departure frequency is such that the bubbles begin to merge with each other, so that at times a vapor film covers the heated surface. Because heat must now be conducted through this film before it can reach the fluid, the heat flux is reduced. However, this region is unstable and corresponds to oscillations between film and nucleate boiling. In region 4, the vapor film blankets the heated surface and film boiling occurs [1,26].

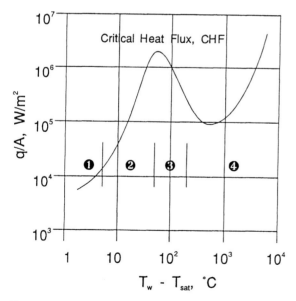

**Figure 7.47** Boiling. (1) Free convection boiling; (2) nucleate boiling; (3) partial film boiling; (4) film boiling.

No single expression exists to represent the complex boiling phenomena. However, for nucleate pool boiling, in which the heated surface is located under the surface of an initially still fluid, Rohsenow's correlation [4] is often used:

$$\frac{C_l \Delta T_x}{h_{fg} Pr_l^s} = C_{sf} \left[ \frac{q/A}{\mu_l h_{fg}} \sqrt{\frac{\sigma}{g(\rho_l - \rho_v)}} \right]^{1/3} \tag{7.138}$$

where subscript $l$ = liquid, subscript $v$ = vapor, $\sigma$ is the surface tension (N/m), $C_l$ is the specific heat of saturated liquid (kJ/kg °C), $\Delta T_x$ is the temperature excess = $T_w - T_{sat}$ (°C), $S = 1$ for water and 1.7 for other fluids, $C_{sf}$ is a fluid surface–dependent coefficient, $\rho$ is density (kg/m³), $g$ is the acceleration due to gravity (m/s²), and $h_{fg}$ is the enthalpy of vaporization (J/kg).

Values of the surface tension $\sigma$ for water are shown in Table 7.8. The fluid surface–dependent coefficient $C_{sf}$ is shown in Table 7.9 for a number of fluids and surfaces.

For water boiling on the outside of submerged surfaces at atmospheric pressure, Holman [26] presents the following empirical relations.

For horizontal surfaces:

$$\frac{q}{A} < 16 \qquad h = 1042(\Delta T_x)^{1/3} \tag{7.139}$$

$$16 < \frac{q}{A} < 240 \qquad h = 5.56(\Delta T_x)^3 \tag{7.140}$$

For vertical surfaces:

$$\frac{q}{A} < 3 \qquad h = 537(\Delta T_x)^{1/7} \tag{7.141}$$

$$3 < \frac{q}{A} < 63 \qquad h = 7.96(\Delta T_x)^3 \tag{7.142}$$

where $\Delta T_x = T_w - T_{sat}$ (°C). $\tag{7.143}$

**Table 7.8** Vapor-Liquid Surface Tension for Water

| Saturation temperature | | Surface tension | |
|---|---|---|---|
| °F | °C | $\sigma \times 10^4 (lb_f/ft)$ | $\sigma(mN/m)$ |
| 32 | 0 | 51.8 | 75.6 |
| 60 | 15.56 | 50.2 | 73.3 |
| 100 | 37.78 | 47.8 | 69.8 |
| 140 | 60 | 45.2 | 66.0 |
| 200 | 93.33 | 41.2 | 60.1 |
| 212 | 100 | 40.3 | 58.8 |
| 320 | 160 | 31.6 | 46.1 |
| 440 | 226.67 | 21.9 | 32.0 |
| 560 | 293.33 | 11.1 | 16.2 |
| 680 | 360 | 1.0 | 1.46 |
| 705.4 | 374.1 | 0 | 0 |

Source: Ref. 201, © Chemical Rubber Publishing Co.

**Table 7.9**  Values of the Coefficient $C_{sf}$ for Various
Liquid-Surface Combinations

| Fluid–heating surface combination | $C_{sf}$ |
|---|---|
| Water-copper [202] | 0.013 |
| Water-platinum [203] | 0.013 |
| Water-brass [204] | 0.0060 |
| Water–emery-polished copper [206] | 0.0128 |
| Water–ground and polished stainless steel [206] | 0.0080 |
| Water–chemically etched stainless steel [206] | 0.0133 |
| Water–mechanically polished stainless steel [206] | 0.0132 |
| Water–emery-polished and paraffin-treated copper [206] | 0.0147 |
| Water–scored copper [206] | 0.0068 |
| Water–Teflon pitted stainless steel [206] | 0.0058 |
| Carbon tetrachloride–copper [202] | 0.013 |
| Carbon tetrachloride–emery-polished copper [206] | 0.0070 |
| Benzene-chromium [205] | 0.010 |
| n-Butyl alcohol–copper [202] | 0.00305 |
| Ethyl alcohol–chromium [205] | 0.027 |
| Isopropyl alcohol–copper [202] | 0.0225 |
| n-Pentane–chromium [205] | 0.015 |
| n-Pentane–emery-polished copper [206] | 0.0154 |
| n-Pentane–emery-polished nickel [206] | 0.0127 |
| n-Pentane–lapped copper [206] | 0.0049 |
| n-Pentane–emery-rubbed copper [206] | 0.0074 |
| 35% $K_2CO_3$–copper [202] | 0.0054 |
| 50% $K_2CO_3$–copper [202] | 0.0027 |

Source: Ref. 26, © McGraw-Hill, Inc.

Boiling and condensation heat transfer are not often encountered in electronic applications. However, two promising techniques—immersion cooling and heat pipes—are the subject of much research. In immersion cooling, electronic components are either directly or indirectly immersed in low boiling point dielectric liquids. References 162–164 describe immersion cooling in detail. In heat pipes, a pipe with wicking material on the inside of the pipe wall and filled with a condensable fluid is used. Heat is added at one end of the pipe, resulting in vapor generation, and heat is removed at the other end of the pipe, resulting in condensation of the vapor and replenishment of the liquid by capillary action. References [1] and [26] discuss heat pipe theory, and Ref. [165] discusses potential applications of heat pipes to the cooling of electronic systems.

For more detailed information on boiling and condensation heat transfer, the interested reader is referred to Kraus and Bar-Cohen [1], Holman [26], Seely and Chu [27], Eckert and Drake [28], and Tong [166].

## 7.7  RADIATION
### 7.7.1  Radiation Fundamentals

Like free convection, radiation heat transfer is always present in any real electronic cooling application, although the effect is small for all but the most severe situations. One area of application in which radiation is significant is satellite or high-vacuum applications.

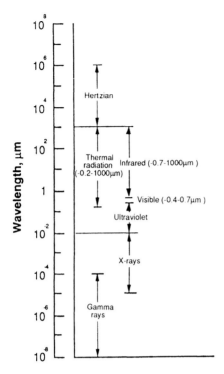

**Figure 7.48** Wavelengths. (From Ref.
199.)

All matter emits electromagnetic radiation at temperatures above absolute zero, with
random frequency, phase, and direction. Radiation is defined as the transfer of energy
by disorganized photon propagation. The photons may be in the wavelength ranges corresponding to the ultraviolet, visible, or infrared regions, all of which are part of the
electromagnetic wave spectrum, as shown in Figure 7.48. Visible light radiation has
wavelengths from approximately 0.4 to 0.7 μm. Infrared radiation has wavelengths from
approximately 0.7 to 1000 μm. When these photons reach another body, they are either
absorbed, reflected, or transmitted through the surface. Three properties are used to
characterize surfaces with regard to their reaction to incident photons:

$\alpha$    absorptivity, the fraction of incident radiation absorbed
$\rho$    reflectivity, the fraction of incident radiation reflected
$\tau$    transmissivity, the fraction of incident radiation transmitted

$$\alpha + \rho + \tau = 1 \tag{7.144}$$

See Figure 7.49 for illustration.

There is heat transfer by radiation between two surfaces at different temperatures
even in the absence of an intervening medium. In fact, heat transfer by radiation occurs
most efficiently in a vacuum. This radiative emission is caused by changes in the electronic
configurations of the constituent atoms or molecules and occurs in liquids and gases as
well as solids.

The Stefan-Boltzmann law defines the maximum flux at which radiation may be
emitted from a surface:

$$q'' - \sigma T_s^4 \quad (\text{W/m}^2) \tag{7.145}$$

where $T_s$ is the absolute temperature of the surface (K) and $\sigma$ is the Stefan-Boltzmann constant, $5.67 \times 10^8$, W/m$^2$ K$^4$. A surface such as this is known as an ideal radiator or blackbody. For a real surface, the heat flux emitted is less than that of a blackbody and is described as

$$q'' = \epsilon \sigma T_s^4 \tag{7.146}$$

where $\epsilon$ is the emissivity. The emissivity value is a measure of how efficiently a surface emits compared to an ideal radiator and may be shown to be numerically equal to the absorptivity of the material.

Some generalizations can be made about emissivity values. The emissivity values for metallic surfaces are generally small and typically are less if the surface finish is polished, and the emissivity values of nonconductors are comparably large. The emissivity values of conductors increase with increasing temperature, while the emissivities of nonconductors may increase or decrease with increasing temperature [23,24].

The peak wavelength of the radiation emitted by any body above absolute zero is dependent on the temperature. The correlation between peak wavelength and temperature is called Planck's spectral distribution of emissive power and is given as

$$\epsilon_{\lambda b} = \frac{C_1}{\lambda^5 e^{c_2/\lambda T} - 1} \tag{7.147}$$

where $C_1$ and $C_2$ are constants whose values are $3.743 \times 10^8$ W $\mu$m$^4$/m$^2$ and $1.4387 \times 10^4$ $\mu$m K, respectively. Here $e_{\lambda b}$ is the radiation energy emitted by a blackbody at wavelength $\lambda$. An object actually emits radiation in a spectrum around its peak wavelength, as shown in Figure 7.50. Equation (7.147) can therefore be integrated from zero to infinity to determine the total radiation, $e_b$, from a blackbody. Integrating Eq. (7.147) from $C_1 = 0$ to $\infty$ produces the Steffan-Boltzmann law

$$e_b = \sigma T^4 \tag{7.148}$$

Radiation directed at an object's surface is reflected, transmitted, or absorbed, and the radiation absorbed is in turn emitted. PCBs are generally made of materials that do not transmit radiation, so that transmission can be neglected. The total radiation heat transferred from the object is therefore

$$e = (\epsilon \sigma T_o^4 + \rho \sigma T_a^4) \tag{7.149}$$

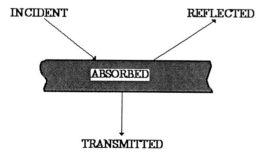

**Figure 7.49**  Radiation.

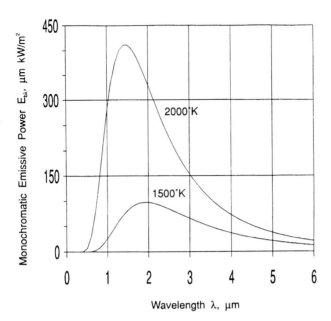

**Figure 7.50**  Blackbody emissive power as a function of wavelength.

where $e$ is the radiation energy being emitted, $T_o$ is the temperature of the object, $T_a$ is
the ambient temperature, $\epsilon$ is the emissivity of the object, and $\rho$ is the reflectivity of the
object. Because transmission is neglected, $\epsilon + \rho = 1$, so that

$$e = \epsilon\sigma T_o^4 + (1 - \epsilon)\sigma T_a^4 \tag{7.150}$$

The first term on the right-hand side of Eq. (7.150) is the radiation energy emitted from
the object and the second term on the right-hand side is the radiation energy from the
ambient air around the object.

These relationships address how energy is emitted from a surface. Determining the
net rate at which radiation is exchanged between surfaces at different temperatures is
more difficult. A special case describes the net exchange between a small surface and a
much larger surface that completely surrounds it, both surfaces being blackbodies. For
these conditions, the net rate of radiation exchange, and thus heat exchange, between
these blackbody surfaces, as shown in Figure 7.51, is

$$q'' = \sigma A_1 f_{12}(T_1^4 - T_2^4) \tag{7.151}$$

where $A_1$ is the surface area of the small inner surface (m), $T_1$ the temperature of the
inner surface (K), $T_2$ the temperature of the outer surface (K), and $f_{12}$ the configuration
factor, that is, the fraction of radiation emitted by surface 1 that is intercepted by surface
2. For the condition of a small body 1 totally enclosed in a larger body 2, $f_{12} = 1$.

For a "gray body" approximation of a real, nonblackbody surface, the rate equation
for the special conditions described above, with a small body 1 totally enclosed in a large
body 2, becomes

$$q'' = A_1 F_{12}\sigma(T_1^4 - T_2^4) \tag{7.152}$$

where $F_{12}$ is the gray body shape factor, such that

$$A_1 F_{12} = \frac{1}{(1 - \epsilon_1)/\epsilon_1 A_1 + 1/A_1 f_{12} + (1 - \epsilon_2)/\epsilon_2 A_2} \tag{7.153}$$

where $A_2$ is the surface area of the large outer surface (m), $\epsilon_1$ the emissivity of surface 1, and $\epsilon_2$ the emissivity of surface 2.

Heat transfer from a surface may occur simultaneously by convection and radiation. The total rate of heat transfer per unit area, then, is described [24] by the sum

$$q'' = q''_{(conv)} + q''_{(rad)} \tag{7.154}$$

$$q'' = h(T_s - T_\infty) + \epsilon\sigma(T_s^4 - T_{sur}^4) \tag{7.155}$$

where $h$ is the convective heat transfer coefficient (W/m$^2$ K), $T_s$ the surface temperature (K), $T_\infty$ the ambient fluid temperature, (K), and $T_{sur}$ the temperature of the surrounding enclosure (K).

Radiation that is emitted from a surface propagates outward in all directions. The directional distribution of the emitted radiant energy is described in terms of radiation intensity. For the surface depicted in Figure 7.52, the rate at which energy is radiated from the surface element $dA$ through the element $dA_n$ can be expressed in terms of the spectral intensity $I_{\lambda,e}$. $I_{\lambda,e}$ is defined as the rate at which radiant energy is emitted at the wavelength $\lambda$ in the ($\theta$, $\phi$) direction, per unit area of the emitting surface normal to this direction, per unit solid angle about this direction, and per unit wavelength interval $d\lambda$ about $\lambda$ [169]. The spectral intensity is based on the area projected in the direction of radiation. This projected area is given by $dA \cos \theta$.

The spectral intensity of the emitted radiation may therefore be expressed as

$$I\lambda_{,e}(\lambda, \theta, \phi) = \frac{dq_\lambda}{dA \cos \theta \, d\omega} \tag{7.156}$$

where $dq_\lambda = dq/d\lambda$ is the spectral heat transfer rate (W/$\mu$m) and $d\omega = dA_n/r^2$ is the solid differential angle.

The emission of radiant energy from a surface can also be described in terms of the spectral emissive power $E_\lambda$, which is defined as the rate at which radiant energy is emitted at the wavelength $\lambda$ in the ($\theta$, $\phi$) direction, per unit area of the emitting surface, per unit solid angle about this direction, and per unit wavelength interval $d\lambda$ about $\lambda$ [169]. The spectral emissive power differs from the intensity in that it is based on the actual surface

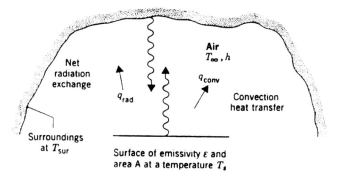

**Figure 7.51** Radiation exchange between a surface and its surroundings. (From Ref. 24, © John Wiley & Sons, Inc., reprinted by permission.)

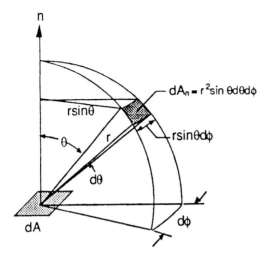

**Figure 7.52** Directional distribution of emitted radiant energy. (From Ref. 199.)

area of the emitting surface rather than the projected surface area. The spectral emissive
power $E_\lambda$ is [199]

$$E_\lambda(\lambda, \theta_1 - \theta_2, \phi_1 - \phi_2) = \int_{\phi_2}^{\phi_1} \int_{\theta_1}^{\theta_1} I_{\lambda,c}(\lambda, \theta, \phi) \cos \theta \sin \theta \, d\theta \, d\phi \qquad (7.157)$$

A rigorous analysis of radiation in complex geometries, such as typical electronic
packages is difficult. The interested reader is referred to references [1, 26, 28, 167–169].
Buller and McNelis [170] have developed an approximate model that avoids the difficulties
of a rigorous analysis of radiation but reasonably predicts radiation heat transfer in some
situations. They found that their model was capable of predicting the radiant heat exchange
to within 5% of the experimentally measured data in (1) vertical natural convection
environments, (2) horizontal natural convection environments, and (3) forced convection
environments in which the air flow rate did not exceed 600 feet per minute (3.1 m/s).

An alternative way to attempt to model and assess the effects of radiant heat loss in
a system is to calculate the ''minimum radiative resistance'' by the relationship

$$R = \frac{1}{4\sigma A T_m^3} \qquad (7.158)$$

where $T_m$ is the mean temperature of the heated surface and surrounding temperatures
(K). It is proposed that the minimum radiative resistance be compared to the other thermal
resistances in the circuit to determine the relative importance of the radiation mode for
heat transport [12].

Note that the simplified gray body approximations do not allow consideration of
variations of emission or other properties with wavelength or temperature, which would
be required for more accurate calculations [24].

In most electronic applications, ignoring thermal radiation does not change the results
within the range of accuracy of the analysis and results in a slight underestimation of heat
transfer coefficients and hence, in terms of reliability, a conservative estimate for the
system.

## 7.7.2 Infrared Thermography

Thermal testing is desirable to determine the accuracy of the calculation methods. There are two main methods of measuring the temperature distribution of powered PCBs. The first is the more obvious method of connecting thermocouples, temperature-sensitive crayons or labels, thermistors, resistance temperature detectors, or semiconductor temperature sensors to the PCB at points of interest. Another method gaining in popularity is the use of infrared thermography. This is an attractive technique because it is nonintrusive and will not change the thermal field while the measurement is being taken. Also, using infrared equipment, the entire field of temperatures can be taken at one time, which is not possible using thermocouples.

Radiation detectors used in thermal vision systems are sensitive to a narrow band of the electromagnetic spectrum. As a result, only the fraction of the total emissive power emitted in this band can be detected. The emissive power for a wavelength range from $\lambda_1 \leq \lambda \leq \lambda_2$ of the detector is given by

$$E_{(\lambda_1 - \lambda_2)}(T) = \int_{\lambda_1}^{\lambda_2} E_{\lambda b}(\lambda, T) \, d\lambda \tag{7.159}$$

Although the emission of radiant energy is a directional quantity, there is a special class of surfaces for which the intensity of the emitted radiation is independent of the direction. A surface of this type is said to be a diffuse emitter. For such a surface

$$I_{\lambda,e}(\lambda, \theta, \phi) = I_{\lambda,e}(\lambda) \tag{7.160}$$

The wavelength at which the emissive power is a maximum for a given temperature is given by Wien's displacement law [169] as

$$\lambda_{\max} = \frac{C_3}{T} \tag{7.161}$$

where $\lambda$ is the wavelength ($\mu$m), $T$ is the absolute temperature (K), and $C_3$ is a constant $= 2897.8 \; \mu$m K.

Gases such as $CO_2$, $O_2$, and $H_2O$ act as wavelength-selective absorbers of radiant energy. Consideration must be given to the attenuation effects caused by the atmosphere. The attenuation effects are negligible for a number of atmospheric "windows," including the wavelength interval 8.0 to 13.0 $\mu$m [171]. By selecting a thermal sensor that is sensitive to one of these wavelength intervals, the problem of atmospheric attenuation can be minimized.

Another consideration is the change in signal versus temperature for the sensor. The greatest percentage change in signal versus temperature occurs at the peak of the blackbody radiation curve [172]. Therefore, by applying Wein's displacement law it can be seen that the wavelength interval from 8 to 13 $\mu$m is best for measuring temperatures ranging from $-50.2$ to $89.1$ °C. If it is not possible to utilize one of the atmospheric windows for infrared measurement, the absorption due to atmospheric gases must be calibrated out of the data.

The blackbody is an idealized construct. Real bodies emit less energy than a blackbody at the same temperature. The ratio

$$\epsilon_{\lambda,\theta}(\lambda, \theta, \phi, T) = \frac{I_{\lambda,\theta}(\lambda, \theta, \phi, T)}{I_{\lambda,b}(\lambda, T)} \tag{7.162}$$

is defined as the spectral directional emissivity. The emissivity is a measure of how efficiently a body emits radian energy. Emissivity values range from zero for a perfect reflector to 1.0 for a blackbody.

The radiative energy flux leaving a given surface is composed of emitted and reflected radiation. This flux, defined as radiosity, accounts for all of the radiant energy leaving the surface. The spectra radiosity $J_\lambda$ is the rate at which radiation leaves a unit area of the surface at the wavelength $\lambda$, per unit wavelength interval $d\lambda$ about $\lambda$. For a gray body, the spectral radiosity is given by

$$J_\lambda(T) = \epsilon_\lambda E_b(T) + \rho_\lambda G \tag{7.163}$$

A thermal vision system sees a target object as an array of discrete picture elements. Each picture element provides a measure of the average radiosity of that portion of the surface. The radiosity of the target surface is a function of the target temperature, the ambient conditions, and the average emissivity for that element. Therefore, it is necessary to determine the unique emissivity value for each element. The resulting array of emissivity values is known as the emissivity map for the target surface. This emissivity map is valid only for a given orientation of the target surface with respect to the thermal vision system. In other words, if the target surface moves with respect to the thermal vision system, the emissivity map is no longer valid.

There are two basic approaches for dealing with unknown emissivities. The first is emissivity equalization, in which the test object is coated with a substance of known emissivity to provide a uniform surface finish. Although this technique is simple and is adequate for some applications, it has the unfortunate effect of altering the thermal characteristics of the test object. Therefore, infrared thermographic measurements made with emissivity equalization are not entirely noninvasive.

The second approach is emissivity compensation. With this approach, no attempt is made to make the target emissivity uniform. Rather, the local emissivity of each part of the target is determined separately. There are three basic techniques for doing this. The first and simplest is to look the emissivity value up in a table of known emissivity values for various materials, surface finishes, and temperatures. The use of this technique is limited to cases in which the properties of the target are well known and published emissivity values are available and accurate.

The second approach is to illuminate the target with a reference source of known temperature and known radiative properties. The radiant energy that is reflected by the target is then measured and the emissivity is calculated from the measured reflectivity. This is the most complicated of the techniques and requires that the testing conditions be very well defined.

The third technique is to heat the test object to a known temperature different from the ambient temperature and measure the radiant energy emitted by the object [176]. This technique is more direct than the reflectivity measurement technique and can be easily implemented [199].

Additional information on infrared thermography and its applications can be found in references [171–184].

## 7.7.3 Numerical Methods

Numerical methods are extensively used to solve problems involving radiation exchange between a gas and a heat transfer surface. However, in most electronic systems, air is the cooling gas used and air can be considered to be transparent to radiation at the temperatures usually encountered in such systems.

Numerical methods are frequently helpful in calculating radiation heat transfer between large numbers of nonblack bodies. Particularly useful for these problems are Monte Carlo methods, in which statistical sampling procedures are used to approximate the radiation heat transfer.

For more detailed information on the use of numerical methods for radiation heat transfer, the reader is referred to Shih [93], Edwards [185], Nice [186], and Howell [187].

## 7.8   HEAT EXCHANGERS

A heat exchanger is a device that permits the exchange of heat between two fluids that are at different temperatures and separated by an intermediate wall, such that mixing does not occur between the two fluids. Heat exchangers are generally classified by the flow arrangement of the fluids and the type of construction.

The fluids may flow in a parallel flow arrangement, in the same or opposite directions, as shown in Figure 7.53, or in a cross-flow manner as shown in Figure 7.54. Figure 7.53 also illustrates the simplest heat exchanger construction, the tubular construction in which the fluids move in concentric tubes. Figure 7.54 shows one type of core for the compact heat exchanger construction. Compact heat exchangers are typically used when one of the fluids is a gas and are characterized by dense arrays of fins and tubes or plates. Figure 7.55 shows four examples; many other core types are used, and the interested reader is referred to Kraus and Bar-Cohen [1] for other examples.

For the simple parallel-flow heat exchanger shown in Figure 7.53A, an energy balance equation can be established for the two fluids, one hot, the other cold, as

$$\dot{m}_h c_{ph}(T_1 - T_2) = \dot{m}_c c_{pc}(t_1 - t_2) \tag{7.164}$$

where $\dot{m}_h$ is the mass flow rate, hot fluid (kg/sec); $\dot{m}_c$ the mass flow rate, cold fluid (kg/sec); $c_{ph}$ the specific heat, hot fluid (kJ/kg °C); $c_{pc}$ the specific heat, cold fluid (kJ/kg °C);

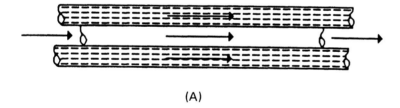

(A)

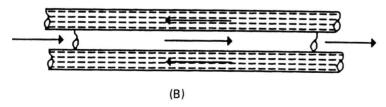

(B)

**Figure 7.53**   Concentric tube heat exchangers. (A) Parallel flow; (B) counterflow.

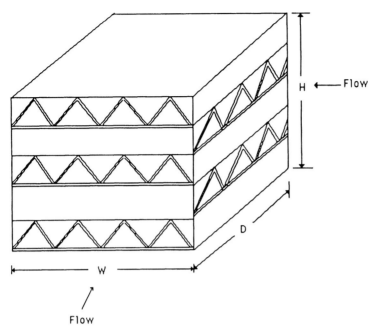

**Figure 7.54**  A two-fluid compact heat exchanger with headers removed.
(From Ref. 1, © McGraw-Hill, Inc.)

$T_1$ the inlet temperature, hot fluid (°C); $T_2$ the outlet temperature, hot fluid (°C); $t_1$ the inlet temperature, cold fluid (°C); and $t_2$ the outlet temperature, cold fluid (°C).

The relationship between $T_1$, $T_2$, $t_1$, and $t_2$ over the length $S$ under examination is as shown in Figure 7.56.

Heat transfer data for compact heat exchangers are shown for individual surfaces, for different fin types, as a plot of the heat transfer factor $j_h$ as a function of the Reynolds number. The heat transfer factor is defined [1] as

$$j_h = (St) \left(\frac{c\mu}{k}\right)^{2/3} \tag{7.165}$$

where St is the Stanton number, $c$ the specific heat (BTU/lb °F), $\mu$ the viscosity (lb/hr ft), and $k$ the thermal conductivity, (BTU/ft °F.

The pressure drop $\Delta p$ in a heat exchanger can be found from

$$\Delta P = \frac{G^2}{2} \left[ (1 + \sigma^2)(v_2 - v_1) + \frac{fSv_m}{A_c} \right] \tag{7.166}$$

where $G$ is the mass velocity, (lb/ft$^2$ hr), $\sigma$ the ratio of free flow area to frontal area (dimensionless), $v_2$ the specific volume of fluid at outlet (ft$^3$/lb), $v_1$ the specific volume of fluid at inlet (ft$^3$/lb), $f$ the friction factor (dimensionless), $S$ the surface area over which flow occurs (ft$^2$), $v_m$ the specific volume of fluid at mean conditions, (ft$^3$/lb), and $A_c$ the free flow area (ft$^2$).

The friction factor $f$ is typically shown as a function of the Reynolds number. There are also pressure losses due to entrance and exit effects, which are also given for a specific exchanger. Kays and London [188] present heat transfer and friction factor data for many different surfaces.

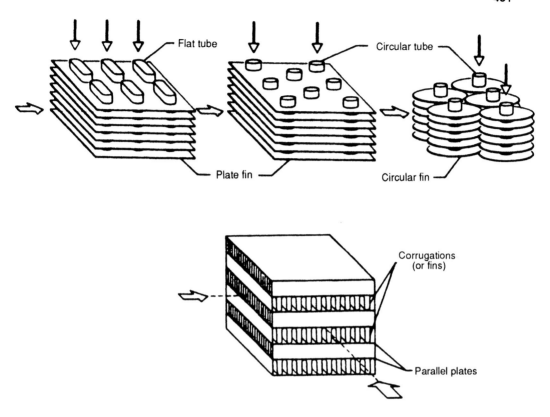

**Figure 7.55**  Compact heat exchanger cores. (From Ref. 24, © John Wiley & Sons, Inc., reprinted by permission.)

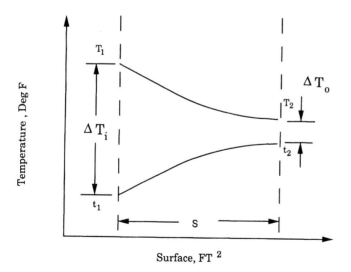

**Figure 7.56**  Simple double-pipe heat exchanger and temperature-length profile for parallel concurrent flow. (From Ref. 61, © 1965 Prentice-Hall, Inc., Englewood Cliffs, NJ, reprinted by permission.)

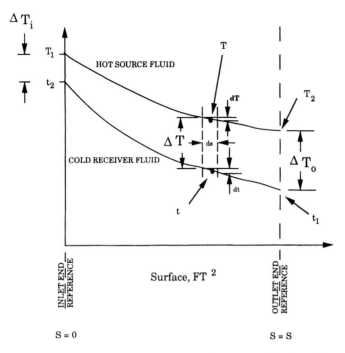

**Figure 7.57** Temperature-length profile for counterflow heat Exchanger. (From Ref. 61, © 1965 Prentice Hall, Inc., Englewood Cliffs, NJ, reprinted by permission.)

For the parallel-flow arrangement shown in Figure 7.53A, the log mean temperature difference (LMTD) can be found [61] to be

$$\Delta T_m = \frac{\Delta T_2 - \Delta T_1}{\ln(\Delta T_2/\Delta T_1)} \tag{7.167}$$

where $\quad \Delta T_1 = T_1 - t_1$ (7.168)
$\qquad\quad \Delta T_2 = T_2 - t_2$ (7.169)

For the counterflow arrangement shown in Figure 7.53B, the temperature-length profile is as shown in Figure 7.57 and the LMTD is found to be

$$\Delta T_m = \frac{(T_2 - t_1) - (T_1 - t_2)}{\ln[(T_2 - t_1)/(T_1 - t_2)]} \tag{7.170}$$

An overall rate equation for the heat exchanger can then be established such that

$$q = UA\,\Delta T_m \tag{7.171}$$

where $U$ is the overall heat transfer coefficient, $A$ the area corresponding to $U$, and $\Delta T_m$ the appropriate mean temperature.

During operation, the interior walls that surround the cooling liquid can develop film or scales due to reactions between the fluid and the wall material, and this material will reduce the heat transfer between the fluids. A fouling resistance $R_f$ is used to describe

**Table 7.10** Representative Fouling Factors

| Fluid | $R_f''$ ($m^2$ K/W) |
|---|---|
| Seawater and treated boiler feedwater (below 50°C) | 0.0001 |
| Seawater and treated boiler feedwater (above 50°C) | 0.0002 |
| River water (below 50°C) | 0.0002–0.0001 |
| Fuel oil | 0.0009 |
| Refrigerating liquids | 0.0002 |
| Steam (non–oil bearing) | 0.0001 |

Source: Ref. 24, © John Wiley & Sons, Inc., reprinted by permission.

this effect. Therefore, for an unfinned, tubular heat exchanger such as is shown in Figure 7.53, the overall heat transfer coefficient is found [24] to be

$$\frac{1}{UA} = \frac{1}{U_i A_i} = \frac{1}{U_o A_o} \tag{7.172}$$

where $i$ is the inner tube surface area $\pi D_i L$ exposed to the fluid.

$$\frac{1}{UA} = \frac{1}{h_i A_i} + \frac{R_{f,i}}{A_i} + \frac{\ln(D_o/D_i)}{2\pi k L} + \frac{R_{f,o}}{A_o} + \frac{1}{h_o A_o} \tag{7.173}$$

For finned exchangers, Eq. (7.172) and (7.173) become [24]

$$\frac{1}{UA} = \frac{1}{U_c A_c} = \frac{1}{U_h A_h} \tag{7.174}$$

where c and h designate the hot and cold fluids, respectively.

$$\frac{1}{UA} = \frac{1}{(\eta_o h A)_c} + \frac{R_{f,c}}{(\eta_o A)_c} + R_w + \frac{R_{f,h}}{(\eta_o A)_h} + \frac{1}{(\eta_o h A)_h} \tag{7.175}$$

where $R_w$ is the conduction resistance for either a plane wall or a cylincrical wall and $\eta$ is the fin efficiency.

Table 7.10 shows representative values for the fouling resistance $R_f$ and Table 7.11 shows representative values for the overall heat transfer coefficient $U$.

**Table 7.11** Representative Values of the Overall Heat Transfer Coefficient

| Fluid combination | $U$(W/$m^2$ K) |
|---|---|
| Water to water | 850–1700 |
| Water to oil | 110–350 |
| Steam condenser (water in tubes) | 1000–6000 |
| Ammonia condenser (water in tubes) | 800–1400 |
| Alcohol condenser (water in tubes) | 250–700 |
| Finned-tube heat exchanger (water in tubes, air in cross flow) | 25–50 |

Source: Ref. 24, © John Wiley & Sons, Inc., reprinted by permission.

Next, a capacity rate $c$ is defined [1] such that

$$c = \dot{m}c_p \tag{7.176}$$

where $\dot{m}$ is the mass flow rate and $c_p$ is the specific heat. There exist both a $c_{hot}$ and a $c_{cold}$. The smaller of $c_{hot}$ and $c_{cold}$ will be referred to as $c_{min}$ and the other as $c_{max}$. A capacity rate ratio $R$ is defined such that

$$R = \frac{c_{min}}{c_{max}} \leq 1 \tag{7.177}$$

The number of transfer units (NTU) is then defined, such that

$$\text{NTU} = \frac{UA}{c_{min}} \tag{7.178}$$

The NTU value is a rough measure of the size of the heat exchanger.
    The exchanger heat transfer effectiveness $\epsilon$ is defined [1] as

$$\epsilon = \frac{c_h(T_1 - T_2)}{c_{min}(T_2 - t_1)} = \frac{c_c(t_2 - t_1)}{c_{min}(T_1 - t_1)} \tag{7.179}$$

$$\epsilon = \frac{c_h(T_1 - T_2)}{c_{min}(T_2 - t_1)} - \frac{c_c(t_2 - t_1)}{c_{min}(T_1 - t_1)}$$

For the counterflow heat exchanger shown in Figure 7.53B, with the cold fluid as $c_{min}$, it can be shown [1] that

$$\epsilon = \frac{t_2 - t_1}{T_1 - t_1} = \frac{1 - e^{-(\text{NTU})(1-R)}}{1 - Re^{-(\text{NTU})(1-R)}} \tag{7.180}$$

If $R = 1$,

$$\epsilon = \frac{\text{NTU}}{\text{NTU} + 1} \tag{7.181}$$

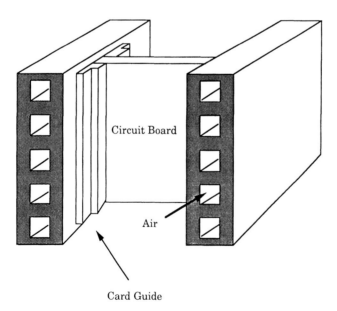

**Figure 7.58**   Compact heat exchangers.

If $R = 0$, it can be shown [1] that

$$\epsilon = 1 - e^{-\text{NTU}} \tag{7.182}$$

The $\epsilon - \text{NTU}$ method is useful for heat exchanger performance calculations. The maximum possible heat exchange in the heat exchanger is

$$q_{\max} = C_{\min}(T_{\max} - t_{\min}) \tag{7.183}$$

$\epsilon$ is a measure of the heat exchanger effectiveness, as

$$\epsilon = \frac{q_{\text{actual}}}{q_{\max}} \tag{7.184}$$

(See Refs. 1, 4, 24, 61).

Figures 7.58 and 7.59 show typical heat exchanger arrangements for electronics

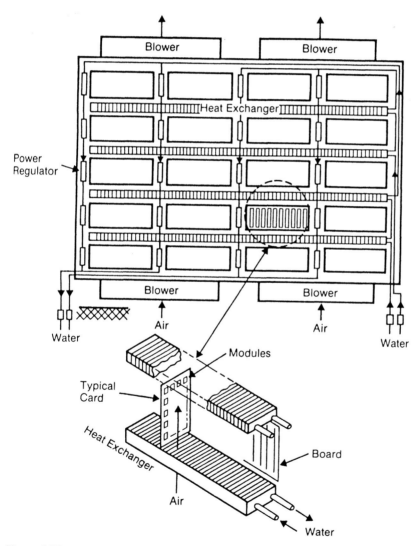

**Figure 7.59** Typical processor gate configuration with water-cooled interboard heat exchanger. (From Ref. 83, courtesy of ISHM).

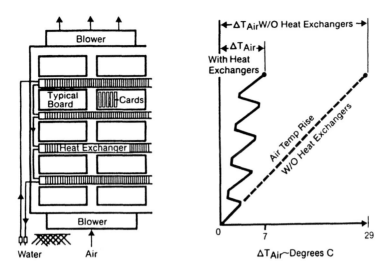

**Figure 7.60** Effect of water-cooled interboard heat exchangers on air temperature distribution throughout a gate. (From Ref. 83, courtesy of ISHM.)

cooling. Figure 7.60 shows the effect of water-cooled heat exchangers on air temperature. For more detailed information on heat exchangers, the interested reader is referred to Kraus and Bar-Cohen [1], Incropera [24], Holman [26], Seely and Chu [27], Kays and London [188], and Schlunder [189].

## 7.9  COMPUTER-AIDED THERMAL DESIGN AND ANALYSIS

Numerous heat transfer software packages are available for computer-aided thermal design and analysis. Reviews of the various software packages are presented by Shuker [190], Noor [191], and Swager [192] and will not be repeated here.

Many programs have been introduced by a wide variety of sources, including

| | | |
|---|---|---|
| CAEDS | CINDA | TRUMP |
| FIDAP | TAP | BETASOFT-R |
| FLUENT | TPS | CALCE Software for Electronics Packaging |
| PHOENICS | TOSS | FLOTRAN |
| THERMAX | TVSSZ | BETASOFT-PC |
| THERMOSTATS | ANSYS | FLOTHERM |
| AUTOTHERM | ABAQUS | PCB THERMAL |
| PROTOTHERM | ADINA | T/SNAP |
| COMPUTHERM | MARC | EE DESIGNER III/E |
| STAP | TAPZ | |
| PADS-THERMAL | FEAP | |

This listing should not be interpreted as an endorsement of these software packages.

## EXERCISES

1. Consider a long cylindrical electronic component 10 mm in radius that reaches a steady-state constant temperature of 100°C, at which point the circuit is switched off. The surrounding temperature is 0°C, the heat transfer coefficient from the component to the ambient is 50 W/m² K, and the thermal conductivity and thermal diffusivity of the component are 12.5 W/m K and $5 \times 10^{-6}$ m²/sec, respectively. Using the charts discussed in Section 7.2.6, find the maximum and minimum temperature of the component after 200 sec of cooling. Compare these temperatures with the temperature predicted by Eq. (7.31).

2. Using King's rule, Eq. (7.99), predict the heat transfer coefficient for a vertical cylinder (diameter = 10 cm, height = 10 cm) that is maintained at 40°C and is exposed to air at 20°C. Compare the calculated results with the correlation of Lienhard [119]:

$$Nu = 0.775Ra^{0.208}$$

3. Suppose a body radiates energy at the rate of 1.0 kW/m² and has an emissivity of 0.7. Find the temperature of the body.

## REFERENCES

1. Kraus, A. and Bar-Cohen, A., *Thermal Analysis and Control of Electronic Equipment*, McGraw-Hill, New York, 1983.
2. Dostal, C. A., sr. ed., *Electronic Materials Handbook*: Vol. 1, *Packaging*, ASM International, Materials Park, Ohio, 1989.
3. Osterman, M. D., and Pecht, M., "Component Placement for Reliability on Conductively Cooled Printed Wiring Boards," *J. Electron. Packag.*, 111:149–156 (1989).
4. Palmer, M. E., Lecture Notes, *Thermal Analysis*, AMSE Training Course, Orlando, Fla., October 1989.
5. Engelmaier, W., "The Use Environments of Electronic Assemblies and Their Impact on Surface Mount Solder Attachment Reliability," *Proc. I-Therm II*, Las Vegas, May 1990, pp. 8–15.
6. Osterman, M. D. and Pecht, M., "Placement for Reliability Based on Physics of Failure Concepts," *1990 IEEE NAECON Conference*, Vol. 3, May 21–25, 1990, pp. 1021–1027.
7. Pecht, M., Lall, P. and Whelan, S. J., "Temperature Dependence of Microelectronic Device Failures," *Qual. Reliab. Eng. Int. J.*, 1990.
8. Westinghouse Electronics Systems Group, Aerospace Divisions, "Reliability Analysis/Assessment of Advanced Technologies," RADC-TR-90-72, Westinghouse Electric Corporation, Baltimore, 1990.
9. Milne, B., "Design Automation Comes to Electronic System Packaging," *Electron. Des.*, February 4, 1988, reprint.
10. Ellison, G. N., "The PCB Thermal Analysis Problem," *Printed Circuit Des.*, October 1987, pp. 27–30.
11. Yao, S. C., "Thermal Modelling of Printed Circuit Boards," *Electron. Packag. Prod.*, February 1989, pp. 94–96.
12. Tummala, R. R. and Rymaszewski, E. J., eds., *Microelectronics Packaging Handbook*, Van Nostrand Reinhold, New York, 1989, pp. 167–200.
13. Andrews, J. A., Mahalingam, L. M. and Berg, H. M., "Thermal Characteristics of 16- and 40-Pin Plastic DIP's," *IEEE Trans. Components, Hybrids, Manuf. Technol. CHMT-4*(4):455–461 (1981).

14. Manno, V. P., Leisk, G. and Azar, K., "Effect of System Orientation and Cooling Mechanism on Thermal Impedances in an Electronic Enclosure," *Proc. Semi-Therm 6*, Phoenix, February 1990, pp. 17–24.

15. Dutta, V. B., "Junction-to-Case Thermal Resistance—Still a Myth?" *Proc. Semi-Therm 4*, San Diego, February 1988, pp. 8–11.

16. Andrews, J. A., "Package Thermal Resistance Model: Dependency on Equipment Design," *IEEE Trans. CHMT-11*(4):528–537 (1988).

17. Bar-Cohen, A., Elperin, T. and Eliasi, R., "$\theta_{jc}$ Characterization of Chip Packages—Justifications Limitations, and Future," *IEEE Trans. Components, Hybrids, Manuf. Technol. CHMT-12*(4):724–731 (1989).

18. Hopkins, T., Cognetti, C. and Tiziani, R., "Designing with Thermal Impedence," *Proc. Semi-Therm 4*, San Diego, February 1988, pp.55–61.

19. Mahalingam, M., "Surface-Mount Plastic Packages—an Assessment of Their Thermal Performance," *IEEE Trans. Components, Hybrids, Manuf. Technol. CHMT-12*(4):745–752 (1989).

20. Swager, A. W., "Circuit Design Requires Thermal Expertise," *EDN*, June 22, 1989, pp. 93–104.

21. Alli, M. M., Mahalingam, M., and Andrews, J. A., "Thermal Characteristics of Plastic Small Outline Transistor (SOT) Packages," *IEEE Trans. Components, Hybrids, Manuf. Technol. CHMT-9*(4):353–363 (1986).

22. Siegal, B. S., "Factors Affecting Semiconductor Device Thermal Resistance Measurements," *Proc. Semi-Therm 4*, San Diego, February 1988, pp. 12–18.

23. Reynolds, W. C. and Perkins, H. C., *Engineering Thermodynamics*, McGraw Hill, New York, 1970.

24. Incropera, F. P. and Dewitt, D. P., *Introduction to Heat Transfer*, Wiley, New York, 1985.

25. Carslaw, H. and Jaeger, J., *Conduction of Heat in Solids*, 2nd ed., Oxford University Press, Fair Lawn, N.J., 1959.

26. Holman, J., *Heat Transfer*, 6th ed., McGraw-Hill, New York, 1986.

27. Seely, J. and Chu, R., *Heat Transfer in Microelectronic Equipment*, Marcel Dekker, New York, 1972.

28. Eckert, E. and Drake, R., *Analysis of Heat and Mass Transfer*, McGraw-Hill, New York, 1972, pp. 153–158.

29. Thorne, C., "Temperature Tables. Part 1. One-Layer Plate, One-Space Variable, Linear," *NAVORD 5562 (NOTS TP 1756)*, U.S. Naval Ordnance Test Station, China Lake, Calif., 1957.

30. Thorne, C. and Morrin, H., "Temperature Tables. Part 2. One-Layer Cylindrical Shell, Internal Heating, One-Space Variable, Linear," *NAVORD 5562 (NOTS TP 2511)*, U.S. Naval Ordnance Test Station, China Lake, Calif., 1960.

31. Thorne, C. and Morrin, H., "Temperature Tables. Part 3. One-Layer Cylindrical Shell, External Heating, One-Space Variable, Linear," *NAVORD 5562 (NOTS TP 2512)*, U.S. Naval Ordnance Test Station, China Lake, Calif., 1960.

32. Thorne, C. and James, M., "Temperature Tables. Part 4. Solid Cylinder, One-Space Variable, Linear," *NAVORD 5562 (NOTS TP 2051)*, U.S. Naval Ordnance Test Station, China Lake, Calif., 1958.

33. Smithson, R. and Thorne, C., "Temperature Tables. Part 5. One-Layer Spherical Shell Segments, Internal Heating, One-Space Variable, Linear," *NAVORD 5562 (NOTS TP 2087)*, U.S. Naval Ordnance Test Station, China Lake, Calif., 1958

34. Smithson, R. and Thorne, C., "Temperature Tables. Part 6. One-Layer Spherical Shell Segments, External Heating, One-Space Variable, Linear," *NAVORD 5562 (NOTS TP 2088)*, U.S. Naval Ordnance Test Station, China Lake, Calif., 1958.

35. Williamson, E. and Adams, L., "Temperature Distribution in Solids During Heating or Cooling," *Phys. Rev. 14*(2):99–114 (1919).

36. Perry, R. and Berggren, W. P., "Transient Heat Conduction in Hollow Cylinders after Sudden Change of Inner-Surface Temperature," *Univ. Calif. Publ. Eng.* 5(3):(1944).

37. Mayer, E., "Heat Flow in Composite Slabs," *J. Am. Rocket Soc.* 22(3):150–158 (1952).

38. Anthony, M., "Temperature Distributions in Slabs with a Linear Temperature Rise at One Surface," Part I, *Proc. General Discussion on Heat Transfer*, Institute of Mechanical Engineers, American Society of Mechanical Engineers, 1951, pp. 250–253.

39. Holter, W. and Grover, J., "Insulation Temperature for the Transient Heating of an Insulated Infinite Metal Slab," *J. Am. Rocket Soc.* 30:907–908 (1960).

40. Grover, J. and Holter, W., "Solution of the Transient Heat-Conduction Equation for an Insulated Infinite Metal Slab," *Jet Propul.* 27:1249–1252 (1957).

41. Brogan, J. and Schneider, P., "Heat Conduction in a Series Composite Wall," *Trans. ASME, J. Heat Transfer, Ser. C* 83:506–508 (1961).

42. Geckler, R., "Transient Radial Heat Conduction in Hollow Circular Cylinders," *Jet Propul.* 25:31–35 (1955).

43. Hatch, J. and Schact, R., "Graphical Presentation of Difference Solutions for Transient Radial Heat Conduction in Hollow Cylinders with Heat Transfer at the Inner Radius and Finite Slabs with Heat Transfer at One Boundary," Report No. NASA TR R-56, National Aeronautics and Space Administration, Washington, D.C., 1960.

44. Newman, A., "The Temperature-Time Relations Resulting from the Electrical Heating of the Face of a Slab," *Trans. AIChE* 30:598–613 (1934).

45. Bergles, A. and Kaye, J., "Solutions of the Heat Conduction Equation with Time Dependent Boundary Conditions," *J. Aerosp. Sci.* 28:251–252 (1961).

46. Robbins, W. H., "Analysis of the Transient Radiation Heat Transfer of an Uncooled Rocket Engine Operating Outside Earth's Atmosphere," NASA TN D-62, National Aeronautics and Space Administration, Washington, D.C., 1959.

47. Kirkpatrick, E. and Stokey, W., "Transient Heat Conduction in Elliptical Plates and Cylinders," *Trans. ASME, J. Heat Transfer, Ser. C* 81:54–60 (1959).

48. Gurney, H. and Lurie, J., "Charts for Estimating Temperature Distributions in Heating and Cooling Solid Shapes," *Ind. Eng. Chem.* 15:1170–1175 (1923).

49. Schneider, P., *Temperature Response Charts*, Wiley, New York, 1963.

50. Heisler, M., "Temperature Charts for Induction and Constant Temperature Heating," *Trans. ASME* 69:227–236 (1947).

51. Kern, D. and Kraus, A., *Extended Surface Heat Transfer*, McGraw-Hill, New York, 1972.

52. Greenwood, A. and Williamson, P., "Contact of Nominally Flat Surfaces," *Proc. R. Soc. London Ser. A* 295:300–319 (1966).

53. Fenech, H., "The Thermal Conductance of Metallic Surfaces in Contact," Sc.D. Dissertation, Massachusetts Institute of Technology, Cambridge, Mass., 1959.

54. Clausing, A., "Transfer at the Interface of Dissimilar Metals: The Influence of Thermal Strain," *Int. J. Heat Mass Transfer* 9:791–798 (1966).

55. Moore, C., "Heat Transfer Across Surfaces in Contact: Studies of Transients in One-Dimensional Composite Systems," Thermal/Fluid Sciences Center Research Report 67–2, Southern Methodist University, Dallas, 1967.

56. Yovanovich, M., "Thermal Contact Conductance in a Vacuum," Sc.D. Dissertation, Massachusetts Institute of Technology, Cambridge, Mass., 1967.

57. Yovanovich, M. and Fenech, H., "Thermal Contact Resistance of Nominally Flat, Rough Surfaces in a Vacuum Environment," AIAA Paper 66–42, 1966.

58. Fenech, H. and Rohsenow, W., "Prediction of Thermal Conductance of Metallic Surfaces in Contact," *J. Heat Transfer* 82:15–24 (1962).

59. Bowden, F. and Tabor, D., *The Friction and Lubrication of Solids*, Part II, Oxford University Press, London, 1966.

60. Shlykov, Y., "Calculating Thermal Contact Resistance of Machined Metal Surfaces," *Teploenergetika* 12(10):79–83 (1965).

61.  Kraus, A., *Cooling Electronic Equipment*, Prentice-Hall, Englewood Cliffs, N.J., 1965, pp. 130–132.
62.  Clausing, A. and Chao, B., "Thermal Conductance in a Vacuum Environment," NASA Report ME-TN-242-1, University of Illinois, Champaign, 1963.
63.  Holm, R., *Electric Contacts Handbook*, Springer-Verlag, Berlin, 1958.
64.  Mikic, B., Yovanovich, M., and Rohsenow, W., "The Effect of Surface Roughness and Waviness upon the Overall Thermal Contact Resistance," EPL Report No. 79361-43, Massachusetts Institute of Technology, Cambridge, Mass. 1966.
65.  Cooper, M., Mikic, B. and Yovanovich, M., "Thermal Contact Resistance," *Int. J. Heat Mass Transfer 12*:279–300 (1969).
66.  Tien, C., "A Correlation for Thermal Contact Conductance of Nominally Flat Surfaces in a Vacuum," Special Publication 302, National Bureau of Standards, Gaithersburg, Md., 1968.
67.  Yovanovich, M. and Burde, S., "Centroidal and Area Average Resistances of Nonsymmetric, Singly Connected Contacts," *AIAA J. 15*(10):1523–1525 (1977).
68.  Yovanovich, M., "New Contact and Gap Correlations for Conforming Rough Surfaces," AIAA Paper 81–1164, 1981.
69.  Yovanovich, M., "Thermal Contact Resistance in Microelectronics," *Proc. Technical Program*, National Electronic Packaging and Production Conference, Anaheim, Calif., 1978.
70.  Yovanovich, M., Hegazy, A. and Antonetti, V., "Experimental Verification of Thermal Conductance Models Based upon Distributed Surface Microhardness," AIAA Paper 83–0532, 1983.
71.  Yovanovich, M., "A Correlation of the Minimum Thermal Resistance at Soldered Joints," *J. Spacecr. Rockets 7*:1013–1014 (1970).
72.  Yovanovich, M., "Effect of Foils upon Joint Resistance: Evidence of Optimum Thickness," in *Thermal Control and Radiation*, Vol. 31, Tien, C., ed., AIAA, Washington, D.C., 1973, pp. 227–245.
73.  Antonetti, V., "On the Use of Metallic Coatings to Enhance Thermal Contact Conductance," Ph.D. Thesis, University of Waterloo, Waterloo, Ontario, 1983.
74.  Cavanaugh, D., "Thermal Considerations of Flip-Chip Relative to Chip-and-Wire Semiconductor Attachment in Hybrid Circuits: An Experimental Approach," *IEEE Trans. Parts, Hybrids, Packag. PHP-12*(4):293–298 (1976).
75.  Fletcher, L., "Thermal Control of Materials for Spacecraft Systems," *Proc. 10th International Symposium on Space Technology and Science*, Tokyo, 1973, pp. 579–586.
76.  Snaith, B., O'Callaghan, P. and Probert, S., "Interstitial Materials for Controlling Thermal Conductances Across Pressed Metallic Contacts," *Appl. Energy 16*:175–191 (1984).
77.  Buchanan, R. and Reeber, M., "Thermal Considerations in the Design of Hybrid Microelectronic Packages," *Solid State Technol.* (1973).
78.  Oktay, S., Dessauer, B. and Horvath, J., "New Internal and External Cooling Enhancements for the Air-Cooled IBM 4381 Module," *IEEE Internal Conference on Computer Design: VLSI in Computers*, Port Chester, N.Y., 1983.
79.  Astrabadi, F., O'Callaghan, P., Snaith, B. and Probert, S., "Prediction of Optimal Interfacial Filler Thickness for Minimum Thermal Contact Resistance," AIAA Paper 81–1166, 1981.
80.  Feldman, K., Hong, Y. and Marjon, P., "Tests on Thermal Joint Compounds to 200°C," AIAA Paper 80–1466, 1980.
81.  Elsby, T., "Thermal Characterization of Epoxy and Alloy Attachments of Hybrid Components," *27th Electronic Components Conference*, Arlington, Va., 1977, pp. 320–323.
82.  Antonetti, V. and Yovanovich, M., "Enhancement of Thermal Contact Conductance by Metallic Coatings: Theory and Experiment," *J. Heat Transfer 107*:513–519 (1985).
83.  Furkay, S., Kilburn, R. and Monti, G., eds., *Thermal Management Concepts in Microelectronic Packaging*, International Society for Hybrid Microelectronics, Silver Spring, Md., 1984.

84. Childres, W. and Peterson, G., "Quantification of Thermal Contact Conductance in Electronic Packages," *Proc. Semi-Therm 5*, San Diego, February 1989, pp. 30–34.

85. Madhusudana, C. and Fletcher, L., "Thermal Contact Conductance: A Review of Recent Literature," Department of Mechanical Engineering Report, Texas A&M University, College Station, 1981.

86. Kadambi, V. and Abuaf, N., "Axisymmetric and Three-Dimensioal Chip Spreader Calculations," AIChE Symposium Series on Heat Transfer, Seattle, 1983.

87. Richardson, P. and Shum, Y., "Use of Finite Element Methods in Solution of Transient Heat Conduction Problems," ASME Paper 69-WA/HT-36, 1969.

88. Emery, A. and Carson, W., "Evaluation of Use of the Finite Element Method in Computation of Temperature," ASME Paper 69-WA/HT-38, 1969.

89. Wilson, E. and Nickell, R., "Application of the Finite Element Method to Heat Conduction Analysis," *Nucl. Eng. Des. 4*:276–286 (1966).

90. Zienkiewicz, O., *The Finite Element Method in Structural and Continuum Mechanics*, McGraw-Hill, New York, 1967.

91. Taylor, C. and Ijam, A., "A Finite Element Numerical Solution of Natural Convection in Enclosed Cavities," *Comput. Methods Appl. Mech. Eng. 19*:429–446 (1979).

92. Shih, T. and Chen, Y., "Comparison of Finite Difference Method and Finite Element Method," in *Numerical Properties and Methodologies in Heat Transfer*, Shih, T., ed., Hemisphere, New York, 1983, pp. 33–54.

93. Shih, T., *Numerical Heat Transfer*, Hemisphere, New York, 1982.

94. Myers, G., *Conduction Heat Transfer*, McGraw-Hill, New York, 1972.

95. Dusinberre, G., *Heat Transfer Calculations by Finite Differences*, International Textbook Company, Scranton, Pa., 1961.

96. Schenck, H., *Fortran Methods in Heat Flow*, Ronald Press, New York, 1963.

97. Leonard, S., "A Convectively Stable, Third-Order Accurate Finite Difference Method for Steady Two-Dimensional Flow and Heat Transfer," in *Numerical Properties and Methodologies in Heat Transfer*, Shih, T., ed., Hemisphere, New York, 1983, pp. 211–226.

98. Ames, W., "A Survey of Finite Difference Schemes for Parabolic Partial Differential Equations," in *Numerical Properties and Methodologies in Heat Transfer*, Shih, T., ed., Hemisphere, New York, 1983, pp. 3–15.

99. Meek, P. and Norbury, J., "An Application of Two-Stage Two-Level Finite Difference Schemes in Nonlinear Heat Diffusion," in *Numerical Properties and Methodologies in Heat Transfer*, Shih, T., ed., Hemisphere, New York, 1983, pp. 55–81.

100. Bar-Cohen, A. and Kraus, A. D., ed., *Advances in Thermal Modeling of Electronic Components and Systems*, Vol. 1, Hemisphere, New York, 1988.

101. Salvadori, M. and Baron, M., *Numerical Methods in Engineering*, Prentice-Hall, Englewood Cliffs, N.J., 1961, pp. 190–293.

102. Vichnevetsky, R., *Computer Methods for Partial Differential Equations*, Vol. 1, Prentice-Hall, Englewood Cliffs, N.J., 1981.

103. Book, D. L., ed., *Finite-Difference Techniques for Vectorized Fluid Dynamics Calculations*, Springer-Verlag, New York, 1981.

104. Spalding, D., "A Novel Finite Difference Formulation for Differential Expressions Involving Both First and Second Derivatives," *Int. J. Numer. Methods Eng. 4*:551–559 (1972).

105. Richtmyer, H., *Difference Methods for Initial Value Problems*, Interscience, New York, 1957.

106. Webb, B. W., "Interaction of Radiation and Free Convection on a Heated Vertical Plate: Experiment and Analysis," *J. Thermophys. 4*(1):117–221 (1990).

107. Eckert, E. and Soehngen, E., "Interferometric Studies on the Stability and Transition to Turbulence of a Free Convection Boundary Layer," *Proc. General Discussion of Heat Transfer ASME-IME*, London, 1951.

108. Gebhart, B., *Heat Transfer*, 2nd ed., McGraw-Hill, New York, 1970, chapter 8.

109. Gebhart, B., "Natural Convection Flow, Instability, and Transition," ASME Paper 69-HT-29, 1969.

110. Mollendorf, J. and Gebhart, B., "An Experimental Study of Vigorous Transient Natural Convection," ASME Paper 70-HT-2, 1970.

111. McAdams, W., *Heat Transmission*, 3rd ed., McGraw-Hill, New York, 1954, chapter 3.

112. Warner, C. and Arpaci, V., "An Experimental Investigation of Turbulent Natural Convection in Air at Low Pressure Along a Vertical Heated Flat Plate," *Int. J. Heat Mass Transfer* *11*:397–409 (1968).

113. Tsuji, T. and Nagano, Y., "Velocity and Temperature Measurements in a Natural Convection Boundary Layer along a Vertical Flate Plate," *J. Exp. Therm. Sci. 2*:208–215 (1989).

114. Morgan, V., "The Overall Convective Heat Transfer from Smooth Circular Cylinders," in *Advances in Heat Transfer*, Vol. 11, Irvine, T. and Hartnett, J., eds., Academic Press, New York, 1975.

115. Yuge, T., "Experiments on Heat Transfer from Spheres Including Combined Natural and Forced Convection," *J. Heat Transfer, Ser. C 82*:214–221 (1962).

116. Amato, W. and Tien, C., "Free Convection Heat Transfer from Isothermal Spheres in Water," *Int. J. Heat Mass Transfer 15*:327–336 (1972).

117. King, W., "A Rule for Multi-Dimensional External Natural Convection," *Mech. Eng. 54*:347–351 (1931).

118. Sparrow, E. M. and Ansari, M. A., "A Refutation of King's Rule for Multi-Dimensional External Natural Convection," *Int. J. Heat Mass Transfer 26*:1357–1364 (1983).

119. Lienhard, J. H., "On the Commonality of Equations for Natural Convection from Immersed Bodies," *Int. J. Heat Mass Transfer 16*:2121–2126 (1973).

120. Bar-Cohen, A. and Rohsenow, W., "Thermally Optimum Arrays of Cards and Fins in Natural Convection," *IEEE Trans. Components, Hybrids, Manuf. Technol. CHMT-6*(2):154–158 (1983).

121. Bar-Cohen, A. and Rohsenow, W., "Thermally Optimum Spacing of Vertical, Natural Convection Cooled, Parallel Plates," *J. Heat Transfer 106*:116–123 (1984).

122. Bar-Cohen, A., "Bounding Relations for Natural Convection Heat Transfer from Vertical Printed Circuit Boards," *Proc. IEEE 73*:1388–1395 (1985).

123. Aung, W., Fletcher, L. and Sernas, V., "Developing Laminar Free Convection Between Vertical Flat Plates with Asymmetric Heating," *Int. J. Heat Mass Transfer 15*:2293–2308 (1972).

124. Wirtz, R. and Sultzman, R., "Experiments on Free Convection Between Vertical Plates with Symmetric Heating," *J. Heat Transfer 104*:501–507 (1982).

125. Johnson, E., "Evaluation of Correlations for Natural Convection Cooling of Electronic Equipment," *ASME HTD 57*:103–111 (1986).

126. Park, K. and Bergles, A., "Natural Convection Heat Transfer Characteristics of Simulated Microelectronic Chips," in *Heat Transfer in Electronic Equipment*, ASME HTD, Vol. 48, 1985, pp. 29–37.

127. Roache, P., *Computational Fluid Dynamics*, Hermosa, Albuquerque, N.M., 1976.

128. Patankar, S., *Numerical Heat Transfer and Fluid Flow*, Hemisphere, New York, 1980.

129. Anderson, D., Tannehill, J. and Pletcher, R., *Computational Fluid Mechanics and Heat Transfer*, Hemisphere, New York, 1983.

130. De Vahl Davis, G. and Jones, I., "Natural Convection in a Square Cavity—a Comparison Exercise," *Proc. 2nd International Conference on Numerical Methods in Thermal Problems*, Venice, 1981.

131. Upson, C., Gresho, P., Sani, R., Chan, S. and Lee, R., "A Thermal Convection Simulation in Three Dimensions by a Modified Finite Element Method," in *Numerical Properties and Methodologies in Heat Transfer*, Shih, T., ed., Hemisphere, New York, 1983, pp. 245–259.

132. Quon, C., "Effects of Grid Distribution on the Computation of High Rayleigh Number

Convection in a Differentially Heated Cavity," in *Numerical Properties and Methodologies in Heat Transfer*, Shih, T., ed., Hemisphere, New York, 1983, pp. 261–281.

133. Quon, C., "High Rayleigh Number Convection in an Enclosure—a Numerical Study," *Phys. Fluids 15*:12–19 (1972).

134. Quon, C., "Free Convection in an Enclosure Revisited," *J. Heat Transfer 99*:340–342 (1977).

135. Mallison, G. and de Vahl Davis, G., "Three-Dimensional Natural Convection in a Box: A Numerical Study," *J. Fluid Mech. 83*:1–31 (1977).

136. Robinson, W., Han, L., Essig, R. and Heddleson, C., "Heat Transfer and Pressure Drop Data for Circular Cylinders in Ducts and Various Arrangements," Research Foundation Report No. 41, Ohio State University, Columbus, 1951.

137. Robinson, W. and Jones, C., "The Design of Arrangements of Prismatic Components for Crossflow Forced Air Cooling," Research Foundation Report No. 47, Ohio State University, Columbus, 1955.

138. Incropera, F. P. and DeWitt, D. P., *Fundamentals of Heat Transfer*, Wiley, New York, 1981.

139. Watson, D., "Thermal Design," in *Reliability and Maintainability of Electronic Systems*, Arsenault, J. E. and Roberts, J. A., eds., Computer Science Press, Rockville, Md., 1980.

140. Dittus, F. and Boelter, L., *Univ. Calif. (Berkeley) Publ. 2*:443–456 (1930).

141. Moffat, R. J., Arvizu, D. E. and Ortega, A., "Cooling Electronic Components: Forced Convection Experiments with an Air-Cooled Array," in *Heat Transfer in Electronic Equipment—1985*, ASME HTD, Vol. 48, pp. 15–27.

142. Arvizu, D. E. and Moffat, R. J., "The Use of Super Position in Calculating Cooling Requirements for Circuit Board Mounted Components," *Proc. 32nd Electronics Components Conference*, IEEE, EIA, CHMT, May 1982.

143. Anderson, A. M. and Moffat, R. J., "Direct Air Cooling of Electronic Components: Reducing Component Temperatures by Controlled Thermal Mixing," in *Symposium on Fundamentals of Forced Convection Heat Transfer*, ASME HTD, Vol. 101, 1988, pp. 9–16.

144. Moffat, R. J. and Anderson, A. M., "Applying Heat Transfer Coefficient Data to Electronics Cooling," *Experimental Methods in Heat Transfer with Emphasis on Air-Cooling of Electronic Components*, Semi-Therm 6, Phoenix, February 1990, pp. 481–502.

145. Ortega, A. and Moffat, R. J., "Heat Transfer from an Array of Simulated Electronic Components: Experimental Results for Free Convection With and Without a Shrouding Wall," in *Heat Transfer in Electronic Equipment—1985*, ASME HTD, Vol. 48, pp.5–15.

146. Moffat, R. J. and Ortega, A., "Buoyancy Induced Forced Convection," *AIAA/ASME Thermophysics and Heat Transfer Conference*, Boston, June 1986.

147. Ortega, A. and Moffat, R. J., "Buoyancy Induced Convection in a Non-Uniformly Heated Array of Cubical Elements on a Vertical Channel Wall," *Experimental Methods in Heat Transfer with Emphasis on Air-Cooling of Electronic Components*, Semi-Therm 6, Phoenix, February 1990, pp. 450–461.

148. Sparrow, E., Neithammer, J. and Chaboki, A., "Heat Transfer and Pressure Drop Characteristics of Arrays of Rectangular Modules Encountered in Electronic Equipment," *Int. J. Heat Mass Transfer 25*:961–973 (1984).

149. Sparrow, E., Yanezoreno, A. and Otis, D., "Convection Heat Transfer Response to Height Differences in an Array of Block Like Electronic Components," *Int. J. Heat Mass Transfer 27*:469–473 (1984).

150. Strassberg, D., "Cooling Devices Take the Heat from SMDs," *EDN*, May 14, 1987, pp. 97–106.

151. Welsh, J., "Handbook of Methods of Cooling Air Force Ground Electronic Equipment," RADC-TR-58-126, Rome Air Development Center—Griffs Air Force Base, Rome, N.Y., 1959.

152. Madison, R., *Fan Engineering*, Buffalo Forge Co., Buffalo, N.Y., 1949.

153. Markstein, H., "Cooling Large Metal Cabinets," *Electron. Packag. Prod.* 27(5):24–27 (1987).

154. Ellison, G., *Thermal Computations for Electronic Equipment*, Van Nostrand Reinhold, New York, 1984.

155. Vogel, G., "Cooling with Blowers," *Electron. Packag. Prod.* 27(1):141–142 (1987).

156. Weghorn, F., "Cooling Equipment," *Electron. Prod.* 12(9):75–76 (1981).

157. Steinberg, D., *Cooling Techniques for Electronic Equipment*, Wiley, New York, 1980.

158. Metais, B. and Eckert, E., "Forced, Mixed, and Free Convection Regimes," *J. Heat Transfer Ser. C 86*:295–306 (1964)

159. Brown, C. and Gauvin, W., "Combined Free and Forced Convection," *Can. J. Chem. Eng.* 43:306–321 (1965).

160. Fand, R. and Keswani, K., "Combined Natural and Forced Convection Heat Transfer from Horizontal Cylinders to Water," *Int. J. Heat Mass Transfer 16*:175–189.

161. Depew, C., Franklin, J. and Ito, C., "Combined Free and Forced Convection in Horizontal, Uniformly Heated Tubes," ASME Paper 75-HT-55, 1975.

162. Bergles, A., Bakhru, N. and Shires, J., "Cooling of High Power Density Computer Components," DSR 70712-60, Massachusetts Institute of Technology, Cambridge, Mass., 1968.

163. Bar-Cohen, A. and Kraus, A. D., *Advances in Thermal Modeling of Electronic Components and Systems*, Vol. 2, ASME Press, New York, 1990.

164. Greene, A. and Wightman, J., "Cooling Electronic Equipment by Direct Evaporation of Liquid Refrigerant," Air Material Command Report PB 136065, Wright-Patterson AFB, Ohio.

165. Basiulis, A. and Hummel, T., "The Application of Heat Pipe Techniques to Electronic Component Cooling," ASME Paper 72-WA/HT-42, 1972.

166. Tong, L., *Boiling Heat Transfer and Two-Phase Flow*, Wiley, New York, 1965.

167. Chapman, A., *Heat Transfer*, Macmillan, New York, 1960.

168. Hottel, H. and Sarofim, *Radiative Transfer*, McGraw-Hill, New York, 1967.

169. Siegel, R. and Howell, J., *Thermal Radiation Heat Transfer*, 2nd ed., McGraw-Hill, New York, 1980.

170. Buller, L. and McNelis, B, "Effects of Radiation on Enhanced Electronic Cooling," *Proc. Semi-Therm Conference*, San Diego, February 1988, pp. 538–544.

171. Beckworth, T., Buck, N. and Marangoni, R., *Mechanical Measurements*, Addison-Wesley, Reading, Mass., 1982.

172. Tenney, A. S., III, "Industrial Radiation Thermometry: Red Hot and Hotter," *Mech. Eng.*, October 1986, pp. 36–41.

173. Bichard, S. and Rogers, L., "A Review of Industrial Applications of Thermography," *Br. J. NDT*, January 1976, pp. 2–11.

174. Boulton, H., "Thermography Systems for PCB Testing Applications," *Electronics*, April 1985, pp. 23–25.

175. Conway, T., "Using Thermal Images to Measure Temperature," *Mech. Eng.*, June 1987, pp. 32–34.

176. Elliot, T., "Infrared Measurement of Surface Temperature Now Hot Option," *Power*, March 1988, pp. 41–48.

177. Kallis, J., Egan, G. and Wirick, M., "Nondestructive Infrared Inspection of Hybrid Microcircuit Substrate-to-Package Thermal Adhesive Bonds," *Trans. Components, Hybrids, Manuf. Technol. CHMT-4*(3):257–260 (1981).

178. Kallis, J., Samuels, A. and Stout, R., "True Temperature Measurement of Electronic Components Using Infrared Thermography," *Proc. 4th Annual International Electronics Packaging Conference*, October 1984, pp. 616–627.

179. Kallis, J., Strattan, L. and Bui, T., "Programs Help Spot Hot Spots," *IEEE Spectrum* 24(3):36–41 (1987).

180. Kaplan, H., "Infrared Thermal Imaging Diagnostics," *Laser Focus*, November 1987, pp. 150–157.
181. Sandor, B., Lohr, D. and Schmid, K., "Nondestructive Testing Using Differential Infrared Thermography," *Materials Evaluation*, April 1987, pp. 392–395.
182. Weight, M., "Thermography Testing of Production PC Boards," *Electron. Packag. Prod. 19*(3):69–74 (1979).
183. Wickersheim, K., "New Thermometry Technique Measures Component Temperature," *Electron. Packag. Prod. 21*(9):121–126.
184. Weight, M., "Thermography Testing of Production PC Boards," *Testing*, October 1981, pp. 69–74.
185. Edwards, D., "Numerical Methods in Radiation Heat Transfer," in *Numerical Properties and Methodologies in Heat Transfer*, Shih, T., ed., Hemisphere, New York, 1983, pp. 479–496.
186. Nice, M., "Application of Finite Elements to Heat Transfer in a Participating Medium," in *Numerical Properties and Methodologies in Heat Transfer*, Shih, T., ed., Hemisphere, New York, 1983, pp. 497–514.
187. Howell, J., "Applications of Monte Carlo to Heat Transfer Problems," in *Advances in Heat Transfer*, Vol. 5, Irvine, T. and Hartnet, J., eds., Hemisphere, New York, 1973, pp. 2–54.
188. Kays, W. and London, A., *Compact Heat Exchangers*, 2nd ed., McGraw-Hill, New York, 1964.
189. Schlunder, E., *Heat Exchanger Design Handbook*, Hemisphere, New York, 1982.
190. Shuker, W., "A Survey of Heat Conduction Computer Programs," *Nucl. Safety 12*:569–582 (1971).
191. Noor, A., "Survey of Computer Programs for Heat Transfer Analysis," in *Finite Elements in Analysis and Design*, Vol. 2, 1986, pp. 259–312.
192. Swager, A., "Thermal Images Provide Reliability Clues," *EDN* February 1, 1990, pp. 47–58.
193. Edwards, A., "TRUMP: A Computer Program for Transient and Steady State Temperature Distributions in Multi-Dimensional Systems," USAEC Report UCRL-14754, Lawrence Radiation Laboratory, Livermore, Calif., 1969.
194. Schauer, D., "FED: A Computer Program to Generate Geometric Input for the Heat Transfer Code TRUMP, "USAEC Report UCRL-50816, Lawrence Radiation Laboratory, Livermore, Calif., 1970.
195. Azar, K., Develle, S. E. and Manno, V. P., "Sensitivity of Circuit Pack Thermal Performance to Convective and Geometric Variation," *IEEE Trans. Components, Hybrids, Manuf. Technol. CHMT-12*(4):732–740 (1989).
196. Moutsoglou, A. and Wong, Y. H., "Convection-Radiation Interaction in Buoyancy-Induced Channel Flow," *J. Thermophys. 3*(2):175–181 (1989).
197. Kumar, R. and Yuan, T., "Recirculating Mixed Convection Flows in Rectangular Cavities," *J. Thermophys. 3*(3):321–329 (1989).
198. Fukuoka, Y. and Ishizuka, M., "An Application of the Thermal Network Method to the Thermal Analysis of Multichip Packages (Proposal of a Simple Thermal Analysis Model)," *Jpn. J. Appl. Phys. 28*(9):1578–1585 (1989).
199. Braunberg, G. C., "Computer Aided Infrared Thermography Using Emissivity Compensated Imaging," M.S. Dissertation, University of Maryland, 1990.
200. *Reliability and Quality Handbook—1990 Edition*, Motorola Semiconductor Product Sector, 1990.
201. *Handbook of Chemistry and Physics,* Chemical Rubber Publishing Co., Cleveland, Ohio, 1960.
202. Piret, E. L., and Isbin, H. S. "Natural Circulation Evaporation Two-Phase Heat Transfer," *Chem. Eng. Prog., 50*:305 (1954).

203. Addoms, J. N., "Heat Transfer at High Rates to Water Boiling Outside Cylinders," Sc.D. thesis, Massachusetts Institute of Technology, Cambridge, Mass., 1948.

204. Cryder, D. S., and Finalbargo, A. C., "Heat Transmission from Metal Surfaces to Boiling Liquids: Effect of Temperature of the Liquid on Film Coefficient," *Trans. AIChE 33*:346 (1937).

205. Cichelli, M. T. and Bonilla, C. F., "Heat Transfer to Liquids Boiling Under Pressure, *Trans. AIChE 41*:755 (1945).

206. Vachon, R. I., Nix, G. H., and Tanger, G. E., "Evaluation of Constants for the Rohsenow Pool-Boiling Correlation," *J. Heat Transfer, 90:*239 (1968).

207. Arsenault, J. E. and Roberts J. A. eds., *Reliability and Maintainability of Electronic Systems,* Computer Science Press, Potomac, Md, 1980.

# 8

# Thermomechanical Analysis and Design

**Abhijit Dasgupta**    *CALCE Center for Electronic Packaging, University of Maryland, College Park, Maryland*

## 8.1  INTRODUCTION

A typical microelectronic packaged assembly consists of a board and several other components manufactured from many different materials. The differences in the coefficients of thermal expansion of these materials result in significant thermomechanical strains. The accompanying strain gradients may be associated with deformations, such as warping of the board and components, and with resulting mechanical stresses due to mismatched thermal expansion in mechanically attached components, during local or global temperature changes. Such temperature changes are commonly encountered during processing operations, such as vapor depositions and soldering operations, and in service conditions such as normal power cycling and global environmental temperature changes.

Expansion mismatch problems can arise at various levels of the assembly. At the smallest length scale, they can be found in the electronic chip or wafer itself, at the interface of oxide or GaAs deposits and the silicon substrate. When packaging the chip, polymeric or ceramic chip carriers are employed. Thermal strains may arise due to thermal mismatches at this interface. This encapsulated (and sometimes hermetically sealed) device is then mounted on a carrier board either by direct soldering or through lead wires. It is at this level that the most significant thermomechanical strains are encountered. Thus fatigue problems are frequently encountered in the solder joints connecting leadless chip carriers to a multilayer printed circuit board (PCB), as well as in conventional solder connections of leaded microelectronic components. Frequently, input ports are present in the encapsulation container for connector pins. These ports are often hermetically sealed with glassy compounds. Fatigue of this seal is often found to lead to loss of functional

477

hermeticity of the device. The main problem at the next higher length scale is the carrier board, which is usually made of laminated fiber-reinforced composites. Thermomechanical stresses can cause damage in the form of microcracking in these brittle, heterogeneous materials. In extreme cases, delaminations can occur between the layers of the composite or between the composite and signal planes. Other problem sources at the board level are the plated through holes in multilayer boards, where failure commonly occurs as a result of the large out-of-plane thermal expansion of the board material.

This chapter will discuss the fundamentals of thermally induced deformations, strains, and stresses and cite some specific examples of thermal deformations and high-temperature fatigue in the context of electronic packaging design. The goal is to develop prediction capabilities for the reliability of microelectronic devices and packaging accessories. Common examples are reliability problems in dies; interconnections such as die attaches, solder joints, wire bonds, and plated through holes; seals in hermetic devices; printed circuit and printed wiring board materials and other components.

## 8.2  CONCEPTS INVOLVED IN THERMOMECHANICAL ANALYSIS AND DESIGN

In order to discuss the damage caused by thermally induced strains (and in the next chapter the damage caused by vibration-induced strains), some basic concepts and definitions in thermoelasticity and fatigue will be reviewed. To simplify the discussion, it will be assumed that on the local level all deformations can be considered small with respect to the critical dimensions of the components. (Large deformations can occur in practice, but their analysis is beyond the scope of this discussion.)

### 8.2.1  Thermoelasticity

Thermoelasticity is the branch of mechanics that deals with the elastic interrelationships between stress, strain, and displacements in a body subjected to loads or displacements, or both, in the presence of a constant or varying temperature field. These elastic relationships are valid only for loadings or displacements which, when removed, allow the body to return to its original geometry. If some permanent set occurs due to inelastic deformations, it may be necessary to use plastic stress-strain relations, time-dependent anelastic stress-strain relations, damage mechanics, or several of these concepts. For the most part in PCB design, plastic and anelastic deformations occur in ductile components such as solders, copper platings in plated through holes, and wire bonds. Irreversible damage in the form of microcracking occurs only in relatively brittle components such as ceramics and laminated composite boards. The inelastic deformations and ultimate failure in ductile materials are described in terms of the local strains, whereas microcracking phenomena in brittle components are better described in terms of the local stresses.

In this section, the concepts of stress, strain, and thermal expansion are briefly reviewed and some simple elastic examples related to PCB design are presented. Inelastic stress-strain relations and damage constitutive laws are beyond the scope of this discussion. It is necessary to point out, however, that stress analysis of all but the simplest geometries and simplest material properties would require computer-assisted numerical methods (such as finite elements) in practice.

## Stress

In the simplest terms, stress is the mechanical load or traction that a body carries per unit cross-sectional area at any given location. In other words, stress is the intensity of the local load distribution that has to be applied across some imaginary cut surface to keep the body in equilibrium.

As an example, consider a simple rod subjected to a tensile loading of magnitude $P$, as shown in Figure 8.1a. Figure 8.1b shows the average uniform load distribution that must be applied across section a–a to keep the body in equilibrium. This stress distribution acts normal to the cut surface and has a magnitude of $P/A$, where $A$ is the cross-sectional area.

If the rod were sectioned along b–b, the required stress distribution to keep the rod in equilibrium would be as shown in Figure 8.1c. The magnitude of the stress distribution is $P/A'$, where $A'$ is the area of this inclined cross section. Rather than showing the stress

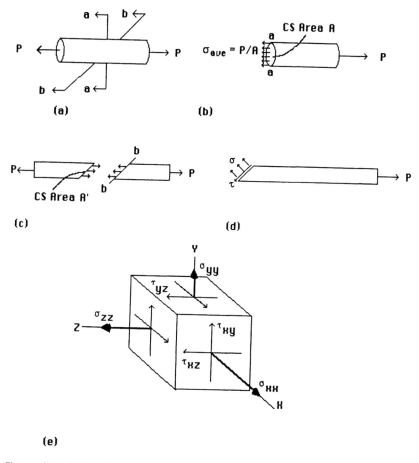

(a)                                            (b)

(c)                                            (d)

(e)

**Figure 8.1** (a) Rod in tension. (b) Averaged normal stress $\sigma$ in section a–a perpendicular to rod axis. (c) Traction on inclined section b–b. (d) Normal ($\sigma$) and shear ($\tau$) stresses. (e) Material element under multiaxial stresses.

distribution acting along the axis of the rod, we could resolve the stress distribution into two components, one acting normal to the cross section and one acting parallel to the cross section. This is shown in Figure 8.1d. The component acting normal to the cut surface is referred to as the normal stress, $\sigma$, and the component acting parallel to the cut surface is the shear stress, $\tau$.

Therefore, all discussions of stress in a body are with implicit reference to some imaginary cut surface. It is common practice to consider cut surfaces that are normal to the coordinate system used to describe the geometry of the body. For one-dimensional problems such as the rod in tension, it is customary to consider a cut surface normal to the axis of the rod. This choice of a cut surface results only in normal stress and no shear stress for the axial loading shown in Figure 8.1a.

For more complex, multidimensional loading situations, normal and shear stresses can coexist at any point along three orthogonal directions, as shown in Figure 8.1e on an infinitesimally small cube. The common notation is to indicate these stresses with the Greek letters $\sigma$ (for normal stress) and $\tau$ (for shear stress), followed by two subscripts. As shown in Figure 8.1e, the first subscript indicates the face on which the stress acts and the second subscript indicates the direction of the stress. Thus, repeated subscripts indicate normal stresses and mixed subscripts indicate shear stresses. It can be shown from moment equilibrium considerations that

$$\tau_{xy} = \tau_{yx}, \qquad \tau_{xz} = \tau_{zx}, \qquad \tau_{yz} = \tau_{zy} \tag{8.1}$$

Thus there are only six independent components of stress at a point. The distribution of these stress components cannot be arbitrary. They are related to each other through a set of coupled partial differential equations, if the point is to be in equilibrium. The reader is referred to any intermediate-level text on the mechanics of deformable bodies [1] for these equations.

Since these stresses are computed with reference to a particular coordinate reference frame, it is obvious that the components of stress will change with coordinate rotations. For any applied loading, it is possible to show the existence of a unique coordinate frame (at some orientation relative to the loading) where the shear stresses vanish completely. Along these coordinate directions, the normal stresses take extreme values and are called the principal stresses. This set of coordinate directions is labeled the principal stress axis. The maximum shear stresses occur in coordinate frames that are oriented at 45° to these principal stress directions. Detailed methods for computing principal stresses and maximum shear stresses may be found in Ref. 1.

Principal stresses are extremely important in failure and reliability modeling because most multiaxial failure theories are formulated in terms of either principal stresses or maximum shear stresses.

## Strain and Stress-Strain Relations

Again considering the rod in tension as an example, if the rod were originally $L$ in length, it would now be of length $(L + \Delta L)$ as a result of the tensile load. This elongation under load is expressed in a nondimensional form and called normal strain $\epsilon$; that is, if the $x$ axis were along the axis of the rod,

$$\epsilon_{xx} = \frac{\Delta L}{L} \tag{8.2}$$

As the rod elongates due to the axial load, there is a contraction in the transverse direction in the material, resulting in a small volume change. The ratio of this transverse contraction to axial elongation is a material property and is called Poisson's ratio, $\nu$.

$$\nu_{yx} = -\frac{\epsilon_{yy}}{\epsilon_{xx}}, \quad \nu_{zx} = -\frac{\epsilon_{zz}}{\epsilon_{xx}} \tag{8.3}$$

Most metals and ceramics used in the electronics industry can be considered to be isotropic materials on the macro scale. That implies that on a length scale much larger than the grain size of the material, all the physical properties—i.e., elastic, thermal, electrical, magnetic, and optical properties—can be considered to be equal in every direction.

In isotropic materials, Poisson's ratio is independent of loading direction and is simply called $\nu$. Poisson's ratio values typically range from about 0.1 (glass-epoxy laminated boards in the plane of the board) to about 0.5 (incompressible, rubberlike materials). Most metals have a $\nu$ of about $\frac{1}{3}$.

In an elastic material, normal stress is proportional to normal strain. The constant of proportionality, Young's modulus, $E$, is a material property and is independent of loading direction in an isotropic material. Thus,

$$\sigma_{xx} = E\epsilon_{xx}$$

or

$$\epsilon_{xx} = \frac{\sigma_{xx}}{E} \tag{8.4}$$

For the more general case of loading in more than one dimension, isotropic stress-strain relations are written as

$$\epsilon_{xx} = \frac{1}{E}[\sigma_{xx} - \nu(\sigma_{yy} + \sigma_{zz})]$$

$$\epsilon_{yy} = \frac{1}{E}[\sigma_{yy} - \nu(\sigma_{xx} + \sigma_{zz})] \tag{8.5}$$

$$\epsilon_{zz} = \frac{1}{E}[\sigma_{zz} - \nu(\sigma_{xx} + \sigma_{yy})]$$

Shear stress causes a distortion of the body with no accompanying volumetric change. As shown in Figure 8.2, shear strain is defined as one half the angular deformation of the body. For small deformations, therefore,

$$\text{Shear strain} \cong \frac{\alpha + \beta}{2} \tag{8.6}$$

where $\alpha$ and $\beta$ are the angles shown in Figure 8.2. This is the mathematical (or tensorial) definition of shear strain and is indicated by $\epsilon_{xy}$, $\epsilon_{yz}$, or $\epsilon_{xz}$ for the shear strain in the $xy$, $yz$, or $xz$ planes, respectively. Obviously, from the definition,

$$\epsilon_{xy} = \epsilon_{yx}, \quad \epsilon_{yz} = \epsilon_{zy}, \quad \epsilon_{zx} = \epsilon_{xz} \tag{8.7}$$

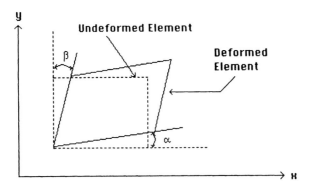

**Figure 8.2**   Definition of shear strain.

The reader is cautioned that there is another commonly used definition of shear strain normally referred to as engineering shear strain, $\gamma$. Referring again to Figure 8.2,

$$\gamma = \alpha + \beta \tag{8.8}$$

Thus the engineering shear strain is twice the mathematical or tensorial shear strain.

For elastic deformations the shear strain is directly proportional to the shear stress. In isotropic materials,

$$\gamma_{xy} = 2\epsilon_{xy} = \frac{\tau_{xy}}{G}$$

and similarly

$$\gamma_{yz} = \frac{\tau_{yz}}{G}, \quad \gamma_{zx} = \frac{\tau_{zx}}{G} \tag{8.9}$$

where $G$ is the constant of proportionality known as the shear modulus. For isotropic materials, material symmetry conditions dictate that there are only two independent elastic constants. Thus it can be shown that $G$ can be expressed in terms of $E$ and $v$ as

$$G = \frac{E}{2(1 + v)} \tag{8.10}$$

As in stresses, strains are also computed with reference to a particular coordinate frame. Hence the discussions in the preceding section regarding coordinate transformations, principal values, and maximum shear values also hold good for multiaxial strain analysis. The principal stress axes and the principal strain axes coincide in isotropic materials. Thus, multiaxial failure theories based on principal stresses predict failure along the same orientation as failure theories based on principal strains.

Materials that have three mutually orthogonal planes of symmetry for their physical properties are said to be orthotropic. The material principal axes are defined by the three normals to the planes of symmetry. Examples of orthotropic materials are wood and fiber-reinforced epoxies (when the fibers are oriented along orthogonal directions). It can be shown [2] that orthotropic materials require nine independent constants for a complete description of their elastic behavior. The elastic stress-strain relationships for orthotropic materials can be expressed in material principal axes (1,2,3) as

$$\epsilon_{11} = \frac{\sigma_{11}}{E_{11}} - \frac{\nu_{12}\sigma_{22}}{E_{22}} - \frac{\nu_{13}\sigma_{33}}{E_{33}}$$

$$\epsilon_{22} = -\frac{\nu_{21}\sigma_{11}}{E_{11}} + \frac{\sigma_{22}}{E_{22}} - \frac{\nu_{23}\sigma_{33}}{E_{33}}$$

$$\epsilon_{33} = -\frac{\nu_{31}\sigma_{11}}{E_{11}} - \frac{\nu_{32}\sigma_{22}}{E_{22}} + \frac{\sigma_{33}}{E_{33}}$$

$$\gamma_{12} = \frac{\tau_{12}}{G_{12}}, \quad \gamma_{23} = \frac{\tau_{23}}{G_{23}}, \quad \gamma_{13} = \frac{\tau_{13}}{G_{13}}$$

$$(8.11)$$

Materials that have no planes of symmetry for their physical properties are said to be anisotropic. Anisotropic materials have 21 independent elastic constants, and their detailed stress-strain relationship is beyond the scope of this discussion.

In the failure analysis of orthotropic and anisotropic materials subjected to mutiaxial stresses, it is important to realize that, unlike in the case of isotropic materials, principal stress axes and principal strain axes do not coincide.

## Thermal Expansion and Strains

An isotropic material will expand uniformly in all directions when heated uniformly. The amount that a material will expand is quantitatively described by the coefficient of thermal expansion (CTE), which has units of microstrain per degree Celsius (or degree Farenheit). Mathematically, this is expressed as

$$\Delta L = L\alpha\,\Delta T \tag{8.12}$$

or normal strain

$$\epsilon = \frac{\Delta L}{L} = \alpha\,\Delta T \tag{8.13}$$

with $\alpha$ being the symbol commonly used for the CTE. It is important to realize that, unlike mechanical strains, thermal strains do not generate transverse contractions along orthogonal directions by the Poisson effect. Neither do unconstrained thermal strains generate stresses in homogeneous materials, unless the temperature gradient is nonuniform [1].

Table 8.1 lists some CTE values for materials commonly used in PCB design. Note that the CTE of silicone elastomers is an order of magnitude greater than the CTE of aluminum. This means that silicone elastomers would have thermal deformations an order of magnitude greater than aluminum when subjected to the same temperature difference. Similarly, epoxy-glass laminate printed wiring boards (PWBs) would have three times the thermal deformation of ceramic chip carriers in the plane of the board and 30 times the deformation of ceramic chip carriers in the thickness direction of the board.

Only one thermal expansion constant is required in addition to the two elastic constants to describe the state of stress and deformation of an isotropic body completely. For an isotropic material subjected to a temperature change $\Delta T$, the normal strains are

$$\epsilon_{xx} = \epsilon_{yy} = \epsilon_{zz} = \alpha\,\Delta T \tag{8.14}$$

There are no thermally induced shear strains in an isotropic material subjected to a uniform temperature gradient.

**Table 8.1** Coefficient of Thermal Expansion of Materials
Used in Electronic Assemblies

| Material | CTE (ppm/°C) |
|---|---|
| Ceramic chip carriers | 5–7 |
| Silicone elastomers | 275–285 |
| Unfilled epoxies | 175–225 |
| Filled epoxies | 65–100 |
| Epoxy-glass laminate (@ room temp.) | |
| x-y axis | 12–20 |
| z axis | 60–100 |
| Polymide-glass (x-y axis) | 11–19 |
| Polymide-Kevlar (x-y axis) | 6–7 |
| Polymide-quartz (x-y axis) | 6–9 |
| Copper-clad Invar | 8 |
| Solder (SnPb) | 25–30 |
| Aluminum | 20–25 |
| Copper | 15–20 |
| Copper | 15–20 |
| Type 400 steels | 5.1–5.6 |
| Kovar | 5.9 |
| Molybdenum | 4.9 |
| Tungsten | 4.3 |
| 96% alumina | 5 |
| Borosilicate glass | 5.0 |
| Single-crystal silicon | 2.3 |
| Quartz | 0.3 |

In orthotropic materials three independent coefficients are needed to describe the thermal expansion along each material principal axis.

$$\epsilon_{11} = \alpha_1 \, \Delta T, \quad \epsilon_{22} = \alpha_2 \, \Delta T, \quad \epsilon_{33} = \alpha_3 \, \Delta T \qquad (8.15)$$

As in isotropic materials, no shear strains are induced in the material principal axis in an orthotropic material subjected to a uniform temperature gradient. However, because the normal strains are not equal in all directions, transformations to axes other than the material principal axis will produce shear strains. This is in direct contrast to the behavior of isotropic materials. Insight into thermomechanical behavior of orthotropic materials is important for proper design of PCBs made from laminated fiber-reinforced composites. Thermally induced strains in fully anisotropic materials are beyond the scope of the current discussion.

In summary, it is important to realize that mechanical strains and stresses can be expected in PCB or PWB assemblies whenever thermal or mechanical deformations are constrained. Constraints arise typically from nonuniform thermal deformations in a component or assembly of components. Such nonuniformities can be expected in deformation fields due to nonlinear temperature gradients in homogeneous materials and/or due to differential expansion in heterogeneous assemblies. Royce [3] has presented an excellent discussion of the common analytical tools for computing nonuniform thermal deformations in microelectronic systems. Analysis of the thermomechanical stress distribution in any

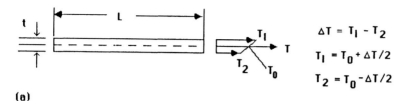

**(a)**

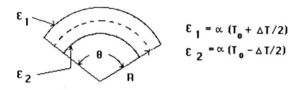

$$\varepsilon_1 = \alpha (T_0 + \Delta T/2)$$
$$\varepsilon_2 = \alpha (T_0 - \Delta T/2)$$

**(b)**

$M \big($ [ ⎯⎯⎯⎯⎯⎯⎯⎯⎯⎯⎯⎯⎯ ] $\big) M = \dfrac{E t^3 \alpha \Delta T}{12 (1-\nu)}$

**Force it flat with a bending moment (M)**

**(c)**

**Figure 8.3**  Example: stresses in plate with a linear thermal gradient. (a) Plate subjected to a linear thermal gradient. (b) Deformed shape of the plate due to the thermal gradient. (c) Plate constrained to lie flat with applied moments.

specific situation depends not only on the geometry of the problem but also on the elastic symmetries of the constituent materials. The examples in the following section illustrate these considerations in the context of PCB/PWB design.

## Examples

The following are some simple examples of thermal stresses and deformations that are encountered in PCB or PWB design. Only simple idealized problems are discussed here in order to present closed-form solutions. For more complex problems, numerical discretization or experimental techniques (such as the finite-element method [4] or photoelasticity) are required.

### Plate with a Linear Thermal Gradient

This situation arises commonly in die materials, substrates, and boards where power dissipation can cause temperature gradients through the thickness direction.

Consider a homogeneous, isotropic plate of sides $L$ and thickness $t$ subjected to a temperature distribution that is uniform in the plane of the plate and linear through its thickness, as shown in Figure 8.3a. The temperature on the upper and lower surfaces of the plate can be expressed as

$$T_1 = T_0 + \frac{\Delta T}{2}$$

$$T_2 = T_0 - \frac{\Delta T}{2}$$

(8.16)

where $T_0$ is the temperature at the midplane of the plate and $\Delta T = (T_1 - T_2)$ is the temperature difference between the top and bottom surfaces of the plate. The resulting thermal strain at the plate's upper surface is

$$\epsilon_1 = \alpha \left( T_0 + \frac{\Delta T}{2} \right)$$

and at the lower surface

$$\epsilon_2 = \alpha \left( T_0 - \frac{\Delta T}{2} \right)$$

(8.17)

The strain differential through the thickness causes a curvature $\kappa$ in the plate. The magnitude of $\kappa$ is independent of direction because the plate is isotropic. Assuming that thickness $t$ is much smaller than length $L$ and that the out-of-plane deformations are small compared to $t$, the curvature $\kappa$ or the radius of curvature $R$ can be expressed in terms of the strain gradient as

$$\kappa = \frac{1}{R} \simeq \frac{\theta}{L} \simeq \frac{\epsilon_1 - \epsilon_2}{t}$$

(8.18)

where $\theta$ is the angle subtended by a length $L$ of the plate. The resulting shape of the plate is shown in Figure 8.3b. If the plate were forced to lay flat with moments applied to the ends, as shown in Figure 8.3c, the total strain (sum of the thermal and mechanical strains) would be $\alpha T_0$ everywhere in the plate. Thus at the upper surface

$$\epsilon_1 = \text{bending strain} + \text{thermal strain} = \alpha T_0$$

(8.19)

Subtracting $\alpha T_0$ from both sides, the induced mechanical bending strains on the top surface can be written as

$$\epsilon_{xx} = \epsilon_{yy} = -\frac{\alpha \Delta T}{2}$$

(8.20)

This gives the induced bending stresses on the upper surface as

$$\sigma_{xx} = \sigma_{yy} = -\frac{E \Delta T \alpha}{2(1 - \nu)}$$

(8.21)

where $E$ is Young's modulus and $\nu$ is Poisson's ratio. Using an approximate bending theory of thin elastic plates, it is possible to compute the moments required per unit length to constrain the plate to remain flat [5]:

$$M_x = M_y = \frac{E t^3 \alpha \Delta T}{12(1 - \nu)}$$

(8.22)

where $M_x$ and $M_y$ are the moments about the $x$ and $y$ axes, respectively.

*Copper-Plated Through Hole in a Fiberglass PWB Subjected
to a Uniform Temperature Change*

Figure 8.4a shows a copper-plated through hole in a PWB being subjected to a uniform
temperature change $\Delta T$. Figure 8.4b shows an approximate idealized model of the assembly
for analysis purposes. The fiberglass and copper have vastly different CTE values in the
axial $z$ direction, giving rise to large thermal stresses. All deformations in the $xy$ plane
are ignored. Under these approximations, this problem can be solved as a one-dimensional
structure. The force equilibrium equation for this assembly is written as

$$F_1 + F_2 = 0 \tag{8.23}$$

where subscript 1 refers to copper and 2 refers to the board. The geometric compatibility
condition for deformations $\delta$ is given as

$$\delta_1 = \delta_2 = \delta \tag{8.24}$$

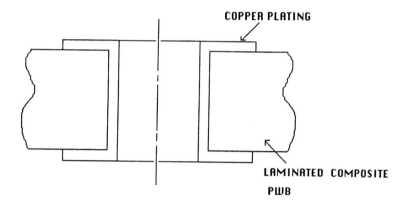

**(a)**

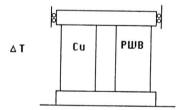

**(b)**

**Figure 8.4** Example: plated through hole in a laminated glass-epoxy board. (a)
Schematic representation of the plated through hole. (b) One-dimensional model
of copper plating and board.

The thermoelastic deformations $\delta_i$ are given by

$$\delta_i = \alpha_i \, \Delta T \, l_i + \frac{F_i l_i}{(E_i A_i)}, \qquad i = 1\text{--}2 \tag{8.25}$$

where $\alpha$ is the CTE, $l$ is the length of each component, $E$ is Young's modulus, and $A$ is the cross-sectional area. Noting that both members have the same length and solving the equilibrium and compatibility equations simultaneously,

$$F_i = \frac{(\alpha_{(3-i)} - \alpha_i)\Delta T}{\sum_i (1/A_i E_i)}, \qquad i = 1,2 \tag{8.26}$$

The normal stress $\sigma$ in each component can now be obtained as

$$\sigma_i = \frac{F_i}{A_i}, \qquad i = 1,2 \tag{8.27}$$

Because the dimensions of the composite board in the $xy$ plane are many orders of magnitude larger than the cross-sectional dimensions of the copper plating, for the purpose of this problem it is necessary at this point to make a reasonable assumption about the effective cross-sectional area of the composite board. Assuming the composite cross section to be infinitely larger than the copper cross section, one obtains the upper bound to the stresses in the copper and lower bound to the stresses in the composite board. Conversely, assuming them to have comparable cross-sectional areas, it is possible to obtain the lower bound to the stresses in the copper and upper bound to the stresses in the board. Typically, in this problem the stresses in the copper are of interest because failure is common in the copper plating. Hence a conservative estimate can be obtained by using the former of the two assumptions listed above. Using standard properties for copper and glass-epoxy composite below the glass transition temperature (see Table 8.1), the upper and lower bound stresses for a temperature change of 10°C are computed as follows.

Upper bound in copper:

$$\sigma_{\text{copper}} = 75.4 \text{ MPa}, \qquad \sigma_{\text{composite}} = 0 \text{ MPa}$$

Lower bound in copper:

$$\sigma_{\text{copper}} = 1.92 \text{ MPa}, \qquad \sigma_{\text{composite}} = -1.92 \text{ MPa}$$

More sophisticated numerical analysis reveals that the upper bound estimate of the stresses in copper is much more realistic than the lower bound value. Clearly, the copper plating experiences large tensile stresses while the composite board is under compression. This is a direct consequence of the large out-of-plane CTE of the composite board. The problem becomes more pronounced at temperatures exceeding the glass transition temperature of the epoxy in the composite material because of the quantum jump in the CTE of the board. The resulting thermal stress often causes premature fatigue failure in the copper plating of through holes in microelectronic assemblies.

*Bonded Assembly of Two Dissimilar Materials Subjected to Uniform Temperature Change*

Consider two plates of dissimilar materials bonded together with a layer of adhesive. Such situations arise commonly (1) when dies are bonded to a substrate or a substrate to a board, (2) in bonded interconnections like leads or wire bonds, or (3) when composite boards are laminated to heat sinks with a compliant adhesive. Both adherend materials and the adhesive are assumed to be homogeneous and isotropic, and the elastic properties of the adhesive are several orders of magnitude smaller than those of the two adherend materials. Figure 8.5a shows a schematic of this assembly being subjected to a uniform temperature change $\Delta T$. The length direction is labeled the $x$ direction, the width direction is out of the plane of the paper, and the direction normal to the plane of the plates is the thickness direction. Assuming generalized plane conditions in the width direction (i.e., stresses and strains do not vary along the width direction), the problem can be solved approximately as a plane problem. Figure 8.5b shows an exploded view of an element of unit width and infinitesimal length $dx$. Ignoring bending deformations in the thickness direction, the in-plane thermal deformation $d\delta$ in the $x$ direction, for the strip $dx$ in length, can be written in each plate as

$$d\delta_i = \frac{(1 - v_i)P_i\, dx}{E_i A_i} + \alpha_i\, \Delta T\, dx, \qquad i = 1,2 \tag{8.28}$$

where $P_i$ are the mechanical tractions in the $x$ direction, per unit width in each plate, as a result of the thermal expansion mismatch and $E_i$, $v_i$, $\alpha_i$, and $A_i$ are Young's modulus, Poisson's ratio, the coefficient of thermal expansion, and the cross-sectional area, respectively, of each plate. It is assumed here that in-plane normal stresses and thermal deformations are uniform through the thickness of the plates.

Assuming plane stress conditions in the thickness direction of the assembly, it can be shown that the thermal expansion mismatch between the two plates will cause only shear deformations (and shear stress) in the adhesive layer. This type of approximation, called the shear lag model, is common in the analysis of heterogeneous media (e.g., fiber-reinforced composites [2]). Figure 8.5b shows the resulting tractions acting on each layer. The equilibrium condition for a unit width can be written in each of the two adherend plates as

$$dP_i + (-1)^i \tau\, dx = 0, \qquad i = 1,2 \tag{8.29}$$

where $\tau$ is the shear stress in the adhesive. The compatible shear strain (and resulting stress) in the adhesive is given by

$$d\gamma = \frac{d\delta_1 - d\delta_2}{b} = \frac{d\tau}{G} \tag{8.30}$$

where $G$ is the shear modulus of the adhesive material and $b$ is the thickness of the adhesive layer. This assumed strain field is uniform in the thickness and width directions and thus reduces the problem further to one dimension. Substituting for $d\delta_i$ from Eq. (8.28) in the compatibility equation (8.30), taking the second derivative with respect to $x$, and using the equilibrium conditions (8.29), one obtains the one-dimensional differential equation

$$\frac{d^2\tau}{dx^2} - A^2\tau = 0 \tag{8.31}$$

where

$$A^2 = \frac{G}{b}\left(\frac{1 - v_1}{E_1 A_1} + \frac{1 - v_2}{E_2 A_2}\right)$$

This second-order, homogeneous, ordinary differential equation with constant coefficients must be solved subject to the following symmetry and boundary conditions on the tractions:

$$\tau = 0 \quad \text{at } x = 0; \qquad P_i = 0, \quad x = \pm l \tag{8.32}$$

The solution to the differential equation (8.31) gives the resulting shear stress distribution along the length ($x$) direction and can be shown to be

$$\tau = \frac{G\,\Delta T(\alpha_1 - \alpha_2)\sinh(Ax)}{bA\,\cosh(Al)} \tag{8.33}$$

The maximum shear stress occurs at the edges and is given by

$$\tau_{\max} = \frac{G\,\Delta T(\alpha_1 - \alpha_2)\tanh(Al)}{bA} \quad \text{at } x = \pm l \tag{8.34}$$

In most practical situations $Al \gg 1$. Thus, $\tanh(Al) \simeq 1$ and the maximum shear stress can be written approximately as

$$\tau_{\max} \simeq \frac{G\,\Delta T(\alpha_1 - \alpha_2)}{bA} \quad \text{at } x = \pm l \tag{8.35}$$

As a particular example, consider a 0.25-in.-square silicon die bonded to an alumina substrate with conductive epoxy as shown in Figure 8.5c. Let the die be 0.02 in. thick,

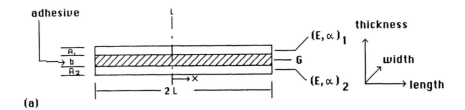

**(a)**

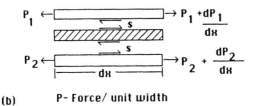

**(b)**      P- Force/ unit width

**Figure 8.5** Example: stresses in a bonded assembly under uniform thermal loading. (a) Two pieces of dissimilar materials joined together with a compliant adhesive. (b) Force balance of joined materials (out-of-plane bending ignored). (c) Diagram showing a die on a substrate. (d) Plot of joint shear stress as a function of position.

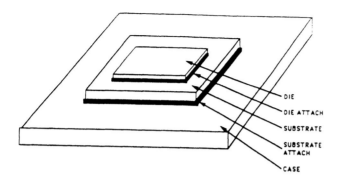

**(c)**

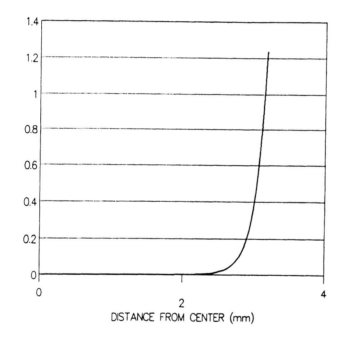

SHEAR STRESS (MPa)

DISTANCE FROM CENTER (mm)

**(d)**

the adhesive layer be 0.0035 in. thick, and the substrate be 0.1 in. thick. In accordance with MIL specification test guidelines, let the assembly be subjected to a temperature cycle of 180°C, ranging from $-55$ to 125°C. If the adhesive shear modulus $G$ is 1.52 GPa, then using standard property values for silicon and alumina, the shear stress in the adhesive can be computed as a function of distance $x$ from the die center using Eq. (8.33).

Figure 8.5d shows that the shear stress is negligible through most of the bond length except for a small boundary layer at the edge ($x = \pm 1$), where it reaches its maximum

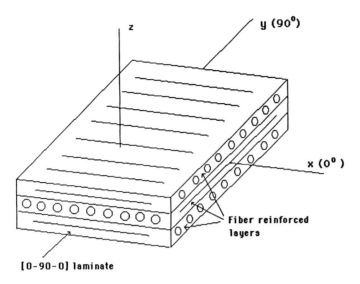

z

y (90°)

x (0°)

Fiber reinforced
layers

[0-90-0] laminate

**Figure 8.6**   Example: stresses in a [0,90,0] orthotropic fiber-reinforced composite laminate under uniform thermal loading.

value. The shear stress gradient is seen to be very large in this boundary layer. Any failure
in the adhesive layer will obviously initiate at this free edge and propagate inward, causing
progressive delamination of the bond. Similar failures can also occur in the attach between
the substrate and the case.

*A Laminated Orthotropic Plate Subjected to Uniform Temperature Change*

Figure 8.6 shows a laminated orthotropic composite board consisting of three fiber-
reinforced layers or laminae. The two outer layers have continuous fibers running in the
0° direction, while the central layer has continuous fibers running in the 90° direction. In
composites terminology, this is a [0/90/0] laminate. Such directionally tailored laminates
are becoming popular as PWB materials for superior anisotropic heat conduction properties.
    When subjected to a uniform temperature change $\Delta T$, the laminate will undergo
overall thermal deformations given by the in-plane thermal strain vector

$$\epsilon_T = \bar{\alpha} \, \Delta T \qquad\qquad (8.36)$$

where $\bar{\alpha}$ is the vector of overall effective coefficients of thermal expansion of the laminate.
As a result of the symmetry about the midplane, the laminate will not experience any
curvatures. In accordance with the assumptions of classical lamination theory, it will be
assumed that the strains in each layer are given by $\epsilon_T$ (Voigt assumption). Thus the in-
plane stress vector in the $i$th layer can be written as [2]

$$\sigma_i = Q_i(\epsilon_T - \alpha_i \, \Delta T) \qquad\qquad (8.37)$$

where $Q_i$ is the stiffness tensor and $\alpha_i$ is the vector of coefficients of thermal expansion
of the $i$th layer.
    Thus even though the laminate is undergoing unconstrained thermal deformations,
each layer still experiences nonzero thermomechanical stresses due to self-equilibrating
anisotropic constraints from adjacent layers. The stress vector can be readily determined

in each layer from the above equation if the $\boldsymbol{\epsilon}_T$ vector can be determined. $\boldsymbol{\epsilon}_T$ is obtained by writing the global equilibrium equations for the laminate; i.e., no external tractions $N$ are acting on the laminate. Thus

$$\left\{\begin{matrix} N_x \\ N_y \\ N_{xy} \end{matrix}\right\} = \left\{\begin{matrix} 0 \\ 0 \\ 0 \end{matrix}\right\} = \sum_i t_i \mathbf{Q}_i (\boldsymbol{\epsilon}_T - \boldsymbol{\alpha}_i \, \Delta T) \tag{8.38}$$

Assuming that all layers have the same thickness $t_i$, the above equation can be rearranged and written for the geometry of Figure 8.6 as

$$\boldsymbol{\epsilon}_T = [2\mathbf{Q}_0 + \mathbf{Q}_{90}]^{-1}[2\mathbf{Q}_0\boldsymbol{\alpha}_0 + \mathbf{Q}_{90}\boldsymbol{\alpha}_{90}]\Delta T \tag{8.39}$$

With $\boldsymbol{\epsilon}_T$ known, $\boldsymbol{\sigma}_i$ can be determined explicitly in each layer using Eq. (8.37).

As a particular example, let the in-plane stiffness tensor of each layer be given as follows. (Msi = millions of pounds per square inch.)

$$\left\{\begin{matrix} \sigma_L \\ \sigma_T \\ \tau_{LT} \end{matrix}\right\} = \begin{bmatrix} 25 & 0.25 & 0 \\ 0.25 & 1.0 & 0 \\ 0 & 0 & 0.5 \end{bmatrix} \left\{\begin{matrix} \epsilon_L \\ \epsilon_T \\ \gamma_{LT} \end{matrix}\right\} \quad \text{(Msi)} \tag{8.40}$$

where subscript L indicates the longitudinal (along fiber) direction and subscript T indicates the direction transverse to the fiber. Further, let the vector of CTE values be given by

$$\{\alpha_L \quad \alpha_T \quad \alpha_{LT}\} = \{2.5 \quad 15.0 \quad 0\} \quad \mu\epsilon/°F \tag{8.41}$$

Then for equal ply thickness, the stress in each layer of the laminate shown in Figure 5.6 can be shown to be

$$\left\{\begin{matrix} \sigma_x \\ \sigma_y \\ \tau_{xy} \end{matrix}\right\}_0 = \left\{\begin{matrix} 5.945 \\ -11.38 \\ 0 \end{matrix}\right\} \Delta T, \qquad \left\{\begin{matrix} \sigma_x \\ \sigma_y \\ \tau_{xy} \end{matrix}\right\}_{90} = \left\{\begin{matrix} -11.89 \\ 22.76 \\ 0 \end{matrix}\right\} \Delta T \tag{8.42}$$

It is noted that in orthotropic laminates uniform thermal gradients do not produce any shear stresses in the material principal axes.

## 8.2.2  Fatigue

It has long been recognized that materials subjected to repetitive or fluctuating loads will fail at load levels lower than that required to cause failure on a single, monotonic application of the load. Such failures are said to be due to material fatigue or are simply called fatigue failures. In microelectronic packages, such fluctuating loads may be caused by environmental temperature cycling, operational power cycling, and vibrational loads.

The monotonic failure stress is called the ultimate strength of the material, and the cyclic stress required to cause failure in $N_f$ cycles is called the fatigue strength for a life of $N_f$ cycles. On a microscopic scale, the fatigue failure surface can be seen to be created by either a single crack or several cracks that have slowly grown under the repetitive load cycling. The origin of these cracks can usually be found at some point of stress concentration. In monolithic materials such stress concentrations can arise at sharp geometric discontinuities associated with cracks, cutouts, or voids and at reentrant corners. They may also be found at the surface of microscopic inhomogeneities such as inclusions. In laminated composite materials, additional stress concentration sites can be found at the

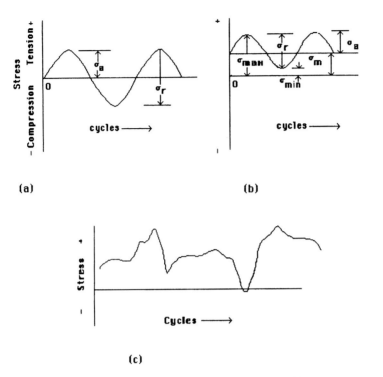

**Figure 8.7**  Fatigue loading. (a) Completely reversed loading.
(b) $\sigma_{mean} \neq 0$. (c) Random loading.

surface of inhomogeneities such as material interfaces, layer interfaces, and free edges
and at discontinuities such as delaminations and debonds.

The three basic parameters that influence fatigue are (1) a maximal tensile stress of
sufficiently high value, (2) a large enough variation or fluctuation in the applied stress,
and (3) a sufficiently large number of applied cycles. In addition, there are a host of other
variables primarily related to the environmental conditions to which the specimen is
subjected.

Figure 8.7 illustrates the types of fluctuating stresses that can cause fatigue failures.
Figure 8.7a illustrates a completely reversed cycle of stress of sinusoidal form. The tensile
stress magnitude is equal to the compressive stress magnitude. Figure 8.7b illustrates
another sinusoidal repeated stress cycle more typical of actual loadings, where the maximum
tensile stress is not equal in magnitude to the maximum compressive stress or minimum
stress. In this repeated cycle the stress can be thought of as a superposition of a mean
stress and an alternating or variable stress amplitude. The mean stress $\sigma_m$ is

$$\sigma_m = \frac{\sigma_{max} + \sigma_{min}}{2} \tag{8.43}$$

and the alternating stress or the stress amplitude $\sigma_a$ is

$$\sigma_a = \frac{\sigma_{max} - \sigma_{min}}{2} \tag{8.44}$$

These two values are commonly used to characterize the fatigue cycle.

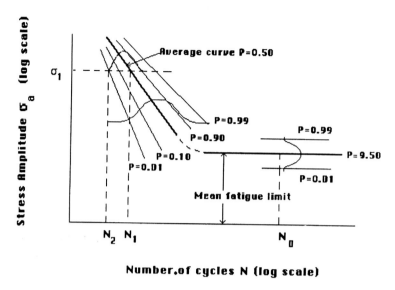

**Figure 8.8**  Fatigue probability distribution.

Figure 8.7c illustrates a complicated stress cycle that might be encountered in a part such as an aircraft wing that is subjected to occasional unpredictable overloads due to wind gusts. Such random loading is common in some electronic equipment and will be discussed in more detail in the next chapter. The greatest fatigue stresses that commonly occur in PCB design are due to the thermally induced strains during power cycling. It has been estimated that 80% of all mechanical board failures are fatigue related due to thermomechanical strains and that the remaining 20% of failures are fatigue related due to vibration-induced strains [6].

Predicting the fatigue life of a part is usually done in terms of probability. The problem is to predict within $c\%$ confidence level when $N\%$ of the samples are going to fail. Fatigue data inherently have a fair amount of scatter, and rarely are enough specimens tested to determine the statistical parameters accurately. The few studies [7] directed toward determining the failure distributions for a given stress cycle suggest that a Gaussian or normal distribution adequately describes the failure when the life is expressed as log $N$, where $N$ is the life in cycles. Figure 8.8 shows a distribution of fatigue life at constant stress along with the curves for constant probability. Note that at the higher stress levels there is decreasing scatter in the fatigue life measurements.

When discussing fatigue there are two distinct classifications: high-cycle fatigue and low-cycle fatigue. Engineers are most commonly used to designing for high-cycle fatigue where the part is subjected to relatively low load levels. The elastic stresses are much greater than the plastic stresses under such conditions. Fatigue failure typically occurs after $10^4$ or more cycles and is a stress-dominated event. Vibrational loads on microelectronic packages typically cause high-cycle fatigue failures. Low-cycle fatigue occurs at higher load levels where the plastic component of strain is comparable to or greater than the elastic strain. Low-cycle fatigue is characterized as a strain-dominated event rather than a stress-dominated event. Temperature and power cycling are the primary causes of low-cycle fatigue failures in microelectronic packages.

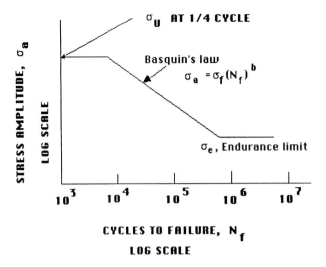

**Figure 8.9** *S-N* plot.

## High-Cycle Fatigue (HCF)

Experimental data for high-cycle fatigue are typically plotted as the log of the alternating stress amplitude, $\sigma_a$, versus the log of the cycles to failure, $N_f$, as shown in Figure 8.9. When plotted in this manner, the data can be approximated with a straight line whose equation is

$$\sigma_a = \sigma_f N_f^b \tag{8.45}$$

This is Basquin's power law, where $b$ and $\sigma_f$ are material constants depending on the surface finish, environment, size, etc. Basquin's law holds for completely reversed cyclic loading. These plots are commonly referred to as *S-N* diagrams (stress amplitude, *S*, versus fatigue life, *N*, diagrams). As shown in the figure, when the alternating stress amplitude, $\sigma_a$, falls below a threshold value, the part appears to have almost infinite life. This type of behavior is encountered in some metals, and the threshold stress value is known as the fatigue limit or endurance limit. A common rule of thumb used by designers in the past rates the endurance limit of low-alloy steels as approximately one half of the ultimate strength and that of aluminum alloys as about one third of the ultimate strength.

When the applied cyclic loading consists of an alternating component superposed on a mean stress $\sigma_m$ (as shown in Figure 8.7b), Basquins's power law requires an appropriate correction

$$\sigma_a = (\sigma_f - \sigma_m)(N_f)^b \tag{8.46}$$

An alternative approach to high-cycle fatigue analysis utilizes fracture mechanics concepts rather than *S-N* diagrams (general background on fracture mechanics can be found in Refs. 8 and 9). With the fracture mechanics approach, fatigue is viewed as the slow propagation of a single dominant crack. When the log of the fatigue crack growth rate, *da/dN*, is plotted against the log of the stress intensity factor range, $\Delta K$, three distinct types of crack propagation behavior are observed as shown in Figure 8.10. Below a threshold $\Delta K$, the fatigue cracks do not propagate. Just above threshold values of $\Delta K$,

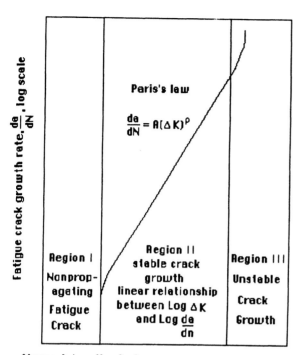

**Figure 8.10** *da/dN* versus $\Delta K$ plot.

crack propagation rates are fairly high. Again, at very high values of $\Delta K$ the cracks propagate in a rapid, unstable fashion. Between these two load levels, the fatigue crack undergoes stable propagation and the log of the fatigue crack growth rate is proportional to log $\Delta K$

$$\frac{da}{dN} = A(\Delta K)^p \tag{8.47}$$

This is Paris' power law for fatigue crack propagation under completely reversed uniaxial (mode I) loading in isotropic materials, where $A$ and $p$ are material constants. It can be shown from crack mechanics [8,9] that $\Delta K$ is related to the applied stress level and to the instantaneous crack size. Thus

$$\Delta K = Y \Delta\sigma \sqrt{\pi a} \tag{8.48}$$

where $Y$ is a calibration factor dependent on specimen geometry, $\Delta\sigma$ is the amplitude of the applied stress level, and $a$ is the crack size.

Thus at a given cyclic stress level, the propagation rate will become unstable when the crack size reaches a critical value. In the presence of mean stresses, Paris' law requires the following correction:

$$\frac{da}{dN} = \frac{A(\Delta K)^p}{(1 - R)(K_{I_c} - K_{Max})}, \mathrm{R} > 0 \tag{8.49}$$

where $R$ is the ratio of the minimum stress level to the maximum stress level. Substituting Eq. (8.48) in (8.47) and integrating, one can obtain an expression for the number of cycles $N_p$ required for a crack to grow from size $a_0$ to $a_f$ under a uniform cyclic stress of amplitude $\Delta\sigma$

$$N_p = \frac{2(a_0^{(2-p)/2} - a_f^{(2-p)/2})}{A(p-2)(\Delta\sigma)^{p}\Pi^{(p/2)}} \tag{8.50}$$

In the presence of multiaxial stresses the crack undergoes mixed-mode propagation and an effective stress intensity factor is required in the above propagation equations.

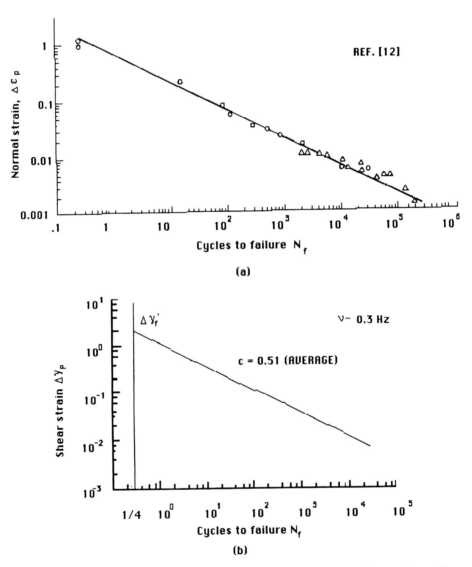

**(a)**

**(b)**

**Figure 8.11**  Plotting the log of the plastic strain amplitude versus the log of the cycles to failure results in a straight line  (low-cycle fatigue).

In the fracture mechanics approach to fatigue analysis, a crack is assumed to be located at the worst possible location, i.e., at the most stressed region and in the worst possible orientation. Paris' power law is then used to compute the number of cycles for the crack to grow to a critical size for catastrophic failure. Alternatively, if the part is required to survive for a specified life period, the crack propagation problem is posed in reverse to set inspection standards. In this case, Paris' power law is used to compute the initial size of a crack that would grow to the critical size under the given loading, within the rated fatigue life. This initial crack size is specified as the inspection criterion. The component is then scheduled for repair or maintenance if any cracks of the specified size or larger are detected in the location of interest.

## Low-Cycle Fatigue (LCF)

Low-cycle fatigue failures occur in less than $10^4$ cycles. Load amplitudes are relatively high and plastic strains are typically much greater than elastic strains. The failure is a strain-dominated event rather than a stress-governed event. Plotting the log of the amplitude of the plastic strain, $\epsilon_p$, versus the log of the cycles to failure, $N_f$, results in a linear fit through the experimental data as shown in Figure 8.11a and b. This is commonly referred to as the Manson-Coffin relation [10,11]

$$\frac{\Delta \epsilon_p}{2} = \epsilon_f (2N_f)^c \tag{8.51}$$

where $\epsilon_f$ and $c$ are material properties. The applied strain amplitude has one half the applied strain range, and note that there are $2N_f$ reversals to failure.

The generalized Manson-Coffin relation takes into account both high-cycle fatigue and low-cycle fatigue and is schematically represented in Figure 8.12. The total strain amplitude can thus be written as

$$\frac{\Delta \epsilon}{2} = \frac{\Delta \epsilon_e}{2} + \frac{\Delta \epsilon_p}{2} \tag{8.52}$$

$$= \left( \frac{\sigma_f}{E} \right) (2N_f)^b + \epsilon_f (2N_f)^c \tag{8.53}$$

where

$\Delta \epsilon/2$ = applied total strain amplitude
$\Delta \epsilon_e/2$ = applied elastic strain = $\Delta \sigma / 2E$
$\Delta \epsilon_p/2$ = applied plastic strain amplitude
$\epsilon_f$ = fatigue ductility coefficient
$N_f$ = number of cycles to failure
$c$ = Manson-Coffin fatigue ductility exponent
$\sigma_f$ = fatigue strength coefficient
$E$ = modulus of elasticity
$b$ = Basquin fatigue strength exponent

The figure and equation show that the plastic and the elastic fatigue failure curves plot as straight lines on a log-log scale. The $y$ intercepts are $\epsilon_f$ (i.e., a measure of the material's ductility) and $\sigma_f/E$ (i.e., a measure of the material's strength), respectively. The plastic curve dominates in the low-cycle fatigue region and the elastic curve dominates in the

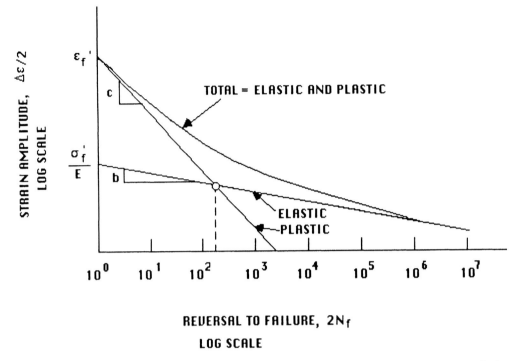

**Figure 8.12**  Schematic representation of generalized Manson-Coffin relation taking into account both high-cycle and low-cycle fatigue.

high-cycle fatigue region. There is a transition region in between, where elastic and plastic strains contribute equally to fatigue damage.

In the presence of a mean stress $\sigma_m$, Eq. (8.53) needs the following modification:

$$\frac{\Delta\epsilon}{2} = \frac{\sigma_f - \sigma_m}{E}(2N_f)^b + \epsilon_f(2N_f)^c \tag{8.54}$$

The above equations have been presented in the context of cyclic uniaxial tensile loading. Solder joints are more commonly subjected to shear loading. Ignoring the elastic strains and working only with the plastic strains for low-cycle fatigue failures, the Manson-Coffin equation can be rewritten for shear loading as

$$\frac{\Delta\gamma_p}{2} = \gamma_f(2N_f)^c \tag{8.55}$$

where $\Delta\gamma_{p/2}$ is the plastic shear strain amplitude. Figure 8.11b illustrates this behavior in the data of Solomon and Harvey [12] for shear strain versus life for 60/40 Pb-Sn solder.

The fracture mechanics approach discussed in the preceding section is not found to be particularly effective in low-cycle fatigue analysis. The concept of a stress intensity factor is restricted to linear elastic situations and does not apply when there is large-scale plasticity at the crack tip. In such situations, an alternative approach is to use one of the many available path-independent conservation integrals [13], such as the famous $J$-integral, to characterize the crack energy release rate. Unfortunately, experimental evidence of

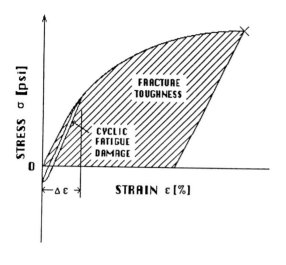

$$\text{CYCLES-TO-FAILURE} \approx \frac{\text{FRACTURE TOUGHNESS}}{\text{CYCLIC FATIGUE DAMAGE}}$$

**Figure 8.13** Stress-strain plot of a material that does not exhibit stress relaxation when subjected to a cyclic strain $\Delta\epsilon$.

correlations between the $J$-integral and the propagation life are scarce and inconclusive at present. It remains to be seen if further research can establish any such correlation.

## High-Temperature Creep Fatigue

Components that are subjected to thermomechanical strains at fairly high temperatures (''hot-worked'') can undergo significant amounts of time-dependent stress relaxation due to inelastic creep deformations. Such time-dependent strains are termed anelastic strains and result, in part, from mechanisms such as diffusion at grain boundaries and within grains, grain boundary sliding, dynamic recrystallization, grain growth, intergranular crack propagation, and dislocation climb. These deformation mechanisms are fundamentally different from dislocation slip mechanisms encountered in time-independent plastic deformation. It is therefore believed that the resulting fatigue damage also differs for these two deformation modes. Several investigators therefore have attempted to modify the conventional Manson-Coffin–type fatigue analysis methods for dealing with anelastic deformations as opposed to plastic deformations.

Stress relaxation phenomena have a critical adverse influence on the fatigue damage of the component during each strain cycle. This is qualitatively illustrated in Figures 8.13 and 8.14. Figure 8.13 shows the stress-strain plot of a material exhibiting no stress relaxation when subjected to a cyclic strain $\Delta\epsilon$. One school of thought postulates that the area within the cyclic strain hysteresis loop is a qualitative measure of the fatigue damage the material has suffered. This damage criterion is based on the contention that fatigue damage can be viewed as a progressive exhaustion of ductility and can be characterized by the inelastic work done during each cycle. Figure 8.14 shows the same figure for a material that exhibits stress relaxation. The large increase in the hysteresis area indicates earlier ductility exhaustion and shorter fatigue life.

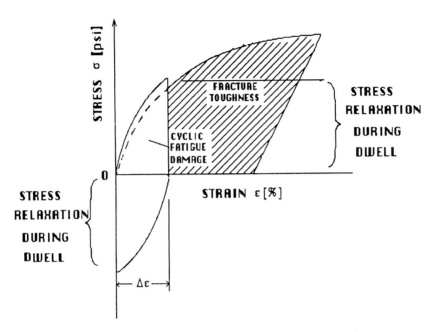

**Figure 8.14**  Stress-strain plot for a material that exhibits stress relaxation.

The amount of stress relaxation is obviously related to the temperature and to the cycle time and any dwell time associated with a particular loading cycle. Thus strain rate or loading frequency is an important variable, in addition to the temperature, in creep fatigue design. Several investigators have attempted to correct the Manson-Coffin strain-life relationship parametrically for frequency and temperature effects, but with limited success in extrapolating down to low service frequencies on the order of one cycle per day.

One such relationship suggested by Coffin [14] applies a correction factor to the fatigue coefficients and exponents in the Manson-Coffin relationship:

$$\Delta\epsilon = C_2[N_f \, v^{(k-1)}]^{-\beta} + \frac{AC_2^n}{E} N_f^{-\beta n} \, v^{[-\beta n(k-1) + k_1]} \tag{8.56}$$

where $C_2$, $A$, $\beta$, $n$, $k$, and $k_1$ are temperature-dependent material fatigue properties and $v$ is the frequency of loading.

An alternative approach, proposed by Halford et al. [15], uses strain-range partitioning techniques in which anelastic strains and plastic strains are recorded separately for each cycle. Figure 8.15a shows the strain partitioning for a typical viscoplastic hysteresis loop. Strain-life curves are generated for each type of strain in laboratory experiments. Sample data are shown in Figure 8.15b. The overall damage is then computed by summing the damage due to the individual strain types from their respective strain-life curves. Details of this method can be found in Ref. 15.

The obvious disadvantages of either of these approaches stem from the fact that additional material properties have to be generated and catalogued from laboratory experiments for each material of interest. In the absence of such information, some investigators use the total strain amplitude in an unmodified Manson-Coffin relationship as a first-order approximation.

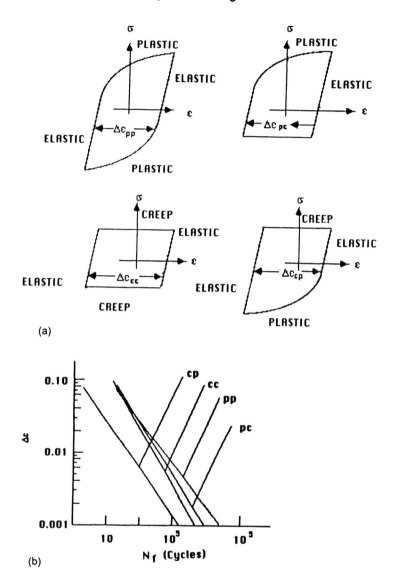

(a)

(b)

**Figure 8.15** (a) Idealized hysteresis loops for the four basic types of inelastic strain range. (b) Partitioned strain range–life relationships for a sample material.

## 8.3  SPECIAL TOPICS IN THERMAL CREEP FATIGUE OF MICROELECTRONIC ASSEMBLIES

The general theoretical considerations in high-temperature fatigue were presented in Section 8.2.2. This section addresses some specific issues and examples that are of importance in the reliability assessment of electronic packages. These topics range from discussions of specific damage mechanisms in different components of the microelectronic assembly to material behavior and testing techniques. Fatigue studies of PWB/PCB components in the past have often approached the problem as an issue in empirical statistics. Consequently, a large fraction of the available fatigue models are problem-specific and dependent on specific loading histories, component geometry, etc. The purpose of this text is to encourage the designer to use more generic, universal, and updated fatigue and fracture models. Several researchers have already attempted such modeling, and a brief review is included in the following section.

It is important to realize that the generalized fatigue models require reliable inputs in the form of accurate stress-strain histories and accurate material properties from controlled fatigue tests. On the one hand, this necessitates the use of sophisticated stress analysis tools, such as the finite-element method, to predict accurately the stress-strain history and the strain partitioning (in terms of elastic, plastic, and anelastic strains) during a complete representative cycle/block of the load history. On the other hand, it is essential to use reliable laboratory test methods to generate accurate sets of thermoviscoplastic constitutive properties as well as fatigue and fracture properties (for each component of the partitioned strains) for all materials commonly used by the electronics industry.

Admittedly, this involves industrywide awareness and a large cooperative experimental effort. It can be argued, however, that without such a concerted commitment to this very important damage problem, the electronics industry is not likely to experience the benefits of the systematic fatigue theories that other industries (such as the automotive, aerospace, and nuclear industries) have successfully adopted and implemented for many years.

### 8.3.1 Common Failure Mechanisms

The most common cause of failure is fatigue in the solder joints of leadless chip carriers due to thermal cycling and to vibrational loads. There are innumerable studies available in the literature dealing with this failure mode. Although most employ empirical or statistical models some have attempted detailed stress analysis in conjunction with a Coffin-Manson–type fatigue model [16–23]. Most of these deal with only thermal loading or only vibrational loading. However, evidence indicates that the superposition of these two loads may be particularly damaging. Most recent modifications of the MIL specifications call for testing under combined thermal and vibrational loads. Analytical procedures for superposing both types of loadings are approximate at best [24], and research is currently in progress to formulate systematic procedures for addressing both types of fatigue loads simultaneously [25]. A very comprehensive literature survey of available fatigue studies may be found in Ref. 26, and the studies are discussed further in Section 8.3.2. Figure 8.16 shows fatigue cracks after 500 cycles in castellated leadless chip carriers due to CTE mismatches between the ceramic chip and the solder material. The substrate in this case is also a ceramic.

In leaded chip carriers the compliance of the lead reduces the stresses in the solder significantly. However, fatigue failures are still observed to initiate at sites of stress concentrations in J-lead and gull-wing lead configurations. The important difference between solder joint failures in leadless and leaded devices is that the average stress level throughout the cycle is lower in leaded devices, even though the total strain range is the same in both joint types. This implies that the inelastic work dissipated within the solder joint is much higher in leadless chip carrier (LCC) packages than in leaded versions. Engelmaier [27] used this difference to explain why LCC solder joints have much poorer fatigue damage tolerance than leaded versions. In this reference, Engelmaier also develops very useful closed-form design equations to predict the actual fatigue energy dissipation and reliability of the solder joints for different package types.

**Figure 8.16** Fatigue cracks in the solder joint of a castellated ceramic LCC device due to thermal cycling. (Courtesy of NASA-GSFC, UNISYS.)

In bonded wire interconnects, the bonds between the wires and the bonding pad and/ or between the bond pad and the substrate often suffer shear fatigue failures due to thermomechanical loading, as shown in Figure 8.17a. Thermal cycling can also cause flexural fatigue failures at the heel of a bonded wire or axial fatigue failures in encapsulated packages with wire bonds. Hu et al. [28] have presented a comprehensive reliability analysis, using the analytical methods of section 8.2.1 on bonded assembly of two dissimilar materials, for wire bond failures. Sometimes the failure may be in the lead seal where a connector comes out of a package, as shown for a ceramic capacitor in Figure 8.17b.

Low-power devices typically use gold wires that are bonded through thermocompression or thermosonic techniques. High-power devices require larger-diameter aluminum wires that are ultrasonically bonded. Thermal bonding techniques are not suitable for aluminum wires because of the formation of intermetallic precipitates. Special care must be taken to avoid vibrational fatigue damage in the wire bonds during ultrasonic bonding (and cleaning and soldering) operations, because the natural frequencies of the bond wires are typically in the 6- to 8-kHz range. Manufacturing defects and material flaws such as voids play dominant roles in governing the failures of these components. Corrosion and diffusion at the bond interface can also lead to premature failures [29–32]. The bond wire diameters typically range from approximately 0.0007 to 0.01 in., and special techniques are often required for measuring the elastic and inelastic properties [33].

Laminated, multilayered, woven-fabric composite boards are often used in PCB designs in an effort to match the coefficients of thermal expansion of the board material with those of the ceramic chip carriers. Details of this matching are discussed in Sections 8.3.3 and 8.3.4. On organic multilayer PWBs, electrical interconnection through the thickness of the board is achieved by using small-diameter through holes of approximately 9 to 13 mils diameter, plated with copper and/or nickel. Unfortunately, the tailoring required to produce favorable in-plane CTE values often leads to extremely high CTE values in the thickness direction. This results in extremely high stresses and early failure in the copper plating of the plated through hole, thereby causing electrical failure of the

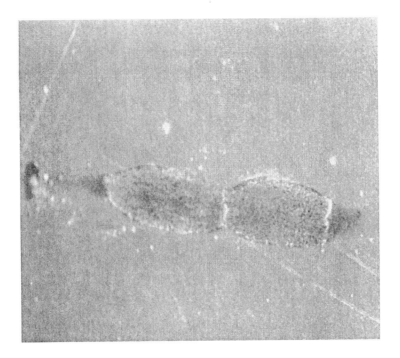

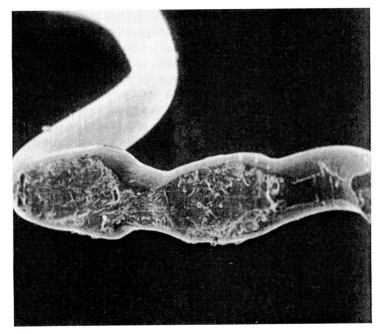

(a)

**Figure 8.17** (a) Shear fatigue failure between an interconnection wire and the bond pad. (b) Brittle fatigue failure in a lead seal. (Courtesy of NASA GSFC, UNISYS.)

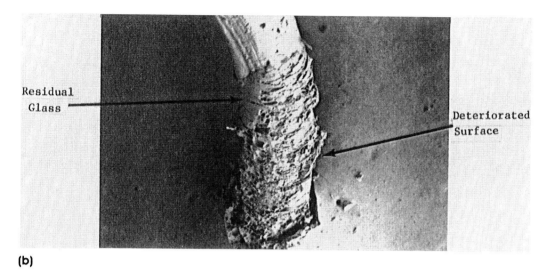

Residual Glass

Deteriorated Surface

(b)

board. This type of failure is enhanced by the soldering technique. Typical methods such as wave soldering, vapor-phase reflow, or infrared reflow subject the component to temperatures around 250°C and higher. This temperature exceeds the glass transition temperature of typical epoxies in the board by almost 100°C, resulting in dramatic increases in the out-of-plane expansions [34]. The damage to the copper plating during this processing cycle is extremely severe [35], and manufacturers are experimenting with strengthening mechanisms, such as an additional layer of nickel at the interface, to overcome this problem.

Fatigue failure problems are also observed in fiber-reinforced composite boards. These boards are typically constructed by reinforcing epoxy or polyimide matrices with woven fabrics made of glass or Kevlar fibers. The large temperature excursion during the soldering process causes damage to the matrix material and generates residual stress fields during the subsequent cool-down. Thermomechanical and vibrational loading during the service life of the component serve to add to the cumulative damage of the material. Damage in these brittle heterogeneous materials takes the form of microcracking in the matrix, debonding at fiber-matrix interface, and sometimes gross delamination between layers of multilayer boards. Distributed microdamage can be statistically characterized through quantitative microscopy and leads to overall degradation of the residual elastic and strength properties of the material. Delamination can lead to macroscopic, localized damage sites, which can then propagate until catastrophic failure occurs in the board. The crucial issue in designing damage-tolerant boards is the optimization of the microarchitecture of the composite to minimize the possibility of a large crack propagating catastrophically through the material. Another concern in board material design is to minimize the out-of-plane CTE value, while optimizing the in-plane CTE, through proper control of the reinforcement architecture and fiber and matrix properties.

Thermomechanical loading generates significant shear stress in the attach between dies and the underlying substrate due to CTE mismatches. This failure mode can be readily

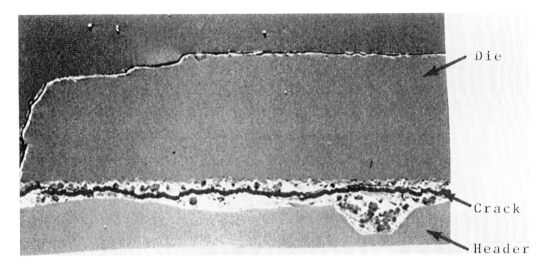

**Figure 8.18**  Fatigue crack in 5Sn-95Pb solder joint between a ceramic LCC and the copper substrate. (Courtesy of NASA-GSFC, UNISYS.)

addressed by the shear lag analysis outlined in section 8.2.1 for bonded assembly of two dissimilar materials. Figure 8.18 shows a fatigue crack that has developed in the 5Sn-95Pb solder attach between a ceramic die and a copper substrate in a power transistor device.

**Figure 8.19**  Thermal fatigue crack in the active area of a ceramic die. (Courtesy of NASA-GSFC, UNISYS.)

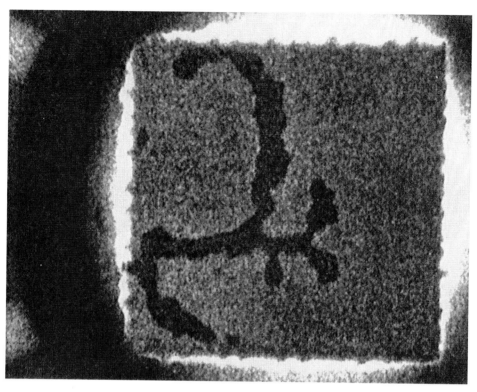

**Figure 8.20**   Void formation in a die attach as revealed by an X-ray radiograph. (Courtesy of NASA GSFC, UNISYS.)

Fatigue damage is also observed in the die itself. Figure 8.19 shows an example of a fatigue crack that has propagated through the active area of a ceramic die. Vertical die cracks often result from excessive voids in the die attach, as shown by X-ray radiography methods in Figure 8.20. The die typically consists of several different materials, such as silicon, silicon dioxide, and gallium arsenide. Most of these are brittle materials and fail by brittle fracture mechanisms [36] under thermomechanical residual and cyclic stresses from manufacturing and operational loads. Fatigue design of these components can, therefore, be successfully addressed by fracture mechanics tools, such as Paris' fatigue crack propagation law given in Eqn. (8.50) [37]. Of course, such an analysis would require fracture and crack propagation data that are almost nonexistent in the current literature for commonly used die materials.

## 8.3.2   Physical and Mechanical Properties of Solder Materials

The properties of common solders are discussed in this section because solder joint fatigue forms a large fraction of the total number of fatigue failures in surface-mounted components.

   In order to understand and predict the failure of solder joints, the physical and mechanical properties of tin-lead (SnPb) solders need to be unambiguously characterized. Much work has been done on the basic properties of SnPb solders and is available in the

**Table 8.2**  Physical Properties of Tin-Lead Bulk Solders

| Tin content (%) | Density (g/cm$^2$) | Electrical conductivity of copper (IACS) (%) | Thermal conductivity (BTU in./sec ft$^2$) | Coefficient of linear thermal expansion (°F × 10$^{-6}$) |
|---|---|---|---|---|
| 0 | 11.34 | 7.9 | 0.067 | 16.3 |
| 5 | 10.80 | 8.1 | 0.068 | 15.8 |
| 10 | 10.50 | 8.2 | 0.069 | 15.5 |
| 20 | 10.04 | 8.7 | 0.072 | 14.7 |
| 30 | 9.66 | 9.3 | 0.078 | 14.2 |
| 40 | 9.28 | 10.1 | 0.084 | 13.7 |
| 50 | 8.90 | 10.90 | 0.090 | 13.1 |
| 60 | 8.52 | 11.50 | 0.096 | 12.0 |
| 70 | 8.17 | 12.50 | — | 11.5 |

literature. Extensive review of the literature is presented in the paper cited earlier by Lau and Rice [26]. Properties of SnPb solders with a wide range of compositions are presented in Table 8.2. As expected, the electrical conductivity and thermal conductivity increase with Sn content, while the density and coefficient of linear expansion decrease with increased Sn content.

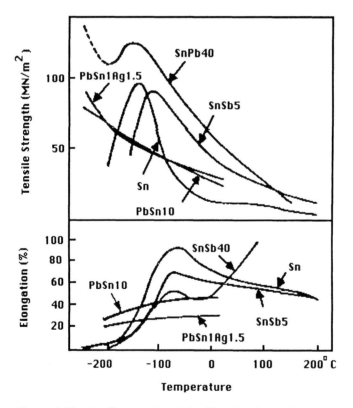

**Figure 8.21**  Tensile strength and ductility of various solders as a function of temperature. (From Ref. 52.)

Although the physical properties are generally well known, the mechanical properties of solder are scattered throughout the literature and are not usually consistent. Generally, the mechanical properties are strongly dependent on temperature, as shown in Figures 8.21 and 8.22 for various typical compositions.

When discussing the fatigue of solder joints in PCB design, it must be realized that solder exhibits a large amount of creep deformation that is uncharacteristic of other common engineering materials at normal operating temperatures. The difference is due to the fact that at room temperature solder is at about two thirds of its melting temperature, expressed on an absolute temperature scale. At such high relative temperatures the solder can be considered to be hot worked. Any deformation of the material, therefore, is accompanied by large time-dependent anelastic strains.

Motivated by Coffin's frequency-correction approach to such creep fatigue problems, various researchers have investigated the changes in mechanical properties of solder materials at different frequencies of cyclic loading and at different temperatures [38–41]. Perhaps the most comprehensive study of the fatigue properties of solder was conducted

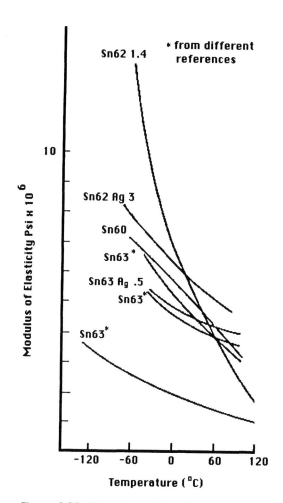

**Figure 8.22**   Young's modulus of various solder alloys as a function of temperature. (From Ref. 20.)

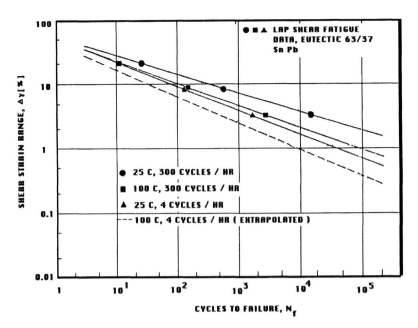

**Figure 8.23**  Wild's data from lap shear fatigue tests on eutectic 63/67 solder.
(From Ref. 42.)

by Wild [42,43]. Figure 8.23 shows Wild's data from lap shear fatigue tests of eutectic
SnPb solder at various temperatures. He observed a straight-line relationship between the
total shear strain range and the cycles to failure on a logarithmic scale. Figure 8.24a
shows the dependence of Young's modulus in 60/40 SnPb solder on frequency and
temperature. These tests were conducted by the dynamic modulus test method for anelastic
materials [44]. Some of the PCB materials also exhibit similar anelastic variations in their
properties. For example, the modulus of FR-4 glass-epoxy is shown as a function of
temperature and loading frequency in Figure 8.24b. The stiffness drops sharply at the
glass transition temperature of the epoxy matrix (approximately 125°C). It is clear therefore
that at operating temperatures and loading conditions, anelastic materials like solder
material can exhibit extremely complex and nonlinear thermoviscoplastic behavior. Other
recent data on anelastic behavior of solders may be found in Refs. 45 and 46. Tribula
and Morris [45] present a very comprehensive ensemble of creep constitutive test data
for near-eutectic solder materials. They illustrate the effects of grain structure and of
adding different alloying elements in small quantities. They concluded that small amounts
of indium and/or cadmium (approximately 2% by weight) can significantly improve the
creep properties of near-eutectic solder material.

   These characteristics have important consequences for PCB design. In actual use,
the dwell times at the extremes of the functional temperature cycles are often long enough
to allow almost complete stress relaxation. As a result, all cyclic strains due to the thermal
expansion mismatch of chip carrier and substrate are converted to inelastic (permanent)
strains. Hence, on each cycle the fatigue damage is maximized. In situations in which
the cycle frequency is too large to allow complete stress relaxation, viscoplastic analyses
may be required to obtain an accurate estimate of the ratio of anelastic, plastic, and elastic

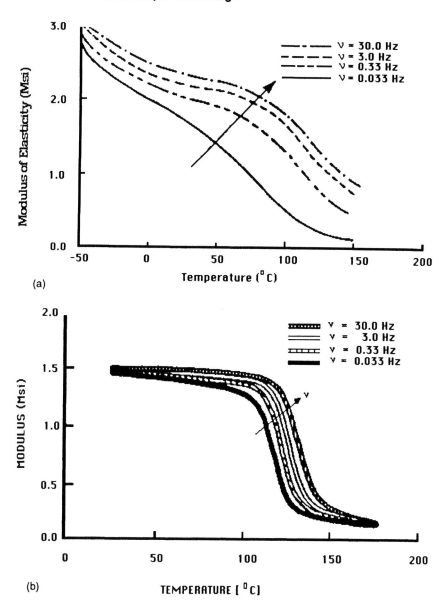

**Figure 8.24** (a) Dependence of 60/40 SnPb solder properties on frequency and temperature. (b) The modulus of FR-4 glass-epoxy as a function of temperature. (From Ref. 44.)

strains. The fatigue analysis may then be conducted according to the methods outlined earlier under High-Temperature Creep Fatigue.

Experimental tests or qualification studies that do not allow complete stress relaxation, due to different temperature or frequency conditions, result in different solder fatigue life assessments. Care must be exercised in comparing test results and in normalizing the data to account for frequency and temperature effects. Further, when planning accelerated

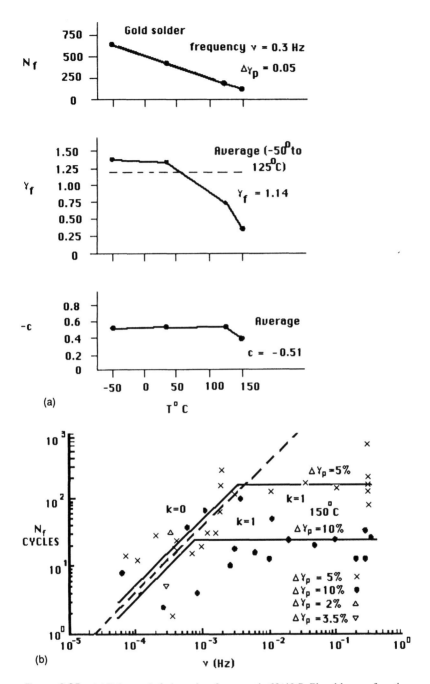

**Figure 8.25** (a) Solomon's fatigue data for eutectic 60/40 SnPb solder as a function of temperature. (From Ref. 12.) (b) Solomon's fatigue data for eutectic SnPb solder as a function of frequency. (From Ref. 12.)

testing, temperature extremes must be carefully chosen to avoid introducing nonrepresentative material behavior such as softening of the substrate epoxy.

Good fatigue data for solder in the open literature are scarce. Different investigators have used different test methods and test parameters, recorded different variables, and interpreted their answers in different (sometimes erroneous) ways. The data are confusing and conflicting at best and misleading and incorrect at worst. It is extremely important to realize that the fatigue behavior of most materials is different in axial and shear modes. Consequently, the experimental data must be chosen carefully to model the actual loading situation encountered in practice. In multiaxial situations, multiaxial fatigue analyses will be required. Some of the most useful information has been developed by Solomon [12, 38], who mechanically strained solder joints under different conditions of temperature and frequency. Most of his experiments were on lap shear specimens, and he presented a relationship between the plastic strain range (unlike Wild [42], who used the total strain range) and cycles to failure, in accordance with the classical Coffin-Manson law for low-cycle fatigue. Figures 8.11a and 8.25a and b show his data for eutectic tin-lead solder, showing that the number of mean cycles to failure for a given shear strain range increases with frequency and decreases with increasing temperature. Typical operating temperatures would fall within his temperature extremes of $-50°C$ and $150°C$, and his lowest test frequencies of $10^{-4}$ Hz are adequate for normal functional frequencies of about 4 cycles per hour. Other comprehensive fatigue studies may be found in Wild [42].

Other attempts at generating Coffin-type frequency-modified fatigue data for solders have met with limited success. The best results have come from corrections to $c$, the fatigue ductility exponent, and only minor corrections to $\epsilon_f$, the fatigue ductility coefficient. Engelmaier [18] determined from Wild's experimental data for 63/37 and 60/40 SnPb solder [42] that the fatigue ductility exponent can be represented by

$$c = -0.442 - 6(10^{-4})T_s + 1.74(10^{-2}) \ln\left(1 + \frac{360}{t_D}\right) \qquad (8.57)$$

where $T_s$ is the mean cyclic solder joint temperature (°C), $t_D$ is the half-cycle dwell time (minutes), and the fatigue ductility coefficient is

$$2\epsilon_f \simeq 0.65 \qquad (8.58)$$

Besides temperature, loading frequency, and dwell time, there are other factors that drastically affect solder fatigue life. Impurities or contaminants can drastically reduce fatigue life. As an example, 6% gold added to a eutectic PbSn solder has been shown to reduce fatigue life by a factor of 4 due to embrittlement from the formation of AuSn intermetallics. Such phenomena are observed when gold from plating diffuses (termed "leaching") into the solder joint. Solder grain size is another parameter that can affect life. As the grain size becomes smaller, plastic strains decrease due to dislocation pile-ups at grain boundaries and anelastic strains decrease due to reduced grain boundary sliding and intergranular cracking. As a result, fatigue life improves with grain refinement. Surface finish, notches, and hostile chemical environments can also limit fatigue life because of local stress concentrations and stress-assisted corrosion.

## 8.3.3 Thermal Cycling Tests

Various testing methods are used to generate fatigue failure data for PCBs. Temperature cycling in environmental chambers produces first-order reliability results quickly and with little effort. It should be clearly understood, however, that this method does not simulate the actual temperature distributions that components and interconnections in a PCB see in service. Some of the better-known specifications for temperature cycling and thermal shock tests are the military specifications MIL-STD-883, Methods 1010 and 1011. These methods define a thermal cycle or shock as the stress a device undergoes when it is temperature cycled in an environmental chamber or when it is alternately switched from a bath of hot liquid to a bath of cold liquid. These methods specify the temperature extremes for cycling and shock tests under several conditions: (1) condition A is between +100 and 0°C; (2) condition B, which was the most commonly adopted test until recently, is between +125 and −55°C; (3) condition C, which is currently most common, is between +65 and 100°C; and (4) condition F is between +22 and −195°C. For example, during screening tests, hybrid components are typically subjected to 10 temperature cycles (Method 1010) under condition C or 15 shock cycles (Method 1011) under condition A. On the other hand, qualification tests under MIL-STD-1772 require 100 temperature cycles (Method 1010) under condition C *and* 15 thermal shocks (Method 1011) under condition A.

These test conditions can be used in simple analytical models to predict fatigue failure. As an example, consider the solder joint of a leadless chip carrier. Because of the differences in expansion between the chip carrier and the substrate in a PCB, the solder joint is subjected to a cyclic strain. A first approximation of the cyclic strain can be obtained by considering only the deformations in the plane of the PCB. Figure 8.26a is a schematic of the chip carrier mounted on the PCB substrate with a solder joint of rectangular cross section and height $h$.

The shear strain in the deformed solder joint due to the differential expansion $\Delta L$, as shown in Figure 8.26b, is

$$\gamma = \frac{\Delta L}{2h} \tag{8.59}$$

The factor of 2 in the denominator accounts for the fact that the expansion mismatch is shared equally by the solder joint at each end. Assuming that the chip carrier and the PCB are much stiffer than the solder joint and ignoring bending deformations (i.e., considering only in-plane displacements),

$$\Delta L = L[(\alpha_s - \alpha_{cc})T_{max} - (\alpha_s - \alpha_{cc})T_{min}] = L \, \Delta\alpha \, \Delta T \tag{8.60}$$

where $\Delta\alpha \,(= \alpha_s - \alpha_{cc})$ is the difference between the thermal coefficients of the PCB substrate and the chip carrier and $\Delta T \,(= T_{max} - T_{min})$ is the difference between the maximum and minimum temperatures in the cycle. Note that there is an underlying assumption that the dwell times at $T_{max}$ and $T_{min}$ are sufficient for complete stress relaxation to occur in the solder joint. Thus

$$\gamma = \frac{L \, \Delta\alpha \, \Delta T}{2h} \tag{8.61}$$

Assuming that the dwell times at $T_{min}$ and at $T_{max}$ are sufficient to allow complete stress relaxation in the solder during each cycle, the mean stress effects can be ignored (experimental evidence suggests that mean strains do not affect fatigue life). Further, ignoring elastic strains in low-cycle fatigue and assuming the anelastic and plastic strains to be equivalent in the context of fatigue damage (i.e., ignoring the frequency modifications to the Coffin-Manson relationship) as a first-order approximation,

$$\frac{\Delta\gamma_P}{2} = \gamma_f(2N_f)^c \tag{8.62}$$

Rearranging Eq. (8.62) and substituting Eq. (8.59),

$$N_f = \frac{1}{2}\left(\frac{\Delta\gamma_P}{2\gamma_f}\right)^{1/c} = \frac{1}{2}\left(\frac{L\,\Delta\alpha\,\Delta T}{4h\gamma_f}\right)^{1/c} \tag{8.63}$$

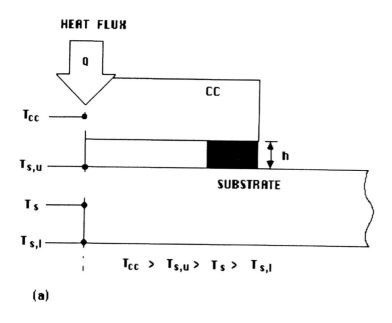

(a)

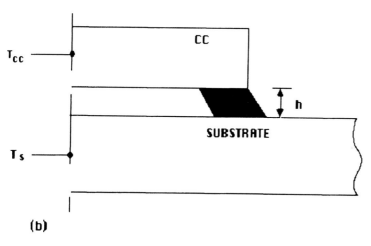

(b)

**Figure 8.26** Shear strain in a deformed solder joint due to differential expansion $\Delta l$. (a) Schematic of a chip carrier mounted on a PCB with a solder joint of rectangular cross section and height $h$. (b) Pure shear deformation (out-of-plane deformations are ignored), $\alpha_s\,\Delta T_s > \alpha_{cc}\,\Delta T_{cc}$.

At 50°C the fatigue ductility exponent, $c$, for eutectic SnPb solder is approximately $-0.5$. Thus

$$N_f = \frac{1}{2}\left(\frac{4h\gamma_f}{L\,\Delta\alpha\,\Delta T}\right)^2 \tag{8.64}$$

It is obvious that fatigue life increases as the square of the height $h$ and is proportional to the solder ductility coefficient $\gamma_f$. However, excessive solder height ($>0.005$ in.) can lead to increased processing difficulties and quality problems. Excessive solder volume can result in bridging between adjacent pads, formation of solder balls, and problems in maintaining uniformity of height. Clearly, therefore, there are practical limitations to the maximum achievable solder height. Similarly, the decrease in the fatigue life is proportional to the square of the chip size $L$, thermal expansion mismatch $\Delta\alpha$, or temperature differential $\Delta T$. This is why solder joint problems become so critical as chip sizes increase. This equation also emphasizes the need to minimize the mismatch between the thermal coefficients of expansion of the board and chip carrier.

### 8.3.4  Power Cycling Tests

Power cycling or functional cycling is the actual thermal cycle that the components see in service. It consists of a power-on transient in which the components go from room temperature, $T_0$, to some steady-state temperature in a matter of minutes. This steady-state condition then lasts for several hours or longer before a power-off (cooling) transient occurs. The power-off condition then lasts for several hours. The transient conditions are difficult to determine either experimentally or analytically.

For comparative purposes, let us consider again the LCC solder joint analyzed in Section 8.3.3. Figure 8.27 shows a schematic of an assumed cyclic temperature history for a chip carrier and substrate. The thermally induced strains in the solder joint due to power cycling can be illustrated by ignoring, for the moment, the bending due to thermal deformations (i.e., considering only the in-plane deformations). Under these simplifications, the thermal deformations are limited only to normal strain, $\epsilon$, multiplied by the length $L$ as seen in Figure 8.28. This is true for the substrate as well as the chip carrier. The total differential expansion at any time $t$ is

$$L\epsilon_{cc}(t) - L\epsilon_s(t) = L\,\Delta\epsilon \tag{8.65}$$

where subscripts cc and s refer to the chip carrier and substrate, respectively. Thus the thermal strain in the chip carrier and the substrate can be written as

$$\epsilon_i(t) = \alpha_i[T_i(t) - T_0] = (\alpha\,\Delta T)_i, \qquad i = 1,2 \tag{8.66}$$

where $T_0$ is the power-off temperature for the $i$th material.

Under steady-state power-on conditions the temperature $T_i$ will be either the maximum chip carrier temperature $T_{cc}$ or the maximum substrate temperature $T_s$. The maximum differential expansion will then be

$$
\begin{aligned}
L\,\Delta\epsilon &= L\,\Delta\alpha\,\Delta T\\
&= L[\alpha_{cc}(T_{cc} - T_0) - \alpha_s(T_s - T_0)]\\
&= L[(\alpha_{cc} - \alpha_s)(T_{cc} - T_0) + \alpha_s(T_{cc} - T_s)]
\end{aligned}
\tag{8.67}
$$

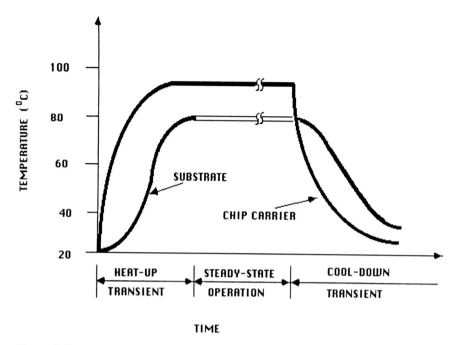

**Figure 8.27**   An assumed cyclic temperature history for a chip carrier and substrate. (From Ref. 47.)

In thermal cycling, thermal strains are less severe because $T_{cc} = T_s$ and the last term in the above equation vanishes. For zero differential expansion in thermal cycling, it can be seen that the coefficient of thermal expansion of the substrate needs to be matched to the chip carrier. Such a substrate is known as a CTE-matched substrate and can be produced by using suitably designed fiber-reinforced composite substrates.

In power cycling or functional cycling the temperatures of the substrate and the chip carrier are not necessarily equal. To get zero differential expansion the thermal coefficient of expansion of the substrate must be chosen so that

$$\alpha_s = \frac{\alpha_{cc}(T_{cc} - T_0)}{T_s - T_0} \tag{8.68}$$

Such a substrate is known as a CTE-tailored substrate [47].

Representative cyclic expansion mismatches for ceramic chip carriers on epoxy-glass, CTE-matched and CTE-tailored expansion substrates are shown in Figure 8.28 for the temperature ranges shown in Figure 8.27. As seen in Figure 8.28, CTE matching does not lead to the minimum strains in the solder joints during actual operation.

Figure 8.28 also illustrates that CTE tailoring provides the minimum strains and maximum reliability for real devices with power dissipation. Typically, $(\alpha_{cc} - \alpha_s)$ should equal 1 to 3 ppm/°C. The practical difficulty encountered in CTE tailoring is that not all devices on the substrate have the same CTE. Further, tailoring for steady-state power-on conditions usually implies a CTE mismatch during transients or during power-off conditions.

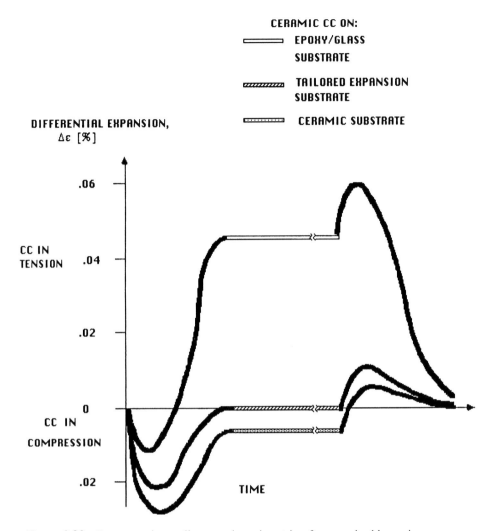

**Figure 8.28**   Representative cyclic expansion mismatches for ceramic chip carriers on epoxy-glass, CTE-matched and CTE-tailored expansion substrates for the temperature history from Figure 8.27. (From Ref. 47.)

The cyclic strains induced in power cycling are actually quite complex. The various components include:

1. In-plane steady-state strains due to the expansion mismatch
2. In-plane transient strains due to the expansion mismatch
3. Strains from warpage due to the expansion mismatch (bimetallic strip)
4. Strains from warpage due to steady-state power dissipation temperature gradients

Figure 8.29a and b show schematic representations of the different components of cyclic strain. Whereas the steady-state in-plane strains are relatively easy to determine, as previously demonstrated, the transient and warpage strains are not easy to determine with analytical or experimental techniques.

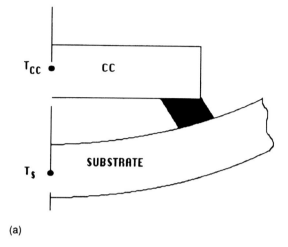

(a)

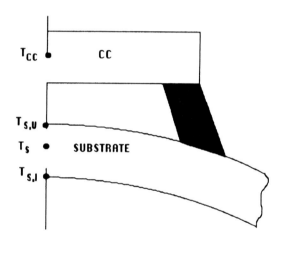

(b)

**Figure 8.29** Schematic of the temperature and strain components due to out-of-plane deformations during power cycling. (a) Warpage due to bimetallic strip effects, $\alpha_s \, \Delta T_s > \alpha_{cc} \, \Delta T_{cc}$. (b) Warpage due to temperature gradient through the thickness of the substrate, $\alpha_s \, \Delta T_{s,u} > \alpha_s \, \Delta T_{s,l}$.

Bimetallic strip–type warpage occurs because the substrate typically expands more than the chip carrier, and it has been extensively studied by Hall et al. [48,49] using strain gauges and holographic interferometry. This warpage of the substrate is concave as viewed from the chip carrier side.

During powered operation of the chip, a temperature gradient exists through the thickness of the substrate because the chip acts as a heat source. The temperature on the surface with the chip carrier is higher than on the opposite surface of the substrate. This thermal gradient through the substrate causes a convex warpage as viewed from the chip carrier side. During powered cycling this warpage approximately offsets the bimetallic-type warpage because they produce curvatures of opposite signs. Care must be exercised when testing only with thermal cycling because the warpage due to the thermal gradient is not present to offset the bimetallic warpage.

In most cases the steady-state expansion mismatch strains are the dominant cause of cyclic fatigue damage because the transients are of such short duration that full stress relaxation cannot occur. On the other hand, it should be noted that failures are typically first noticed during transient periods due to higher elastic stress concentrations. It is the out-of-plane warpage that breaks the continuity across the joint.

Assuming an ideal joint geometry with uniform shear strain distribution and neglecting warpage and transient strains, the shear strain for a solder joint of height $h$ can be expressed as

$$\Delta\gamma = \frac{L[(\alpha_s - \alpha_{cc})(T_{cc} - T_0) - \alpha_s(T_{cc} - T_s)]}{2h} \tag{8.69}$$

where $L$ is the chip carrier length and $h$ the height of the solder joint between the attachment pads. The equation differs from the strain expression developed for thermal cycling because the chip carrier and substrate are not at the same temperature during power cycling.

The underlying assumptions in this equation are all interrelated and somewhat nonconservative. The idealized joint is assumed to be cylindrical with fillets at both ends to reduce the strain concentrations and the effects of intermetallic compounds. In reality, the joints typically have a larger cross section in the center, subjecting the ends of the joint near the bonded surfaces to higher strains. Failures typically occur at or near these bonded surfaces due to the higher strains and brittle intermetallic compounds such as $Cu_6Sn_5$, $Cu_3Sn$, and $AuSn_4$. With large solder joint heights, the joint is not subjected to pure shear. Tensile stresses due to warpage also play an important role. The effect of castellations is not completely clear at this point, but some preliminary experimental evidence indicates that they may hurt the performance.

To account for nonideal factors such as cyclic warpage, cyclic transients, nonideal joint geometry, brittle intermetallics, and inaccuracies and uncertainties in the parameters $\alpha$, $h$, and $T$, an empirical factor, $K$, has been introduced by Engelmaier [50] as follows:

$$\Delta\gamma = \frac{KL[(\alpha_s - \alpha_{cc})(T_{cc} - T_0) - \alpha_s(T_{cc} - T_s)]}{2h} \tag{8.70}$$

This empirical factor $K$ is typically in the range $1.1 < K < 1.6$ and can be thought of as a measure of the merit of the particular attachment technology. Depending on the attachment method, steady-state thermal expansion mismatch can underestimate the effective strain in a solder joint by 10 to 60% [50].

Using the Manson-Coffin relation

$$\frac{\Delta\gamma_p}{2} = \gamma_f(2N_f)^c \tag{8.71}$$

and making the same set of assumptions as in thermal cycling,

$$N_f = \frac{1}{2}\left\{\frac{KL[(\alpha_s - \alpha_{cc})(T_{cc} - T_0) - \alpha_s(T_{cc} - T_s)]}{4h\gamma_f}\right\}^{1/c} \tag{8.72}$$

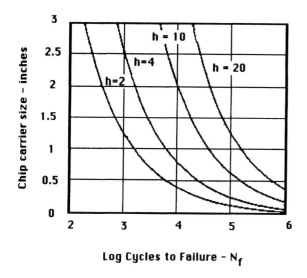

**Figure 8.30**   Fatigue life as a function of chip carrier size for various solder joint heights.

The influence of the various parameters, with the exception of the empirical factor $K$, on the cycles to failure is shown in Figures 8.30 to 8.32 for a typical chip and substrate. Figure 8.30 clearly shows the increased fatigue life with increased solder joint height. Figure 8.31 illustrates that even CTE-matched ceramic substrates can experience fatigue failures due to thermal gradients. The need to use CTE-tailored substrates rather than CTE-matched substrates is further illustrated in Figure 8.32, where the fatigue life increases asymptotically as the CTE approaches the tailored value.

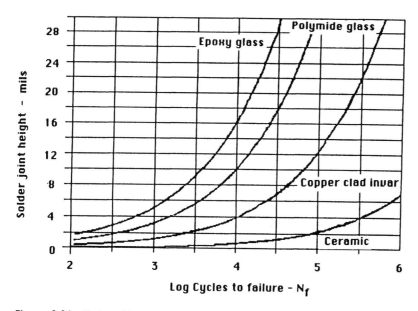

**Figure 8.31**   Fatigue life as a function of joint height for various board materials.

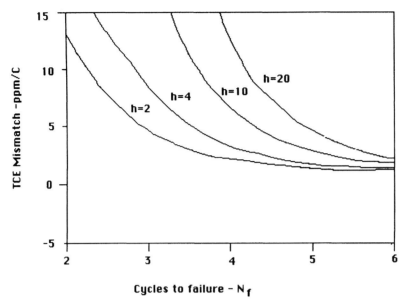

**Figure 8.32**   Fatigue life as a function of CTE mismatch for various solder heights.

### 8.3.5   Experimental Reliability Testing
## Testing Concerns

The concerns in experimental reliability testing of solder joint connections reflect the variabilities and uncertainties in the mechanical properties of the solder and substrate and the stresses developed in the connection joint. There are strengths and weaknesses in the various testing approaches, and these must be understood. Inappropriate cyclic testing can lead to erroneous and extraneous failure modes, misleading results, and false conclusions.

Testing parameters require careful consideration. If the cyclic stress levels are too high, failures can occur in the metallization pads or solder joints which are stress-induced and not representative of low-cycle strain–induced fatigue failures. Test frequencies that are too high can produce results unrepresentative of functional service because the dwell times are too short for stress relaxation. This can lead to a disproportional influence of second-order effects such as out-of-plane bending and warpage. Test temperature extremes that are too high or too low can result in unrepresentative material behavior such as modulus changes that will not occur in actual service conditions.

The test method itself requires careful implementation. Incorrect test methods, such as thermal cycling using CTE-matched components, will not produce fatigue failures within reasonable test durations because there is not enough thermal expansion mismatch under these idealized laboratory conditions of zero thermal gradients. Power cycling is therefore a superior and more meaningful test method.

Another concern is inappropriate test conclusions. Care must be exercised to ensure that test models are adequately representative of the actual product. Care must also be exercised when using thermal cycling or mechanical cycling, since the out-of-plane bending and warpage is different from that encountered in powered functional cycling.

## Testing Methods

When designing a reliability assessment experiment the most important parameter is the test method itself. Basically, there are three different methods for applying cyclic strains to solder joints: (1) powered functional cycling, (2) thermal cycling, and (3) mechanical cycling. The methods have their respective advantages and disadvantages.

Powered functional cycling most closely matches the actual service condition. The same support structure should be modeled as closely as possible. The same environmental conditions, such as cooling air flow, should be duplicated. With such care this approach will yield highly accurate reliability information, but this method is the most difficult, expensive, and time-consuming of the three. The actual service conditions can be varied for test acceleration purposes. If this is done, the frequency is the best parameter to vary, but only with great care and thought about the resulting influence.

Temperature cycling in an environment chamber is an excellent test method for first-order reliability assessment. The method is quick and requires little effort. The method is not good for products with closely matched thermal coefficients of expansion because second-order effects, such as out-of-plane warpage (which are the dominant failure modes in such boards), are not duplicated in this test method. Thus it is important to guard against nonrepresentative failure modes and against test results that are unduly influenced by nonrepresentative cyclic behavior.

Mechanical cycling with some sort of bending apparatus is the quickest method for gross comparisons between connection technologies. All the caveats listed under temperature cycling apply here. The most significant limitation of mechanically induced inelastic strains is that the ratio of plastic to anelastic strains tends to be higher than in thermal loading.

## Acceleration Methods

The ability to accelerate the wear-out reliability test is of very real importance, but extreme care must be exercised with full knowledge of the ramifications. One of the easiest and obvious ways to accelerate the test is by increasing the cyclic frequency. However, in the case of viscoplastic materials at high temperature, e.g., solder, due to the need for sufficient dwell times at the cyclic extremes and the typical time spent in transition between these extremes, powered functional and temperature cycling tests are normally limited to frequencies of no more than 100 cycles per day. Mechanical cycling probably can be raised to about 1000 cycles per day without significant deviations from the solder fatigue model [12,50].

Another way to accelerate the tests is to increase the cyclic strain range. This can be done by the obvious method of raising the upper temperature and/or lowering the lower temperature in temperature cycling. Care must be exercised to avoid significantly changing the material properties of the constituents by doing this. Further, it is important to ensure that the failure mechanism itself does not change from one mode to another due to the acceleration. Some not so obvious ways in solder joints, for example, are to increase the CTE mismatch, increase the chip size, or decrease the solder joint height. All three methods increase the imposed cyclic strain and thus accelerate the test.

Test acceleration is a complex task and the specific method is dependent on the parameters in the failure model, the nature of the failure mechanism, and the properties of the materials involved. In all cases, a valid failure model should first be formulated, based on the physics and mechanics of the failure mode, before any acceleration schemes can be examined [51].

## EXERCISES

1. During wave-soldering operations, the temperature difference between the top and bottom surfaces of a PCB may reach up to 100°C. Assume the thickness of the board to be 0.062 inches.
    (i) Compute the curvature of the board (assume a linear temp gradient through the thickness and ignore variations of board properties with temperature).
    (ii) What harmful effects do you anticipate due to this curvature?
    (iii) Compute the bending stresses on the top and bottom surfaces if the board is constrained to remain flat with clamps.
    (iv) what moment should the clamps exert on the board?

    Glass-polyimide properties in the plane of the board:
    $\alpha = 19 \, \mu$ strain/°C
    $E = 14 \, \text{GPa}$
    $\nu = .16$

2. A planar-pack device, measuring $0.5'' \times 0.5''$ is attached to a glass-polyimide PCB with a 0.008″ thick layer of conductive epoxy. According to MIL HDBK 217 this assembly is required to withstand temperature cycling between −55°C and 125°C. What is the maximum shear stress generated in the adhesive during such cycling? Assume the joint to be stress-free at room temperature (20°C) and assume all materials to behave linear elastic. Properties:

|           | E        | $\nu$ | $\alpha$             |
|-----------|----------|-------|----------------------|
| Board     | 14 GPa   | .16   | 19 $\mu$ strain/°C   |
| Epoxy     | 3.45 GPa | .37   | 70 $\mu$ strain/°C   |
| Component | 138 GPa  | 0.3   | 5.9 $\mu$ strain/°C  |

## REFERENCES

1. Timoshenko, S. and Goodier, J., *Theory of Elasticity*, 3rd ed., McGraw-Hill, New York, 1970.
2. Jones, R. M., *Mechanics of Composite Materials*, McGraw-Hill, New York, 1975.
3. Royce, B. S., "Differential Thermal Expansion in Microelectronic Systems," *IEEE Trans. Components, Hybrids, Manuf. Technol. CHMT-11*(4): 454–463, Dec (1988).
4. Cook, R. D., Malkus, D. S. and Plesha, M. E., *Concepts and Applications of Finite Element Analysis*, 3rd ed., Wiley, New York, 1989.
5. Langhaar, H. L., *Energy Methods in Applied Mechanics*, Wiley, New York, 1962.
6. Markstein, H. W., "Designing Electronics for High Vibration and Shock," *Electron. Packag. Prod.*, April 1987, pp. 40–43.
7. Little, R. E. and Jebe, E. H., *Statistical Design of Fatigue Experiments*, Wiley, New York, 1975.
8. Broek, D., *Elementary Engineering Fracture Mechanics*, Sijthoff & Noordhoff, Alphen aan den Rijn, The Netherlands, 1978.
9. Lawn, B. R. and Wilshaw, T. R., *Fracture of Brittle Solids*, Cambridge University Press, 1975.
10. Coffin, L. F., Jr., "A Study of the Effects of Cyclic Thermal Stresses on a Ductile Metal," *Trans. ASME 76*:931–950 (1954).

11. Manson, S. S., "Fatigue: A Complex Subject—Some Simple Approximations," *Exp. Mech.* 5:193–226 (1965).

12. Solomon, H. D., "Fatigue of 60/40 Solder," *IEEE Trans. Components, Hybrids, Manuf. Technol.* CHMT-9(4):423–432 (1986).

13. Atluri, S., "Path Independent Integrals in Finite Elasticity and Inelasticity with Body Forces, Inertia and Arbitrary Crack-Face Conditions," *Eng. Fract. Mech.* 16(3):341–364 (1982).

14. Coffin, L. F., Jr., "Fatigue at High Temperatures," ASTM STP 520, American Society for Testing and Materials, Philadelphia, 1973.

15. Halford, G. R., Hirschberg, M. H. and Manson, S. S., "Temperature Effects on the Strain-Range Partitioning Approach for Creep Fatigue Analysis," ASTM STP 520, American Society for Testing and Materials, Philadelphia, 1973, pp. 658–667.

16. Merrell, L. J., "A Methodology for Analysis of Fatigue in Solder Joints," Sandia Report SC-RR-71-0326, August 1971.

17. Hagge, J. C., "Predicting Fatigue Life of Leadless Chip Carriers Using Manson-Coffin Equations," *International Electronics Packaging Conference*, November 1982, pp. 200–208.

18. Engelmaier, W., "Fatigue Life of Leadless Chip Carrier Solder Joints During Power Cycling," *IEEE Trans. Components, Hybrids, Manuf. Technol.* CHMT-6(3):232–237, Sept. (1983).

19. Waller, D. L., Fox, L. R. and Hannemann, R. J., "Analysis of Surface Mount Thermal and Thermal Stress Performance," *IEEE Trans. Components, Hybrids, Manuf. Technol.* CHMT-6(3):257–266, Sept. (1983).

20. Clatterbaugh, C. V. and Charles, H. K., Jr., "Thermomechanical Behavior of Soldered Interconnects for Surface Mounting: A Comparison of Theory and Experiment," *Proc., IEEE Electronics Components Conference*, 1985, pp. 60–72.

21. Sherry, W. M., Erich, J. S., Bartschat, M. K. and Prinz, F. B., "Analytical and Experimental Analysis of LCCC Solder Joint Fatigue Life," IEEE, CHMT, Vol. 8, No. 4, Dec 1985, pp. 417–426.

22. Taylor, J. R. and Pedder, D. J., "Joint Strength and Thermal Fatigue in Chip Carrier Assembly," *Proc. ISHM Conference*, 1982, pp. 209–214.

23. Smeby, J. E., "Solder Joint Behavior in HCC/PWB Interconnections," *IEEE Trans. Components, Hybrids, Manuf. Technol.* CHMT-8(3): (Sept. 1985), 391–396.

24. Desaulnier, W. E., LaFlamme, T. E., Ammerman, W. B. and Binder, M. C., "LCC Solder Joint Fatigue Analysis Procedure," *Proc. Institute of Environmental Sciences*, 1986, pp. 6–19.

25. Barker, D., Vodzak, J., Dasgupta, A. and Pecht, M., "Solder Joint Fatigue Under Combined Thermal and Vibrational Loads: A Generalized Strain vs. Life Approach," *ASME J. Electron. Packag.*, Vol. 112, No. 2, pp. 129–134, June 1990.

26. Lau, J. H. and Rice, D. W., "Solder Joint Fatigue in Surface Mount Technology: State of the Art," *Solid State Technol.* 28(10):91–104 (1985).

27. Engelmaier, W., "Surface Mount Attachment Reliability of Clip-Leaded Ceramic Chip Carriers on Fr-4 Circuit Boards," *IEPS J.* 9(4): Jan, 1988, pp. 3–11.

28. Hu, J.-M., Pecht, M. and Dasgupta, A., "A Probabilistic Approach for Predicting Thermal Fatigue Life of Wire Bonding in Microelectronics," *ASME Trans. Electron. Packag.*, submitted (1990).

29. Ching, T. B. and Schroen, W. H., "Bond Pad Structure Reliability," *Proc., IEEE/IRPS*, 1988, pp. 64–70.

30. Koch, T., Richling, W., Whitlock, and Hall, D., "A Bond Failure Mechanism," *Proc., IEEE/IRPS*, 1986, pp. 55–60.

31. Shukla, R. and Deo, J. S., "Reliability Hazards of Silver-Aluminum Substrate Bonds in MOS Devices," *Proc., IEEE/IRPS*, 1982, pp. 122–127.

32. Koyma, H., Shiozaki, H., Okumura, I., Mizogashira, S., Higuchi, H. and Ajiki, T., "A New Bond Failure Wire Crater in Surface Mount Devices," *Proc., IEEE/IRPS*, 1988, pp. 59–63.

33. Hannula, S.-P., Wanagel, J. and Li, C.-Y., "Evaluation of Mechanical Properties of Thin

Wires for Electrical Interconnections," *Proc., IEEE Electronic Components Conference*, 1983, pp. 181–188.

34. Boey, W. K. and Walker, R. J., "Wave Soldering of Surface Mount Components," presented at *IEPS Fifth Annual International Electronic Packaging Conference*, Orlando, Fla., October 1985, pp. 25–29.

35. Wright, E. A. and Wolverton, W. M., "The Effect of the Solder Reflow Method and Joint Design on the Thermal Fatigue Life of Leadless Chip Carrier Solder Joint," *Proc. IEEE Electronic Components Conference*, 1984, pp. 149–155.

36. Van Kessel, C. G. M., Gee, S. A. and Murphy, J. J., "The Quality of Die-Attachment and Its Relationship to Stresses and Vertical Die Cracking," *IEEE Trans., Components, Hybrids, Manufacturing Technology* 6(4):414–420 (1983).

37. Pecht, M., Dasgupta, A., et al., "Reliability Assessment of Advanced Technologies," Final Research Report submitted to Westinghouse Defense Electronics, CALCE Center for Electronic Packaging, University of Maryland, College Park, 1989.

38. Solomon, H. D., "The Influence of Hold Time and Fatigue Cycle Wave Shape on the Low-Cycle Fatigue of 60/40 Solder," *Proc. IEEE Electronic Components Conference*, 1988, pp. 7–12.

39. Lambert, R. G., "Some Aspects of Solder Joint Fatigue During Thermal Cycling," *Proc. Institute of Environmental Sciences*, 1987, pp. 67–72.

40. Norris, K. C. and Landzberg, A. H., "Reliability of Controlled Collapse Interconnections," *IBM J. Res. Dev. 13*(3): (1969).

41. Stone, D., Hannula, S.-P. and Li, C.-Y., "The Effects of Service and Material Variables on the Fatigue Behavior of Solder Joints During the Thermal Cycle," *Proc. IEEE Electronic Components Conference*, 1985, pp. 46–51.

42. Wild, R. N., "Some Fatigue Properties of Solders and Solder Joints," IBM Technical Report No. 73Z00421, January 1973.

43. Lake, J. K. and Wild, R. N., "Some Factors Affecting Leadless Chip Carrier Solder Joint Fatigue Life," IBM Technical Report No. 83A80132, 1983.

44. Engelmaier, W., "Is Present-Day Accelerated Cycling Adequate for Surface Mount Attachment Reliability Evaluation?" *IPC-TP-653, Proc. IPC, Institute for Interconnecting and Packaging Electronic Circuits*, Lincolnwood, Ill., September 1986.

45. Tribula, D. and Morris, J. W., Jr., "Creep in Shear of Experimental Solder Joints," *ASME J. Electronic Packaging 112*(2):87–93 (June 1990).

46. Frost, H. J., Howard, R. T., Lavery, P. R. and Lutender, S. D., "Creep and Tensile Behavior of Lead-Rich Lead-Tin Solder Alloys," *IEEE Trans. Components, Hybrids, Manufacturing Technology, CHMT-11*(4):371–379 (Dec 1988).

47. Engelmaier, W., "Effects of Power Cycling on Leadless Chip Carrier Mounting Reliability and Technology," *Electron. Packag. Prod.*, April 1983, pp. 58–63.

48. Hall, P. M., Dudderar, T. D. and Argyle, J. F., "Thermal Deformations Observed in Leadless Ceramic Chip Carriers Surface Mounted to Printed Wiring Boards," *IEEE Trans. Components, Hybrids, Manuf. Technol. CHMT-6*(3):544–557 (1983).

49. Hall, P. M., "Forces, Moments, and Displacements During Thermal Chamber Cycling of Leadless Ceramic Chip Carriers Soldered to Printed Boards," *IEEE Trans. Components, Hybrids, Manuf. Technol. CHMT-7*(4):314–327 (1984).

50. Engelmaier, W., "Functional Cycles and Surface Mounting Reliability," Surface Mount Technology, Technical Monograph Series 6984-02, ISHM, Silver Spring, Md., 1984, pp. 87–114.

51. Hu, T.-M., Pecht, M., Barker, D. and Dasgupta, A., "The Role of Failure Mechanism Identification in Accelerated Testing, *Proc., Annual Reliability and Maintainability Symposium*, January 1991.

52. Ainsworth, P. A., "The Formation of Soft Soldered Joints," *Met. Mater.*, November 1971, pp. 374–379.

# 9

# Design for Vibration and Shock

**Donald B. Barker**  *CALCE Center for Electronic Packaging, University of Maryland, College Park, Maryland*

## 9.1  INTRODUCTION

Fatigue failures are very common in electronic systems subjected to vibrations. These failures are usually in the form of broken wires, broken component leads, cracked castings, cracked welds, and loose screws. Surface mount components are more prone to fatigue failure than dual in-line packages (DIPs) and discrete components soldered in plated through holes because of the lack of the very compliant pin and lead wires. It is obvious that the majority of engineers designing electronic systems would benefit from a better understanding of the basic fundamentals of vibration. In many cases they do not take the time and spend the effort to analyze the electronic support structure to determine the critical dynamic load paths. Most of the engineering time and money goes into the electronic circuitry. This seems logical because the purpose of an electronic system is to work electrically. However, unless an electronic system is designed to function in its dynamic environment, there can be many structural problems that may require an extensive amount of redesign and retesting.

Electronic systems are actually subjected to three different phases of a dynamic or vibrational environment. The first phase is during manufacture and screening tests, the second is during transportation from the assembly plant to the customer, and the final phase is in the operational environment. Depending on the product, the severest vibrational environment may be any one of the three phases.

Even well-designed products may be built with defective components. To keep the reliability of the product up in the field, defects must be found before the product leaves the plant. Manufacturing screening tests are used for both commercial and military systems to determine and, if necessary, eliminate these defects [1,2]. The screening tests are adapted to specific products and typically consist of a random vibration with the power spectral density of the exposure varied with frequency (random vibrations and power spectral density are discussed in Section 9.2.5). Usually the product under test is positioned on a vibration table with the axis of motion perpendicular to the printed circuit boards. Additional tests with the axis of motion positioned orthogonally are also used for complex systems. Malfunctions in the operation of the product while it is exposed to random vibrations indicate a latent defect that can be corrected before shipment.

In addition to vibration, the manufacturing screening may include temperature cycling. The temperature range in the cycle is usually from $-55$ to $+125°C$. Typically, the product is exposed to the thermal cycling separately from the vibrational testing, but military specifications are beginning to consider simultaneous thermal and vibrational tests.

The shipping environment varies from product to product and even from shipment to shipment. Differences may be due to type of transport vehicle, placement within vehicle, vehicle operator, routing, road conditions, weather, drops and tosses during handling, etc. The American Society for Testing and Materials (ASTM) has characterized the shock and vibration exposures that may be encountered, and these are presented in Table 9.1 [3]. Drop height is related to product weight. A lighter product is likely to be dropped from a higher position. Assurance level is related to number of units shipped and their value. The vibrational environment is again specified in terms of a random vibration spectrum.

The product operational shock and vibration environment clearly depends on application. Commercial products for the home and office have an innocuous environment, but products for the factory floor and vehicles clearly have environments including shock, vibration, and temperature fluctuations. Military applications typically have the severest environments, and the specifications are defined in the procurement documentation (e.g., see Refs. 4–7).

Many vibration problems can be minimized by avoiding coincident resonances that magnify acceleration $G$ forces very rapidly. There are many fine texts on vibration, but very little is published that directly relates to the basic methods for analyzing today's sophisticated electronic hardware. This chapter attempts to acquaint the engineer with vibration and shock design procedures. It begins with a brief review of terminology and vibration of one-degree-of-freedom systems before applying the equations and concepts to selected electronic components.

**Table 9.1**   Shock Environments (from ASTM Std D41-69)

| Shipping weight lb. (kg) | Drop height, in. (mm), assurance level | | |
| --- | --- | --- | --- |
| | I | II | III |
| 0–20 (0–9.1) | 24 (610) | 15 (381) | 9 (229) |
| 20–40 (9.1–18.1) | 21 (533) | 13 (330) | 8 (203) |
| 40–60 (18.1–27.2) | 18 (457) | 12 (305) | 7 (178) |
| 60–80 (27.2–36.3) | 15 (381) | 10 (254) | 6 (152) |
| 80–100 (36.3–45.4) | 12 (305) | 9 (229) | 5 (127) |
| 100–200 (45.4–90.7) | 10 (254) | 7 (178) | 4 (102) |

## 9.1.1  Basic Definitions

In order to discuss vibration and shock loading, some basic definitions and terminology must be reviewed. Vibration in its simplest form is the oscillatory back-and-forth motion of some structure. If this back-and-forth motion is repeated with all the same characteristics, the motion is periodic. This motion can be quite complex, but as long as it is repeated it is periodic. If the motion never repeats, then the motion is random. Simple harmonic motion is the simplest form of periodic motion and is usually represented as a sine wave plot of displacement versus time as shown in Figure 9.1. The equation for the displacement is

$$x(t) = A \sin(\omega t) \tag{9.1}$$

where $A$ is the amplitude of displacement and $\omega$ is the circular frequency in radians per second (or other unit of time).

The period, $\tau$, of periodic motion is the time required before the motion repeats itself and is usually expressed in seconds. The reciprocal of period is frequency, $f$, and has units of cycles per second or hertz (Hz).

$$f = \frac{1}{\tau} \quad \text{(Hertz, cycles/sec)} \tag{9.2}$$

The relationship between frequency and circular frequency is best illustrated as in Figure 9.2. The motion of the mass on a spring can be represented in the complex plane by looking at a vector of length $Y_0$ rotating about the origin with a circular frequency of $\omega$ radians per second. The motion can be mathematically expressed by taking the imaginary part

$$y = Y_0 \sin(\omega t) \tag{9.3}$$

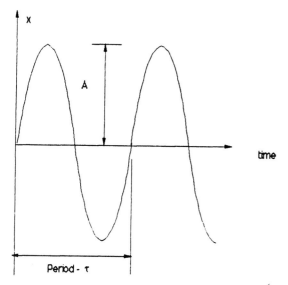

**Figure 9.1**  Simple harmonic motion, $x(t) = A \sin\left(\dfrac{2\pi}{\tau} t\right)$.

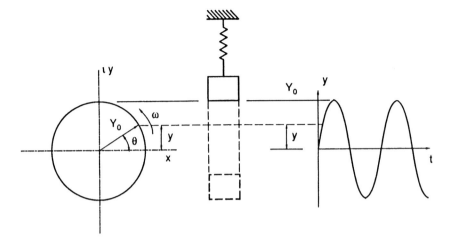

**Figure 9.2**  Simple harmonic motion as seen in the complex plane and $y$–$t$ plane.

where

$$\omega = 2\pi f = \frac{2\pi}{\tau} \tag{9.4}$$

The number of degrees of freedom in a system indicates the number of coordinates needed to describe the deformation. Some simple single-degree-of-freedom systems are

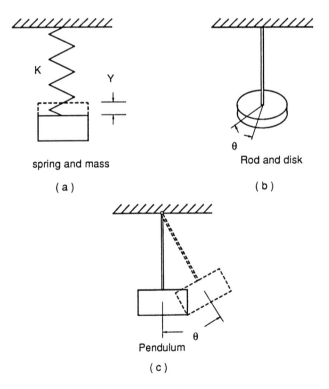

**Figure 9.3**  Single-degree-of-freedom systems.

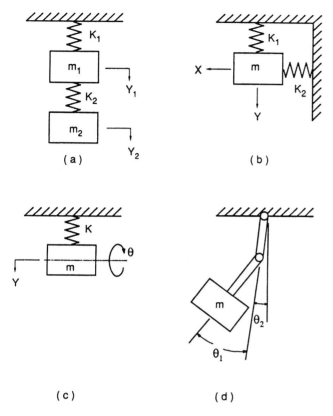

**Figure 9.4**  Two-degree-of-freedom systems.

shown in Figure 9.3. Each of these systems has only one possible type of motion and can be described with a single coordinate variable. Figure 9.4 shows some two-degree-of-freedom systems, for which two coordinate variables are required to describe the motion. The most general element or rigid body has six degrees of freedom—displacement and rotation about each coordinate axis—as shown in Figure 9.5.

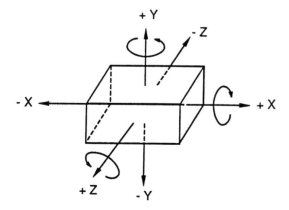

**Figure 9.5**  General element has six degrees of freedom.

**Figure 9.6**  A simply supported beam showing several degrees of freedom.

When a body is subjected to some excitation, there are unique shapes to which it
can deform. Each of these shapes is called a mode. Take, for example, a simply supported
beam as shown in Figure 9.6. A continuous beam has an infinite number of degrees of
freedom. The first harmonic mode, or simplest deformation shape the body can take, is
shown in Figure 9.7a. The next more complicated shape or second mode is shown in
Figure 9.7b. Also shown in Figure 9.7b is a node in the center of the beam. A node is
a point that always has zero deformation during the vibrational motion. Figure 9.8 shows
the first four modes of a circular membrane and the first three modes of a square plate
with the nodal lines shown as dashed lines.

## 9.2  ONE-DEGREE-OF-FREEDOM SYSTEM

The vibratory response of an actual mechanical system is often quite complex, with several
time-dependent motions occurring simultaneously. Fortunately, one motion usually dominates and is essentially independent of the other, much smaller motions. In these instances,

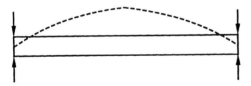

( a )  First harmonic mode

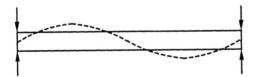

( b )  Second harmonic mode

**Figure 9.7**  First and second harmonic modes
for a simply supported beam.

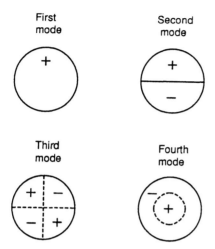

First four harmonic modes for a circular membrane

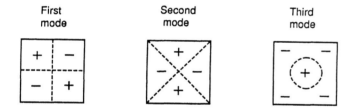

First three harmonic modes of a square plate

**Figure 9.8** First four harmonic modes of a circular plate and a square plate.

the mechanical system can be treated as a single-degree-of-freedom system without introducing significant error.

A single-degree-of-freedom system is represented in Figure 9.9 with a spring, mass, and damper affixed to a rigid support and the deformation measured by the single coordinate variable $x$. Care must be exercised to use proper units for the mass, $m$. Remember that weight, $W$, has units of force.

$$W = mg \tag{9.5a}$$

or, solving for mass,

$$m = \frac{W}{g} \tag{9.5b}$$

where $g$ is the gravitational constant, 32.2 ft/sec$^2$ or 9.8 m/sec$^2$.

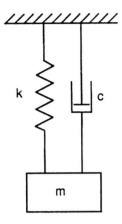

**Figure 9.9** Spring-mass-damper single-degree-of-freedom system.

Usually the spring force is taken to be directly proportional to spring deformation. The constant of proportionality is the spring constant, $k$, and has units of force/length. The damping coefficient, $c$, represents energy loss—for instance, due to friction. Normally, viscous damping is assumed, damping proportional to velocity, primarily for simplification of mathematics.

## 9.2.1 Undamped Spring-Mass System

Taking as a first example a simple undamped spring-mass system, the equation of motion can be written directly using the free body diagram as shown in Figure 9.10. From Newton's law of motion, force equals mass times acceleration, and noting that the spring force $F_s$ acts in the negative coordinate direction

$$-F_s = \frac{d^2x}{dt^2} = m\ddot{x} \tag{9.6}$$

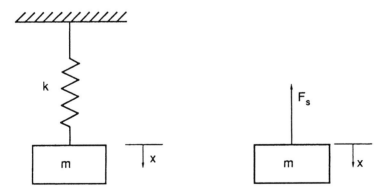

**Figure 9.10** Simple one-degree-of-freedom spring-mass system and free body diagram.

where $\dot{x}$ represents differentiation of the position variable $x$ with respect to time. Substituting for the spring force,

$$-kx = m\ddot{x} \tag{9.7}$$

or

$$m\ddot{x} + kx = 0$$
$$\ddot{x} + \left(\frac{k}{m}\right)x = 0 \tag{9.8}$$

This is an ordinary differential equation (ODE) with constant coefficients. Trying a solution of the form

$$x(t) = Ce^{st} \tag{9.9}$$

and substituting into the equation yields

$$Ce^{st}\left(s^2 + \frac{k}{m}\right) = 0 \tag{9.10}$$

The equation for the nontrivial solution is thus

$$\left(s^2 + \frac{k}{m}\right) = 0 \tag{9.11}$$

which gives us two solutions,

$$s_1 = +i\sqrt{\frac{k}{m}} \quad \text{and} \quad s_2 = -i\sqrt{\frac{k}{m}} \tag{9.12}$$

The natural frequency of the system is

$$\omega = \sqrt{\frac{k}{m}} \tag{9.13}$$

The displacement is

$$x = C_1 e^{i\omega t} + C_2 e^{-i\omega t} \tag{9.14}$$

commonly written as

$$x = A\cos(\omega t) + B\sin(\omega t) \tag{9.15}$$

where the constants $C_1$ and $C_2$ or $A$ and $B$ are determined from the initial boundary conditions.

The two constants $A$ and $B$ can be arbitrarily replaced with $C$ and $\phi$ such that

$$A = C\cos(\phi) \quad \text{and} \quad B = C\sin(\phi) \tag{9.16}$$

Thus

$$\begin{aligned} x &= A\cos(\omega t) + B\sin(\omega t) \\ &= C\cos(\phi)\cos(\omega t) + C\sin(\phi)\sin(\omega t) \\ &= C\sin(\omega t + \phi) \end{aligned} \tag{9.17}$$

where $C$ is the amplitude of the harmonic motion and $\phi$ is the phase angle. In terms of $A$ and $B$,

$$C = \sqrt{A^2 + B^2}$$

and

$$\phi = \tan^{-1}\left(\frac{B}{A}\right)$$

The natural circular frequency

$$\omega = \sqrt{\frac{k}{m}} \quad \text{radians/sec} \tag{9.18}$$

and

$$f = \frac{\omega}{2\pi} \quad \text{Hz (cycles/sec)} \tag{9.19}$$

This result indicates that the vibratory motion given by the displacement $x$ [Eq. (9.17)] will continue indefinitely. This is not realistic because some damping is always present and will cause the amplitude of the oscillation to decay with time.

## 9.2.2 Damped Single-Degree-of-Freedom System, Spring-Mass Damper

The equation of motion for a damped single-degree-of-freedom system (shown in Figure 9.9) is

$$m(\ddot{x}) + c(\dot{x}) + k(x) = 0 \tag{9.20a}$$

or

$$(\ddot{x}) + \frac{c}{m}(\dot{x}) + \frac{k}{m}(x) = 0 \tag{9.20b}$$

Trying a solution for this ODE of the same form as done for Eq. (9.9)

$$x(t) = Ce^{st} \tag{9.21}$$

yields an equation for the nontrivial solution

$$s^2 + \left(\frac{c}{m}\right)s + \frac{k}{m} = 0 \tag{9.22}$$

which gives

$$s_{1,2} \frac{-c/m \pm \sqrt{(c/m)^2 - 4k/m}}{2} \tag{9.23a}$$

or

$$s_{1,2} = \frac{-c}{2m} \pm \sqrt{\left(\frac{c}{2m}\right)^2 - \frac{k}{m}} \tag{9.23b}$$

There is a special significance if the term under the square root radical is zero, i.e.,

$$\frac{c}{2m} = \sqrt{\frac{k}{m}} = \omega \tag{9.24}$$

where $\omega$ is the undamped natural frequency. When this condition occurs, the displacement

$$x = Ce^{-ct/2m} \tag{9.25}$$

is an exponential decay to zero with no oscillations. This is known as the critically damped condition, and the critical damping factor is

$$c_c = 2m\omega \tag{9.26}$$

Rewriting the characteristic equation in terms of the natural frequency and the critical damping factor,

$$s_{1,2} = -\left(\frac{c}{c_c}\right)\omega \pm \omega\sqrt{\left(\frac{c}{c_c}\right)^2 - 1} \tag{9.27a}$$

or

$$s_{1,2} = \omega(-\zeta \pm \sqrt{\zeta^2 - 1}) \tag{9.27b}$$

where

$$\zeta = \frac{c}{c_c}, \quad \text{the damping factor} \tag{9.28}$$

· Case 1: Underdamped, $\zeta < 1$ or $c < c_c = 2m\omega$

In this condition the roots of the characteristic equation are imaginary

$$s_{1,2} = \omega[-\zeta \pm i\sqrt{1 - \zeta^2}] \tag{9.29}$$

and the displacement is

$$x(t) = e^{-\zeta\omega t}[A \cos(\sqrt{1 - \zeta^2}\,\omega t) + B \sin(\sqrt{1 - \zeta^2}\,\omega t)] \tag{9.30}$$

Alternatively, the displacements can be more easily visualized as

$$x(t) = e^{-\zeta\omega t}C \sin(\omega_d t + \phi) \tag{9.31}$$

where

$$\omega_d = \omega\sqrt{1 - \zeta^2} = \text{damped circular frequency} \tag{9.32}$$

and

$$\phi = \text{phase angle}$$

Note that the damped circular frequency is always less than the natural frequency $\omega$. The displacement is a harmonic function decaying with time as shown in Figure 9.11.

· Case 2: Critically damped, $\zeta = 1$ or $c = c_c = 2m\omega$

For this case the roots of the characteristic equation are real and the displacement is

$$x(t) = (A + Bt)e^{-\omega t} \tag{9.33}$$

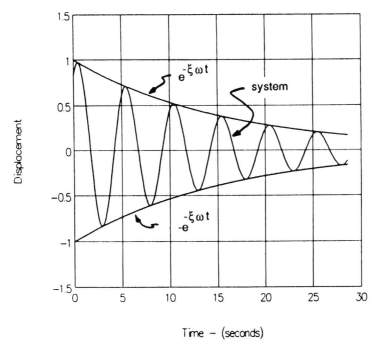

**Figure 9.11**  Free vibrational response for an underdamped system.

which is a product of a linear function of time and a decaying exponential. The exact form of the displacement is a function of the initial conditions, but it decays exponentially with time without oscillations as shown in Figure 9.12.

· Case 3: Overdamped, $\zeta > 1$ or $c > c_c = 2m\omega$

In this case both roots of the characteristic equation are real,

$$s_{1,2} = \omega[-\zeta \pm \sqrt{\zeta^2 - 1}] \tag{9.34}$$

Since

$$\sqrt{\zeta^2 - 1} < \zeta$$

both $s_1$ and $s_2$ are negative. The displacement is thus the sum of two decaying exponentials, which gives a nonoscillatory motion as shown in Figure 9.13.

For problems of main interest in this chapter the amount of damping associated with mechanical systems is usually small; thus critical and overdamped motions (Cases 2 and 3) are rare and not usually considered in the study of vibration. The concern is the underdamped case, in which vibrations occur that can damage electronic components because of either excessive displacements or forces generated during the oscillatory motion. Free vibration is not a major concern in the design of electronic systems. Forced vibration, in which a periodic external force is imposed on the system, is. With forced vibrations, the forces constraining the system and the rigid body displacements of the system can be amplified.

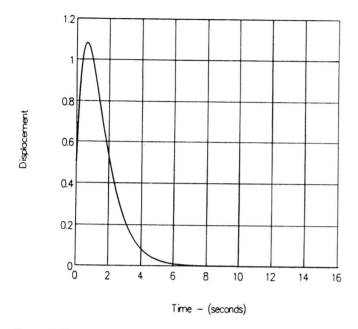

**Figure 9.12**  Free vibrational response for a critically damped system.

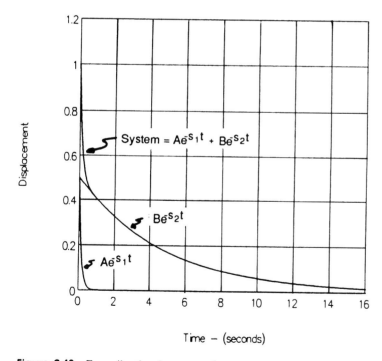

**Figure 9.13**  Free vibrational response for an overdamped system (two decaying exponentials).

### 9.2.3  Forced Vibrations (Harmonic Varying Force)

The differential equation for the free body subjected to a harmonic varying force as shown in Figure 9.14 is

$$\ddot{x} + \left(\frac{c}{m}\right)\dot{x} + \left(\frac{k}{m}\right)x = F_0 \sin(\omega t) \tag{9.35}$$

(Only a harmonic varying force is considered here because through a Fourier series any forcing function can be represented by a superposition of a series of harmonic functions.) The solution of this ODE with constant coefficients consists of the complementary solution (homogeneous solution) plus the particular solution. As discussed in the last section, the homogeneous solution dies out with time, leaving only the particular solution to be considered.

It is easiest to work with complex notation, where

$$e^{i\omega t} = \cos(\omega t) + i \sin(\omega t) \tag{9.36}$$

and then Eq. (9.35) can be written as

$$\ddot{x} + \left(\frac{c}{m}\right)\dot{x} + \left(\frac{k}{m}\right)x = F_0 e^{i\omega t} \tag{9.37}$$

where we are interested in the imaginary solution. Trying

$$x = D e^{i\omega t} \tag{9.38}$$

yields the equation

$$-m\omega^2 D + ic\omega D + kD = F_0 \tag{9.39}$$

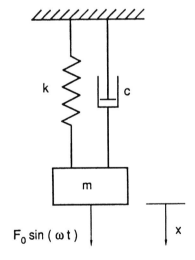

**Figure 9.14**  Simple single-degree-of-freedom system with forced excitation.

Solving for $D$,

$$D = \frac{F_0}{k - m\omega^2 + ic\omega} \tag{9.40}$$

and the particular solution

$$x_p(t) = \frac{F_0}{k - m\omega^2 + ic\omega} e^{i\omega t} \tag{9.41}$$

Noting that

$$x + iy = \sqrt{x^2 + y^2} e^{i\theta} = re^{i\theta} \tag{9.42}$$

and rewriting the particular solution

$$x_p(t) = \frac{F_0 e^{i\omega t}}{\sqrt{(k - m\omega^2)^2 + (c\omega)^2} \, e^{i\theta}} \tag{9.43}$$

where

$$\theta = \tan^{-1}\left(\frac{c\omega}{k - m\omega^2}\right) \tag{9.44}$$

Taking only the imaginary part,

$$x_p(t) = \frac{F_0 \sin(\omega t - \theta)}{\sqrt{(k - m\omega^2)^2 + (c\omega)^2}} \tag{9.45}$$

or

$$x_p(t) = X \sin(\omega t - \theta) \tag{9.46a}$$

where

$$X = \frac{F_0}{\sqrt{(k - m\omega^2)^2 + (c\omega)^2}} \tag{9.46b}$$

Rewriting this relationship,

$$X = \frac{F_0/k}{\sqrt{(1 - \omega^2/k/m))^2 + ((c\omega/k)^2}} \tag{9.47}$$

and remembering that $\omega_n = \sqrt{k/m}$

$$X = \frac{x_{\text{static}}}{\sqrt{(1 - r^2)^2 + (2\zeta r)^2}} \tag{9.48}$$

where

$$x_{\text{static}} = \text{static deflection, } F_0/k$$

$$r = \frac{\omega}{\omega_n}, \text{ frequency ratio} \tag{9.49a}$$

$$\zeta = \frac{c}{c_c} = \frac{c}{2m\omega_n}, \text{ damping factor} \tag{9.49b}$$

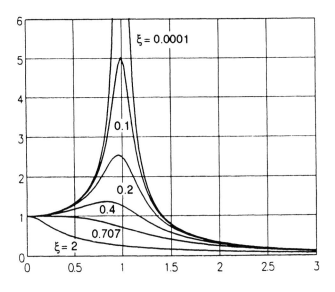

**Figure 9.15**   Dynamic magnification factor as a function of frequency ratio and damping factor.

The dynamic magnification factor, $D$, is defined as the ratio of the actual dynamic displacement to the displacement that would occur under static loading

$$D = \frac{x}{x_{\text{static}}} = \frac{1}{\sqrt{(1 - r^2)^2 + (2\zeta r)^2}} \tag{9.50}$$

A plot of $D$ is shown in Figure 9.15 as a function of the frequency ratio. Note that for any frequency ratio the magnification factor decreases as the damping factor, $\zeta$, increases.

The infinite magnification factor for no damping, $\zeta = 0$, at resonance frequency $r = \omega/\omega_n = 1$, is expected. (In fact, the finite dynamic displacement of a real system is one means of measuring the damping factor.) Otherwise, the maximum magnification factor for a fixed damping ratio can be calculated from

$$\frac{dD}{dr} = 0 \frac{r(1 - r^2 - 2\zeta^2)}{[(1 - r^2)^2 + (2\zeta r)^2]^{3/2}} \tag{9.51}$$

This equation is satisfied at $r = 0$, i.e., $\omega = 0$, a static system and $D = 1$. The other end point at $r = \infty$ yields $D = 0$. When $r = \infty$, the forcing frequency is so large that the system cannot respond. After examining the end points, the solution for a local maximum is

$$(1 - r^2 - 2\zeta^2) = 0 \tag{9.52a}$$

or

$$r = \sqrt{1 - 2\zeta} \tag{9.52b}$$

provided that $\zeta \le 0.707$. For such values of $\zeta$ there is a local maximum

$$D_{max} = \frac{1}{2\zeta\sqrt{1-\zeta^2}} \tag{9.53}$$

Unfortunately, the damping ratio in most electronic assemblies is small and values of $D_{max} > 10$ are common.

Thus it is advisable to design the electronic assembly with a natural frequency that is higher than the driving frequency. For the worst case, damping factor $= 0$, the magnification factor $D$ can be kept below 2 if the frequency ratio $r$ is less than 0.7. If it is not possible to avoid the resonance region, $0.9 < r < 1.1$, use of a vibration isolation system to protect the system from the detrimental influence of the excessive amplitudes may be required.

The displacement solution, Eq. (9.45), can now be written in terms of the dynamic magnification factor as

$$x_p(t) = Dx_{static}\sin(\omega t - \theta) \tag{9.54}$$

The phase angle $\theta$, as given in (9.44), is shown in Figure 9.16 as a function of the frequency ratio.

The transmissibility, $Q$, of a system is the ratio of support motion to the actual system motion. As an example, take the simple system shown in Figure 9.17, where the support has some motion. The equation of motion for the system is

$$m(\ddot{x}) + c(\dot{x} - \dot{x}_s) + k(x - x_s) = 0 \tag{9.55a}$$

or

$$m(\ddot{x}) + c(\dot{x}) + kx = kx_0\sin(\omega t) + c\omega x_0\cos(\omega t) \tag{9.55b}$$

$$m(\ddot{x}) + c(\dot{x}) + kx = F_0\sin(\omega t + \beta)$$

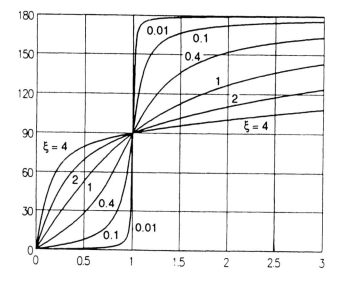

Frequency Ratio

**Figure 9.16** Phase angle as a function of frequency ratio and damping factor.

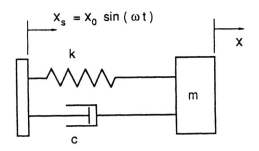

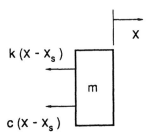

**Figure 9.17**   Simple system in which support has some motion.

where

$$F_0 = x_0 \sqrt{k^2 + (c\omega)^2}$$ (9.56)
$$= x_0 k \sqrt{1 + (2r\zeta)^2}$$

$$\beta = \tan^{-1}\left(\frac{c\omega}{k}\right) = \tan^{-1}(2r\zeta)$$ (9.57)

Using Eqs. (9.45) and (9.49), the particular solution to (9.55c) can be immediately written as

$$x_p(t) = \frac{F_0/k \, \sin(\omega t + \beta - \theta)}{\sqrt{1 - r^2)^2 + (2\zeta r)^2}}$$ (9.58)

Combining with Eq. (9.56),

$$\frac{X_p(t)}{x_0} = \frac{\sqrt{1 + (2r\zeta)^2}}{\sqrt{(1 - r^2)^2 + (2\zeta r)^2}} \sin(\omega t + \beta - \theta)$$ (9.59)

Using the definition of transmissibility as

$$Q = \frac{\text{amplitude of the system motion}}{\text{amplitude of the support motion}}$$ (9.60)

$$Q = \frac{X}{x_0} = \frac{\sqrt{1 + (2r\zeta)^2}}{\sqrt{(1 - r^2)^2 + (2\zeta r)^2}}$$ (9.61)

Alternatively, we could have used

$$Q = \frac{\text{amplitude of transmitted force}}{\text{amplitude of applied force}}$$ (9.62)

and come up with the same results.

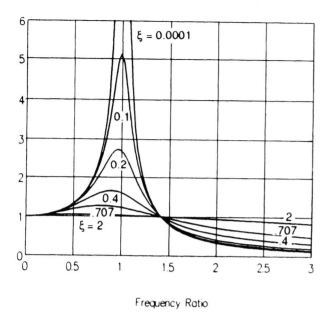

**Figure 9.18** Transmissibility as a function of frequency ratio and damping factor.

Figure 9.18 is a plot of the transmissibility as a function of the frequency ratio. Note that for all the damping factors, the lines all pass through the same point at $r = \sqrt{2}$. For $r < \sqrt{2}$, increases in the damping factor decrease the transmissibility and increase the effectiveness of vibration isolation. For $r > \sqrt{2}$ just the opposite occurs; the greater the damping factor, the less effective the vibration isolation.

Since damping in electronic assemblies is usually very low, one can anticipate large amplification of vibrating amplitude and forces transmitted from the support structure if the system is driven at a frequency near the natural frequency $\omega$. Clearly, extended exposure of the system to resonance conditions must be avoided to prevent damage due to fatigue failure of solder joints, lead wires, and fasteners. If it is not possible to design with $\omega_n > 2\omega$, then isolation of the electronic system from the disturbing force may be required.

The idea in vibration isolation is to place a spring and dashpot between the mass of interest and the surrounding support structure or base. Isolation is accomplished by designing a suspension system with a frequency ratio $r > 3$, which gives a transmissibility of less then 0.1 for systems with small damping. There are many commercially available spring-dashpot assemblies, called isolation mounts, for lightweight printed circuit boards to large mounts for cabinets.

## 9.2.4 Random Vibrations

Periodic and harmonic motions are characteristic of deterministic vibrations because an instantaneous value for displacement, velocity, or acceleration can be predicted at any time. A random motion never develops a repeatable pattern, so it has no period, and the amplitude at any moment in the future cannot be predicted. Thus a random signal is called a nondeterministic phenomenon. This type of motion may exhibit either narrow-band or broadband characteristics. A narrow-band disturbance transmits loads that occupy a narrow range of frequencies, whereas a broadband disturbance occupies a wide range of frequencies, usually encompassing the major structural resonances of equipment.

In addition to never developing a repeatable pattern of amplitude versus time, a random signal can theoretically have amplitude spikes that approach infinity. This combination of uncertain moment-to-moment amplitude with the possibility of unlimited excursions makes the task of working with a random signal appear formidable, if not impossible. But rather than trying to predict the instantaneous amplitude of a random signal, we work with the probabilities of the signal occurring within certain bounds.

A random signal can be understood if we can judge its intensity as well as its frequency content. This is accomplished by measuring its standard deviation and spectral density, respectively.

A random signal is stationary if its standard deviation and its spectral density do not vary with time. A nonstationary random signal would show a change in standard deviation and/or spectral density with time. An ergodic random signal is a stationary random signal whose standard deviation and spectral density are identical to those of a group of other random signals. In other words, a single record of the random signal is sufficient to obtain a statistical description of the function.

## Amplitude Considerations in a Random Signal

The distribution of peaks in a random signal may be characterized as Gaussian, which means that the motion can be defined mathematically in terms of a mean, standard deviation, and moments. The mean value is defined as

$$\bar{y} = \frac{1}{T} \int_0^T y(t)\, dt \tag{9.63}$$

and the mean square value as

$$\overline{y^2} = \frac{1}{T} \int_0^T y^2(t)\, dt \tag{9.64}$$

Both the mean and mean square values provide measurements for the average value of the random function $y(t)$. The measure of how widely the function $y(t)$ differs from the average is given by its variance, namely

$$\sigma_y^2 = \frac{1}{T} \int_0^T [y(t) - \bar{y}]^2\, dt \tag{9.65}$$

When the expression under the integrand is expanded and integrated, we find that

$$\sigma_y^2 = \overline{y^2} - (\bar{y})^2 \tag{9.66}$$

which means that the variance can be calculated by the mean square minus the square of the mean. Most commonly in random vibrations the mean value is zero, in which case the variance is equal to the mean square value.

$$\sigma_y^2 = \overline{y^2} \quad \text{for } \bar{y} = 0 \tag{9.67}$$

The root mean square value is defined as

$$y_{\text{rms}} = \sqrt{\overline{y^2}} \tag{9.68}$$

The standard deviation $\sigma_y$ of $y(t)$ is the square root of the variance

$$\sigma_y = \sqrt{\overline{y^2} - (\overline{y})^2}$$ (9.69)

and when the mean is zero, the standard deviation is equal to the root mean square

$$\sigma_y = y_{rms} \quad \text{for } \overline{y} = 0$$ (9.70)

For a stationary random signal with a zero mean, a measured rms quantity is sufficient to characterize the signal. Reference can immediately be made to statistical tables that list the areas under the Gaussian probably density curve. This gives us the means of defining, analyzing, and understanding all aspects of the intensity or magnitude, characteristics of the random signal.

As an example, the small table below lists the occurrence of a peak signal between certain bounds.

| Magnitude within (multiple of rms quantity) | Occurrence (% of time) |
|:---:|:---:|
| $\mp\sigma$ | 68.27 |
| $\pm 2\sigma$ | 95.54 |
| $\pm 3\sigma$ | 99.74 |
| $\pm 4\sigma$ | 99.96 |

For fatigue design of electronic components mounted on PCBs it is the $3\sigma$ values that need to be considered [8, 9]. When an rms acceleration, displacement, or stress is under consideration, its value should immediately be multiplied by 3 to obtain the $3\sigma$ values. This means that a random vibration level of 10 $g$ rms requires a design that will be capable of withstanding accelerations levels in excess of $3 \times 10$ or 30 $g$, not 100% of the time but only approximately 0.3% of the time.

## Frequency Considerations in a Random Signal

If a random signal is processed by a spectrum analyzer and the fraction of the mean square value $\Delta\overline{y^2}$ is recorded for frequencies in a narrow range $\Delta f$, the power spectral density (PSD) for $f \pm \Delta f/2$ becomes

$$PSD(f) = \frac{\Delta\overline{y^2}}{\Delta f}$$ (9.71)

As $\Delta f$ approaches zero, the spectrum becomes continuous and may be written as

$$PSD(f) = \frac{d\overline{y^2}}{df}$$ (9.72a)

or

$$d\overline{y^2} = PSD(f)\, df$$ (9.72b)

The overall mean square value may be obtained by integrating over the limits of the spectrum

$$\overline{y^2} = \int_{f_1}^{f_2} \text{PSD}(f)\, df \tag{9.73}$$

### 9.2.5  Shock Loading

Any sudden change that affects the position, velocity, acceleration, or forces applied to the system may be considered as shock. Shock can occur during assembly, handling, or transport of the equipment. Shock can also result when two adjacent vibrating assemblies come in contact with each other due to excessive displacements. Near-miss explosions and load due to wind gusts are also examples of shock-producing environments.

Shock can develop significant internal forces that diminish rapidly. After the application of shock loading, motion consists of damped harmonic vibration. Characteristics that differentiate shock from static loading involve the time required for the response to reach maximum values of displacement, velocity, and acceleration. The intensity of the response to the applied load depends on how close the fundamental frequencies of the assembly are to the frequency of the loading.

The evaluation of the response of an assembly can be simplified by assuming a simple single-degree-of-freedom system. The limiting response of the system is related to the fundamental frequency, $f_n$, and the duration of the loading pulse, $T_p$. If the product of $f_n T_p$ is less than $1/3$, then the loading is an impact loading and the response will generate greater displacements than an equivalent static loading. If $f_n T_p$ is greater than 3, the loading can be considered static. When the shock parameter $1/3 < f_n T_p < 3$, a solution of the differential equations of motion for the applied loading is required (see Ref. 9).

### 9.3  MODELING PRINTED WIRING BOARDS AS RECTANGULAR PLATES

The previous section on the response of single-degree-of-freedom systems was presented because we will approximate the response of a PWB as that of a single-degree-of-freedom system. It is true that in reality a PWB is a multi-degree-of-freedom system, but the fundamental mode is of primary importance. It is the fundamental mode that has the large displacements that cause fatigue damage to solder joints, component wires, and pin connections.

Printed wiring boards come in various sizes and the predominant shape is rectangular. This rectangular shape is dictated by the ease of service for plug-in type boards. Non-rectangular boards are used for specialized applications in which space is at a premium and every available inch is used. The concepts presented in this section are applicable to all PWBs, but the discussion will consider rectangular boards. Determination of the PWB fundamental natural frequencies is the first and most critical step in determining the fatigue life of components mounted on the board. The PWB loading, dynamic displacements, stresses, and ultimately fatigue lives can be calculated once the natural frequencies are known. The fundamental natural frequency is the critical value, since it is associated with the highest displacements and stresses. Most of the vibrational fatigue damage occurs at this fundamental resonance mode.

Assuming that the PWB acts like a single-degree-of-freedom system, there is a simple relationship between the maximum displacement of the board and the acceleration. This can be determined by starting with the assumed harmonic equation of motion and differentiating it twice:

$$Z = Z_0 \sin(\omega t) \tag{9.74a}$$

$$\frac{dZ}{dt} = \omega z_0 \cos(\omega t) \tag{9.74b}$$

$$\frac{d^2 Z}{dt^2} = -\omega^2 z_0 \sin(\omega t) \tag{9.74c}$$

The maximum acceleration in terms of the maximum displacement amplitude, $Z_0$, is

$$a_{max} = \omega^2 Z_0 \tag{9.75}$$

or expressed in terms of $g$'s

$$G_{max} = \frac{a_{max}}{g} = \frac{\omega^2 Z_0}{g} = \frac{(2\pi f)^2 Z_0}{9.8} \tag{9.76}$$

Solving for maximum displacement at the board's natural frequency, $f_n$,

$$Z_0 = \frac{9.8 G_{max}}{f_n^2} = \frac{9.8 G_{in} Q}{f_n^2} \tag{9.77}$$

where $G_{in}$ is the input acceleration to the PWB and $Q$ the transmissibility of the PWB at its natural frequency (for PWB natural frequencies between 200 and 400 Hz, Steinberg [10] has shown that a good approximation is $Q = \sqrt{f_n}$).

By only slightly increasing the natural frequency of the board, the maximum deflection amplitude, $Z_0$, is rapidly decreased, reducing the stresses and strains on the components. Thus it is in the designer's best interest to keep the natural frequencies as high as practically possible.

The natural frequency is dependent on PWB size, shape, and material; how it is supported; and the location, orientation, and mounting method of the components attached to it. Obviously, size, or board dimensions, influences the natural frequency, with larger PWBs having lower natural frequencies. Typically a PWB is constructed from thin layers of composite materials such as fiberglass-epoxy. With woven composite materials there are different material properties in orthogonal directions. When such layers are used, the stiffness of the board becomes dependent on the fiber orientation of each layer and on the manner in which the layers are sequenced. In addition, metallic layers, such as aluminum, are also sometimes used to increase heat transfer. All of these material influences affect the stiffness and thus the natural frequency of the board.

PWB shape is usually governed by the space available and by the ease of maintenance. Most plug-in PWBs are rectangular in shape but with edge and corner cutouts. Odd-shaped boards are usually limited to special applications with extreme space restrictions. The effect of shape on the natural frequency can be large but depends very strongly on boundary conditions.

## 9.3.1   Edge or Boundary Conditions

The manner in which the PWB is supported in the electronic assembly is important in determining the response to vibration and shock. PWBs can be supported in many different ways, depending on such factors as environment, weight, maintainability, accessibility, and cost. Loose fits permit easy connector engagement but are not desirable in a vibration or shock environment. Connectors and board edge guides that firmly grip the PWB are desirable to reduce deflections, translations, and edge rotations. Such grips increase a board's natural frequency and reduce deflections that cause failures of component connections.

The greatest effect on the natural frequency of the board is due to the manner in which it is supported. Supports restrict the out-of-plane and/or rotational motion of the PWB edge. Classically, the support or boundary conditions are considered along an entire edge as either free, simply supported, or clamped. A simple support restricts the out-of-plane movement of the edge, and both the out-of-plane movement and the rotational movement are restricted by a clamped support. As the natural frequency is inversely proportional to the square root of displacements or deflections, the natural frequency is much higher if the edge is clamped rather than simply supported or unsupported.

For high natural frequencies, one would wish to have clamped supports an all edges of the PWB. This is not practical for plug-in PWBs. In general, the normal supports in high-vibration environments are wedge-lock–type edge retainers along two sides of the board. Unfortunately, even though the wedge-lock retainers are normally considered to provide clamped-type support, they do not completely restrict the rotational movement of the edge. The wedge-locks provide a support between clamped and simple.

Bircher-type edge guides, another common edge support, are springlike supports. They allow not only rotational movement but also translational movement of the edge, i.e., a support between simple and free. The same is true for one-part edge card connectors and two-part pin-in-socket connectors due to the contact between the mating surfaces.

In reality, some edge conditions change depending on the magnitude of loading. Typically, the higher the *g* loading, the more translation can result in what would at first appear to be a simple support. Rotations typically begin to occur in all but the most rigid clamped supports. Thus, unless there is a great deal of confidence in knowing the edge conditions, discretion must be exercised in the effort expended in calculating the PWB's natural frequency. Edge supports and connectors commonly used with PWBs cannnot be represented by the classical supports (clamped, simple, free). Edge supports and connectors must be represented as elastic supports. It is meaningless to spend a great amount of effort in a complex finite-element modeling of the board if the boundary conditions are suspect.

## 9.3.2  Calculating Natural Frequencies

The natural frequency is the most important parameter for vibration and shock design of PWBs. Most of the vibrational damage occurs at this fundamental resonance mode, where the displacements and stresses are greatest. This was illustrated in Figure 9.15 for single-degree-of-freedom systems with small damping factors.

There are several ways to calculate approximately the natural frequency of a rectangular PWB. The easiest method is to treat the PWB as a uniform flat vibrating plate and use the Raleigh method [11]. In this method a deflection curve is assumed that satisfies the geometric boundary conditions for a particular plate. This assumed deflection curve is then used to obtain the strain and kinetic energy of the plate. If no energy is dissipated, the strain energy equals the kinetic energy and the approximate natural frequency of the plate can be determined.

The Rayleigh method results in a natural frequency that is slightly higher than the true natural frequency for a given set of conditions, unless the exact deflection curve is used. If the exact deflection curve is used, the exact natural frequency is found. The natural frequencies of various rectangular plates as determined by the Rayleigh method are given in Table 9.2.

**Table 9.2**  Fundamental Frequencies of Various Uniformly Loaded Rectangular Plates

| Case | Edge configuration | | | | $C_o$ | $P_o$ |
|------|-------|-------|-------|-------|-------|-------|
|      | $A_1$ | $A_2$ | $B_1$ | $B_2$ | | |
| 1  | F | F | F | F | $\pi/2$    | $[2.08(ab)^{-2}]^{1/2}$ |
| 2  | F | F | F | C | 0.55       | $a^{-2}$ |
| 3  | F | F | C | C | 3.55       | $a^{-2}$ |
| 4  | F | F | C | S | 2.45       | $a^{-2}$ |
| 5  | F | F | S | S | $\pi/2$    | $a^{-2}$ |
| 6  | F | C | C | F | $\pi/5.42$ | $[a^{-4} + 3.2(ab)^{-2} + b^{-4}]^{1/2}$ |
| 7  | F | S | S | F | $\pi/11$   | $(a^{-2} + b^{-2})$ |
| 8  | F | S | C | F | $\pi/2$    | $[0.138a^{-4} + 0.251(ab)^{-2}]^{1/2}$ |
| 9  | F | S | C | C | $\pi/2$    | $(0.25b^{-2} + 2.25a^{-2})$ |
| 10 | F | C | S | S | $\pi/2$    | $[a^{-4} + 0.608(ab)^{-2} + 0.1266b^{-4}]^{1/2}$ |
| 11 | F | C | C | C | $\pi/3$    | $[12a^{-4} + 2.25(ab)^{-2} + 0.31b^{-4}]^{1/2}$ |
| 12 | F | S | S | S | $\pi/2$    | $[0.927a^{-4} + 0.65(ab)^{-2} + 0.31b^{-4}]^{1/2}$ |
| 13 | F | C | C | S | $\pi/2$    | $[2.44a^{-4} + 0.707(ab)^{-2} + 0.127b^{-4}]^{1/2}$ |
| 14 | F | S | C | S | $\pi/2$    | $[2.56a^{-4} + 0.57(ab)^{-2}]^{1/2}$ |
| 15 | S | S | S | S | $\pi/2$    | $(a^{-2} + b^{-2})$ |
| 16 | C | C | C | C | $\pi/1.5$  | $[2.91a^{-4} + 1.663(ab)^{-2} + 2.91b^{-4}]^{1/2}$ |
| 17 | S | S | C | C | $\pi/3.53$ | $[16a^{-4} + 7.7(ab)^{-2} + 3b^{-4}]^{1/2}$ |
| 18 | C | C | C | S | $\pi/2$    | $[2.45a^{-4} + 2.9(ab)^{-2} + 5.13b^{-4}]^{1/2}$ |
| 19 | S | C | C | S | $\pi/2$    | $[2.45a^{-4} + 2.68(ab)^{-2} + 2.45b^{-4}]^{1/2}$ |
| 20 | S | S | C | S | $\pi/2$    | $[2.43a^{-4} + 2.33(ab)^{-2} + 0.985b^{-4}]^{1/2}$ |

$F$ = Free edge

$S$ = Simple support

$C$ = Clamped support

$$f_n = C_o P_o \sqrt{\frac{Dab}{M}} \text{ Hz}$$

$$D = \frac{Et^3}{12(1 - v^2)}$$

$$M = W/g \text{ (mass)}$$

Source: Ref. 2

The following example demonstrates how sensitive the natural frequency calculation is to the various boundary or edge conditions.

*Example 9.1*  Calculate the natural frequencies of a 5 × 8 in. G-10 board $\frac{1}{16}$ in. thick, assuming a uniform total weight of board and components to be $\frac{1}{2}$ pound. Using the equations from Table 9.2 with the associated case numbers.

$$D = \frac{Et^3}{12(1 - t^2)} = \frac{2E6(\frac{1}{16})^3}{12(1 - 0.12^2)} = 169.1 \text{ lb in.}$$

$$M = \frac{W}{g} = \frac{0.5}{386.4} = 1.294E - 3 \text{ lb sec}^2/\text{in.}$$

$a = 8$ in.

$b = 5$ in.

$$f_n = C_0 P_0 \sqrt{\frac{Dab}{M}} \text{ Hz}$$

Case 18—all four edges champed:

$$C_0 = \frac{p}{1.5}$$

$$P_0 = (2.91a^{-4} + 1.663(ab)^{-2} + 2.91b^{-4})^{1/2}$$

$$= 80.04E - 3$$

$$f_n = 383.3 \text{ Hz}$$

Case 15—all four edges simply supported:

$$C_0 = \frac{p}{2}$$

$$P_0 = (a^{-2} + b^{-2}$$

$$= 55.63E - 3$$

$$f_n = 199.8 \text{ Hz}$$

Case 1—all four edges free (not possible, but for comparison purposes):

$$C_0 = \frac{p}{2}$$

$$P_0 = (2.08(ab)^{-2})^{-1/2}$$

$$= 36.06E - 3$$

$$f_n = 129.5 \text{ Hz}$$

Case 10—simple along long sides and clamped along one short side with the other being free (card inserted in guides with connection along one edge). For this case

$$C_0 = \frac{p}{2}$$

$$P_0 = (a^{-4} + 0.608(ab)^{-2} + 0.1266b^{-4})^{1/2}$$

$$= 44.84E - 3$$

$$f_n = 161.0 \text{ Hz}$$

When the Rayleigh method is used, the PWB is assumed to be a homogeneous isotropic plate rather than a multilayer composite. Board shape is assumed to be a perfect rectangle with no edge or corner cutouts. Boundary conditions are assumed to be the classical conditions of simple, clamped, or free along the entire edge. The mass of components, if considered, is treated only as uniformly distributed over the entire board. Any stiffening effect of components is completely ignored.

Table 9.3 lists the measured and calculated natural frequencies of a PWB that was held along three sides. The measurements were made for the unpopulated board and after the board was populated with 14-pin DIPs. The Rayleigh method was used assuming three sides clamped and three sides free. For the populated case the mass of the components was assumed evenly distributed over the entire board. This example clearly points out the magnitude of potential error (plus or minus 50%) in calculating the natural frequency by the Rayleigh method when the boundary conditions are not accurately known or do

**Table 9.3** Experimentally Measured Natural Frequency Versus Calculated Natural Frequency (Rayleigh analysis)

|  | Measured | Analytic (FSSS) | Analytic (FCCC) |
|---|---|---|---|
| Unpopulated | 335 Hz | 210 Hz ($-45\%$) | 490 Hz ($55\%$) |
| Populated | 200 Hz | 147 Hz ($-30\%$) | 320 Hz ($70\%$) |

not behave as classical boundary conditions. More accuracy can be obtained by modeling the boundary conditions as elastic supports with translational and rotational spring constants.

The components can also be better modeled as finite masses attached to the board at various locations. There are several theoretical solutions for handling plates with concentrated masses, but the mathematics are too complex for practical solutions. Finite-element techniques can be practically used when better accuracies are required.

The attachment of the components to the PWB reinforces or increases the stiffness of the PWB in the immediate region of the component. The stiffening and mass effects are functions of component size, thickness, material, and attachment technology. The location and orientation of the component are also important factors. If the component is mounted with its longer side in the direction of maximum curvature of the board, its stiffening effect will be more than if it is oriented in the other direction. This stiffening effect is also emphasized if the component is located in the region of maximum curvature. This is difficult to model with anything but finite-element techniques.

Figure 9.19 is an example showing the effect on a PWB's natural frequency of slowly adding components. The example board was suspended by light threads to simulate free boundary conditions on all edges. The board was slowly populated in a uniform pattern with 14-pin DIPS as natural frequency measurements were made. The lower line in the figure is the calculated natural frequency of the PWB using the Rayleigh method and assuming the mass of the components is uniformly distributed. Two boards were slowly populated, one with the components mounted parallel to the long axis of the board and one with the components mounted transverse. Note that the longitudinally mounted components stiffened the board more than the transversely mounted components. The local reinforcement of the longitudinally mounted components almost completely canceled the effect of the increased mass of the assembled board.

The Rayleigh method should be used only with a complete understanding of its limitations. Under these conditions, it is very useful for initial quick approximations and ''what if'' calculations to determine initial trends and potential problem. It is particularly useful in the early stages of design, but it should not be used for any detailed analysis or whenever the design is close to being finalized.

## Finite-Element Analysis

Finite-element analysis is one of the most powerful tools for modern design engineers. It provides the designer with an analysis tool to accurately model a PWB without having to build a prototype and subject it to vibration testing. Any shape and material can be modeled, immediately overcoming the initial restrictions of the Rayleigh method. In addition, the finite-element method has the capabilities to model elastic boundary conditions as well as the stiffness and mass influences of the components.

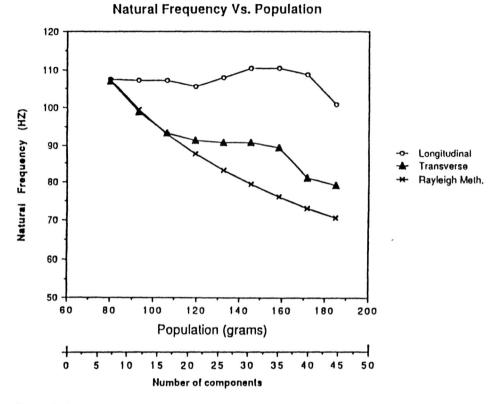

**Figure 9.19**   Effect of mounting 14-pin DIPs on natural frequency of a PWB.

Proper use of a finite-element analysis program is not an easy task and requires a good theoretical background. Although a large number of general-purpose software packages have been developed, such as NASTRAN, ANSYS, SAP, MARC, ABAQUS, and ADINA, they are not automated and intended for the novice. The user must give a long list of commands before a realistic model can be developed for analysis. These commands are confusing for a designer with limited knowledge of the method. Small errors in the modeling phase, which are not easy to detect, can give unrealistic results. Only through experience with particular codes can users start to trust finite-element results.

### 9.3.3   Estimating Transmissibility

For calculating the PWB response it is necessary to know the transmissibilities developed by the boards. Remember that the transmissibility (Section 9.2.3 and Figure 9.18) is defined as

$$Q = \frac{\text{maximum amplitude of board}}{\text{maximum amplitude of supporting edges}} \qquad (9.60)$$

Obviously, the best source of information on transmissibility for various types of PWBs is old test data, but in the design process of new PWBs it becomes necessary to estimate the response characteristics.

The factors that must be considered are all related to the damping characteristic of the board. The damping characteristic is a measure of the amount of energy lost during the vibrating condition. (Note in Figure 9.18 at the natural frequency, $r = 1$, how sensitive $Q$ is to the damping factor, $f$.) When more energy is lost, or transformed to heat, there is less energy remaining and the transmissibility is lower, hence the dynamic loads and stresses are also lower.

The greatest energy losses are due to hysteresis and friction. Hysteresis losses are generally due to internal strains that are developed during bending deflections of the board. Friction losses are generally due to relative motion between high-pressure interfaces, such as mounting surfaces, stiffening ribs, and edge guides. These energy losses are greatest when the deflections are greatest and smallest when the deflections are smallest. Since higher frequencies have smaller deflections, they will also have less damping. This means higher frequencies will usually have higher transmissibilities at resonant conditions.

Terms such as "low" and "high" resonant frequency are only relative. In general, low resonant frequencies for PWBs are resonance frequencies less than 100 Hz. The term high resonant frequency applies to PWBs whose resonant frequency is above about 400 Hz. Most PWBs have fundamental resonant frequencies between 200 and 300 Hz.

Test data of Steinberg [10] show that in general transmissibility is related to the natural frequency by the equation

$$Q = A \sqrt{f_n} \tag{9.78}$$

where

$$A \sim \frac{1}{2} \quad \text{for } f_n < 100 \text{ Hz}$$

$$A \sim 1 \quad \text{for } 200 < f_n < 300 \text{ Hz}$$

$$A \sim 2 \quad \text{for } f_n > 400 \text{ Hz}$$

This relationship is commonly simplified by taking $A = 1$ for all preliminary calculations.

## 9.3.4 Stresses and Deflections

If, as a first approximation, the PWB can be considered to be a homogeneous isotropic uniformly loaded plate, the elastic stresses and displacements of the PWB can be calculated from classical rectangular plate solutions [12]. Considering the PWB as a uniformly loaded plate ignores the local stiffening of the PWB directly under the mounted component. This local stiffening is minimal for leaded components such as resisters, diodes, and capacitors but can be significant for large surface-mounted leadless chip carriers (LCC's). This stiffening will result in lower displacements and stresses, except for the possibility of local stress concentrations due to the transition regions between stiffened regions of the PWB and nonstiffened regions. The assumption of isotropy is also not correct for fiber-reinforced PWBs; they are more typically othrotropic. Nevertheless, the classical plate solutions give insight into the stresses and displacements to be expected in vibrational deformation.

When the out-of-plane displacement of a plate, whether static or due to vibratory excitation, is greater than about one half the plate thickness, small-displacement linear elastic solutions break down. With these larger displacements, in-plane stretching contributes significantly to the plate stress and quickly becomes dominant over the bending stress.

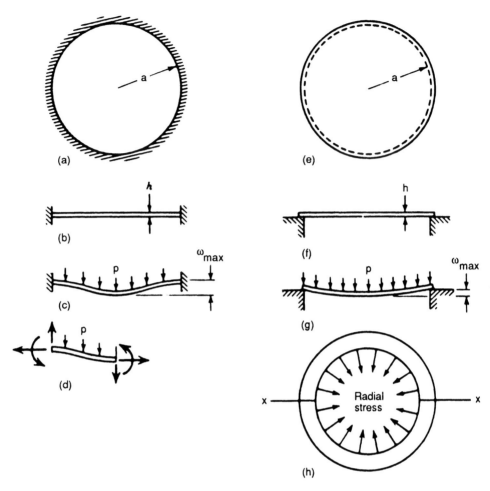

**Figure 9.20**   Thin plates having large deflections in which direct tension is significant. (From Ref. 13.)

The magnitude of this in-plane stretching or membrane stress is a function of the boundary conditions. This is easily visualized by considering two different plates, one clamped along its edges to prevent any translation or rotation and the other simply supported along its edges allowing only rotation (Figure 9.20). For equal maximum displacements, the deflected surface length of the clamped plate is greater than that of the simply supported plate, resulting in higher membrane stresses.

A plate with no edge supports also generates membrane or in-plane stretching stresses. This can be visualized in Figure 9.20h by looking at a strip along the outer edges of a deflected plate with free boundary conditions. Because of the plate deflection the outer edges of the plate try to move inward but the surrounding material in the outer strip resists this motion. Thus tensile stresses are generated in the direction toward the middle of the plate and compressive stresses are generated tangent to the edges of the plate. These compressive stresses cause buckling of the plate. You can easily demonstrate this phenomenon by supporting a piece of paper on the fingers on one hand while gently loading

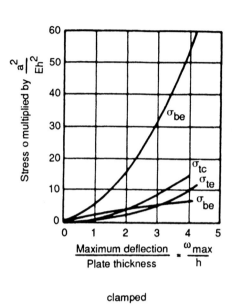
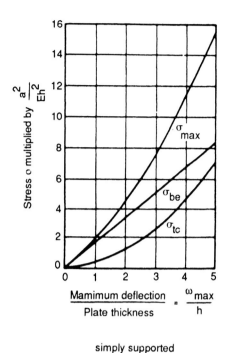

clamped

simply supported

**Figure 9.21** Membrane or tension stresses and bending stresses at the center of the plate, $\sigma_{tc}$ and $\sigma_{bc}$, and the tension and bending stresses at the edge of the plate, $\sigma_{te}$ and $\sigma_{be}$, for the clamped and simply supported circular plates. (From Ref. 13.)

the middle of the paper with a finger from the other hand. The ripplies generated in the paper are caused by the compressive buckling stresses. Obviously, compared to the clamped and simply supported cases, the free-edge condition generates the smallest membrane stresses.

Figure 9.21 shows the membrane tension stress and the bending stress at the center of the plate, $\sigma_{tc}$ and $\sigma_{bc}$, and the membrane and bending stresses at the edge of the plate, $\sigma_{te}$ and $\sigma_{be}$, as a function of the plate displacement for the two conditions of clamped edges and simply supported edges. These illustrations are actually for circular plates, for which the mathematics are simpler, but are representative of the stresses for rectangular plates. The exact quantitative values are not important, but illustrate the nonlinear behavior when plates undergo large out-of-plane displacements. Note that the maximum stress for the clamped-edge condition is at the edge of the plate and not at the center of the plate, as might be expected.

To gain a feel for the magnitudes of stresses in a PWB, two cases from the example presented in Section 9.3.2 will be continued.

*Example 9.2* The natural frequencies of a 5 × 8 in. G-10 board $\frac{1}{16}$ in. thick assuming a uniform total weight of board and components to be $\frac{1}{2}$ pound were found to be:

· 383.3 Hz for clamped edges
· 199.8 Hz for simply supported edges.

Assume that the example board is subjected to a 10 $g$ loading. Since the natural frequencies of the example PWB boards are between 200 and 400 Hz, the transmissibility can be approximated by the relation

$$Q = \sqrt{f_n}$$

Thus the maximum board acceleration is

$$G_{max} = QG_{in}$$
$$= \sqrt{f_n}\, G_{in}$$
$$= \sqrt{383.3}\; 10$$
$$= 195\; g \text{ for the clamped edges}$$
$$= 141\; g \text{ for the simply supported edges}$$

In Eq. (9.77) it was shown that the maximum displacement for a single-degree-of-freedom system is

$$Z_0 = \frac{9.8G_{max}}{f_n^2} = \frac{9.8G_{in}Q}{f_n^2} \tag{9.77}$$

where $a_0$ is the amplitude of the acceleration. Thus

· $Z_0 = 0.013$ in. for the clamped edges
· $Z_0 = 0.036$ in. for the simply supported

Note that the deflection of the simply supported plate is about three times the deflection of the clamped supported plate.

From Westergaard's solutions for the moments and stresses in slabs [12], as summarized by Boresi et al. [13], the tensile stresses for the clamped-edge plate are:

$$\sigma_{max} = 1879 \text{ psi}$$

at the center edge of the short side and

$$\sigma = 830 \text{ psi}$$

in the center of the plate in the direction of the short side. For the simply supported plate

$$\sigma_{max} = 2212 \text{ psi}$$

at the center of the plate in the direction of the short side.

As noted earlier, the maximum stress in the clamped plate does not occur at the center of the plate but at the edges. The edge stress is a little over two times the stress in the center of the plate. Note that the maximum stress in the simply supported plate is slightly greater than the maximum stress in the clamped-edge plate. The magnitude of the board stresses, as calculated in this example and as typically occur in practice, is so low that there is rarely any worry of fatigue failure of the board material.

The fact that the board stresses are so low and that there is no concern about failure of the board does not reflect what happens to components mounted on the surface of the board. Using the example board dimensions and stresses, assume that over the central 1 in. of the board, in the short-dimension direction, the stress in the outer fiber of the board is about 2000 psi or a strain of

$$\epsilon_x = \frac{1}{E}(\sigma_x - \sigma_y) \sim \frac{2000}{2,000,000} = 1000 \ \mu\text{in./in.} \tag{9.79}$$

assuming the stress in the perpendicular direction, $\sigma_y$, is small. The strain in the copper traces can be logically assumed to be equal to the surface strains in the PWB. Thus the copper traces see a stress of

$$\sigma = E\epsilon = (16E6)(1000E - 6) = 16,000 \ \text{psi} \tag{9.80}$$

which is significantly higher than the PWB stress. It is easy to reach high enough stress levels in the copper traces to have a fatigue problem.

Components mounted on the PWB surface can experience significant strains. The surface elongation or change in length in the direction of the 2000 psi PWB stress is

$$\Delta L = L\epsilon = L\frac{\sigma}{E} = \frac{1(2000)}{2,000,000}$$
$$= 0.001 \ \text{in.} \tag{9.81}$$

Assume that a 1 in. square LCC, which can be assumed to be rigid compared to the G-10 board, is mounted in the center of the board with a solder height of 20 mils. One half of this elongation of the board with respect to the rigid chip carrier must be taken up as shear strain in the solder joints on each side of the LCC. The shear strain in the solder joints can be approximated by

$$\gamma = \frac{\Delta L/2}{h} = \frac{0.0005}{0.020} = 2.5\% \ \text{strain} \tag{9.82}$$

a nontrivial amount and clearly beyond the elastic limit of the solder—about 7000 psi and 0.14% strain. With this amount of plastic strain, the low-cycle fatigue equations as presented in the previous chapter apply. The exact number of cycles to failure is difficult to predict, but it is definitely between 1 and about 100. Since the natural frequency of the example PWB is 200 Hz (cycles per second), failure will occur within 1 sec of application of the vibratory 10 $g$ load.

Some of the assumptions in the foregoing example are not quite correct, and the implications need to be discussed. The assumption that the chip carrier was rigid compared to the G-10 board is pretty good in that the elastic modulus of alumina is about 38 million psi, whereas the elastic modulus of G-10 is about 2 million. With the board being 0.1 in. thick and assuming the chip carrier is nominally 0.1 in. thick, this gives the stiffness of the chip carrier relative to the board as about 20 to 1. Since the chip carrier is so stiff, when mounted on the PWB it increases the local stiffness of the assembly. This increase in local stiffness is caused by the solder joints trying to prevent board elongation. Because of this increase in stiffness, the natural frequency calculation is incorrect and will result in a higher value. This higher natural frequency will result in a lower maximum deflection, which will in turn lower the stress calculation. Thus the local increase in stiffness has a compounding effect in our calculations, all leading to a lower solder joint strain.

The exact amount of local stiffening of the board and its influence on the natural frequency calculation and influence on the stresses and strains within the board are an area of current research. It also should be pointed out that the edge conditions of the board must be known precisely. Without precise knowledge of the boundary conditions, a sophisticated analysis must be tempered.

The purpose of this section and example is to point out that it is not the stresses in the PWB that are important, but the outer fiber strains. The stresses are so low that failure of the board material is rare and occurs only in the severest of loadings. It is the strains that must be accounted for. The strains are a measure of the elongation of the board. Any components mounted on the board must be able to accommodate this elongation.

It is also necessary to note that for vibratory loadings with frequencies on the order of 200 to 400 Hz, the components must be able to withstand hundreds of thousands of cycles for reasonable lives. At a frequency of 300 Hz and a life of 1000 hr, this corresponds to

(300 cycles/sec)(1000 hr)(60 sec/hr) = 18,000,000 cycles

The failure mechanism is high-cycle fatigue, which is dominated by the elastic strains. Figure 9.22 is an *S-N* plot for solder showing that the elastic stresses in the solder joints need to be less than about 2000 psi for long lives. Good fatigue data for solder at these high frequencies are not readily available.

## 9.3.5  Rib Stiffeners

If the stiffness of a PWB can be increased without a significant weight increase, the natural frequency will increase and the deflection at the center of the board will decrease rapidly. A decrease in the deflection means a decrease in connection stresses. One simple method

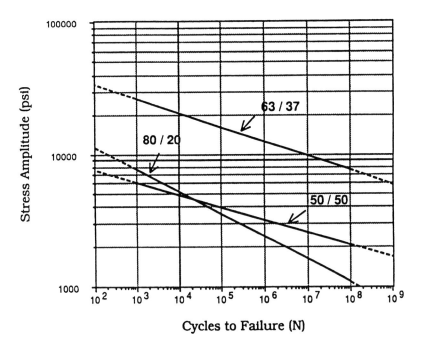

**Figure 9.22**   *S-N* plot for solder.

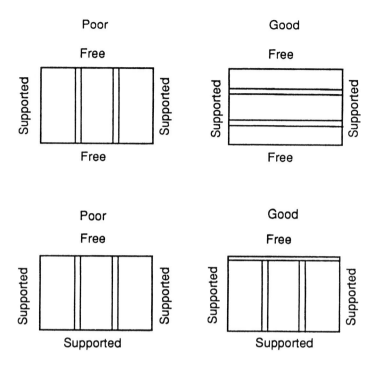

**Figure 9.23**   Adding ribs to effectively stiffen a flat plate.

for increasing the stiffness of the PWB is to add ribs. Ribs can be made of thin steel, copper, or brass and soldered directly to the copper cladding on the PWB. Such ribs form a very stiff section because of the high modulus of elasticity. These ribs can be undercut to allow the printed circuits to run under them without causing shorts.

To properly increase the stiffness of a PWB or plate, the ribs must be properly used. In order to be effective, they must carry a load directly to the supports. Consider the two cases in Figure 9.23, which emphasizes this point. The two ribs at the left in Figure 9.23 do nothing to stiffen the plate.

The natural frequency of rib-stiffened plates can be calculated by using equations similar to those presented in Table 9.2. For example, for a rectangular plate with simple supports along its edges, the natural frequency equation [14] is

$$f_n = \frac{\pi}{2} \sqrt{\frac{1}{\rho}\left(\frac{D_x}{a^4} + \frac{4D_{xy}}{a^2b^2} + \frac{D_x}{b^4}\right)} \qquad (9.83)$$

where $D_x$ and $D_y$ are the bending stiffness along the $x$ and $y$ axes, respectively, and $D_{xy}$ is the torsional stiffness.

## 9.4   FATIGUE LIFE CALCULATION FOR COMPONENTS MOUNTED ON PWBs

Two approaches can be taken to calculate the life of solder joints. The first approach uses a strictly empirical equation originally developed by Steinberg and based on the maximum

PWB displacement. The second approach directly determines the stress within the solder joint and then uses a high-cycle fatigue relation to determine life.

## 9.4.1 Empirical Maximum Deflection Calculations

Through extensive testing and design experience with PWB assemblies, Steinberg [8,10,15,16] has developed an empirical equation for the maximum dynamic single amplitude displacement at the center of the board, $d$. Steinberg's first equation, presented in his oft-referenced book *Vibration Analysis for Electronic Equipment*, first published in 1973 [10], was simply

$$d = 0.003b \qquad (9.84)$$

where $b$ is the short side of the rectangular board. His latest equation [10] has more parameters and begins to reflect the sophistication of modern PWB design.

$$d = \frac{0.00022B}{ct\sqrt{L}} \qquad (9.85)$$

where  $B$ = length of the PWB edge parallel to the component located at the center of the board (worst case), in.
$L$ = length of component, in.
$t$ = thickness of PWB, in.
$c$ = 1.0 for standard DIP
= 1.26 for DIP with side brazed leads
= 1.0 for pin-grid array with four rows of pins (one row extending along the perimeter of each edge)
= 2.25 for LCC

Remember that these are only empirical equations. The latest equation is rated for 10 million stress reversals under harmonic vibration (sinusoidal), and about 20 million stress reversals can be achieved under random vibration.

Assuming that the PWB acts like a single-degree-of-freedom system and combining the equations

$$Z_0 = \frac{9.8G_{max}}{f_n^2} = \frac{9.8G_{in}Q}{f_n^2} \qquad (9.77)$$

and

$$Q = A\sqrt{f_n} \qquad (9.78)$$

and solving for the natural frequency

$$f_n = \left(\frac{9.8G_{in}tc\sqrt{L}}{0.00022B}\right)^{0.66} \qquad (9.86)$$

This is the minimum natural frequency for a PWB for achieving a fatigue life of 10 million stress reversals in a simple harmonic vibration (sinusoidal) and about 20 million stress reversals under random vibration. Any natural frequency lower than this value would result in a greater deflection and resultant stresses that would ultimately lead to an early failure. But it must be remembered that the underlying assumption in the development of the above equation is that the PWB is simply supported on all four sides.

This purely empirical equation can be used as a first approximation for solder joint life prediction. Steinberg assumes that the stresses in the solder joints can be directly related to the out-of-plane displacement of the board. This assumption is only valid with simply supported boundary conditions on all board edges. In reality, the solder joint stresses are directly proportional to the radius of curvature directly under the component. This more accurate approach will be discussed later. A further implied assumption is that failures are due to solder joint high-cycle fatigue and the failures can be described by the relationship,

$$\sigma N_f^b = \text{constant} \qquad (9.87)$$

where $\sigma$ is the solder joint maximum stress amplitude, $N_f$ is cycles for failure, and $b$ and constant are material properties. That is, failures can be represented by a straight line on a log-log plot of stress versus cycles to failure.

With the implied assumption of stress being directly proportional to the out-of-plane displacement, the high-cycle fatigue relationship can be rewritten in terms of the out-of-plane displacement $Z$ as

$$ZN_f^b = \text{constant} \qquad (9.88)$$

This relationship can be used to determine the life of a board for any out-of-plane displacement $Z_2$, i.e.,

$$Z_1 N_1^b = Z_2 N_2^b \qquad (9.89)$$

or solving for life $N_2$

$$N_2 = N_1 \left( \frac{Z_1}{Z_2} \right)^{1/b} \qquad (9.90)$$

where $Z_1$ is the displacement from Steinberg's equation (9.85), $N_1 = 10,000,000$ cycles for sinusoidal and $20,000,000$ cycles for random vibration, and $Z_2$ is the maximum board displacement amplitude from

$$Z = \frac{9.8 G_{\text{in}} Q}{f_n^2} \qquad (9.77)$$

The life of components also depends on where they are placed. Assuming a simply supported board, components mounted in the center of the board have a shorter life than components mounted farther away. This variation in life is a function of the radius of curvature under the components and is included in the relationship

$$N_2 = N_1 \left( \frac{Z_1}{Z_2 \sin(\pi x) \sin(\pi y)} \right)^{1/b} \qquad (9.91)$$

where $x$ and $y$ are the nondimensionalized board coordinates of the component center.

Random vibration response equations are usually discussed in terms of the rms accelerations. When the acceleration distribution is Gaussian, which it is most of the time, then the maximum accelerations and stresses will be within 3 standard deviations 99.73% of the time. In engineering terms, this means that any acceleration or displacement should be multiplied by 3 to obtain the highest response occurring 99.73% of the time. This means that a random vibration level of 10 $g$ rms requires a design that will be capable of withstanding acceleration levels of $3 \times 10$ or $30\ g$. Thus to account for the $3\sigma$ extremes for random vibration

$$Z_0 = 3\left(\frac{9.8G_{\mathrm{rms}}}{f_n^2}\right) \tag{9.92}$$

Near resonance, the acceleration response of a single-degree-of-freedom system, subjected to a random vibration input that is flat white noise,

$$G_{\mathrm{rms}} = \sqrt{\frac{\pi}{2}PSD\,f_n\,Q} \tag{9.93}$$

where $G_{\mathrm{rms}}$ is the $1\sigma$ rms acceleration response and PSD is the power spectral density input at $f_n$, $g^2/\mathrm{Hz}$. The maximum displacement then becomes

$$Z_0 = \frac{36.8\sqrt{PSD}}{f_n^{1.25}} \tag{9.94}$$

which is then used in the life prediction as explained above.

Steinberg uses the simple rule of thumb that in a shock environment of less than a few thousand total cycles, the allowable stress levels can be up to six times higher than for normal vibration. Therefore, the maximum displacement equation for $d$ is multiplied by 6. The maximum shock displacement of the PWB is

$$d = \frac{0.00132B}{ct\sqrt{L}} \tag{9.95}$$

The maximum shock displacement at the center of the board can be expressed as

$$Z = \frac{9.8G_{\mathrm{in}}A}{f_n^2} \tag{9.96}$$

where $A$ is the shock amplification factor as found in shock spectra tables; typical values are from 0.5 to 1.5. Again, this is then used in the life prediction as explained for the harmonically excited board.

## Solder Joint Stress Calculations

More accurate fatigue life estimation can be accomplished by calculating the stresses in the solder joints and using a solder fatigue life relationship based on stress amplitude. A detailed finite-element analysis could be used to determine the dynamic stresses in all the solder joints, but because of the size and number of components on a typical PWB the problem quickly becomes unmanageable even for a supercomputer. The problem can be simplified by using matrix structural analysis concepts typically used by civil engineers. Each component lead can be represented with a stiffness matrix. Knowing the relative displacements between the component and the board, the stiffness matrix can be used to give the force transmitted across the solder joint. Implied in the use of this concept is the assumption that the deformation of the component case is negligible with respect to the deformation of the board.

In the simplest implementation of the concept, the component can be assumed to be a rigid case mounted above the board with linear springs as shown in Figure 9.24. It is assumed that all forces and rotational moments are negligible compared to the out-of-plane forces. The second-order dynamic displacements are also assumed to be negligible.

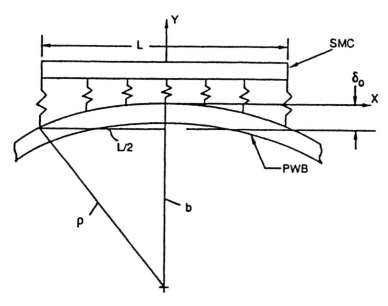

**Figure 9.24**   Component is modeled as a rigid body suspended above board with springs.

The PWB deformation for the fundamental mode can be expressed in terms of the radius of curvature, as shown in Figure 9.25. This information is directly obtainable from the natural frequency finite-element analysis. The change in each spring length can then be determined as a function of the radius of curvature for a specific component geometry by using the equilibrium equation. For each component

$$\sum_{i=1}^{m} F_{yi} = 0 = k_i(\delta_i - \delta_0) \tag{9.97}$$

where   $k_i$ = spring stiffness
$\delta_0$ = spring's initial length measured from an arbitrary reference
$\delta_i$ = spring's deflected length measured from the same reference
$m$ = number of solder joints for the component

This procedure can be used for both leaded and leadless components. For leaded components the spring representing the attachment is actually two springs in series, one for the solder and one for the lead. Once the spring deflections have been determined, the average maximum stress amplitude in each solder joint can be calculated by dividing the force by the cross-sectional area of the solder. The normal high-cycle fatigue relation

$$\sigma N_f^b = constant \tag{9.87}$$

can then be used to determine the fatigue life of the solder joint.

## Combined Vibration and Thermal Load

For an accurate account of the total damage to which a solder joint is subjected by simultaneous vibration and thermal cycles, the effects of vibration strain and thermal strain should be superposed appropriately. One of the simplest and most widely used

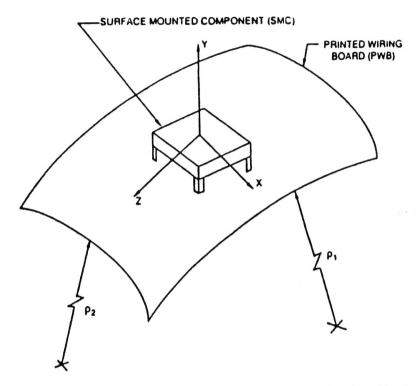

**Figure 9.25** Stress in component leads and solder joints is a function of local radius of curvature.

superposition schemes is Miner's rule [17]. Miner assumes that every structural member has a useful fatigue life and that every stress cycle uses up a part of this life. When enough damage has been accumulated due to stress/strain cycling, the effective life is exhausted and the member fails. Effective superposition is possible once the strain histories from vibrational and thermal loading are quantified.

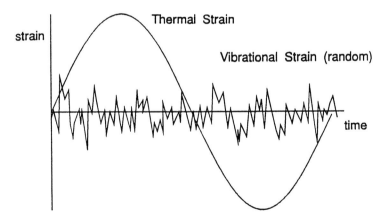

**Figure 9.26** Schematic representation of vibrational and thermal strain.

Most electronic devices are subjected to random vibrations as shown schematically in Figure 9.26. The associated stress or strain amplitudes in the solder joints can be assumed to have a Gaussian distribution about the rms amplitude near the PWB's natural frequency. For properly designed electronics, vibration strain amplitudes that exceed the $3\sigma$ values are considered to cause fatigue [8]. (Some feel that the $3\sigma$ value is not conservative enough and use $2.2\sigma$ [16].) Strain amplitudes above the $3\sigma$ value occur only 0.135% of the time in the Gaussian distribution, and thus the PWB has to be designed to handle only 0.135% of the total number of vibration cycles.

The strain produced by thermal cycles is caused by a mismatch in the steady-state thermal coefficients of expansion between the SMC and the PWB. By predicting the shear strain, $\Delta\gamma$, in the solder joint caused by both vibration and thermal cycles, the fatigue life can be predicted by utilizing the generalized Coffin-Manson equation [18]

$$\frac{\Delta\epsilon_T}{2} = \frac{\Delta\epsilon_e}{2} + \frac{\Delta\epsilon_p}{2} = \frac{\sigma_f}{E}(2N_f)^b + \epsilon_f(2N_f)^c \tag{9.98}$$

where $\dfrac{\Delta\epsilon_T}{2}$ = total strain amplitude

$\dfrac{\Delta\epsilon_e}{2}$ = elastic strain amplitude

$\dfrac{\Delta\epsilon_p}{2}$ = plastic strain amplitude

$\epsilon_f$ = fatigue ductility coefficient

$c$ = fatigue ductility exponent

$\sigma_f$ = fatigue strength coefficient

$b$ = fatigue strength exponent

$E$ = modulus of elasticity

Utilizing Miner's relationship, the ratio $R$ (of cycles experienced, $n_i$, to the fatigue life, $N_i$, at a specific $i$th strain level) is used to define the percent of the solder joint life that is exhausted.

$$R = \sum \frac{n_i}{N_i} \tag{9.99}$$

(Failure is assumed to occur when $R = 1$, but commonly failure is assumed to occur at a more conservative value such as 0.7 for electronic equipment, or lower when loss of life is at stake.) For the solder joint application, the subscripts th for thermal and v for vibration are used, and the fatigue-cycle ratio, $R_n$, becomes

$$R_n = \frac{n_{th}}{N_{th}} + \frac{n_v}{N_v} \tag{9.100}$$

The thermal strain damage, $D_{th}$, is given by

$$D_{th} = \frac{n_{th}}{N_{th}} \tag{9.101}$$

Because the vibration cycles occur at a much higher frequency than the thermal cycles, the ratio of vibrational frequency to thermal frequency, $f_v/f_{th}$, is used to determine the vibration strain damage, $D_v$, for $n_{th}$ thermal cycles or blocks

$$D_v = \frac{n_{th} f_v/f_{th}}{N_v} \tag{9.102}$$

The total damage or the total percent of the solder joint life exhausted by the combined loading, $D_{total}$, is given by

$$D_{total} = D_v + D_{th} = n_{th} \left( \frac{f_v/f_{th}}{N_v} + \frac{1}{N_{th}} \right) \tag{9.103}$$

The estimated solder joint fatigue life, $N_f$, for solder subjected to both thermal and vibrational strain cycles is given by

$$N_f = \frac{1}{D_{total}} \tag{9.104}$$

where $N_f$ is the number of thermal cycles or load blocks.

Low-cycle fatigue models, such as the modified Coffin-Manson model for solder joints of surface-mounted components, generally predict the fatigue life assuming the strain is completely inelastic [19]

$$N_f = \frac{1}{2} \left( \frac{\Delta\gamma}{2\gamma_f} \right)^{1/c} \tag{9.105}$$

where   $N_f$ = mean cycles to failure
   $\gamma_f$ = fatigue ductility coefficient
   $c$ = fatigue ductility exponent in shear
   $\Delta\gamma$ = shear strain range

It is usually assumed that the strains are primarily due to creep, anelastic. The fatigue life predictions from such low-cycle thermal fatigue models are found to be inaccurate when compared to actual lives of electronic assemblies [19–22]. A more elaborate and presumably more accurate procedure for predicting solder fatigue life uses the generalized Coffin-Manson equation (9.98), where the measured thermal strain is partitioned into inelastic and elastic portions. However, when compared to solder joint life experienced by actual electronic assemblies, this procedure also proves to be inaccurate.

Fatigue damage caused by high-cycle vibrations can help to explain the inaccuracies of the thermal fatigue models. Fatigue damage caused by vibration can be superposed with the damage caused by thermal loading utilizing Miner's rule. The fatigue damage caused by vibrational strain can be predicted two ways: by using an elastic strain fatigue model described by Eq. (9.87), or by partitioning the measured strain into inelastic and elastic portions and using the generalized Coffin-Manson equation (9.98). The more elaborate procedure of utilizing the generalized Coffin-Manson equation (9.98) presumably will provide the more accurate results because the strain is correctly divided into elastic and inelastic components. As a first-order approximation, the fatigue behavior for plastic and anelastic (creep) strains is assumed to be the same.

To develop an understanding of how solder joint life predictions will vary when vibrational loads are superposed on thermal cycling, three separate cases are examined. In each case typical minimum and maximum extremes of vibration strains experienced by electronic equipment are combined with various magnitudes of thermal strains.

Two approaches are used to determine how the high-frequency vibrational strain can influence the predicted solder joint life. In the first approach, the thermally induced strains are assumed to be completely inelastic and the vibrational strains are assumed to be completely elastic. Fatigue lives $N_v$ and $N_{th}$ are predicted from Eqs. (9.105) and (9.87), respectively. With these assumptions the fatigue life of the solder joint can be determined using Eq. (9.104). In the following discussions this method will be referred to as approach I.

In the second approach, both the thermal and vibrational strains are recognized to contain elastic and inelastic components and fatigue lives are calculated from the total strain, utilizing Eq. (9.98) in conjunction with Miner's rule as given in Eq. (9.104). This method will be referred to as the total strain approach or approach II.

A surface-mounted component is subjected to combined thermal and vibrational strains. The thermal load has a frequency of one cycle per day (block) and produces a shear strain of 6%. The natural frequency of the PWB is 200 Hz and the electronic assembly is subjected to random vibration. The strain amplitude in the solder joints can be assumed to have a Gaussian distribution about the rms amplitude near the PWB's natural frequency. Strain amplitudes that exceed the $3\sigma$ values are considered to cause the fatigue damage [8]. Strain amplitudes above $3\sigma$ occur only 0.135% of the time; thus the PWB must be designed to handle vibrational fatigue for only 0.135% of its life. Assuming 10 hr of operation per day, the PWB is subjected to 9720 vibrational cycles per block (per one thermal cycle).

$$\left(\frac{200\ cycles}{sec}\right)\left(\frac{3600\ sec}{hr}\right)\left(\frac{10\ hr}{block}\right)(0.00135) = 9720\ \frac{cycles}{block}$$

Different vibration strain amplitude levels are considered up to 1% shear strain, and the results are shown in Figure 9.27a.

For the second and third cases the solder joint is subjected to the same environment except that thermal shear strains of 3% and 1.5% are introduced. The results are shown in Figure 9.27b and c, respectively.

For all three cases the fatigue life is predicted using four different models and the results are summarized in Table 9.4. Columns three and four, labeled approach I, are based on the assumption that the life can be predicted by assuming the simple power law fatigue relation is valid. In column three, labeled $(N_f)_{th}$, the life is calculated using only the inelastic thermal strain, modified Coffin-Manson equation (9.105). Column four, labeled $(N_f)_c$, is calculated by using Miner's rule to combine thermal and vibrational damages, but assumes that the thermal strain is totally inelastic and that the vibrational strain is totally elastic, Eqs. (9.105) and (9.87), respectively. Columns five and six, labeled approach II, contain both the thermal strain–only model prediction, $(N_f)_{th}$, and the combined model utilizing Miners rule, $(N_f)_c$, calculated by utilizing the generalized Coffin-Manson equation (9.98).

Figure 9.27 summarizes the results of the three examples selected to represent a range of typical thermal and vibrational environments for electronic equipment. The predicted fatigue life is shown for the approach of assuming that the thermal strain is completely inelastic and that the vibrational strain is completely elastic, approach I, and shown with a dashed line in the figure. The predicted fatigue life is also shown for the more general total strain approach, where the elastic and inelastic strain components of the thermal and vibrational strain are used to predict the fatigue life, approach II, and shown with a solid line in the figure.

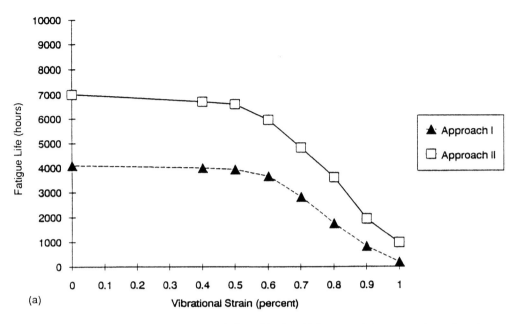

**Figure 9.27**   Vibrational effects on thermal fatigue life, in which approach I assumes thermal strains are only inelastic and vibrational strains are only elastic and approach II recognizes that thermal and vibrational strains have elastic and inelastic components, total strain approach: (a) 6% thermal strain, (b) 3% thermal strain, and (c) 1.5% thermal strain.

For the case of 6% thermal strain, shown in Figure 9.27a, vibrational loads affect the fatigue life prediction for vibrational strain levels over 0.5%. For a thermal strain level of 3%, shown in Figure 9.27b, vibrational strain levels over 0.5% are shown again to influence the fatigue life prediction. When the thermal strain level of 1.5% is examined, Figure 9.27c, vibrational strain levels above 0.4% are shown to influence the fatigue life prediction. For all three thermal strain levels, vibrational strains greater than 0.7% significantly reduce the fatigue life prediction.

Figure 9.28 incorporates the results of all three examples and shows the variance in the influence of vibrational strain in combined loading problems for different thermal strain levels. Figures 9.28 points out that as the thermal strain level decreases, the effect of vibrational strain on fatigue life increases. In addition, as the vibrational strain is increased above 0.4% there is a dramatic drop in fatigue life and the influence of vibrational strain must be accounted for.

The two different approaches used to predict the fatigue life produce dramatically different results. Assuming that the thermal strain is totally inelastic and that the vibrational strain is totally elastic significantly underestimates the life prediction made by the more general approach, which recognizes that both the thermal and vibrational strains are composed of inelastic and elastic components. Table 9.4 illustrates that the order of magnitude in difference between the two fatigue life predictions increases as the thermal strain is reduced. Table 9.4 also reveals that when the amount of vibrational strain decreases the difference between the two approaches decreases.

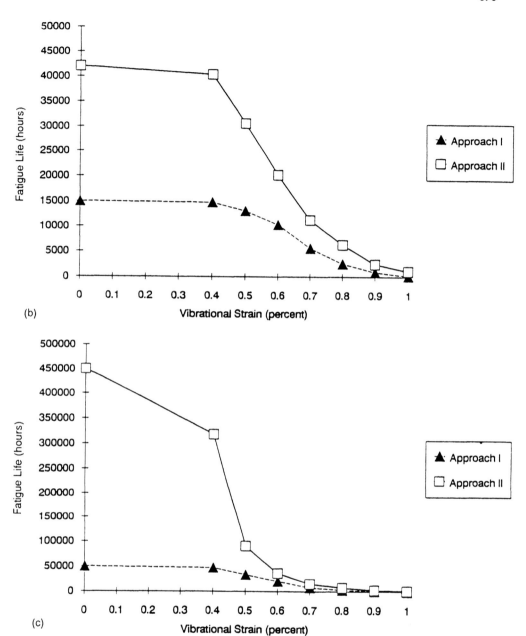

(b)

(c)

In the transition region from $10^3$ to $10^4$ cycles, both elastic and inelastic strains contribute a significant amount of fatigue damage. In particular, this suggests that high-frequency vibrations will shorten the life of solder joints and therefore cannot be neglected in a solder joint reliability prediction. There needs to be a generic unified method to properly account for the thermal as well as the vibrational strains.

**Table 9.4**  Solder Fatigue Life Predictions

| $(\Delta\gamma)_{th}$ (%) | $(\Delta\gamma)_v$ (%) | Approach I | | Approach II | |
|---|---|---|---|---|---|
| | | $(N_f)_{th}$ (hr) | $(N_f)_c$ (hr) | $(N_f)_{th}$ (hr) | $(N_f)_c$ (hr) |
| 6 | 1.0 | 4,100 | 186 | 7,000 | 970 |
| 6 | 0.9 | 4,100 | 823 | 7,000 | 1,930 |
| 6 | 0.8 | 4,100 | 1,760 | 7,000 | 3,600 |
| 6 | 0.7 | 4,100 | 2,820 | 7,000 | 4,820 |
| 6 | 0.6 | 4,100 | 3,660 | 7,000 | 5,940 |
| 6 | 0.5 | 4,100 | 3,940 | 7,000 | 6,590 |
| 3 | 1.0 | 15,000 | 192 | 42,000 | 1,100 |
| 3 | 0.9 | 15,000 | 964 | 42,000 | 2,510 |
| 3 | 0.8 | 15,000 | 2,560 | 42,000 | 6,290 |
| 3 | 0.7 | 15,000 | 5,640 | 42,000 | 11,300 |
| 3 | 0.6 | 15,000 | 10,400 | 42,000 | 20,230 |
| 3 | 0.5 | 15,000 | 13,100 | 42,000 | 30,600 |
| 3 | 0.4 | 15,000 | 14,800 | 42,000 | 40,400 |
| 1.5 | 1.0 | 50,000 | 194 | 450,000 | 1,130 |
| 1.5 | 0.9 | 50,000 | 1,010 | 450,000 | 2,650 |
| 1.5 | 0.8 | 50,000 | 2,910 | 450,000 | 7,280 |
| 1.5 | 0.7 | 50,000 | 7,660 | 450,000 | 14,900 |
| 1.5 | 0.6 | 50,000 | 20,200 | 450,000 | 36,000 |
| 1.5 | 0.5 | 50,000 | 33,700 | 450,000 | 90,300 |
| 1.5 | 0.4 | 50,000 | 47,700 | 450,000 | 318,000 |

The three examples with typical ranges of thermal and vibrational strains presented show that the fatigue damage caused by vibrations will have a significant effect on the reliability of solder joints. The use of a simple low-cycle power law fatigue model based on thermal strains will overestimate the actual life if the assembly is subjected to random vibration. The increased accuracy gained by superposing the damage accumulated from high-cycle vibrational strains is crucial in determining the overall reliability of electronic assemblies.

Superposition results obtained by assuming the thermal strain is totally inelastic and the vibrational strain totally elastic are incorrect. When solder joint life predictions are required in the transition region between low-cycle fatigue and high-cycle fatigue (1000 to 10,000 cycles), the increased accuracy of a generalized approach using total strain concepts and the generalized Coffin-Manson equation is required.

Close examination of the solder life predictions by superposing the thermal and vibrational strains utilizing the generalized Coffin-Manson equation and Miner's rule reveals the high sensitivity of the life predictions to solder properties. Consistent and accurate solder property data are required to model solder joint behavior accurately. The use of Miner's linear superposition rule to superimpose vibrational and thermal fatigue damage caused by both inelastic and elastic strains provides a sensitive and accurate model for predicting the reliability of electronic assemblies. Higher-order superposition models can be applied to provide a higher degree of accuracy only when more accurate solder properties are available.

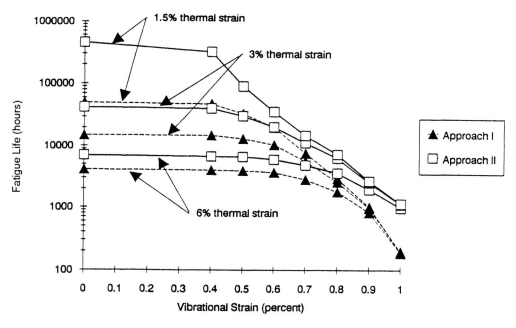

**Figure 9.28**   Vibrational effects on thermal fatigue life.

## EXERCISES

1. What is the natural frequency of vertical motion for a 10-lb weight suspended by a 10-ft length of $\frac{1}{16}$-in.-diameter steel wire? If this weight experiences a maximum acceleration loading of 2 $g$, what is the maximum velocity and displacement of the weight?

2. The octave rule of thumb recommends that the natural frequency of a PWB is always at least twice as large as the natural frequency of the chassis. Explain the rationale for this in terms of the dynamic magnification factor and/or transmissibility.

3. What is the maximum displacement of a PWB that has a natural frequency of 225 Hz and is subjected to a maximum acceleration of 10 $g$?

4. Compare the natural frequency of a 3 × 5 in. G-10 PWB 0.40 in. thick with that of a PWB with all dimensions doubled. Assume the PWB is simply support along the long sides and clamped along one of the short sides.

5. Using a finite-element program, compare the natural frequency of a simple rectangular board that has two clamped edges with that of the same board in which the clamped edge condition occurs over only one half the length of the edge.

6. Using a finite-element program, install rotational spring elements along two edges of a rectangular plate to simulate elastic supports. Create a plot of natural frequency versus rotational spring constant and show the transition from the classical boundary conditions of simply supported to clamped.

7. What is the maximum safe displacement of a 5 × 8 in. PWB 0.60 in. thick that has 1 in. square leadless chip carriers mounted on it and will be subjected to simple harmonic vibration?

8. Assume that the PWB in Exercise 7 is subjected to 15 $g$. What is the life of a 1 in. long dip mounted one third of the way in from the edges and parallel to the long axis of the board?

9. Assume that the PWB in Exercise 7 is subjected a uniform PSD of 10 $g^2$/Hz. What is the life of a 1 in. square leadless chip carrier mounted one third of the way in from the edges?

10. Model a 1 in. square LCC as shown in Figure 9.24 with springs suspended between the rigid chip carrier and the PWB. Assume the springs are evenly spaced on 0.05-in. centers around the entire perimeter of the chip carrier. Plot the equilibrium spring extension for the corner spring and the spring in the center of a side as a function of PWB radius of curvature. Assume the radius of curvature is infinity in one direction and vary the other.

11. If a PWB with a natural frequency of 250 Hz was found to have solder joint fatigue failures after 10,000 cycles when subjected to a 20 $g$ sinusoidal load, what would be the expected increase in life if the natural frequency of the PWB could be increased to 300 Hz by using different card guides?

12. Calculate the natural frequency of a 5 × 8 in. G-10 PWB 0.060 in. thick assuming that it is simply supported along the two short sides and subjected to a 10 $g$ sinusoidal load. Now calculate the natural frequency of the PWB assuming the edge guides are 0.080-in. slots. Hint: assume the PWB translates uniformly at constant velocity across the gap from contacting one side of the edge guide to the other side.

13. Derive the function expressing the system damping as a function of the ratio of displacement amplitudes measured at two points with a known number of intervening cycles.

14. What is the minimum spacing between two PWBs that have a natural frequency of 250 Hz and will be subjected to random vibration with a uniform PSD of 15 $g^2$/Hz?

15. Calculate the increase in fatigue life by (a) increasing the natural frequency of the PWB by a factor of 2 and (b) decreasing the $g$ loading by a factor of 2.

16. A 40-pin DIP is mounted at the center of a 6 × 8 in. PWB with a thickness of 0.080 in. The DIP will be mounted parallel to the long edge of the board. Determine the minimum desired natural frequency of the board when it will be subjected to a maximum acceleration level of 10 $g$. What is the fatigue life at this resonant frequency?

17. A PWB has a natural frequency of 196 Hz, with a transmissibility of 14 at the center of the board. It is desired to keep the maximum dynamic deflection to less than 0.020 in. What should the natural frequency of the PWB be for a 5 $g$ peak sinusoidal input?

18. Determine the natural frequency of a rectangular G-10 PWB of 8 × 6 × 0.060 in. with all four edges assumed to be simply supported.

19. The term decibel, as it is used in random vibration to measure PSD ratios $P_2/P_1$, is defined as

$$\text{Number of decibels (dB)} = 10 \log_{10} \frac{P_2}{P_1}$$

If the PSD doubles for a double in the frequency, how many decibels per octave is this?

20. Determine an expression for calculating the damping factor by measuring the static and dynamic displacement of a system at its natural frequency.

## REFERENCES

1.  Navy Manufacturing Screening Program, NACMAT P-9492, Navy Department, May 1979.
2.  Tuskin, W., "Recipe for Reliability: Shake and Bake," *IEEE Spectrum*, December 1986, pp.37–42.
3.  ASTM Standard D41.
4.  Mil-Standard-810D, Environmental Test Methods and Engineering Guidelines, July 19, 1983.
5.  Mil-Standard-167–1.
6.  Mil-Standard-1540B (USAF), Test Requirements of Space Vehicles, October 10, 1982.
7.  Mil-Standard-901C.
8.  Markstein, H. W., "Designing Electronics for High Vibration and Shock," *Electron. Packag. Prod.*, April 1987, pp.40–43.
9.  Sloan, J. L., *Design and Packaging of Electronic Equipment*, Van Nostrand Reinhold, New York, 1985.
10. Steinberg, D. S., *Vibration Analysis for Electronic Equipment*, Wiley, New York, 1973, 2nd ed., 1988.
11. Rayleigh, J. W. S., *The Theory of Sound*, Dover, New York, 1945.
12. Westergaard, H. M., "Moments and Stresses in Slabs," *Proc. Am. Concrete Inst. 17*(1921).
13. Boresi, A. P., Sidebottom, O. M., Seely, F. B. and Smith, J. O., *Advanced Strength of Materials*, 3rd ed., Wiley, New York, 1978.
14. Timoshenko, S. and Krieger, S. W., *Theory of Plates and Shells*, McGraw-Hill, New York, 1959.
15. Steinberg, D. S., "Stress Screening with Random Vibration," *Proc. Institute of Environmental Sciences*, 1980.
16. Steinberg, D. S., "Design Guidelines for Random Vibration," Designing Electronic Equipment for Random Vibration Environments, *Proc. Institute of Environmental Sciences*, 1982, pp.13–16.
17. Miner, M. A., "Cumulative Damage in Fatigue," *J. Appl. Mech.*, September 1945.
18. Manson, S. S., "Fatigue: A Complex Subject—Some Simple Approximations," *Exp. Mech.* 5:193–226 (1965).
19. Engelmaier, W., "Fatigue Life of Leadless Chip Carrier Solder Joints During Power Cycling," *IEEE Trans. Components, Hybrids, Manuf. Technol. CHMT-6*(3) (1985).
20. Lau, J. H. and Rice, D. W., "Solder Joint Fatigue in Surface Mount Technology: State of the Art," *Solid State Technol.*, October 1985, pp. 91–102.
21. Solomon, H. D., "Fatigue of 60/40 Slder," *IEEE Trans. Components, Hybrids, Manuf. Technol. CHMT*(4):423–432 (1986).
22. Wild, R. N., "Fatigue Properties of Solder Joints," Welding Research Supplement, November 1972, pp. 521s–526s.

# 10

# Humidity and Corrosion Analysis and Design

**Wing C. Ko and Michael Pecht**   *CALCE Center for Electronic Packaging, University of Maryland, College Park, Maryland*

## 10.1  HUMIDITY

### 10.1.1  Introduction

Electronic packages can be affected by many environments, including temperature, vibration, altitude, and humidity. One study [1] suggests that humidity alone constitutes up to 20% of the environmental stresses that lead to failure, and, together with temperature, is responsible for more than half of device failures.

Humidity is the amount of water vapor present in air. The air–water vapor mixtures follow Dalton's law of partial pressures, which states that in a given mixture of gases or vapors, each gas or vapor exerts the same pressure it would exert if it existed alone in the same space and at the same temperature as those in the mixture. The relative humidity (RH) is the ratio of the partial pressure of water vapor in the air to the pressure of saturated water vapor at the temperature of the air.

Typical problems that can result from exposure to a humid environment include the following [2]:

1. Swelling of materials due to moisture absorption
2. Changes in mechanical properties, including loss of physical strength
3. Degradation of electrical and thermal properties in insulating materials
4. Electrical shorts due to condensation
5. Binding of moving parts due to corrosion or fouling of lubricants
6. Oxidation and/or galvanic corrosion of metals
7. Accelerated chemical reactions
8. Chemical or electrochemical breakdown of organic surface coatings
9. Deterioration of electrical components

The high mobility of the water molecules in the vapor phase provides for easy ingress into a sealed unit, through either a defective seam or a selectively permeable gasket too tight for a liquid water molecule. The penetration, or permeation, of moisture is governed by two engineering principles. One is Dalton's gas law, which states that gases tend to stabilize themselves with respect to their own vapor pressure independently of other gases or vapors present. The other is that most materials are hygroscopic; that is, they can absorb moisture. It is often assumed that as long as hermetically sealed units are pressurized with dry gas, such as nitrogen, and a positive pressure condition is maintained, the units will remain dry. If the material used as a seal is selectively permeable to water vapor or is hygroscopic, water vapor will penetrate the "sealed" unit to reach an equilibrium vapor pressure.

In addition to permeation, water vapor may be entrapped within the cavity of hermetic parts at sealing or released by package materials after sealing [3]. At a constant temperature, most truly hermetic package cavities will contain a moisture level in equilibrium with the cavity walls. Temperature excursions will shift the equilibrium. In particular, an extreme drop in temperature can cause the sealed cavity to attain its dew point and condensation can form on inner surfaces of the cavity, including the integrated circuit chip. Integrated circuits encapsulated in a molding compound are more susceptible to moisture ingress than hermetically sealed circuits, because plastic molding compounds can contain contaminants and have been known to offer poor resistance to contaminant ingress from the ambient.

## 10.1.2 Hermeticity

Moisture by itself should not cause electronic problems when trapped in a microelectronic package, because it is a poor electrical conductor. However, water can dissolve salts and other polar molecules to form an electrolyte, which, together with the metal conductors and the potential difference between them, can create leakage paths as well as corrosion problems. This is the primary reason why electronic packages are usually categorized as either hermetic or nonhermetic [4].

If the seals are intact, hermetic packages are those made of metals, ceramics, or glasses, and nonhermetic packages are those encapsulated with polymers. High-reliability electronic packages, such as those used in military and space applications, are usually hermetic. In addition, MIL-STD-883 requires that all such packages contain no more than 5000 ppm water vapor when analyzed at 100°C by a mass spectrometer. To better understand the purpose of this limit, one needs to examine the phase diagram of water at various pressures and temperatures (Figure 10.1).

Once a hermetic package is sealed, it contains a fixed amount of moisture corresponding to the partial pressure of water contained in its initial internal atmosphere, which is generally below 100 ppm in a dry box. Further equilibrium will take place because of the positive contribution from absorbed moisture from the walls and other pores and surfaces as well as from outgassing products from such materials as epoxy adhesives and flux residues.

Generally, not enough water is trapped in the package at room temperature to intersect the gas-liquid line; therefore, no condensation will occur. If this initial moisture content is above line 1 in Figure 10.1, then cooling the package to $T_1$°C will initiate the formation of water. The minimum concentration for a hermetic package at 22°C is 6500 ppm [5]. Below this amount it is thought that no water can form, regardless of the temperature of the package, because moisture will change into ice directly (line 2, Figure 10.1). Therefore,

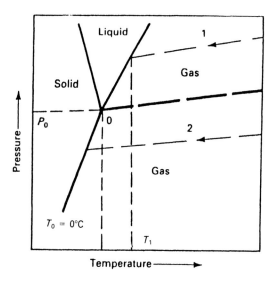

**Figure 10.1**  Phase diagram of water at various pressures and temperatures.

no ionic conduction facilitated by the formation of an aqueous solution can take place. In practice, supercooled water can form at temperatures below 0°C because water can remain liquid in micropores within the package, in effect lowering the triple point of water to well below 0°C. As a result, the minimum concentration of moisture was set at 5000 ppm to ensure that no liquid phase will form, even at temperatures lower than 0°C [4].

Research [6] indicates that the absolute minimum amount of moisture required to raise the surface conductance to a significant level is three monolayers of condensed water molecules. Table 10.1 summarizes the amount of internal vapor that can result in three-monolayer coverage [7]. Figure 10.2 shows the data graphically. Note that the larger the internal volume, the less moisture is required to produce the three monolayers. This is because with the ever-diminishing area-to-volume ratio there is less package surface area competing for water molecules, and thus they are more likely to reside on the chip. For single integrated circuit chips in their relatively small (<0.01 to 0.1 cm³) packages, the present MIL-SPEC internal water vapor limit of 5000 ppm is in good agreement with the

**Table 10.1**  Amount of Internal Moisture to Reach Critical Moisture Content

| Volume cm$^3$ | Surface area (cm$^2$) | Ratio of area to volume | Water thickness (cm) | No. of monolayers | Moisture content |
|---|---|---|---|---|---|
| 0.001 | 0.08 | 80 | | | 13,000 |
| 0.01 | 0.38 | 38 | | | 6,200 |
| 0.1 | 1.7 | 17 | | | 2,700 |
| 1 | 8.2 | 8.2 | $1.2 \times 10^{-7}$ | 3 | 1,300 |
| 10 | 38 | 3.8 | | | 600 |
| 100 | 175 | 1.75 | | | 280 |

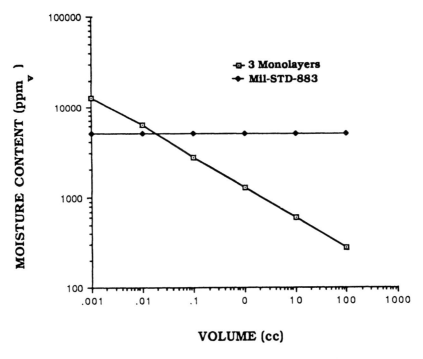

**Figure 10.2**  Amount of internal vapor that can result in three-monolayer coverage.

three-monolayer model. However, for hybrid assemblies in the large ($>1.0\,\text{cm}^3$) packages, MIL-STD-883 Method 1018.2 allows too much moisture in the package.

No packaging material is truly hermetic to moisture. The permeability to moisture of glasses, ceramics, and metals, however, is very low and is orders of magnitude lower than that of any plastic material. Figure 10.3 illustrates the time scales for moisture penetration for these materials. Although polymeric sealed devices can be designed to pass even fine leak tests, moisture will move through the seal in hours [8].

Hermetic packages are considered to be those made of metals, ceramics, and glasses. The common feature of these packages is the use of a lid or a cap to seal in the semiconductor device mounted on a suitable substrate. The leads entering the package must also be hermetically sealed. In metal packages the individual leads are sealed into the metal platform or "header" by separate glass seals. In ceramic packages the leads are formed by thick-film-screen techniques.

## 10.1.3  Hermeticity Testing

Hermetically sealed packages are tested to detect package sealing defects such as improperly formed seals, improperly sealed lids, and cracks and holes in the seals. Early detection and correction of leaks in hermetic packages will keep potential corrosion-related defects from becoming reliability problems.

There are two separate hermeticity test sequences: fine leak and gross leak tests. This is necessary because different leak rates exist, and one method alone is not sufficient. Gross leak test methods do not have the sensitivity to detect fine leaks. The fine leak test is usually performed first because fluids used in the gross leak test methods may contain large enough particles to plug fine leak holes.

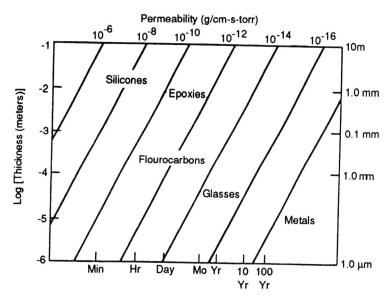

**Figure 10.3**  Time for moisture to permeate various sealant materials in one defined geometry. (From Ref. 8.)

## Gross Leak Test

The gross leak test detects devices with leak rates greater than $1 \times 10^{-5}$ atm cm$^3$/sec. The fluorocarbon bubble test is the most widely accepted gross leak test. It is safe and nondestructive because the test fluids are chemically inert. Gross leak testing is performed by depressurizing devices in a vacuum chamber. After a vacuum period, a detector fluid is poured over the devices in the chamber. While the devices are immersed in the detector fluid within the vacuum chamber, the chamber pressure is increased to a range of 30 to 75 psig and maintained there for several hours. The devices are then removed from the chamber while still immersed in the low-boiling fluorocarbon, the surfaces are dried for a couple of minutes, and then the devices are resubmerged in a detecting fluorocarbon fluid of a different density at 125°C. Bubble emission is the means of detection. Any devices emitting bubbles are rejected. This test method is effective and economical but is not applicable to noncavity or molded plastic parts.

The fine leak test is used to detect devices with leak rates between $1 \times 10^{-6}$ and $1 \times 10^{-8}$ atm cm$^3$/sec. Two methods currently in use for fine leak detection are the helium and radioisotope (krypton-85) gas methods.

### Helium Leak Test

Helium, the lightest chemically inert gas that is nontoxic, nonflammable, and nonhazardous, is an ideal tracer because of its detectability by the mass spectrometer. In the helium leak test, devices are pressurized in a pure helium pressure chamber, usually at 60 psig for 2 hr. The devices are then placed in a mass spectrometer leak detector, which draws a vacuum to force out helium from a leaky package. Helium is detected by the mass spectrometer, which produces a measured leak rate reading. This test method is effective and relatively inexpensive but is not applicable to noncavity or molded plastic parts.

*Krypton-85 Tracer Gas Test*

The krypton-85 tracer gas test requires pressurization of devices in the detectable gas, which in this case is dry nitrogen mixed with approximately 1% krypton-85 gas. After pressurization, the devices are taken to a detector (scintillation counter). The leak rate may then be measured by detecting gamma rays from the krypton-85 that was forced into the leaky package during the pressurization period. This test is faster and less costly than the helium method but is again not applicable to noncavity or molded plastic parts.

## Other Tests

Various methods are used to evaluate the resistance of electronic packaging to moisture intrusion and its deteriorating effects. Hermetic devices intended for military use are subjected to a moisture resistance test, such as Test Method 1004 of MIL-STD-883. This test is designed to evaluate, in an accelerated manner, the resistance of packages to high-humidity, high-heat conditions typical of tropical environments. The packages are subjected to 10 cycles, each consisting of temperature cycles of 25 to 65°C in an 80 to 100% relative humidity environment. The duration of the test is typically 240 hr. The test also includes low-temperature subcycles to −10°C to assess the effects of freezing moisture on packages. Temperature-humidity chambers preprogrammed to perform the required cycles automatically are widely available. Evaluations after the test include electrical measurements and a visual inspection for corrosion on the package exterior. The visual inspection and failure criteria are similar to those for the salt atmosphere test [9].

The standard humidity stress test for nonhermetic packages is the 85/85 test. A typical procedure is described in Test Method A100 of JEDEC-STD-22. Devices are placed under electrical bias in a temperature-humidity chamber at 85°C and 85% RH. The test duration is typically 1000 hr.

The unbiased autoclave, or pressure cooker, test is used to evaluate, in a more accelerated manner, the moisture resistance of nonhermetic packages. Because relationships between test failure mechanisms and actual field failures have not been established for this test, it has been found to be useful only for rough qualitative comparisons of different production lots or for qualification testing [10,11]. Difficulties with test reproducibility have also been encountered [12]. For testing in accordance with Test Method A102 of JEDEC-STD-22, the chamber is maintained at a pressure of 206 kPa absolute and at 100% RH. The temperature of the water in the chamber, which is established by the pressure and humidity conditions, is generally 121°C. Typical test durations range from 24 to 96 hr.

A test for evaluating the reliability of nonhermetic devices in a humid environment is the highly accelerated temperature and humidity stress test (HAST). Severe conditions of temperature, humidity, and bias are used to accelerate the ingress of moisture through encapsulant or seals, thus significantly reducing required test times [13]. Results have been found to be consistent with those of standard 85°C/85% testing [14]. A standard procedure for HAST testing is Test Method A110 of JEDEC-STD-22-A110 [15]. Four different conditions of temperature, humidity, and test duration, as well as provision for continuous or cycled bias, are incorporated into the test procedure. The applied temperatures range from 110 to 140°C, with test durations ranging from 25 to 200 hr. The relative humidity is maintained at 85%. Acceptance after completion of all the humidity stress tests of nonhermetic packages is based on compliance with electrical parametric limits and the demonstration of electrical functionality under nominal and worst-case conditions.

## 10.1.4   Impact of Moisture on Plastic Integrated Circuit Packages

Soldering of plastic surface mount components (PSMCs), including J-bend and gull-wing leaded packages such as PLCCs, SOICs, and PQFPs, is often performed in an environment that exposes the PSMC to extreme temperature conditions. Several PSMC manufacturers have discovered that certain solder reflow processes such as vapor phase, infrared, and hot air and hot plate can cause defects in the plastic packages [16].

PSMC cracking can occur when the package is heated such that moisture trapped inside the PSMC expands due to outgassing and then collapses to its original dimension, leaving cracks in the plastic shell. Cracks that propagate from the surface of the package to the die provide an open air trail allowing flux and other contaminants to reach the die pad. The failure mechanism may be wire bond pad or metallization corrosion, which may cause long-term wearout-type failure. Wire bond damage, such as wire breakage or bond cratering, may also occur. Plastic components that are socketed rather than soldered are not at risk for cracking. Localized solder reflow processes that concentrate heat on the package leads are also safe. These processes include hot bar reflow, laser soldering, and hand soldering.

Package cracking can be internal or external and is difficult to detect nondestructively. The most common mode is bottom cracking (Figure 10.4). Cracks may start at the edge of the die, or lead frame paddle. Delamination of molding compound from the chip or lead frame surface may occur. Study of the package cracking problem was summarized by the IPC [16]. Wire bond degradation due to cratering has also been recognized in the literature [16]. There are as yet no published reports of field failures due to cracked SMT modules resulting in moisture-induced wire bond damage. Moisture-induced wire bond damage is commonly found in temperature cycling, temperature and humidity cycling, and pressure cooker tests.

Any plastic package that reaches an internal body temperature greater than 210°C is at risk for package cracking. The higher the package temperature, the greater the risk. High package heating occurs primarily during solder reflow processes that heat the entire package, such as vapor phase, infrared, and hot air flow. In addition, solder wave processes that immerse the package body in the solder can result in package cracking.

### Failure Mechanism

The reflow solder temperature is above the glass transition temperature (approximately 140 to 160°C) of the mold compound, the temperature at which the coefficient of expansion increases sharply. During solder reflow, the mismatch between the coefficients of expansion of the mold compound, lead frame, and silicon die is maximum. If at the same time the plastic has absorbed excessive moisture, the resulting stress caused by moisture flashing into steam and the plastic expansion may be sufficient to cause delamination between the mold compound and the lead frame and/or die.

Many plastic packages absorb moisture over time, starting from the moment they leave the cure oven after molding. The probability of package cracking is heavily dependent on the amount of moisture that has been absorbed by the package and has diffused to the die paddle-epoxy interface. Because it is impossible to know exactly how much water is at the interface when a moisture concentration gradient exists, package cracking guidelines have been based on the total amount of water absorbed by the package at saturation at a given relative humidity condition. The amount of absorbed moisture is expressed as a

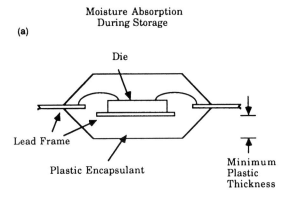

## CRACK GENERATION DURING SOLDER

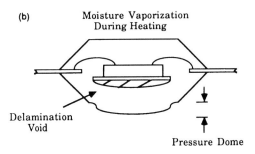

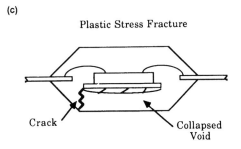

**Figure 10.4**  Schematic diagram of die cracking due to soldering.

percentage of the dry package weight and can be determined by baking a few packages for 48 hr at 125°C and recording the weight loss. The calculation is

$$\text{Weight percent moisture} = 100 \times \frac{\text{wet wt.} - \text{dry wt}}{\text{dry wt}} \qquad (10.1)$$

An absorbed moisture limit of 0.12% by weight has been used as a guideline for deciding whether a moisture-sensitive package needs to be baked before board mount.

## Guidelines

In order to eliminate the risk of moisture-induced package cracking of PSMCs, it is necessary to change the level of susceptibility to cracking at a specific moisture content of the package. The immediate method of achieving this goal is by ensuring that the PSMCs are "dry" at the time they are exposed to solder reflow process conditions. As of this writing, there are two ways to supply dry components to the manufacturing floor. The components can be baked to desorb moisture immediately prior to solder reflow or they can be stored in a desiccant environment until use.

In order to obtain PSMC components that will not be moisture sensitive to package cracking, new component package materials and designs are being investigated. Areas of study include molding compound modifications, lead frame surface finish evaluation, lead frame design changes, and the effects of molding processes on sensitivity to package cracking. Epoxy molding compound formulation changes that are being pursued include increased fracture toughness, lowered moisture absorption rate and saturation limits, and increased modulus and glass transition temperature ($T_g$). Lead frame design changes include the use of "paddle locks" and geometric modifications to lengthen the critical distance between leads and paddle. Molding processes are under closer scrutiny to determine methods of decreasing process-induced epoxy void size and distribution. Roughening the surface of the lead frame is being studied as a method of providing a mechanical lock to increase epoxy–lead frame adhesion to prevent delamination [16].

## 10.2  CORROSION

### 10.2.1  Introduction

Corrosion can occur during manufacturing, storage, shipping, and service. Failure of electronic equipment due to corrosion is dependent on the package type, corroding material, fabrication and assembly processes, and environmental conditions. The package and environmental conditions control moisture ingress into the package. The properties of the corroding material, contaminant and condensed moisture, control the rate of the corrosion process.

Corrosion is broadly defined as material deterioration caused by chemical or electrochemical attack. Although direct chemical attack can occur with most materials, electrochemical attack usually occurs only with metals. Typical corrosion sites in a package are shown in Figures 10.5 to 10.8. In electronic packages, the three most common corrosion mechanisms are uniform, galvanic, and pitting corrosion.

## Uniform Corrosion

Uniform corrosion is defined as a heterogeneous chemical reaction that occurs at a metal-electrolyte interface and involves the metal as one of the reactants. The corrosion process occurs uniformly over the surface of a material. The process can be either time dependent or self-limiting. If the reaction products are soluble, such as Al $(OH)_3$ or Al $(OH)_2Cl$, the corrosion process will continue monotonically with time until all the material is corroded. If the corrosion products do not dissolve readily in the corrodent, the process becomes a self-limiting phenomenon, in the sense that the current magnitude decreases as the corrosion product film thickness increases. In such cases, the corroded material is

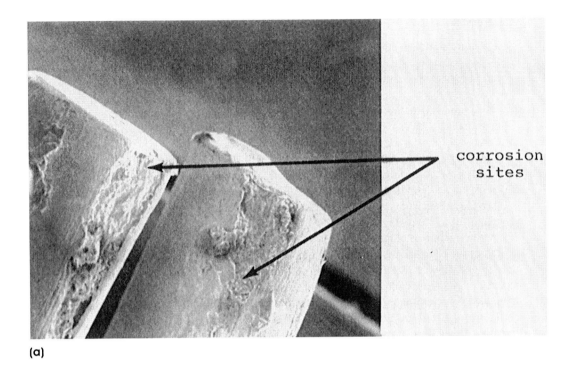

(a)

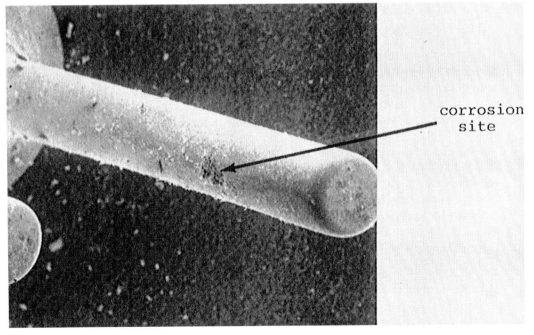

(b)

**Figure 10.5** Scanning electron micrographs of (a) extensive corrosion damage at the end of the lead and (b contamination in the upper part of the lead. (Courtesy of Unisys—NASA.)

nickel

gold

Kovar

**Figure 10.6**   Scanning electron micrograph of extensive corrosion at the end of the lead. (Courtesy of Unisys—NASA.)

typically in the form of an oxide that adheres to the corrosion surface and acts as a protective layer to retard further corrosion. The rate of corrosion will depend on the stability of this corroded layer. For example, aluminum oxide will protect aluminum from corrosion in a normal environment. However, in the presence of water and ionic contaminants such as sodium and chlorine, this oxide layer can be broken down [17].

Most commonly, uniform attack occurs on metal surfaces that are homogeneous in chemical composition or have homogeneous microstructure. Typical examples include corrosion of bare bond pads and unpassivated metallization (Figures 10.9 to 10.12).

Plating

Corrosion
Site

Copper
Cladding

Lead

Iron
Core

Header

**Figure 10.7**   Appearance of corrosion site beneath the plating of a randomly selected sample. (Courtesy of Unisys—NASA.)

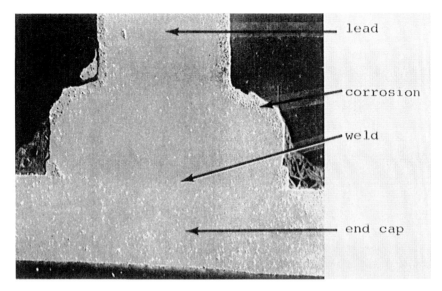

**Figure 10.8** Scanning electron micrograph of lead weld area cross section of Victoreen MOX-400 resistor. (Courtesy of Unisys—NASA.)

## Galvanic Corrosion

Galvanic corrosion occurs when dissimilar metal structures, such as molybdenum-gold metallization or the aluminum metallization–gold bond wire interface, are subjected to corroding environments. Corrosion occurs at exposed regions such as bonding pads and can proceed along the passivation-metal interface (Figures 10.13 and 10.14). An electrical potential difference will usually exist between two dissimilar metals exposed to a corrosive solution. When these two metals are electrically connected, the more active metal will become the anode in the resulting corrosion cell, and its corrosion rate will be increased. The extent of this increase in corrosion will depend on several geometric and electrical factors. A high resistance in the electrical connection between the dissimilar metals will tend to decrease the rate of attack. On the other hand, if a large area of the more noble metal is connected to a smaller specimen of the more active metal, attack of the more active metal will be accelerated.

The conductivity of the corrosion medium will affect both the rate and the distribution of galvanic attack. In solutions of high conductivity, the corrosion of the more active alloy will be dispersed over a relatively large area. In solutions having low conductivity, most of the galvanic attack will occur near the point of electrical contact between the dissimilar metals. The latter situation is usually the case, for example, under atmospheric corrosion conditions [17].

Aluminum is the material most often used for die metallization, and corrosion of aluminum is one of the most common failure mechanisms in microelectronic devices. The aluminum corrosion reaction is written in three parts. The electrolysis of water, which produces hydroxyl and hydrogen ions, is given by an anodic reaction specifying the oxidation of water:

$$2H_2O \rightarrow O_2\uparrow + 4H^+ + 4e^- \tag{10.2}$$

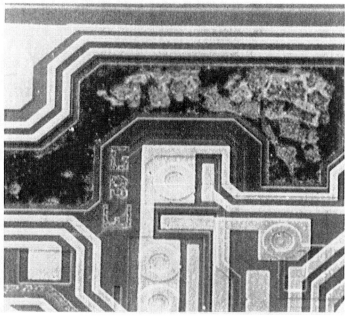

(a)

(b)

**Figure 10.9**   Scanning electron micrographs of metallization corrosion in SN14. (a) ×300; (b) ×480. (Courtesy of Unisys—NASA.)

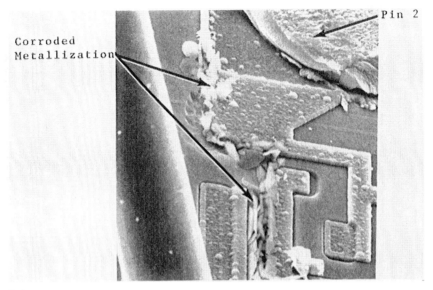

(a)

(b)

**Figure 10.10** Scanning electron micrographs of corrosion on the die. (a) ×550; (b) ×800. (Courtesy of Unisys—NASA.)

(a)

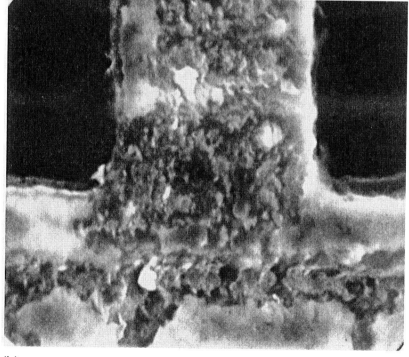

(b)

**Figure 10.11** Scanning electron micrographs of metallization corrosion in SN15. (a) ×1000; (b) ×4400. (Courtesy of Unisys—NASA.)

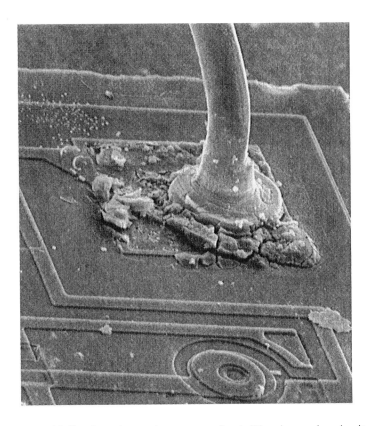

**Figure 10.12**  Corrosion products surrounding ball bond on a microcircuit die. (Courtesy of Unisys—NASA.)

and a cathodic reaction specifying reduction of hydrogen

$$2H^+ + 2e^- \rightarrow H_2\uparrow \qquad (10.3)$$

The total reaction is typically a cathodic corrosion product

$$2Al + 6H^+ \rightarrow 2Al^{3+} + 3H_2\uparrow \qquad (10.4)$$

$$2Al^+ + 6H_2O \rightarrow 2Al(OH)_3 + 3H_2\uparrow \qquad (10.5)$$

Aluminum corrosion can occur at both anodic and cathodic sites. In general, anodic corrosion of aluminum occurs either in the presence of an impressed electronic field or when a material of a different galvanic potential and an electrolyte are present. The presence of an electrolytic solution with conductivity $>10^{-7}$ mho will significantly increase the corrosion rate of many metals, including aluminum.

With the use of chlorine-based dry etches for very large scale integration (VLSI) metallization, the following reactions with aluminum are also likely:

$$6HCl + 2Al \rightarrow 2AlCl_3 + 3H_2 \qquad (10.6)$$

$$AlCl_3 + 3H_2O \rightarrow Al(OH)_3 + 3HCl \qquad (10.7)$$

$$2Al(OH)_3 + aging \rightarrow Al_2O_3 + 3H_2O \qquad (10.8)$$

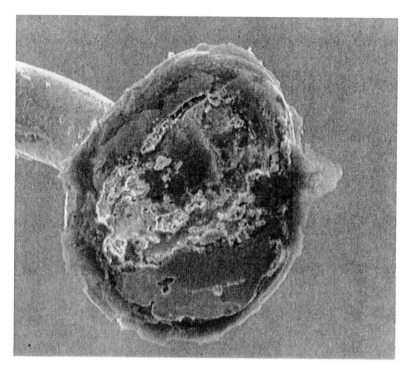

**Figure 10.13**  Underside of ball bond after lifting off of die. (Courtesy of Unisys—NASA.)

Equation (10.6) governs the corrosion process for aluminum in an HCl environment. If the moisture content in a package is sufficient, the reaction may continue until the exposed aluminum is completely reacted.

Since aluminum is easily corroded, gold is often substituted. However, when highly reactive ions such as $Cl^-$, $Na^+$, or $K^+$ are present, even gold corrodes. For example, corrosion occurs when there is generation of $AuCl_3$, which diffuses or migrates to the cathode, where it starts dendritic growth back to the anode (Figure 10.15). Gold metallization can also corrode in the absence of highly reactive ions, given enough humidity, high voltage, and small spacing between conductors. The process is the anodic oxidation of gold to form $Au(OH)_3$.

## Pitting Corrosion

In pitting corrosion, attack is highly localized at specific areas that develop into pits. Active metals such as aluminum, as well as alloys that depend on aluminum-rich passive oxide films for resistance to corrosion, are prone to this form of attack. Pits usually appear at well-defined boundaries at the surface, but pit growth can often change direction as penetration progresses [17]. When solid corrosion products are produced, the actual corrosion cavity may be obscured but the phenomenon can still be recognized from the well-defined nature of the corrosion product accumulations.

Pitting corrosion is usually the result of localized, autocatalytic corrosion cell action, in the sense that the corrosion conditions produced within the pit accelerate the corrosion

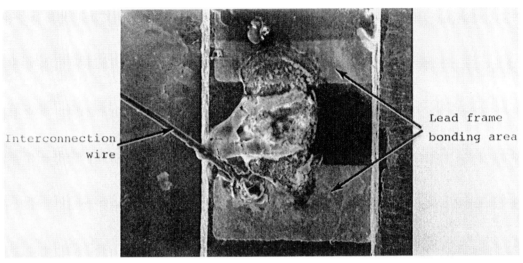

**(a)**

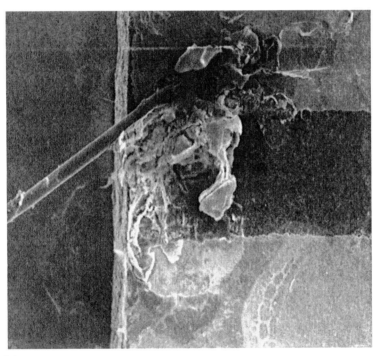

**(b)**

**Figure 10.14**  Scanning electron micrographs of LCC 5 showing corrosion products between bonding pads. (a) ×110; (b) ×180. (Courtesy of Unisys—NASA.)

**Figure 10.15**   Scanning electron micrograph of dendritic growth in SN 14. (Courtesy of Unisys—NASA.)

process. As an example of how such autocatalysis works, assume that there exists a weak spot in an oxide film in an oxygenated solution of sodium chloride covering the aluminum surface so that the corrosion process initiates at this point. The local accumulation of $Al^{3+}$ ions will lead to a local increase in acidity due to the hydrolysis of these ions. That is, the hydrolysis of aluminum ions gives an overall anodic reaction.

If the cathodic oxygen reduction occurs at a region removed from the anodic reaction, the localized corrosion of the aluminum will produce an accumulation of acid. This acid destroys the protective oxide film and increases the rate of attack. In addition, the accumulation of a positive charge in solution will cause the migration of $Cl^-$ ions to achieve solution neutrality. This increased $Cl^-$ concentration can then further increase the rate of attack. Because the oxygen concentration within the pit is low, the cathodic oxygen-reduction reaction occurs at the mouth of the pit, thus limiting its lateral growth. In most cases pits tend to be randomly distributed and of varying depth and size [17].

## 10.2.2   Factors Affecting the Rate Of Corrosion

There are several mechanisms by which a package can be contaminated and corrosion initiated. Contaminants may be sealed within a package, depending on the properties of the structural materials and fabrication methodology, or may permeate into the package. When corroding contaminants are coupled with the appropriate environmental conditions, such as temperature, temperature cycling, relative humidity, electrical bias (either applied or galvanic), and ionic contamination levels, corrosion occurs.

Corrosion failures in microelectronic devices are generally accelerated by high moisture levels in the die cavity, ionic contamination from lead glass seals, improper phosphorus levels, high negative potential on the die surface, and low operating temperatures. Device failures associated with corrosion include opens in lead wires; shorting due to flaking, peeling, dendrite growth, and migration; and migration and parameter degradation due to electrical leakages.

## Properties of the Structural Materials

Corrosion of metallization and bonding materials occurs predominantly on aluminum subjected to chlorine or other halogen contaminants. Depending on the stability and inertness of a material, the rate of corrosion is determined. For example, gold has a more stable atomic structure than aluminum, and in the corrosion process the gold atoms will react much slower (approximately 10 times slower) than the aluminum atoms. However, as the component dimensions are miniaturized and the current densities are increased, even gold will corrode in the presence of complex builders such as $Cl^-$, $Na^+$, or $K^+$ ions [18].

To minimize moisture access to the metallization layer in a package, a passivation layer is often applied. A high-quality, contaminant-free passivation layer can extend the time to failure by as much as three orders of magnitude compared to an unprotected counterpart [19]. The passivation layer is frequently a glass, silicon nitride, silicon dioxide, or other dielectric layer deposited on the device surface. Plasma-deposited silicon nitride is used as a passivating layer because it is an effective barrier for alkali ions that may be left on the device surface after the final wash. However, silicon nitride can give rise to a change in the surface potential, which can introduce charge trapping and storage effects and lead to interface conduction due to the activation of traps. These effects can contribute to corrosion as well as device electrical performance degradation [20].

Failures due to corrosion of the metallization are frequently associated with defects in the passivation, such as cracks and pinholes. These defects act as sites for the entrapment of contaminants and moisture, which are instrumental in corrosion. The chemical vapor deposition (CVD) process deposits films that are in tensile stress and can result in localized lifting or cracking of the passivating layer in regions of poor adhesion. This mechanism creates sites at which contaminants can penetrate protected metallization patterns and initiate corrosive attack.

The phosphorus content of the silicon dioxide layers used in passivation is an important factor in the reliability of the devices. Phosphorus is added both to provide mechanical stability of the glass coating and as a getter for mobile sodium ions. However, too little phosphorus causes cracks in the passivation layer due to tensile stresses in the film, and too much phosphorus can enhance aluminum corrosion. Again, cracking of the passivation increases the susceptibility of the underlying aluminum to corrosion by impurities.

## Fabrication Methodology

Fabrication methodologies affect the amount of contaminants inherent in the package, the potential for contaminant ingression, and the rate at which corrosive damage will lead to a failure.

Throughout the fabrication process, wafers are cleaned with various acids and solvents, which can function as contaminants of integrated circuits and thus affect wafer ecology (Figure 10.16). Wafers are also exposed to industrial and specialty gases at every process step. As dry processing becomes more prevalent, gases are now one of the major raw materials employed in the manufacture of microelectronic packages.

Another factor to consider during the fabrication process is the recontamination of baked-out parts during handling. That is, even properly baked-out parts can be recontaminated by the reexposure to moisture at ambient environment. Thomas [21] found that a dry surface will adsorb many monolayers of moisture during exposure in less than 1 sec. Therefore, baked-out parts should be handled carefully.

**Figure 10.16** Interior of LCC 5. (Courtesy of Unisys—NASA.)

Die attach materials can also affect the contaminant level in a package when improper fabrication techniques are employed. For example, when epoxies are used as an adhesive, they can act as a source of ionic contamination because the epoxide resin itself is typically a mixture of hydrolyzable chlorine, bromine, and sodium.

The lid-sealing process also affects the final moisture content of the package through the silicon used in the chip-bonding eutectic. The amount of silicon available is dependent on the temperature and duration of the sealing process. As the sealing time and therefore the temperature duration increase, the amount of silicon diffused from the eutectic bond increases and the silicon produced will react with the water vapor to produce silicon oxide. Hence the level of moisture decreases.

## Package Permeability and Leakage

In 1966, Eisenberg et al. [22] discovered that the low-temperature reliability of the 2N1132 (*pnp*) mesa transistor was related to moisture ingress and subsequent corrosion of the integrated circuit metallization. Other research [21] suggested that nichrome resistors with glass passivation were sensitive to moisture contamination and that acceptable reliability could be achieved by limiting moisture content to less than 500 ppm, since corrosion-related failures were observed with moisture contents greater than 1000 ppm. Subsequent research [23–26] supported these early results, and it was concluded that integrated circuits should be encapsulated to protect them against environmental conditions such as moisture. However, even with the utilization of hermetically sealed packages, corrosion still remains one of the critical concerns associated with the reliability of microelectronic devices.

The materials used to fabricate an electronic package will determine the inherent resistance of the package to corrosion, its ability to minimize the moisture ingress rate, and the amount of impurities that enter it. Microelectronic packages provide protection to the internal die and interconnection circuitry and also provide a mating surface to the external circuitry. The permeability and leakage of a package determine the ingress rate of contaminants from the ambient environment.

Package enclosures are referred to as hermetic or nonhermetic, according to the package's ability to resist moisture intrusion. The ability to resist the ingress of moisture can differ by four orders of magnitude between hermetic and nonhermetic packages [18]. Hermetic packages are enclosed with inorganic moisture-resistant material such as metal, glass, and ceramic. Nonhermetic packages are made of organic materials, such as plastics, that are permeable to moisture. Plastic packages are typically molded from silicone, phenolic, or epoxy. Because of their cost, plastic packages have become very popular. However, they offer reduced resistance to ingress of moisture and have not been acceptable for military application. Efforts are being made to improve their moisture and contaminant protection characteristics. With new encapsulating packaging materials, moisture ingress by permeation and diffusion is playing a smaller role than moisture flow through cracks and defects in the package.

Moisture is often trapped in the cavity of a hermetically sealed microelectronic during the assembly process. In addition, materials such as polymers used for the die attachments, gold-plated surfaces, and even the plastic used for the package inherently contain adsorbed moisture. Moisture can also enter a package by permeation through polymers used in the seals and the package material, or by leakage through cracks or small voids between the plastic material and lead frame in molded plastic packages. Furthermore, temperature and power cycling may induce crack growth in package materials, destroying the hermetic seal [20].

For a hermetic package, moisture ingress by diffusion through the housing material is insignificant. The quantity of the permeated or leaked contaminants is dependent on the construction processes, materials, and operating stresses. For example, the oxide coating on the lead frame conductor must be sufficiently thick to allow a good chemical bond to the sealing glass, and the design of the seal and the residual stress present in the seal must be able to withstand the rigors of thermal shock.

All packages leak at some rate. A package is defined as hermetic if its leak rate is less than a specified value. For military microelectronic packages this value ranges from $< 1 \times 10^{-8}$ atm $cm^3$/sec for package volume less than 0.01 $cm^3$ to $< 5 \times 10^{-7}$ atm $cm^3$/sec for package volume greater than 0.4 $cm^3$, as defined in MIL-STD-883, Method 1014 [27]. The leak rate is dependent on the shape of the leak path. In general, the leak rate decreases as the cross section decreases or as it becomes more elongated and the flow changes from viscous to molecular. Generally, leaks into packages are due to stress cracks that have elongated rather than having circular cross sections. The differential pressure of the leaking medium and the properties of leaking medium, including viscosity, density, and molecular mean path, also affect the leak rate.


## Environmental Conditions

The environmental conditions (relative humidity and temperature) determine the rate of moisture ingress and the moisture induction time for the package. For corrosion to occur inside a package, moisture has to ingress into the package to support the transfer of ions.

Therefore, the time to corrosion failure of a package is dependent on the time for the moisture ingress, or induction time, and the time for the corrosion process.

As the temperature increases, the rate of moisture ingress increases and the induction time is reduced. However, corrosion will not occur if the temperature surrounding a potential corroding material is high enough to prevent the formation of a medium (i.e., moisture condensation for ionic transfer). Thus the nonoperating environment of the package can affect the corrosion process more severely than the operating environment, especially when the active device generates a high temperature on the die.

## Relative Humidity and Moisture Content

Uhlig [28] stated that a critical relative humidity exists below which corrosion is negligible. He found that an electrolytic layer (say of condensed water) must reach a critical thickness in order to dissolve contaminants and support ionic conduction. The absolute minimum, or threshold, moisture content has been described as one that provides three monolayers of condensed water molecules [6]. The time to reach the threshold moisture content (induction time) depends primarily on the internal volume and leak rate of the package and the environment. During operation, new leak paths can start and old paths may grow, which can decrease the estimated induction time.

After the critical moisture content is reached inside a package, corrosion will start once the nonoperating sealed package is exposed to a temperature below the dew point, so that the moisture inside the package condenses and combines with any ionic contaminant present to provide a conductive path between adjacent metallic conductors. The conductive path serves as a medium for the transfer of ions in the corrosion process. When the package is operating, the heat dissipated by the chip will typically elevate the temperature inside the package above the dew point. Consequently, the electrolyte will evaporate and will no longer provide an electrolytic path between conductors.

The inside temperature functions as a "switch" for the corrosion process. The thermal-corrosion activation function, $f*(T)$, has a value of 0 when the inside temperature is above saturation temperature or below the freezing point and has a value of 1 when the inside package temperature is below the saturation temperature. Hence, the function can be expressed as

$$f*(T) = \begin{cases} 0 & \text{for } \begin{cases} T_{inside} > T_{saturation} \\ T_{inside} < T_{freezing} \end{cases} \\ 1 & \text{for } T_{inside} < T_{saturation} \end{cases} \qquad (10.9)$$

## Temperature and Temperature/Power Cycling

The permeation and corrosion rates are affected by the surrounding environment—in particular, temperature combined with moisture. Chemical reaction rates are enhanced at elevated temperatures. Thus the rate of corrosion will increase accordingly due to a higher rate of electron transfer. Temperature changes can also induce leakage of moisture into otherwise hermetic packages or increase the permeability of molded packages. On the other hand, localized high temperature can inhibit moisture condensation leading to corrosion. The primary components of the microelectronic package in terms of moisture ingress are the case and the lid and lead seals.

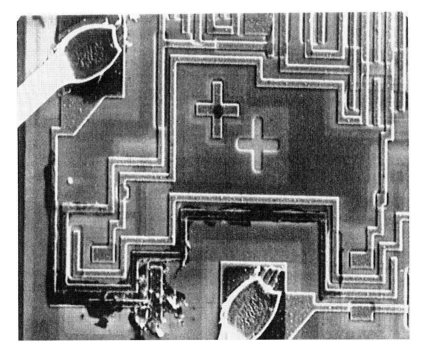

(a)

(b)

**Figure 10.17**  Scanning electron micrographs of organic contamination on die surface in SN 2. (a) ×200; (b) ×780. (Courtesy of Unisys—NASA.)

A microelectronic package undergoes numerous duty cycles in its life, and the application and removal of power will subject the package to thermal cycling. Because mated parts of a package often have different coefficients of thermal expansion, expansion and contraction of the material can increase device cracking at regions of maximum stress and provide a path for moisture ingress.

*Electrical Bias*

When sufficient condensed moisture is present to act as an electrolyte and when a bias is applied, ions are carried to the anode or cathode, depending on the impurities involved. The bias may be simply a galvanic bias or may be the varying voltage levels on different conductors caused by normal operation of the integrated circuits. Ionic contamination will accelerate the corrosion process in the presence of mobile ions in the electrolyte.

*Ionic Contamination*

Abbott [29] found that parts per billion levels of selected pollutants are sufficient under proper environmental conditions to accelerate corrosion reactions in electronic equipment. Corrosive agents include chlorides, fluorides, hydrogen sulfide ($H_2S$), sulfur dioxide ($SO_2$), nitrogen compounds such as ammonia ($NH_3$), and other airborne contaminants [30]. Sources of chlorine ions in a package include the chlorine-based dry etches used for microcircuit metallization or, if epoxy is used in the package, the outgassing of the surrounding epoxy. Sodium ions can be produced from the epoxy or glass used in the package.

Contaminants can also be introduced by the use of forced ventilation cooling in an electronic enclosure, with air generally drawn from the surrounding atmosphere. In sites with aggressive atmospheres, this type of cooling can accelerate corrosion when the circulating contaminated air comes in intimate contact with sensitive electronics (Figures 10.17 and 10.18) [29].

## 10.2.3 Review of Corrosion Models

Various deterministic and empirical models have been developed to predict the corrosion failure rate of a microelectronic package. Some of the more recent and representative models are discussed below.

The Motorola Discrete and Special Technologies Group [31] proposed an empirical model that incorporates temperature and humidity effects in the time to failure, $\tau$. The model is based on an Arrhenius-Eyring formulation in terms of temperature and relative humidity. The form of the model is

$$\tau = \tau_B \, AF \tag{10.10}$$

where $\tau_B$ is a base time to failure and

$$AF = \exp\left[\frac{E_a}{K}\left(\frac{1}{T} - \frac{1}{T_R}\right) + B\left(\frac{1}{RH} - \frac{1}{RH_R}\right)\right] \tag{10.11}$$

is an acceleration factor, $E_A$ is a thermal activation energy (eV), $K$ is Boltzmann's constant ($8.62 \times 10^{-5}$ eV/K), $T$ is the operating temperature (K), $T_R$ is a reference temperature (K), $B$ is a humidity activation energy (%), RH is the operating relative humidity (%), and $RH_R$ is a reference humidity (%).

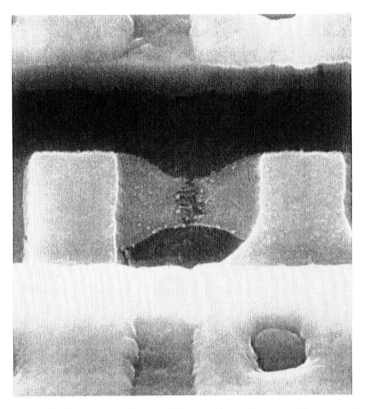

**Figure 10.18** Fused link caused by ionic contamination. (Courtesy of Unisys—NASA.)

ITT/Honeywell [32] proposed a failure rate model that incorporates an effective relative humidity caused by an averaged or effective device operational history. The time-to-failure distribution for the corrosion associated with moisture is modeled with a log-normal distribution (which is not a wear-out distribution as observed from experimental studies) with a mean life of

$$t_{50} = Ae^{\Delta H/KT} e^{296/RH_{eff}} \tag{10.12}$$

where $\Delta H$ is the activation energy (0.2 eV) and $RH_{eff}$ is an effective relative humidity. $RH_{eff}$ is a function of the component junction temperature, $T_j$, and the ambient relative humidity, RH,

$$RH_{eff} = DC(RH) e^{5230(1/T_j - 1/T_A)} + (1 - DC)(RH) \tag{10.13}$$

where DC is the duty cycle (% operating time) and $T_A$ is the normalizing temperature. In this model, moisture ingress was neglected and thus the associated corrosion life for "hermetic" packages was set to zero. Equations (10.12) and (10.13) are plotted in Figure 10.19 to demonstrate the effect of duty cycles and relative humidity at different temperature.

Slota and Levitz [33] proposed a corrosion model that addresses the temperature, humidity, and duty cycle dependences of plastic-encapsulated PROM packages. The model

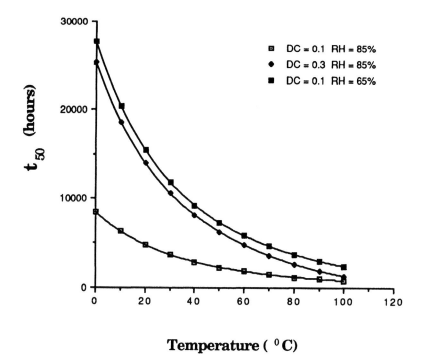

**Figure 10.19**  Effects of duty cycles and relative humidity at different temperatures based on the IIT/Honeywell model [31].

was based on empirical equations with constants determined from experimental data. The model has the same form as Eq. (10.11) but

$$AF = (AF_T)(AF_{RH})(AF_{DC}) \tag{10.14}$$

where

$$AF_T = \exp\left[\frac{\Delta H}{K}\left(\frac{1}{T} - \frac{1}{T_R}\right)\right] \tag{10.15}$$

$$AF_{RH} = e^{-c(RH^2 - RH_R^2)} \tag{10.16}$$

$$AF_{DC} = \left(\frac{DC}{DC_R}\right)^{0.5} \tag{10.17}$$

Here the subscript R means the reference condition.

Howard [34] proposed a model for metallization corrosion using Ohm's and Faraday's laws. The model was developed to account for the corrosion process in terms of the transfer of ions. The leakage current between the conductors in Howard's model (Figure 10.20) is

$$i = \frac{V}{R} = \frac{V}{\rho S/A} = \frac{VL}{(\rho/z)S} \tag{10.18}$$

where $S$ is the separation of conductors connected by the electrolyte and $A = Lz$ is the cross section of current flow given in terms of the electrolyte thickness $z$ and the length

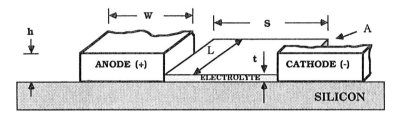

**Figure 10.20** Schematic diagram of die metallization.

$L$ of the conductor edge exposed to the electrolyte and perpendicular to $S$. The term $\rho/z$ is the sheet resistance of the electrolyte, and the number of squares is given by $S/L$. The quantity of material corroded is given by Faraday's law with the introduction of Eq. (10.19),

$$\frac{w^2 h n d F}{M} = \int_\tau i \, d\tau = \frac{V\tau}{\rho/z} \tag{10.19}$$

where $M$ is the atomic weight of the metal conductor of density $d$, width $w$, height $h$, and chemical valence $n$. The time to corrosion failure is modeled as

$$\tau = \frac{w^2 h n d F}{MV} \frac{\rho}{z} \tag{10.20}$$

## 10.2.4 Development of a Corrosion Model

The time to corrosion failure $\tau$ is modeled as the sum of an induction time $\tau_I$ and a time for the corrosion attack and failure $\tau_c$ by

$$\tau = \tau_I + \tau_c \tag{10.21}$$

To show the effect of induction time on the time to corrosion failure, a Weibull model was fit to the data obtained by Berg and Paulson [35] in accelerated corrosion tests of encapsulated and unencapsulated TO-5 packages at 85°C and 85% RH. Figure 10.21 shows the experimental data for the encapsulated package with a fit of a Weibull distribution with $\gamma = 135$, $\beta = 2.1$, and $\eta = 410$, where $\gamma$ is a location parameter, $\beta$ a shape parameter, and $\eta$ a scale parameter. Thus there exists a period of time ($\gamma = 135$ hr) before any significant failure occurs. Here $\eta$ is approximately 15 to 20% of the time for which 90% of the failures occur. Hence the induction time plays a significant role in predicting the time to corrosion failure. For the unencapsulated package $\eta = 0$.

Because the predominant mechanisms of moisture ingress for hermetic and nonhermetic epoxy-type packages are different, they must be treated separately. For a nonhermetic package, moisture ingress is primarily by permeation. The permeation equation has the form

$$\frac{\delta P}{\delta t} = D^2 P \tag{10.22}$$

where $D$ is the permeation constant, $\nabla = \partial^2/\partial x^2$, and $P$ is the partial pressure. With the boundary conditions that at time $t = 0$, the partial pressure $P_{t=0} = 0$; a time $t = \infty$, the

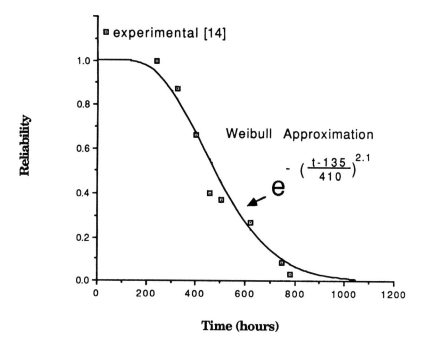

**Figure 10.21**   Reliability of an encapsulated package in a corrosive environment.

partial pressure $P_{t=\infty} = P_{out}$; and at the boundary $x = L$, $\partial P / \partial x = 0$ for all $t$, Eq. (10.22) yields

$$P_{in} = P_{out}(1 - e^{-\pi^2 Dt/4L^2})  \qquad (10.23)$$

where $P_{in}$ is the inside partial pressure. The time required for a nonhermetic package to reach an external vapor pressure is

$$\tau_{I_{nh}} = -\frac{4L^2}{\pi^2 D} \ln\left(1 - \frac{P_{in}}{P_{out}}\right)  \qquad (10.24)$$

For a hermetic package of known internal volume, the time to reach the critical moisture content is obtained from DerMarderosian's data [36], which we interpreted and plotted in Figure 10.22. As an example of how to use Figure 10.22, consider a hermetic package with an internal volume of 0.1 cm³. If the package passes the MIL-STD-883 leakage screening test, it will have a leak rate of $1 \times 10^{-7}$ atm cm³/sec or less. Referring to Figure 10.22. the package will have an induction time of 2570 hr, assuming the leak rate is exactly $1 \times 10^{-7}$ atm cm³/sec. This is a worst-case estimate valid at the time of the test but may prove to be an underestimate if leak paths grow as the package is subjected to further screening, installation, and operation.

As the basis of model development for the time $\tau_c$ to corrosion failure, we modify Howard's [34] model for two parallel traces (Figure 10.20) where corrosion at the anode proceeds until a length of the electrode is corroded.

$$\tau_c = \frac{k_1 k_2 k_3}{k_4} \frac{w^2 hndf}{4MV} \frac{\rho}{z}  \qquad (10.25)$$

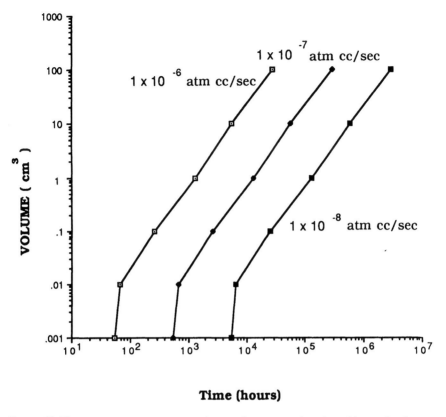

**Figure 10.22**  Time to reach three monolayers of water as a function of internal volume
and leak rate.


Here we incorporated additional parameters to address the corroding materials being
used in constructing the metallization and passivation layers, temperature cycling due to
the mission profile, and operating conditions. We also utilized $w/2$ as the trace width to
failure dimension to be consistent with MIL-STD-883C, which states that the maximum
corrosion width is 50% of the actual width, rather than a complete open as specified by
Howard. In fact, the effects of electromigration will act in conjunction with corrosion to
shorten the time to an ''open'' condition on the trace (Figure 10.23).

The time to failure, $\tau$, given by Eq. (10.25) is dependent on the geometry and
composition of the corroded electrode and can be calculated from design parameters. The
time to corrosion also depends on the sheet resistivity, thickness of the electrolyte, and
value of the thermal-corrosion function $f^*(T)$ given in Eq. (10.9). The resistivity of the
electrolyte, $\rho$, is calculated based on the amount of contaminants in an average indoor
environment [37] and is presented in Table 10.2.

The physical properties index, $k_1$, is a measure of the resistivity to corrosion (Table
10.3), normalized to that of gold. Following Schneider [18], because aluminum corrodes
approximately 10 times faster than gold, $k_1$ is set to 0.1.

The coating integrity index, $k_2$, factor accounts for the existence and integrity of a
passivation layer covering the die metallization (Table 10.4). The presence of a passivation

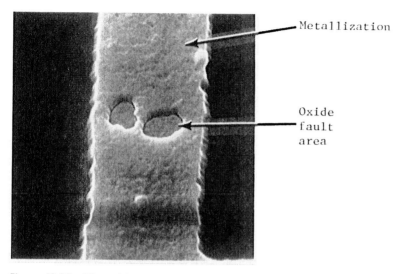

**Figure 10.23** View of the oxide faults under the aluminum metallization. (Courtesy of Unisys—NASA.)

**Table 10.2** Resistivity of Electrolyte in Different Environments

| Environment | $\rho$ (ohm-cm) |
|---|---|
| Normal | $7.3 \times 10^6$ |
| Corrosive | $2.3 \times 10^6$ |

**Table 10.3** Physical and Chemical Properties of Metallization Materials

| Material | Physical property index $(k_1)$ | Atomic weight $(M)$ | Density $(d)$ | Valence $(n)$ |
|---|---|---|---|---|
| Aluminum | 0.1 | 27 | 2.7 | 3 |
| Copper | 0.7 | 64 | 8.9 | 2 |
| Gold | 1.0 | 197 | 19.32 | 3 |

**Table 10.4** Coating Integrity Index

| Coating type | Coating integrity index $(k_2)$ |
|---|---|
| No coating | 1 |
| Partially bonded | 10–50 |
| Completely bonded | 100 |

layer will reduce the corrosion rate by minimizing moisture permeation and the formation of an electrolytic layer. However, pitting corrosion does occur and can also be incorporated by the $k_2$ factor. Sbar [19] has demonstrated experimentally that a completely bonded passivation layer will reduce the metallization corrosion rate by two to three orders of magnitude because voids and pinholes, which allow accumulation of electrolyte in contact with the metallization and promote corrosion, will be covered.

A value of $k_2 = 1$ is assigned to unpassivated metallization, indicating that under this condition corrosion proceeds without protection. A conservative value of $k_2 = 100$ is assigned to represent a defect-free, completely bonded passivation layer. Between these two limits $k_2$ will typically assume values varying from 10 to 50 to account for varying defect levels. For conservatism, a default value of 10 is assigned when a passivation layer exists and the defect level is unknown. Further investigation should be conducted to develop definitions for passivation layer defect types and magnitudes and associated values of $k_2$ between the two limits.

The mission profile correction factor, $k_3$, is introduced to represent the influence of temperature and power cycling, to reflect the reduction in the corrosion rate when the internal package temperature is increased on the die. The mission profile correction factor is calculated by the approximation

$$k_3 = \frac{1}{(1 - DC)^{10DC - 1}} \tag{10.26}$$

where DC is the duty cycle (% operating time). For a nonoperating package, DC = 0, $k_3 = 1$, and the actual time to failure decreases to the base time. For a continuously

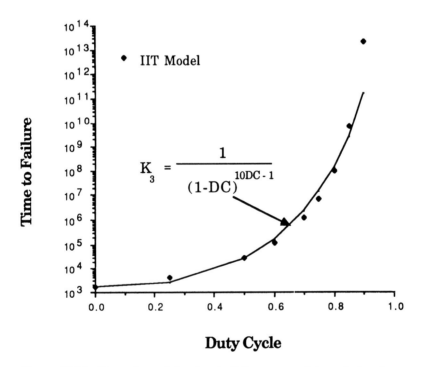

**Figure 10.24** Comparison of the time to failure versus duty cycle based on IIT/Honeywell study [31].

operating package, DC $= 1$, $k_3 = \infty$, and the time to corrosion failure becomes infinite. This approach is similar to that of the ITT/Honeywell study [32], as shown in Figure 10.24. In contrast, from Eqs. (10.12) and (10.13), with DC $= 1$, both $RH_{eff}$ and $t_{50}$ will have a finite value. In our mode, $\tau$ will be infinite, assuming the heat generated by an operating package will forbid the corrosion process as discussed above.

The environment correction factor, $k_4$, is utilized to determine the time to failure for various temperature and humidity conditions. Peck [38] evaluated many temperature-humidity correlation relationships [39–43] and proposed a modified Arrhenius model based on analysis of published data from 61 tests conducted over the period from 1970 to 1985. Peck subsequently discarded the less reliable data prior to 1979, evaluated new data, and proposed the following acceleration model based on 90 tests:

$$k_4 = \frac{(RH_R)^n \exp(E_a/KT_R)}{(RH)^n \exp(E_a/KT)} \tag{10.27}$$

where $RH_R$ is a reference relative humidity (%), $E_a$ is an activation energy (eV), $K$ is Boltzmann's constant (eV/K), $n$ is a material constant, and $T_R$ is a reference temperature (K).

Peck [38] found that $n = -3.0$ and $E_a = 0.87$ eV provided excellent correlation between the test data and the predicted acceleration factor over the range of 25% $<$ RH $<$ 100% and $T <$ 150°C. By using a reference relative humidity $RH_R = 85\%$ and a reference temperature of $T_R = 85°C$, Eq. (10.27) gives the relationship.

$$k_4 = \frac{2.9 \times 10^6}{(RH)^{-3} \exp(10{,}096/(T + 273))} \tag{10.28}$$

where $T$ and RH are the operating environment temperature (°C) and relative humidity (%), respectively. Equation (10.28) is plotted in Figure 10.25.

## 10.2.5 Corrosion Testing

Environmental tests are performed on packages to detect flaws in materials or workmanship that may lead to device failure in subsequent assembly process or field use. Many of the environmental tests, such as salt atmosphere and moisture resistance, provide accelerated forms of anticipated environments and are therefore applied as destructive tests to assess package quality and reliability [9].

Electronic devices intended for use in a corrosive environment, such as military or automotive electronics, are subjected to a salt fog test to assess the resistance of the package to corrosion (Figure 10.26). Several versions of the salt fog test are currently in use for testing electronic devices. All such tests evolved from the original salt fog test proposed in 1914 [44]. The test was originally proposed as an accelerated test for simulating the effects of a seacoast atmosphere on test samples. The salt fog tests used for electronic devices are listed in Table 10.5. The basic procedures of these tests are similar, but because of differences in specific test parameters, such as salt concentration and evaluation criteria, the tests are not interchangeable. Studies have shown that the results of salt fog tests do not correlate well with service life, primarily because the tests do not reproduce all the factors that may be involved in marine service. For this reason, the salt fog test is considered to be most suitable as an arbitrary performance test for closely related materials [45–50].

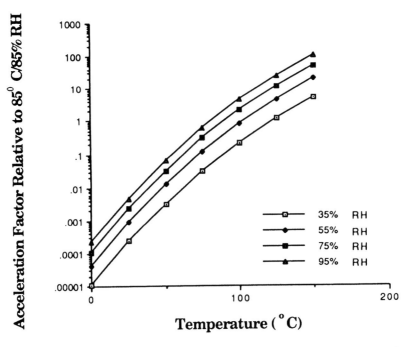

**Figure 10.25** Environment correction factor as a function of temperature and humidity.

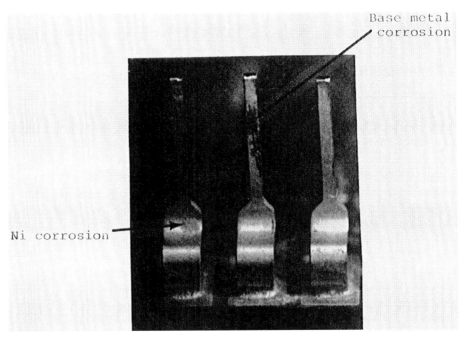

**Figure 10.26** Macrograph showing corroded leads on the MDC281 after 96 hr in salt spray. (Courtesy of Unisys—NASA.)

**Table 10.5**  Salt Fog Tests Used for Testing Electronic Components

| Salt fog test method | Specification |
| --- | --- |
| Method 101, salt spray (corrosion) | MIL-STD-202, "Test Methods for Electronic and Electrical Component Parts" |
| Method 1009, salt atmosphere (corrosion) | MIL-STD-883, "Test Methods and Procedures for Microelectronics" |
| Method 1041, salt atmosphere (corrosion) | MIL-STD-750, "Test Methods for Semiconductor Devices" |
| Method A107, salt atmosphere | JEDEC-STD-22, "Test Methods and Procedures for Solid State Devices Used in Transportation/Automotive Applications" |

The basic principle of the test is that a salt fog generated from a salt solution is permitted to fall by gravity onto test specimens inside an enclosed chamber. A typical salt fog apparatus consists of an exposure chamber, a humidifying tower, a reservoir for salt solution, atomizing nozzles, heaters and controls for maintaining the required temperature, and racks or screens for specimen support and positioning. The exposure portion of the chamber is maintained within a few degree of 35°C and at a relative humidity of approximately 95 to 98%. The specified test duration for electronic devices is generally 24 or 48 hr. A visual examination is performed after the exposure period to assess the extent of corrosion damage to the package exterior. For microelectronic devices tested in accordance with MIL-STD-883, a hermeticity test is also performed after the salt fog test to determine whether corrosion has penetrated through the package case or lid or along lead-package material interfaces to the package interior.

# EXERCISES

1. Define relative humidity.
2. List five typical problems that a package will undergo when exposed to a warm and humid environment.
3. Describe the cracking phenomenon of a PSMC under an extreme temperature condition.
4. Suggest a way to reduce moisture-induced package cracking of PSMCs.
5. What are the three most common forms of corrosion in integrated circuits?
6. Describe the occurrence of galvanic corrosion.
7. List four major factors that affect the rate of corrosion.
8. Justify the usage of organic/inorganic epoxies in the manufacturing processes.
9. List the path of moisture intrusion for both hermetic and nonhermetic packages.
10. What are the potential failure sites at which corrosion can occur inside a package?
11. Justify the application of passivation materials over the metallization and bonding pads. Does a passivation layer always improve the corrosion resistance?
12. Comment on the reliability of two identical packages, one operating 20 hr a day and the other operating 1 hr a day. Suggest the failure mechanism most likely to occur for each package.

13. Most military avionic equipment operates on an average of 1 hr per day. What extra precautions should be taken in terms of corrosion prevention?
14. Does the result of an accelerated test always predict the response of a package in real life?
15. Will an electronic package in space corrode? Justify your answer.

## REFERENCES

1. Summary Chart of 1984/1987 Failure Analysis Memos, Westinghouse Electric Corporation, 1989.
2. Nash, R. V., Jr., "MIL-STD-810D and Humidity," *Proc. Institute of Environmental Sciences*, 1988.
3. Lowry, R. K., "Microcircuit Corrosion and Moisture Control," *Microcontamination*, May 1985, p. 63.
4. Uy, O. M. and Benson, R. C., "Package Sealing and Passivation Coatings," in *Electronic Materials Handbook*, Vol. 1, ASM International, Materials Park, Ohio, 1989, p. 243.
5. Kane, D. and Domingos, H., "Nondestructive Moisture Measurement in Microelectronics," RADC-TR-87-210, Rome Air Development Center, November 1987.
6. Cvijanovich, G. B., "Conductivities and Electrolytic Properties of Adsorbed Layers of Water," *Proc. NBS/RADC Workshop, Moisture Measurement Technology for Hermetic Semiconductor Devices, II*, Gaithersburg, Md., 1980, pp. 149–164.
7. Der Marderosian, A., "Permissible Leak Rates and Moisture Ingress," in *Hermeticity Tutorial Notes*, 22nd Annual International Reliability Physics Symposium (IRPS), Las Vegas, IEEE, 1984.
8. Doyle, K. B., "Some Experiences and Conclusions Using Solder and Welded Packages for Hermetic Thick Film Hybrids," *Microelectron. Reliab. 14*:303–307 (1977).
9. *Electronic Materials Handbook*, ASM International, Materials Park, Ohio, 1989, p. 935.
10. "Test Methods and Procedures for Solid State Devices Used in Transportation/Automotive Applications," JEDEC-STD-22, Joint Electronic Devices Engineering Council
11. Gavin, F. S., "Application Notes on Autoclave Testing," *Eval. Eng.*, February 1986, pp. 108–112.
12. Toi, K., Yamamoto, T. and Yoshida, H., "Problems in Pressure Cooker Test," *Proc. 5th International Conference on Reliability and Maintainability*, France, 1986.
13. Gavin, F. S., "Accelerated Stress Testing Reduces Test Times and Costs," *Electron. Test*, December 1985, pp. 31–34.
14. Gunn, J. E., Camenga, R. E. and Malik, S. K., "Rapid Assessment of the Humidity Dependence of IC Failure Modes by Use of HAST," *Proc. International Reliability Physcis Symposium*, IEEE, 1983, pp. 66–72.
15. "Highly-Accelerated Temperature and Humidity Stress Test (HAST)," JEDEC-STD-22-A110, Joint Electronic Devices Engineering Council
16. "Technical Report on Impact of Moisture on Plastic I/C Package Cracking," Institute for Interconnecting and Packaging Electronic Circuits, January 1989.
17. Cocks, F. H., *Manual of Industrial Corrosion Standards and Control*, American Society for Testing and Materials, Philadelphia, 1973.
18. Schneider, G., "Non-Metal Hermetic Encapsulation of a Hybrid Circuit," *Microelectron. Reliab. Br. 28*(1): (1988).
19. Sbar, N. S., "Bias Humidity Performance of Encapsulated and Unencapsulated Ti-Pd-Au Thin Film Conductors in an Environment Contaminated with Cl," *Proc. 1976 26th Electronics Components Conference*, San Francisco, 1976, pp. 277–284.
20. Hnatek, E. R., *Integrated Circuit Quality and Reliability*, Marcel Dekker, New York, 1987.

21. Thomas, R., "Moisture, Myths and Microcircuits," *IEEE Trans. Parts, Hybrids Packag. PHP-12*(3):169 (1976).
22. Eisenberg, P. H., et al., "Effects of Ambient Gases and Vapors at Low Temperature on Solid State Devices," presented at the 7th New York Conference on Electronic Reliability, IEEE, 1966.
23. Crawford, W. M. and Weigand, B. L., "Contamination of Relay Internal Ambients," *Proc. 15th Annual Relay Conference*, 1956, pp. 1–8.
24. Cannon, D. L. and Trapp, O. D., "The Effects of Cleanness on Integrated Circuit Reliability," *6th Annual Proc. Reliability Physics*, 1968, pp. 68–69.
25. Young, M. R. P. and Peterman, D. A., "Reliability Engineering," *Microelectron. Reliab.* 7:91–103 (1968).
26. Thomas, R. W., "IC Packages and Hermetically Sealed-in Contaminants," *Government Microcircuits Applications Conference Digest*, 1972, pp. 31–36.
27. MIL-STD-883C, Method 1014, 1983.
28. Uhlig, H. H., *Corrosion and Corrosion Control*, Wiley, New York, 1963, p. 146.
29. Abbott, W. H., "Field Versus Laboratory Experience in the Evaluation of Electronic Components and Materials," Paper presented at Corrosion/83, Anaheim, Calif., National Association of Corrosion Engineers, April 1983.
30. Baboian, R., "Corrosion—a National Problem," *ASTM Stand. News*, March 1988, p. 37.
31. Motorola, "TMOS Power MOSFET Reliability Report," Motorola Literature Distribution, Phoenix, 1988.
32. IIT/Honeywell, "VHSIC/VHSIC-LIKE Reliability Prediction Modeling," Contract No. F30602-86-C-0261, December 1988.
33. Slota, P. and Levitz, P. J., "A Model for a New Failure Mechanism—Nichrome Corrosion in Plastic Encapsulated PROMs," *IEEE Proc. 33rd Electronic Components Conference*, 1983.
34. Howard, R. T., "Electrochemical Model for Corrosion of Conductors on Ceramic Substrates," *IEEE Trans. Components, Hybrids, Manuf. Technol. CHMT-4*(4): (1981).
35. Berg, H. M. and Paulson, W. M., "Chip Corrosion in Plastic Packages," *IEEE Symp. Microelectron. Reliab.* 20:247 (1980).
36. DerMarderosian, A., "Hermeticity and Moisture Ingress," Raytheon, Sudbury, Mass., 1988.
37. Commizzoli, R. B., Franenthal, R. R., Milner, P. C. and Sinclair, J. D., "Corrosion of Electronic Materials and Devices," *Science 234*:340–345 (1986).
38. Peck, D. S., "Comprehensive Model for Humidity Testing Correlations," *Proc. IRPS*, 1986, pp. 44–50.
39. Reich, B. and Hakim, E., "Environmental Factors Governing Field Reliability of Plastic Transistors and Integrated Circuits," *Proc. IRPS*, 1972, pp. 82–87.
40. Peck, D. S. and Zierdt, C. H., Jr., "Temperature-Humidity Acceleration of Metal-Electrolysis Failure in Semiconductor Devices," *Proc. IRPS*, 1973, pp. 146–152.
41. Larson, R. W., "The Accelerated Testing of Plastic Encapsulated Semiconductor Components," *Proc. IRPS*, 1974, pp. 243–249.
42. Larson, R. W., "A Review of the Status of Plastic Encapsulated Semiconductor Reliability," *Br. Telecom Technol. J.* 2(2): (1984).
43. Hallberg, O., "Accelerated Factors for Temperature-Humidity Testing of Al-Metallized Semiconductors," SINTOM, Copenhagen, 1979.
44. Capp, J. A., "A Rational Test for Metal Protective Coatings," *Proc. ASTM 14*(II):474 (1914).
45. MIL-STD-202, "Test Methods for Electronic and Electrical Component Parts."
46. Uhlig, H. H., *The Corrosion Handbook*, Wiley, New York, 1948.
47. "Standard Method of Salt Spray (Fog) Testing," B 117-85, in *Annual Book of ASTM Standards*, American Society for Testing and Materials, Philadelphia
48. Durbin, C. O., "Accelerated Corrosion-Tests," in *Electroplating Engineering Handbook*, 3rd ed., Graham, A. K., ed., 1971.

49.  Lee, T. S. and Money, K. L., "Difficulties in Developing Tests to Simulate Corrosion in Marine Environments," *Mater. Perform.* *3*(8):28–33 (1984).
50.  Altmayer, F., "Critical Aspects of the Salt Spray Test," *Plat. Surf. Finish.*, September 1985, pp. 36–40.

## SUGGESTED READING

*CRC Handbook of Chemistry and Physics*, 59th ed., Chemical Rubber Publishing Co., Cleveland, 1978–1979, p. D-205.

Davy, J. G., "Model Calculations for Maximum Allowable Leak Rates of Hermetic Devices," *J. Vac. Sci. Technol.*, *12*(1): (1975).

Graham, J. W., "Gas Leakage in Sealed Systems," *Chem. Eng.*, May 1964, pp. 169–174.

Olberg, R. C. and Bozarth, J. L., "Factors Contributing to the Corrosion of the Aluminum Metal on Semiconductor Devices Packaged in Plastics," *Microelectron. Reliab.* *15*:60

Paulson, W. M. and Kirk, R. W., "The Effects of Phosphorus-Doped Passivation Glass on the Corrosion of Aluminum," *Proc. International Reliability Physics Symposium*, 1974, p. 172.

"Permeability Data for Aerospace Applications," IIT Research Institute, Chicago, 1968.

Stojadinovic, N. D., "Failure Physics of Integrated Circuits—a Review," Reliability Physics Symposium, 1981.

Traeger, R. K., "Non-Hermeticity of Polymer Lid Sealants," *IEEE Trans. Parts, Hybrids Packag.* *PHP-13*(2): (1977).

Zamanzadeh, M., et al., "Electrochemical Examination of Dendritic Failure in Electronic Devices," *Proc. Symposium on Multilevel Metallization, Interconnection and Contact Technologies*, Vol. 87–4, Electrochemical Society, 1987, pp. 173–174.

# 11

# Design for Reliability

**Michael Pecht**     *CALCE Center for Electronic Packaging, University of Maryland, College Park, Maryland*

## 11.1  INTRODUCTION

Reliability engineering is the aspect of the product development process concerned with the operational readiness, successful performance over time, maintenance, and service requirements of systems, subsystems, and parts. The reliability of a product can be established only if reliability is designed in (failure-related design), manufacturing processes and testing are in control, and maintenance procedures are properly established and carried out. Improvement in reliability occurs when reliability information, obtained from test and field data, is utilized to adjust design, manufacturing, and maintenance parameters, thereby reducing or eliminating potential failure causes. Test and field data are obtained from inspection, failure analysis, manufacturing, field service, and customers. Figure 11.1 shows the interaction of reliability tasks within the framework of product development.

Design for reliability is a process that includes the following operations:

- Defining the mission operating conditions, environments, and maintenance conditions
- Defining the system (or product) and its components in terms of its functional, physical, and operational characteristics
- Characterizing the materials, and the manufacturing and assembly processes
- Determining the potential modes, sites, mechanisms of failure, and corrective actions
- Assigning probabilities of survival
- Qualifying and controlling the manufacturing and assembly processes

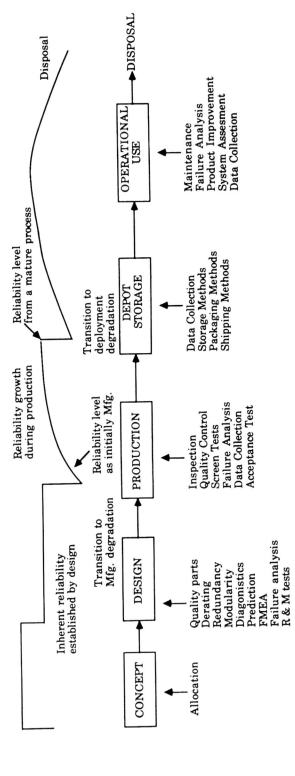

**Figure 11.1**  Product development and reliability tasks. (From Ref. 53.)

· Computing the system maintainability
· Refining the design to achieve performance, reliability, and maintainability characteristics that meet cost and schedule requirements effectively

The design for reliability process involves establishing a system-level goal for reliability, such as probability of mission completion, product warranty, or dispatch reliability, often stated in terms of, or related to, a decomposition of the reliability goal to subsystems. Each interacting subsystem thus has its individual goal for reliability design, and each designer can make cost-benefit trades to optimize the subsystem. The requirements differ with the application: in a locomotive weight and volume generally may not be terribly important, but in a spacecraft they would be. Dollar cost in a military fighter airplane could be less important than high reliability, but in an automobile the opposite may be true, and so forth.

Generally, the more complex a system is, the more difficult, time consuming, and thus expensive it will be to locate and repair a fault. Figure 11.2 shows the cost to locate and repair one fault as a function of where and when that fault is detected; e.g., the cost to find and correct a fault in a product in the field may be 1000 to 10,000 times the cost to find that fault at the component level prior to product assembly. If the design does not adequately address issues of reliability early in the product development cycle, then severe maintenance, repair, and cost penalties could be incurred.

Electronic components have, for the most part, very low rates of failure (typically less than 1 failure per million hours) and are continuously improving. However, reliability remains an important issue for modern electronic equipment because of rapid technology changes, high component densities, and complexities of the manufacturing and assembly processes. Table 11.1 illustrates the effect of complexity on system reliability and shows that 1000 series components with 99% reliability each yield a system reliability of only

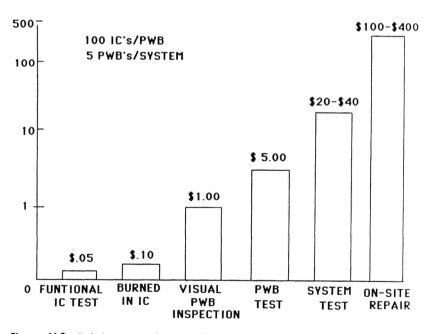

**Figure 11.2**   Relative cost to locate and repair one fault.

**Table 11.1**   Effect of Complexity on System Reliability

| No. of components in series | System reliability for individual component reliability of | | | |
|---|---|---|---|---|
|  | 99.999% | 99.99% | 99.9% | 99.0% |
| 10 | 99.99% | 99.900% | 99.004% | 90.438% |
| 100 | 99.900% | 99.005% | 90.479% | 36.603% |
| 250 | 99.750% | 97.531% | 77.870% | 8.106% |
| 500 | 99.501% | 95.123% | 60.638% | 0.657% |
| 1000 | 99.005% | 90.483% | 36.770% | 0.004% |

0.004%. For such a system, if the desired system reliability is that 1 out of 10 systems operates successfully, then the component reliability must be at least 99.99%. That is, only 1 in 10,000 can fail during the product life. Figure 11.3 shows the typical technology product life cycle where usage or sales is depicted as a function of time. The problem of achieving a specification standard, based on reliability data, is apparent for rapidly changing technologies when the learning and decay curves follow each other closely in time.

   To affect the reliability of electronics, the designer must identify and understand all the potential failure sites, mechanisms, and causes of failure. Electrical, thermal, mechanical, and systems management and production control all aid in reducing failures. Investigating and assessing the physics of failure processes, the mechanisms, and the various manufacturing, assembly, storage, operational, and environmental conditions is the foundation of good reliability analysis. Such analysis provides information on the causes of failure and points to the mechanisms that will dominate under operational conditions.

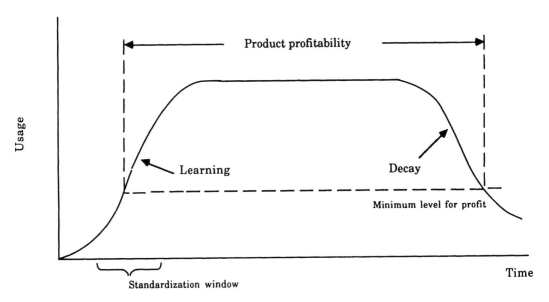

**Figure 11.3**   Technology curve.

## 11.2 CLASSIFICATION OF FAILURES

Failure is defined as loss of the ability of the product to perform a required operation in a specified environment. This definition includes catastrophic failures, which imply the end of life of the part or product, as well as degradation failures, wherein an important parameter drifts significantly to cause improper functioning. Often, a parameter drift or an intermittent loss of function will eventually lead to a complete failure.

Failures are typically grouped into one of three categories depending on when the failure occurs during a product's operating life. The categories are infant mortality, useful life, and wearout failure.

Infant mortality failures occur early in a product's operating life. The definition is that the rate of failures decreases with age. The causes of early failures include

- Poor workmanship
- Poor manufacturing, assembly, storage, and transportation
- Poor quality control (flawed materials, contamination)
- Insufficient burn-in and debugging
- Improper start-up

The associated rate of failure during the "useful" life of the product is assumed to be constant. This has been a widely used approximation for many electronic devices. The causes of failures include

- Inadequate derating, higher than expected random loads, and/or lower than expected strength characteristics
- Abuse or misapplication

Wearout failures occur late in the operating life of a product. The definition is that the rate of failure increases with age. The causes of wearout failures include

- Aging
- Wear, degradation
- Creep
- Fatigue
- Poor service, maintenance, or repair

If a plot of the rate of failure of a product versus its operating life is constructed from data taken from a large sample of identical products placed in operation at time $t = 0$, the resulting curve often has a shape such that it is referred to as a "bathtub" curve (Figure 11.4). The high initial rate of failure represents early failures, the central portion of the curve represents the useful life of the product, and the increasing rate of failure with advanced age is due to wearout failures. The useful life or useful operating period provides the basis for failure data, which are typically published in reliability product manuals.

It is important to realize that the bathtub curve will be different for unrepairable and repairable products. From a probabilistic view for unrepairable products, the ordinate can be interpreted as the number of failures per given period of time per number of products operating at the beginning of the time period (i.e., number of products at risk). For repairable products, the ordinate can be interpreted as the number of failures per given period of time per total number of products put on test at time 0.

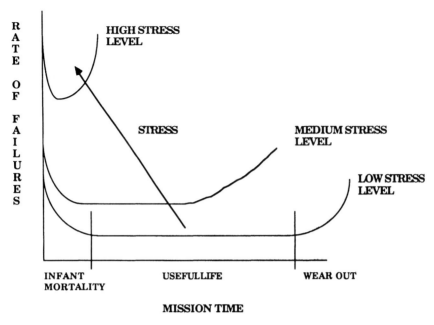

**Figure 11.4** The bathtub curve.

Electronic products should never be used unless infant mortality failures have been precipitated and it is ensured that wearout failures will not be encountered during the expected life. Often a product is called reliable when only a small percentage of the products surviving infant mortality have failed up to the beginning of wearout. Table 11.2 shows typical electronic device reliability objectives prevalent in the 1980s. In this table the initial failure period is further decomposed into two early failure periods to separate out start-up failures from those representative of other early failure causes.

## 11.3  RELIABILITY THEORY

This section presents some of the fundamental definitions and mathematical theory for reliability. The focus is on the reliability function, the probability distribution function, the hazard rate, the conditional reliability function, and the mean time to failure. Here we define a part as an item that is not subject to disassembly and is thus discarded the first time it fails; a socket as a circuit or equipment position that at any given time holds

**Table 11.2**  Typical Device Reliability Objectives

| Failure period | Time period | Failure goal |
|---|---|---|
| Infant mortality | 0–1000 hours | 0.02% |
| Early life | 0–1 year | 100 fr/$10^9$ hours |
| Random | 1–20 years | 100 fr/$10^9$ hours |
| Wearout | >20 years | MTTF > 20 years |

a part; a system as a collection of two or more sockets and their associated parts, inter-connected to perform a function; a nonrepairable system as a system that is discarded the first time it fails to perform satisfactorily; and a repairable system as a system that can be restored to performing all of its required functions by any method other than total system replacement.

## 11.3.1 Reliability

The reliability of a product is the probability that the product will perform satisfactorily for a given time at a desired confidence level under specified operating and environmental conditions. For a constant sample size, $n_0$, of identical products which are tested or being monitored, it may be observed at time $t$ that $n_f$ have failed and that the remaining $n_s$ are still operating satisfactorily. Thus

$$n_s(t) + n_f(t) = n_0 \tag{11.1}$$

The reliability $R(t)$ of the product at time $t$ is estimated by the ratio of operating products per sample size,

$$R(t) = \frac{n_s(t)}{n_0} \tag{11.2}$$

Similarly, the unreliability or probability of failure $Q(t)$ of the product at time $t$ is estimated by the equation

$$Q(t) = \frac{n_f(t)}{n_0} = 1 - R(t) \tag{11.3}$$

The accuracy of a reliability estimate at a given time is improved for a larger sample size $n_0$. The requirement of a large test sample is analogous to the conditions required in experimental measurements of probabilities associated with coin tossing and dice rolling.

## 11.3.2 Probability Density Function

A plot of the ratio of the number of product failures $n_f(t)$ per total number of products, as a function of the operating life, is called the failure probability density function $f(t)$ curve (Figure 11.5). The plot starts at $t = 0$ with $n_f(t \le 0) = 0$. The probability density function is expressed by

$$f(t) = \frac{1}{n_0} \frac{d[n_f(t)]}{dt} = \frac{d[Q(t)]}{dt} \tag{11.4}$$

Integrating both sides of this equation gives

$$\frac{n_f(t)}{n_0} = \int_0^t f(\tau) \, d\tau \tag{11.5}$$

where $\tau$ is just a variable of integration. Since the total probability of failures must equal 1, the function $n_f(t)$ is appropriately normalized. That is,

$$\int_0^\infty f(t) \, dt = 1 \tag{11.6}$$

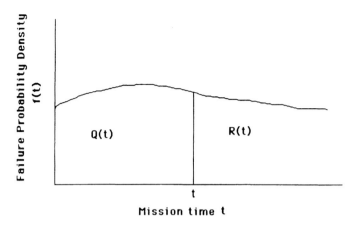

**Figure 11.5** The failure probability density distribution.

From the $f(t)$ curve and Eq. (11.5), the area under the curve to the left of the time line is the total percentage of failures and is called the cumulative failure probability distribution (or unreliability) function for a continuous random variable.

$$Q(t) = \int_0^t f(\tau)\, d\tau \tag{11.7}$$

Similarly, the percentage of products that have not failed up to time $t$ is represented by the area under the curve to the right of $t$ by

$$R(t) = \int_t^\infty f(\tau)\, d\tau \tag{11.8}$$

### 11.3.3 Hazard Rate

The average hazard rate $\lambda_a$, estimated at time $t_i$, is the failure probability of a part (i.e., nonrepairable item) in a given time period (from $t_i$ to $t_i + \Delta t$), assuming that the part did not fail at time $t_i$. More formally, this is an estimate of the force of mortality (FOM). It is mathematically expressed by the equation

$$\lambda_a = \frac{1}{n_{bp}(t_i)} \frac{\Delta n_f}{\Delta t} \tag{11.9}$$

where $n_{bp}(t_i)$ is the number of parts monitored or tested that have not failed at the beginning of the time period, $\Delta n_f$ is the number of failures in the sample time period, and $\Delta t$ is the size of the sampling period. As $n_{bp}$ becomes large and as the sampling time period goes to zero, the average hazard rate estimate approaches the hazard rate at time $t_i$.

The hazard rate $\lambda(t)$ provides a relationship between the mission time of a part and the failure frequency at that time. The hazard rate is defined as the number of failures per unit time per number of operational parts left. It is thus a relative rate of failures. This can be derived from Eq. (11.9) by letting $\Delta t$ go to zero such that

$$\lambda(t) = \frac{1}{n_s} \frac{d[n_f(t)]}{dt} \tag{11.10}$$

Combining this equation with (11.1), (11.2), and (11.3) gives a relation for the hazard rate in terms of the reliability,

$$\lambda(t) = \frac{-1}{R(t)} \frac{d[R(t)]}{dt} \tag{11.11}$$

Integrating Eq. (11.11) over a mission time $\tau$ from 0 to $t$, noting that $R(t = 0) = 1$, and taking the exponential of each side gives

$$R(t) = \exp\left(-\int_0^t \lambda(\tau)\, d\tau\right) \tag{11.12}$$

The hazard rate can also be expressed as the ratio of the failure probability density function to the reliability by combining Eqs. (11.2) and (11.6) with (11.10),

$$\lambda(t) = \frac{f(t)}{R(t)} \tag{11.13}$$

For a repairable system, Eq. (11.10) is modified by changing $n_s$ to the initial number of systems, $n_0$. This absolute rate equation is often called the rate of occurrence of failures (ROCOF).

### 11.3.4 Conditional Reliability

The conditional reliability function $R(t,T)$ is defined as the probability of operating for the time interval $t$, given that the nonrepairable system has operated for a time $T$ to the beginning of the interval. It can be expressed as the ratio of the reliability at time $t + T$ to the reliability at a mission duration $T$, where $T$ is the age of the system at the beginning of the mission. That is,

$$R(t,T) = \frac{R(t + T)}{R(T)} \tag{11.14}$$

which is the conditional probability that the system will complete a mission of duration $t$, given that it had survived a period $T$ up to the start of the mission. It is important to note that the conditional reliability of a system with a constant rate of failure is independent of $T$; i.e., the reliability for a mission time $t$ is independent of previous mission times. However, for a product with a decreasing hazard rate, the conditional reliability will improve as the previous mission time $T$ increases, and vice versa for a product with an increasing hazard rate.

### 11.3.5 Mean Time to Failure

The mean time to failure (MTTF) is defined as

$$\mathrm{MTTF} = \int_0^\infty tf(t)\, dt \tag{11.15}$$

Using Eq. (11.8), it can be shown that the MTTF, given above as an expected value or first moment of the failure probability density function, is equivalent to

$$\text{MTTF} = \int_0^\infty R(t)\, dt \tag{11.16}$$

The MTTF should be used only when the failure distribution function is specified because the reliability at a given MTTF depends on the distribution function. In fact, different failure distributions can have the same MTTF while having different reliabilities.

Other measures can be used to characterize the failure probability and related terms. In particular, the median $M$ of the probability distribution is the point in time at which the area under the distribution is divided in half (i.e., 50% reliability). That is,

$$\int_0^M f(t)\, dt = 0.5$$

Since $M$ occurs as a limit, determining an explicit relation for $M$ is often extremely complex.

## 11.4 THREE USEFUL DISTRIBUTIONS

This section presents one discrete distribution (binomial) and two common continuous distributions (Weibull and exponential). The binomial is extremely useful in analyzing system redundancies, while the Weibull and exponential are commonly used to model the failure characteristics of parts. Before discussing these distributions, some other continuous distributions are briefly described below.

The normal distribution is one of the fundamental distributions used in statistics. It is used in applications in which the observed frequency of occurrence approaches a bell-shaped or normal distribution. This is the case with many physical measurements such as product dimensions and average temperatures.

The gamma distribution is a fundamental statistical distribution that is used when the associated variables are bounded on one side. It can be used to model a distribution of time required for $n$ independent events to occur at a constant rate. It is frequently used in reliability engineering, queuing theory, and industrial applications including determining the time between inventory restocking, the time to failure for a system with standby products, and the distribution of time between recalibration of maintained instruments. The exponential distribution is a special case of this distribution.

The beta distribution is a commonly used statistical distribution employed when the variables are bounded on both sides. In particular, it is used to model distributions of daily percent yield in a manufacturing process, elapsed times to a task completion, and the proportion of a population located between lowest and highest values in a sample.

The log-normal distribution permits representation of random variables whose logarithm follows a normal distribution. It is often used when the value of an observed variable is a random proportion of the previously observed value. Other applications include the life distribution of some transistor types, the distribution of sizes from a breakage process, and the distribution of various biological phenomena.

The Rayleigh distribution models the radial error when the errors in two mutually perpendicular axes are independent and normally distributed around zero with equal variances. Applications include bomb-sighting problems and the amplitude of a noise envelope when a linear detector is used. The Rayleigh distribution is a special case of the Weibull distribution.

The extreme value distribution is a limiting model for the distribution of the maximum or minimum of values selected from an exponential-type distribution such as the normal, gamma, or exponential distribution. It has applications in the breaking strength of some materials, in capacitor breakdown voltages, and in extinction times of bacteria.

## 11.4.1 The Binomial Distribution

The binomial distribution is a discrete distribution applicable in situations in which there are only two mutually exclusive outcomes for each trial. For example, for a roll of a die, the probability is $\frac{1}{6}$ that a specified number will occur (success) and $\frac{5}{6}$ that it will not occur (failure). The probability density function $f(k)$ for the binomial distribution gives the probability of exactly $k$ successes. It is defined by

$$f(k) = \binom{m}{k} p^k q^{m-k}, \qquad 0 \le p \le 1, q = 1 - p, k = 1, 2, \ldots, m \qquad (11.17)$$

where $p$ is the probability of success, $q$ is the probability of failure, $m$ is the number of independent trials, $k$ is the number of successes in $m$ trials, and $\binom{m}{k}$ is the combinatorial formula, defined by

$$\binom{m}{k} = \frac{m!}{k!(m-k)!} \qquad (11.18)$$

and where ! is the symbol for factorial. Other notations for $\binom{m}{k}$ include $c(m,k)$ and $m^c k$.

For a discrete distribution, the cumulative distribution function $F(k)$ gives the probability of $k$ or fewer successes in $m$ trials. It is defined in terms of the discrete probability density function $f(i)$,

$$F(k) = \sum_{i=0}^{k} f(i) \qquad (11.19)$$

*Example 11.1*   Consider the selection of four resistors from a large lot in which 10% of the resistors are defective. What are the probabilities of (a) no defects, (b) one defect, (c) two defects, and (d) two or fewer defects?
    *Solution*   Using Eqs. (11.17) and (11.19) with $p = 0.1$ and thus $q = 0.9$ gives

(a) $f(0) = \binom{4}{0} (0.1)^0 (0.9)^4 = 0.6561$

(b) $f(1) = \binom{4}{1} (0.1)^1 (0.9)^3 = 0.2916$

(c) $f(2) = \binom{4}{2} (0.1)^2 (0.9)^2 = 0.0486$

(d) $F(2) = f(0) + f(1) + f(2) = 0.9963$

It can easily be shown that since $(p + q) = 1$,

$$(p + q)^j = F(j) = 1 \tag{11.20}$$

The coefficients of the expansion of Eq. (11.20) are represented by the probability of failure corresponding to the probability density function. For example, for three identical components

$$(p + q)^3 = p^3 + 3p^2q + 3pq^2 + q^3 = 1$$

The four terms can also be written by Eq. (11.17) for $f(k)$ where $f(0) = p^3, f(1) = 3p^2q$, $f(2) = 3pq^2$, and $f(3) = q^3$.

The binomial expansion is also useful when there are products with different success and failure probabilities. The formula for the binomial expansion in this case is

$$\prod_{i=1}^{m} (p_i + q_i) = 1 \tag{11.21}$$

where $i$ pertains to the $i$th component in a system consisting of $m$ components. For a system of three different components, the expansion takes the form

$$(p_1 + q_1)(p_2 + q_2)(p_3 + q_3) = p_1p_2p_3 + (p_1p_2q_3 + p_1q_2p_3$$
$$+ q_1p_2p_3) + (p_1q_2q_3 + q_1p_2q_3 + q_1q_2p_3) + q_1q_2q_3 = 1$$

where the first term on the right gives the probability of success of all three components, the second term (in parentheses) gives the probability of success of any two components, the third term (in parentheses) gives the probability of success of one component, and the last term gives the probability of all components failing.

*Example 11.2*   A computer component is composed of three active and identical microprocessors in parallel. All microprocessors fail independently. At least one microprocessor must operate normally for the computer component to function successfully. The probability of success of each microprocessor for a given mission time is $0.95$. Determine the failure probability of the computer component assuming it can only be in an operating or failed state.

*Solution*   The probability of a failure of each microprocessor is given by

$$q = 1 - p = 1 - 0.95 = .05$$

Using Eq. (11.17) with $q = 0.05$, $p = 0.95$, $m = 3$, and $k = 0$ gives

$$f(0) = \binom{3}{0} (0.95)^0 (0.05)^3 = 1.25 \times 10^{-4}$$

Thus the probability of the computer component failing is $1.25 \times 10^{-4}$.

The expression for the mean, $\mu$, of the binomial distribution is

$$\mu = mp \tag{11.22}$$

*Example 11.3*   An electronics company manufacturers a particular type of electronic relay. From a sample of 75, the probability of a defective relay is $0.075$. Calculate the mean for the distribution of defective relays in this sample.

*Solution*   Substituting the given information into Eq. (11.22) gives the mean time to failure

$$\mu = (75)(0.075) = 5.63$$

which means that in a sample of 75, between 5 and 6 can be expected to be defective.

## 11.4.2 The Weibull Distribution

The Weibull distribution is a continuous distribution developed in 1951 by W. Weibull. It is widely used for reliability analyses because it is a general time-to-failure distribution with which a wide diversity of hazard rate curves can be modeled and because it is an extreme value distribution for variables bounded on the left. It has applications in life distributions for microelectronic devices, capacitors, and relays. The exponential and Rayleigh distributions are special cases of the Weibull.

The three-parameter Weibull probability distribution function is expressed by

$$f_w(t) = \frac{\beta}{\eta}\left(\frac{t-\gamma}{\eta}\right)^{\beta-1} e^{-[(t-\gamma)/\eta]^\beta} \tag{11.23}$$

where $\beta > 0$ is the shape parameter, $\eta > 0$ is the scale parameter, and $\gamma$ is the location parameter. Often the Weibull distribution is formulated as a two-parameter distribution with $\gamma = 0$.

The shape factor can be utilized as a guide to the type of failure that has occurred or is predicted. Early life is modeled with $0 < \beta < 1$ and indicates that the hazard rate is a decreasing function of time. With $\beta = 1$, the useful life of a part having a constant hazard rate is modeled. If, in addition to $\beta = 1$, $\gamma = 0$ (failures begin at $t = 0$), then it can be shown that the Weibull distribution reduces to the exponential distribution. The wearout life of a part when the hazard rate is an increasing function can be modeled with $\beta > 1$. Figure 11.6 shows the effects of $\beta$ on the probability density function curve with $\eta = 1$ and $\gamma = 0$. Figure 11.7 shows the effect of $\beta$ on the hazard rate curve with $\eta = 1$ and $\gamma = 0$. Note that for $1 < \beta < 2$ the hazard rate curve is concave and for $\beta > 2$ the curve is convex. For $\beta = 2$ the hazard rate curve is a straight line.

The scale parameter $\eta$ has the effect of scaling the mission time axis while keeping the total area under the distribution equal to unity. Thus for $\gamma$ and $\beta$ fixed, an increase in $\eta$ will stretch the distribution to the right while maintaining its starting location and shape (although there will be a decrease in the amplitude). Figure 11.8 shows the effects of $\eta$ on the probability density function for the case of $\beta = 2$ and $\gamma = 0$.

The location parameter estimates the earliest time to failure and locates the distribution along the mission time axis. For $\gamma = 0$, the distribution starts at $t = 0$. However, if, for example, $\gamma = -250$ hr, this implies that failures could have started to occur 250 hr before operation, possibly as a result of early failure phenomena such as transportation or storage problems. With $\gamma > 0$, this implies that the product has a failure-free period equal to $\gamma$. Figure 11.9 shows the effects of $\gamma$ on the probability density function curve with $\beta = 2$ and $\eta = 1$. Note that if $\gamma$ is positive, the distribution starts to the right of the $t = 0$ origin. If $\gamma$ is negative, the distribution starts to the left of the $t = 0$ origin.

The reliability function can be determined from Eq. (11.8) with the introduction of Eq. (11.23),

$$R(t) = e^{-[(t-\gamma)/n]^\beta} \tag{11.24}$$

It can be shown that for a mission of duration $t = \gamma + \eta$, starting at an age of zero, the reliability $R(t) = 36.8\%$ regardless of the value of $\beta$. Thus for any Weibull failure probability density function, 36.8% of the components survive for $t = \gamma + \eta$.

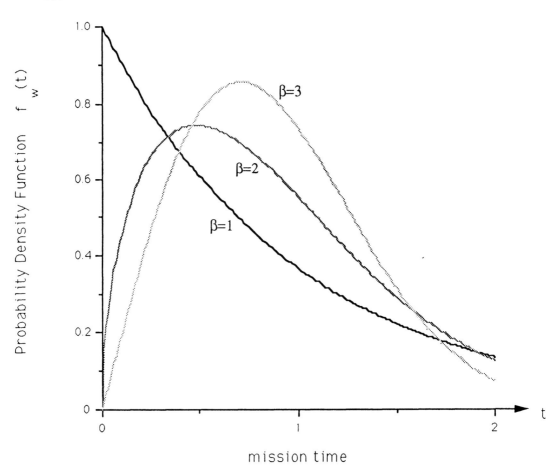

**Figure 11.6** Effects of shape parameter $\beta$ on the Weibull probability density distribution ($\eta = 1$, $\gamma = 0$). (From Ref. 64.)

To obtain the mission life (starting at $t = 0$) of a product with a specified reliability $R$, Eq. (11.24) is solved in terms of $t$,

$$t = \gamma + \eta(-\ln R)^{1/\beta} \tag{11.25}$$

To obtain the median life for which half of the products will fail, let $R = 0.50$ in Eq. (11.25).

The hazard rate function $\lambda(t)$ can be derived from Eqs. (11.11), (11.23), and (11.24),

$$\lambda(t) = \frac{\beta}{\eta}\left(\frac{t - \gamma}{\eta}\right)^{\beta - 1} \tag{11.26}$$

The conditional reliability function can be developed by introducing Eq. (11.24) into (11.14) to give

$$R(t,T) = \exp\left\{-\left[\frac{(t + T - \gamma)^{\beta}}{\eta}\right] - \left[\frac{(T - \gamma)^{\beta}}{\eta}\right]\right\} \tag{11.27}$$

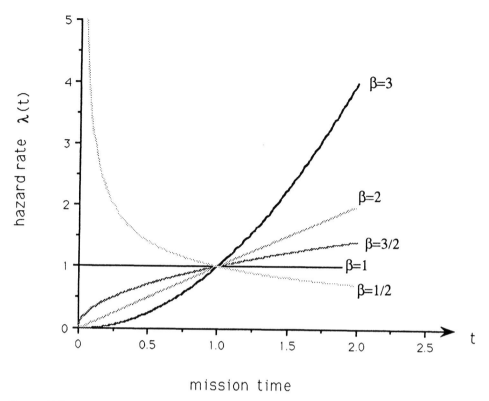

**Figure 11.7** Effects of shape parameter $\beta$ on the Weibull hazard rate curve ($\eta = 1, \gamma = 0$). (From Ref. 64.)

This equation gives the reliability for a new mission of duration $t$ for which $T$ hours of operation were previously accumulated up to the beginning of this new mission. Unlike the exponential distribution, the Weibull distribution is dependent on both the age at the beginning of the mission and the mission duration.

From Eq. (11.24), it can be shown that for $\beta = 1$ and $\gamma = 0$ the Weibull distribution reduces to the single-parameter exponential distribution. In particular, the probability density function, the reliability, and the hazard rate, respectively, simplify to

$$f(t) = \frac{1}{\eta} e^{-t/\eta}$$

$$R(t) = e^{-t/\eta}$$

$$\lambda(t) = \frac{1}{\eta}$$

When these results are compared to the exponential case, it is observed that $\eta = 1/\lambda$ is the mean life. If $\beta = 1$ and $\gamma > 0$, then the Weibull distribution is similar to the exponential distribution but with a delay term that can be considered to be a period within which no failures occur, i.e., a minimum or failure-free operation life. If $\beta = 1$ and $\gamma < 0$, then failures can occur before $t = 0$, i.e., failures occurring in storage or during transport.

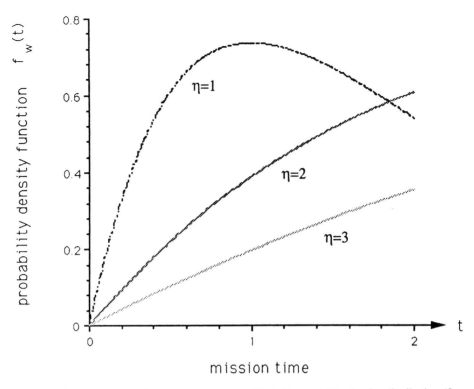

**Figure 11.8**   Effects of scale parameter η on the Weibull probability density distribution (β = 2, γ = 0). (From Ref. 64.)

The Weibull distribution reduces to the Rayleigh distribution when β = 2. The probability density function, reliability, and hazard rate are, respectively,

$$f(t) = \frac{2}{\eta}\left(\frac{t-\gamma}{\eta}\right)\exp\left[-\left(\frac{t-\gamma}{\eta}\right)^2\right]$$

$$R(t) = \exp\left[-\left(\frac{t-\gamma}{\eta}\right)^2\right]$$

$$\lambda(t) = \frac{2}{\eta}\left(\frac{t-\gamma}{\eta}\right)$$

As previously noted, there exists a linear relationship between the hazard rate and the mission time when β = 2. When γ = 0, the straight line passes through the origin with slope $2/\eta^2$.

The MTTF for the Weibull distribution is obtained using Eqs. (11.16) and (11.24),

$$\text{MTTF} = \int_0^\infty \exp\left[-\left(\frac{t-\gamma}{\eta}\right)^\beta\right]dt = \gamma + \eta\Gamma\left(\frac{1}{\beta}+1\right) \tag{11.28}$$

where Γ(1/β + 1) is the gamma function of the value 1/β + 1.

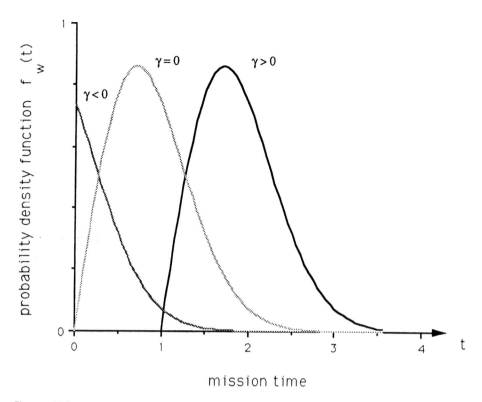

**Figure 11.9** Effects of location parameter $\gamma$ on the Weibull probability density distribution ($\beta = 2$, $\eta = 1$). (From Ref. 64.)

The variance of the Weibull distribution with $\gamma = 0$ is

$$\sigma^2 = \eta^2 \Gamma\left(\frac{\beta + 2}{\beta}\right) - \left[\eta\Gamma\left(\frac{1}{\beta} + 1\right)\right]^2 \qquad (11.29)$$

*Example 11.4*  Assuming the failure times of a lot of capacitors can be described by the Weibull distribution with estimated values of $\eta = 1000$ hr, $\gamma = 0$, and $\beta = 2$, determine the reliability of a capacitor that must operate for a 100-hr mission starting at time $t = 0$. Also determine the MTTF.

*Solution*  From Eqs. (11.24) and (11.28), respectively,

$$R(100) = e^{-(100/1000)^2} = 0.990$$

and

$$\text{MTTF} = 1000\Gamma(\tfrac{1}{2} + 1) = 1000\Gamma(1.5) = 886 \text{ hr}$$

The parameters for the Weibull distribution can be obtained from field data or reliability tests by suitably plotting the accumulated data using a special Weibull probability paper. This paper utilizes the fact that

$$1 - Q(t) = R(t) = \exp\left[-\left(\frac{t - \gamma}{\eta}\right)^\beta\right]$$

Taking the natural logarithm on both sides of the equation twice gives

$$\ln\{\ln[1 - Q(t)]^{-1}\} = \beta \ln(t - \gamma) - \beta \ln \eta$$

which is a linear equation of the form

$$y = \beta x + \text{constant}$$

if $\ln\{\ln[1 - Q(t)]^{-1}\}$ is plotted with respect to $\ln(t - \gamma)$ on linear versus linear graph paper. The Weibull paper is a modified ln-ln versus ln graph paper where the value of $Q(t)$ from which $\ln\{\ln[1 - Q(t)]^{-1}\}$ is determined becomes the ordinate value. Thus an engineer can work with $Q(t)$ and $t - \gamma$ directly.

Since the graphical method requires plotting the probability of failure or unreliability $Q(t)$ versus the time-to-failure data, an estimate of $Q(t)$ must be obtained. One estimate is the median rank (MR), which is found by setting the binomial distribution (Section 11.4.1) equal to 0.5 and solving for $Q$.

As an example, Table 11.3 gives some representative data for a reliability test for 10 identical parts in which 6 parts failed within the test duration of 600 hr. The unreliability was determined from median rank values. The data are plotted using the median rank (50% confidence level) and parameters are estimated from the curve. The unreliability versus time-to-failure data are then plotted (Figure 11.10) using Weibull probability paper. In Figure 11.10, the unreliability ranges, and thus the reliability, from 1% to 99%. If a reliability higher than 99% is required, then a Weibull probability paper with a percent range lower than 1% must be used.

Once all the data have been plotted, a line is drawn through the points that best represent the data. If the line is approximately straight, as is the case with the given data, the location parameter $\gamma = 0$. The case of $\gamma \neq 0$ will be discussed later. The shape parameter ($\beta$) can be found from the Weibull slope indicator at the top left of the Weibull probability paper. Once a straight line is drawn through the data, a line parallel to it and passing through the origin mark of the Weibull slope indicator must be drawn. For this example, the shape parameter is estimated to be $\beta = 0.66$. The value of the scale parameter ($\eta$) can be estimated by drawing a vertical line from the 63.2% unreliability point on the line. For this case, $\eta = 850$ hr approximately. It is now possible to write out the equation for the reliability and use it for analysis. However, for the most part, the curve can be used to determine the reliability values quite efficiently.

**Table 11.3**

| Sample number | Time to failure (hr) | Unreliability (%) based on median rank |
|---------------|----------------------|----------------------------------------|
| 1 | 14 | 6.70 |
| 2 | 58 | 16.23 |
| 3 | 130 | 25.86 |
| 4 | 245 | 35.51 |
| 5 | 382 | 45.17 |
| 6 | 563 | 54.83 |
| 7 | — | |
| 8 | — | |
| 9 | — | |
| 10 | — | |

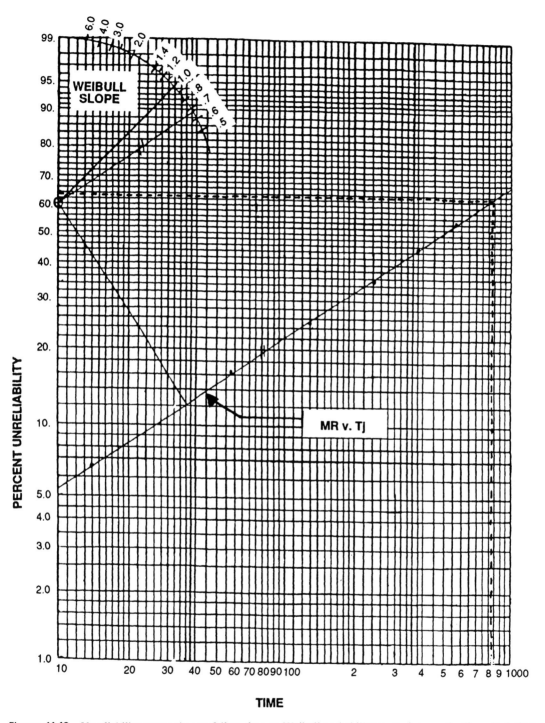

**Figure 11.10**   Unreliability versus time to failure data on Weibull probability paper for $\gamma = 0$. (From Ref. 64.)

*Example 11.5*   Using Figure 11.10 representing data from Table 11.3, determine
(1) the unreliability and reliability for a mission duration of 50 hr, (2) the reliability
for a new mission of 50 hr starting after the end of the previous 50-hr mission,
and (3) the longest mission duration that will provide a reliability of 95% assuming
the mission starts at an age of 50 hr.

   *Solution*   (1) From Figure 11.10, the unreliability estimate for a mission
time of 50 hr can be read directly from the curve. The value is $Q(50) = 15\%$.
Thus, the reliability for this mission time is $R(50) = 1 - Q(50) = 85\%$. (2)
The reliability for a new 50-hr mission starting with an age of 50 hr is

$$R(50,50) = \frac{R(50 + 50)}{R(50)} = \frac{R(100)}{R(50)} = \frac{0.78}{0.85} = 91.7\%$$

where $R(100) = 1 - Q(100)$ was taken directly from the curve. (3) For a mission
time $t$ that starts after a 50-hr mission and must have a reliability of 95%,

$$R(t,50) = \frac{R(t + 50)}{R(50)} = \frac{R(t + 50)}{0.85} = 0.95$$

or

$$R(t + 50) = 0.95 \times 0.85 = 0.808$$

To obtain this reliability, the unreliability is 0.192 or 19.2%. From the curve,
the time to obtain this unreliability is about 75 hr. Thus $50 + t = 75$ gives a
maximum new mission time of 25 hr in order to maintain a reliability of 95%.

   When the plotted data on the Weibull probability paper do not lie in a straight line,
it is possible that either $\gamma \neq 0$ or the Weibull distribution cannot be used to model the
data. The basic methodology consists of determining whether a time factor $\gamma$ exists such
that, when it is subtracted from the failure times and plotted, the resulting plot is a straight
line. Usually this value is obtained by successive approximations starting with the value
of the first failure time, $t_1$. If the plot obtained from $t_j - t_1$ is concave, this implies that
the time factor is greater than $t_1$, which is impossible because a failure has already occurred.
Thus, for this case there is no Weibull distribution that is capable of representing the data.
If the original curve is concave and the $t_j - t_1$ curve is convex, then the time factor lies
between 0 and $t_1$. If both curves are convex, then there exists a time factor that is less
than zero.

   As an example, Table 11.4 gives some representative data for a reliability test of 10
identical parts in which 6 failed within the test duration of 200 hr. From Figure 11.11,
the plot of the data on the Weibull probability paper does not lie in a straight line. It can
also be seen from Figure 11.11 that the plot of the reliability versus $t_j - t_1$ is convex,
which implies that the data can be represented by the Weibull distribution. Since the
original curve is concave, the time factor must lie between 0 and $t_1 = 50$. Then, by
picking values for the time factors within the range, the time factor that gives a straight
line can be found. In this case the estimate for the time factor $\gamma = 30$ is shown in Figure
11.11. The shape parameter is determined by drawing a parallel line through the Weibull
slope scale such that $\beta = 1.17$. The scale factor is determined by drawing a vertical line
from the curve at an unreliability of 63.2% such that $\eta = 200$. The Weibull equations
can then be obtained or reliability analysis can be accomplished using the plot.

Table 11.4

| Sample number | Time to failure (hr) | $t_j - t_1$ ($t_j - 50$) | $t_j - \gamma$ ($t_j - 30$) | Unreliability (%) based on median rank |
|---|---|---|---|---|
| 1 | 50 | 0 | 20 | 6.70 |
| 2 | 75 | 25 | 45 | 16.23 |
| 3 | 105 | 55 | 75 | 25.86 |
| 4 | 131 | 81 | 101 | 35.51 |
| 5 | 160 | 110 | 130 | 45.17 |
| 6 | 190 | 140 | 160 | 54.83 |
| 7 | — | | | |
| 8 | — | | | |
| 9 | — | | | |
| 10 | — | | | |

*Example 11.6* Using Figure 11.11, determine (1) the unreliability and reliability for a mission duration of 50 hr, (2) the reliability for a new mission of 50 hr starting after the end of the previous 50 hr mission, and (3) the longest mission duration that will provide a reliability of 90% assuming the mission starts at an age of 50 hr.

*Solution* (1) From Figure 11.11, the unreliability estimate for a mission time of 50 hr can be read directly from the curve using a time value of $50 - 30 = 20$. From this, the unreliability is 7%, and the reliability is thus 93%. (2) The reliability for a new 50-hr mission starting with an age of 50 hr is

$$R(50,50) = \frac{R(50 + 50)}{R(50)} + \frac{R(100)}{R(50)} = \frac{0.74}{0.93} = 79.5\%$$

where $R(100) = 1 - Q(100)$ was taken directly from the curve, using the time value of $100 - 30 = 70$. (3) For a mission time $t$ that starts after a 50-hr mission and must have a reliability of 90%,

$$R(t,50) = \frac{R(t + 50)}{R(50)} = \frac{R(t + 50)}{0.93} = 0.90$$

or

$$R(t + 50) = (0.90)(0.93) = 0.837$$

To obtain this reliability, the unreliability is 0.163 or 16.3%. From the curve, the time to obtain this unreliability is about 43 hr. Thus $50 + t - 30 = 43$ gives a maximum new mission time of 23 hr in order to maintain a reliability of 90%.

## 11.4.3 The Exponential Distribution

The exponential distribution is the most widely used statistical distribution in reliability analysis, with wide applicability in electronic systems. It is a single-parameter distribution,

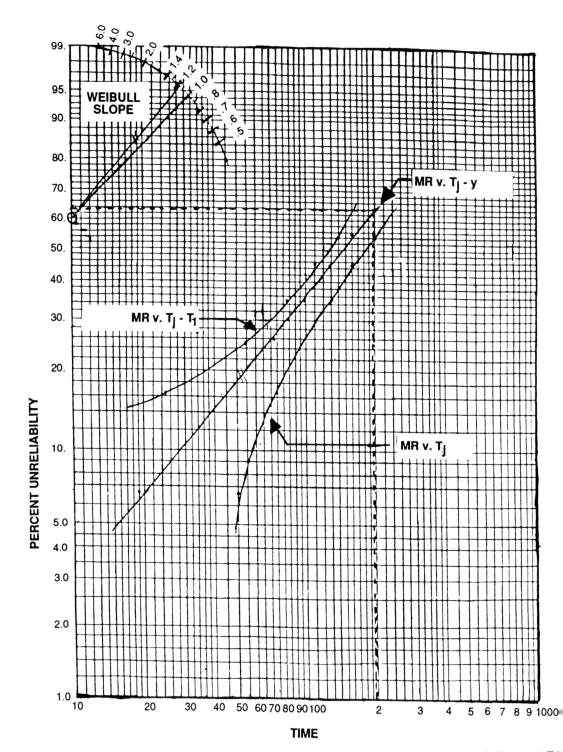

**Figure 11.11** Unreliability versus time to failure data on Weibull probability paper for $\gamma \neq 0$ $(0 < \gamma < T_1$. (From Ref. 64.)

which makes it very convenient to use in modeling failures, such as chance (random) failures, that are defined by a constant hazard rate.

The exponential distribution models the time between independent events that occur at a constant rate. It has been applied in modeling chance failures in reliability engineering, the life distribution of complex nonredundant systems, and the useful life of products exposed to both initial burn-in and preventive maintenance for product wearout. The exponential distribution is a special case of both the gamma and Weibull distributions.

The exponential distribution has the form

$$f(t) = \lambda_0 e^{-\lambda_0 t} u(t) \tag{11.30}$$

where $u(t)$ is the Heaviside step function and $\lambda_0$ is a positive real number, often called the chance hazard rate. The parameter $\lambda_0$ is typically an unknown that must be defined. The step function ensures that the function is valid only for $t \geq 0$.

Once $\lambda_0$ is known, the reliability can be determined from the probability density function,

$$R(t) = \int_t^\infty f(\tau) \, d\tau = \int_t^\infty \lambda_0 e^{-\lambda_0 \tau} \, d\tau = e^{-\lambda_0 \tau} \tag{11.31}$$

The hazard rate (force of mortality) is

$$\lambda(t) = \frac{f(t)}{R(t)} = \frac{1}{e^{-\lambda_0 t}} (\lambda_0 e^{-\lambda_0 t}) = \lambda_0 \tag{11.32}$$

The conditional reliability function is used in determining the reliability of additional missions of duration $t$ after the part has already accumulated $T$ hours of operation. Introducing Eq. (11.14) into Eq. (11.31) gives

$$R(t,T) = \frac{e^{-\lambda_0 (t+T)}}{e^{-\lambda_0 T}} = e^{-\lambda_0 T} \tag{11.33}$$

Equation (11.33) shows that previous missions do not affect the future reliability for the exponential distribution. This result stems from the fact that the hazard rate is constant. The mean time to failure (MTTF) can be determined from the general MTTF equation (11.16) using the reliability equation (11.31),

$$\text{MTTF} = \int_0^\infty e^{-\lambda_0 t} \, dt = \frac{1}{\lambda_0} \tag{11.34}$$

Since the MTTF is indirectly proportional to the failure risk, the reliability can be expressed in terms of the MTTF as

$$R(t) = e^{-t/\text{MTTF}} \tag{11.35}$$

Thus, for a mission time of $t = \text{MTTF}$, the reliability can be calculated from Eq. (11.35) to give $R(\text{MTTF}) = 0.368$. That is, only 37% of the parts will still be surviving for a mission time equal to the MTTF. At 10% MTTF, the reliability $R(0.1\text{MTTF}) = 0.90$, and at 1% MTTF the reliability $R(0.01\text{MTTF}) = 0.99$.

*Example 11.7*   Consider electronic parts that obey chance failure. If the MTTF is 100,000 hr, then 10% of the part failures will occur when $t = 0.1 \times \text{MTTF} = 10,000$ hr.

## 11.5  FAILURE MECHANISMS

The study of failure and failure mechanisms in electronic equipment presents ever-changing challenges as problems become solved and new ones arise. Fortunately, design, packaging, and manufacturing engineers are not necessarily faced with starting from the beginning with every new product, but can rely on an understanding of the basic principles of process control, physics of failure, and reliability engineering to aid in the development of new products and the improvement of existing products.

Failures can be classified by the failure location or failure site, the consequence or mode by which a failure is observed (i.e., open, short, parameter drift), and the failure mechanism (i.e., electrical overstress, corrosion, fatigue). Design for reliability incorporates the investigation of all potential failure mechanisms, with identification of where and when they may occur and their effect on and criticality for the operation of the product over the required useful life. As an example, Table 11.5 lists many of the failure sites,

**Table 11.5**  Microelectronic Failure Sites, Mechanisms, and Modes

| Failure site/ manufacturing process | Failure mechanism | Failure mode |
|---|---|---|
| Slice | Dislocations and stacking faults | Degradation of junction characteristics |
| | Nonuniform resistivity | Unpredictable component values |
| | Irregular surface | Improper electrical performance, shorts and/or opens |
| | Cracks, chips, scratches (general handling damage) | Opens, possible shorts in subsequent metallization |
| | Contamination | Degradation of junction characteristics |
| Passivation | Cracks and pinholes | Electrical breakdown in oxide layer between metallization and substrate; shorts caused by faulty oxide diffusion mask |
| | Nonuniform thickness | Low breakdown and increased leakage in the oxide layer |
| Mask | Scratches, nicks, blemishes in the photomask | Opens and/or shorts |
| | Misalignment | Opens and/or shorts |
| | Irregularities in photoresist patterns (line widths, spaces, pinholes) | Performance degradation caused by parameter drift, opens, or shorts |
| Etch | Improper removal of oxide | Opens and/or shorts or intermittents |
| | Undercutting | Shorts and/or opens in metallization |
| | Spotting (etch splash) | Potential shorts |
| | Contamination (photoresist, chemical residue) | Low breakdown; increased leakage |
| Diffusions | Improper control of doping profiles | Performance degradation resulting from unstable and faulty passive and active components |
| Metallization | Scratched or smeared metallization (handling damage) | Opens, near opens, shorts, near shorts |
| | Thin metallization to insufficient deposition or oxide steps | Opens and/or high-resistance intraconnections |

Table 11.5  Continued

| Failure site/ manufacturing process | Failure mechanism | Failure mode |
|---|---|---|
| | Oxide contamination–material incompatibility | Open metallization to poor adhesion |
| | Corrosion (chemical residue) | Opens in metallization |
| | Misalignment and contaminated contact areas | High contact resistance or opens |
| | Improper alloying temperature or time | Open metallization, poor adhesion, or shorts |
| Die separation | Improper die separation resulting in cracked or chipped dice | Opens and potential opens |
| Die bond | Voids between header and die | Performance degradation caused by overheating |
| | Overspreading and/or loose particles of eutectic solder | Shorts or intermittent shorts |
| | Poor die-to-header bond | Cracked or lifted die |
| | Material mismatch | Lifted or cracked die |
| Wire bond | Overbonding and underbonding | Wire weakened and breaks or is intermittent; lifted bond; open |
| | Material incompatibility or contaminated bonding pad | Lifted lead bond |
| | Plague formation | Open bonds |
| | Insufficient bonding pad area or spacings | Opens or shorted bonds |
| | Improper bonding procedure or control | Opens, shorts, or intermittent operation |
| | Improper bond alignment | Open and/or shorts |
| | Cracked or chipped die | Open |
| | Excessive loops, sags, or lead length | Shorts to case, substrate, or other leads |
| | Nicks, cuts, and abrasions on leads | Broken leads causing opens or shorts |
| | Unremoved pigtails | Shorts or intermittent shorts |
| Final seal | Poor hermetic seal | Performance degradation; shorts or opens caused by chemical corrosion or moisture |
| | Incorrect atmosphere sealed in package | Performance degradation caused by inversion and channeling |
| | Broken or bent external leads | Open circuit |
| | Cracks, voids in kovar-to-glass seals | Shorts and/or opens in the metallization caused by a leak |
| | Electrolytic growth of metals or metallic compounds across glass seals between leads and metal case | Intermittent shorts |
| | Loose conducting particles in package | Intermittent shorts |
| | Improper marking | Completely inoperative |

associated manufacturing processes, failure mechanisms, and failure modes for micro-electronic devices. Table 11.6 lists similar information for printed circuit and wiring board assemblies. Table 11.7 lists typical failure modes and the approximate percentages for some common microelectronic devices.

The classification of failures by sites is complex because of the various ways in which electronics can be constructed. A common failure site classification for electronic components is bulk, interface, oxide, metallization, and package. A common failure classification for circuit or wiring board assemblies is laminate, routing tracks, via and through holes, and component-to-board interconnects (i.e., solder joints). Connectors are typically classified independently of the components and assemblies.

In order to address failures most generically, failures must be identified as to the failure mechanisms and the predominant stresses (i.e., electrical, chemical, thermal, mechanical, and radiation) that induce failure, with the understanding that a given failure mechanism can be applied to many sites. The failure mechanism is the process by which the specific combination of physical, electrical, chemical, and mechanical stresses induces failure.

Examples of electrical stress failure mechanisms include electrical overstress (EOS), electromagnetic interference (EMI), and electrostatic discharge (ESD) associated with poor design and operation. ESD also arises because of poor handling. Electrical failures occur predominantly in devices, although track failures on a printed wiring or circuit board can arise.

Examples of chemical stress failure mechanisms include corrosion- and diffusion-related failures resulting from contamination, mismatched materials, and radiation. Chemical failures can occur within a packaged component, on the boards, and on all interconnects and connectors.

Examples of mechanical stress failure mechanisms include failures by fatigue, fracture, and misalignment and general degradation resulting from viscoelastic effects, such as creep. Mechanical failures can arise in every level of electronic equipment, from the device to the component to the system package, interconnects, and interfaces.

**Table 11.6**  Typical Printed Wiring Board Failure

| Failure site | Failure description | Failure mode | Failure mechanism |
|---|---|---|---|
| Plated through hole | Circumferential barrel crack | Electrical inter-mittent/open | Thermal shock Vibration–induced fatigue Thermal cycle–induced fatigue |
| | Inner layer conductor separation | Electrical inter-mittent/open | Vibration–induced fatigue Thermal cycle–induced fatigue Moisture adsorption–induced expansion |
| Surface layer conductor | Loss of conductor material | Electrical open | Corrosion |
| | Transverse conductor crack | Electrical inter-mittent/open | Vibration–induced fatigue Thermal cycle–induced fatigue |
| | Small cross-section metallic paths be-tween adjacent conductors | Electrical leak-age/short | Dendritic growth |

**Table 11.7** Typical Device Failure Modes

| Type | Main failure modes |
|---|---|
| Microcircuits | |
|   Digital logic | Outputs SA 0 or SA 1 |
| | No function |
|   Linear | Parameter drift |
| | No output |
| | Hard over output |
| Transistors | |
| | Low gain |
| | Open circuit |
| | Short circuit |
| | High-leakage collector base |
| Diodes | |
|   Rectifier, general purpose | Short circuit |
| | Open circuit |
| | High reverse current |
| Resistors | |
|   Film, fixed | Open circuit |
| | Parameter change |
|   Composition, fixed | Open circuit |
| | Parameter change |
|   Variables | Open circuit |
| | Intermittent |
| | Noisy |
| | Parameter change |
| Relays | |
| | No transfer |
| | Intermittent |
| | Short circuit |
| Capacitors | |
|   Fixed | Short circuit |
| | Open circuit |
| | Excessive leakage |
| | Parameter change |

Source: Ref. 53.

It is difficult to predict generically the dominant failure mechanism for any given structure unless the exact geometry; material specifications; manufacturing, assembly, and operating processes; and environmental conditions are known. A single failure mechanism cannot be applied to all equipment. Each new technology and manufacturing process must be researched to determine the potential and actual mechanisms of failure. Some failure mechanisms are dominant only under certain conditions. Furthermore, new materials and structures must be characterized and evaluated with respect to their potential to cause failure. Figure 11.12 illustrates a methodology for assessing materials and structures and their applicability in a product.

A common mistake in reliability engineering is to assume that an electronic system can be separated into parts (perhaps based on the independence of the facilities in which

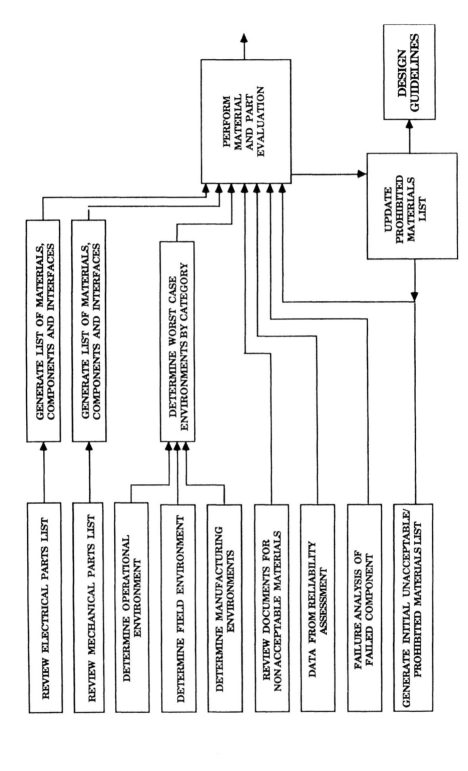

**Figure 11.12(a)** Materials characterization methodology. (Courtesy of Allied Signal-Bendix.)

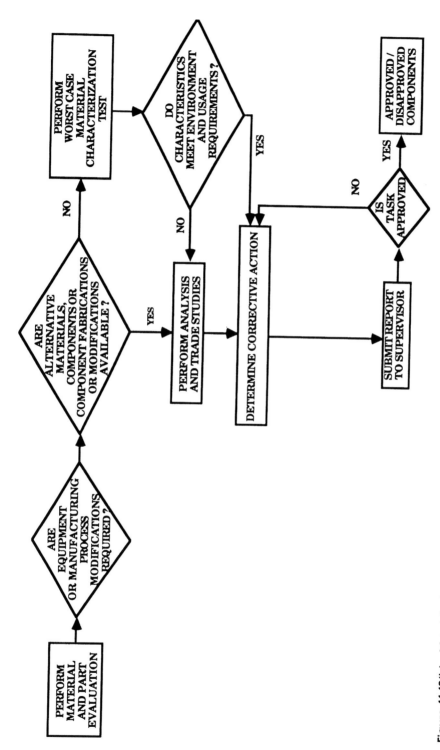

**Figure 11.12(b)**   Materials characterization methodology. (Courtesy of Allied Signal-Bendix.)

the individual parts were manufactured) and that the reliability of the system can be determined based on the reliability of the parts. The problem with this assumption is that often the dominant failures arise due to interactions of parts and stresses in assembly, handling, and operational use. A classical example is the problem of solder joints, where the interaction between component and board expansions due to power and temperature cycling can induce solder cracking and electrical failure.

Often, conditions and failure mechanisms at one site of a product will affect other sites and failure mechanisms. As an example of the interaction of conditions of stress, corrosion of wire bonds is associated with hermeticity, passivation and encapsulation integrity, seal integrity, and manufacturing and environmental contaminant conditions. As such, the prevention of corrosion of wire bonds could perhaps focus on making sure that seals do not fatigue crack so that moisture cannot ingress into the product and subsequently induce corrosion. As an example of the interaction of two failure mechanisms, even a small amount of corrosion attack on die metallization can accelerate metallization failure due to electromigration. Finally, as an example of the interaction of assembly conditions in failure mechanisms, microelectronic interconnection wires can typically withstand vibrations greater than 10 kHz. However, ultrasonic cleaning of an assembled printed wiring board assembly can induce high-cycle fatigue of wire bonds and induce failures in a microelectronic package.

Electronic equipment failures can be accelerated by defects in the crystal; chemically deposited epitaxial, metallization, and passivation layers; flaws in the oxide die and bonding materials; cracks and voids in the package encapsulant, seals, and leads; poor solder joints; cracked board connectors, etc. Failures can be further accelerated by various operational and environmental conditions, such as temperature, vibration, and humidity, and the gradients and history of these conditions. In many cases, coupling of stresses can result in further acceleration of failures.

Manufacturers set specifications on the stresses (temperature, current, voltage, fan-out, etc.) for which the parts or equipment will operate successfully. These specifications are typically minimum and maximum values and should be used as base guidelines for equipment operation. Occasionally, parts are subject to unreasonable usage conditions or greater than expected stresses occur in nature, which cause the parts to fail. To satisfy the customer, the manufacturer strives for part durability, defined by the ability of the

**Table 11.8** Typical Failure Mechanisms in MOS and Bipolar Devices

| Failure mechanisms | Type |
| --- | --- |
| Oxide defects | Infant/random |
| Silicon defects | Infant/random |
| Mask defects | Random |
| Refresh | Random |
| Surface charge | Wearout |
| Electromigration | Wearout |
| Slow trapping | Wearout |
| Contamination | Wearout/infant |
| Microcracks | Random |
| Hot electron injection | Wearout |

## Yield vs. Time, Texas Instruments devices

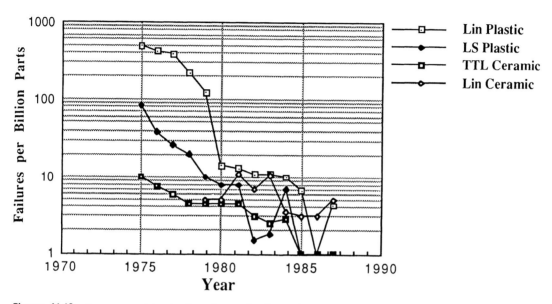

**Figure 11.13** Improvement in plastic and ceramic electronic component yield over time. (From Ref. 30.)

parts to withstand overstress. Durability also helps the manufacturer, since overstress can arise in manufacturing, assembly, and handling (Tables 11.5 and 11.6). Manufacturers thus provide specification margins to test parts under more stringent conditions than their data sheet specification limits and to provide some margin of safety.

Some technologies, such as many of the bipolar transistor–transistor logic (TTL) devices, have matured and been effectively controlled to the point that very few failures occur in the useful life of the equipment. This is true of metal oxide semiconductor (MOS) devices as well, as seen in Table 11.8. However, equipment utilizing more recent advances in technology and manufacturing processes often exhibits failure mechanisms that are not as well understood or controlled. For example, the reliability of CMOS devices was initially low in the late 1970s, but by the late 1980s it was comparable with that of the more mature bipolar TTL devices. As another example, Figure 11.13 shows the improvement in yield of plastic and ceramic electronic components over time.

## 11.5.1 Electrical Stress Failures

Manufacturers and equipment users typically claim that electrical overstress and electrostatic discharge equally account for between 60 and 80% of operational failures in the field. Although the failure stresses are different, in many cases the failure mechanisms often cannot be distinguished because the symptoms are shared by both failure mechanisms.

Protective circuitry is often incorporated in devices and equipment to accommodate excessive electrical stress. Device protection techniques typically focus on the usage of resistor diode networks to protect high-impedance gates in microelectronic devices. However, this circuitry is susceptible to the same failure mechanisms as the device being protected, so the circuitry depends on the margin of safety allowed by the component and equipment design. Furthermore, the striving for faster equipment has narrowed the margin of safety to the point that protection may not be effective.

## Electrical Overstress

Electrical overstress occurs when a device is subject to a voltage, current, or power that is larger than designed for. The overstress is associated with the generation of a hot spot at a semiconductor junction or at the drain in MOS devices. As the junction gets hotter, an increase in current is required to accommodate the additional stress, which in turn increases the temperature of the junction. Failure typically occurs when the melting point of the die is reached and the metallization open-circuits. In MOS devices, the failure mechanism is typically gate oxide breakdown due to high electric fields in the oxide.

High electric fields may also be generated by EOS (and ESD) that exceeds the oxide breakdown field strength, especially if poor yields result in uneven oxide growth. However, failures in MOS devices due to oxide breakdown can for the most part be eliminated by a voltage screen 50% above rated (i.e., 7.5 V for a 5-V rated device). On the other hand, with decreasing device size and the manufacturing of ultra-high-density circuits, the integrity and evenness of the oxide, especially with respect to microcracks and process defects, must remain a high priority.

**Table 11.9** Typical Charge Sources

| Object and process | Material or activity |
| --- | --- |
| Work surfaces | Waxed, painted, or varnished surfaces |
| | Common vinyl or plastics |
| Floors | Sealed concrete |
| | Waxed, finished wood |
| | Common vinyl tile or sheeting |
| Clothes | Common clean-room smocks |
| | Common synthetic personnel garments |
| | Nonconductive shoes |
| | Virgin cotton |
| Chairs | Finished wood |
| | Vinyl |
| | Fiberglass |
| Packaging and handling | Foam, plastic bags, wraps, envelopes |
| | Plastic trays, boxes, vials, parts bins |
| Assembly, cleaning, test, and repair areas | Spray cleaners |
| | Plastic solder suckers |
| | Solder irons with ungrounded tips |
| | Solvent brushes (synthetic bristles) |
| | Cleaning or drying |
| | Temperature chambers |
| | Cryogenic sprays |
| | Heat guns and blowers |
| | Sandblasting |
| | Electrostatic copiers |

Source: Ref. 53.

**Table 11.10** Dependence of Electrostatic Voltage on Humidity

| Methods of electrostatic generation | Electrostatic voltage | |
|---|---|---|
| | 10 to 20% relative humidity | 65 to 90% relative humidity |
| Walking across carpet | 35,000 | 1,500 |
| Walking over vinyl floor | 12,000 | 250 |
| Fabricating wafer | 2,000–8,000 | 100–2,000 |
| Picking up a polyethylene bag | 20,000 | 800 |
| Working at bench on a chair padded with polyurethane foam | 18,000 | 1,500 |

Source: Ref. 53.

## Electrostatic Discharge

Electrostatic discharge, or triboelectric charge, is the charge transferred when two insulators are rubbed together. The charge magnitude is a function of the separation, time of contact, rate of separation, humidity, and triboelectric materials. Although the ESD current is generally small, the magnitude of charge in a typical operating environment ranges from 100 V to 20 kV. Table 11.9 shows common charge sources. Table 11.10 shows the dependence of static voltage on humidity as well as the magnitude of charge generated by various common actions. Table 11.11 provides the ranges and sensitivity to voltage for various component types.

ESD failures occur when the potential across an electronic component is sufficient to induce dielectric breakdown, junction short circuit, or cracking between isolation regions. In extreme cases, cracking and pitting of the die can also arise.

In general, MOSFETs are the most susceptible to damage, with total breakdown for $n$-channel devices occurring around 0.1 to 2.5 kV. For example, the gates on Motorola's devices are about 800 angstroms thick and break down at a gate-source potential ($V_{gs}$) of about 100 V. For bipolar devices such as TTL, breakdown of the reverse-biased junction as a result of ionization and subsequent Joule heating occurs around 1000 V.

In some cases, rather than a catastrophic failure, ESD results in a reduction in the saturation drain current of NMOS gate oxides or concentrated cracking and melting of silicon of GaAs, which in turn causes voiding. In these cases, ESD must be treated as a cumulative failure process because a series of pulses will cumulate damage.

ESD prevention combines inhibition and removal of static charge generation and removal of any existing charge with device protection. ESD inhibition procedures include using carpets and equipment made or sprayed with antistatic solutions, packaging components in conductive or antistatic packages, and maintaining the environmental relative humidity above 65%.

Charge dissipation procedures are dependent on the materials of concern. If the part is a conductor, discharging can be accomplished simply by grounding. Insulator materials must be placed in contact with a grounded conductor to remove charge. If a part is in an insulated package, ionized air may be required. General precautions for ESD prevention are given in Table 11.12.

**Table 11.11** ESD Ranges and Sensitivity to Voltage

Class 1: Sensitivity range: 0 to 1000 V

Metal oxide semiconductor (MOS) devices including C, D, N, P, V, and other MOS technology without protective circuitry, or protective circuitry having Class 1 sensitivity.

Surface acoustic wave (SAW) devices

Operational amplifiers (OP AMP) with unprotected MOS capacitors

Junction field effect transistors (JFETs) (Ref.: similar to MIL-STD-701: Junction field effect, transistors and junction field effect transistors, dual unitized)

Silicon-controlled rectifier (SCR) with $I_0 < 0.175$ amperes at 100°C ambient temperature (Ref.: similar to MIL-STD-701: thyristors (silicon-controlled rectifiers))

Precision voltage regulator microcircuits: line or load voltage regulation $< 0.5\%$

Microwave and ultra-high-frequency semiconductors and microcircuits: frequency $> 1$ gigahertz

Thin-film resistors (type RN) with tolerance of $< 0.1\%$; power $> 0.05$ watt

Thin-film resistor (type RN) with tolerance of $> 0.1\%$; power $< 0.05$ watt

Large-scale integrated (LSI) microcircuits including microprocessors and memories without protective circuitry, or protective circuity having Class 1 sensitivity (Note: LSI devices usually have two or three layers of circuitry with metallization crossovers and small geometry active elements)

Hybrids utilizing Class 1 parts.

Class 2: Sensitivity range: 1000 to 4000 V

MOS devices or devices containing MOS constituents including C, D, N, P, V, or other MOS technology with protective circuitry having Class 2 sensitivity.

Schottky diodes (Ref: similar to MIL-STD-701: Silicon switching diodes)

Precision resistor networks (type RZ)

High-speed emitter-coupled logic (ECL) microcircuits with propagation delay $< 1$ nanosecond

Class 2: Sensitivity range: 3000 to 4000 V

Transistor-transistor logic (TTL) microcircuits (Schottky, low power, high speed, and standard)

Operational amplifiers (OP AMP) with MOS capacitors with protective circuitry having Class 2 sensitivity

LSI with input protection having Class 2 sensitivity

Hybrids utilizing Class 2 parts

Class 3: Sensitivity range: 4000 to 15000 V

Lower-power chopper resistors (Ref.: Similar to MIL-STD-701: Silicon low-power chopper transistors)

Resistor chips

Small signal diodes with power $< 1$ watt excluding zeners (Ref.: Similar to MIL-STD-701: Silicon axial leaded power rectifiers, silicon power diodes listed in order of maximum DC output current), fast-recovery diodes

Low-power silicon transistors with power $< 5$ watts at 25°C (Ref.: Similar to MIL-STD-701: Silicon switching diodes, thyristors (bidirectional triodes), silicon *pnp* low-power transistors ($P_c < 5$ watts at $T_A = 25°C$), silicon RF transistors)

All other microcircuits not included in Class 1 or Class 2

Piezoelectric crystals

Hybrids utilizing Class 3 parts

Source: Ref. 53.

**Table 11.12**   General Precautions for ESD Prevention

Utilize work benches with grounded metallic tips, conductive floor mats and containers, and ionized air.

Use grounded wrist straps in contact with skin.

Use a ground strap on the tip of all soldering equipment.

Ground all components prior to handling.

Place complete assemblies in antistatic bags if they are to be stored or transported.

Do not exceed maximum rated specifications.

Never insert or remove an IC with power applied.

Connect all low-impedance equipment to MOS devices only after power-up and disconnect before shutdown.

## 11.5.2   Thermal-Mechanical Stress Failures

Many electronic equipment failures can be traced to thermomechanical causes. Failure mechanisms commonly encountered are

1. Overstress failures
   a. brittle fracture
   b. excessive deformation
   c. catastrophic yielding and failure
   d. creep rupture in viscoplastic materials
   e. mechanical instability under compression
   f. electrical or thermal overload
2. Cumulative damage
   a. low-cycle fatigue
   b. high-cycle fatigue
   c. fatigue crack propagation
   d. creep-fatigue interactions
   e. excessive deformation due to ratcheting
   f. mechanical wear

The failure mechanisms listed above must be applied to study failures in components, component interconnects, wiring and circuit boards, assemblies, and systems. Once the failure mechanisms have been identified, sites that are prone to such mechanisms can be identified. For example, for a microelectronic component composed of a die, die attachment, lead frame assembly, interconnects, encapsulation, and enclosure assembly, the failure sites that need to be addressed are included in Table 11.13. For higher levels of packaging which includes the component-board interconnect, board and assembly of boards, connections and connectors, the failure sites which need to be addressed are included in Table 11.14.

Depending on the structural geometry, loads, and materials used, some of the failure mechanisms become active and critical at a given failure site. The result is a wide spectrum of different types of potential failures. Often there may be more than one active failure mechanism, and the designer needs to identify the most critical mechanism or mechanisms that will play the limiting role in the life of the component, device, or system. Table 11.15 presents a set of potential reliability problems for electronic package and assembly processes.

**Table 11.13**  Potential Failure Sites in IC Packages

Brittle die including silicon and gallium arsenide
Hybrid passive elements
Attachment adhesives including gold-silicon eutectics, silver-glass, polymers,
    solders, etc. used for components and substrates
Substrates
Interconnection wires including gold, aluminum, copper, and palladium
Bonds between wires and bond pads (Al, Au, Ag, Ni, Cr, Cu)
Tape-automated bonds (TAB)
Flip-chip bonds
Leads
Lids
Seals including lid and lead seals
Case, molding, or header
Lead and case finish

Regardless of whether the failure is to a component, a board, the interconnects, or the connectors, the physical dimensions, material properties, and "stresses" can never be treated as constants. The designer must obtain data on the variation of values and express the mean value with the associated variance. Physical dimensions and tolerances must also be considered with respect to their effects on the dominant failure mechanisms.

Beside dimensional variances, natural material property variabilities must be considered. Variabilities must be included in the material constitutive and damage model coefficients such as moduli, thermal coefficients of expansion, orthotropic behavior, fracture toughness, fatigue properties, and material irregularities. Chapter 12 discusses materials and their properties.

Similarly, assembly and manufacturing variabilities must be considered. All assembly processes impose transient stresses on the equipment, as well as introducing variabilities that need to be accounted for in the life cycle evaluation process. Assembly conditions must be assessed in terms of the assembly of the components onto the board, boards into boxes, etc. For example, wave soldering methods and the associated thermal profile can affect plated through hole reliability. Another example is the effect of ultrasonic cleaning and the associated vibrations on the reliability of microelectronic wire bonds.

Stress conditions used in the failure models must include the operating conditions and the testing conditions to which the component is subjected. These include temperature cycling, thermal shock, burn-in screening, stabilization baking, constant acceleration tests, mechanical shock, vibration, moisture resistance, salt atmosphere, and insulation resistance.

**Table 11.14**  Potential Failure Sites in Higher-Level Packaging

Bulk and constituent board materials
Board lamination
Plated through holes and vias
Routing tracks and pads
Component-to-board interconnect (adhesive, solder, joint, solder bump,
    tape-automated bonds)
Connectors

**Table 11.15** Potential Reliability Problems for Electronic Package and Assembly Process

Loss of substrate adhesion from the case or header in microelectronic packages or multichip carriers due to thermomechanical stresses, temperature cycling, power cycling, and poor adhesion

Loss of die adhesion at land pads due to thermomechanical stresses, temperature cycling, power cycling, and poor adhesion

Bond wire failure due to axial fatigue, bond pad fatigue, corrosion, and thermomechanical corrosion stresses

Loss of wire/TAB/flip-chip bond adhesion due to chemical/thermomechanical stresses, cleaning/ soldering and lid sealing operations, and poor adhesion

Fractured and fatigued external leads due to vibration and shock loading screening and burn-in stresses, corrosion, and thermomechanical stresses

Loss of land pad adhesion in all surface mount components due to chemical and thermo-mechanical stresses.

Hermeticity failure from poor bake stabilization, leaks through lid seals, lead seals, glass seal fracture

Corrosion of package exterior through welds and pinholes

Current leakage through or on the surface of insulating external package components due to dielectric breakdown and electromigration from metallization

Loss of die function due to propagation of fatigue crack through active elements under thermo-mechanical cycling, resulting from inspection and test requirements, manufacturing and assembly conditions, and thermal-mechanical stresses

Cracked substrate from thermomechanical stress

## 11.5.3 Corrosion Stress Failures

Corrosion-induced failure mechanisms have been steadily decreased over the years as contamination in the manufacturing processes has been reduced and as methods for package sealing and inspection have improved. However, as material dimensions have decreased, even parts per billion levels of pollutants are sufficient to contaminate circuitry and accelerate corrosion reactions in electronic equipment. Contamination caused by mobile ions, especially Na, Cl, and K, is perhaps most common because of the abundant nature of these ions in the manufacturing process materials, packaging materials, environment, atmosphere, sweat, and breath. Sources of process contaminants include solder flux residues, residual electroplating and other processing chemicals, sulfur and other reactive substances from storage container materials, and vaporized contaminants from adhesives. In fact, the single largest source of contamination at the printed circuit board assembly level is solder flux residue. Solder fluxes are used to remove oxides, tarnish films, and other impurities from the surface during soldering, while simultaneously lowering the surface tension of the molten solder to allow rapid flow and adherence to the metal surfaces. In general, fluxes leave corrosive residues.

Within each package type, the failure sites that must be addressed include metallization on die, wire bond pads on die and lead frame, component leads, routing tracks on printed circuit and wiring boards, connectors and other electrical interfaces, and electromechanical moving parts. Guidelines for corrosion control are given in Chapter 10.

## 11.5.4 Storage Stress Failures

Often, reliability engineers assume that if electronic equipment is stored in a dry environment at a constant low temperature, even poor-quality electronics may show only slow degradation. However, in typical storage environments, where the relative humidity oc-

**Table 11.16**  Failures Encountered During Storage

| Component | Failures |
|---|---|
| Batteries | Dry batteries have limited shelf life. They become unusable at temperatures above 350°C. The output of storage batteries drops at very low temperatures. |
| Capacitors | Moisture permeates solid dielectrics and increases losses, which may lead to breakdown. Moisture on plates of an air capacitor changes the capacitance. |
| Coils | Moisture causes changes in inductance. Moisture swells phenolic forms. Wax coverings soften at high temperatures. |
| Connectors | Corrosion causes poor electrical contact and seizure of mating members. Moisture causes shorting at the ends. |
| Relays and solenoids | Corrosion of the metallic parts causes malfunction. Dust and sand damage the contacts. Fungi grow on the coils. |
| Resistors | Fixed composition resistors drift, and these resistors are not suitable at temperatures above 85°C. Enameled and cement-coated resistors can have small pinholes that bleed moisture, accounting for eventual breakdown. Precision wire-wound fixed resistors fail rapidly when exposed to high humidity and temperatures above 125°C. |
| Diodes, transistors | Plastic-encapsulated devices often provide a poor hermetic seal, resulting in shorts, or opens caused by chemical corrosion or moisture. |
| Motors, blowers, and dynamotors | Swelling and rupture of plastic parts and corrosion of metal parts. Moisture abortion and fungus growth on coils. |
| Plugs, jacks, dial-lamp sockets, etc. | Corrosion and dirt produce high-resistance contacts. Plastic insulation absorbs moisture. |
| Switches | Metal parts corrode and plastic bodies and wafers warp owing to moisture absorption. |
| Transformers | Windings corrode, causing short or open circuits. |

Source: Ref. 53.

casionally exceeds 80%, even the better-quality electronics can exhibit degradation. In general, the failure mechanisms encountered during storage are associated with contamination and corrosion, although temperature changes can induce mechanical-type failures such as creep and fatigue. Table 11.16 shows some of the dominant failures encountered during storage. A difficulty in failure analysis of storage reliability is in determining the exact source of failure, especially if the equipment is occasionally handled.

## 11.5.5  Radiation Stress Failures

In microelectronic devices, radiation stress failures are caused by both external radiation (gamma, cosmic, and X-rays) and radiation internal to the device and packaging materials (alpha particles and beta radiation). Internal radiation comes from trace impurities of radioactive elements such as uranium and thorium present in many packaging materials, including alumina, epoxy, and package lids. Table 11.17 shows the susceptibility of various semiconductor devices to various radiation environments.

In semiconductor devices, radiation can generate hole-electron pairs that can induce low-frequency noise, increased leakage current across $p$-$n$ junctions, and reduce current gain in bipolar devices. In MOS devices, the dominant effects include shifts in the threshold

**Table 11.17** Radiation Susceptibility of Various Semiconductors

| Radiation environment | Semiconductor technology | | | | | | | | |
|---|---|---|---|---|---|---|---|---|---|
| | Discrete bipolar transistors and J-FET's | Silicon-controlled rectifiers | TTL | Low-power Schottky TTL | Analog integrated circuit | C-MOS | n-MOS | Light-emitting diodes | Isoplanar II ECL |
| Neutrons (c/nm$^2$) | $10^{10}$–$10^{12}$ | $10^{10}$–$10^{12}$ | $10^{14}$ | $10^{14}$ | $10^{13}$ | $10^{15}$ | $10^{15}$ | $10^{13}$ | $>10^{15}$ |
| Ionizing radiation | | | | | | | | | |
| Total dose (rads (Si)) | $>10^4$ | $10^4$ | $10^6$ | $10^6$ | $10^4$–$10^5$ | $10^3$–$10^4$ | $10^3$ | $>10^5$ | $10^7$ |
| Transient dose rate (rads (Si)/s) (upset or saturation) | — | $10^3$ | $10^7$ | $5 \times 10^7$ | $10^6$ | $10^7$ | $10^5$ | — | $>10^8$ |
| Transient dose rate (rads (Si)/s) (survival) | $10^{10}$ | $10^{10}$ | $>10^{10}$ | $>10^{10}$ | $>10^{10}$ | $10^9$ | $10^{10}$ | $>10^{10}$ | $10^{11}$ |
| Dormant total dose (zero bias) | $>10^4$ | $10^4$ | $10^6$ | $10^6$ | $10^5$ | $10^6$ | $10^4$ | $>10^5$ | $>10^7$ |

Source: Ref. 71.

voltage, reduction in transconductance, and activation of parasitics. Alpha particle radiation in particular can disturb stored data in dynamic memory, thus generating a recoverable or soft error.

When external radiation is a concern, shielding is required. When internal radiation is a concern, especially for high-reliability devices, material selection and quality control are the major methods of improvement.

### 11.5.6  Particulate Contamination Stress Failures

Contamination failures include both particulate (i.e., loose particles or debris) and ionic contamination. As discussed in Section 11.5.3, ionic contamination is a concern at the device level and can lead to performance degradation and corrosion. Particulate contamination can occur at all packaging levels. The dominant failure mechanism of particulate contamination is shorting between electrical connection paths, although contaminants can inhibit electrical contact in switches and can block heat transfer paths. Occasionally, degradation of performance may arise prior to catastrophic failure.

On the component level, sealed (hermetic packages are the most susceptible to particulate contamination. Loose conducting particles can result from manufacturing processes in the form of bond wire pieces, solder pieces (balls) from wire bonds and seals, particles of silicon from scribing, and breakages within the packaged device. For higher levels of packaging, particulate contamination can occur anywhere debris can enter and induce a short, cause a switch not to make contact, or inhibit a function (i.e., block a cooling path).

The effect of particulate contaminants on device performance is unpredictable and difficult to detect. In many cases, it is possible for components to pass most of the high-reliability screening tests without being detected. The most effective tests for particulate contamination include X-radiography and particle impact noise detection (PIND). In X-radiography, comparison images are made of a package before and after shaking to observe for differences due to loose particles. In the PIND test, sensitive acoustic sensors are used to detect small particle impacts within the package as the package is shaken. However, screening is useless if loose particles form after screening or during field operation.

## 11.6  RELIABILITY ISSUES IN ELECTRONICS DESIGN AND MANUFACTURE

The goal of design for reliability and the reliability program is to improve operational readiness and mission success of the major end item; reduce item demand for maintenance, manpower, and logistic support; provide essential management information; and hold on its own impact an overall program cost and schedule. Reliability programs must therefore include an appropriate mix of reliability engineering and accounting tasks, depending on the life cycle phase. These tasks must be planned, integrated, and accomplished in conjunction with other design, development, and manufacturing functions. Thus, the reliability engineering tasks focus on prevention, detection, and correction of reliability design deficiencies, weak parts, and workmanship defects. An efficient reliability program stresses early investment in reliability engineering tasks to avoid subsequent cost and schedule delays.

The reliability program begins during the concept development and design stage and continues through manufacture, assembly, and test and during the useful life of the product. When a product is designed with consideration for reliability, various criteria for reliability assessment, growth and demonstration testing, maintainability, supportability, and logistic cost studies will also be incorporated into the design. Figure 11.14 illustrates the phases and flow of reliability analysis and testing during product development. Design for reliability during the concept stages of design requires that:

· Target reliabilities are specified taking into consideration stress and environmental constraints, the use of new technologies and their risks, and the possibility of designing redundancy into the product.
· Target maintenance parameters are specified taking into consideration the potential for built-in self-test capabilities and modular design for easy repair by minimally skilled maintenance personnel using a minimal amount of support equipment.
· A design framework exists within which detailed design is developed and specified.

After the concept phase of design, the major categories of the reliability program are design verification, qualification procedures, manufacturing reliability monitors, and fundamental reliability analysis. Throughout the program, all engineers who are part of the product development must meet to review the design, manufacture, assembly, test, and usage of the product.

Design verification is employed to ensure that product reliability and functionality, quality, performance, and yield are considered in the design. As part of design verification, the packaging team must ensure that packages meet strict reliability goals before package qualification is initiated. At the same time, the process development team works with manufacturing to ensure that new processes do not violate any reliability standards.

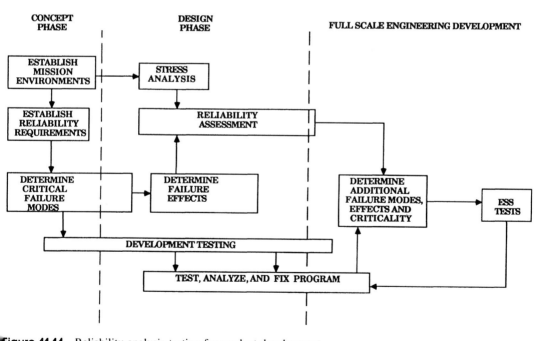

**Figure 11.14**  Reliability analysis testing for product development.

Qualification procedures are used predominantly to ascertain the major characteristics of new technologies or products or to evaluate changes in existing technologies or products. Qualification must be conducted upon the introduction of a new product technology or process, equipment changes or modifications, or initiation or transfer to a new manufacturing site. All changes require qualification, but the degree of qualification depends on the nature of the change and therefore ranges from detailed qualification to simply reviewing engineering data. However, it must be understood that device and equipment parameter changes can occur as a result of even the most seemingly trivial changes in the manufacturing and assembly processes.

Reliability monitors are conducted on products from each assembly plant associated with the product production process. Monitors are conducted on each representative manufacturing process technology, each generic product and package technology, and each assembly plant. The monitor requires subjecting the product to electrical, mechanical, thermal, chemical, and general environmental tests similar to those used to qualify the product. Test results are continuously reported and problems are identified. If the rate of failure is greater than an acceptable limit, a review board meeting must be called to review the data, determine the appropriate corrective action, and control the problem or issue until it is resolved. It is important to note that even if the problem or potential problem does not have current implications, it could become critical as geometries, materials, and technologies change.

Fundamental reliability analysis pertains to the understanding and study of the physics of existing and potential failure mechanisms; circuit, layout, and operating dependencies; reliability physics and geometry to forecast both environmental and electrical accelerating factors; failure models of each failure mechanism; and the key physical and electrical parameters and their limits for process control. Many mechanisms affect product reliability, and therefore each mechanism must be investigated and characterized.

Reliability accounting tasks focus on the provision of information essential to acquisition, operation, and support management, including properly defined inputs for estimates of operational effectiveness and ownership cost. An efficient reliability program provides this information while ensuring that cost and schedule investments are visible and controlled.

Up-front design for reliability has been shown to improve product performance over time, reduce the design time, and reduce life cycle costs. The performance over time of electronic equipment can be increased through improved design practices, selection of preferred parts, judicious derating, addition of fault tolerance, quality control, and proper stress screening and burn-in practices. Several of these techniques are discussed in the following sections.

## 11.6.1  Reliability Allocation

Reliability allocation is a technique employed in the conceptual phase of design to assess potential system reliability requirements by assigning reliability constraints to the subsystems of the system. Allocation aids in establishing reliability guidelines and goals that can be used to assess feasibility and consider fundamental trade-offs.

The mathematical foundation for reliability allocation involves solving the inequality

$$f(\hat{R}_1, \hat{R}_2, \ldots, \hat{R}_n) \geq \hat{R} \tag{11.36}$$

where $\hat{R}_i$ is the allocated reliability for the $i$th unit, $\hat{R}$ is the system reliability requirement, and $f$ is a function that defines the system reliability in terms of the unit reliabilities.

For a system of $n$ failure-independent subsystems in series, each with a constant hazard rate and mission time equal to that of the system, the system reliability $R_s(t)$ is

$$R_s(t) = \prod_{i=1}^{n} R_i(t) = \exp(-\Sigma \lambda_i t) = \exp(-\lambda_s t) \qquad (11.37)$$

where $R_i$ and $\lambda_i$ are the reliability and hazard rate of the $i$th subsystem, respectively, and $\lambda_s$ is an estimated system hazard rate.

The allocated hazard rate $\hat{\lambda}_i$ for each subsystem must result in a system hazard rate $\hat{\lambda}_s$ such that

$$\hat{\lambda}_s = \sum_{i=1}^{n} \hat{\lambda}_i \qquad (11.38)$$

The subsystem hazard rate allocation is then given by

$$\hat{\lambda}_i = \frac{\hat{\lambda}}{\lambda_s} \lambda_i \qquad (11.39)$$

where $\lambda$ is the required hazard rate for the system.

*Example 11.8*   Consider a series system composed of four independent subsystems with respective estimated constant hazard rates of $\lambda_1 = 0.0002$, $\lambda_2 = 0.0007$, $\lambda_3 = 0.0005$, and $\lambda_4 = 0.0008$ failure/hr. If the required hazard rate for the system is $\lambda = 0.004$, determine the hazard rate to be allowed to each system.

*Solution*   Using Eq. (11.37) for the system reliability gives

$$\lambda_s = \Sigma \lambda_i = 0.0002 + 0.0007 + 0.0005 + 0.0008 = 0.0022 \text{ failure/hr}$$

Then, from Eqs. (11.38) and (11.39),

$$\hat{\lambda}_1 = \lambda_1 \frac{\hat{\lambda}}{\lambda_s} = \frac{(0.0002)\,(0.004)}{0.0022} = 0.00036$$

$$\hat{\lambda}_2 = \lambda_2 \frac{\hat{\lambda}}{\lambda_s} = \frac{(0.0007)\,(0.004)}{0.0022} = 0.00127$$

$$\hat{\lambda}_3 = \lambda_3 \frac{\hat{\lambda}}{\lambda_s} = \frac{(0.0005)\,(0.004)}{0.0022} = 0.00091$$

$$\hat{\lambda}_4 = \lambda_4 \frac{\hat{\lambda}}{\lambda_s} = \frac{(0.0008)\,(0.004)}{0.0022} = 0.00145$$

Other more complex allocation schemes might recognize that some components are preferred over others and selectivity set allocation goals to minimize cost, space, or test time. Other schemes might be based on the inherent and/or operational availability and as such recognize the need for redundant systems and maintenance concepts.

Allocation of availability involves solving Eq. (11.36) in terms of a system availability, but with the $i$th unit availability replacing the $i$th unit reliability. Figure 11.15 shows the complex dependences to determine availability.

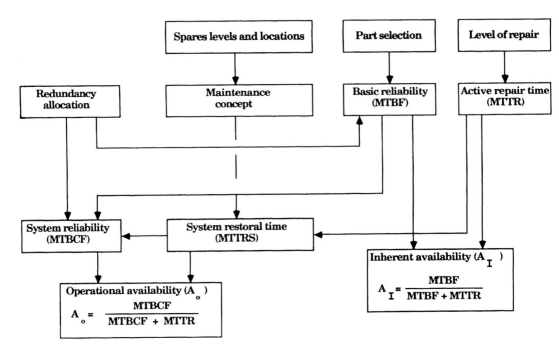

**Figure 11.15**  Reliability, availability, maintainability dependence tree for exponentially distributed system.

## 11.6.2  Preferred Parts Selection and Qualification

Preferred parts selection is a method whereby parts (i.e., subsystems, components, or circuits) are chosen from a data base of parts that have been tested, generally based on field data, to demonstrate acceptable reliability for the operational and environmental stresses of the proposed mission.

The use of preferred parts is frequently mandated on military and government programs. The designer can select proven components and subsystems from companies that produce products with high quality, unless custom or semicustom components are to be produced. NASA, the Department of Defense, and many companies maintain preferred parts lists for components and tested circuits that are known to perform in desired environments. Selection of potentially superior but less proven parts (in terms of the accumulated test time) must be justified and approved. Such approval might be based on accelerated testing to establish the reliability of the new parts as quickly as possible. This requires understanding the failure mechanisms of the parts and the interaction of the parts within the system.

Parts qualification is a process by which a manufacturer acquires a manufacturing line certification and qualification for a given part or product. A vendor part qualifies by following the requirements, procedures, and tests detailed in numerous military standards and specifications. For each device type requiring qualification, procedures that must be followed include process line certification, such as manufacturer audits, product assurance program plan, materials, and construction baseline; inspection during manufacture, such as equipment, review processes and techniques, audit plans, and records; screening tests, such as nondestructive bond pull, internal visual, stabilization bake, temperature cycle, burn-in, particle impact noise detection, leak test, final electrical, and external visual; and a host of quality conformance tests.

Certification is the recognition of evidence by the qualifying activity that the manufacturing line is capable of producing parts of high quality that are compliant with requirements. For military microelectronic parts, MIL-I-38535, "General Specification for Integrated Circuit Manufacturing," is the specification manual. Here, qualification is the actual demonstration of the certified manufacturing line capabilities by producing microelectronics compliant with MIL-I-38535 and the device specification. Satisfying these procedures takes from 7 to 25 months. The minimum number of functioning devices needed for all testing is typically 200, half of which are destroyed in test. Typical package qualification requirements are given in Table 11.18.

In order to modernize the procedure of qualification of parts, in particular high-quality, high-reliability microelectronics, an approach to generic qualification is often utilized. The concept is to introduce a single methodology for both commercial and military products that permits certification of design, fabrication, assembly, packaging, and test. The assumption is that high-quality and high-reliability microelectronics can be developed if the processes are properly monitored and controlled at each stage of the manufacturing and assembly process, without excessive end-of-line testing. This involves improved statistical process control, field failure return programs, corrective action procedures, and quality improvement programs.

## 11.6.3   Identification of Failure Mechanisms

A key to the success of design for reliability is the determination of the failure sites, mechanisms, and modes of failure and the development of knowledge of the manufacturing,

**Table 11.18**   Typical Package Qualification Requirements

| Test | Conditions | Duration |
|---|---|---|
| High temperature | 150°C | 1000 hours |
| Humidity | 85°C/85% RH | 1000 hours |
| Temperature cycle | $-55°C/+125°C$ | 1000 cycles |
| Thermal shock | $-55°C/+125°C$ | 200 cycles |
| Vibration | 0.002–2 kHz, 20 $G$ max. | 3 axes |
| Mechanical shock | 500 $G$ | 6 axes |
| Salt | 35°C | 24 hours |
| Solvents | Chemicals | 1 cycle |
| Die shear | 2500 grams | To destruction |
| Wire pull | 4 grams | To destruction |
| Lead frame testing | Various | To destruction |
| Nonvolatile storage—hot | 75°C | 4 hours |
|  | 95°C | 4 hours |
|  | 100°C | 4 hours |
| Nonvolatile storage—cold | $-20°C$ | 4 hours |
|  | $-40°C$ | 4 hours |
|  | $-55°C$ | 4 hours |
| Benchmark monitor | Full data sheet with various worst-case patterns | 6 weeks |

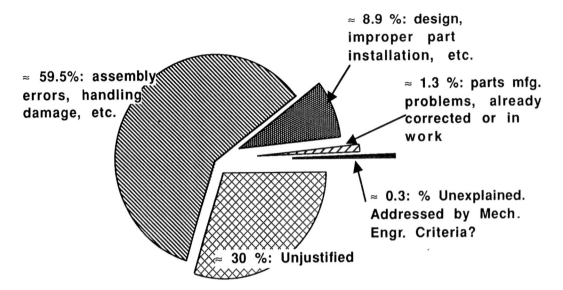

≈ 8.9 %: design, improper part installation, etc.

≈ 59.5%: assembly errors, handling damage, etc.

≈ 1.3 %: parts mfg. problems, already corrected or in work

≈ 0.3: % Unexplained. Addressed by Mech. Engr. Criteria?

≈ 30 %: Unjustified

**Figure 11.16**  Summary of diagnosis results, complex computer-type boxes, based on total removals. (From Ref. 30.)

assembly, storage, operational, and environmental conditions. The designer must utilize this information when assessing the reliability of electronic devices, their interconnects, and the complete interacting product. Figure 11.16 shows a diagnosis summary of failures of computer-type boxes, based on total removals.

Failure mechanisms and potential failure sites are identified and analyzed subject to manufacture, assembly, test, storage, and field use under various mechanical, thermal, electrical, chemical, and other "stresses." The goal is to encompass the mean, variations, and, where appropriate, bounds (minimum and maximum) on package dimensions, materials and their properties, I/O terminations, manufacturing/assembly/storage processes, and service conditions. Occasionally, sensors are placed in the operating environment to determine the actual on-board vibration, temperature, and humidity stress history. Potential failure modes are then identified to aid in determining the consequence of a failure in terms of system operation, as well as to aid in determining test and maintenance procedures. Section 11.5 focuses on some of the major physics of failure mechanisms for electronics.

### 11.6.4  Derating and Stress Management

Derating and stress management are methods of increasing a product's life by decreasing applied thermal, electrical, mechanical, and other stresses. The premise is that the greater the stress level, the more accelerated the failure mechanisms; and the lower the stress level, the longer the product will operate successfully. Thus, derating is similar to placing a safety factor on the operating parameters that can induce degradation of the product. When the product is operated under conservative stresses that are below designer/manufacturer-specified maximum rated stress levels, the reliability should be improved. However, there is often a minimum stress below which the circuit will not function properly

| Factors | | Score |
|---|---|---|
| Reliability Challenge | • For *proven design*, achievable with standard parts/circuits | 1 |
| | • For high reliability requirements, *special design features* needed | 2 |
| | • For new design challenging the state-of-the-art, *new concept* | 3 |
| System Repair | • For *easily accessible*, quickly and economically repaired systems | 1 |
| | • For high repair cost, limited access, *high skill levels required*, very low downtimes allowable | 2 |
| | • For *nonaccessible repair*, or economically unjustifiable repairs | 3 |
| Safety | • For *routine safety* program, no expected problems | 1 |
| | • For potential system or equipment *high cost damage* | 2 |
| | • For potential *jeopardization of life* of personnel | 3 |
| Size, Weight | • For no significant design limitation, *standard practices* | 1 |
| | • For *special design features* needed, difficult requirements | 2 |
| | • For new concepts needed, severe design limitations | 3 |
| Life Cycle | • For *economical repairs*, no unusual spare part costs expected | 1 |
| | • For potentially *high repair cost* or unique high cost spares | 2 |
| | • For systems that may require *complete substitution* | 3 |

**Instuctions:** Select score for each factor, sum and determine derating level or parameter.

| Derating Level | Total Score |
|---|---|
| I | 11–15 |
| II | 7–10 |
| III | 6 or less |

**Figure 11.17** RADC part level derating classes.

or at which the increased complexity required to allow the lower stress level will not offer an advantage in reliability or cost effectiveness.

Manufacturers' ratings or users' procurement specifications are generally used to determine a maximum rating value. These published values are often conservative estimates of the actual stress level of the product based on the design, the manufacturing uniformity and repeatability, and the desired margin of safety.

The choice of an appropriate derating criterion depends on the dominant failure mechanisms, the application, the criticality of the part or subsystem within the system, and the required safety of the mission. In order to address different derating environments, derating guidelines for electronic components have been based on classes of criticality. Figure 11.17 shows typical derating classes from the RADC derating guidelines. NASA and the U.S. Navy, as well as most industries, have their own derating guidelines.

The derating factors specify the maximum recommended stress limits based on the ratio of the actual applied stress to the maximum rated stress. The designer must be aware of the difference between the specified part environment and operating conditions and the actual part environment and operating conditions. For unlisted parts or new devices,

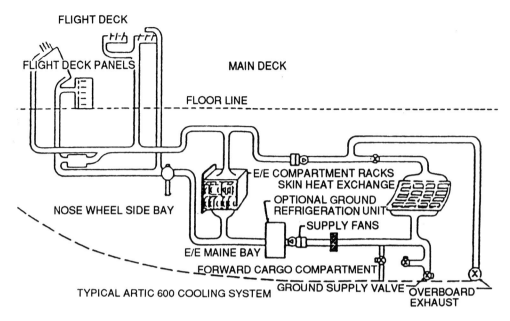

FLIGHT DECK

FLIGHT DECK PANELS

MAIN DECK

FLOOR LINE

E/E COMPARTMENT RACKS
SKIN HEAT EXCHANGE

OPTIONAL GROUND
REFRIGERATION UNIT

SUPPLY FANS

NOSE WHEEL SIDE BAY

E/E MAINE BAY

FORWARD CARGO COMPARTMENT

TYPICAL ARTIC 600 COOLING SYSTEM   GROUND SUPPLY VALVE   OVERBOARD
                                                                                                           EXHAUST

**Figure 11.18**   Typical cooling system weight for electronic equipment on contemporary commercial airplanes. (From Ref. 30.)

derating factors may be determined based on suspected failure mechanisms to increase the margin of safety.

The methodology behind derating depends on the part type. For example, semiconductors are often derated by maintaining the power of dissipation below a rated level. Capacitors are derated by maintaining the applied voltage below the rated part voltage. Resistors are derated by maintaining the ratio of operating power to rated power below a fixed value.

Derating must be conducted simultaneously with thermal, thermal-mechanical, and thermal-electrical stress management. Stress management involves determining and reducing stresses on critical parts and in critical locations.

Thermal stress management addresses the failure mechanisms that are sensitive to temperature, temperature changes, temperature gradients, and the overall temperature history. Selection of a cooling technology and associated parameters for cooling, thermal modeling and simulation, and testing are addressed in this task.

For electronics design, once the heat dissipation per unit heat transfer area is estimated and a need for cooling is recognized, the designer identifies appropriate cooling mechanisms based on cost, size, weight, and reliability criteria. Cooling parameters can then be bounded by examining the limiting temperatures dictated by inherent performance and reliability requirements. Once the design parameters associated with a cooling mechanism have been bounded, the designer can conduct trade-off studies. Analysis and trade-offs must be extended to the assembly, box, and system. Figure 11.18 depicts the on-board cooling system for electronic equipment of large commercial airplanes.

As an indication of the importance attached to reliable operation of avionics as demonstrated by the willingness to accept heavy complex avionics cooling systems to improve or attain reliability, Figure 11.19 illustrates the growth of active cooling systems

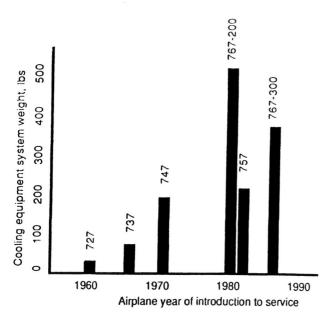

**Figure 11.19**  Schematic of avionics cooling system, typical practice. (From Ref. 30.)

over the history of commercial jet aviation. Commensurate with increased weight, as the active cooling systems have grown in complexity they consume more electrical power, resulting in larger airplane generators, electrical feeders, etc. Furthermore, a result of active cooling provisions is that when active cooling is provided for a box as a reliability enhancement measure, that box can then be dependent on operation of its cooling system because the box may overheat without its coolant flow.

The ability to simulate or measure the thermal behavior of the complete electronic equipment, from the assembly of the board to the component temperatures, is key to thermal management and reliability. Only when the complete thermal conditions and the temperature-dependent failure mechanisms are known can analysis be accurate. Figures 11.20 and 11.21 are analysis plots of an electronic circuit board. It is noted that thermal hot spots may not necessarily reflect components with high unreliability because different components' reliability may depend on temperature in different ways.

As an example of vibrational stress management within the design process, the fundamental mode of vibration can be calculated early in the design process based on the board dimensions, a uniformly distributed weight, and simplified boundary conditions. This gives the designer a baseline resonant mode shape. Then, given the external transmissibility and the applied loading conditions, the dynamic bending stresses can be calculated and the expected areas of maximum deformation can be located. This estimate can be improved when components are placed and boundary conditions are applied which more closely reflect those of the actual circuit board fixture within the electronic assembly. Components can be appropriately placed, or support mountings, ribs, and stiffeners can be incorporated into the design to reduce stress amplitudes.

Stresses arising from coupled thermal, vibrational, and shock loading, covering the complete range of potential conditions based on field data and predicted events, must also

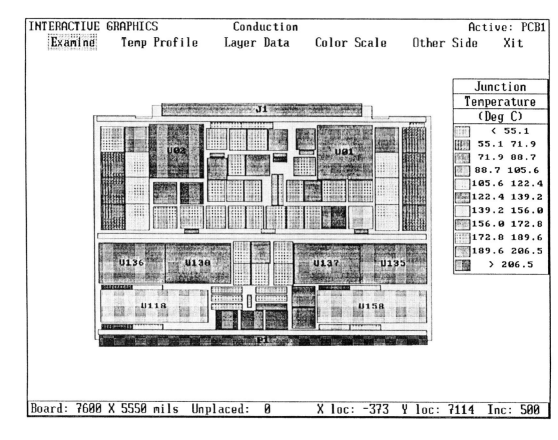

**Figure 11.20**  Typical temperature analysis plot of an electronic circuit board.

be evaluated and reduced if a failure mechanism is assumed to occur within the expected life of the product. Loading conditions that arise from stress coupling must also be considered. Examples of stress coupling include cyclic deformations due to in-plane thermal expansion mismatch, cyclic and transient thermal loading conditions, and loading due to temperature- and time-dependent nonlinear material and metallurgical properties.

Areas showing high thermal gradients and high deformation gradients are especially important because failure critical effects such as thermal mismatch, solder joint fatigue, and stresses on thin vias can occur at these locations. Figure 11.22 is a solder joint fatigue tradeoff screen. The figure presents the cycles to failure for a given solder joint height for different substrate materials, and, as such, suggests two methods to reduce stresses in solder joints.

Stress management also involves testing. For example, image processing and pattern recognition techniques coupled with infrared thermography are often employed to locate faulty nonpowered or shorted (hot spot) components and to determine board and cabinet cooling conditions. Table 11.19 lists several measurement techniques and equipment used to evaluate microelectronic piece parts. Such experiments are useful as an aid in reliability modeling of components and assemblies and for determining step-stress testing.

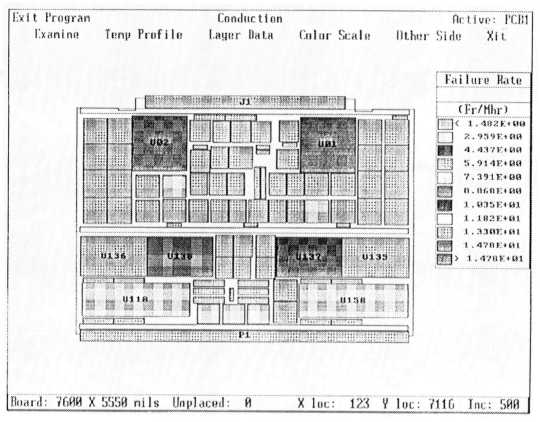

**Figure 11.21**  Typical failure rate analysis plot.

## 11.6.5  Reliability Modeling

Reliability modeling is a means of evaluating possible design configurations and calculating system reliability based on subsystem (part) reliabilities. The approach involves determination of the system reliability block diagram, development of the mathematical models for individual path reliability, and subsequent system analysis using probability theory and part-failure information.

Reliability block diagrams are drawn to aid in the development of the reliability models and determine the paths for successful operation. It is often assumed that the parts that make up a system are independent and can be described in discrete terms as either failed or operational. The block diagram for the system describes the effect of subsystem reliability on system reliability. Probability theory is applied to each part, then to each subsystem, until the whole system is analyzed. Failure information is then introduced to compute the system's reliability.

The architecture of a block diagram for a system depends on how the subsystems (parts) function in terms of reliability. For example, if the reliability concern of a pipe is blockage (i.e., closure), then two pipes connected functionally ''in series'' would be represented by Figure 11.23. The same figure could also be used to represent an electrical ''open'' of two resistors.

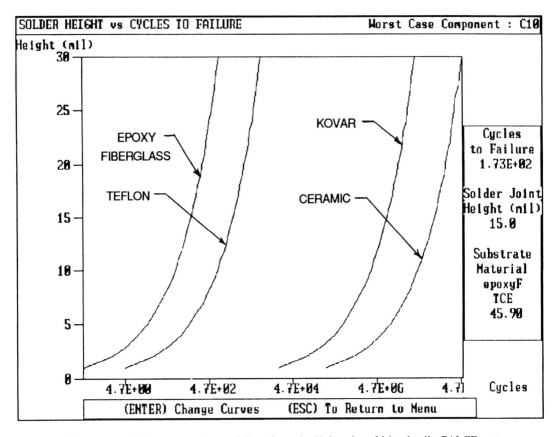

**Figure 11.22**   Solder height versus cycles to failure from the University of Maryland's CALCE system.

## Series Reliability

If parts in a system are connected in a manner (not necessarily physically) such that system failure would occur as a result of the failure of any of the parts, then those parts are considered to be connected in series. The success of the system can thus be no better than that given by the lowest probability of successful operation of any part within the series system.

A system of $n$ parts is assumed to be in series if failure of the system occurs when any component fails. The reliability of the series system, $R_{ss}$, is therefore the product of the reliabilities of the individual parts,

$$R_{ss} = \prod_{i=1}^{n} R_i \tag{11.40}$$

where $R_i$ is the reliability of the $i$th part. This is another way to state that system failure is a function of the weakest link in the system (i.e., $R_{ss} \le \min\{R_i\}$).

*Example 11.9*   A computer system consists of two different circuit boards that operate in series. Assuming that failures are governed by a constant hazard rate distribution for each board, determine the system hazard rate, the system reliability

**Table 11.19**  Typical Measurements and Equipment Used for Evaluating Piece Parts

| Part | Parameter | Equipment |
|---|---|---|
| All | Visual inspection, dimensions | Magnifying glass, microscope<br>Micrometer, caliper<br>Dial indicator, comparator<br>Toolmaker's microscope |
| Cerdip caps and bases | Alpha radiation | Alpha scintillation counter<br>Gas flow proportional counter |
|  | Moisture in glass | Moisture evolution analyser |
|  | UV light transmittance | UV intensity monitor |
|  | Silver-ceramic adhesion | Stud pull tester |
|  | Die adhesion | Die shear tester |
|  | Grain size, % porosity | Scanning electron microscope |
| Cerdip packages | Ceramic strength | Dynamic mechanical analyzer |
|  | Gold plating thickness | Beta backscatter, X-ray fluorescence |
|  | Nickel thickness | Cross section |
|  | Lead thickness | Digital ohmmeter |
| Ceramic and plastic frames | Lead pull | Lead pull tester |
|  | Lead fatigue | Lead fatigue tester |
|  | Lead coplanarity, lead spacing, die pad tilt . . . | Comparator, toolmaker's microscope |
|  | Lead fatigue | Lead fatigue tester |
|  | Plating thickness | Beta backscatter, X-ray fluorescence |
|  | Plating quality | Tape adhesion test |
|  | Bondability | Bond pull tester |
| Molding epoxies | Alpha radiation | Gas flow proportional counter |
|  | Glass transition | Thermal mechanical analyzer |
|  | Thermal expansion coefficient |  |
|  | Spiral flow | Molding press |
|  | Filer content | Muffle furnace, analytical balance |
|  | Chloride (Cl) content | Chromatograph |
|  | pH | pH meter |

for a 1000-hr mission, the system mean time to failure $MTTF_{ss}$, and the system failure probability density function. The hazard rates of the boards are

$\lambda_1 = 0.65\%$ fr/1000 = 6.5 fr/$10^6$ hr

$\lambda_2 = 2.60\%$ fr/1000 = 26.0 fr/$10^6$ hr

*Solution*  For a constant hazard rate, the reliability $R_i$ for the $i$th circuit board has the form

$$R_i = \exp\left(-\int_0^t \lambda_i(t)\, dt\right) = e^{-\lambda_i t} \tag{11.41}$$

The reliability $R_{ss}$ of the series system from Eq. (11.40) is

$$R_{ss} = \exp\left(-\sum_{i=1}^n \lambda_i t\right) = e^{-\lambda_{ss} t} \tag{11.42}$$

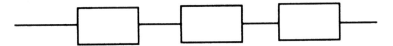

**Figure 11.23** Series system.

where the hazard rate $\lambda_{ss}$ of the system is

$$\lambda_{ss} = \sum_{i=1}^{n} \lambda_i \qquad (11.43)$$

Introducing the values gives,

$$\lambda_{ss} = 6.5 \times 10^{-6} + 26 \times 10^{-6} = 32.5 \times 10^{-6} = 32.5 \text{ fr}/10^6 \text{ hr}$$

The reliability for a 1000-hr mission time is thus

$$R_{ss}(1000) = e^{-(32.5 \times 10^{-6})(1000)} = 0.96802$$

The mean time to failure for the system is

$$\text{MTTF}_{ss} = \int_0^\infty R_{ss}(t)\, dt = \int_0^\infty e^{-\lambda_{ss}t}\, dt = \frac{1}{\lambda} \qquad (11.44)$$

Thus for this problem

$$\text{MTTF}_{ss} = \frac{1}{\lambda_{ss}} = 30{,}770 \text{ hr}$$

This means that a time equal to the mean time to failure, approximately 63.2% of the parts would have failed.

The failure probability density distribution, $f(t)$, is

$$f_{ss}(t) = \frac{d[R_{ss}(t)]}{dt} = -\frac{d[e^{-\lambda_{ss}t}]}{dt} = \lambda_{ss} e^{-\lambda_{ss}t}$$

which is the general form of the exponential probability density distribution. For this case,

$$f_{ss}(t) = 32.5 \times 10^{-6} e^{-32.5 \times 10^{-6}t}$$

*Example 11.10* An electronic system consisting of 100 circuit boards connected in series is required to operate for 8 years with a reliability of 0.9999. Determine the approximate subsystem (board) reliability needed to meet the specified system reliability.

*Solution* From Eq. (11.40), when all the subsystem reliabilities are equal

$$R_i = (R_{ss})^{1/n} = (0.9999)^{1/100} = 0.999999$$

## Parallel Reliability

Redundancy implies that if one part fails, there is a parallel or redundant part that will continue to do the work. That is, redundancy suggests that there exist backup paths for system success. Reliability for such a system is defined as the probability that at least one part is operable between the input and output, although some systems may require that more than one path is operable.

For a basic parallel system in which all components are assumed to be active, $n$ paths for success are possible. For the system to operate successfully, at least one part must operate without failure for the duration of the mission. Conversely, such a system fails only when all the parts in the system fail. This gives rise to the mathematical expression for the system unreliability $Q_{ps}$ in terms of the part unreliabilities $Q_i$,

$$Q_{ps} = \prod_{i=1}^{n} Q_i = \prod_{i=1}^{n} (1 - R_i) \tag{11.45}$$

The system reliability can thus be expressed as

$$R_{ps} = 1 - Q_{ps} = 1 - \prod_{i=1}^{n} (1 - R_i) \tag{11.46}$$

*Example 11.11* Consider a computer system consisting of the same two circuit boards with the same distributions as given by Example 11.9, but connected in parallel. Determine the system reliability for a 1000-hr mission, the system mean time to failure $\text{MTTF}_{ps}$, the system failure probability density function, and the system hazard rate.

*Solution* For a constant hazard rate, the reliability $R_i$ for the $i$th part has the form

$$R_i = \exp\left[ -\int_0^t \lambda_i(t)\, dt \right] = e^{-\lambda_i t} \tag{11.47}$$

The reliability $R_{ps}$ of the parallel system can be obtained from Eq. (11.46) to give

$$R_{ps} = 1 - \prod_{i=1}^{2} (1 - e^{-\lambda_i t}) = e^{-\lambda_1 t} + e^{-\lambda_2 t} - e^{-(\lambda_1 + \lambda_2)t} \tag{11.48}$$

Introducing the part values, the reliability for a mission time of 1000 hr is

$$R_{ps}(1000) = 0.99352 + 0.97434 - 0.96802 = 0.99983$$

This is much greater than $R_{ss}(1000) = 0.96802$ for the series case. Parallel systems will always generate better reliability than series systems because parallel systems provide more than one part to do the job, and if one fails there is another to take its place. In fact, the reliability of an active parallel system will always be greater than that of the part with the highest reliability.

The failure probability density function is

$$f_{ps}(t) = \frac{-d[R_{ps}(t)]}{dt} = \lambda_1 e^{-\lambda_1 t} + \lambda_2 e^{-\lambda_2 t} - (\lambda_1 + \lambda_2)e^{-(\lambda_1 + \lambda_2)t} \tag{11.49}$$

Note that this is not an exponential distribution. The system hazard rate can next be determined as

$$\lambda_{ps} = \frac{f_{ps}(t)}{R_{ps}(t)} = \frac{\lambda_1 e^{-\lambda_1 t} + \lambda_2 e^{-\lambda_2 t} - (\lambda_1 + \lambda_2)e^{-(\lambda_1 + \lambda_2)t}}{e^{-\lambda_1 t} + e^{-\lambda_2 t} - e^{-(\lambda_1 + \lambda_2)t}} \tag{11.50}$$

It should be noted that even though the part hazard rates are constant, the hazard rate $\lambda_{ps}$ of the system is dependent on the mission time.

The mean time to failure for the parallel system is

$$\text{MTTF}_{ps} = \int_0^\infty R_{ps}(t)\,dt = \frac{1}{\lambda_1} + \frac{1}{\lambda_2} - \frac{1}{(\lambda_1 + \lambda_2)} = 161{,}540\ \text{hr}$$

This value is much larger than the $\text{MTTF}_{ss} = 30{,}770$ hr observed for the series case. It is also important to note that for a system consisting of parallel parts

$$\lambda_{ps} \neq \frac{1}{\text{MTTF}_{ps}} \tag{11.51}$$

unlike the case of the series system, even though both systems were assumed to have parts with constant hazard rates.

The reliability $R_s$ of a system in which at least $r$ out of $n$ possible parts must operate if the system is to operate is given by

$$R_s = \sum_{i=r}^n \binom{n}{i} R^i (1 - R)^{n-i} \tag{11.52}$$

where $R$ is the part reliability and is assumed equal for all components and

$$\binom{n}{i} = \frac{n!}{i!(n-i)!} \tag{11.53}$$

Examples of $r$ out of $n$ redundancy occur in memory structures in which a certain minimum number of bytes are necessary to support an outcome, and in products involving high risk to humans, such as a space shuttle, where a minimum number of similar computer outcomes are required for flight.

## Mixed Series-Parallel Reliability

For a series-parallel system such as that shown in Figure 11.24 in block diagram form, the reliability for each parallel unit $i$ is

$$R_i = 1 - \prod_{j=1}^n Q_{ij} \tag{11.54}$$

where $n$ is the number of parallel parts $j$ in the unit. For $m$ units in series, the system reliability $R_T$ is thus

$$R_T = \prod_{i=1}^m R_i = \prod_{i=1}^m \left( 1 - \prod_{j=1}^n Q_{ij} \right) \tag{11.55}$$

If all the parts are identical such that $Q_{ij} = Q$, then Eq. (11.55) can be simplified to

$$R_T = (1 - Q^n)^m \tag{11.56}$$

Figure 11.25 shows the relationship of the system reliability $R$ to part reliabilities for various degrees of redundancy. The curve for $n = 1$ represents the nonredundant system. The greatest increase in reliability occurs for a single increase in redundancy to $n = 2$ and for low values of part reliability. For values of part reliability greater than 0.8, increasing

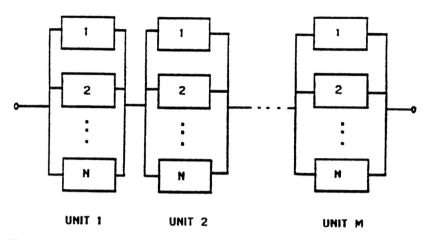

**Figure 11.24** Series-parallel system.

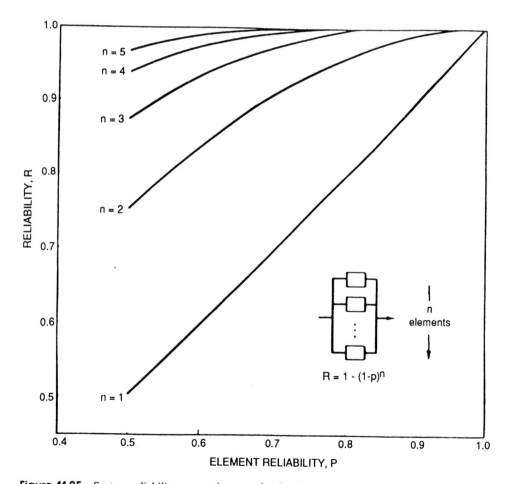

**Figure 11.25** System reliability versus degrees of redundancy. (From Ref. 27.)

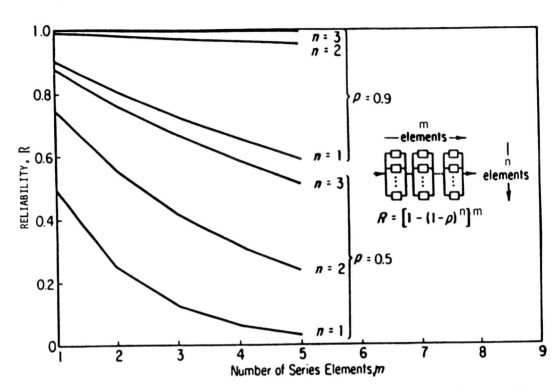

**Figure 11.26** System reliability versus number of series components in a series-parallel system. (From Ref. 27.)

the redundancy to more than three parts does not increase the reliability; for reliabilities greater than 0.9 there is little advantage to having over two parts.

Figure 11.26 is a plot of Eq. (11.56) for the total system reliability versus the number of series parts in a series-parallel configuration. Referring again to Figure 11.24, $n$ is the number of parallel parts in each unit and $m$ is the number of units in series. The effect of parallel redundancy increases with $m$; larger numbers of units in series correspond to greater parallel redundancy effects. For $n = 1$ the system reduces to a nonredundant series configuration. For system reliabilities greater than 0.8, values of $m$ less than 3 usually give little improvement in the system reliability.

For the case of a parallel-series configuration (diagrammed in Figure 11.27), the reliability of the series parts for each path $i$ is calculated first.

$$R_i = \prod_{j=1}^{n} R_{ij} \tag{11.57}$$

where $n$ is the total number of parts $j$ in path $i$. Then taking the reliability for each of the $m$ paths in parallel gives a total system reliability $R_T$ of

$$R_T = 1 - \prod_{i=1}^{m}(1 - R_i) = 1 - \prod_{i=1}^{m}\left(1 - \prod_{j=1}^{n} R_{ij}\right) \tag{11.58}$$

If all the paths are identical with identical parts such as $R_{ij} = R$, then Eq. (11.58) can be simplified to give

$$R_T = 1 - (1 - R^n)^m \tag{11.59}$$

Another type of parallel redundant system is the voting redundant system. A voting redundant system consists of three or more parts in active redundancy. A comparator samples the outputs of the parts and switches off parts that do not agree with the majority. The system can operate successfully if at least two (i.e., a majority) of the parts are operating properly. The reliability of the three-part voting redundant system is calculated as

$$R_{SY} = R_A R_B + R_A R_C + R_B R_C - 2R_A R_B R_C$$

*Example 11.12* Compute the reliability of a voting redundant system if $R_A = .90$, $R_B = .92$, $R_C = .98$, and $R_{COM} = .99$

*Solution* If the reliability of the comparator is taken into account, then the reliability is

$$R_{SY} = (R_A R_B + R_A R_C + R_B R_C - 2R_A R_B R_C)R_{COM}$$
$$R = [(.9)(.92) + (.9)(.98) + (.92)(.98) - 2(.9)(.92)(.98)](.99)$$
$$= .979$$

## Complex System Analysis

System architecture that cannot be resolved into combinations of parallel and series configurations are called complex systems. Although there are many approaches to solving complex configurations, there is no generally efficient method. An example of a complex system is given in Figure 11.28. One method of determining the reliability of this system is by enumeration, whereby all $2^4 = 16$ possible configurations are examined to see whether the system can operate in those configurations. Table 11.20 lists all possible states along with those that are operable.

Using the data from Table 11.20 and letting $a = P(A)$, $1 - a = P(\overline{A})$, $b = P(B)$, etc., the system reliability is

$$
\begin{aligned}
R = {} & abcd + abc(1 - d) + ab(1 - c)d && + ab(1 - c)(1 - d) \\
& + a(1 - b)cd + a(1 - b)c(1 - d) + a(1 - b)(1 - c)d \\
& + (1 - a)bcd + (1 - a)(1 - b)cd + (1 - a)(1 - b)c(1 - d) \\
& + (1 - a)bc(1 - d) = c + ab + ad - abc - acd - abd + abcd
\end{aligned}
$$

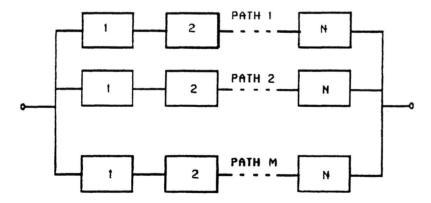

**Figure 11.27**   General parallel-series configuration.

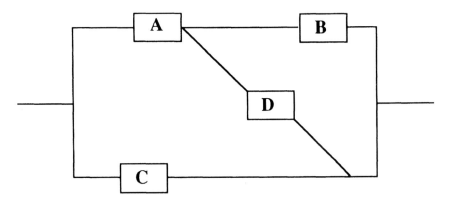

**Figure 11.28**   Complex system.

As another example, consider the complex system given in Figure 11.29. To solve this problem by ennumeration regions $2^5 = 32$ possible configurations. Instead, we can utilize a method called the key compound method, whereby a "key" part is selected as a foundation to forming conditional probability statements. Selecting component $C$ as the key and using the decomposition rules of probability gives

$$
\begin{aligned}
R &= P(AD \text{ or } BE \text{ or } AE \text{ or } BD)P(C) + P(AD \text{ or } BE)\overline{P}(C) \\
&= (ad + be + ae + bd + abde - ade - abd - abe - edb)c \\
&\quad + (ad + be - abde)(1 - c) \\
&= ad + be + ace + bcd - acde - abcd - abce - bcde - abde + 2abcde
\end{aligned}
$$

In general, calculation of system reliability for even a moderately large complex system is extremely time consuming. As a result, a number of researchers have proposed techniques to determine the bounds on reliability using minimal paths and minimal cut methods. Basically, a minimal path is the minimal set of operating parts that ensures system operation and a minimal cut is a minimal set of failed parts that ensures system failure.

## Reliability of Stand-by Systems

A standby system consists of parallel redundancies, except that only one part is active and the rest of the parts wait idle until required to function. A monitor detects when the

**Table 11.20**   Complex System Analysis Chart

| Operating states | | Nonoperating states |
|---|---|---|
| $A\,B\,C\,D$   $\overline{A}\,B\,C\,D$ | | $A\,\overline{B}\,\overline{C}\,\overline{D}$ |
| $A\,B\,C\,\overline{D}$   $\overline{A}\,B\,C\,\overline{D}$ | | $\overline{A}\,B\,\overline{C}\,D$ |
| $A\,B\,\overline{C}\,D$   $\overline{A}\,\overline{B}\,C\,D$ | | $\overline{A}\,B\,\overline{C}\,\overline{D}$ |
| $A\,B\,\overline{C}\,\overline{D}$   $\overline{A}\,\overline{B}\,C\,\overline{D}$ | | $\overline{A}\,\overline{B}\,\overline{C}\,D$ |
| $A\,\overline{B}\,C\,D$ | | $\overline{A}\,\overline{B}\,\overline{C}\,\overline{D}$ |
| $A\,\overline{B}\,C\,\overline{D}$ | | $\overline{A}\,\overline{B}\,\overline{C}\,D$ |
| $A\,\overline{B}\,\overline{C}\,D$ | | |

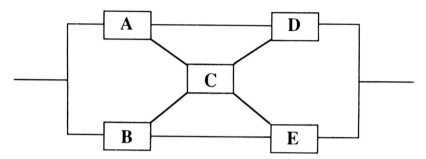

**Figure 11.29**   More complex system.

active part has failed and signals the switch to activate or connect a standby part. A switch controls which one of the redundant parts is active.

To incorporate time dependencies in the model, we present the case of two parts in a standby redundancy and neglect the switch and monitor. The probability that the primary part fails during time $t_1$ is

$$P(E_1) = f_1(t_1)\, dt_1 \tag{11.60}$$

The probability that the secondary part fails after time $t_1$ is

$$P(E_2) = f_2\,(t - t_1)\, dt \tag{11.61}$$

The system failure in time $dt$ is

$$f_s(t)\, dt = \int_{t_1 = 0}^{t} f_1(t_1)\, dt_1\, f_2(t - t_1)\, dt \tag{11.62}$$

Thus the system reliability is

$$R_s(t) = \int_{t}^{\infty} f_s(\tau)\, d\tau = \int_{t}^{\infty} \int_{0}^{t} f_1(t_1) f_2(\tau - t_1)\, dt_1\, d\tau \tag{11.63}$$

*Example 11.13*   Assume that the failure probability density functions for the parts are

$$f_1(t) = \lambda_1 e^{-\lambda_1 t}$$

and

$$f_2(t_1) = \lambda_2 e^{-\lambda_2 (t_1)}$$

Determine the system reliability.

*Solution*   The system probability of failure is obtained by the product of $f_1$ and $f_2$, since the system fails only if both parts fail:

$$f_1(t_1) f_2(t - t_1) = \lambda_1 \lambda_2 e^{-\lambda_1 t_1} e^{-\lambda_2 (t - t_1)}$$

Integrating with respect to $t_1$ to obtain the joint density function in terms of $t$,

$$f_s(t) = \int_{t_1 = 0}^{t} f_1(t_1) f_2(t_1)\, dt_1 = \lambda_1 \lambda_2 \left( \frac{e^{-\lambda_1 t}}{\lambda_2 - \lambda_1} + \frac{e^{-\lambda_2 t}}{\lambda_1 - \lambda_2} \right)$$

The reliability is given by the integration of the joint probability density function

$$R(t) = \int_t^\infty f(t)\, dt = e^{-\lambda_1 t} + \frac{\lambda_1}{\lambda_1 - \lambda_2}(e^{-\lambda_1 t} - e^{-\lambda_2 t})$$

The first term is the reliability of the primary part and the second term is the added reliability resulting from the standby part. We can also express the reliability as

$$R(t) = \frac{\lambda_1 e^{-\lambda_1 t}}{\lambda_1 - \lambda_2} + \frac{\lambda_1 e^{-\lambda_2 t}}{\lambda_2 - \lambda_1}$$

In general, the reliability for $n$ parts of which one is primary is

$$R_s(t) = \sum_{i=1}^n e^{\lambda_i t} \prod_{\substack{j=1 \\ j \neq i}} \frac{\lambda_j}{\lambda_j - \lambda_i} \tag{11.64}$$

## Redundancy Trade-offs and Optimization

Although redundancy methods can be effective in increasing overall system reliability, implementing redundancy involves designing for alternative signal paths to connect the redundant subsystems or components, meeting space and weight requirements, and conforming to acceptable power requirements, heat dissipation, and fan-out. Thus, it is necessary to trade off the "cost" of the improvement in reliability with the "cost" of the constraints.

Redundancy may be justified when the increased cost due to additional parts and connections is compensated for by increased reliability. Redundancy may not be justified when one cannot satisfactorily answer questions such as how reliable the new interconnections are, whether there is any interaction among the redundant parts, whether if one part fails the other(s) can carry the load properly, how the parts will be switched in a standby system, and whether spurious results can occur.

Achieving the desired reliability while minimizing the forementioned factors is of paramount importance to the reliability engineer. By concentrating on redundancies for parts of the system that are more likely to fail, optimum reliability can be achieved.

*Example 11.14* Consider two memory subsystems, A and B, with known reliabilities after a certain time period of .95 and .80, respectively. The two subsystems must be connected in series so that the overall system reliability at the given time period is .90. Design a system so that this can be achieved using a minimum number of subsystems.

*Solution* The simplest series system would consist of two subsystems, subsystem A in series with subsystem B. The overall reliability of this system is

$$R_{SY} = R_A R_B = .95(.80) = .76$$

which is less than the required system reliability of .90.

The next simplest series system consists of three subsystems, either subsystem A in series with two B subsystems in parallel or two A subsystems in parallel and in series with one subsystem B. Since the reliability of B is less than the reliability of A, the redundancy should be another subsystem B. The overall reliability of a subsystem A in series with two B subsystems in parallel is calculated from the following:

$$R_{SY} = R_A(1 - \Pi\, Q_B) = .95\,[1 - (1 - .8)^2] = .912$$

which is greater than the required system reliability of .9. By using one A subsystem and two B subsystems, the overall system reliability requirements are met using only three subsystems.

Generally, it is necessary to allocate redundant components subject to resource constraints so as to satisfy some measure of system performance. In general, this kind of problem belongs to the class of nonlinear integer or mix-integer constrained optimization problems when the constraint functions are nonlinear but the solution is an integer type.

Tillman et al. [1] present an excellent review of solution procedures to maximize system reliability subject to resource constraints, such as cost, weight, or power. Luss [2] developed a procedure for redundancy selection involving two stages of searching. In phase I an approximate continuous pseudosolution is obtained, and in phase II the procedure uses a direct search in the neighborhood of the pseudosolution to obtain an optimum integer solution.

Aggarwal et al. [3] presented a heuristic solution approach to redundancy selection whereby the component that has the highest value of a sensitivity function (i.e., reliability) with respect to a previous vector is added to the system until there is a violation of constraints.

McLeavey [4] introduced a branch-and-bound method to obtain the optimal integer point for reliability for parallel redundancy. McLeavey and McLeavey [5] compared the algorithms of Luss [2], Aggarwal [3], and Ghare-Taylor (McLeavey version [4] with 40 tests). Luss's [2] algorithm could not guarantee the optimal solution (in 37.5% of the problems, the optimum was not reached), but it has the advantage of handling nonlinear constraints. The Ghare-Taylor [4] algorithm gave an optimal solution for all the problems but required linear constraints. Aggarwal's method [3] used less computer time but couldn't reach an optimum for 55% of the problems.

Hwang and Lee [6] formulated an algorithm using branch-and-bound methods and the Hwang-Masud [7] nonlinear goal programming method to solve integer constrained problems encountered in system reliability optimization. Kuo and Lin [8] incorporated the Lagrange multiplier method and the branch-and-bound technique to solve redundancy allocation problems.

Kuo and Lin [8] suggested that branch-and-bound is a better iterative method than the heuristic method in terms of giving an accurate solution. One disadvantage of the branch-and-bound method is that it cannot easily be applied to a nonlinear constrained integer problem because the programming effectivity depends on the choices of branching nodes and, on the other hand, the bound condition is not satisfied in most cases.

## 11.6.6  Accelerated Testing

Many electronic devices demonstrate very high reliability when operating within their intended normal use environment. Problems in investigating failure modes and measuring reliability for systems where long life is required can thus occur because a very long test period under the actual operating conditions is necessary to obtain sufficient data to determine the actual failure modes and failure distributions. One approach to the problem of obtaining meaningful test data for high-reliability devices is accelerated testing, sometimes called accelerated stress testing or accelerated stress life testing. Accelerated testing involves measuring the performance of the test device at stressed conditions of load, stress, temperature, voltage, chemical attack, humidity, etc. that are more severe than the normal operating level in order to induce failures within a reduced time period.

An accelerated test is intended to shorten the normal time to failure of a specific failure mechanism. However, in engineering practice, accelerated tests have often been conducted without knowledge of the failure mechanism involved or whether the test precisely accelerated the same mechanism as that observed in normal operating conditions.

Accelerated tests have been used to aid in screening; accelerated failure mode, site, and mechanism identification; and accelerated reliability demonstration. Accelerated screening is an approach used to remove causes of microcircuit failure resulting from surface defects due to inadequate manufacturing and assembly processes. Accelerated failure mode tests attempt to identify the main failure modes of a component or a system under certain environmental conditions so that the failures can be predicted and corrected. Accelerated tests for reliability demonstration are used to provide quantitative reliability prediction for a product in normal use conditions. However, it is generally recognized that the main difficulty associated with accelerated testing lies in using test data to predict reliability, the mean life, or other parameters under normal use conditions.

The electronics industry has traditionally addressed the issue of reliability prediction as a constant-failure-rate problem by assuming that a burn-in period has been completed and that wearout will not occur within the design life of the device. Some acceleration models have also been derived based on this assumption. However, many of the failures in products, including electronic packages, are due to mechanical failure mechanisms such as fracture, fatigue, and corrosion. These failures are primarily wearout phenomena and hence cannot be characterized by a constant-failure-rate process with respect to accelerated testing. As such, it is possible that the accelerated test may conceal the actual wearout failure mechanism and suggest an incorrect reliability prediction. Furthermore, it has been found in the reinvestigation of the bathtub curve that even in the region of the hazard function between burn-in and wearout, the failure rate is not constant. Then the acceleration models based on a constant failure rate may generally be inapplicable. Often, the main parameter that affects the failure or operation life of a product, such as temperature, is called the acceleration parameter. The acceleration factor indicates the ratio of the life under the accelerated condition and that under the normal use condition.

In the past few decades, tests on electronic devices have been carried out under accelerated conditions. The acceleration parameters that may be varied to cause early failure in a device include temperature, humidity, voltage, and mechanical stresses.

## Tests Carried Out at Increased Temperature

Temperature is known to vary the rates of many physical and chemical reactions. Because failure mechanisms in a device are basically physical or chemical processes, temperature is often used as an acceleration parameter in life testing.

Most frequently, the dependence of a failure mechanism on temperature is modeled by the law of Arrhenius. According to this law, the reaction rate, $r$, is

$$r = r_0 \exp\left(\frac{-E_a}{kT}\right) \tag{11.65}$$

where $E_a$ is the activation energy for the failure mechanism under consideration, $k$ is Boltzmann's constant $(= 8.63 \times 10^{-5} \text{ eV/K})$, $T$ is the absolute temperature, and $r_0$ is a constant.

The Arrhenius law associates with each failure mechanism an activation energy which is a measure of the extent to which the reaction depends on temperature. Further, the Arrhenius law can be used if only one failure mechanism occurs, so that $E_a$ has a constant value.

If a component fails at time $t_1$ at temperature $T_1$ and at time $t_2$ at temperature $T_2$, then

$$\frac{t_1}{t_2} = \exp\left[\frac{E_a}{k}\left(\frac{1}{T_1} - \frac{1}{T_2}\right)\right], \qquad T_1 > T_2 \tag{11.66}$$

where $t_1/t_2$ is the acceleration factor. If $T_1$ is greater than $T_2$, then as $T_1$ increases with respect to $T_2$ the ratio $t_1/t_2$ decreases. This means that the test time $t_1$ at temperature $T_1$ is decreasing with respect to the test time $t_2$ at temperature $T_2$.

Figure 11.30 shows a plot of the inverse of junction temperature versus the time to failure of the device on a log scale [9]. From Eq. (11.65), the time to failure, $t_f$, is

$$t_f = A_0 \exp\left(\frac{E_a}{kT}\right)$$

Taking the logarithm of terms on both sides of the equation gives

$$\ln(t_f) = \ln(A) + [E_a/k][1/T] \tag{11.67}$$

The above relationship illustrates that the natural logarithm of time to failure varies linearly with the inverse of the absolute temperature for a given failure mechanism. The failure

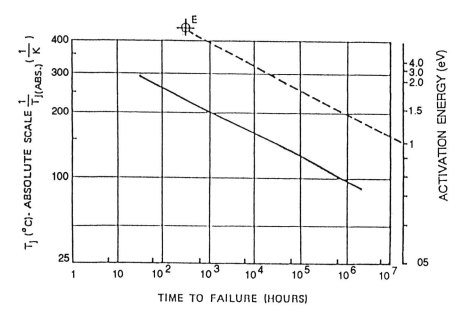

**Figure 11.30** Arrhenius plot: inverse of junction temperature versus time to failure. (From Ref. 9.)

**Table 11.21** Activation Energies for Various Failure Mechanisms of Silicon Devices

| Sensitive element | Failure mechanism | Relevant factors[a] | Accelerating stress[a] | Typical activation energy | Reference |
|---|---|---|---|---|---|
| Dielectric and interfaces Si-dielectric | Accumulation of surface charges | T,E/V | T,V | 1.0–1.05 eV (leakage-type failures) | 54,55 |
| | Breakdown | T,E/V | E,T | 0.3–0.6 eV (0.36 eV for MOS gate structures) | 56 |
| | Oxide pinholes | T,E/V | E,T | 1.0 eV (dielectric breakdown of bipolar devices) | 57 |
| | Ionic contamination | T | T | 1.0 eV | 9 |
| | Hot carrier trapping in oxide | E/V,T | E,T | −0.06 eV | 11 |
| | Charge loss | E,T | E,T | 0.8 eV (MOS) EPROM | 58 |
| Metallization | Electromigration | T,J; Grain size | T,J | 1.0 eV (large grain Al, glassivated) | 59 |
| | | | | 0.5 eV (small grain Al) | 60 |
| | | | | 0.7 eV (small grain Cu-Al) | |
| | Contact electromigration Si in Al Al at sidewalls | T,J | T,J | 0.9 eV 0.8–1.4 eV | 11 |
| | Contact metal migration through barrier layer | T | T | 1.8 eV | 11 |
| | Corrosion Chemical Galvanic Electrolytic | Contamination | H,E/V,T | 0.3–0.7 eV (Al) 0.6–0.7 eV (for electrolysis) 0.65 eV (for corrosion mechanism) E/V may have thresholds | 61 |
| Metal interfaces | Intermetallic compounds (Au, Al) | T, Impurities Bond strength | T | 1.0 eV (for open wires or high-resistance bonds at the pad bond due to Au-Al intermetallics) | 62 |
| Wafer fabrication and assembly-related defects | Metal scratches | T,V | T,V | 0.5–0.7 eV 0.7 eV for assembly-related defects | 63 |

[a] H = humidity, V = voltage, E = electric field, T = temperature, J = current density.

mechanism is characterized by an activation energy, $E_a$ which is obtained from the slope of the curve described by Eq. (11.67), as shown in Figure 11.30. The activation energy is found by drawing a line from the reference point E parallel to the curve obtained. The intercept of this line (dashed line in Figure 11.30) on the right hand side scale gives the activation energy of the failure mechanism. Table 11.21 lists activation energies for various failure mechanisms.

Figure 11.31 [9] plots the logarithm of the time to failure versus the cumulative percentage of failures transferred on a normal scale. The straight-line plot suggests that the failure distribution in time follows a lognormal law. This also agrees with experimental results by Stitch et al [10]. When the plots obtained for different temperatures are parallel to each other, the distribution of failures in time is log-normal for all the tests. Thus, the change in test temperature changes only the average life of the distribution.

Any lot of devices that is tested may include a ''freak'' population that fails very early [10]. This part of the test sample is characterized by different failure mechanisms resulting from workmanship and intrinsic material defects and should be separated from the main population by the use of a screening test before carrying out an accelerated test.

To determine the limits of stress that can be applied to a device that would not cause atypical failures, a step stress test is often conducted. Progressively increasing stress (temperature stress in this case) is applied to the device for constant time intervals. The number of failures that occur at each stress level are recorded and the failure modes are investigated. The stress step at which devices start failing with atypical failure modes sets the stress limit at which an accelerated test can be carried out. The step stress test enables us to test a device at different stress levels using the same sample, which makes it fast and inexpensive. However, the estimate of the maximum allowable stress on the device is not very precise because the test is based on the assumption that failures at a particular stress step are independent of the preceding steps. In practice, the effects of temperature

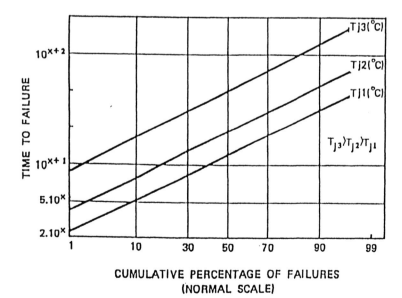

**Figure 11.31**  Time to failure versus cumulative percentage of failures. (From Ref. 9.)

stresses are cumulative and the data obtained are only approximate. Nevertheless, it is a useful method of determining limits of stresses that can be used to conduct more accurate and more time-consuming stress tests.

A study of failure mechanisms and the parameters that affect them is essential for designing an accelerated test. Each test is designed to accelerate a particular failure mechanism in a device. Therefore, one must be able to identify the failure type and stipulate parameters that would accelerate that failure mechanism. An excellent discussion of various failure mechanisms of semiconductor devices is provided in [11].

## Accelerated Humidity Tests

Peck and Zierdt [12] have identified three main failure mechanisms in which the presence of water vapor plays a necessary part:

1.  Charge separation on the surface of MOS structures: Moisture on the surface of field effect devices can provide mobility to insulator surface charges, extending the gate potential over the nearby surfaces. In certain structures, or with processing defects, parasitic gates become operable, causing malfunction of the device. This effect will not cause cumulative degradation with time but will occur as soon as sufficient humidity and suitable fields are available.

2.  Corrosion of metallization: In the presence of ionic contamination, moisture, and metals with different electrochemical potentials, a galvanic couple is formed resulting in the corrosion of the metal with higher electrochemical potential. A typical example is a galvanic cell leading to the corrosion of the aluminum metallization near a gold wire bond in the presence of humidity and ionic contamination (usually as a residue from processing).

3.  Electrolytic conduction between electrically biased metallization paths: The rate of transfer of metal from one electrode to another, across a conductive surface, will depend on (among other things) the electrolytic current flow. The conductivity of the surface is a function of the amount of moisture on the surface. The degree of metal transfer shows up as a leakage current (and eventually a short) resulting from a developing metal–metal compound film in the transfer path. The time required for failure will depend on the sensitivity of the circuit to interpath leakage, the spacing between metal traces, the voltage applied, the metals involved, and the character of the insulating surface. This failure mechanism has been extensively studied by Peck and Zierdt and the results of their investigations have been described in [12].

Various test procedures have been devised to accelerate humidity-related failures in semiconductor devices. Two popular tests are the temperature, humidity, and bias (THB) 85/85 test and the highly accelerated stress test (HAST).

The THB 85/85 test is carried out at 85°C and 85% relative humidity with the device operating under reverse bias conditions. Because of low current, the power dissipated by the device is low and the package retains a high level of humidity.

HAST was developed to reduce the test time required by the THB 85/85 test, which is usually in excess of a thousand hours. This test employs high temperatures (100 to 175°C) at controlled humidity levels (relative humidity of 50 to 85%) and electrical bias as stress factors. The use of saturated vapor has not been recommended [13] because it results in undesirable electrolytic action through condensing droplets between the external lead wires when devices are tested under bias. This has necessitated the use of unsaturated water vapor pressure as an aging stress. So that the vapor will remain unsaturated at high temperatures, i.e., temperatures above the boiling point of water, tests are carried out in a pressurized vessel.

Gunn et al. [14] used the HAST technique to determine corrosion kinetics of a commercially available, plastic-encapsulated, bipolar integrated circuit. These devices had a prior history of aluminum land corrosion under the effects of temperature, humidity, and voltage stress. The test samples were randomly selected from a typical manufacturing lot. To minimize power consumption and to prevent localized heating (resulting in local reductions of relative humidity), the output power transistors were held in the reverse bias mode. The study found that chlorine-induced corrosion was the dominant failure mechanism in three of the four test conditions. This failure mode is identical to that observed in a THB 85/85 test for the device. Failures due to ionic contamination were found in the fourth case, which used a 150°C HAST cell. This failure mode is typical of a 150°C standard high-temperature reverse bias (HTRB) test for the device. Thus, no new failure mode was observed in HAST. This implies that HAST can be used to increase the acceleration factors considerably over the conventional THB tests.

*Cyclic Biased THB Test*

The THB test uses reverse bias conditions to ensure that localized heating does not cause a reduction in relative humidity. However, it is extremely difficult to put all the circuit elements under reverse bias. Therefore, for ICs the THB test is generally performed under actual operating conditions.

Ajiki et al. [15] have suggested a new cyclic biased THB test for power-dissipating ICs. It was found that in the case of plastic-encapsulated devices the device failure related to moisture absorption mainly comes from corrosion of the aluminum metallization. The chemical reaction of aluminum with hydroxyl and hydrogen ions produced by the electrolysis of water aided by the applied bias leads to the formation of aluminum hydroxide and alumina. This finally results in an open circuit (failure). Thus, the corrosion mechanism is accelerated by both water vapor and the applied bias. The application of forward bias, however, results in an increase in the junction temperature, which causes the test sample to emit moisture. As an effective operating test method during which high humidity levels can be attained, Ajika et al. have suggested an intermittent operating life test in which moisture absorption occurs during the off-bias state and electrolysis of the metallization takes place when the bias is applied.

Figure 11.32 [15] shows the dependence of the mean time to failure on the time ratio, which is defined as [on/(on + off)] × 100%. The test was carried out at 85°C and a relative humidity of 85%. The mean time to failure was observed to decrease with an increase in time ratio, reach a minimum, and thereafter increase with an increase in time ratio. For this device, the highest acceleration factor would be achieved at time ratios of 20 to 40%.

Figure 11.33 [15] shows the variation of MTTF with time ratio for devices with different power dissipations (0.6, 0.9, and 2.0 W). The test was conducted at an ambient of 85°C and 85% relative humidity. The variation in MTTF with time ratio is much larger for the higher power dissipating device, and the selectable range of time ratios giving minimum MTTF becomes narrower as the power dissipated increases.

Figure 11.34 [15] shows the effect of cycle time on the cumulative failure rate for a time ratio of 1:7. The experiments were conducted at 85°C and 85% relative humidity for a device dissipating 2 W. It is seen that higher acceleration factors are achieved with shorter cycle times for devices dissipating high power.

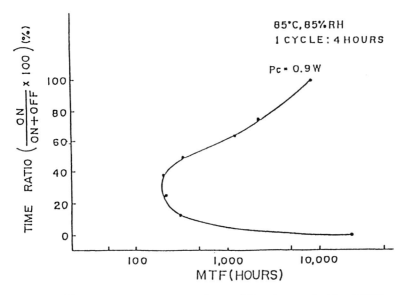

**Figure 11.32**  Time ratio versus mean time to failure in a cyclic biased THB test. (From Ref. 15.)

For the 600-mil row spacing DILs that were used for the experiment, the following time ratios were recommended as the best conditions for the cyclic bias test [15]:

· Power dissipation under 100 mW    steady state on
· Power dissipation 100 mW to 1 W    on/off = 1:2
· Power dissipation 1 W to 1.5 W    on/off = 1:4
· Power dissipation over 2 W    on/off = 1:7

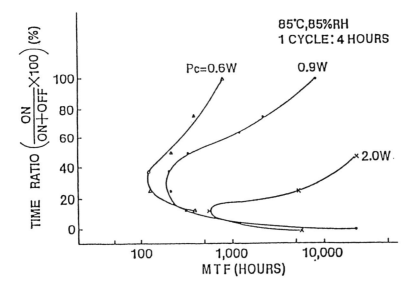

**Figure 11.33**  Time ratio versus mean time to failure in cyclic biased THB test for various dissipated powers. (From Ref. 15.)

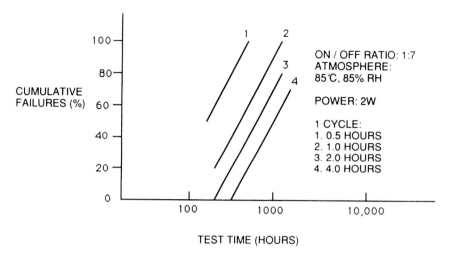

**Figure 11.34** Influence of cycle time on cumulative failure rate percent in a cyclic biased THB test. (From Ref. 15.)

## Voltage-Accelerated Tests

Voltage, in conjunction with temperature, is an agent of many surface degradation processes. At low power dissipation, the voltage-activated failure mechanisms of the cellular diode dominate, while at high temperatures the failure of the emitter diode caused by thermal runaway becomes equally important [16]. Thus, the failure rate for a device depends on both the collector base voltage, $V_{cb}$, and the junction temperature, $T_j$. So that voltage-accelerated failures will predominate, ambient temperature is maintained low ($\sim 25°C$) while conducting operating tests. Also, higher voltage stresses (near the breakdown voltage of the device) induce multiple failure modes that depend differently on voltage variations. Thus, voltage is a poor acceleration factor.

The main reference model used for voltage accelerated tests is the Eyring model [9], which relates failure rate, $\lambda$, of the device to the device junction temperature, $T_j$, and the applied collector base voltage, $V_{cb}$, as follows:

$$\lambda = AT_j \left[ \exp\left(\frac{-B}{kT_j}\right) \right] \left[ \exp\left(CV_{cb} + \frac{DV_{cb}}{kT_j}\right) \right] \tag{11.68}$$

where $k$ is Boltzmann's constant and $A, B, C, D$ are constants to be determined experimentally.

Kemeny [16] has suggested a model that does not have this requirement. As per this model:

$$\lambda = \left[ \exp\left(C_0 - \frac{E_a}{kT_j}\right) \right] \left[ \exp\left(C_1 \frac{V_{cb}}{V_{cbmax}}\right) \right] \tag{11.69}$$

where $E_a$ is the activation energy, $k$ is Boltzmann's constant, $T_j$ is the junction temperature, $V_{cb}$ is the collector base voltage, $V_{cbmax}$ is the maximum allowable collector base voltage, and $C_0$ and $C_1$ are constants.

The first part of the relationship deals entirely with the dependence of failure rate, $\lambda$, on the junction temperature, $T_j$, and is indeed the Arrhenius relationship. The second

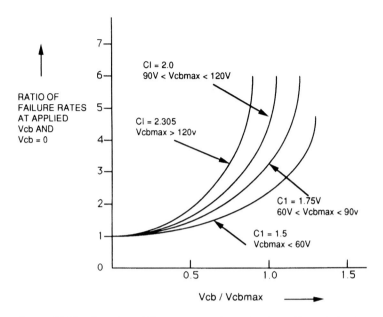

**Figure 11.35**   Transistor failure rate as voltage function. (From Ref. 16.)

exponential term represents the dependence of $\lambda$ on applied voltage, $V_{cb}$. In fact if $V_{cb}$ is set equal to zero, the Kemeny equation reduces to the Arrhenius equation. The value of the constant $C_1$ has been found to be 1.5 [16]. This value is valid for low- to medium-voltage devices, up to $V_{cbmax} = 60$ V. For higher-voltage devices, $C_1$ should be chosen between 1.5 and 2.305, depending on the voltage rating of the device (refer to Figure 11.35). Having determined $C_1$, for a failure mechanism with known activation energy, the Kemeny equation can be used to find $C_0$ by conducting a single life test. It has been reported by Kemeny [16] that the upper junction temperature limit to be used was 200 to 230°C for germanium mesa devices and beyond 300°C for silicon planar devices. Above these limits, atypical failure mechanisms occurred, defeating the purpose of accelerated tests.

## Tests Using Increased Mechanical Stresses

Electronic components are required to withstand both constant and cyclic mechanical stresses. The failure mechanisms of greatest significance are those related to fatigue damage and creep.

### Accelerated Fatigue Tests

Electronic devices suffer fatigue damage when they have to withstand cyclic loads such as those caused by vibrations and thermal cycling. The *S-N* curve is used to plan and interpret the results of an accelerated fatigue life test. Figure 11.36 [17] shows the dependence of fatigue strength on the number of stress cycles for cold-drawn copper. Tests are accelerated by increasing the stress level beyond those expected during normal usage. For Ni, by increasing the stress level from 15 to 20 ksi, the number of cycles to failure is reduced from $10^7$ to $10^6$.

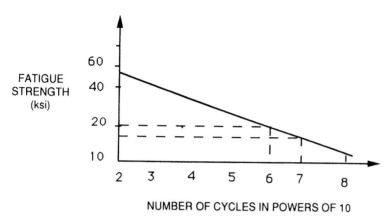

**Figure 11.36**  Fatigue strength versus life plot for cold-drawn copper. (From Ref. 17.)

*Accelerated Creep Tests*

For polymers, which are extensively used in electronic device packaging, the stress, $\sigma(t)$, after a given time $t$ is the product of the strain, $\epsilon$, and the relaxation modulus of the polymer, $E(t)$. Tests are designed based on relationships of superposition. For example, the plots of relaxation modulus versus test time at different temperatures (Figure 11.37)

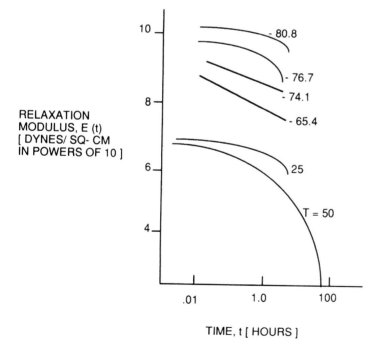

**Figure 11.37**  Relaxation modulus versus time curve for polyisobutylene. (From Ref. 65.)

[18] of some polymers can be shifted to lie on a master curve by using a shift factor, $a_T$. Let an arbitrary reference temperature, $T_s$, be taken to fix a curve. If $t_s$ is the time at a point on that curve with a particular relaxation modulus and $t$ is the time at a point with the same relaxation modulus on a curve at a different temperature, $T$, then the amount of shift required to superpose the two curves is a displacement of $\log(t_s) - \log(t)$ along the log time axis, i.e., the $x$ axis. The shift factor $a_T$ is defined as

$$\log(a_T) = \log(t_s) - \log(t) = \log\left(\frac{t_s}{t}\right) \tag{11.70}$$

Williams, Landel, and Ferry [18] have proposed that $a_T$ can be expressed as

$$\log(a_T) = \frac{-C_1(T - T_s)}{C_2 + (T - T_s)} \tag{11.71}$$

where $C_1$ and $C_2$ are constants and $T_s$ is the reference temperature. This relationship, called the WLF equation, was developed empirically and relates test time to test temperature, $T$. It holds very well for a wide range of polymers in the vicinity of the glass transition temperature, $T_g$. If the reference temperature is taken as the glass transition temperature, Eq. (11.71) may be written as

$$\log(a_T) = \frac{-C_1(T - T_g)}{C_2 + (T - T_g)} \tag{11.72}$$

For this case, values assigned to the constants $C_1$ and $C_2$ were 17.4 and 51.6 K, respectively [18]. The shift factor $a_T$ is a measure of the acceleration factor achieved.

### Crack Initiation, Propagation, and Fracture

Failures can be due to flaws that vary from gross drilled hole misalignment, unsoldered joints, and chemical contaminants to crystalline imperfections. In electronic devices, the alternating stress induced by both thermal cycling and mechanical vibrations is a dominant stress, and fatigue is a major failure mechanism. The fatigue fracture failure mechanism is generally classified into two stages, crack initiation and crack propagation, for which the micromechanical phenomena are different. In general, under small stress cycles, fatigue life is dominated by crack initiation. Under a large stress range, the life is dominated by crack propagation. Accelerated testing may shift fatigue failure from crack initiation–dominated phenomena to crack propagation–dominated phenomena because the stress amplitude is elevated.

Once a crack is initiated, there is typically a sharp notch that gives rise to a high stress concentration. If there is plastic deformation of the material at the high stress, the material near the crack tip will flow and bring about a rounded plastic zone ahead of the tip. The crack may grow slowly within the plastic zone but can extend at a high rate in the elastically deformed material when it moves out of the plastic zone. Thus, the total crack growth rate depends on the size of the plastic zone. According to fracture mechanics, the plastic zone is proportional to the square of fracture toughness and is inversely proportional to the square of yield flow stress. In typical temperature-accelerated tests, the material properties that affect plastic zone size are often changed. In particular, the fracture toughness is increased, the yield point is decreased, and the fracture process is shifted from brittle fracture to ductile fracture when the temperature is elevated. Thus, the crack growth rate in temperature-accelerated testing may be lower than that of the normal service condition, which gives unconservative reliability prediction. Furthermore, the critical crack size to fast extension is large in acceleration tests because of the increase in fracture toughness at high temperature. This may also increase the fatigue life. An accelerated test may conceal the sudden brittle fracture that occurs under normal service conditions.

*Low-Cycle and High-Cycle Phenomena*

The stress-accelerated test may shift fatigue failure from high-cycle to low-cycle phenomena. In fatigue analysis, low-cycle fatigue and high-cycle fatigue have different statistical analytical models and different material properties. If accelerated tests do shift the fatigue from high-cycle to low-cycle, the use of conventional acceleration factors is questionable and more laboratory investigation is necessary.

*Random Vibration*

The accelerated test under random stress is usually conducted by elevating the power spectral density (PSD) function of stress, displacement, or acceleration of the device. However, the value of the PSD function is not explicitly related to the fatigue life as is the *S-N* curve. Only for some very special cases, such as narrow-band, zero-mean Gaussian stress processes, is there a relation between PSD and fatigue life. If the purpose of the vibration test is a reliability measurement (i.e., random fatigue testing), the failure mechanism must be considered in the test design. Because the fatigue damage is sensitive to the sequence of random stress amplitude, the possibility of accelerating the damage parameters in the frequency domain is questionable and it may be that no unique acceleration factor exists for this case. In addition, for random fatigue accelerated tests, the exciting signals can be displacement or acceleration if only the relationship between displacement and stress or between acceleration and stress is linear.

## 11.6.7  Screening

Screening is the process by which defective parts due to improper manufacturing and assembly are detected and eliminated from a production batch. The principle involves inducing latent defect–type failures through the application of short-term stresses only in a population of parts that are already "weak," without reducing the reliability in the population of "strong" parts. The assumption is that the weak population (e.g., weak wire bonds, weak die attach, weak solder joints) can be eliminated, leaving a high-reliability population.

Effective screens test all the parts throughout the manufacturing process and are tuned to the specific failure mechanisms of concern. The concept of 100% screening assumes that infant mortality–type failures are discovered such that the parts begin operational life with minimal defects. Although "defects-free" parts that result from a "mature" production history may not require screening, this is generally not the case for rapidly changing designs and production methods.

Screening is used under the assumption that latent defects can be precipitated by the appropriate testing. However, in some cases, screening can accelerate failure mechanisms that may have been associated with the main population of product failures, so that the life of the product will be reduced. An understanding of dominant failure mechanisms and their distribution is thus important. In general, screen tests are useless if the failure mechanisms for which one is screening have not been identified. On the other hand, once the failure mechanisms and the associated screen tests have been identified, 100% screening of a batch is desirable.

The number of infant mortality failures that occur can be associated with the maturity of the part technology and the quality/yield of the manufacturing and assembly processes. Table 11.22 lists various environmental stresses, their effects, and improvement techniques in electronic equipment. Table 11.23 lists stress screening environments, applications, and trade-offs in both subsystems and systems. The fundamental screening processes for electronic components include the following:

1.  High-temperature burn-in: In this screen, electrically activated components are subject to a minimum of 48 hr at temperatures of 125°C for military components and 70°C for commercial components. Burn-in is typically conducted to eliminate component failures due to nonspecific manufacturing defects. This is a poor practice.

**Table 11.22**  Stresses, Effects, and Improvement Techniques

| Stress | Effects | Reliability improvement techniques |
|---|---|---|
| Humidity | Penetrates porous substances and causes leakage paths between electrical conductors; causes oxidation that leads to corrosion; moisture causes swelling in materials such as gaskets; excessive loss of humidity causes embrittlement and granulation. | Hermetic sealing, moisture—resistant material, dehumidifiers, protective coatings |
| Salt atmosphere and spray | Salt combined with water is a good conductor that can lower insulation resistance; causes galvanic corrosion of metals; chemical corrosion of metals is accelerated. | Nonmetal protective covers, reduced use of dissimilar metals in contact, hermetic sealing, dehumidifiers |
| Electromagnetic radiation | Causes spurious and erroneous signals from electrical and electronic equipment and components; may cause complete disruption of normal electrical and electronic equipment such as communication and measuring systems. | Shielding, material selection, part type selection |
| Nuclear/cosmic radiation | Causes heating and thermal aging; can alter chemical, physical, and electrical properties of materials; can produce gases and secondary radiation; can cause oxidation and discoloration of surfaces; damages electrical and electronic components, especially semiconductors. | Shielding, component selection, nuclear hardening |
| Sand and dust | Finely finished surfaces are scratched and abraded; friction between surfaces may be increased; lubricants can be contaminated; clogging of orifices, etc.; materials may be worn, cracked or chipped; abrasion, contaminates insulation, corona paths. | Air-filtering, hermetic sealing |

**Table 11.22**  Continued

| Stress | Effects | Reliability improvement techniques |
|---|---|---|
| Low pressure (high altitude) | Structures such as containers and tanks are overstressed and can explode or fracture; seals may leak; air bubbles in material may explode, causing damage; internal heating may increase due to lack of cooling medium; insulations may suffer arcing breakdown; ozone may be formed; outgassing is more likely. | Increased mechanical strength of containers, pressurization, alternate liquids (low volatility), improved insulation, improved heat transfer methods |
| High temperature | Parameters of resistance, inductance, capacitance, power factor, dielectric constant, etc. will vary; insulation may soften; moving parts may jam due to expansion; finishes may blister; devices suffer thermal aging; oxidation and other chemical reactions are enhanced; viscosity reduction and evaporation of lubricants are problems; structural overloads may occur due to physical expansions. | Heat dissipation devices, cooling systems, thermal insulation, heat-withstanding materials |
| Low temperature | Plastics and rubber lose flexibility and become brittle; electrical constants vary; ice formation occurs when moisture is present; lubricants and gels increase viscosity; high heat losses; finishes may crack; structures may be overloaded due to physical contraction. | Heating devices, thermal insulation, cold-withstanding materials |
| Thermal shock | Materials may be instantaneously overstressed, causing cracks and mechanical failures; electrical properties may be permanently altered. Crazing, delamination, ruptured seals. | Combination of techniques for high and low temperatures |
| Shock | Mechanical structures may be overstressed, causing weakening or collapse; items may be ripped from their mounts; mechanical functions may be impaired. | Strengthened members, reduced inertia and moments, shock-absorbing mounts |
| Vibration | Mechanical strength may deteriorate due to fatigue or overstress; electrical signals may be erroneously modulated; materials and structure may be cracked, displaced, or shaken loose from mounts; mechanical functions may be impaired; finishes may be scoured by other surfaces; wear may be increased. | Stiffening control of resource |

Source: Ref. 53.

**Table 11.23**  Typical Stress Screening Applications

| Thermal cycling | Application | Trade-offs |
|---|---|---|
| Module level | | |
| Temperature range | Max: −55 to 125°C (180°C)<br>Nom: −40 to +95°C (135°C)<br>Min: −40 C to +75°C<br>(115°C) | In-house failure rates may in some cases be increased at the next assembly level; hence, equipment behavior under proposed stress screening environment should be evaluated prior to implementation. |
| Temperature Rate | Max: 20°C/min<br>Nom: 15°C/min<br>Min: 5°C/min | Temperature rates of change are as measured by thermocouple on components mounted on modules. |
| No. of cycles | Max: 40<br>Nom: 30<br>Min: 20 | Power-ON screening may be continued to early production until latent design problems are exposed and production problems, processes, and test procedures are proven. |
| Power | Power ON (Development phase)<br>Power OFF (Production phase) | Power-Off screening is considerably cheaper and is effective on mature production hardware. |
| System level | | |
| Temperature range | Max: −55 to 125°C (180°C)<br>Nom: −40 to 95°C (135°C)<br>Min: −40 to 75°C (115°C) | In-house failure rate may in some cases be increased at next assembly level; hence equipment behavior under proposed stress screening environment should be evaluated prior to implementation. |
| Temperature Rate | Max: 20°C/min<br>Nom: 15°C/min<br>Min: 5°C/min | Higher temperature rates may require open-unit exposure with higher air flow rate to overcome slower temperature response of higher mass. |
| No. of cycles | Max: 12<br>Nom: 10<br>Min: 8 | Functional testing at high and low temperature increases failure detectability. |
| Power | Power ON | |

Source: Ref. 53.

2.  High-temperature storage bake: In this screen, nonelectrically activated components are subjected to a temperature of 250°C for hermetic components or 150°C for plastic-encapsulated components. The purpose of the bake is to determine instability mechanisms, gross contamination, and moisture and metallization problems.

3.  High-temperature reverse bias test: In this screen, elevated voltages are coupled with high temperatures.

4. Overstress and voltage step stressing: In this screen, high continuous electrical overstress or a step electrical stress (i.e., voltage) is coupled with high temperature. The major application has been to precipitate gate oxide failures in MOS devices.

5. Thermal shocks: Thermal shock tests are used to accelerate thermal-mechanical stress problems. The part is subject to rapidly induced temperature extremes (i.e., by hot-oil drenching) in order to precipitate poor bonding (i.e., die to substrate) and material defects that can lead to fracture.

6. Temperature cycling: This type of screen is also used to precipitate structural and bonding anomalies. Parts are placed in a chamber and subjected to a time-varying temperature. Figure 11.38 shows a typical temperature cycling profile for ambient cooled and supplementary cooled equipment. For military electronics, the temperature cycle range is approximately $-65$ to $150°C$.

7. Mechanical shock tests: Package integrity problems can also be precipitated by mechanical shock and vibration screens. Such tests are used to determine bond defects, weak seals, and cracks.

8. Centrifugal spin tests: This screen is used to detect package integrity by the application of centrifugal loading, often in the vicinity of 30,000 $g$.

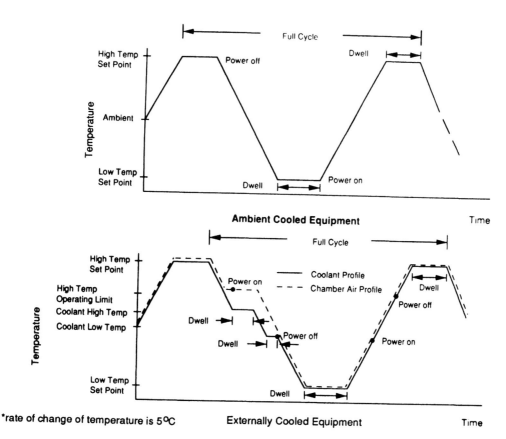

**Figure 11.38** Temperature cycling for ambient and externally cooled equipment. (From Ref. 53, p. 10-5.)

9.    Package leak tests: These screens are applied to detect package hermeticity. Both fine and gross tests are used. The gross leak test requires immersion of the component in hot fluorocarbon with inspection for bubbles. The fine leak test requires immersion in a radioactive gas.

10.    Humidity tests: These screens typically consist of coupling high temperature and humidity. The common test is an 85°C/85% RH test. The purpose is to indicate the resistance to moisture ingress.

11.    Particle impact noise detection tests: This is a shake test in which acoustic monitors detect stray particles, often as small as 25 μm in diameter.

NAVMAT P-9492 is one of the governing publications for environmental stress screening. This publication provides guidance for the use of temperature cycling and random vibration as manufacturing screens for defects in both parts and workmanship. NAVMAT P-9492 is not a specification. In general, the stress screening process is one that is unique to the product being screened and to the people and machines producing the product. Thus, a ''cookbook'' cannot exist on how to stress screen any particular product.

Finally, when designing a screening program for a product, there should be recognition of the cost. Trade-offs involve the costs of planning and conducting the screens and the costs associated with warranty and selling price.

## 11.6.8   Reliability and Prediction Models

Reliability prediction is a methodology for estimating an item's ability to meet specified reliability requirements. A mission reliability prediction estimates the probability that an item will perform its required functions during the mission. A basic reliability prediction estimates the demand for maintenance and logistic support caused by an item's unreliability. When used properly, prediction approaches can provide a useful tool for identifying areas wherein special emphasis or attention is needed and for comparing the ownership cost-effectiveness of various design configurations. To be beneficial for design, models must address the failure modes of the devices as a function of the history-dependent material properties, the manufacturing and assembly processes, and the test and loading (environment) conditions. Table 11.24 presents potential uses of reliability predictions.

Lumped-parameter statistical approaches to failure prediction are often applied to electronic parts and systems because of the large number of failure variables, the difficulty of testing the effects of the variables independently, and the lack of failure-specific constitutive equations with which to write equations to model modes of failure. In many cases, these predictions are based on experience data from similar parts that are used in a similar situation.

Systematic reliability assessments of electronic components using lumped-parameter statistics are represented in well-known and widely used handbooks such as MIL-HDBK-217E [19], British Telecom [20], CNET [21], NTT [22], and Bellcore [23]. Although the failure models used in these handbooks incorporate some physical and operating parameters, the models are not directly derived from physics or mechanics of failure processes. As such, these models do not give the designer any insight into or control over the actual failure mechanisms. Difficulties in data collection, multiparameter curve-fitting techniques, and translation of system-level ''interaction'' failures to component failures further reduce the utility of the models as a design tool. In addition, for ease of modeling,

**Table 11.24**  Uses of Reliability Prediction Models

Establishment of firm reliability requirements in planning documents, preliminary design specifications, and requests for proposals, as well as determination of the feasibility of a proposed reliability requirement

Comparison of an established reliability requirement with state-of-the-art feasibility, and guidance in budget and schedule decisions.

Provide a basis for uniform proposal preparation and evaluation and ultimate contractor selection

Evaluation of potential reliability through predictions submitted in technical proposals and reports in precontract transactions

Identification and ranking of potential problem areas and the suggestion of possible solutions

Allocation of reliability requirements among the subsystems and lower-level items

Evaluation of the choice of proposed parts, materials, units, and processes

Conditional evaluation of the design for prototype fabrication during the development phase

Provide a basis for trade-off analysis

Source: Ref. 53.

the lumped-parameter models typically assume a constant hazard rate for the components. This is conceptually problematic, since the hazard rates of many of the component failure mechanisms are known to be generally modelable only by a Weibull or similar distribution.

Mechanistic approaches based on actual physics of failure mechanisms have the advantage of being useful for prediction of new structures, materials, and technologies. However, mechanistic models are typically of a deterministic nature and thus do not account for the probabilistic nature of materials, manufacturing processes, and the environment. Lumped-parameter statistical methods have the advantage over mechanistic methods of being able to describe effects not easily modeled, such as application, quality, and learning, and to describe interactions between multiple effects.

The mechanistic approach to failure modeling involves determining the specific failure mechanisms in terms of stress-strain, fracture, or fatigue models which are tied to vibration, shock, and thermal loading conditions. This can aid the designer in predicting and improving the causes of failure. For example, as a result of a mechanical analysis of the solder joint fatigue, the number of power cycles can be predicted and the materials constituting the component package and the substrate can be subsequently matched. In addition, trade-offs can be considered. For example, even though ceramic boards (substrates), especially thick films, can be manufactured with a thermal coefficient matching those of hermetically sealed dual in-line packages (DIPs) and flatpacks, such boards are expensive and have a very high dielectric constant, which can cause electrical problems. An alternative solution to the thermal mismatch problem is to use compliant solder tails to avoid cracking resulting from thermally induced fatigue and fracture. The disadvantages of using solder tails include the additional use of board space and the extended component profile.

In the RADC effort "Reliability Analysis/Assessment of Advanced Technologies" [24], the Westinghouse/University of Maryland team was tasked with updating the microelectronics section of MIL-HDBK-217E. It was apparent early in that study that coupled statistical and deterministic methods were required, possibly in the form of design guidelines. To provide a realistic aid to reliability prediction that can be proactive, the method models must incorporate parameters associated with the package materials, dimensions, tolerances, manufacturing, assembly, test, and service processes and stresses due to op-

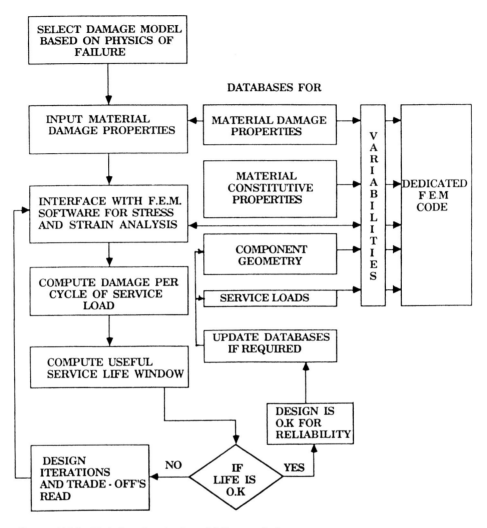

**Figure 11.39**   Task flow for physics of failure analysis.

erating and nonoperating conditions, and thus describe and simulate "stress" interactions between the components and environment. Figure 11.39 shows the requirements and the flow of tasks and information necessary to conduct a proper reliability prediction based on both statistical and deterministic methods.

The approach to developing guidelines includes identification and analysis of package-related failure mechanisms, analysis of the manufacturing and assembly variabilities that affect reliability, development of failure guideline models, and guideline verification. The goal is to establish design guideline models that can be utilized to predict package failure trade-offs and parametric and sensitivity analyses. Of concern, though, is that error bounds be well understood, especially when applying nonlinear or nonmonotonic functions. Furthermore, the effect of inaccurate data introduced into trade-off routines that apply multivariable optimization techniques must be understood.

## 11.6.9  Layout

Once the parts and interconnections have been chosen, the layout of an electronic system can be initiated. The layout process is frequently conducted on the basis of space and interconnection considerations. Layout is a critical part of the electronics design process [25–32].

Dancer et al. [33] have examined several optimization schemes for placement for reliability with emphasis on computational accuracy and speed. Mayer [34] has presented layout techniques incorporating thermal and cost factors in the design of avionics. In Osterman and Pecht [35], a methodology was developed for placement of components for reliability and routability. The reliability goals was to reduce the total system failure rate, and the routing goal was to minimize the total interconnection wirelength. A force-directed approach was developed whereby springs represented both the interconnectivity and the sensitivity of the components to a ''thermal'' location in the system. The scheme was to then minimize the total energy of the system. A more detailed discussion is given in Chapter 6.

## 11.6.10  Failure Mode, Effects and Criticality Analysis

A failure mode, effects, and criticality analysis (FMECA) is an evaluation process for analyzing and assessing the potential failures in a system or equipment design. The objective is to determine the effect of failures on system operation, identify the failures critical to operational success and personnel safety, and rank each potential failure according to the effects on other portions of the system, the probability of the failure occurring, and the criticality of the failure mode [36].

FMECAs began in the 1960s and were used in the Apollo space program to evaluate the reliability and criticality of the complex systems required for spaceflight [37]. Today FMECAs are used early in the design and testing phases by both military and commercial electronics manufacturers to avoid costly modifications by ferreting out latent design and operational deficiencies in components and subsystems and to ensure a high level of achieved reliability before the initiation of full-scale production [38]. In most military procurements today the FMECA is also a required part of the logistic support analysis (LSA), used also to determine the maintenance tasks required for each failure, the repair parts material list, and the personnel requirements for support of a failure. The military places major significance on the ability of the FMECA to contribute positively to the overall logistics of major military systems [39].

To be useful for LSAs and improving the design reliability, FMECAs must be performed early and often during the system development with constant feedback to the design groups. This requires a good understanding of the system by the reliability engineer to determine the failure modes and effects accurately. Application of the criticality factors must be based on the failure effects with respect to the system definition. A poor understanding of the system design and operation can lead to incorrect failure mode severity classifications, resulting in inaccurate FMECAs.

The major specifications covering the FMECA are MIL-STD-1629A, ''Procedures for Performing a Failure Mode, Effects and Criticality Analysis,'' and Society of Automotive Engineering report ARP-926A, ''Fault/Failure Analysis Procedure.'' The reader is referred to these documents for greater detail on FMECA preparation.

The FMECA should initially be performed during the early stages of the design and development cycle to evaluate systematically and document, by item failure mode analysis, the potential problem areas of a design that could result in system failures which could affect mission success and system performance and/or threaten personnel and system safety. The analysis is oriented toward the system only and does not generally cover the effects of human-system interactions. The basic objective is to determine the ways in which the system and equipment can fail (i.e., the failure mode) and map out how the failure mode affects the system, including the criticality of the particular failure mode in relation to system operation and potential for personnel injury [40]. In general, these objectives are accomplished by itemizing and evaluating the components and subsystems in terms of potential failures, based on knowledge of the component/subsystem function. The FMECA analysis includes

1. Item identification and function
2. Failure mode description
3. Effects of failure mode and possible causes
4. Probability of occurrence
5. Criticality of failure

When the FMECA is performed early during the design and development cycle of a system or product, some of the potential benefits of the FMECA include [36]

1. Determining need for redundancy, fail-safe design features, further derating, and/ or design simplification.
2. Determining need to select more reliable materials, parts, devices, and/or components.
3. Identifying single failure points. These are system failures that may result in loss of life or system and are not compensated for by redundancy or alternate system operation.
4. Identifying critical items for design review, configuration control, and traceability.
5. Providing the logic model required to estimate quantitatively the probability of anomalous system conditions.
6. Disclosing safety hazard and liability problem areas.
7. Ensuring that test programs are responsive to identified failure modes and safety hazards.
8. Pinpointing key areas for concentrating quality, inspection, and manufacturing process controls.
9. Establishing data-recording requirements and monitoring frequency for testing, checkout, and mission one.
10. Supporting logistics planning and maintainability analysis. Providing information for selection of preventive maintenance schedules and development of troubleshooting guide, built-in test equipment, and suitable test periods.
11. Identifying circuits for worst-case analysis. (Failure modes involving parameter drifts frequently require worst-case analysis.)
12. Supporting the need for built-in test and isolation for critical system circuits.

Normally the criteria for the FMECA are to accomplish the objectives of the analysis in the most cost-effective manner within given constraints. The principal constraints include the budget and schedule for the FMECA as well as the status of the system design. Variations in design complexity and available design data will generally dictate the analysis approach to be used, provided the contractual requirements or work statement requirements are met [36]. The reliability engineer must be cognizant of the contractual or self-imposed work statement requirements that may impose specific restrictions on the FMECA. Failure to do so could result in costly wasted efforts. The constraints usually determine the FMECA approach, either hardware or functional. Regardless of the approach selected, the reliability engineer must work closely with the system designers so that he or she has a detailed understanding of the entire system function and the interrelationships between the various system functions. At the most detailed level a complete understanding of the design down to the component level may be required.

The functional approach to FMECA is normally used when specific hardware details are not yet known but system definition has been identified. As an example, consider the system functional diagram for a typical ground-based radar system as illustrated in Figure 11.40. The dashed-line boxes represent the basic functional subsystem breakdowns, such as antenna, transmitter, receiver, and signal processor, to highlight a few. Detailed hardware requirements for the subsystem are normally not represented on this block level diagram, and often the detailed hardware has not yet been identified. The functional approach is used when the system complexity is such that a system level–down analysis is the more practical approach. However, because it is a function-oriented approach it is not as detailed as the hardware approach presented in the next section, and as a result certain failure modes can be overlooked with this approach [36].

The documentation of the functional approach to FMECA normally begins with the initial system indenture level and proceeds downward through the lower indenture levels. The indenture levels are determined from the system information flow diagrams. "Flow diagrams are employed as a mechanism for portraying system design requirements in a pictorial manner, illustrating series-parallel relationships, the hierarchy of system functions, and functional and hardware interfaces. The flow diagrams are designated as top level, first level, second level and so on. The top level shows gross operational and mission requirements. The first and second level diagrams represent progressive expansion of the individual functions of the preceding level" [40]. An example of this indenture level number and schematic approach is illustrated in Figure 11.41. Normally this documentation is prepared down to the level necessary to establish the hardware, software, facilities, personnel, and data requirements of the system. The functions identified in each block are numbered to ensure continuity of the functions and provide traceability throughout the system [40]. This system identification methodology permits a natural extension beyond the functional approach to the hardware approach, which is presented in the next discussion.

The hardware approach to FMECA is normally practical only when the reliability engineer has access to schematics, drawings, and other engineering and design data normally available once the system has matured beyond the functional design stage. This approach is normally initiated in a part level–up fashion. However, it can be initiated at any indenture level and move in either direction [41]. Essentially, the hardware approach is an extension of the functional approach down to the component level. In electronic systems the component level is usually associated with discrete electronic components making up printed wiring board assemblies (PWAs). PWAs can be line replaceable units (LRUs) or subassemblies of a chassis that has been designated an LRU in place of the PWAs. A chassis level LRU usually represents a subsystem. The LRU is normally defined by the customer and is based on maintenance philosophy.

Components can also refer to building blocks for mechanical systems. Consider, for example, the ground-based radar system that was introduced previously. Here the antenna is a subassembly, but it is highly mechanical in nature. The antenna is built up from waveguides, gimbals, servomotors, gear assemblies, etc. These would all represent subassemblies and finally components of the antenna.

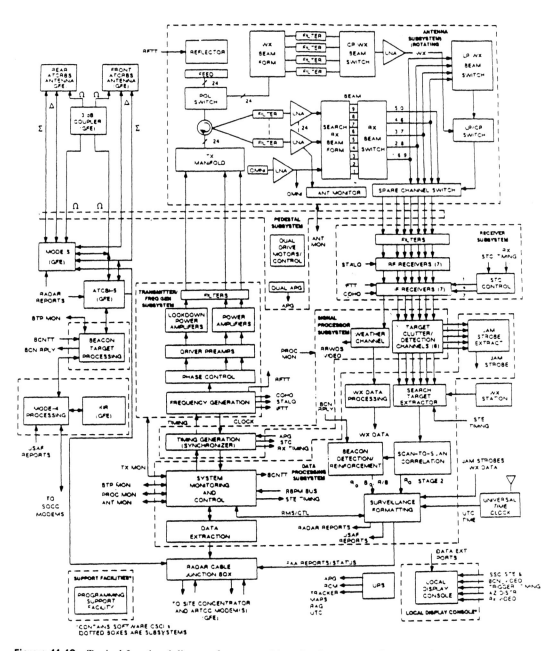

**Figure 11.40**  Typical functional diagram for a ground-based radar system. (Courtesy of Westinghouse Electric Corporation.)

The level of detail refers to the system hardware level at which the failures are postulated. In relation to the functional approach, the failures are associated with the input and output requirements of the particular subsystem under evaluation. The hardware approach level of detail would include failures that could occur at the individual component level. The level of detail for the FMECA may be prescribed in the contract or the self-

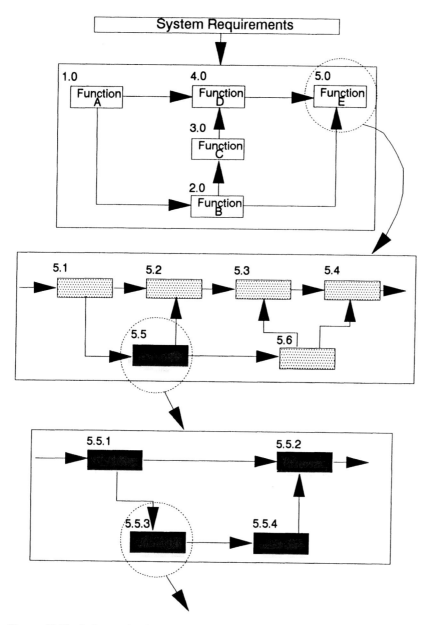

**Figure 11.41** Indenture level numbering for system documentation. (From Ref. 40.)

imposed work statement [36]. During the conceptual stage the level of detail will be driven by the design status maturity. The benefit of a lower-level-of-detail FMECA performed early in the design development cycle is the uncovering of costly design mistakes. Critical nonredundant subsystems may require a hardware approach level of detail to ensure the overall system reliability. It may not be possible to add redundancy to a critical subsystem because of overall cost constraints or system weight and space limitations. Redundant systems may require a lower level of detail because the redundancy will increase the system reliability.

The level-of-detail arguments presented here are ones the reliability engineer should address before beginning the analysis, but the most important consideration is to be certain that the level of detail for the FMECA is performed within the contract requirements [36].

Figure 11.42 is a typical FMECA worksheet, which contains the information required to document and compute the criticality of the failures. The general guidelines that should be followed in performing the analysis are as follows.

1.   Select the approach for the analysis, either functional or hardware.
2.   Define the system utilizing block diagrams and indenture level numbering.
3.   Define the failure modes and the failure mechanism, indicating any environmental stress factors if known. The four basic failure modes are premature operation, failure to operate at a prescribed time, failure to cease operation at a prescribed time, and failure during operation.
4.   Define the system or mission requirements. Using our example, consider the system requirements for the ground-based radar system. Stated simply, the system is to provide full capability for air traffic surveillance and detection within the azimuth, height, and range limitations specified in the radar specifications and to disseminate detected air traffic location parameters to a specified control center within its required availability value.
5.   Define the failure severity classification for the criticality analysis. These are assigned during the analysis to provide a qualitative measure of the worst potential consequences of the failure occurrence. The criticality classifications are as follows [41]:

   · Category I—Catastrophic. A failure that may cause death or weapon system loss (aircraft, tank, missile, ship, etc.)
   · Category II—Critical. A failure that may cause severe injury, major property damage, or major system damage that will result in mission loss.
   · Category III—Marginal. A failure that may cause minor injury, minor property damage, or minor system damage that will result in delay or loss of availability or mission degradation.
   · Category IV—Minor. A failure that is not serious enough to cause injury, property damage, or system damage but that will result in unscheduled maintenance or repair.

6.   Generate required corrective action recommendations for the design or manufacturing process and evaluate new procedures and their effectiveness.
7.   Compile and document the results of FMECA, including the recommendations for system reliability improvements.

To aid in the overall understanding of the analysis, a typical worksheet for a ground-based radar subsystem will be utilized and is illustrated in Figure 11.43. Items 1 and 2 on the FMECA worksheet refer to the subsystem under examination and the indenture level covered by this worksheet, respectively. Here, the subsystem is the transmitter and the indenture level is 2.1 for the lookdown combiner. This indenture level represents an LRU in the subsystem. When one compares this information to the system block diagram, it provides a quick reference to the portion of the system represented by the worksheet as well as the indenture level. Item 4 defines the failure rate data source for this portion of the analysis. Items 3 and 5 to 10 represent further data for the FMECA documentation and are self-explanatory or company-specific documentation information.

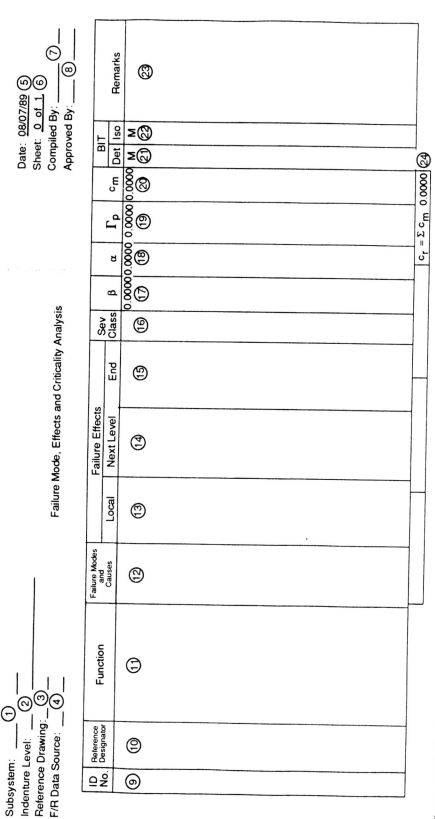

**Figure 11.42** Typical FMECA blank worksheet. (Courtesy of Westinghouse Electric Corporation.)

Subsystem: TRANSMITTER

Identure Level: 2:1 Lookdown Combiner.

Reference Drawing:

F/R Data Source: MIL-HDBK-217E

Date: 03/01/90

Sheet: 1 of 1

Compiled By:

Approved By:

**Failure Mode, Effects and Criticality Analysis**

| ID No. | Reference Designator | Function | Failure Modes and Causes | Failure Effects | | | Sev Class | $\beta$ | $\alpha$ | $r_p$ | $c_m$ | BIT | | Remarks |
|---|---|---|---|---|---|---|---|---|---|---|---|---|---|---|
| | | | | Local | Next Level | End | | | | | | Det | Iso | |
| 30 | B30a | Combines the two outputs from the lookdown Power Panel and sends the signal to the Lookdown Harmonic/Filter Coupler. | Electrical failure- no output. | Power signal is not combined in module. | Reduced power signal is sent to lookdown mode coupler/filter. | No transmitter operation in lookdown mode. | III | 0.5000 | 0.1000 | 0.5258 | 0.0263 | Y | Y | There is one lookdown combiner. |
| | | | Electrical failure- erroneous output. | Erroneous power generation in lookdown combiner. | Erroneous power signal sent to lookdown mode coupler/filter. | Degraded transmitter operation in lookdown mode. | III | 0.5000 | 0.8980 | 0.5258 | 0.2361 | N | N | |
| | | | Input connector fails to transfer signal- dirty, broken terminal. | Lookdown signal does not enter combiner. No additional power combining. | Reduced power available for lookdown channel. | Degraded transmitter operation in lookdown mode. | III | 0.5000 | 0.0010 | 0.5258 | 0.0003 | Y | Y | |
| | | | Output connector fails to transfer signal- dirty, broken terminal. | Amplified lookdown power signal does not exit module. | Significant loss of lookdown power signal. | Same as above. | III | 0.5000 | 0.0010 | 0.5258 | 0.0003 | Y | Y | |

$C_r = \Sigma\,C_m$          0.2629

**Figure 11.43**  FMECA worksheet for ground-based radar system. (Courtesy of Westinghouse Electric Corporation.)

The function of the indenture level under analysis is described in column 11 of the worksheet. For example, the function of the lookdown combiner for the radar is to combine the two outputs from the lookdown power panel and send the signal to the lookdown harmonic/filter coupler. The failure modes and potential causes of a failure for the described function are entered in column 12. For example, one failure mode for the lookdown combiner is described as an electrical failure with no output. The effects of this failure are covered in the local, next level, and end or subsystem level categories (columns 13, 14, and 15, respectively). The local effect is a lack of power signal combination in the module, with a next level effect of reduced power sent to the lookdown mode coupler/ filter. The result is the transmitter will not operate in the lookdown mode.

The severity classification of the particular failure mode is entered in column 16. This classification is a critical portion of the analysis because it tends to be subjective. The reliability engineer determines from the functional requirements the severity classification of the particular failure mode. Without proper understanding of the mission and functional requirements, the engineer may enter an inappropriate value in this column. As an example, consider the operation of a weapons system radar in a tactical fighter aircraft. During a hostile engagement with enemy aircraft, loss of the antennae for tracking and arming would normally be considered a severity class I because loss of aircraft and life could result from the antennae failure. But if one were to evaluate the severity class incorrectly based on further conditional requirements, the reliability engineer could label the antennae failure as severity class III if the aircraft is not engaged in combat. This is an extreme example, but it serves to point out that the severity classification must ultimately be referenced back to the stated mission system requirements to be useful.

Column 17 represents the failure effect probability ($\beta$), which is a conditional probability that the failure effect will result in the identified criticality classification, given that the failure mode occurs. For this FMECA $\beta$ is defined in the following manner. For severity class I failures are defined earlier, $\beta = 1.0$. For severity classes II, III, and IV, $\beta$ will have values of 0.75, 0.50, and 0.25, respectively. These values of $\beta$ must be clearly defined at the beginning of the FMECA.

The failure mode ratio ($\alpha$) represents the fraction of the part failure rate (column 19) related to the particular failure mode under consideration for the LRU or indenture level. If all potential failure modes are included, this column should sum to 1.0. The multipliers are usually obtained from field data, which are constantly compiled by organizations such as Rome Air Development Center (RADC) and are available in the technical report RADC-TR-84-244.

The failure mode criticality number $C_m$ is defined as

$$C_m = \alpha\beta\Gamma_p$$

where $\Gamma_p$ is the failure rate obtained from the failure predictions methodology. The criticality number is entered in column 20 on the FMECA worksheet. There is a different criticality number for each severity classification. For a particular severity classification and mission phase, the sum of the criticality numbers is given by the overall criticality, $C_r$ (column 24). If the system has built-in test capability, circuitry may exist to aid in the detection of a particular failure mode and isolation of that failure mode to a particular LRU or indenture level. This information must also be included in the FMECA worksheet (columns 21 and 22).

The criticality matrix permits identifying and comparing each failure mode to all other failure modes with respect to criticality. By assigning a reference designator to the failure

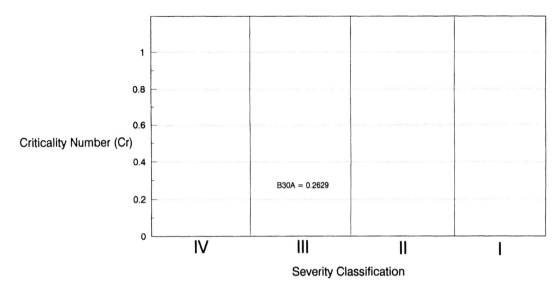

**Figure 11.44**  Example criticality matrix. (From Ref. 41.)

mode and resulting criticality for a particular LRU or indenture level, the results for the
FMECA can be displayed graphically as illustrated in Figure 11.44. The reference des-
ignators, which may be part of the system documentation, serve as points on the plot.
As indicated in the figure, the overall criticality $C_r$ is plotted as a function of the severity
classification for each failure mode. The result is a criticality distribution as a function
of severity classification [41]. The $C_r$ for the example worksheet has been plotted to serve
as an illustration.

The criticality matrix provides an overall reporting method for quickly determining
the parts of the system that may require corrective actions. Catastrophic and Critical
(severity classes I and II, respectively) are the highest-priority categories for evaluation.
To reduce the criticality, a decision may be reached to add redundancy to the system at
the offending indenture level. In sophisticated aerospace, missile, and weapon systems,
the final decision to add redundancy must be carefully weighed. The increase in reliability
and decrease in severity of the failure mode must be traded against the increased weight,
cost, maintainability, and logistic support cost of the redundant system [42].

The most severe criticality levels should be the first to be considered, but as we have
seen from the previous discussion the addition of redundancy can have other system
impacts. However, the critical matrix may also point out areas of the system that have
inherently poor reliability, which may be addressed and corrected at the design stage.
This is why it is extremely important that the FMECA begin early in the system development
stage with continuous feedback to the design engineers. Failure to provide the constant
feedback may result in a system with several high-criticality areas that may not be addressed
because of the cost associated with making the design change.

The FMECA can be effective in the development of a system, but it is not without
its drawbacks. The reliability engineer must have a good understanding of the system in
order to perform an accurate FMECA. Without fundamental system background knowledge, incorrect failure modes and rates may be predicted, resulting in incorrect system
effects and criticality. The criticality factors are subjective because they are assigned by

(a)

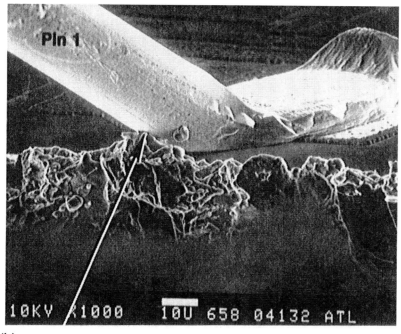

(b)

**Figure 11.45** Failures in an RF amplifier module. (Courtesy of Westinghouse Electric Corporation.)

**(a)**

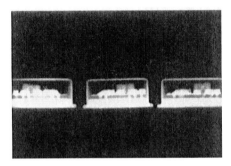

**(b)**

**Figure 11.46**  X-ray inspection to determine part quality: (a) X-ray of amplifiers to assure proper spacing, (b) X-ray of mixer package. (Courtesy of Westinghouse Electric Corporation.)

the engineer performing the analysis; as a result, these factors are left to individual interpretation. The interpretations are not always correct.

## 11.6.11  Reliability Growth

When complex equipment is designed with innovative technology or advanced production methods, the prototypes often have unforeseen design deficiencies which lower the expected reliability of the equipment. In order to meet reliability requirements, a reliability development test program is instituted. The process of achieving reliability goals by improving system designs without compromising performance is called Reliability Growth. The objective is that faulty and weak aspects of the design are located and corrected through the development testing program which employs a test, analyze, and fix (TAAF) process. According to the TAAF process, the detection of failures during test instigates redesign of the prototype, and the corrective action is verified with further testing. The TAAF process can be illustrated by a feedback loop, as seen in Figure 11.47.

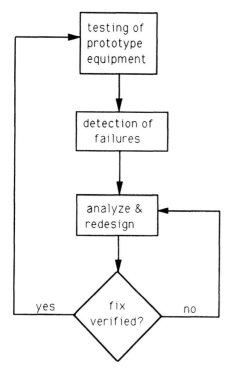

**Figure 11.47** The test, analyze, and fix (TAAF) process. (From Ref. 70.)

The length of a typical reliability development program ranges from five to twenty-five multiples of the required mean time between failures (MTBF). This translates to about twice the number of testing hours in calendar time. Reliability growth can be measured in other units, such as miles or cycles. The testing program requires two or three prototype equipment. Before the prototypes are officially placed under test, they must pass environmental stress screening. Also, the performance of field environment simulation instruments that will be used to test the prototypes must be verified. Once officially under test, any anomalies in the prototypes' behavior are evaluated and classified as relevant or nonrelevant failures. Any anomaly in the system operating behavior that could not be expected to occur in the field application is classified as non-relevant. Nonrelevant failures can be caused by installation damage, accident or mishandling, failures of the test facility, or equipment failures due to externally applied overstress exceeding the amount approved for testing. Relevant failures include any operational anomalies not classified as nonrelevant, regardless of whether the cause of failure is verified or unverified. Momentary cessation of equipment operation, termed an intermittent failure, and parameter drift are examples of relevant failures. Relevant failures are investigated and may result in design or production modifications. Some contracts require that the causes of all relevant failures must be determined.

If an anomaly is classified as a relevant failure, it is further classified as chargeable or nonchargeable. Nonchargeable failures are failures that result from another failure, called a dependent failure, failures of parts whose specified life expectancy has been exceeded, or failures that are induced by equipment furnished by the customer. Independent failures are defined as failures that are neither a consequence of another failure of the equipment under test nor any part of the testing facility. Chargeable failures include intermittent failures, independent failures of the equipment design, equipment manufacturing, part design, part manufacturing and failures that result from contractor furnished equipment (CFE) operating, maintenance or repair procedures. Failures which are determined to have the same cause, failure mode, and failure mechanism and occur at the same environmental conditions are only counted chargeable once. Chargeable failures can be used to determine the system MTBF. A schematic of failure classification is illustrated in Figure 11.48.

Test time as well as failures is categorized as relevant or nonrelevant. The test time accumulated between failures when the equipment is officially under test is termed relevant test time. Time spent troubleshooting equipment failures and verifying repairs is termed nonrelevant test time. Only the accumulated relevant test time is used to determine the improvement in system reliability, that is the system MTBF. Also, for contracts which must comply with military standards, each prototype system or equipment under test must operate at least one half of the average relevant test time of all the equipment under test. For example, for a test of 1500 hours on two systems, neither system can be tested for less than 750/2 or 375 hours.

The relevant test time consists of a series of cycles that combine the environmental stresses the equipment will experience in the field. Depending on the field application,

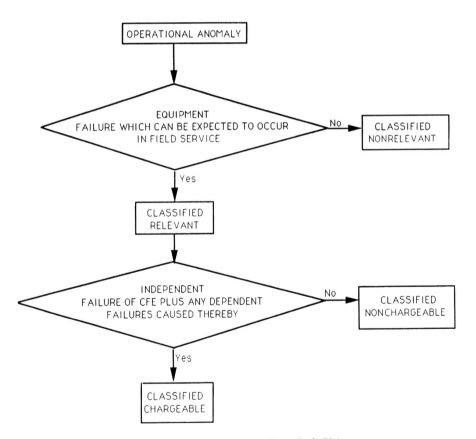

**Figure 11.48** Classification of equipment failure. (From Ref. 70.)

the test cycle may include electrical, thermal, moisture and vibration stresses. The environmental conditions in the test are often the worst case stresses the equipment could encounter in the field. An example of a typical test plan cycle is illustrated in Figure 11.49. Operational checks on the equipment are conducted at regular intervals during each test cycle, and performance checks are conducted less frequently, perhaps every 100 power-on hours during testing. The performance checks are conducted at room temperature and consist of the operational check plus additional verification of equipment behavior such as precision and measurement repeatability.

Other types of testing, such as functional and safety testing, should be performed concurrently with growth testing, since design changes stemming from other test results may affect reliability. Test data shared from among the different types of tests provides optimum utilization of this information and may provide deeper insight into the equipment behavior.

As the system improves with appropriate design modifications resulting from the TAAF process, the subsequent times between failures increases. The failure time data obtained from reliability growth testing is used to determine the improvement in the system MTBF. Reliability growth is depicted as a function of test time. The equipment MTBF is plotted versus test time to illustrate growth. The MTBF is found by dividing the

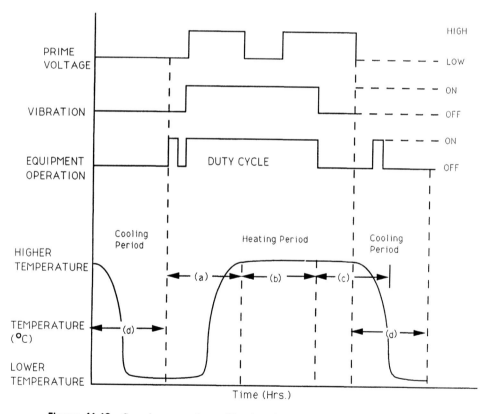

**Figure 11.49** Sample test cycle profile of environmental stresses: (a) time from chamber to reach stabilization at higher temperature, (b) time of equipment operation at higher temperature, (c) optional hot soak and hot start-up checkout, (d) equipment off (can be operated if required.) (From Ref. 70.)

cumulative relevant test time by the cumulative number of chargeable equipment failures. This formula for calculating MTBF was originally implemented by Duane [44], who recognized a general trend in the improvement of the cumulative failure rate of various systems under reliability development. The systems included hydro-mechanical devices, aircraft generators, and an aircraft jet engine. The cumulative number of failures plotted on log-log paper versus the cumulative operating hours nearly produced a straight line for all of the systems. Figure 11.50 illustrates their similarity in growth trends. Progressively fewer failures occurred during the test program as design improvements were incorporated into the systems. This phenomenon is mathematically modelled by

$$\lambda_\Sigma = \frac{\Sigma F}{t} = Kt^{-\alpha} \tag{11.73}$$

where $\lambda_\Sigma$ is the cumulative failure rate, $\Sigma F$ is the cumulative number of failures, $t$ is the cumulative operating hours, $K$ is a constant indicating an initial failure rate and $\alpha$ is the growth rate. The growth rate must be between zero and one to model a decreasing failure rate. Growth rates are normally bound by a range of 0.3 to 0.7 maximum.

A current or instantaneous failure rate, $\lambda_I$, can be found from the derivative of the cumulative number of failures, $\Sigma F$.

$$\lambda_I = \frac{d(\Sigma F)}{dt} = (1 - \alpha)Kt^{-\alpha} \tag{11.74}$$

The Duane model is useful for constructing curves for the predicted or planned growth in order to map the progress of reliability growth. For scheduling purposes, a planned growth curve is constructed from the Duane model to predict the improved MTBF based on the allotted test time and a growth rate from similar previous systems. The reliability

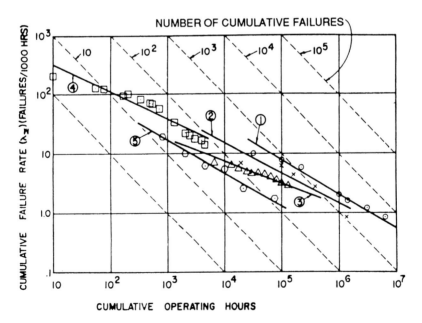

**Figure 11.50** Reliability growth trends for various systems under test development at General Electric. (From Ref. 44, © 1964 IEEE.)

can be assessed at any point during testing to determine whether the system is improving on schedule and that resources are allocated appropriately.

Another popular model, called the Army Materiel Systems Analysis Activity (AMSAA) model [45], utilizes the Weibull intensity function and the Poisson distribution. It is explained as follows. Let $DM_1 < DM_2 < \cdots < DM_k$ represent the cumulative test times when design modifications are made during the TAAF process. Between design modifications, the failure rate can be assumed constant, as illustrated in Figure 11.51. Let $\lambda_i$ represent the failure rate during the i-th time period between modifications $(DM_i - DM_{i-1})$. Based on the constant failure rate assumption, the number of failures $N_i$ during the i-th time period can be estimated by the Poisson distribution

$$\text{Prob } \{N_i = n\} = \frac{[\lambda_i(DM_i - DM_{i-1})]^n \, e^{-\lambda_i(DM_i - DM_{i-1})}}{n!} \tag{11.75}$$

where n is an integer. The mean number of failures $\Theta(t)$ for the i-th time period is equal to $\lambda_i (DM_i - DM_{i-1})$.

Let $t$ represent the cumulative test time and let $N(t)$ represent the total number of system failures by time $t$. If $t$ is in the first interval, then $N(t)$ has the Poisson distribution with mean $\Theta(t) = \lambda_1 t$. If $t$ is in the second interval, then $N(t)$ is the number of failures $N_1$ in the first interval plus the number of failures in the second interval between $DM_1$ and $t$. Thus, in the second interval, $N(t)$ has the mean $\Theta(t) = \lambda_1 N_1 + \lambda_2(t - DM_1)$.

When the failure rate is assumed constant $(\lambda_0)$ over a test interval, that is between design modifications, then $N(t)$ is said to follow a homogeneous Poisson process with mean of the form $\lambda_0 t$. When the failure rates change with time, that is, over test time with design modifications, then $N(t)$ is said to follow a nonhomogeneous Poisson process with the mean value function

$$\Theta(t) = \int_0^t v(x) \, dx \tag{11.76}$$

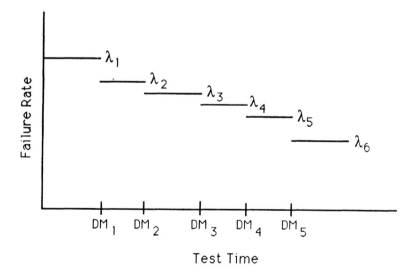

**Figure 11.51** Assumed constant failure rates between design modifications. (From Ref. 43.)

where the intensity function $v(x) = \lambda_i$ and $DM_{i-1} < x < DM_i$. Thus, for any $t$,

$$\text{Prob}\{N(t) = n\} = \frac{[\Theta(t)]^n \, e^{-\Theta(t)}}{n!} \tag{11.77}$$

where $n$ is an integer. As $\Delta t$ approaches zero, $v(t) \, \Delta t$ approximates the probability of a system failure in the time interval $(t, t + \Delta t)$. The intensity function is approximated by a continuous parametric function so that the test data can be compiled and the parameters estimated. The AMSAA model assumes that the intensity function can be approximated by a parametric function defined as

$$v(t) = \lambda \beta t^{\beta - 1} \tag{11.78}$$

This is the Weibull intensity function with $\beta > 0$, $\lambda > 0$ and $t > 0$. Note that the AMSAA model assumes a Poisson process with the Weibull intensity function, and not the Weibull distribution. Crow [45] recognized equation 11.78 can model various processes, including reliability growth. The parameter of interest is $\beta$, because 1-$\beta$ is the reliability growth rate.

From the parametric assumption, the mean number of failures by time $t$, $\Theta(t)$, is defined as

$$\Theta(t) = \lambda \, t^\beta \tag{11.79}$$

Also, the cumulative failure rate, $\omega(t)$, is defined as

$$\omega(t) = \frac{N(t)}{t} \tag{11.80}$$

If the cumulative failure rate, $\omega(t)$, is linear with respect to time on a log-log scale, then the growth is analogous to that modelled by Duane. Then the parameters $\beta$ and $\lambda$ (or $\alpha$ and $K$) can be found graphically on full logarithmic paper or determined statistically using estimation theory. For graphical estimation, a straight line is fitted to the plot of the cumulative failure rate (or mean number of failures) versus the cumulative test time on log-log paper. By taking the logarithm of the cumulative failure rate (or mean number of failures), the relationship between the parameters and the slope and ordinate intercept is apparent. For statistical estimation, the Maximum Likelihood Estimators provide point approximations of the parameters. Confidence limits can also be calculated, and goodness of fit tests can be implemented.

Once the parameters are found, a curve can be drawn to predict the future behavior of the product. If the system is not modified after time $t_0$, then failures are assumed to continue at the constant rate

$$\lambda_0 = v(t_0) = \lambda \beta t_0^{\beta - 1} \tag{11.81}$$

according to the exponential distribution.

The Duane and AMSAA reliability growth models are examples of continuous models, which were developed for repairable systems in which reliability is measured in terms of times between failure. Other models, called discrete models, measure reliability growth under different circumstances of failure, that is for systems that are described as either failing or operating when called into service, also known as one-shot systems. Examples of one-shot systems are missiles and space shuttles. Other reliability growth models have been developed which model systems under assumptions which are tailored to the development program and to specific circumstances. References to other models include Lloyd and Lipow [66], Rosner [67], Wolman [68], and Cox and Lewis [69].

## 11.6.12 Design Guidelines, Standards, and Handbooks

The need for consistent reliability programs prompted the military to set requirements for reliability in the late 1950s. The requirements were set in documents known as military standards (MIL-STD) and military handbooks (MIL-HDBK). These documents, as well as similar documents developed by professional societies and industry, provide guidelines for reliability testing and evaluation, measurement, prediction, and improvement. Many of these documents have been revised several times since their first issue.

Standards have been developed to establish requirements in all areas of the reliability program, with emphasis given to overall management and specific taskings. Standards also give guidance to procuring activities when setting requirements for contracting activities.

Handbooks typically support the standards with technical requirements. The handbooks often give guidance in fulfilling the technical requirements by supplying equations and other necessary information. Some of the general military standards are briefly discussed below. Appendix C gives a detailed listing of military and industry standards, guidelines, and handbooks for electronics.

MIL-STD-785, entitled "Reliability Program for Systems and Equipment Development and Production," provides the general requirements and specific tasks for reliability programs during the development, production, and initial deployment of systems and equipment. The standard defines system reliability parameters in units of measurement directly related to operational readiness, mission success, demand for maintenance power, and demand for logistic support, as applicable to the type of system. The standard states that the reliability program shall improve operational readiness and mission success of the major end item, reduce item demand for maintenance manpower and logistic support, provide essential management information, and manage the overall program cost and schedule. Furthermore, each reliability program shall include an appropriate mix of reliability engineering and accounting tasks, depending on the life cycle phase. These tasks shall be planned, integrated, and accomplished in conjunction with other design, development, and manufacturing functions.

MIL-STD-756, entitled "Reliability Modeling and Prediction," provides procedures and rules for the preparation of mission reliability models for electronic, electrical, electromechanical, mechanical, and ordinance systems and equipments. The complexity of these items may range from a complete weapon system to the simplest subdivision of a system. Failure distribution information is not given in this standard, so it is not really a prediction tool, except in the sense that, given part reliabilities, a model can be developed to calculate system reliability. The primary value of reliability modeling is as a design tool to provide relative measures of item reliability to design decisions. The standard includes tasking sections on reliability modeling and prediction for both basic and mission reliability and methods including conventional probability, truth tables, logic diagrams, and Monte Carlo simulation.

MIL-STD-781, entitled "Reliability Design Qualification and Production Acceptance Tests: Exponential Distribution," establishes criteria for the reliability qualification tests and reliability acceptance tests for equipment whose failure rate distribution is exponential. The requirements include test conditions, procedures, and various fixed-length and sequential test plans with respective accept/reject criteria. When testing in accordance with this standard, the minimum test program normally consists of preproduction reliability acceptance tests or equipment production reliability acceptance tests. Production reliability acceptance tests are a series of tests conducted during the production run to ascertain whether the equipment continues to meet specified performance and reliability requirements under specified environmental conditions. This test is normally performed on each lot produced.

MIL-STD-790, entitled "Reliability Assurance Program for Electronic Parts Specifications," was developed to provide guidelines to ensure the manufacturer's reliability assurance program. The standard requires that the reliability assurance program be an integral part of all phases of the design and development of a part. A description of the necessary documentation required for the program approval is found in the general and detailed requirements sections.

MIL-STD-2068, entitled "Reliability Development Tests," establishes requirements and procedures for a reliability development test to implement the requirements set in MIL-STD-785. The purpose of the test is to provide reliability growth and to promote standardized reliability improvement. The standard is not applicable to "single shot" and "passive use" devices. The standard requires that tests should be conducted to provide engineering information on failures under natural and induced environments in order to determine and improve the reliability of the system.

MIL-STD-1629, entitled "Procedures for Performing a Failure Mode, Effects, and Criticality Analysis," establishes requirements and procedures for performing a failure mode, effects, and criticality analysis (FMECA) to evaluate systematically and document, by item failure mode analysis, the potential impact of each functional or hardware failure on mission success, personnel and system safety, system performance, maintainability, and maintenance requirements. Each potential failure is ranked by the severity of its effect in order that appropriate corrective actions may be taken to eliminate or control high-risk items. The standard applies to the program activity phase of demonstration and validation and full-scale engineering development, e.g., design, research and development, and test and evaluation.

### 11.6.13   Quality Assurance and Audits

Quality pertains to the product characteristics that enable the product to satisfy a given need. Quality assurance involves ensuring that manufactured parts have the characteristics required to satisfy the need.

Often quality and reliability are distinguished in terms of the type of failures, how they were initiated, and when they occurred. Quality is tied to yield failures in the sense that poor workmanship, assembly, and handling affect process yields, part yields, acceptance test yields, and functional test yields. Reliability, on the other hand, is tied to the fraction of the total number of parts that initially meet specifications but subsequently fail to meet specification requirements in operation. According to this definition, a field failure associated with some aspect of equipment that had never been operational is classified as a quality/yield problem. However, there should be a much tighter mixture of quality and reliability, because it is possible to design for reliability in such a manner that quality-related problems cannot occur.

Quality is typically addressed by generating quality specifications and controlling part conformance with specifications through the application of inspections and tests. However, with increased complexities in parts come complexities in the ability to inspect and test the parts. The electronics industry has realized that concentration must be on quality control in the manufacturing and assembly processes, rather than on testing after completion. The goal is to remove flaws and defects resulting from poor processing. Typical gauges used to measure quality are the defect levels obtained from samples at key inspection points in the manufacture, assembly, and test processes and the lot reject rate report, which summarizes the lot rejection rates based on pass/fail of some sample acceptance criteria.

Improvements in manufacturing yields and subsequent quality come from sources that include improved on-line process control, reduced defect levels in both materials and processes, improved incoming material and part quality, increased attention to quality issues such as workmanship and test, increased automation where appropriate, and improved product design. However, the solution to quality/yield problems can often be complex. For example, failures in an RF amplifier module (Figure 11.45) initiated analysis that included X-ray inspection to determine part quality (Figure 11.46). It was determined that a lead wire from the RF coil was occasionally shorting to the case of the packaged amplifier because of greater than expected temperatures and greater than expected thermal expansion. The problem could be solved by either inspecting all components via an extensive quality control and screening program or redesigning to provide more tolerance room. The latter approach is ultimately more cost effective but may require up-front costs for redesign and production.

Parts that fail screening or burn-in applied by the manufacturer are considered yield losses. With high product complexity and increasingly high dependence on material and process parameters, inspection methods are costly and features are often too small to be observed visually. Even electrical inspection and test is expensive and time consuming. Design for defect tolerance is thus becoming an alternative.

## EXERCISES

1. Determine the mission profile and the operating and environmental stresses that must be considered in the design of an ignition system for a lawn mower, a car, a boat, and a plane.
2. Discuss the trade-offs between cost, reliability, and maintainability for a lawn mower engine, a car engine, a plane engine, and an unmanned space shuttle engine.
3. Discuss the trade-offs between cost, reliability, weight, and size for the engine, a car engine control electronics, and the potential use of redundancy for a lawn mower, a plane, a car, and a space shuttle.
4. Prove the following relation using Eq. (11.9):

$$\text{MTTF} = \int_0^\infty tf(t)\, dt = \int_0^\infty R(t)\, dt$$

5. If all the products in a reliability test fail by time $T$ according to the probability density distribution $f(t) = t$, what percentage of $n_0$ has failed at time $T/2$?
6. In one reliability test, there is a 50% chance that two of five parts fail. What is the part reliability assuming the parts are identical?
7. What is the significance of the Weibull $\eta$ if $\beta = 1$ and $\gamma = 0$?

8. What is the MTTF for products that can be modeled with the Weibull distribution $\beta = 1$, $\gamma = 10$, and $\eta = 5$?

9. Three parts are operating in parallel, of which two are required for success. If the parts have reliabilities of 0.9, 0.8, and 0.7, what is the reliability of the redundant assembly?

10. Prove that the Weibull reliability function gives a reliability of 36.8% for a time $t = \gamma + \eta$.

11. From Figure 11.10, unreliability versus time to failure data ($\gamma = 0$), what is the longest mission duration that will provide a reliability of 90% if the mission starts at an age of 200 hr?

12. Plot the following sets of data in Tables 11.25 and 11.26 from reliability tests on

**Table 11.25** Reliability Test Data #1

| j | $T_j$ | %Q |
|---|---|---|
| 1 | 136 | 12.9 |
| 2 | 150 | 31.4 |
| 3 | 190 | 50.0 |
| 4 | 300 | 68.6 |
| 5 | 500 | 87.1 |

**Table 11.26** Reliability Test Data #2

| j | $T_j$ | %Q |
|---|---|---|
| 1 | 5000 | 12.9 |
| 2 | 5500 | 31.4 |
| 3 | 6000 | 50.0 |
| 4 | 8000 | 68.6 |
| 5 | 10000 | 87.1 |

Weibull probability paper. Determine if the data can be modeled by the Weibull distribution and the Weibull parameters $\gamma$, $\eta$, and $\beta$ if so.

13. Consider a mixed parallel-series system in which there are three different series components (a,b,c) on path 1 and two different series components (d,e) on path 2. Determine the reliability formula for the system.

14. Consider three elements, A, B, and C, whose reliabilities are 0.95, 0.80, and 0.85, respectively. Maximize the reliability of a series-parallel system of elements A, B, and C using only five elements.

15. Discuss which combinations of physical, mechanical, electrical, and chemical stresses induce each of the failures in Table 11.5.

16. Discuss how the failure mechanisms and modes in Table 11.5 may affect each other, if one mechanism may cause another or counteract the effects.

17. Draw a reliability block diagram for the avionics cooling system in Figure 11.18.

## REFERENCES

1. Tillman, F. A., Hwang, C. L., and Kuo, W., "Optimization Techniques for System Reliability with Redundancy—a Review," *IEEE Trans. Reliab. R-26*:148–155 (1977).

2. Luss, R., "Optimization of System Reliability by a New Nonlinear Integer Programming Procedure," *IEEE Trans. Reliab. R-24*(1):14–16 (1975).

3. Aggarwal, K. K., Gupta, J. S. and Misra, K. B., "A New Heuristic Criterion for Solving a Redundancy Optimization Problem," *IEEE Trans. Reliab. R-24*(1):86–87 (1975).

4. McLeavey, D. W., "Numerical Investigation of Parallel Redundancy in Series Systems," *Oper. Res. 22*:1110–1177 (1974).

5. McLeavey, D. W. and McLeavey, J. A., "Optimization of System Reliability by Branch-and-Bound," *IEEE Trans. Reliab. R-25*(5):327–328 (1976).

6. Hwang, C. L. and Lee, H. B., "Nonlinear Integer Goal Programming Applied to Optimal System Reliability," *IEEE Trans. Reliab. R-33*(5):431–438 (1984).

7. Hwang, C. L. and Masud, A. S., *Multiple Objective Decision Making Methods and Applications*, Springer-Verlag, New York, 1979.

8. Kuo, W. and Lin, H.-H., "Reliability Optimization with the Lagrange-Multiplier and Branch-and-Bound Technique," *IEEE Trans. Reliab. R-36*(5):624–630 (1987).

9. Boccaletti, G., Borri, F. R., D'Espinosa, G., Fioravanti, G. and Ghio, E., "Accelerated Tests," in *Microelectronic Reliability*, Vol. 2, Poollino, E. (ed.), Artech House, Norwood, MA (1989).

10. Stitch, M., Johnson, G. M., Kirk, B. P. and Brauer, J. B., "Microcircuit Accelerated Testing Using High Temperature Operating Tests," *IEEE Trans. Reliab. R-24*(4):238–250 (1975).

11. Amerasekera, E. A. and Campbell, D. S., *Failure Mechanisms in Semiconductor Devices*, Wiley, New York, 1987.

12. Peck, D. S. and Zierdt, C. H., "Temperature Humidity Acceleration of Metal-Electrolysis Failure in Semiconductor Devices," *Reliab. Phys.* 11th Annual Proceedings, Reliability Physics, Las Vegas, Nevada, April, 1973, pp. 146–152 (1973).

13. Sinnadurai, F. N., "The Accelerated Aging of Plastic Encapsulated Semiconductor Devices in Environments Containing a High Vapor Pressure of Water," *Microelectron. Reliab. 13*:23–27 (1974). February

14. Gunn, J. E., Malik, S. K. and Mazumdar, P. M., "Highly Accelerated Temperature and Humidity Stress Test Technique (HAST)," *Proc. IEEE*, 1981, pp. 48–51. 19th Annual Proceedings Reliability Physics, Orlando, Fla. April, 1981, pp. 48–51.

15. Ajiki, T., Sugimoto, M., Higuchi, H. and Kumada, S., "A New Cyclic Biased T.H.B. Test for Power Dissipating ICs," *Reliab. Phys. 00*:118–126 (1979). 17th Annual Proceedings Reliability Physics San Francisco, CA April, 1979 pp. 118–126.

16. Kemeny, A. P., "Experiments Concerning the Life Testing of Semiconductor Devices," *Microelectron. Reliab. 10*(3):169–193 (1971). June

17. Dally, J. W., *Packaging of Electronic Systems: A Mechanical Engineering Approach*, McGraw-Hill, New York, 1990.

18. Christensen, R. M., *Theory of Viscoelasticity*, Academic Press, Orlando, Fla., 1982.

19. *Reliability Prediction of Electronic Equipment*, Military Handbook 217E, October 21, 1987.

20. *Handbook of Reliability Data for Electronic Components Used in Telecommunications Systems*, British Telecom Handbook HRD3, Issue 3, January 1984.

21. *Recueil de Données de Fiabilitie du CNET*, Vol. 1: *Partie Composants Électroniques*, Centre National d'Études des Télécommunications, 1983.

22. "Standard Reliability Table for Semiconductor Devices," Nippon Telegraph and Telephone Public Corporation, December 1982.

23. "Reliability Prediction Procedure for Electronic Equipment," Bellcore Technical Advisory TA-000-23620-84-01, Issue 1, 1984.

24. "Reliability Analysis/Assessment of Advanced Technologies," RADC Report RADC-TR-89-177, 1989.

25. Brownlee, K. A. *Statistical Theory and Methodology in Science and Engineering*, 2nd ed., Wiley, New York, 1965.

26. Bazovsky, I., *Reliability Theory and Practice*, Prentice-Hall, Englewood Cliffs, N.J., 1965.

27. Von Alven, W. H. (ed.), *Reliability Engineering*, Prentice-Hall, Englewood Cliffs, N.J., 1964.

28. Pieruschka, E., *Principles of Reliability*, Prentice-Hall, Englewood Cliffs, N.J., 1963.
29. Pecht, M., "The University of Maryland CALCE/RAMCAD for Electronics," *IEEE Trans. Reliab. R-36*:501 (1987).
30. Leonard, C., "Mechanical Engineering Issues and Electronic Equipment Reliability: Incurred Costs without Compensating Benefits," *Intersociety Conference on Thermal Phenomena*, 1990, pp. 1–6.
31. Usell, R. J. and Rao, B. S., "Designing-in and Building-in Reliability: An Operational Definition with Examples from Electronic Packaging Products," *Proc. International Electronic Packaging Society Conference*, October 29–31, 1984, pp. 540–547.
32. Howard, R. T., "Packaging Reliability: How to Define and Measure It," *IEEE Trans. Components, Manuf. Technol., 00*:376–384 (1982).
33. Dancer, D., Pecht, M. and Palmer, M., "Reliability Optimization Scheme for Convection Cooled Printed Circuit Boards," accepted for publication, *IEEE Trans. Reliab.* in press.
34. Mayer, A., "Computer Aided Thermal Design of Avionics for Optimal Reliability and Minimum Life Cycle Cost," AFFDL-TR-7848, March 1978.
35. Osterman, M. and Pecht, M., "Placement for Reliability and Routability of Convectively Cooled PWBs," *IEEE Trans. Computer-Aided Des. Integrated Circuits Syst.*, in press (1990).
36. ARP 926A, "Fault/Failure Analysis Procedure," Society of Automotive Engineers, Warrendale, Pa., November 15, 1979, p. 5.
37. Burgess, J. A., "Spotting Trouble Before It Happens," *Mach. Des.*, September 17, 1967.
38. Arnzen, H. E., "Failure Mode and Effect Analysis: A Powerful Engineering Tool for Component and System Optimization," *Ann. Reliab. Maintainab.*, 1966, p. 355.
39. Harris, J. R., Jr., Team Leader—Logistics Review Group, Naval Air Systems Command, Crystal City, Va., telephone interview.
40. Blanchard, B. S. and Fabrycky, W. J., *Systems Engineering and Analysis*, Prentice-Hall, Englewood Cliffs, N.J., 1981, p. 353.
41. MIL-STD 1629A, "Procedures for Performing a Failure Mode, Effects and Criticality Analysis," Engineering Specifications and Standards Department (Code 93), Naval Air Engineering Center, Lakehurst, N.J., November 24, 1980.
42. Ball, L. W., "Failure Mode and Effect Analysis: A Powerful Engineering Tool for Component and System Optimization," *Ann. Reliab. Maintainab.*, Fifth Reliability and Maintainability Conference, July 18–20, 1966, p. 368.
43. MIL-HDBK 189, "Reliability Growth Management," Department of Defense, February 13, 1981.
44. Duane, J. T., "Learning Curve Approach to Reliability Monitoring," *IEEE Trans. Aerosp.*, 1964, pp. 563–566.
45. Crow, L. H., "Reliability Analysis for Complex, Repairable Systems," in *Reliability and Biometry Statistical Analysis of Lifelength*, Proschan, F. and Serfling, R. J. (eds.), Society for Industrial and Applied Mathematics, Philadelphia, 1974, pp. 579–410.
46. Bassin, W. M., "Increasing Hazard Functions and Overhaul Policy," *Proc. Annual Symposium on Reliability*, New York, 1969, pp. 173–178.
47. Crow, L. H., "On Tracking Reliability Growth," *Proc. 1975 Annual Reliability and Maintainability Symposium*, Washington, D.C., 1975, pp. 438–443.
48. Crow, L. H., "Confidence Interval Procedures for Reliability Growth Analysis," U.S. Army Material Systems Analysis Activity, Technical Report 197, Aberdeen Proving Ground, Md., AD-A044788, 1977.
49. Englehardt, M. and Bain, L. J., "Prediction Intervals for the Weibull Process," *Technometrics 20*:167–169 (1978).
50. Finkelstein, J. M., "Confidence Bounds on the Parameters of the Weibull Process," *Technometrics 18*:115–117 (1976).
51. Kempthorne, O. and Folks, L., *Probability, Statistics, and Data Analysis*, Iowa State University Press, Ames, 1971.

52. Lee, L. and Lee, S. K., "Some Results on Inference for the Weibull Process," *Technometrics* 20:41–45 (1978).
53. Wilbur, J. W. and Fuqua, N. B., "A Primer for DoD Reliability, Maintainability and Safety Standards," IIT Research Institute (RAC) RADC, Griffiths Air Force Base, Rome, N.Y., 1988.
54. Fitch, W. T., Greer, P. and Lycoudes, N., "Data to Support 0.001%/1000 Hours for Plastic I/C's"
55. Peck, D. S., "New Concerns about Integrated Circuit Reliability," Reliability Physics Symposium, 1978.
56. Domangue, E., Rivera, R. and Shepard, C., "Reliability Prediction Using Large MOS Capacitors," Reliability Physics Symposium, 1984.
57. Hokari, Y. et al., IEDM Technical Digest, 1982.
58. Gear, G., "FAMOS PROM Reliability Studies," Reliability Physics Symposium, 1976.
59. Nanda, Vangurd, Gj-P and J. R. Black, "Electromigration of Al-Si Alloy Films," Reliability Physics Symposium, 1978.
60. Black, J. R., "Current Limitation of Thin Film Conductor," Reliability Physics Symposium, 1982.
61. Lycoudes, N.E., "The Reliability of Plastic Microcircuits in Moist Environments," Solid State Technology, 1978.
62. Fitch, W. T., "Operating Life vs. Junction Temperatures for Plastic Encapsulated I/C (1.5 Mil Au Wire)," unpublished report.
63. Howes, M. G. and Morgan, D. V., *Reliability and Degradation, Semiconductor Devices and Circuits*, Wiley, New York, 1981.
64. Kececioglu, D., "The Weibull Distribution and Its Applications to Reliability Engineering," University of Arizona, Tucson, 1985.
65. Young, R. J., *Introduction to Polymers*, Chapman & Hall, London, 1981.
66. Lloyd, D. K. and Lipow, M. R., *Reliability: Management, Methods and Mathematics*, Prentice Hall, Englewood Cliffs, N.J., 1962.
67. Rosner, N., "Systems Analysis—Nonlinear Estimation Techniques," *Proceedings National Symposium on Reliability and Quality Control*, New York, 1961, pp. 203–207.
68. Wolman, W., "Problems in System Reliability Analysis," *Statistical Theory of Reliability*, ed. M. Zelen, The University of Wisconsin Press, 1963, pp. 149–160.
69. Cox, D. R. and Lewis, P. A. W., *The Statistical Analysis of Series of Events*, John Wiley & Sons, New York, 1966.
70. MIL-STD-781D (Reliability Design Qualification and Production Acceptance Tests: Exponential Distribution).
71. Myers, D. K., "What Happens to Semiconductors in a Nuclear Environment?" in "Microelectronics Interconnection and Packaging," *Electronics Magazine*, J. Lyman, ed., 1980:283, table 1.

## SUGGESTED READING

Abbott, W. H., "Field Versus Laboratory Experience in the Evaluation of Electronic Components and Material," paper presented at Corrosion/83, Anaheim, Calif., National Association of Corrosion Engineers, April 1983.
Abuaf, N. and Kadambi, V., "Thermal Investigation of Power Chip Packages: Effect of Voids and Cracks," *IEPS: 6th Annual International Electronics Packaging Conference*, November 1986, p. 821.
Altoz, F. et al., "A Method for Reliability—Thermal Optimization," *Proc. 1982 Annual Reliability and Maintainability Symposium*. January 1982, pp. 303–308.
Amstadter, B. L., *Reliability Mathematics*, McGraw-Hill, New York, 1971.

Anderson, R. T., *Reliability Design Handbook* (Cat. RDH-376), IIT Research Institute (RAC) RADC, Griffiths Air Force Base, Rome, N.Y., 1976.

Baboian, R., "Corrosion—a National Problem," *ASTM Stand. News*, March 1988, p. 37.

Bain, L. J., *Statistical Analysis of Reliability and Life Testing Models: Theory and Methods*, Marcel Dekker, New York, 1978.

Barlow, R. E. and Proschan, F., *Statistical Theory of Reliability and Life Testing Probability Models*, Holt, Reinhart & Winston, New York, 1975.

Berg, H. M. and Paulson, W. M., "Chip Corrosion in Plastic Packages," *IEEE Symp. Microelectron. Reliab. 20*:247 (1980).

Billington, R., et al., *Power System Reliability Calculations*, MIT Press, Cambridge, Mass., 1973.

Blanchard, B. S., *Design to Manage Life Cycle Cost*, M/A Press, Portland, 1978.

Blanchard, B. S., *Logistics Engineering and Management*, 2nd ed., Prentice-Hall, Englewood Cliffs, N.J., 1981.

Bompas-Smith, J., *Mechanical Survival: The Use of Reliability Data*, McGraw-Hill, New York, 1973.

Bourne, A. J. and Greene, A. E., *Reliability Technology*, Wiley Interscience, New York, 1972.

Braunberg, G., Naft, J., Madison, L. and Pecht, M., "Computer Aided Life Cycle Engineering," 1988 Annual Reliability and Maintainability Symposium, January 1988.

Breipchu, A. M., *Probabilistic Systems Analysis*, Wiley, New York, 1970.

Brockland, W. R., *Statistical Assessment of the Life Characteristic, a Biographical Account*, Hafner, Collier-Distribution Center, Riverside, N.J., 1974.

Brook, R. H. W., *Reliability Concepts in Engineering Manufacture*, Wiley, New York, 1972.

Caflen, R. H., *A Practical Approach to Reliability*, Business Books Limited, 1972.

Cannon, D. L. and Trapp, O. D., "The Effects of Cleanness of Integrated Circuit Reliability," *6th Annual Proc. Reliability Physics*, 1968, pp. 68–79.

Carruepa, E. R., et al., "Assuring Product Integrity," D. C. Heath & Co., Lexington, MA, 1975.

Carter, A. C. S., *Mechanical Reliability*, Halsted, New York, 1973.

Cluley, J. C., *Electronic Equipment Reliability*, Macmillan, London, 1974.

Cocks, F. H., *Manual of Industrial Corrosion Standards and Control*, American Society for Testing and Materials, Philadelphia, 1973.

Commizzoli, R. B., Franenthal, R. R., Milner, P. C. and Sinclair, J. D., "Corrosion of Electronic Materials and Devices," *Science 234*:340–345 (1986).

Crawford, W. M. and Weigand, B. L., "Contamination of Relay Internal Ambients," *Proc. 15th Annual Relay Conference*, 1956, pp. 1–8.

Crook, D. L., "Method of Determining Reliability Screens for Time Dependent Dielectric Breakdown," Reliability Physics Symposium, 1979.

Cunningham, C. E. and Cox, W., *Applied Maintainability Engineering*, Wiley, New York, 1972.

Cvijanovich, G. B., "Conductivities and Electrolytic Properties of Adsorbed Layers of Water," *Proc. NBS/RADC Workshop, Moisture Measurement Technology for Hermetic Semiconductor Devices*, II, Gaithersburg, Md., 1980, pp. 149–164.

DerMarderosian, A., "Hermeticity and Moisture Ingress," Raytheon, Sudbury, Mass., 1988.

Dhillon, B. S., "The Analysis of the Reliability of Multi-State Device Networks," University of Windsor, Windsor, Ontario, 1975.

Edwards, D. R., et al., "Shear Stress Evaluation of Plastic Packages," *Proc. 37th ECC*, May 1987, p. 84.

Eisenberg, P. H., et al., "Effects of Ambient Gases and Vapors at Low Temperature on Solid State Devices," presented at the 7th New York Conference on Electronic Reliability, IEEE, 1966.

Enrick, N. L., *Quality Control & Reliability*, Indus. Pub., Edison, N.J., 1972.

Gordon, F., *Maintenance Engineering: Organization and Management*, Halsted-Wiley, New York, 1973.

Green, A. E. and Bourne, A. J., *Reliability Technology*, Wiley Interscience, London, 1972.

Grouchko, D., *Operations Research & Reliability*, Gordon & Breach, New York, 1972.

Gyrna, F. M. and Juran, J. M., *Quality Planning and Analysis*, McGraw-Hill, New York, 1970.

Haasl, D. F., et al., "Fault Tree Handbook," NUREG-0492, Division of Technical Information and Document Control, U.S. Nuclear Regulatory Commission, Washington, D.C., 1981.

Hahn, G. J. and Shapiro, S. S., *Statistical Models in Engineering*, Wiley, New York, 1967.

Hallberg, O., "Accelerated Factors for Temperature-Humidity Testing of Al-Metallized Semiconductors," SINTOM, Copenhagen, 1979.

Henley, E. J. and Kumamoto, H., *Reliability Engineering and Risk Assessment*, Prentice-Hall, Englewood Cliffs, N.J., 1981.

Hnatek, E. R., *Integrated Circuit Quality and Reliability*, Marcel Dekker, New York, 1987.

Howard, R. T., "Electrochemical Model for Corrosion of Conductors on Ceramic Substrates," *IEEE Trans. Components, Hybrids, Manuf. Technol.* CHMT-4(4):520–525 (1981).

Hu, J. M., Pecht, M. G., Barker, D. B. and Dasgupta, A., "The Role of Failure Mechanism Identification in Accelerated Testing," submitted for presentation at the Annual Reliability and Maintainability Symposium, Orlando, Fla., January 24–31, 1991.

Ireson, G. (ed.), *Reliability Handbook*, McGraw-Hill, New York, 1966.

Jardine, A. K. S., *Operational Research in Maintenance*, Halsted-Wiley, New York, 1970.

Jelen, F. C., *Cost and Optimization Engineering*, McGraw-Hill, New York, 1970.

Jensen, F. and Petersen, N. R., *Burn-In: An Engineering Approach to the Design and Analysis of Burn-In Procedures*, Wiley, New York, 1982.

Jowett, C. E., *Reliable Electronic Assembly Production*, Tab Books, Blue Ridge Summit, Pa., 1971.

Kapur, K. C. and Lamberson, L. R., *Reliability in Engineering Design*, Wiley, New York, 1977.

Kivenson, G., *Durability and Reliability in Engineering Design*, Hayden Rock Co., Rochelle Park, N.J., 1971.

Kozlov, B. A. and Ushakov, *Reliability Handbook*, Rosenblatt, L. (transl.), Holt, Rinehart & Winston, New York, 1970.

Larson, R. W., "The Accelerated Testing of Plastic Encapsulated Semiconductor Components," *Proc. International Reliability Physics Symposium*, 1974, pp. 243–249.

Larson, R. W., "A Review of the Status of Plastic Encapsulated Semiconductor Reliability," *Br. Telecom Technol. J.* 2(2):(1984).

Lawless, J. F. *Statistical Models and Methods for Lifetime Data*, Wiley, New York, 1982.

Lipson, C. and Sheth, M. J., *Statistical Design and Analysis of Engineering Experiments*, McGraw-Hill, New York, 1973.

Mann, N. R., and Shafer, E., *Methods for Statistical Analysis of Reliability and Life Data*, Wiley, New York, 1974.

Mann, N. R., Schafer, R. E. and Singpurwalla, N. D., *Methods for Statistical Analysis of Reliability and Life Data*, Wiley, New York, 1974.

Motorola, "TMOS Power MOSFET Reliability Report," Motorola Literature Distribution, Phoenix, 1988.

Myers, G. J., *Reliable Software Through Composite Design*, Petrocelli Books, New York, 1975.

Myers, G. J., *Software Reliability Principles and Practice*, Wiley Interscience, New York, 1976.

Nishimura, A., et al., "Life Estimation for IC Plastic Packages Under Temperature Cycling Based on Fracture Mechanics," *Proc. ASM 3rd Conference on Electronic Packaging, Material, and Processing and Corrosion in Microelectronics*, 1987, p. 90.

Nixon, F., *Managing Costs to Achieve Quality and Reliability*, McGraw-Hill, New York, 1971.

O'Connor, D. T., *Practical Reliability Engineering*, Wiley-Heyden, 1983, reprinted March 1984.

Osterman, M. and Pecht, M., "Placement of Integrated Circuits for Reliability on Conductively Cooled Printed Wiring Boards," *Trans. ASME J. Electron. Packag.*, June 1989, pp. 149–156.

Pearne, N. K., "Future Trends in IC Packaging," *19th Annual Connectors and Interconnection Technology Symposium Proc.*, October 1986, p. 321.

Pecht, M. and Ko, W. C., "A Corrosion Rate Equation for Microelectronic Die Metallization," ISHM, in press.

Pecht, M., Palmer, M., Schenke, W. and Porter, R., "An Investigation into PCB Component Placement Tradeoffs," *IEEE Trans. Reliab. R-36*:524 (1987).

Pecht, M., Azarm, S. and Praharaj, S., "Optimal Redundancy Allocation For Electronic Equipment," *Proc. 34th Annual Institute of Environmental Sciences Conference*, May 1, 1988.

Peck, D. S., "Comprehensive Model for Humidity Testing Correlation," *Proc. International Reliability Physics Symposium*, 1986, pp. 44–50.

Peck, D. S. and Zierdt, C. H., Jr., "Temperature-Humidity Acceleration of Metal—Electrolysis Failure in Semiconductor Devices," *Proc. International Reliability Physics Symposium*, 1973, pp. 146–152.

Pound, R., "Thermal Management Balances Cooling and Size Demands," *Electron. Packag. Prod. 25*(9):112 (1985).

*Probability Charts for Decision Making*, Industrial Press, New York, 1971.

Pronikov, A. S., *Dependability and Durability of Engineering Products*, Halsted-Wiley, New York, 1973.

Rau, J. G., *Optimization and Probability in Systems Engineering*, Van Nostrand, New York, 1970.

Reich, B. and Hakim, E., "Environmental Factors Governing Field Reliability of Plastic Transistors and Integrated Circuits," *Proc. International Reliability Physics Symposium*, 1972, pp. 82–87.

Robertson, S. R., "A Finite Element Analysis of the Thermal Behaviour Contacts," *IEEE Trans. Components, Hybrids, Manuf. Technol. CHMT-5*(1):3–10 (1982).

Sbar, N. S., "Bias Humidity Performance of Encapsulated and Unencapsulated Ti-Pd-Au Thin Film Conductors in an Environment Contaminated with $Cl_2$," *Proc. 1976 26th Electronics Components Conference*, San Francisco, 1976, pp. 277–284.

Schneider, G., "Non-Metal Hermetic Encapsulation of a Hybrid Circuit," *Microelectron. and Reliab. 28*(1):75–92 (1988).

Shooman, M. L., *Probabilistic Reliability—an Engineering Approach*, McGraw-Hill, New York, 1968.

Shooman, M., *Software Engineering—Reliability, Design, Management*, McGraw-Hill, New York, 1980.

Slota, P. and Levitz, P. J., "A Model for a New Failure Mechanism—Nichrome Corrosion in Plastic Encapsulated PROMs," *IEEE Proc. 33rd Electronic Components Conference*, 1983.

Smith, C. J., *Reliability Engineering*, Barnes and Noble, Scranton, Pa., 1972.

Smith, D. J. and Babb, A. H., *Maintainability Engineering*, Pitman, New York, 1973.

Smoliar, S. W., "Operational Requirements Accommodation in Distributed System Design," *IEEE Trans Software Eng.*, November 1981, p. 531.

Thomas, R. W., "IC Packages and Hermetically Sealed-in Contaminants," in *Gov. Microcircuits Applications Conference Digest*, 1972, pp. 31–36.

Thomas, R., "Moisture, Myths and Microcircuits," *IEEE Trans. Parts, Hybrids Packag. PHP-12*(3):169 (1976).

Tillman, F., et al., "System Reliability Subject to Multiple Non-Linear Constraints," *IEEE Trans. Reliab. R-17*:153 (1968).

Uhlig, H. H., *Corrosion and Corrosion Control*, Wiley, New York, 1963, p. 146.

Young, M. R. P. and Peterman, D. A., "Reliability Engineering," *Microelectron. Reliab. 7*:91–103 (1968).

Zave, P. and Schell, W., "Salient Features of an Executable Specification Language and Its Environment," *IEEE Trans. Software Eng.*, February 1986, p. 312.

# 12

# Electronic Materials and Properties

**Jillian Y. Evans**   *NASA, Goddard Space Flight Center, Greenbelt, Maryland*

**John W. Evans**   *UNISYS Corporation, Goddard Space Flight Center,*
*Greenbelt, Maryland*

## 12.1  INTRODUCTION

Electronic packaging involves developing systems to house individual active and passive electronic devices and systems to interconnect those devices to form a functioning electronic product. A wide variety of metals, ceramics, and polymers are employed. As a result, the packaging engineer must often consult numerous references to obtain design information or data necessary to compare performance and reliability. This chapter is intended to reduce this effort by providing tabulated electrical, mechanical, and thermal data for many of the materials commonly used at various levels of packaging.

Electronic packaging design consists of several levels. Individual devices are manufactured, mounted on substrates, and packaged in metal or ceramic enclosures. For example, integrated circuits are enclosed in hermetic ceramic dual in-line or leadless packages. The packages may include more than one device, such as in a multichip module or thick-film hybrid. The packaged parts are then interconnected by joining them to a printed circuit. Individual circuit boards are joined together by backplanes in metal enclosures to form a functioning electronic system.

The hierarchical nature of the process was considered in the development of this chapter. Rather than categorizing materials as ceramic, polymer, and metals in the traditional manner, the data are tabulated for the elements of the packaging process. Thermal, mechanical, and electrical data and applications information are tabulated for semicon-

727

ductor materials, substrates, joining materials, encapsulants, enclosures, and contact/ conductor materials. A separate section on corrosion behavior is also included.

A considerable body of data has been compiled from various references on electronic packaging and material properties. The sources of the data are cited and tabulated with the data. All of the property data are at room temperature unless otherwise noted. Photographs are used to illustrate examples of the applications of various materials, and important aspects of materials behavior are discussed.

## 12.2  SEMICONDUCTOR MATERIALS

A variety of materials are utilized for their semiconducting properties. A partial list of semiconducting materials is shown in Table 12.1. Silicon is the most widely used semiconductor and most familiar to the packaging engineer. It is used in bulk for simple devices such as diodes and transistors and as a substrate for the growth of epitaxial Si films in which complex very large scale integrated circuits are created. A single-crystal sapphire

**Table 12.1**  Semiconductors and Their Band Gaps

| Materials | Band Gap (eV) | |
|---|---|---|
|  | 0K | 300K |
| Diamond (C) | 5.48 | 5.47 |
| Si | 1.17 | 1.12 |
| Ge | 0.74 | 0.66 |
| $\alpha$Sn | 0.092 | 0.08 |
| InP | 1.42 | 1.35 |
| InAs | 0.42 | 0.36 |
| InSb | 0.24 | 0.17 |
| GaP | 2.35 | 2.26 |
| GaAs | 1.52 | 1.42 |
| GaSb | 0.81 | 0.72 |
| AlSb | 1.70 | 1.58 |
| SiC | 3.03 | 3.0 |
| Te | 0.33 | — |
| CaO | 1.09 | — |
| PbS | 0.286 | 0.41 |
| PbSe | 0.145 | 0.27 |
| PbTe | 0.187 | 0.31 |
| CdS | 2.58 | 2.42 |
| CdSe | 1.85 | 1.70 |
| CdTe | 1.61 | 1.56 |
| ZnO | 3.44 | — |
| ZnS | 3.85 | 3.68 |
| SnTe | 0.3 | — |
| AgC | 3.25 | — |
| AgI | 3.02 | — |
| $Cu_2O$ | 2.172 | — |
| $TiO_2$ | 3.03 | — |

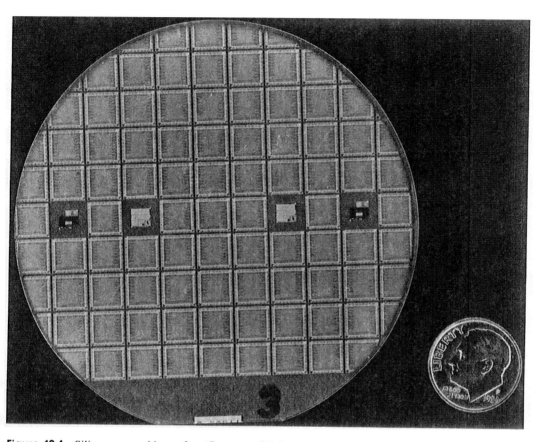

**Figure 12.1**  Silicon on sapphire wafer. (Courtesy of E-Systems.)

wafer on which epitaxial silicon has been deposited to create large-scale devices is shown in Figure 12.1.

GaAs is seeing increasing use, particularly in high-frequency applications. Its continued development as a technologically important material will depend on successful development of the processes for economically manufacturing integrated circuits [117]. A GaAs wafer is shown in Figure 12.2.

Other semiconducting compounds (III-V and II-VI compounds) are used for the unique optical behavior that arises from their structure in such applications as detectors. As optics and electronics see increasing integration, compound semiconductors may see increased applications. Table 12.2 summarizes the uses of some of the important semiconducting materials.

## 12.2.1  Electrons in Semiconductors

Materials are classified as semiconductors because of their unique electrical behavior. In defining a semiconductor it is useful to begin by thinking about a metal. Metals are often thought of as having free electrons that readily travel through the structure of the metal,

**Figure 12.2**  GaAs wafer. (Courtesy of E-Systems.)

giving rise to the excellent electrical conductivity exhibited by metals. However, this is a somewhat unsatisfying picture in many respects.

In reality, the theory of quantum mechanics clearly shows that electrons can occupy only certain allowed energy levels. In solids, these allowable levels become allowable bands. In the case of metals, the valence band is not completely filled and there are allowed energy states that an electron can occupy immediately adjacent to the occupied states within the valance band. Thus, when a metal is subjected to an electric field, these electrons are easily excited into the unfilled states, allowing the metal to be conductive.

In semiconductors and insulators, the valence band is filled and there are no immediately available states for the electrons to occupy. For conduction to occur, the electrons

**Table 12.2**  Applications of Semiconductor Materials

| Material | Chemical Symbol | Applications |
|---|---|---|
| Silicon | Si | basic semiconductor material, diodes, transistors, rectifiers, integrated circuits, solar cells, near-infrared detectors |
| Germanium | Ge | low forward voltage drop devices, photodetectors |
| Silicon-Carbide | SiC | varistors, diodes, rectifiers |
| Gallium-Arsenide | GaAs | high-speed, high-frequency applications, field-effect transistors, varactor diodes, light-emitting diodes, lasers, high-speed integrated circuits |
| Cadmium-Sulfide | CdS | optical sensors, solar cells |
| Indium-Antimonide | InSb | infrared detectors |
| Lead-Sulfide | PbS | rf detectors, infrared detectors |
| III-V, II-VI Compounds | — | optical applications, detectors |

must be excited across an energy gap into the next allowable set of energy states, or into the *conduction band*. The distinction between semiconductors and insulators is made on the basis of the size of the gap. A gap on the order of 1 to 4 eV usually defines a semiconductor [16, 98]. A useful number of electrons can be thermally excited across a gap of this magnitude, allowing conduction to occur. In insulators, the gap is large enough that electrons are effectively prevented from crossing it. The conductivity of a semiconductor arising only from thermally excited electrons is known as intrinsic conductivity.

Semiconductors that are most useful for devices are those in which the conductivity can be modified by the controlled introduction of impurities (doping) by available processing technologies. Consequently, silicon has become the most important semiconducting material. The modified conductivity or extrinsic behavior of semiconductors is dependent on the concentration and species of the dopant.

The electronic structure of semiconductors also gives rise to mechanical and thermal properties important to the packaging engineer. In contrast to metals, semiconductors tend to be very brittle or lack ductility. In addition, they do not exhibit good thermal conductivity relative to many metals. Mechanical, thermal, and electrical properties are summarized in Tables 12.3 through 12.6.

**Table 12.3**  Mechanical Properties of Semiconductors

| Material | Flexural[a] Strength MPa | Elastic[a] Modulus GPa | Poisson Ratio | Density kg/m$^3$ | Hardness kg/mm$^2$ | Source |
|---|---|---|---|---|---|---|
| Si | 62 | 109-190 | 0.28 | 2330 | 850-950 | 14, 20, 25, 64, 85, 95, 120, 112 |
| Ge | – | 103[b] | 0.28 | 5360 | 750 | 95, 102 |
| Te | – | 41.3 | – | 6240 | – | 100, 102 |
| Diamond | 71.4 | – | – | 3510 | 8800 | 14, 25, 64, 102, 103, 120 |
| SiC | 69 | 655 | 0.19 | – | 2200-2950 | 95 |
| InP | – | – | – | 4787 | 535 | 64, 102 |
| InSb | – | – | – | 5775 | 220 | 64 |
| AlAs | – | – | – | 3810 | – | 64 |
| AlP | – | – | – | 2850 | – | 64 |
| AlSb | – | – | – | 4218 | 360 | 64 |
| GaAs | 150 | 84.95 | 0.31 | 5316 | 535-750 | 53, 64, 120, 112 |
| GaP | – | – | – | – | 950 | 64 |
| GaSb | – | – | – | 5619 | 450 | 64 |
| CdS | – | – | – | 4820 | 55 | 64, 102 |
| CdSe | – | – | – | 5810 | – | 102 |
| CdTe | – | – | – | 6200 | – | 102 |
| ZnS | – | – | – | 3980 | 180 | 64, 102 |
| ZnSe | – | – | – | 5420 | – | 102 |
| ZnTe | – | – | – | 6340 | – | 102 |
| PbS | – | – | – | 7500 | – | 102 |
| PbSe | – | – | – | 8100 | – | 102 |
| PbTe | – | – | – | 8164 | – | 102 |

b. < 100 >                          a. Crystallographic directions not reported unless noted.

**Table 12.4** Elastic Constants of Cubic Semiconductor Crystals

| Material | $C_{11}$ (GPa) | $C_{12}$ (GPa) | $C_{44}$ (GPa) | Source |
|---|---|---|---|---|
| Diamond (C) | 950 | 390 | 430 | 98 |
| Ge | 128.9 - 129.2 | 47.9 - 48.3 | 67.0 - 67.1 | 64, 98 |
| Si | 165.7 - 167.4 | 63.9 - 65.2 | 79.6 | 64, 98 |
| GaAs | 11.92 | 5.986 - 6.0 | 5.38 | 64, 98 |
| GaSb | 8.85 | 40.4 | 43.3 | 64, 98 |
| InSb | 67.2 | 36.7 | 30.2 | 64, 98 |

## 12.2.2 Crystal Anisotropy

Semiconductors are used primarily as single crystals. Unlike polycrystalline metals and ceramics, single-crystal semiconductor materials are generally not isotropic. The properties of the material will vary with the orientation of the crystal. Of particular importance to the packaging engineer are the elastic properties, which may be used to relate the stresses

**Table 12.5** Electrical Properties of Semiconductors

| Material | Resistivity ohm-m | Dielectric[a] Constant $\epsilon_\infty$ | Source |
|---|---|---|---|
| Si | 640 | 11.9 | 13, 14, 21, 25, 98, 120, 112 |
| Ge | 8.9 | 16.0 | 98, 100 |
| Te | 20 | – | 98, 100 |
| Diamond | $10^{14}$ | 5.7 | 98, 100, 120 |
| SiC | – | 10.0 | 98 |
| InP | – | 12.4 | 98 |
| InAs | – | 14.6 | 98 |
| InSb | – | 17.7 | 98 |
| AlAs | – | 8.5 | 98 |
| AlP | – | 11.6 | 98 |
| AlSb | – | 14.4 | 98 |
| GaAs | 4000 | 13.1 | 53, 98, 120, 112 |
| GaP | – | 11.1 | 98 |
| GaSb | – | 15.7 | 98 |
| CdS | – | 5.4 | 98 |
| CdSe | – | 10.0 | 98 |
| CdTe | – | 10.2 | 98 |
| ZnS | – | 5.2 | 98 |
| ZnSe | – | 5.9 | 98 |
| ZnTe | – | 9.0 | 98 |
| PbS | – | 17.0 | 98 |
| PbSe | – | 24.0 | 98 |
| PbTe | – | 30.0 | 98 |

a. High (optical) frequency limit.

**Table 12.6** Thermal Properties of Semiconductors

| Material | Conductivity W/m-K | Coefficient of Expansion ppm/°C | Heat Capacity J/kg-K | Melting Point K | Source |
|---|---|---|---|---|---|
| Si | 148 | 2.3 - 2.6 | 712 | 1685 | 13, 14, 20, 25, 64, 101, 112 |
| Ge | 64 | 5.7 - 6.1 | 335 | 958 | 64, 95, 100 |
| Te | – | 16.8 | 197 | 723 | 100 |
| Diamond | 2000 - 2300 | 1.2 | – | > 3823 | 64, 102, 103 |
| SiC | 283 | 4.5 - 4.9 | 670 | – | 64, 95 |
| InP | 67 | 4.5 | – | 1344 | 64, 101, 102 |
| InAs | – | 5.3 | – | 1216 | 64, 102 |
| InSb | 16 | 5.0 | – | 808 | 64, 102 |
| AlAs | – | – | – | > 1873 | 64 |
| AlP | – | – | – | > 1773 | 64 |
| AlSb | 46 | – | – | 1323 | 64 |
| GaAs | 44 - 58 | 5.72 | 322 | 1511 | 53, 64, 95, 112 |
| GaP | 79 | 5.3 | – | 1738 | 64 |
| GaSb | 33 | 6.9 | – | 985 | 64 |
| CdS | – | – | – | 2023[b] | 102 |
| CdSe | – | – | – | 1623 | 102 |
| CdTe | – | – | – | 1365 | 101, 102 |
| ZnS | – | – | – | 2122[c] | 95 |
| ZnSe | – | – | – | 2271 | 95 |
| ZnTe | – | – | – | 1511 | 102 |
| PbS | – | – | – | 1386 | 95, 102 |
| PbSe | – | – | – | 1338 | 102 |
| PbTe | – | – | – | 1190 | 102 |

a. high purity
b. 100 atm, sublimes @ 1253K
c. 150 atm, sublimes @ 1460K

and strains induced in single-crystal semiconductors. Isotropic materials are simply described by the elastic modulus or Young's modulus and Poisson's ratio. However, in the general case, the elastic behavior of a solid is much more complex.

The elastic behavior of any solid is described by a fourth-order tensor consisting of 81 independent components [98, 104]. This can be reduced to 21 independent constants through thermodynamic and static considerations. Fortunately, many important semiconductors crystals have cubic symmetry, which further reduces the number of independent constants to 3. For a cubic crystal, the elastic constants are referred to as $C_{11}$, $C_{12}$, and $C_{44}$. The subscripts are standard notation and refer to the orthogonal coordinate system assigned to the crystallographic axes [104]. The elastic constants of semiconductors with cubic symmetry are shown in Table 12.4.

In the case of an isotropic solid, $C_{11} = C_{44}$ [98, 104, 105] and the elastic constants are related to the elastic modulus and Poisson's ratio by the following equations:

$$C_{11} + 2C_{12} = \frac{1}{S_{11} + 2S_{12}}$$

$$C_{11} - C_{12} = \frac{1}{S_{11} - S_{12}}$$

$$C_{44} = \frac{1}{S_{44}}$$

$$S_{11} = \frac{1}{E}$$

$$S_{12} = \frac{v}{E}$$

$$S_{44} = \frac{1}{G} = \frac{2(1 + v)}{E}$$

The constants $S_{11}$, $S_{12}$, and $S_{44}$ are known as the elastic compliances, $E$ is the elastic modulus, and $v$ is the Poisson's ratio.

## 12.2.3 Semiconductor Crystal Growth

Single-crystal semiconductors are manufactured from the melt or by a variety of vapor deposition processes. Growth from the melt is primarily by the Czochralski (CZ) process or by modifications of the process to accommodate material characteristics and inherent process problems [101]. The process begins with a seed crystal that is essentially free of dislocations. The seed is immersed in the melt and slowly withdrawn with rotation. The crystal grows by solidification of the melt, producing a single crystal up to 6 inches in diameter, depending on the material and process parameters. Growth of crystals with volatile constituents is done through liquid encapsulation of the melt, and the distribution of impurities may be controlled through magnetic stabilization of the melt.

Alternative processes include the float zoning and gradient freeze methods. Float zoning begins with a polysilicon rod. The rod is supported vertically, and a localized melt is created and moved down the length of the rod. Very high purity silicon crystals are grown by this technique [64, 101]. In the gradient freeze technique the container holding the melt is passed through an abrupt temperature change, resulting in solidification of the melt. This technique used with a horizontal boat (horizontal Bridgeman method) is used for the growth of many III-V compounds [101].

## 12.3   CHIP CARRIER AND SUBSTRATE MATERIALS

A chip carrier may be defined as a housing for a semiconductor device or a substrate to which a semiconductor device is attached. A variety of ceramic materials are used and serve as hermetic enclosures or substrates. Alumina and beryllia are among the most common. Other materials serve specialized purposes, including various metals such as molybdenum, copper, and tungsten and copper-tungsten powder composites.

Ceramic insulating materials offer many advantages for packages and substrates. In general, ceramics have low dielectric constants, allowing them to accommodate circuitry without excessive parasitic capacitance. In addition, ceramics have high resistivity and high dielectric strengths. They have low coefficients of thermal expansion (CTEs), making them a good match for mounting silicon and other semiconductors. On the other hand, ceramic materials are brittle and tend to have low toughness, which makes ceramic packages susceptible to shock loads. With a few exceptions, ceramics also tend to be poor thermal conductors.

In general, ceramic materials are processed from finely divided particles into the desired package or substrate configuration. The processes and factors affecting them that lead to a finished ceramic product are complex. However, some general observations can be made. An overview of alumina processes is shown in Figure 12.3 [112]. Ceramic powders are generally mixed with small amounts of glass binders and agglomerated by dry pressing into the final shape or mixed with organic vehicles, cast, and processed into green tape [37, 112, 118]. Green ceramic can be cut to size and stacked to the desired thickness and configuration. Dry-pressed ceramic or green tape is fired or sintered to produce the finished product. The process of sintering occurs through diffusion of the ceramic constituents between particles. Considerable shrinkage occurs with a corresponding reduction in porosity.

The properties of the finished package or substrate are dependent on the constituents of the ceramic and the processing. The uses and properties of selected ceramic substrates are presented in Tables 12.7 through 12.10. Silicate materials, such as steatite, forsterite, and cordierite, are used for a variety of passive device applications such as resistor bodies. Steatite and forsterite are low-loss materials, making them candidates for high-frequency applications. Silicates, however, have low strength and shock resistance relative to other ceramic materials.

Alumina is the most widely used ceramic material for substrates and chip carriers. Alumina has many desirable properties for electronic applications. Alumina is refractory and has relatively good dielectric properties coupled with high strength and excellent shock resistance. However, alumina is a relatively poor thermal conductor. A typical cofired alumina substrate used for surface-mounted devices is shown in Figure 12.4. J-lead ceramic packages manufactured from alumina and hermetically sealed with a leaded glass are shown in Figure 12.5. A typical alumina hybrid substrate is shown in Figure 12.6.

Good thermal conductivity is desirable in electronic applications. Size reduction has led to increased power densities, presenting a challenge to the packaging engineer to develop designs that can rapidly dissipate heat. Beryllia is generally selected for power

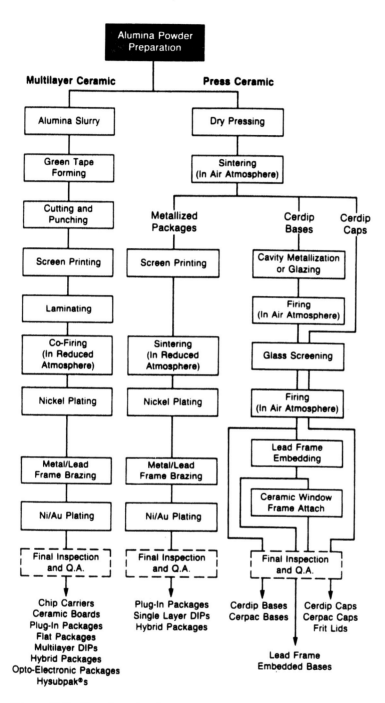

**Figure 12.3**  Flowchart showing the manufacturing steps involved in producing alumina packages and substrates. (Courtesy of Kyocera America.)

**Table 12.7**  Application and Composition of Chip Carriers and Substrates

| Material | Composition | Applications |
|---|---|---|
| Beryllia | BeO | high thermal conductivity, low expansion, power hybrids, HF power transistors, (toxic dust) |
| Silicon carbide | SiC | high thermal conductivity, multichip module substrates |
| Aluminum nitride | AlN | high thermal conductivity, substitute for BeO, multichip modules, power devices, hybrid substrates |
| Alumina | $Al_2O_3$ | low expansion, integrated circuit packages, hybrid substrates |
| Diamond | C | high thermal conductivity IMPATT diodes, passivation for integrated circuits |
| Steatite | $Mg_3Si_4O_{10}(OH)_2$ | thin film resistor substrates, high frequency applications |
| Cordierite | $2MgO \cdot 2Al_2O_3 \cdot 5SiO_2$ | high-frequency applications, resistor bodies |
| Silicon nitride | $Si_3N_4$ | multichip substrates |
| Titanates | $BaTiO_3$ | capacitor dielectrics |
| Forsterite | $2MgO \cdot SiO_2$ | high frequency applications, ceramic-to-metal seals |
| Mullite | $3Al_2O_3 \cdot 2SiO_2$ | cofired substrates, integrated circuit packages |
| Molybdenum | Mo | low expansion metal, power transistor pedestals, diode headers |
| Copper-tungsten (CuW) | 10% Cu-90%W | low expansion, high conductivity, powder metallurgy material, power transistors, hybrids |
| Tungsten | W | low expansion metal, package metallization |

dissipation applications [28]. However, developments in ceramic technology have produced alternative thermally conducting substrate materials, including aluminum nitride and silicon carbide [19, 23, 31]. A comparison of the thermal conductivities of various substrate materials over temperature is shown in Figure 12.7.

Single-crystal silicon has also been employed as substrate material [3, 17, 18, 19]. As a substrate, silicon affords relatively good thermal conductivity. Silicon also matches the CTE of the devices mounted to it. Silicon, however, has a relatively high dielectric constant and is very brittle. Silicon, aluminum nitride, and silicon carbide are particularly well suited for use in high-density multichip module applications as thin-film substrates

Table 12.8  Mechanical Properties of Chip Carriers and Substrates

| Material | Tensile Strength MPa | Compressive Strength MPa | Flexural Strength MPa | Elastic Modulus GPa | Density Kg/m³ | Hardness[a] | Impact Strength Joule | Source |
|---|---|---|---|---|---|---|---|---|
| **Ceramics** | | | | | | | | |
| BeO | 230 | – | 250 | 345 | 3000 | 100K | – | 14, 21, 29, 35, 85 |
| SiC | 17.24 | 490 | 440-460 | 412 | 3160 | 2800K | 65 | 14, 15, 16, 21, 72 |
| AIN | – | 392-441 | 360-410 | 310-343 | 3260 | 1200K | – | 21, 52, 72, 85 |
| Si | – | – | 580 | 190 | 2330 | 850K | – | 14, 16, 21, 25, 60, 64 |
| Si$_3$N$_4$ | 96.5-965 | – | 275-932 | 314 | 2400 | – | – | 21, 35, 72, 85, 14 |
| SiO$_2$ | 96.5-386 | – | 30-100 | 69 | – | – | – | 21, 35, 72, 85 |
| Al$_2$O$_3$ 85% | 124.11 | 1620 | 290 | 221 | 3970 | 9MH | 8.5-8.8 | 14, 15, 16, 34 |
| Al$_2$O$_3$ 90% | 137.9 | 2413 | 317 | 269 | 3970 | 9MH | 8.8 | 14, 15, 16, 34 |
| Al$_2$O$_3$ 92% | 127.56 | 1931 | 321 | 290 | 3970 | – | 8.8-9.2 | 14, 16, 34 |
| Al$_2$O$_3$ 95% | 127.4-193 | 2069-2413 | 310-338 | 296-317 | 3970 | 9MH | 8.8-10.3 | 14, 15, 16, 34 |
| Al$_2$O$_3$ 96% | 127.4 | 2344 | 317 | 310.3 | 3970 | 2000K | 9.5 | 14, 16, 34 |
| Al$_2$O$_3$ 99% | 206.9 | 2586 | 345 | 345 | 3970 | 9MH | 9.5 | 14, 15, 16, 34 |
| Al$_2$O$_3$ 99.5% | 206.9 | 2620 | 345 | 379 | 3970 | – | 8.1 | 14, 16, 34 |
| Al$_2$O$_3$ 99.8% | 206.9 | 2758 | 414 | 386 | 3970 | 93.5RA | 9.5 | 14, 15, 16, 34 |
| Diamond | – | – | 71.4 | – | 3500 | 700K | – | 14, 25, 64 |
| Steatite | 55.2-69 | 448-896 | 110-165 | 90-103 | 2500-2700 | – | 0.4-0.5 | 32, 64 |
| Forsterite | 55.2-69 | 414-690 | 124-138 | 90-103 | 2700-2900 | – | 0.04-0.05 | 32, 64 |
| Titanate | 27.6-69 | 276-827 | 69-152 | 69-103 | 3500-5500 | – | 0.4-0.7 | 32, 64 |
| Cordierite | 55.2-69 | 138-310.3 | 10.3-48 | 13.8-34 | 1600-2100 | – | 0.3-0.34 | 21, 32, 64 |
| Mullite | – | – | 125-275 | 175 | – | – | – | 85, 108 |
| **Metals** | | | | | | | | |
| Mo | 655 | – | – | 324 | 10240 | – | – | 14, 66 |
| CuW (10/90) | 489.5 | – | 1062 | 331 | 17300 | 427K | – | 33, 57, 85, 125 |
| W | 310-1517 | – | 28.3 | 345 | 19300 | 485K | – | 14, 16 |
| Kovar | 552 | – | – | 138 | 8360 | 68RB | – | 9, 51, 53, 58, 59, 65, 85 |

a: K = Knoop; MH = Moh; RA = Rockwell A; RB = Rockwell B

**Table 12.9**  Electrical Properties of Chip Carriers and Substrates

| Material | Resistivity ohm-m | Dielectric Constant @ 1MHz | Dielectric Strength V/mil | Dielectric Loss $10^{-4}$@ 1MHz | Dielectric Permitivity | Source |
|---|---|---|---|---|---|---|
| **Ceramics** | | | | | | |
| BeO | $10^{11}$-$10^{12}$ | 7.0-8.9 | 19.7 | 4-7 | 6.6-6.8 | 21, 23, 29, 31, 40, 75, 85, 123 |
| SiC | $> 10^{11}$ | 20-42 | 1 | 500 | 42 | 21, 23, 31, 40, 72, 75 |
| AlN | $> 10^{11}$ | 8.5-10 | 14 | 5-10 | 8.9 | 21, 23, 29, 31, 72, 75, 85, 123 |
| Si | 640 | – | – | – | 12 | 13, 14, 21, 25 |
| $Si_3N_4$ | $10^{10}$ | 6-10 | 5000 | – | 5-8 | 21, 35, 72, 75, 85 |
| $SiO_2$ | $>10^{14}$ | 3.5-4 | 5000 | – | 3.8 | 21, 35, 72 |
| $Al_2O_3$ | $> 10^{12}$ | 9.9-10 | 8.3 | 3-5 | 8.5-9.4 | 16, 21, 23, 31, 40, 54, 75, 79, 121, 123 |
| Diamond | $> 10^4$ | – | – | – | – | 85 |
| Steatite | $10^{13}$-$10^{15a}$ | 5.5-7.5 | 200-350 | 8-35 | – | 16, 31, 32 |
| Forsterite | $10^{13}$-$10^{15a}$ | 6.2 | 200-300 | 3 | – | 31, 32 |
| Titanate | $10^{8}$-$10^{15a}$ | 15-10000 | 50-300 | 2-500 | – | 31, 32 |
| Cordierite | $10^{12}$-$10^{14a}$ | 4.5-5.5 | 40-100 | 40-100 | 5 | 21, 31, 32 |
| Mullite | $> 10^{12}$ | 6.8 | – | – | – | 85, 108 |
| **Metals** | | | | | | |
| Mo | $5.2 \times 10^{-8}$ | – | – | – | – | 11 |
| CuW (10/90) | $6.6 \times 10^{-8}$ | – | – | – | – | 125 |
| W | $5.5 \times 10^{-8}$ | – | – | – | – | 11, 13, 16 |
| Kovar | $50 \times 10^{-8}$ | – | – | – | – | 11, 13, 52, 112 |

a = in Ohm/cm$^3$ as reported in source 32.

[17, 18, 19, 40]. As shown in Figure 12.7, aluminum nitride and silicon carbide closely match the CTE of silicon over temperature. A complex multichip module employing a silicon substrate is shown in Figure 12.8.

In certain applications where electrical conductivity is required, metals are selected as chip carrier materials. Copper, molybdenum, tungsten, and Kovar may be used for this purpose. The large difference in CTE between silicon and copper requires the use of compliant high-temperature solder to bond the die to the substrate. Under power cycling conditions, however, solder will fatigue, presenting a potential reliability problem. Low-expansion refractory metals, such as molybdenum and tungsten, have relatively poor thermal conductivity in comparison to copper. Powder metallurgy composites of tungsten and copper offer an alternative when higher thermal conductivity is desired.

## 12.4  PRINTED CIRCUIT LAMINATES

The printed circuit board typically consists of a copper circuit pattern created on a copper-clad composite laminate by lithography and electroplating technologies [64, 65, 85, 110]. The type of laminate is selected on the basis of electrical, mechanical, and thermal characteristics. A variety of laminate materials are used in electronic packaging for printed circuitry. Selected laminate types and their applications are shown in Table 12.11. The thermal, electrical, and mechanical properties of selected laminates are shown in Tables 12.12 to 12.14. Each laminate type has its own unique set of problems and the packaging engineer must compromise.

**Table 12.10**   Thermal Properties of Chip Carriers and Substrates

| Material | Conductivity W/m-K | Coeff. of expansion ppm/°C | Heat Capacity J/Kg-K | Max. use Temp. °C | Melting Point K | Source |
|---|---|---|---|---|---|---|
| **Ceramics** | | | | | | |
| BeO | 150-300 | 6.8-7.5 | 1047-2093 | – | 2725 | 14, 17, 19, 21, 23, 31, 40, 61, 75 |
| SiC | 120-270 | 3.5-4.6 | 675 | > 1000 | 3100 | 14, 16, 17, 19, 21, 23, 40, 61, 72, 75 |
| AlN | 170-260 | 4.3-4.7 | 745 | > 1000 | 2677 | 17, 19, 21, 23, 52, 72, 75 |
| Si | 125-148 | 2.33 | 712 | – | 1685 | 13, 14, 16, 20, 21, 25, 56, 60, 64, 87, 90 |
| $Si_3N_4$ | 25-35 | 2.8-3.2 | 691 | > 1000 | 2173 | 21, 23, 72, 75, 14, 52 |
| $SiO_2$ | 1.5 | 0.6 | – | > 800 | – | 21, 72 |
| $Al_2O_3$ | 15-33 | 4.3-7.4 | 765 | 1600 | 2323 | 4, 14, 21, 23, 40, 54, 56, 57, 60, 75, 79 |
| Quartz | 43[a] | 1.0 | 816-1193 | – | N/A | 61 |
| Diamond | 2000-2300 | 1.0 | 509 | – | N/A | 14, 23, 25, 61 |
| Steatite | 2.1-2.5 | 8.6-10.5 | – | 1000-1100 | – | 32, 61 |
| Forsterite | 2.1-4.2 | 11 | 118087 | 1000-1100 | – | 32, 61 |
| Titanate | 3.3-4.2 | 7-10 | – | – | – | 32 |
| Cordierite | 1.3-1.7 | 2.5-3 | 770 | 1250 | – | 21, 23, 32, 61 |
| Mullite | 5.0 | 4.2 | – | – | – | 57, 125 |
| **Metals** | | | | | | |
| Mo | 138 | 3-5.5 | 251 | – | 2894 | 11, 14, 17, 18, 19, 66 |
| CuW (10/90) | 209.3 | 6.0 | 209 | – | – | 57, 109 |
| W | 174 | 4.5 | 132 | – | 3660 | 11, 14, 16, 25, 54, 57 |
| Kovar | 15.5-17 | 5.87 | 439 | – | 1450 | 9, 11, 17, 85, 94 |

a. Parallel to C-axis

**Figure 12.4**   Cofired alumina substrates. (Courtesy of Unisys NASA/GSFC.)

**Figure 12.5**   Alumina J-lead surface mount chip carriers. (Courtesy of Unisys NASA/GSFC.)

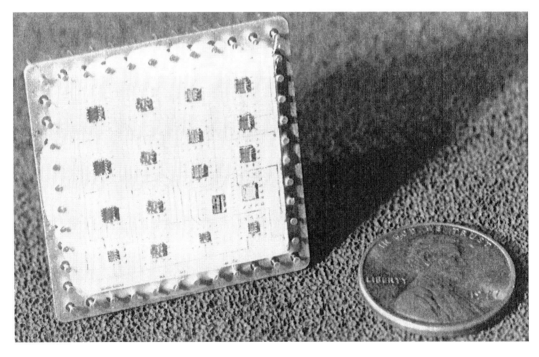

**Figure 12.6**   Typical hybrid with an alumina substrate. (Courtesy of E-Systems.)

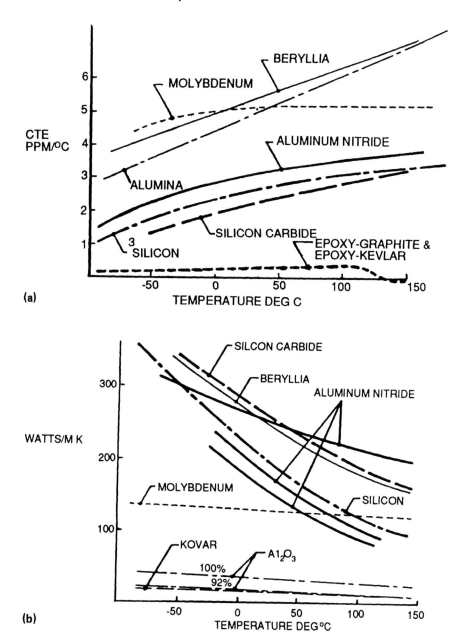

**Figure 12.7** (a) Thermal expansion of typical packaging materials; (b) thermal conductivity of typical packaging materials.

Epoxy-glass laminates are among the most widely used circuit board materials [64, 85]. Typical circuit board laminates employ resins based on diglycidyl ether of tetrabromobisphenol A (DGEBA) epoxy with a woven "E-glass" reinforcement [85]. The relatively low dielectric constant and the processability of the glass-epoxy system make it ideal for general usage. These laminates are chemically resistant and able to withstand

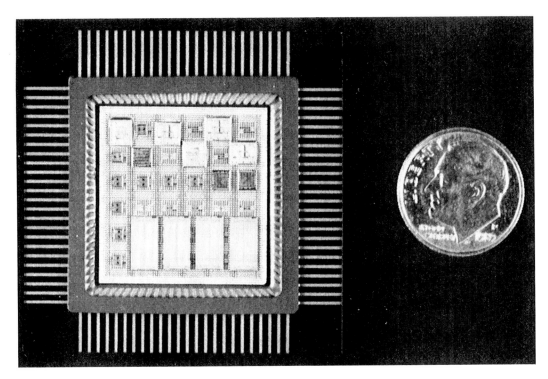

**Figure 12.8**  Multichip module (MCM) employing a silicon substrate in an alumina package. (Courtesy of E-Systems.)

electroplating processes without sustaining damage. They are easily machined and laminated together to create multilayer structures. However, standard NEMA laminates, such as G-10 and FR-4, have a high CTE, which presents problems for utilization of these materials for applications such as surface mounting. In addition, the glass transition temperature is about 125°C.

Polyimide- and PTFE-based systems are selected for their superior thermal or electrical characteristics. Polyimide is very heat resistant and will not sustain damage from most assembly processes. Thus, the polyimide-glass system is less affected by local applications of heat by soldering irons during rework and touch-up. Assembled polyimide printed wiring boards are shown in Figure 12.9. These boards are mixed surface mount and through hole technology and require a significant amount of processing, making polyimide a good choice. Polyimide films are also utilized for flexible printed circuits. An example is shown in Figure 12.10. Polyimide is expensive and tends to be difficult to machine [110].

PTFE (Teflon) systems are ideal for microwave applications. They have a low dielectric constant (2.1 to 2.2 at 1 MHz), which reduces propagation delays for very high speed circuitry. PTFE systems tend to have a very high CTE. Fortunately, this does not always preclude their application because common nonwoven glassfiber–reinforced materials are not very stiff and have a low in-plane modulus of elasticity [12, 64, 110].

With the advent of surface mounting, several new materials have been developed in an effort to "tailor" the coefficient of thermal expansion to that of ceramic chip carrier materials. Epoxy with aramid fiber (Kevlar) reinforcement is one of the more popular

**Table 12.11**  Laminate Applications

| Materials | Applications |
|---|---|
| Polyimide-glass | good repairability, high temperature, high glass transition temperature |
| Epoxy-glass (G-10, FR-4) | most common PWB material, low-cost, FR4-fire retardant |
| PTFE—Glass | low dielectric material, high frequency circuitry |
| Epoxy-aramid | low expansion, surface mount laminate |
| Epoxy-quartz | low expansion, surface mount laminate |
| Polyimide-aramid | low expansion, surface mount laminate |
| Polyimide-quartz | low expansion, surface mount laminate |
| Metal composites (Cu/Invar/Cu) (Ni/Mo/Ni) (Cu/Mo/Cu) | low expansion for SMT applications, clad with polyimide-glass |

systems [85, 110]. The superior properties of epoxy are retained, while the aramid fiber reinforcement substantially reduces the CTE. Aramid reinforcement is also used with polyimide resins. Quartz reinforcement also serves to reduce the CTE. A more complex method of reducing the CTE is to laminate polyimide to a constraining metal core such as copper-Invar-copper, copper-molybdenum-copper, or nickel-molybdenum-nickel. However, metal core boards tend to be very difficult to process through conventional printed circuit manufacturing operations.

As indicated by the discussion already presented, the properties of a laminate are derived from the reinforcement and the resin. Laminates are generally orthotropic with respect to most properties. The volume fraction of the resin ($V_{rsn}$) of the laminate is the key to determining the properties of the system. The dielectric constant can be estimated within ±15% from the following equation [85]:

$$\kappa_{lam} = \kappa_{rsn}V_{rsn} + \kappa_{rnf}V_{rnf} \tag{12.1}$$

$\kappa_{rsn}$ = dielectric constant of the resin

$\kappa_{rnf}$ = dielectric constant of the reinforcement

where

$$V_{rsn} = 1 - V_{rnf} \tag{12.2}$$

and

$V_{rsn}$ = volume fraction of resin

$V_{rnf}$ = volume fraction of reinforcement

A typical laminate may have 37 to 73% resin content, depending on the resin type

Table 12.12  Mechanical Properties of Printed Circuit Laminates

| Material | Tensile Strength MPa | Yield Strength MPa | Elongation % | Flexural Strength MPa | Impact Strength Joule | Elastic Modulus GPa | Density kg/m³ | Typical Resin Content % | Source |
|---|---|---|---|---|---|---|---|---|---|
| Polymer Composites | | | | | | | | | |
| Polyimide Glass | 345 | – | – | 503 | – | 19.3-20.6 | 2214 | 55 | 6, 39, 47, 64, 110 |
| Epoxy Glass[a] | 276 | – | – | 310-379 | 7.4-9.5 | 17.2 | 1938 | – | 6, 39, 47, 65, 64 |
| PTFE[c] Glass, non-woven | 38-52 | – | – | 69-103 | 1.8-3.4 | 1.38 | 2150 | 73 | 6, 39, 47, 64, 85 |
| Epoxy Aramid | – | – | – | – | – | 30.3 | – | 60 | 39, 47, 110 |
| Epoxy Quartz | – | – | – | – | – | – | – | 64 | 110 |
| Polyimide Aramid | 207 | – | – | – | – | 20.7-27.6 | 2214 | – | 6, 39, 47 |
| Polyimide Quartz | – | – | – | – | – | – | 1938 | 46 | 6, 110 |
| Epoxy-Cordierite | – | – | – | 117-176 | 0.3-1.1 | – | 1800-1900 | 30 | 112 |
| PTFE[c] Glass, Woven | 68-103 | – | – | 76-117 | 5.4-8.1 | – | 2200 | 73 | 65, 85, 110 |
| Modified Epoxy Aramid | – | – | – | – | – | – | – | 61 | 110 |
| PTFE[c] Quartz | – | – | – | – | – | – | – | 67 | 110 |
| Polyimide | 34.5-96.5 | – | 6-7 | 48-96.5 | 1.1-1.5 | 3.2-5.2 | – | 100 | 1, 12 |
| Metal Composites | | | | | | | | | |
| Cu/Invar/Cu (20/60/20) | 310-414 | 170-270 | 36 | – | – | 131-135 | 8430 | N/A | 6, 111 |
| Cu/Invar/Cu (12.5/75/12.5) | 380-480 | 240-340 | – | – | – | 140 | 8330 | N/A | 111 |
| Cu/Mo/Cu | – | – | 50 | – | – | 260-280 | 44288 | N/A | 110, 111 |
| Ni/Mo/Ni | 621 | 552 | 50 | – | – | 310 | 10241.6 | N/A | 6 |

a.  FR-4, G-10    c.  PTFE = Polytetrafluoroethylene

**Table 12.13**  Electrical Properties of Printed Circuit Laminates

| Material | Resistivity ohm-m | Dielectric Constant @ 1 MHz | Dielectric Strength V/mil | Dissipation Factor @ 1 MHz | Source |
|---|---|---|---|---|---|
| **Polymer Composites** | | | | | |
| Polyimide Glass | $10^{12}$-$10^{13}$ | 3.9-4.8 | 140-300 | 0.00445 | 24, 39, 47, 64 |
| Epoxy Glass[a] | $10^{10}$ | 4.5-5.2 | 45-400 | 0.01-0.025 | 24, 39, 47, 64, 81, 111 |
| Modified Epoxy[b] Glass | – | 4.1-4.2 | 1200-1400 | 0.028-0.03 | 85 |
| PTFE[c] Glass, Non-woven | $10^8$ | 2.1-2.2 | 2000 | 0.0005-0.001 | 24, 39, 47, 64, 65, 111 |
| PTFE[c] Glass, Woven | – | 2.5-2.8 | – | 0.002-0.006 | 85, 110 |
| Epoxy Aramid | $10^{14}$ | 3.7-3.9 | – | 0.0097 | 39, 47, 110, 111 |
| Polyimide Aramid | $10^{10}$-$10^{11}$ | 3.6 | – | – | 6, 39, 47 |
| Polyimide Quartz | $10^8$-$10^9$ | 3.8-4.0 | – | – | 34, 47, 85, 111, 112 |
| Modified Epoxy Aramid | – | 4.0 | – | 0.0024 | 110 |
| PTFE[c] Quartz | – | 2.4 | – | 0.00042 | 110 |
| Polyimide (Flex) | $10^{16}$ | 3.5-3.9 | 310-560 | 0.004-0 011 | 1, 12, 72 |
| Epoxy Cordierite | – | 4.0 | – | – | 112 |
| **Metal Composites** | | | | | |
| Cu/Invar/Cu (20/60/20) | $4.3$-$5.5 \times 10^{-8}$ | N/A | N/A | N/A | 6,111 |
| Cu/Invar/Cu (12.5/75/12.5) | $7 \times 10^{-8}$ | N/A | N/A | N/A | 111 |
| Cu/Mo/Cu | $3.5 \times 10^{-8}$ | N/A | N/A | N/A | 111 |
| Ni/Mo/Ni | $6 \times 10^{-8}$ | N/A | N/A | N/A | 6 |

a. FR-4, G-10    b. Polyfunctional FR-4    c. PTFE = Polytetrafluoroethylene

and desired properties [85, 110]. E-glass has a dielectric constant of 5.80 at 1 MHz, while epoxy resins for printed circuit applications will have dielectric constants ranging from 3.5 to 3.6 at 1 MHz. The benefit for electrical performance of decreasing the fiber volume fraction or increasing the resin content is apparent.

The effects of resin content or fiber volume fraction on the mechanical properties of epoxy-glass laminates have been investigated by numerical techniques [126]. As Figures 12.11 through 12.13 show, increasing the fiber volume fraction or decreasing the resin content has a very significant effect on the CTE and laminate modulus. In this case, the out-of-plane expansion will dramatically increase as the resin content is increased or fiber volume fraction is decreased. This will affect the reliability of the plated through holes used in the finished board. These calculations are based on the resin and reinforcement properties shown in Table 12.15.

The mechanical properties of laminated multilayer structures, such as constrained metal core boards, depend on the properties of the layers used and their relative thicknesses. The following equation can be used to estimate the in-plane CTE of laminated structures under the assumption that warping is negligible and the materials obey Hooke's law:

$$\alpha_{lam} = \frac{\sum_{i=1}^{n} \alpha_i E_i t_i}{\sum_{i=1}^{n} E_i t_i} \tag{12.3}$$

**Table 12.14**  Thermal Properties of Printed Circuit Laminates

| Material | Conductivity W/m-K | CTE x,y Dir. ppm/°C | CTE z Dir. ppm/°C | Heat Capacity J/Kg-K | Water Absorption % | Max. Use Temperature °C | Glass Transition Temp. °C | Source |
|---|---|---|---|---|---|---|---|---|
| **Polymer Composites** | | | | | | | | |
| Polyimide Glass | 0.35 | 12-16 | 40-60 | – | 0.35 | 215-280 | 250-260 | 39, 47, 81, 112 |
| Epoxy Glass[a] | 0.16-0.2 | 14-18 | 180 | 878.6 | 0.1 | 130-160 | 125-135 | 26, 39, 47, 93, 64, 111 |
| Modified Epoxy[b] Glass | – | 14-16 | – | – | – | – | 140-150 | 85 |
| PTFE[e] Glass, Non-woven | 0.1-0.26 | 20 | – | 962.3 | 1.1 | 230-260 | – | 39, 47, 65, 111 |
| PFTE[e] Glass, Woven | 419-837 | 10-25 | – | 836.8 | 0.02 | 248 | – | 65, 110 |
| Epoxy Aramid | 0.12 | 6-8 | 66 | – | 0.85 | – | 125 | 27, 39, 47, 111 |
| Epoxy Quartz | – | 6-13 | 62 | – | – | – | 125 | 27, 39, 47, 111 |
| Polyimide Aramid | 0.28 | 5-8 | 83 | – | 1.5 | – | 250 | 6, 27, 39, 47, 111 |
| Polyimide Quartz | 0.35 | 6-12 | 35 | – | 0.5 | – | 188-250 | 39, 47, 111 |
| Modified Epoxy Aramid | – | 5.5-5.6 | 100 | – | – | – | 137 | 110 |
| PFTE[e] Quartz | – | 7.5-9.4 | 88 | – | – | – | 19[d] | 110 |
| Polyimide (Flex) | 4.3-11.8 | 45-50 | – | 1130-1298 | 0.24-0.47 | 260-315 | – | 1, 12 |
| Epoxy-Cordierite | 0.9-1.3 | 3.3-3.8 | – | – | – | – | – | 112 |
| **Metal Composites** | | | | | | | | |
| Cu/Invar/Cu (20/60/20) | 15-18[c] | 5.3-5.5 | 16 | 4590 | N/A | – | N/A | 6, 111 |
| Cu/Invar/Cu (12.5/75/12.5) | 14[c] | 4.4 | – | 4840 | N/A | – | N/A | 111 |
| Cu/Mo/Cu | 90-174 | 2-6 | – | – | N/A | – | N/A | 110, 111 |
| Ni/Mo/Ni | 129.8[c] | 5.2-6 | 5.2-6.0 | – | N/A | – | N/A | 6 |

a.  FR-4, G-10    b.  Polyfunctional FR-4    c.  z-direction    d.  Polymorphic phase transformation    e.  PTFE = Polytetrafluoroethylene

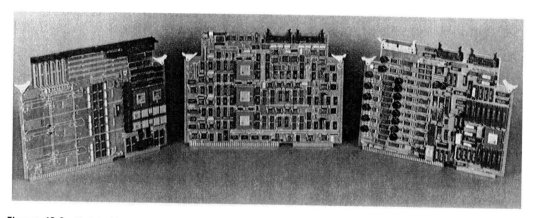

**Figure 12.9** Polyimide printed wiring boards populated with surface mount and through hole components. (Courtesy of E-Systems.)

where

$\alpha_{lam}$ = CTE of laminated structure

$\alpha_i$ = CTE of $i$th layer

$E_i$ = elastic modulus of $i$th layer

$t_i$ = thickness of $i$th layer

$n$ = number of layers

The assumption of negligible warping is valid only for symmetrical structures.

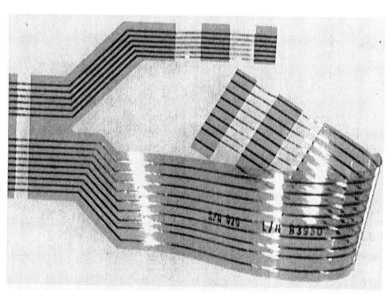

**Figure 12.10** Polyimide flex circuit. (Courtesy of E-Systems. Photo courtesy of B. Munoz, Unisys.)

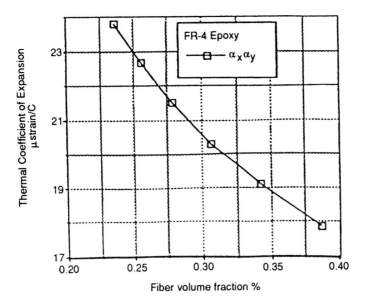

**Figure 12.11**  CTE versus fiber volume fraction of epoxy-glass laminates.

## 12.5  ENCLOSURE MATERIALS FOR HYBRIDS AND SYSTEMS

Substrates bearing semiconductor devices and circuitry are housed in metal packages. Likewise, printed wiring assemblies, consisting of numerous active and passive devices,

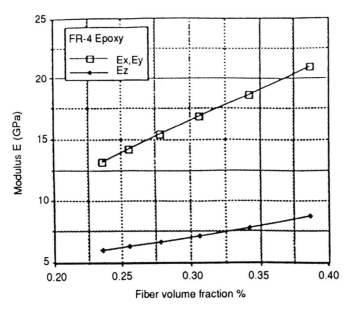

**Figure 12.12**  Extensional modulus versus fiber volume fraction.

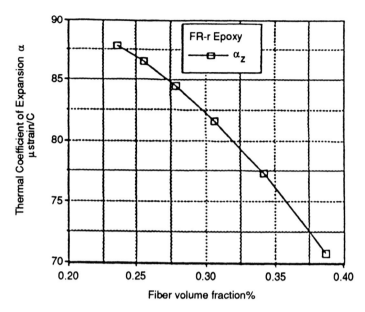

**Figure 12.13**   TCE versus fiber volume fraction in the z direction
of epoxyglass.

**Table 12.15**   Reinforcement and Resin Properties for Epoxy-Glass Laminates

| E glass | Epoxy |
|---------|-------|
| E = 72.4 GPa | E = 3.45 GPa |
| $\nu$ = 0.22 | $\nu$ = 0.37 |
| $\alpha$ = 5.4 x $10^{-6}$/°C | $\alpha$ = 69 x $10^{-6}$/°C |

must be housed in a chassis. Hybrid substrates, power semiconductors, and various passive devices may be enclosed in hermetic packages manufactured from iron, copper, and nickel alloys. A typical hybrid enclosure housing a gallium arsenide monolithic microwave IC (MMIC) is shown in Figure 12.14.

Chassis structures for military or aerospace uses are generally manufactured from aluminum alloys. They offer suitable weight-to-strength ratios for many applications in electronic packaging. Aluminum alloys may be welded, brazed, or cast. A typical aluminum chassis and housing designed for rack mounting is shown in Figure 12.15. A typical rack system consisting of several standard chassis is shown in Figure 12.16. When weight is not a consideration, as with chassis and electronic cabinets for ground uses, low-carbon steels may be more economical. Properties of selected alloys used for chassis, cabinets, and chassis hardware, as well as hybrid enclosures, are shown in Tables 12.16 through 12.19.

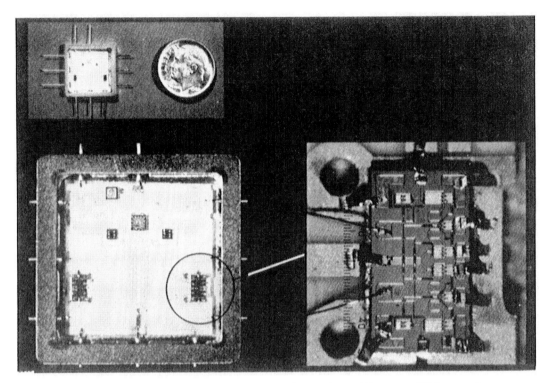

**Figure 12.14**   GaAs MMIC device housed in a copper alloy package. (Courtesy of E-Systems.)

## 12.6  JOINING AND BONDING MATERIALS

Joining and bonding materials are very important in electronic packaging. They are often critical to the overall reliability of the electronic system. They appear in all levels of packaging. Semiconductor devices must be brazed, soldered, or adhesive bonded to their substrates. Individually packaged devices must be soldered to printed circuit boards. Enclosures are often sealed by soldering or brazing, and substrates are joined to their enclosures by soldering, brazing, or adhesive bonding.

### 12.6.1  Common Factors

The formation of a reliable joint depends on good wetting between the joining material and workpieces. Wetting must take place to allow the joining material to interact and bond with the workpiece surface. This applies to adhesive bonding as well as metallurgical processes. Wetting is also necessary for capillary action to take place. Capillary action allows solders and brazing materials to fill a joint gap.

The mechanics of wetting depends on the surface tensions or surface energies of the materials involved in the joining process [41]. The concept of wetting can be illustrated by considering a droplet of a joining alloy or adhesive on a solid surface, as shown in Figure 12.17. At thermodynamic equilibrium there is a balance between the forces involved as expressed by the following equation:

$$\gamma_{13} = \gamma_{12} \cos\theta + \gamma_{23} \qquad (12.4)$$

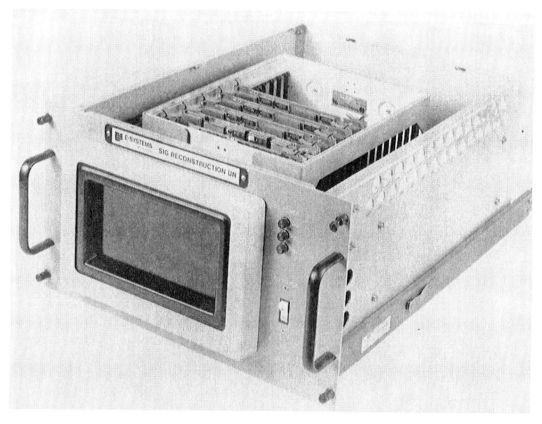

**Figure 12.15** Typical standard rack-mount aluminum chassis. (Courtesy of E-Systems.)

The dihedral angle, $\theta$, is a measure of the degree of wetting. A small dihedral angle is desirable in the joining process and indicates good wetting and hence a reliable joint.

Several factors affect the degree of wetting during bonding. The principal factor is the surface condition of the base materials [41, 44, 70, 73]. Surface cleanliness is essential, since the presence of soils will impede the wetting process and prevent bonding. In the case of soldering and brazing materials, surface oxides or in some cases sulfides on the base metal must be reduced. Fluxes are used for this purpose. However, a suitable flux must be selected for the severity of the surface oxidation and types of base metals being joined [41, 73].

Another common factor in joining is the effect of joint geometry on the properties of the joint. For example, joint thickness can significantly affect the mechanical properties of the joint [41, 73, 85]. As Figure 12.18 indicates, the optimum shear strength for eutectic solder on copper is approximately 34 MPa at a thickness of 0.076 mm [41].

Similar behavior is exhibited by braze joints of Cu-Ag-Zn brazing alloys. As Figure 12.19 shows, the tensile strength of a braze joint is also dependent on thickness. The optimum joint strength for brazed stainless steel is 930 MPa at a joint thickness of 0.038 mm [73]. Polymer adhesive joints also exhibit this behavior. In the case of die bonding, the optimum shear strength occurs at a bond line thickness of 25 $\mu$m [85]. These effects must be considered in the design of a joint.

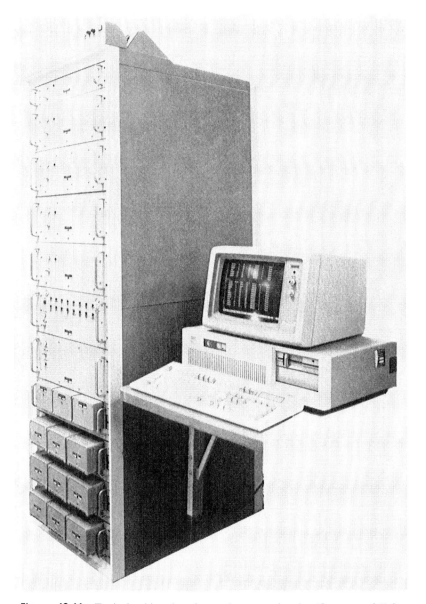

**Figure 12.16**  Typical cabinet housing rack-mount chassis. (Courtesy of E-Systems.)

## 12.6.2  Soldering and Brazing Alloys

Applications of selected solders are shown in Table 12.20. The nominal composition of various alloys is shown in Table 12.21. The mechanical, electrical, and thermal properties of selected solder alloys are surveyed in Tables 12.22 through 12.24. Creep properties of some tin-lead and tin-lead-silver alloys are shown in Table 12.25, and the melting range and composition of several silver brazing alloys are shown in Table 12.26.

**Table 12.16**  Materials for Enclosure Applications

| Materials | Alloy Designations | Applications |
|---|---|---|
| Aluminum alloys | 6061, 5052, 2024, 380, 356, 308, 208 | low density, aerospace chassis, housings |
| Low carbon steels | 1010, 1020 | ground applications, housings, headers for transistors |
| Stainless steel | 304, 304L, 316 | corrosion resistant, fasteners, chassis components |
| Copper alloys | C11000 | hybrid enclosures |
| Nickel alloys | Kovar, Ni42, Ni200 | hybrid enclosures, integrated circuit and transistor packages |
| Molybdenum | – | multichip module package components |

Tin-lead alloys are the most commonly used soldering materials in the electronics industry. A wide variety of alloys are available covering nearly the entire spectrum of the Sn-Pb phase diagram. 63%Sn-37%Pb (alloy Sn63) is generally referred to as the eutectic. However, recently the actual eutectic composition has been shown to be 61.9% Sn [85]. Near-eutectic alloys (Sn60, Sn63) are most commonly used for joining components to circuit boards and are well suited to machine soldering and mass soldering methods. High-lead solders (Sn5, Sn10) are generally used when a higher melting point material is required, as in joining a power transistor die to a copper header. The tin-lead phase diagram is shown in Figure 12.20.

The mechanical properties of tin-lead alloys depend on several factors. Measured properties are dependent on strain rate, temperature, and microstructure [41]. Bulk properties may differ significantly from joint properties. Actual joint properties will also depend on the materials being joined because different type and amounts of intermetallic compound will form [41, 44, 85]. For example, joints that bond copper alloys may differ from joints that bond nickel materials, in part because of the kinetics of intermetallic compound formation. Joint properties will also be affected by the amount of time the base metals are in contact with the molten solder. Soldering time affects intermetallic compound formation as well as the amount of base material dissolved by the solder.

Dissolution of some base materials by the joining alloy, such as gold in tin-lead, has a severe effect on the properties. As shown in Figure 12.21, the ductility of solder is severely reduced by the presence of a relatively small amount of gold. This effect is known as gold embrittlement. Generally, soldering gold with tin-lead alloys is avoided for this reason.

Tin-lead alloys are used at high temperatures relative to their melting point. Consequently, they are subject to creep or stress relaxation [41, 43, 44, 68]. The interaction of creep with cyclic loading (fatigue) is the dominant wearout mechanism in surface mount assemblies [4, 7, 26, 27]. The process of stress relaxation during temperature cycling of these assemblies results in the maximum amount of cyclic damage to the surface-mounted joint. Careful consideration must be given to this behavior in the design of surface-mounted assemblies. Figure 12.22 shows creep-fatigue damage in alloy Sn63 surface-mounted joints on an alumina substrate. In this case, the primary source of strains in the joint was

Table 12.17   Mechanical Properties of Enclosure Materials

| Material | Tensile Strength MPa | Yield Strength MPa | Shear Strength MPa | Elongation % | Elastic Modulus GPa | Endurance Limit MPa | Density Kg/m³ | Hardness[a] | Source |
|---|---|---|---|---|---|---|---|---|---|
| Al 6061-T6 | 310.3 | 276 | 207 | 12 | 69 | 96.5 | 2713 | 95B | 1, 16, 25, 60 |
| Al 6061-T4 | 241.3 | 145 | 165 | 22 | 69 | 90 | 2713 | 65B | 1, 16, 25, 60 |
| Al 5052-H34 | 262 | 214 | 145 | 10 | 69 | 124 | 2685 | 68B | 1, 16, 25, 60 |
| Al 5052-H38 | 290 | 256 | 165 | 7 | 69 | 138 | 2685 | 77B | 1, 16, 25, 60 |
| Al 2024-T4 | 469 | 324 | 290 | 20 | 69 | 138 | 2768 | 120B | 1, 16, 25, 60 |
| Al 2024-T6 | 476 | 393 | 283 | 10 | 69 | 124 | 2768 | 125B | 1, 14, 16, 60 |
| Al 380 | 324 | 172.4 | – | 3 | 69 | – | 2685 | 80B | 1, 16, 60 |
| Al 356-T6 | 227.5-256 | 165-186 | 138-207 | 3.5-5 | 69 | – | 2685 | 90B | 1, 16, 60 |
| Al-308 | 193 | 110 | 152 | 2 | 69 | – | 2796 | 70B | 1, 16, 60 |
| Al 208 | 145 | 96.5 | 117 | 2.5 | 69 | – | 2796 | 55B | 1, 16, 60 |
| Kovar | 552 | 343 | – | – | 138 | – | 8360 | 68B | 9, 51, 53, 58, 59, 65, 66 |
| W | 310-1517 | 103-1517 | – | – | 379-407 | – | 19238 | 350-480V | 1, 14, 16 |
| Ni 200 | 345-758.5 | 69-690 | – | 3-60 | 204-207 | – | 8885 | 75-230B | 1, 14, 16 |
| Ni42/Fe58 | 565 | 276 | – | 30 | 145 | – | 8110 | 76RB | 53, 59, 62, 65, 85 |
| ETPCu C11000[b] | 220.6-241.3 | 69-75.8 | 152-165 | 45-55 | 117 | 75.8 | 8858 | 40-45RF | 1, 60 |
| Steel 1020 | 379 | 207 | – | 25 | 207 | – | 7870 | 111B | 16, 62 |
| Steel 1010 | 324 | 179 | – | 28 | 207 | – | 7860 | 95B | 16, 53, 59, 62 |
| Stainless Steel 304 | 579 | 289 | – | 55 | 193 | – | 8027 | 79-143RB | 1, 58 |
| Mo | 655 | 552 | – | – | 324 | – | 10240 | 270V | 14, 66 |

a.   B = Brinnell,  V = Vicker,  RB = Rockwell B,  RF = Rockwell F        b.   Annealed

**Table 12.18** Electrical Properties of Enclosure Materials

| Material | Conductivity micro mho-cm | Resistivity $10^{-8}$ ohm-m | Source |
|---|---|---|---|
| Al Pure | 0.372 | 2.8 | 11, 13, 16, 40 |
| Al 6061 | 0.263 | 3.8 | 1 |
| Al 5052 | 0.204 | 4.9 | 1 |
| Al 2024 | 0.174-0.29 | 3.45-5.75 | 1 |
| Al 380 | 0.147 | 6.8 | 1 |
| Al 356-T6 | 0.244 | 4.1 | 1 |
| Al 308 | 0.217 | 4.6 | 1 |
| Al 208 | 0.182 | 5.5 | 1 |
| Kovar | 0.024 | 50 | 11, 13, 52, 112 |
| W | 0.177 | 5.5 | 5, 11, 13, 72 |
| Ni 200 | 0.105-0.118 | 8.5-9.5 | 1, 11, 13 |
| Ni42/Fe58 | 0.021 | 48.4 | 53 |
| ETP Cu C11000 | 0.588 | 1.7 | 1, 5 |
| Steel 1020 | 0.059 | 16.9 | 16 |
| Steel 1010 | 0.059 | 16.9 | 16, 53 |
| Stainless Steel 304 | 0.014 | 72 | 1 |
| Mo | 0.192 | 5.2 | 1 |

**Table 12.19** Thermal Properties of Enclosure Materials

| Material | Conductivity W/m-K | Coefficient of Expansion ppm/°C | Heat Capacity J/Kg-K | Melting Point K | Source |
|---|---|---|---|---|---|
| Al Pure | 237 | 25 | 903 | 933 | 14, 16, 25, 60, 85, 94 |
| Al 6061 | 180 | 23.4-25 | 963 | 855-922 | 1, 6, 85 |
| Al 5052 | 138 | 23.8 | 921 | 866-922 | 1, 85 |
| Al 2024-T6 | 189 | 23.2-24.7 | 921 | 775-911 | 1, 14, 16, 25 |
| Al 195 | 168 | 25 | 883 | — | 14, 16, 25 |
| Al 380 | 100 | 21.1 | — | 866-811 | 1 |
| Al 356-T6 | 159 | 21.4 | — | 886-830 | 1, 85 |
| Al 308 | 142 | 21.4 | — | 886-794 | 1 |
| Al 208 | 121 | 22 | — | 922-844 | 1 |
| Kovar | 16.3 | 5.87 | 439 | 1450 | 9, 11, 17, 18, 52, 53, 85 |
| W | 174 | 4.5 | 132 | 3660 | 1, 14, 16, 18 |
| Ni | 90.7 | 15.3 | 444 | 1708-1719 | 1, 14, 16 |
| Ni42/Fe58 | 129 | 4.45 | 477 | — | 53, 62, 65 |
| ETP Cu C11000 | 391 | 17.6 | 385 | 1338-1356 | 1 |
| Steel 1020 | 50 | 11.7 | | | 16, 85 |
| Steel 1010 | 63.9 | 11.3 | 434 | | 16, 53, 62, 85 |
| Stainless Steel 304 | 16.3 | 17.2 | 502 | 1672-1727 | 1 |
| Mo | 138 | 5.04 | 251 | 2894 | 1, 14, 68, 85 |

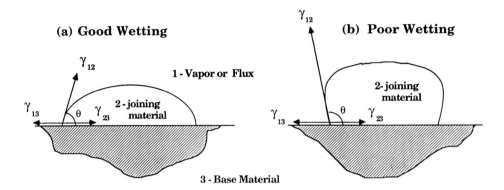

**Figure 12.17**   Wetting of a base material by a droplet of joining material. θ, the dihedral angle, is a measure of the degree of wetting. The vectors shown represent the forces developed as a result of the surface tension, γ.

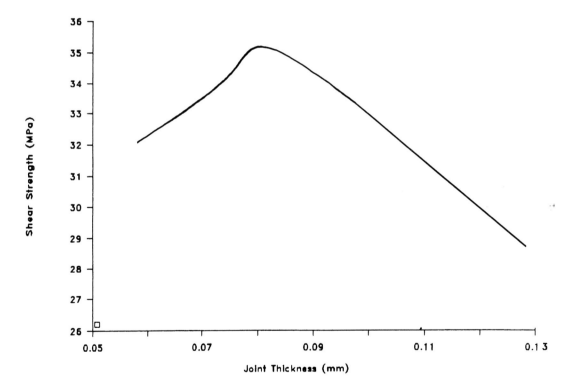

**Figure 12.18**   Effect of joint thickness on the shear strength of solder joints.

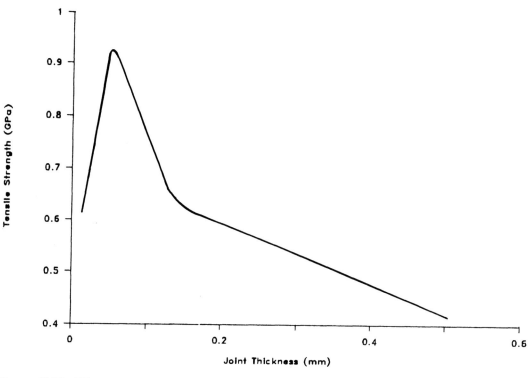

**Figure 12.19**   Effect of joint thickness on silver braze joints.

**Table 12.20**   Applications for Joining Alloys

| Alloy Types | Alloys | Applications |
|---|---|---|
| Tin-lead | Sn63, Sn62, Sn60[a] | soldering components to printed circuitry |
| Lead, lead-silver, Lead-antimony | Sn10, Sn5[a] Pb80, Sb5[a] Ag1.5, Ag 2.5, Ag 5.5[a] | high temperature soldering applications lead attachments, internal construction of passive devices, die attach |
| Gold-silicon | 97% Au-3% Si | eutectic alloy, die attach |
| Gold-tin, Gold-germanium | 80% Au-20% Sn 88% Au-12% Ge | eutectic alloy, lid sealing, substrate attachment |
| Silver-copper Silver-copper-zinc | | substrate attachments, lead attachments |
| Indium-silver Indium-lead Indium-tin | 50% Pb-50% In 50% Sn-50% In 50% Ag-50% In 75% Pb-25% In 95% Pb-5% In | soldering to gold substrates, low melting point applications, directly wet ceramics and glasses |

a.  QQ-S-571  designation.

**Table 12.21**   Nominal Chemical Composition of Soldering Alloys

| Alloy | Sn, % | Pb, % | Sb, % | Ag, % | Au, % | Si, % | In, % | Ge, % | Source |
|---|---|---|---|---|---|---|---|---|---|
| Sn96[a] | Rem. | 0.1 Max | — | 3.6-4.4 | — | — | — | — | 41, 42, 77, 85 |
| Sn63[a] | 62.5-63.5 | Rem. | 0.2-0.5 | 0.015 Max | — | — | — | — | 41, 42, 46, 77, 85 |
| Sn62[a] | 61.5-62.5 | Rem. | 0.2-0.5 | 1.75-2.2 | — | — | — | — | 41, 42, 46, 77, 85 |
| Sn60 | 59.5-61.5 | Rem. | 0.2-0.5 | 0.015 Max | — | — | — | — | 41, 42, 46, 77, 85 |
| Sn50 | 49.5-51.5 | Rem. | 0.2-0.5 | 0.015 Max | — | — | — | — | 41, 42, 77, 85 |
| Sn10 | 9.0-11.0 | Rem. | 0.2 Max | 1.70-2.4 | — | — | — | — | 41, 42, 77, 85 |
| Sn5 | 4.5-5.5 | Rem. | 0.5 Max | 0.015 Max | — | — | — | — | 41, 42, 77, 85 |
| Pb80 | Rem. | 78.5-80.5 | 0.2-0.5 | 0.015 Max. | — | — | — | — | 41, 42, 77, 85 |
| Sb5 | 94.0 Min. | 0.2 Max | 4.0-6.0 | 0.015 Max. | — | — | — | — | 41, 42, 77, 85 |
| Ag 1.5[a] | 0.75-1.25 | Rem. | 0.4 Max. | 1.30-1.7 | — | — | — | — | 41, 42, 77, 85 |
| Ag 2.5[a] | 0.25 Max. | Rem. | 0.4 Max. | 2.30-2.7 | — | — | — | — | 41, 42, 77, 85 |
| Ag 5.5 | 0.25 Max. | Rem. | 0.4 Max. | 5.0-6.0 | — | — | — | — | 41, 42, 77, 85 |
| AuSn[a] | 20 | — | — | — | 80 | — | — | — | 38, 72, 85 |
| AuSi[a] | — | — | — | — | 97 | 3 | — | — | 38, 72, 85 |
| AuGe[a] | — | — | — | — | 88 | — | — | 12 | 85 |
| PbIn50 | — | 50 | — | — | — | — | 50 | — | 85 |
| SnIn50 | 50 | — | — | — | — | — | 50 | — | 41 |
| AgIn90 | — | — | — | 10 | — | — | 90 | — | 41 |
| PbIn25 | — | 75 | — | — | — | — | 25 | — | 41 |
| PbIn5 | — | 95 | — | — | — | — | 5 | — | 41 |

a.   Eutectic Alloy

**Table 12.22**   Mechanical Properties of Soldering Alloys[a]

| Alloy | Tensile Strength MPa | Yield Strength MPa | Shear Strength MPa | Density Kg/m$^3$ | Elonga-ation % | Elastic Modulus GPa | Brinnell Hardness | Source |
|---|---|---|---|---|---|---|---|---|
| Sn96 | 36.3-57.6 | 48.8 | 32.1 | 7290 | 0.69 | – | 14.8 | 43, 45, 64 |
| Sn63 | 35.4-42.2 | 16.1 | 28.5 | 8470 | 1.38 | 14.9 | 17.0 | 10, 43, 45, 64 |
| Sn62 | 31.0-59 | 17.7 | 27.6-37.9 | 8500 | – | – | 15.4 | 43, 45, 46, 64 |
| Sn60 | 18.6-28.0 | 14.2 | 24.1-33 | 8520 | 5.3 | – | 16.0 | 38, 41, 43, 64 |
| Sn50 | – | – | 24.2 | 8890 | – | – | 14.0 | 64 |
| SnIn50 | 11.8 | – | 11.2 | – | 83 | – | 4.9 | 41 |
| AgIn90 | 11.3 | – | 11.0 | – | 55 | – | 2.7 | 41 |
| PbIn25 | 37.5 | – | 24.2 | – | 47.5 | – | 10.2 | 41 |
| PbIn5 | 29.8 | – | 22.2 | – | 52 | – | 5.9 | 41 |
| Sn10 | 19.7-24.3 | 13.9 | 19.3 | 10500 | 18.3 | – | | 43, 45, 64 |
| Sn5 | 23.2 | 13.3 | – | 10990 | 26.0 | 23.5 | – | 43, 45, 52 |
| Sb5 | 56.2 | 38.1 | 31.9 | 7250 | 1.06 | – | 15.0 | 43, 45, 64, 72 |
| Ag 1.5 | 38.5 | 29.9 | 21.0 | 11100 | 1.15 | – | – | 43, 45, 64 |
| Ag 2.5 | 26.5 | 16.5 | 17.9 | – | 12.8 | – | – | 43, 45, 64 |
| AuSn | 198 | – | 185 | – | 0.0 | – | 165 | 85 |
| AuSi | 255-304 | – | – | 1568 | – | 69.5 | – | 53, 72 |
| AuGe | 233 | – | 220 | – | 0.9 | – | 139 | 85 |
| PbIn50 | 32.2 | – | 18.5 | – | 55 | – | 9.6 | 41, 85 |

a. Properties vary depending on strain rates, temperature and microstructure.

**Table 12.23**   Electrical Properties of Soldering Alloys

| Alloy | Conductivity % IACS | Resistivity ohm-m | Source |
|---|---|---|---|
| Sn96 | 12.6 | – | 85 |
| Sn63 | 11.5-11.9 | $1.45 \times 10^{-7}$ | 85 |
| Sn62 | 14.0 | – | 85, 94 |
| Sn60 | 11.5 | $1.7 \times 10^{-7}$ | 85 |
| Sn50 | 10.9 | – | 85 |
| Sn10 | 8.2 | – | 85 |
| Sn5 | – | $2.14 \times 10^{-7}$ | 52 |
| Sb5 | 10.9-11.9 | – | 41, 85 |
| AuSn | 4.8 | – | 85 |
| AuSi | – | $7.75 \times 10^{-6}$ | 59, 85 |
| AuGe | 6.0 | – | 85 |
| PbIn50 | 5.1-6.0 | – | 41 |
| SnIn50 | 11.7 | – | 41 |
| AgIn90 | 22.1 | – | 41 |
| PbIn25 | 4.6 | – | 41 |
| PbIn5 | 5.1 | – | 41 |

**Table 12.24**  Thermal Properties of Soldering Alloys

| Alloy | Conductivity W/m-k | Coefficient of Expansion ppm/°C | Heat Capacity J/kg K | Surface Tension[a] Dyne/cm | Melting Range Solidus Deg. °C | Liquidus Deg. °C | Source |
|---|---|---|---|---|---|---|---|
| Sn96 | — | 29.3 | — | — | 221 | 221 | 42, 77, 85 |
| Sn63 | 50.6 | 24.7 | — | 490 | 183 | 183 | 41, 42, 77, 85 |
| Sn62 | — | 27.0 | — | — | 179 | 179 | 42, 77, 85 |
| Sn60 | 50-65 | 23.9 | 150-176 | 480 | 183 | 191 | 42, 44, 77, 85 |
| Sn50 | 46.5 | 23.4 | — | 476 | — | — | 85 |
| Sn10 | 36.0 | 28.0 | — | 485-505 | 268 | 290 | 42, 77, 85, 45 |
| Sn5 | 35.5 | 28.7 | — | — | 308 | 312 | 41, 42, 52, 77 |
| Sb5 | — | — | — | — | 235 | 240 | 41, 42, 77 |
| Ag 1.5 | — | — | — | — | 309 | 309 | 42, 77 |
| Ag 2.5 | — | — | — | — | 304 | 304 | 42, 77 |
| Ag 5.5 | — | — | — | — | 304 | 380 | 42, 77 |
| AuSn | 251 | 16.0 | — | — | 280 | 280 | 38, 72 |
| AuSi | 293 (94.1-5.9) | 10-12.9 | — | — | — | — | 59, 72 |
| AuGe | — | 13.0 | — | — | 356 | 356 | 85 |
| PbIn50 | — | 26.3 | — | — | 180 | 210 | 85 |
| SnIn50 | — | — | — | — | 117 | 125 | 41 |
| AgIn90 | — | — | — | — | 141 | 237 | 41 |
| PbIn25 | — | — | — | — | 226 | 264 | 41 |

a. 280 °C

**Table 12.25**  Creep Behavior of Solder Alloys

| Alloy Designation | Creep Resistance Ranking | Stress for Failure @ 1000 h (MPa) 20°C | 100°C |
|---|---|---|---|
| Sn96 | 1 | 14 | 6 |
| Sn63 | 2 | — | — |
| Sn62 | 1 | 5 | 2 |
| Sn60 | 3 | 3 | 1 |
| Sn10 | 2 | — | — |
| Sn5 | 1 – 2 | — | — |
| Sb5 | 1 | 11 | 4 |
| Ag 1.5 | 2 | 5 | 3 |

1 – High
2 – Moderate
3 – Low

**Table 12.26**  Brazing Alloys

| Listing No. | ASTM Designation | Composition — Ag | Cu | Zn | Cd | P | Sn | Ni | Au | Melting Range — Solidus (°C) | Liquidus (°C) | Source |
|---|---|---|---|---|---|---|---|---|---|---|---|---|
| 1 | — | 10 | 52 | 38 | — | — | — | — | — | 788 | 852 | 100 |
| 2 | — | 20 | 45 | 35 | — | — | — | — | — | 777 | 816 | 73, 100 |
| 3 | — | 20 | 45 | 30 | 5 | — | — | — | — | 616 | 816 | 73, 100 |
| 4 | — | 45 | 30 | 25 | — | — | — | — | — | 677 | 743 | 100 |
| 5 | — | 50 | 34 | 16 | — | — | — | — | — | 671 | 774 | 73, 100 |
| 6 | — | 65 | 20 | 15 | — | — | — | — | — | 671 | 707 | 73, 100 |
| 7 | — | 70 | 20 | 10 | — | — | — | — | — | 723 | 754 | 73, 100 |
| 8 | — | 80 | 16 | 4 | — | — | — | — | — | 721 | 810 | 100 |
| 9 | BAg-6 | 50 | 15.5 | 16.5 | 18 | — | — | — | — | 625 | 635 | 73, 100 |
| 10 | — | 15 | 80 | — | — | 5 | — | — | — | 640 | 802 | 1, 73, 100 |
| 11 | — | 30 | 38 | 32 | — | — | — | — | — | 743 | 765 | 100 |
| 12 | — | 40 | 36 | 24 | — | — | — | — | — | 721 | 785 | 73, 100 |
| 13 | — | 60 | 25 | 15 | — | — | — | — | — | 682 | 718 | 73, 100 |
| 14 | BAg-8 | 72 | 28 | 15 | — | — | — | — | — | 780 | 780 | 1, 73, 100 |
| 15 | BAg-7 | 56 | 22 | 17 | — | — | 5 | — | — | 630 | 649 | 1, 73, 100 |
| 16 | BAg-3 | 50 | 15.5 | 16.5 | 18 | — | — | 3 | — | 630 | 690 | 1, 73 |
| 17 | BAg-1 | 45 | 15 | 16 | 24 | — | — | — | — | 605 | 620 | 1, 73 |
| 18 | — | 5 | 89 | — | — | 6 | — | — | — | 640 | 802 | 73 |
| 19 | — | 7 | 85 | — | — | — | 8 | — | — | 665 | 985 | 73 |
| 20 | — | 25 | 40 | 33 | — | — | 2 | — | — | 690 | 780 | 73 |
| 21 | — | 50 | 20 | 28 | — | — | — | 2 | — | 660 | 705 | 73 |
| 22 | — | 63 | 28.5 | — | — | — | 6 | 2.5 | — | 690 | 800 | 73 |
| 23 | BAu-4 | — | — | — | — | — | — | 18 | 82 | 950 | 950 | 1, 73 |

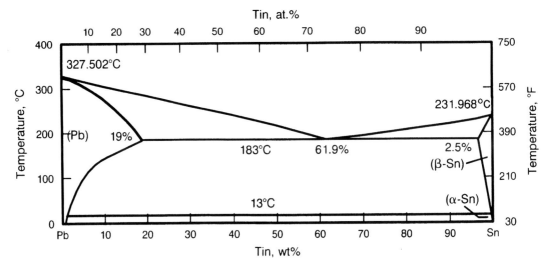

**Figure 12.20**  Tin-lead phase diagram.

the difference in the coefficients of thermal expansion of the solder and the alumina ceramic. The relative creep resistance of selected solders is shown in Table 12.25.

Tin-lead-silver, tin-lead-antimony, and indium alloys are used for various specialized applications. Ag1.5, Ag2.5, and Ag5.5 are high melting point silver-bearing alloys. Sn62 has a nominal composition of 62%Sn-36%Pb-2%Ag with a melting point near the tin-lead eutectic. The silver content tends to reduce the dissolution of silver and tends to improve the creep resistance over that of near-eutectic tin-lead alloys. Indium alloys have a reduced tendency to dissolve gold. They also have the ability to wet glasses and ceramics.

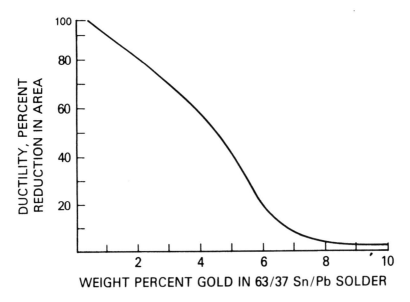

**Figure 12.21**  Effect of gold content from base metal dissolution on the ductility of eutectic solder.

**Figure 12.22**   Creep-fatigue damage in surface mount solder joints after thermal cycling. (Courtesy of Unisys NASA/GSFC.)

Various gold alloys may be selected for applications requiring a very high melting point and high strength. Gold-germanium and gold-tin are used for such applications as substrate attachment and lid sealing, and gold-silicon is widely used as a die attach alloy. These alloys have a high elastic modulus and low ductility compared to other solders and are often referred to as hard solders.

Brazing alloys have melting temperatures in excess of 450°C [73]. Many alloys are available over a wide range of melting temperatures. Several alloys are listed in Table 12.26. Brazing is used for applications such as joining copper headers to package flanges in power transistor packages or joining leads to cofired ceramic packages. Very high strengths can be achieved in brazed joints. The tensile strength of brazed joints will approach the tensile strength of the base metal [73]. For example, properly brazed copper test pieces will exhibit tensile strengths in excess of 200 MPa and brazed 1020 steel test pieces will exhibit tensile strengths approaching 450 MPa.

## 12.6.3.  Adhesives

Adhesives are made from many different types of polymers and come in several forms [64, 70, 113, 124]. Commonly used adhesives in electronic packaging include polyurethanes, silicones, polyimides, acrylics, cyanoacrylates, epoxies, and modified epoxies. These materials may be employed as one- or two-component liquids, pastes, or films. Fillers may also be utilized to enhance or modify certain properties, such as the coefficient of thermal expansion or electrical resistivity [12]. Adhesives are used for such applications as bonding surface-mounted components to printed circuit boards, bonding semiconductor devices to substrates, and for structural joints in chassis and housings. The mechanical, electrical, and thermal properties of various adhesives are summarized in Tables 12.27, 12.28, and 12.29. Additional discussion and property data for various resins are also presented in Section 12.7.

**Table 12.27**  Mechanical Properties of Adhesives

| Material | Tensile Strength MPa | Shear Strength MPa | Elong-ation % | Elastic Modulus GPa | Specific Gravity | Hardness[a] | Source |
|---|---|---|---|---|---|---|---|
| Silicone | 10.3 | – | 100-800 | 2.21 | 1.02-1.2 | 20-90A | 1, 3, 12, 85 |
| Urethane | 5.5-55 | 15.5 | 250-800 | .172-34.5 | 1.1-1.6 | 10A-80D | 1, 12, 65 |
| Acrylic | 12.4-13.8 | – | 100-400 | .69-10.34 | 1.09 | 40-90A | 1 |
| Epoxy Silicone | – | 11.7 | – | – | – | – | |
| Epoxy Novolac | 55-82.7 | 26.2 | 2-5 | 2.76-3.45 | 1.2 | – | 1 |
| Epoxy Bisphenol A | 43-85 | – | 4.4-11.0 | 2.7-3.3 | 1.15 | 106RM | 1, 85 |
| Epoxy Elec. Cond. | 3.4-34 | – | – | – | – | – | 69, 85 |
| Polyimide Modified | – | – | – | 0.275 | – | – | 85 |
| Polyimide | – | 16.5 | – | 3.0 | – | – | 65, 85 |
| Epoxy Polyamide | – | 41 | – | – | – | – | 65 |
| Epoxy Polyurethane[b] | 34 | – | 10 | – | – | – | 65 |

a. A = Shore A;  D = Shore D;  RM = Rockwell M
b. 50:50 composition

As a joining process, adhesive bonding offers several advantages. Many dissimilar materials can be joined, including metals, ceramics, and polymers. The adhesive can be cured at relatively low temperatures, ranging from room temperature for a two-part epoxy to 177°C for an epoxy-polyamide adhesive [70]. A structural joint can be designed with a large bond area that will provide high strength, toughness, and fatigue resistance. Adhesive joints insulate dissimilar metals, preventing galvanic interaction. Most adhesives, however, rapidly lose significant strength with increasing temperature [64, 70].

Epoxies are the most widely used adhesive materials. In general, epoxies offer relatively high strength but do not have good flexibility. Silver particles may be added to impart electrical conductivity and enhance thermal conductivity. A variety of epoxy resins

**Table 12.28**  Electrical Properties of Adhesives

| Material | Resistivity ohm-m | Dielectric Constant @ 1MHz | Dielectric Strength V/mil | Dissipation Factor @1MHz | Source |
|---|---|---|---|---|---|
| Silicone | $10^{13}$-$10^{15}$ | 2.9-4.0 | 400-700 | .001-.002 | 1, 64, 67, 3 |
| Polyurethane | $3 \times 10^{8}$ | 5.9-8.5 | 330-700 | .05-.06[b] | 1, 64, 67 |
| Acrylic | $7 \times 10^{11}$ | – | – | – | 1 |
| Epoxy Novolac | $10^{13}$-$10^{16}$ | 3.4-3.6 | – | 0.016 | 1, 85 |
| Epoxy Phenolic | $6.1 \times 10^{14}$ | 3.4 | 400 | 0.32 | 1 |
| Epoxy, electrically conductive | $10^{-6}$ | – | – | – | 69 |
| Epoxy Bisphenol A | $19^{14}$-$10^{16}$ | 3.2-3.8 | – | 0.013-0.024 | 85 |
| Epoxy Polyamide[a] | $10^{12}$ | – | – | – | 65 |

a. 50:50 composition      b. 1 KHz

**Table 12.29** Thermal Properties of Adhesives

| Material | Conductivity W/m-k | Coefficient of Expansion ppm/°C | Heat Capacity J/kg K | Water[a] Absorption % | Maximum Temperature °C | Source |
|---|---|---|---|---|---|---|
| Silicone, RTV | 6.4-7.5 | 262 | – | 0.08-0.3 | 260 | 1, 3, 12 |
| Polyurethane | 1.9-4.6 | 90-450 | – | – | 65.6 | 1, 12 |
| Acrylic | – | – | – | – | 93.3 | 1, 12 |
| Epoxy Silicone | – | – | – | – | 260 | 1, 12 |
| Epoxy Phenolic | 25-74.7 | 33 | 1674-2093 | 0.02 | 87.8 | 1 |
| Epoxy Elec. Cond. | .17-1.5 | – | – | – | – | 36, 50 |
| Cyanoacrylate | – | – | – | – | 82.2 | 1 |
| Polyimide | – | 50 | – | – | – | 85 |
| Polyimide Mod. | – | 73 | – | – | – | 85 |

and blends or "alloys" may be used in various joining applications. The addition of polyurethane to epoxy increases the flexibility of the adhesive. Epoxy-polyamides or epoxy-nylons offer high strength that is retained to very low temperatures ($-240°C$) [64]. However, the strength will decrease rapidly as the temperature increases. Epoxy-polyamides are also sensitive to moisture. Epoxy-silicones have a relatively low shear strength at room temperature but retain their strength with increasing temperature. They can withstand temperatures to 450°C for short durations while retaining 80% of their shear strength. [64].

Polyurethane resins also serve as adhesives. Polyurethane adhesives will maintain their strength and toughness to very low temperatures, which makes them suitable for cryogenic uses. However, the strength of polyurethanes decreases rapidly above room temperature. Polyurethane adhesives are limited to applications below about 65°C [12].

Silicone room temperature vulcanizing (RTV) materials are generally used as coatings or sealants but may be used in low-strength adhesive applications. Silicones have high flexibility and may be used at high temperatures.

Polyimide-based adhesives are useful to temperatures up to 315°C. Their strength and temperature resistance make polyimides an excellent choice for applications such as die bonding. The conductivity of polyimides can also be modified with silver particles.

## 12.7 ENCAPSULATING AND COATING MATERIALS

A wide variety of thermosetting resins and thermoplastic materials are used in electronic packaging for coating and encapsulating [34, 64, 85]. The uses of selected materials and the range of properties associated with them are summarized in Tables 12.30 through 12.33.

Organic coatings and encapsulants serve the purpose of providing protection for the assembly or device from subsequent processing or from the intended service environment. They may also serve specialized functions, such as facilitating heat transfer from components or damping mechanical vibrations. Coatings and encapsulants can provide effective protection while maintaining electrical isolation. The packaging engineer must select a suitable material for the application by balancing the protection achieved by the use of a particular material against such issues as repairability, applied thickness, and sensitivity of the parts to thermomechanical loads applied via the coating or encapsulant.

**Table 12.30**  Applications of Encapsulating and Coating Materials

| Material (Resin Types) | Typical Applications |
|---|---|
| Silicones | printed circuit conformal coating, vibration dampening applications, high voltage insulation, potting applications, junction coatings |
| Epoxies | molded encapsulation of semiconductors, molded encapsulation of capacitors and resistors, potting applications, adhesives, printed circuit conformal coating, solder mask |
| Polyimides | flexible circuits, wire insulation, multichip dielectrics, TAB applications, junction coating, passivation |
| Polyamides | wire insulation |
| Polyurethanes | printed circuit conformal coating, potting applications |
| Fluorocarbons | wire insulation, high voltage insulation applications, high frequency applications |
| Acrylics | printed circuit conformal coating, encapsulation, adhesives |
| Dialyll phthalate | connector blocks, molded encapsulation |
| Polyvinyl chlorides | wire insulation |
| Parylene | conformal coating, multichip dielectric, passivation |

Several different materials are used for conformal coatings. Printed wiring assemblies utilize many different materials, including acrylics, polyurethane, silicones, epoxies, fluorocarbons, and parylene. With the exception of parylene, these materials, as conformal coatings, are generally applied as liquids and cured to their final state using elevated temperature or ultraviolent curing. Parylene is applied using vapor deposition [91, 99]. Conformal coatings are generally applied in a thin layer ranging from 1 to 8 mils for most materials [85].

The selection of a particular conformal coating is based on performance and manufacturing criteria. For example, acrylics are easily repairable, whereas epoxies are not. However, epoxies may provide better protection against moisture. Fluorocarbons are generally selected for their low dielectric constant and low loss, and silicones are selected for their high dielectric strength, flexibility, and stability.

Coatings are also used for other purposes. Polyimide and silicone are used as for their ability to resist elevated temperature in such applications as junction coatings for power devices. Polyimide, fluorocarbons, and polyamide are also utilized as wire insulation.

**Table 12.31**  Mechanical Properties of Encapsulating and Coating Materials

| Material | Tensile Strength MPa | Comp. Strength MPa | Flexural Strength MPa | Impact Strength Joule | Elongation % | Elastic Modulus GPa | Density kg/m³ | Specific Gravity | Hardness[a] | Source |
|---|---|---|---|---|---|---|---|---|---|---|
| **Thermosets** | | | | | | | | | | |
| Silicone, Coating | 1.7-6.2 | – | – | – | 50-1200 | – | – | – | 20-70A | 85 |
| Silicone, Molding | 17-34 | 69-124 | 45-59 | 0.4-0.54 | 8 | 9.7-17 | 1661 | 1.7-2.0 | 70-95RM | 12, 35, 65 |
| Silicone Ribber, RTV | 4.5 | – | – | – | 100 | – | – | 1.17 | 45-60A | 35, 84 |
| Silicone flex foam | .035-.1 | – | – | – | – | – | 112-144 | – | – | 12 |
| Epoxy, Cast, Bisphenol | 69-207 | – | 138-179 | 0.7 | 1-4 | 20.7 | 1206 | 1.1-1.4 | 100-108RM | 1, 12, 35, 36 |
| Epoxy, Mold, Bisphenol[e] | 96-207 | 172-207 | 138-179 | 0.54 | 3 | 21 | – | 1.5-2.0 | 100RM | 1, 12, 35 |
| Epoxy, Cast Novolac | 55-82.7 | 117-131 | 75.8-110 | 0.68 | 2-5 | 2.8-3.45 | – | 1.2 | 107-112RM | 1, 12, 65 |
| Epoxy, Mold, Novolac[e] | 36-36.5 | 152-179 | 69-83 | 0.4-0.7 | – | – | – | 1.7 | 94-96D | 1 |
| Polyimide | 34.5-96.5 | 88-165.5 | 48-96.5 | 1.1-1.5 | 6-7 | 3.2-5.2 | – | 1.4 | 97-122RM | 1 |
| Polyimide Films | 103-172 | – | – | – | 15-70 | 1.4-2.4 | 1400-1600 | 1.42 | – | 1, 12 |
| Polyurethane rigid foam | .35 | .17-.21 | – | – | – | – | 32-640 | – | – | 12, 85 |
| Polyurethane | 35 | – | – | – | 400 | – | – | – | – | 12 |
| Polyurethane coating | 6.9 | – | – | – | 25 | – | – | – | – | 35 |
| Polystyrene foam | 0.14-1.52 | .069-1.4 | 0.17-2.28 | 0.4-0.7 | – | – | 16-160 | 1.09 | 35D | 84 |
| Diallyl Phthalate[e] | 27.6-48 | – | 55-124 | 0.34-0.54 | – | 8.3-14 | – | – | – | 12 |
| Urea Filled | 34-89.6 | 172-276 | 69-110 | – | .5-1 | 6.9-10 | 22.4-320.4 | 1.4-1.9 | 100-110RM | 12 |
| Urea Unfilled | 3.5 | 138 | – | – | 300 | .04-.07 | 96-160 | 1.4-1.5 | 110-120RM | 12, 35 |
| **Thermoplastics** | | | | | | | | | | |
| Acrylic[b] | 60-76 | 96.5-117 | 96.5-117 | 0.4-0.54 | 3-6 | 2.4-3.1 | – | 1.18-1.19 | 84-97RM | 35, 36 |
| Acrylic coating | 17.2 | – | – | – | 88 | 0.7 | – | – | – | 65 |
| Polyamide | 77-90 | 34-90 | 100.7 | 1.2-2.7 | 60-300 | 2.8-3.3 | – | 1.13-1.15 | 118-123RR | 85 |
| Polyamide Films | 75.8 | – | – | – | 7 | – | – | 0.8-1.0 | – | 65 |
| PTFE[c] | 14-34 | 28.83 | – | – | 75-400 | .34-.69 | – | – | – | 12 |
| Parylene | 45-75 | – | – | – | 30-200 | 2.4-3.2 | 1139 | 2.13-2.18 | 60-65D | 61 |
| FEP[d] | 18.6-21.4 | – | – | – | 250-330 | 0.34-.56 | 1100-1400 | 2.14-2.17 | 80-85RR | 35, 85 |
| Polyvinyl Chloride | 20 | – | – | – | 100-360 | 6.9-20.7 | – | 1.2-1.5 | 80D | 1, 12, 35 |

a: A = Shore A; D = Shore D; RM = Rockwell M; RR = Rockwell R.   b: Casting/liquid Resins   c: PTFE = Polytetrafluoroethylene

d: FEP = Fluorinated Ethylene Propylene   e: Glass or mineral filled

**Table 12.32**  Electrical Properties of Encapsulating and Coating Materials

| Material | Resistivity ohm-m | Dielectric Constant @ 1MHz | Dielectric Strength V/mil | Dissipation Factor @ 1MHz | Source |
|---|---|---|---|---|---|
| **Thermosets** | | | | | |
| Silicone, Coating | $10^{12}$-$10^{14}$ | 2.6-2.7 | 500-2000 | 0.009-0.001 | 84, 85 |
| Silicone, Molding | $10^{11}$-$10^{13}$ | 3.4-6.3 | 300-465 | 0.002 | 65 |
| Silicone Rubber, RTV | $2 \times 10^{13}$ | 3-4 | 550-600 | 0.001-0.005 | 64, 84 |
| Silicone flex foam | – | – | – | – | – |
| Epoxy, cast, Bisphenol | $10^{13}$ | 4.2 | 320-450 | 0.006 | 12, 64, 65 |
| Epoxy, mold, Bisphenol | $10^{13}$ | 4.7 | 370-450 | 0.02 | 12, 64, 65 |
| Epoxy, cast, Novolac | $2 \times 10^{12}$ | 3.5 | – | 0.029 | 1 |
| Epoxy, mold, Novolac | $5 \times 10^{12}$ | 4.3-4.8 | 280-400 | – | 1 |
| Polyimide | $10^{16}$ | 3.5-3.9 | 310-560 | 0.004-0.011 | 1, 12, 72 |
| Polyimide Films | $10^{16}$ | 3.5[b] | 6000-7000 | .003 | 12, 85 |
| Polyurethane | $10^{9}$ | 3.5 | 350 | 0.04 | 35 |
| Polyurethane coating | $10^{12}$ | 4.2-5.2 | 3500 | 0.05-0.07 | 85 |
| Polystyrene foam | – | 2.55 | 6 | – | 13 |
| Diallyl Phthalate | – | 4.5-6 | – | .04-.08 | 65 |
| Urea Filled | .5-5 $\times 10^{9}$ | 6.7-6.9 | 330-370 | .029-.03 | 65 |
| Urea Unfilled | $2 \times 10^{9}$ | 5.2 | 400 | 0.016 | 64 |
| **Thermoplastics** | | | | | |
| Acrylic[a] | $10^{12}$ | 3.3-3.9[b] | 400 | .04-.05[b] | 65 |
| Acrylic coating | $10^{13}$ | 2.2-3.2 | 800-3500 | 0.02-0.04 | 85 |
| Polyamide | $10^{12}$-$10^{13}$ | 3.6-4.0[b] | 300-400 | 0.014[b] | 65 |
| Polyamide Films | $10^{16}$ | 2.2-3.3[b] | 700 | .008[b] | 12 |
| PTFE[c] | $10^{16}$ | 2.1[b] | 400 | .0001[b] | 65 |
| Parylene | $10^{13}$-$10^{16}$ | 2.6-3.5 | 5500-6500 | .01-.03 | 12, 85, 3 |
| FEP[d] | $10^{16}$ | 2.1 | 500 | 0.0002 | 12, 3 |
| Polyvinyl Chloride | $10^{9}$-$10^{13}$ | 2.8 | 400 | 0.006 | 1, 12, 35 |

a.  casting/liquid resins      b. @ 60Hz      c.  PTFE = polytetrafluoroethylene
d.  FEP = fluorinated ethylene propylene

Encapsulants or embedding and potting materials include polyurethanes, silicones, and epoxies. Much of the information regarding embedding resins will also apply to coatings.

Epoxies offer many benefits and are widely employed as encapsulants for electronic devices and assemblies. In general, they have relatively low shrinkage, good mechanical strength, and excellent adhesion and are resistant to many chemical processes and application environments. A wide variety of resin types and curing agents offer a range of properties and processing options. Properties such as the coefficient of expansion can be modified with various fillers. This is illustrated in Figure 12.23. Common filler materials include glass and alumina and other mineral or ceramic particulates.

Table 12.33 Thermal Properties of Encapsulating and Coating Materials

| Material | Conductivity W/m-K | Coeff. of expansion ppm/°C | Heat Capacity J/kg-K | Heat Distortion. Temp. °C | Temperature Resis. °C | Water Absorption % | Source |
|---|---|---|---|---|---|---|---|
| **Thermosets** | | | | | | | |
| Silicone, Coating | 0.12 | – | 1464 | – | 200 | .1-.4 | 84 |
| Silicone, Molding | 2-3.6 | 70-150 | 1004-1255 | 171-482 | 315.6 | .1-.2 | 12, 65 |
| Silicone Rubber, RTV | 0.38 | 400 | 1339-1464 | <RT | 93-149 | .12 | 12, 64, 84 |
| Silicone flex foam | 0.54-0.6 | – | – | – | 316 | – | 12 |
| Epoxy, Cast, Bisphenol | 0.17-0.2 | 40-65 | 1046 | 46-260 | 121-260 | .12 | 1, 12, 64, 65 |
| Epoxy, Mold, Bisphenol | 0.4-0.8 | 20-40 | 834-1255 | 71-288 | 121-260 | .05-.095 | 1, 12, 64, 65 |
| Epoxy, Cast, Novolac | – | – | – | 149-204 | 232 | – | 1 |
| Epoxy, Mold, Novolac | – | 30-40 | – | 149-218 | 232-260 | .11-.2 | 1 |
| Polyimide | 4.3-11.8 | 45-50 | 1130-1298 | 360 | 260-315.6 | .24-.27 | 1, 12 |
| Polyimide Films | 0.26 | 20-40 | 18000 | – | 300-400 | 1-3 | 12, 85 |
| Polyurethane rigid foam | 0.19-0.29 | 97 | – | – | 149-177 | 10 | 12 |
| Polyurethane | – | 180-250 | – | 65 | 95 | 0.4 | 35, 85 |
| Polyurethane coating | – | – | – | – | 120 | – | 84 |
| Polystyrene foam | 0.42 | 70 | – | – | 85 | 0.01-1 | 12 |
| Diallyl Phthalate | 8.7-12 | 10-40 | .4 | 163-260 | 204-232 | .2-.5 | 1, 12 |
| Urea Filled | 0.3-0.4 | 25-50 | .4 | 127-143 | 93.3 | .4-.8 | 12 |
| Urea Unfilled | 0.2 | 150 | – | <RT | – | .65 | 12, 64 |
| **Thermoplastics** | | | | | | | |
| Acrylic[a] | 0.06 | 65-105 | – | 75-92 | 54-90.6 | .3 | 12, 65 |
| Acrylic coating | 0.15 | – | – | – | – | – | 85 |
| Polyamide | 0.24 | 81 | .4 | 93.3 | 121 | 1.3 | 12, 65 |
| Polyamide Films | – | – | – | – | – | 1.5 | 12 |
| PTFE[b] | 0.25 | 55 | – | 51 | 204 | 0.01 | 65 |
| Parylene Coating | 0.12 | 35-69 | 1000-1300 | – | – | 0.1 | 85 |
| FEP | 2511.6 | 83-90 | 1172 | – | 400 | 0.01-.05 | 12, 3 |
| Polyvinyl Chloride | 1.45-2.1 | – | – | – | – | 0.15-0.8 | 1, 35 |

a. casting/liquid resins   b. PTFE = Polytetrafluoroethylene   c. FEP = Fluorinated ethylene propylene

d. = Water absorption test methods vary.

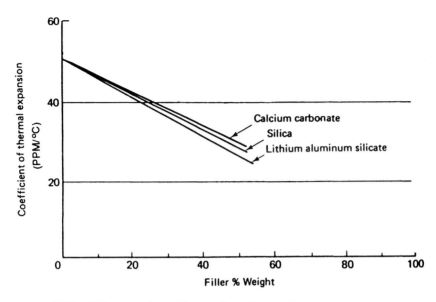

**Figure 12.23**   Effect of various fillers on thermal expansion.

Epoxies are relatively rigid, even when modified by alloying with polyurethane or other less rigid materials. As such, they can transfer considerable stress to encapsulated components, particularly at low temperatures [20]. Epoxies must be employed with caution in environments that include thermal shock or wide temperature excursions.

A widely used category of silicones is the RTV elastomers. These materials offer flexibility down to $-45$ to $-55°C$ and are resistant to temperatures as high as $200°C$ [85]. Silicone encapsulants have excellent electrical properties, including a relatively low dielectric constant, small dielectric loss, and high dielectric strength, which are stable over temperature. These materials do not have good adhesion and are not very resistant to tearing. Silicones are also available as rigid transfer molding compounds and as "gels."

Polyurethanes also serve as encapsulants. Polyurethane resins are flexible, yet offer excellent toughness, abrasion resistance, and tear resistance and good adhesion in comparison to other flexible materials. Polyurethanes can be used in applications requiring exposure to $130°C$. They retain flexibility down to $-40°C$. The electrical properties are not stable with temperature or frequency [85]. Consequently, polyurethane resins are not suitable for microwave applications.

## 12.8   CONDUCTORS AND CONTACT MATERIALS

In the electronic package, at both the system and the device level, a wide range of metals must be selected to serve as conductors. Copper is one of the most desirable conductors because of its excellent electrical conductivity. However, limitations imposed by processing considerations, mechanical requirements, or environmental concerns result in many other metals serving as conductors [11, 64, 85, 112]. Typical constraints involve differential thermal expansion in a glass-to-metal sealing processes, corrosion resistance, wear resistance, or mechanical stress levels.

Alternative conductors range from gold and gold-cobalt alloys to silver and tungsten. Electroplated materials are utilized for applications such as printed circuits and contact finishes. A contact pin, for example, may be manufactured from phosphor bronze. However, in order to retain good contact resistance and impart wear resistance, the pin will require a gold-cobalt alloy (hard gold) electroplated finish.

Applications and mechanical, electrical, and thermal properties of conductor materials are shown in Tables 12.34 to 12.37. Properties of both wrought metals and electroplated materials are presented, because considerable differences in properties can exist [109]. Data for electroplated alloys are given for commonly used plating baths. Little or no mechanical or thermal data are available for thick-film conductors. However, the electrical conductivity of several of these materials is shown in Table 12.38.

**Table 12.34**  Applications of Conductors and Contact Materials

| Material | Alloys | Typical Applications |
|---|---|---|
| Copper alloys | C11100, C19400, C10100, C19500, C10400 | lead materials, lead frames |
| Electroplated copper | | printed wiring |
| Beryllium copper alloys | C17000, C17200, C17400 | contact springs |
| Brass alloys | C36000, C27000, C26000 | contact pins |
| Phosphor-bronze alloys | C51000, C52400 | contact pins |
| Electroplated gold | 99.9% Au-0.1% Co; 99% Au-1% Co | finish for contact surfaces |
| Gold alloys | 99.99% Au | bond wires |
| Silver alloys | | contact materials in relays, leads |
| Electroplated silver | | finish for high frequency components |
| Tungsten[a] | W powder/binder | high temperature Co-fired metallization |
| Molybdenum[a] | Mo or Mo/Mn powder/binder | Co-fired metallization |
| Electroplated Rhodium | | finish for mating contact surfaces, high wear applications |
| Aluminum | Al/1% Si; Al/1% Mg | bond wires |
| Kovar | 29% Ni/17% Co/54% Fe | pins for glass sealed feed-throughs lead frames, glass sealing applications |
| Alloy 42 | 42% Ni/58% Fe | lead frames; brazed leads |

[a]powder with glass/ceramic binder in as fired condition

**Table 12.35**  Mechanical Properties of Conductors and Contact Materials

| Material | Tensile Strength MPa | Yield Strength MPa | Shear Strength MPa | Elonga-tion % | Elastic Modulus GPa | Density $Kg/m^3$ | Hardness | Source |
|---|---|---|---|---|---|---|---|---|
| **Wrought Alloys** | | | | | | | | |
| OFE Cu C10100 (A) | 221-241 | 69-76 | 152-165 | 45-55 | 110.3 | 8941.5 | 40-45RF | 1, 14, 16, 20, 51 |
| OFE Cu C10100 (H) | 310-379 | 276-345 | 179-200 | 6-20 | 110.3 | 8941.5 | 85-95RF | 1, 14, 16, 20 |
| Be Cu C17400 (A) | 469 | 221-331 | 345-414 | 45, 35 | 110.3 | 8249.5 | 60RB | 1, 16 |
| Be Cu C17400 (H) | 1255[f] | 483-1103 | 621-690 | 2-25 | 110.3 | 8249.5 | 80-99RB | 1, 16 |
| Brass C26000 (A) | 303 | 75.8 | – | 66, 64 | 110.3 | 8526.2 | 54RF | 1, 14 |
| Brass C26000 (H) | 524 | 434 | 303 | 8 | 110.3 | 8526.2 | 82RB | 1, 14 |
| Brass C27000 (A) | 317 | 96.5 | 221 | 65, 60 | 103.4 | 8470.9 | 58RF | 1 |
| Brass C27000 (H) | 510 | 414 | 297 | 8,8 | 103.4 | 8470.9 | 80RB | 1 |
| Bronze C51000 (A) | 324 | 131 | – | 64, 58 | 110.3 | 8858.4 | 75RF | 1, 14 |
| Bronze C51000 (H) | 559 | 517 | – | 10 | 110.3 | 8858.4 | 92RB | 1, 14 |
| Gold (A) | 131 | – | – | 45 | 82.7 | 19322.4 | 25V | 1, 16, 51, 52 |
| Gold (CW) | 221 | 207 | – | 4 | 82.7 | 19322.4 | – | 1, 51, 52 |
| Silver (A) | 152 | 55.2 | – | 48 | 75.8 | 10270.2 | 26V | 1, 16, 60, 69 |
| Silver (CW) | 345 | 303 | – | 2.5 | 75.8 | 10270.2 | – | 1, 16, 60, 69 |
| Tungsten | 310-1517 | 103-1517 | – | – | 406.8 | 19239.4 | 480V | 1, 14, 16 |
| Kovar | 552 | 343 | – | – | 138 | 8360 | 68B | 9, 51, 53, 58, 59, 65 |
| Ni42/Fe58 | 565 | 276 | – | 30 | 145 | 8110 | 76RB | 53, 59, 62, 65 |
| Molybdenum | 655 | 552 | – | – | 324 | 10240 | 270V | 14, 66 |
| Al, High Purity | – | – | – | – | 61.8 | – | – | 72 |
| **Electrodeposits, plating bath type** | | | | | | | | |
| Cu, cyanide[b] | 441-618 | – | – | 4-18 | 108-117 | 8860-8930 | 131-159V | 109 |
| Cu, Pyrophosphate | 280 | – | – | 35 | 117 | 8860-8930 | 92 | 109 |
| Cu, sulfate | 137-265 | – | – | 8-41 | 108 | 8860-8930 | 48-64V | 109 |
| Ni, sulfamate | 549-824 | 314-540 | – | 7-14 | 146-156 | 8930 | 140-650 | 109 |
| Au-0.1% Co, cyanide | – | – | – | – | – | 17800 | 137-196 | 109 |
| Au-1% Co, cyanide | – | – | – | – | – | – | 194-238 | 109 |
| Au, cyanide | 152-213 | 88-128 | – | 3.5-6.3 | – | 18900-19270 | 43-82 | 109 |
| Ag, cyanide | 235-333 | – | – | 12-19 | – | 9200-10500 | 40-185 | 109 |
| Sn, sulfate | – | – | – | – | – | – | 15-30 | 109 |
| Sn, fluoborate | – | – | – | – | – | – | 9-12 | 109 |
| Rh, sulfate | – | – | – | – | 317 | 12370 | 819-1039 | 1, 109 |
| Electroless Ni[d] | 441-771 | – | – | 3-6 | 117-196 | – | 500-800[e] | 109 |

a.  A = Annealed, H = Hard        b. High strength deposits. Also attainable in sulfate baths.
c.  V = Vickers $(Kg/mm^2)$; RF = Rockwell F; RB = Rockwell B; all other values reported as $Kg/mm^2$
d.  8 - 9% P    e. As deposited. Heat treatment increases hardness.       f. Heat treated

## 12.9  CORROSION

Materials selection and design must carefully consider the environment intended for the product. Electronic devices and systems used in military, aerospace, and many automotive applications must include corrosion protection. In addition, medical implants, such as pacemakers, must also provide protection from corrosion. Hence, corrosion is a primary concern of the electronic packaging engineer [36, 116].

### 12.9.1  Corrosion Cells

Corrosion begins with the formation of an electrochemical cell [97]. A cell consists of an anode and cathode immersed in an electrolyte. A potential between the anode and cathode results in the flow of electrons from the anode to the cathode and the transfer of metallic ions from the anode to the electrolyte. Figure 12.24 diagrammatically illustrates this process. The end result is deterioration of the anode and failure of the system.

The wide variety of materials and processes used in the electronics industry may lead to the formation of several types of cells in a corrosive environment. One of the most common types of cells is the dissimilar electrode or galvanic cell. Two dissimilar metals in contact will result in corrosion of the anode in the presence of an electrolyte.

**Table 12.36**  Electrical Properties of Conductors and Contact Materials

| Material | Conductivity %IACS | Resistivity $10^{-9}$ ohm-m | Source |
|---|---|---|---|
| **Wrought Alloys** | | | |
| OFE Cu. C10100 | 101 | 17 | 1, 13, 16, 51, 122 |
| BeCu. C17400 | 22 | 16.7 | 1, 72 |
| Brass C26000 | 28 | 62 | 1, 16 |
| Brass C27000 | 27 | 60 | 1, 16 |
| Bronze C51000 | 15 | 100 | 1, 16 |
| Gold | – | 21.9-24.4 | 1, 16, 51, 59, 72, 122 |
| Silver | – | 16 | 13, 16, 69, 72 |
| Tungsten | – | 55 | 5, 11, 13, 72 |
| Kovar | – | 500 | 11, 13, 52, 112 |
| Ni42/Fe58 | – | 48.4 | 53 |
| Molybdenum | – | 48 | 11, 72 |
| Al. Pure | – | 26.5 | 122 |
| **Electrodeposits, bath type** | | | |
| Cu, cyanide | – | 17.5-20.2 | 109 |
| Cu, pyrophosphate | – | 17.0-17.4 | 109 |
| Cu, sulfate | – | 17.0-18.2 | 109 |
| Ni, sulfamate | – | 86.0 | 109 |
| Au, Co 0.1% cyanide | – | 28.5-48.0 | 109 |
| Au, Co 1%, cyanide | – | 31.5-97.0 | 109 |
| Au, cyanide | – | 22.0-35.0 | 109 |
| Ag, cyanide | – | 16-19 | 109 |
| Sn, sulfate | – | 110 | 109 |
| Sn, fluoborate | – | 110 | 109 |
| Rh, sulfate | – | 85 | 109 |
| Electroless Ni | – | 300-1100 | 109 |

In microelectronic applications, gold wires bonded to aluminum bond pads form a potentially destructive couple. A small amount of moisture present in the hermetic package can result in severe corrosion of the aluminum bond pad [116]. An additional example is shown in Figure 12.25. In this case, zinc inclusions in the axial leaded glass diode package formed a galvanic couple with the molybdenum header during humidity testing. The corrosion products formed the unique morphology shown in Figure 12.25b.

Differences in the distribution of stresses can also result in the formation of an electrochemical cell. For this reason, the grain boundaries in some metals are preferentially attacked. Also, differences in oxygen concentration between two regions can form a cell. An unprotected crevice will form a cell, resulting in corrosion of a metal structure at the site of the crevice. This process is enhanced by entrapped residues or dirt in the crevice site. This can occur in welded aluminum chassis in which the joints are spot welded and subsequently processed by electroplating or chromate finishing [36]. Entrapped residues can form a highly corrosive electrolyte, severely damaging the structure in a moist environment.

**Table 12.37**  Thermal Properties of Conductors and Contact Materials

| Material | Conductivity W/m-K | Coefficient of expansion ppm/°C | Heat Capacity J/kg-K | Temperature Resistance °C | Melting Temperature K | Source |
|---|---|---|---|---|---|---|
| Wrought Alloys | | | | | | |
| OFE Cu C10100 | 380-403 | 17 | 385 | 1084 | 1356 | 1, 14, 16, 20, 51, 56, 72, 85, 22, 123 |
| BeCu C17400 | 107.3-130 | 17.8 | 418.6 | — | 1139-1255 | 1, 85 |
| Brass C26000 | 121 | 20 | 377 | — | 1188-1228 | 1, 14, 16, 85 |
| Brass C27000 | 115.9 | 20.3 | 377 | — | 1178-1205 | 1, 85 |
| Bronze C51000 | 69 | 17.8 | 377 | — | 1228-1322 | 1, 14, 16, 72 |
| Gold | 318 | 14.2 | 129 | 1064 | 1336 | 1, 16, 51, 52, 59, 72, 85, 22 |
| Silver | 419 | 18 | 235 | 962 | 1235 | 1, 14, 16, 60, 72 |
| Tungsten | 167-174 | 4.4 | 132 | 3387 | 3660 | 1, 14, 16, 72 |
| Kovar | 16.3 | 5.87 | 439 | — | 1450 | 9, 11, 17 |
| Ni42/Fe58 | 129 | 4.45 | 477 | — | — | 53, 62, 65 |
| Molybdenum | 138 | 5.04 | 251 | 2610 | 2894 | 1, 14, 68, 72 |
| Al. Pure | 237 | 25 | 903 | 660 | 993 | 14, 16, 25, 60, 72, 85, 22 |
| Electrodeposits, bath type | | | | | | |
| Cu, cyanide | — | 16.7 | — | — | — | 109 |
| Cu, pyrophosphate | — | 16.7 | — | — | — | 109 |
| Cu, sulfate | — | 17.1-18.9 | — | — | — | 109 |
| Ni, sulfamate | 82-108 | 13.6-17.0 | 453 | — | — | 109 |
| Electroless Ni | 4.3-5.7 | 13.5-14.5 | — | — | — | 109 |
| Rh, sulfate | 150 | 8.3 | 247 | — | — | 1, 109 |

**Table 12.38**  Conductivity of Thick-Film Conductors

| Material | Resistivity ohms/sq. | Source |
|----------|----------------------|--------|
| Gold (Au) | 0.005-0.01 | 65 |
| Pt-Au | 0.05-0.1 | 65 |
| Pd-Au | 0.05-0.1 | 65 |
| Pd-Ag | 0.01 | 65 |
| Silver (Ag) | 0.005-0.1 | 65 |

When confronted with a potentially corrosive environment, the packaging engineer must design a system that will not be susceptible to the formation of destructive electrochemical cells. Only compatible metal couples should be utilized. In addition, metals that are subject to general atmospheric attack must be protected by electroplating or coating. In general, unprotected crevices should be avoided and adequate drainage provided as appropriate. Circuit cards should be mounted vertically and connectors should be mounted only on the sides of the chassis. Unavoidable couples or potentially destructive cells must be isolated from potential electrolytes such as moisture.

## 12.9.2  Predicting Susceptibility to Corrosion

The galvanic series can be a useful tool in assessing the relative behavior of potential galvanic couples [16, 36, 97]. The series is shown in Table 12.39. The chemical reactions are the partial reactions that occur when the metal is anodic. The standard oxidation potential, $E_0$, is given in volts relative to a hydrogen electrode. The metals are listed from most active (most anodic) to least active or most noble (cathodic). In a destructive couple the more active member will be anodic and suffer deterioration.

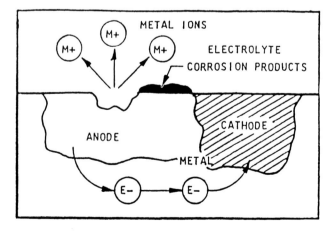

**Figure 12.24**  Diagrammatic representation of a corrosion cell.

**(a)**

**(b)**

**Figure 12.25**   (a) Diode package that corroded in an accelerated humidity test due to galvanic interaction between zinc and molybdenum. (Courtesy of Rockwell International.) (b) Morphology of corrosion products shown in (a). (Courtesy of Rockwell International.)

**Table 12.39**  The Galvanic Series

| Electrode Reaction | Standard Oxidation Potential, $E°(V)$, 25°C |
|---|---|
| $Li = Li^+ + e^-$ | 3.05 |
| $K = K^+ + e^-$ | 2.93 |
| $Ca = Ca^{++} + 2e^-$ | 2.87 |
| $Na = Na^+ + e^-$ | 2.71 |
| $Mg = Mg^{++} + 2e^-$ | 2.37 |
| $Be = Be^{++} + 2e^-$ | 1.85 |
| $U = U^{+3} + 3e^-$ | 1.80 |
| $Hf = Hf^{+4} + 4e^-$ | 1.70 |
| $Al = Al^{+3} + 3e^-$ | 1.66 |
| $Ti = Ti^{++} + 2e^-$ | 1.63 |
| $Zr = Zr^{+4} + 4e^-$ | 1.53 |
| $Mn = Mn^{++} + 2e^-$ | 1.18 |
| $Nb = Mb^{+3} + 3e^-$ | 1.1 |
| $Zn = Zn^{++} + 2e^-$ | 0.763 |
| $Cr = Cr^{+3} + 3e^-$ | 0.74 |
| $Ga = Ga^{+3} + 3e^-$ | 0.53 |
| $Fe = Fe^{++} + 2e^-$ | 0.440 |
| $Cd = Cd^{++} + 2e^-$ | 0.403 |
| $In = In^{+3} + 3e^-$ | 0.342 |
| $Tl = Tl^+ + e^-$ | 0.336 |
| $Co = Co^{++} + 2e^-$ | 0.277 |
| $Ni = Ni^{++} + 2e^-$ | 0.250 |
| $Mo = Mo^{+3} + 3e^-$ | 0.2 |
| $Sn = Sn^{++} + 2e^-$ | 0.136 |
| $Pb = Pb^{++} + 2e^-$ | 0.126 |
| $H_2 = 2H^+ + 2e^-$ | 0.000 |
| $Cu = Cu^{++} + 2e^-$ | −0.337 |
| $Cu = Cu^+ + e^-$ | −0.521 |
| $2HG = Hg_2^{++} + 2e^-$ | −0.789 |
| $Ag = Ag^+ + e^-$ | −0.800 |
| $Pd = Pd^{++} + 2e^-$ | −0.987 |
| $Hg = Hg^{++} + 2e^-$ | −0.854 |
| $Pt = Pt^{++} + 2e^-$ | −1.2 |
| $Au = Au^{+3} + 3e^-$ | −1.50 |

**Table 12.40**  Galvanic Series in Seawater
**Active (Read down)**

Magnesium
Magnesium alloys
Zinc
Aluminum 5052H
Aluminum 3004
Aluminum 3003
Aluminum 1100
Aluminum 6053T
Alclad
Cadmium
Aluminum 2017T
Aluminum 2024T
Mild steel
Wrought iron
Cast iron
Ni-Resist
13% Chromium stainless steel, type 410 (active)
50-50 Lead-tin solder
18-8 stainless steel, type 304 (active)
18-8, 3% Mo stainless steel, type 316 (active)
Lead
Tin
Muntz metal
Manganese bronze
Naval brass
Nickel (active)
76% Ni-16% Cr-7% Fe (Inconel) (active)
Yellow brass
Aluminum bronze
Red brass
Copper
Silicon bronze
5% Zn-20% Ni, Bal. Cu (Ambrac)
70% Cu-30% Ni
88% Cu-2% Zn-10% Sn (composition G-bronze)
88% Cu-3% Zn-6.5% Sn-1.5% Pb (composition M-bronze)
Nickel (passive)
76% Ni-16% Cr-7% Fe (Inconel) (passive)
70% Ni-30% Cu (Monel)
Titanium
18-8 stainless steel, type 304 (passive)
18-8, 3% Mo stainless steel, type 316 (passive)

**Noble (Read up)**

**Table 12.41**  Compatible Couples (MIL-F-14072B)

| Group No. | Metallurgical Category | Compatible Couples (see note below) |
|---|---|---|
| 1 | Gold, Solid and Plated; Gold-Platinum Alloys; Wrought Platinum | |
| 2 | Rhodium-Plated or Silver-Plated Copper | |
| 3 | Silver, Solid or Plated; High Silver Alloys | |
| 4 | Nickel, Solid or Plated; Monel Metal, High-Nickel-Copper Alloys | |
| 5 | Copper, Solid or Plated; Low Brasses or Bronzes; Silver Solder; German Silver; High Copper-Nickel Alloys; Nickel-Chromium Alloys; Austenitic Corrosion-Resistant Steels | |
| 6 | Commercial Yellow Brasses and Bronzes | |
| 7 | High Brasses and Bronzes; Naval Brass; Muntz Metal | |
| 8 | 18-Percent Chromium-Type Corrosion-Resistant Steels | |
| 9 | Chromium, Plated; Tin, Plated; 12-Percent Chromium-Type Corrosion-Resistant Steels | |
| 10 | Tin-Plate; Terneplate; Tin-Lead Solder | |
| 11 | Lead, Solid or Plated; High Lead Alloys | |
| 12 | Aluminum, Wrought Alloys of the Duralumin Type | |
| 13 | Iron, Wrought, Gray, or Malleable; Plain Carbon and Low Alloy Steels, Armco Iron | |
| 14 | Aluminum, Wrought Alloys other than Duralumin Type; Aluminum, Cast Alloys of the Silicon Type | |
| 15 | Aluminum, Cast Alloys other than Silicon Type; Cadmium, Plated and Chromated | |
| 16 | Hot-Dip-Zinc Plate; Galvanized Steel | |
| 17 | Zinc, Wrought; Zinc-Base Die-Casting Alloys; Zinc, Plated | |
| 18 | Magnesium and Magnesium-Base Alloys, Cast or Wrought | |

NOTE:  O = Indicates the most cathodic members of the series
       * = Indicates an anodic member
       Arrows indicate the anodic direction

**Table 12.42**  Metals Requiring Protection from Atmospheric Exposure

| Metals | Withstands Exposure[1] | Requires Protection |
|---|---|---|
| Aluminum | | X |
| Aluminum alloys | | X |
| Brasses | | X |
| Copper | | X |
| Copper alloys | | X |
| Kovar | | X |
| Alloy 42 (Ni Fe) | | X |
| Gold plating | X | |
| Molybdenum | | X |
| Nickel alloys | X | |
| Nickel plating | X | |
| Rhodium plating | X | |
| Silver plating | | X |
| Tin plating | X | |
| Tin-lead alloys | X | |
| Tungsten | | X |
| Low carbon steels | | X |
| 18-8 Stainless steels | X | |

[1]Including salt air and high humidity

The tendency for a destructive cell to form is measured by the relative distance between two metals in the series as determined by the following equation:

$$\Delta E_0 = E_0(\text{anode}) - E_0(\text{cathode})$$

The higher the value of $\Delta E_0$ or cell electromotive force (EMF) corresponds to an increased tendency to form a destructive galvanic couple.

The galvanic series in seawater shown in Table 12.40 can provide a better indication of the behavior of materials in more severe environments and can also be used to assess galvanic compatibility. As before, the metals are listed from most active (anodic) to least active (noble). Materials near the top of the list are more likely to suffer general attack, and the farther apart the metals are on the series, the more likely a detrimental couple will occur.

Table 12.41 lists various metallic couples that are considered acceptable for general military ground equipment. This table is part of MIL-F-14072 and can be useful in selecting materials that will minimize galvanic activity for many applications. Acceptable couples were determined from combinations of alloys that have a small EMF in seawater at room temperature.

Permissible couples in accordance with the table are indicated by the arrows at the right of the alloy category. Members of the alloy groups interconnected by arrows form acceptable couples. The most noble member of a group of alloy categories forming acceptable couples is indicated by an open circle ($\bigcirc$), while an asterisk (*) indicates an anodic member. Groups that are not connected by a continuous line of arrows do not form acceptable couples.

Many materials used in electronic assemblies require protection from general atmospheric attack. Table 12.42 lists many common packaging materials and indicates which materials will require protection.

## EXERCISES

1. Assume that single-crystal silicon is isotropic. Estimate its elastic modulus from the elastic constants. Use your estimate to determine the stresses in a simply supported beam of the configuration shown below. The beam experiences a maximum deflection of 0.05 mm. Comment on the accuracy of your calculation. How can greater accuracy be achieved?

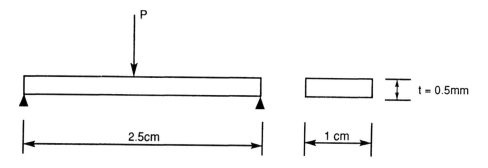

2. Discuss the effects of increasing fiber volume on the mechanical properties of glass-epoxy laminates.
3. A large power hybrid design requires a ceramic substrate material with a high thermal conductivity. Suggest three possibilities. Compare the mechanical and dielectric properties of your selections.
4. The behavior of solder under cyclic loading is known to be a complex interaction between creep and fatigue. Based on this information, select an alternative solder to SnPb eutectic for this type of application. What are potential drawbacks or additional advantages of your selection? Review the electrical and thermal properties.

5. Estimate the dielectric constant at 1 MHz of a glass-epoxy laminate with a fiber volume fraction of 0.3.
6. Discuss the advantages and disadvantages of using Kovar over ETP copper for the power hybrid package enclosure for the application discussed in Exercise 12.3. Carefully consider the thermal properties. Refer to the sketch below.

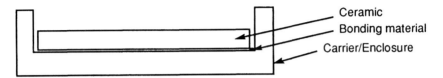

7. Epoxy has been selected for a conformal coating application on modules intended for use under the hood of an automobile. Discuss potential disadvantages. Suggest an alternative.
8. Kovar has been selected for a hybrid enclosure. The package will be exposed to a humid seacoast environment Will Kovar survive long-term exposure to this environment? If not, suggest an appropriate electroplated finish.
9. Zinc-plated steel fasteners have been selected for use in a wrought aluminum alloy chassis. Is this acceptable for a seacoast environment? Discuss your answer.
10. List potential applications of indium alloy solders.
11. Discuss the effect of thickness of proper braze joint design.
12. Estimate the coefficient of thermal expansion of the system shown in the sketch. Assume no warping.

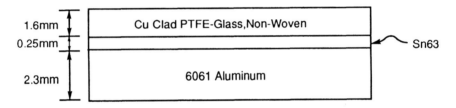

# REFERENCES

1. *Materials Selector 1990, Materials Engineering*, Penton Publications, December 1989.
2. Sadhir, R. K. and Saunders, H. E., "Protective Thin Film Coatings by Plasma Polymerization," *IEPS Proc.*, October 1984, pp. 789–803.
3. Pommer, D. and Chiechi, J., "Multi-Chip Modules Without Thin Film Wafer Processing," *IEPS Proc.*, September 1989, pp. 211–229.
4. Smeby, J. M., "Behavior of Solder Materials in Surface Mount Applications," *Proc. ASM 2nd Electronic Packaging Materials and Processes Conference*, October 1985, pp. 97–107.
5. Kraus, J. D. and Carver, K. R., *Electromagnetics*, 2nd ed., McGraw-Hill, New York, 1973.
6. "Advanced Packaging and Interconnection of Electronic Components, IPC Guidelines, "Institute for Interconnecting and Packaging Electronic Circuits, August 1985.
7. Engelmaier, W., Functional Cycles and Surface Mounting Attachment Reliability," *Surface Mount Technology*, pp. 87–114, ISHM Technical Monograph Series 6984-002, 1984.
8. Amick, P. J. and Buckley, C. L., "Thermal Analysis of Chip Carrier Compatible Substrates," *IEPS Proc.*, October 1984, pp. 34–39.

9. Kokini, K., "Thermal Shock Testing of Glass-to-Metal Seals in Microelectronics: Effect of Package Lid," *IEPS Proc.*, October 1984, pp. 637–651.

10. Dogra, K. S., "A Bismuth Tin Alloy for Hermetic Seals," *IEPS Proc.*, October 1984, pp. 631–636.

11. Kohl, W., "Materials and Process for Electron Tubes," in *Handbook of Materials and Techniques for Vacuum Devices*, Reinhold, New York, 1967.

12. Grzegorczyk, D. and Feineman, G. *Handbook of Plastic Electronics*, Reston Publishing, Reston, Va., 1974.

13. Tipler, P. A., *Physics*, Worth, New York, 1978.

14. Incropera, F. P. and Dewitt, D. P., *Fundamentals of Heat Transfer*, Wiley, New York, 1981.

15. "Guide to Selecting Engineered Materials, Advanced Materials and Processes," ASM International Publication, Special Issue, 1989.

16. Van Vlack, V., *Elements of Materials Science and Engineering*, 4th ed., Addison-Wesley, Reading, Mass., 1979.

17. Hagge, J. K., "Ultrareliable HWSI with Aluminum Nitride Packaging," NEPCON, March 1989.

18. Hagge, J. K., "Mechanical Considerations for Reliable Interfaces in Next Generation Electronics Packaging," NAECON, May 1989.

19. Hagge, J. K., "Ultra-Reliable Packaging for Silicon-on-Silicon WSI," *IEEE Trans. Components, Hybrids, Manuf. Technol.*, 12(2):170–179 (1989).

20. Glaser, J. C. and Juaire, M. P., "Thermal and Structural Analysis of a PLCC Device for Surface Mount Process," *ASME Trans. J. 178* (1989).

21. Rabinovich, E. M., "Ceramic Materials for Electronic Packaging," *ASME Trans. J. Electron. Packag. 111*:183–191 (1989).

22. 1987/1988 Electrionics Desk Manual, Vol. 33, No. 10, Lake Publishing, Libertyville, Illinois, June 1987.

23. Suryanarayana, D., "Thermally Conductive Ceramics for Electronic Packaging," *ASME Trans. J. Electron. Packag. 111*:192–198 (1989).

24. Leishman, S., "Finstrate: A New Concept in IC Substrates," *IPC Tech. Rev.*, April/May 1984, pp. 12–17.

25. Peterson, K. E., "Silicon as a Mechanical Material," *IEEE Proc.* 70(5):420–456 (1982).

26. Engelmaier, W., "Effects of Power Cycling on Leadless Chip Carrier Mounting Reliability and Technology," *Electron. Packag. Prod.*, April 1983, pp. 58–63.

27. Dance, F. J., "Can PWB's Meet the Surface Mount Challenge? Part 2," *Electronics*, June 1983, pp. 35–38.

28. "Properties of High Performance Copper Alloys," Brushwellman Engineered Materials pamphlet, Brushwellman Customer Technical Service Dept., Cleveland, Ohio.

29. "Design Guidelines for Multilayer Cofired Ceramic Modules, Ceramic Systems," Table 1, p. 2. General Ceramics, Inc., Haskell, New Jersey.

30. Simon, R. M., "Conductive Plastics Fight EMI and ESD: Improved Heat Transfer, Too," *Circuits Manu.*, March 1983, pp. 48–50.

31. McPhillips, R. B., "Advanced Ceramic Materials for High Thermal Conductivity Substrate Applications," *Hybrid Circuit Technol.*, August 1988, pp. 21–23.

32. Kingery, W. D., Bowen, H. K. and Uhlmann, D. R., *Introduction to Ceramics*, 2nd ed., Wiley, New York, 1976.

33. Saito, S., *Fine Ceramics*, Elsevier, New York, 1985.

34. "Guide to Selecting Engineering Materials for Wear Resistance," *Advanced Materials and Processing*, special issue, June 1987.

35. Wong, C. P., "An Overview of Integrated Circuit Device Encapsulants," *Trans. ASME J. Electron. Packag. 111*:97–107 (1989).

36. "Preferred Materials and Processes," Rockwell International, 1986.

37. Shigley, J. E. and Mitchell, L. D., *Mechanical Engineering Design*, 4th ed., McGraw-Hill, New York, 1983.

38. Roos-Kozel, B., "Solder Pastes," *Surface Mount Technology*, pp. 115–170, ISHM Technical Monograph Series 6984-002, 1984.

39. "Guidelines for Surface Mounting and Interconnecting Chip Carriers," Institute for Interconnecting and Packaging Electronic Circuits, November 1983.

40. Okutani, K., Otsuka, K., Sahara, K. and Satoh, K., "Packaging Design of a SIC Ceramic Multi-Chip RAM Module," *IEPS Proc.*, October 1984, pp. 299–304.

41. Manko, H. H., *Solders and Soldering*, 2nd ed., McGraw-Hill, New York, 1979.

42. Federal Specification QQ-S-571E.

43. Hwang, J. S. and Vargas, R. M., "Solder Joint Reliability—Can Solder Creep?" *ISHM Proc.*, October 1989, pp. 513–24.

44. Wassink, R. J., *Soldering in Electronics*, 2nd ed., Electrochemical Publications, Ayr, Scotland, 1989.

45. Unpublished JPL document.

46. Moy, P. H., "Solder Application," *Surface Mount Technology*, pp. 171–184, ISHM Technical Monograph Series 6984-002, 1984.

47. "Surface Mount Land Patterns (Configurations and Design Rules)," Institute for Interconnecting and Packaging Electronic Circuits, ANSI/IPC-SM-782, April 1987.

48. Fennimore, J. E. and Gerard, J. P., "Package and Interconnect Development for Future VLSI/VHSIC Devices," *IEPS Proc.*, October 1984, pp. 217–228.

49. Tessier, T. G., Turlik, I., Adema, G. M., Sivan, D., Yung, E. K. and Berry, M. J., "Process Considerations in Fabricating Thin Film Multichip Modules," *IEPS Proc.*, September 1989, pp. 294–313.

50. Ohtomo, S., Kamamura, M. and Nishigaki, S., "High Density Hybrid IC Substrate by Low Temperature Fireable Ceramic (LFC) and Cu Conductor," *IEPS Proc.*, September 1989, pp. 454–460.

51. Kurtz, J., Cousens, D. and Dufour, M., "Ball Bonding with Copper Wire," *IEPS Proc.*, October 1984, pp. 688–698.

52. Desai, P. D., El-Rahaiby, S. K., Bogaard, R. H., Chen, J. C. F., Li, H. H. and Ho, C. Y., "Thermal, Physical, Electrical, and Mechanical Properties of Selected Packaging Materials, Part 6," Second Quarterly Technical Report on SRC Contract 88-IJ-122, October–December 1988.

53. Desai, P. D., El-Rahaiby, S. K., Bogaard, R. H., Chen, J. C. F., Li, H. H. and Ho, C. Y., "Thermal, Physical, Electrical, and Mechanical Properties of Selected Packaging Materials, Part 5," First Quarterly Technical Report on SRC Contract 88-IJ-122, July–September 1988.

54. "Design Guide for Press Fit Rigid Printed Board Backplanes," Institute for Interconnecting and Packaging Electronic Circuits, ANSI/IPC-D-422, September 1982.

55. Livingston, S. and Piacente, R., "An Experience with Cu/Invar/Cu and LCC's in Avionic Equipment," *IEPS Proc.*, October 1984, pp. 770–785.

56. Nagurny, N. J., "Thermal Modeling and Testing of Surface Mounted Components on Conduction Cooled Military Standard Electronic Modules," *IEPS Proc.*, September 1989, pp. 375–384.

57. Rogers, T., "High Lead Count Package Alternatives," *IEPS Proc.*, October 1984, pp. 711–716.

58. Smith, W. F., *Structure and Properties of Engineering Alloys*, McGraw-Hill, New York, 1981.

59. Desai, P. D., El-Rahaiby, S. K., Bogaard, R. H., Chen, J. C. F., Li, H. H. and Ho, C. Y., "Thermal, Physical, Electrical, and Mechanical Properties of Selected Packaging Materials, Part 7," Third Quarterly Technical Report on SRC Contract 88-IJ-122, January–March 1989.

60. Walter, K. and Dershem, S. M., "An Evaluation of the Process Variables Affecting Bond Reliability and Measurement in Silver-Glass Die Attach," *IEPS Proc.*, September 1989, pp. 602–611.

61. "Properties of Solids," *GE Heat Transfer and Fluid Flow Data Book*, General Electric Co., Schenectady, New York, 1977.

62. Gray, F., "Substrates for Chip Carrier Interconnections," *Surface Mount Technology*, pp. 59–85, ISHM Technical Monograph Series 6984-002, 1984.

63. Ng, C. K., "Thermomechanical Stresses of a Standard Electronics Module Assembly," *IEPS Proc.*, September 1989, pp. 530–554.

64. Harper, C. A., *Handbook of Materials and Processes for Electronics*, McGraw-Hill, New York, 1970.

65. Harper, C. A., *Handbook of Electronic Packaging*, McGraw-Hill, New York, 1969.

66. Le Goff, Y. and Viret, P., "Macromodules for Military and Space Equipment Design, Technology and Trend," *IEPS Proc.*, September 1989, pp. 55–65.

67. Olson, R., "Parylene Conformal Coatings for PCB Application," *IEPS Proc.*, October 1984, pp. 804–809.

68. Reed-Hill, R. E., *Physical Metallurgy Principles*, 2nd ed., Litton Educational Publishing, Brooks/Cole Engineering Division, Monterey, California, 1973.

69. Lutz, M. A. and Cole, R. L., "High Performance Electrically Conductive Silicone Adhesives," *IEPS Proc.*, September 1989, pp. 612–624.

70. Degarmo, E. P., Black, J. T. and Kosher, R. A., *Materials and Processes in Manufacturing*, 6th ed., Macmillan, New York, 1984.

71. Herrington, T. L., Ferrier, G. G. and Smith, H. L., "Low Temperature Silver Glass Die Attach Material," *IEPS Proc.*, September 1989, pp. 640–655.

72. Wen, S., "Electronics Packaging Materials," in *Advances in Electronic Materials*, Wessels, B. W. and Chin, G. Y. eds., ASM International, Materials Park, Ohio, September 1984, pp. 263–280.

73. "Fundamentals of Silver Brazing," Handy and Harman, Bulletin 666, 1982.

74. Chen, C. F., "Characteristics of High-Thermal-Conductivity Aluminum Nitride Ceramics," *IEPS Proc.*, September 1989, pp. 1291–1304.

75. Copeland, D. W. and Powell, D. O., "Aluminum Nitride Metallized Ceramic Substrate Performance, *6th Annual IEEE Semiconductor Thermal and Temperature Measurement Symposium Proc.*, February 1990, pp. 104–107.

76. Berson, B., "Strategies for Microwave and Millimeter Wave Packaging Today," *IEPS Proc.*, September 1989, pp. 697–720.

77. "Soldering of Electronic Products," Lead Industries Association, Brochure 50-4-78.

78. Sucec, J., *Heat Transfer*, W. M. C. Brown, Dubuque, Iowa, 1985.

79. Hazzard, H., "Multilayer Ceramic Arrays Solve Problems in High-Density Packaging," *Electronic Packaging and Production*, pp. 214–218, February 1986.

80. Zilley, E. L., Jordan, R. C., Baker, J. A. and Goetz, D. P., "Resin Fracture Properties and PCB Reliability," *IEPS Proc.*, September 1989, pp. 1276–1290.

81. McShanne, M., Lin, P. and Wilson, H., "A Unique Low Cost Pin Grid Array Package with Heat-Spreader," *IEPS Proc.*, September 1989, pp. 199–207.

82. Desai, P. D., El-Rahaiby, S. K., Bogaard, R. H., Chen, J. C. F., Li, H. H. and Ho, C. Y., "Thermal, Physical, Electrical, and Mechanical Properties of Selected Packaging Materials, Part 8," First Quarterly Technical Report on SRC Contract 88-IJ-122, April–June 1989.

83. Eaton, R. M., Pennington, D. C., Sparrow, J. A. and Colton, F. J., "GaAs IC Packaging—Single and Multichip Modules in the 0–20 GHz Range," *IEPS Proc.*, September 1989, pp. 1203–1211.

84. "Materials for High Technology Applications," Dow Corning Product Guide, Form 10–008E–89, Dow Corning Corp., Midland, Michigan, 1989.

85. *Electronic Materials Handbook.* Vol. 1., *Packaging*, ASM International, Materials Park, Ohio, 1989.

86. Aakalu, N. G., "Thermal Induced Stress Molding in a Switcher Transformer," *IEPS Proc.*, September 1989, pp. 334–341.

87. Burkhart, A., "New Epoxies for Hybrid Circuit Applications," *Hybrid Circuit Technol.*, October 1988, pp. 35–40.

88. Ozmat, B. and Guy, M. and Hastings, R., "High Reliability Surface Mount Technology Module for Avionics Applications," *IEPS Proc.*, September 1989.

89. Morrel, R., *Ceramics in Modern Engineering, Engineering Applications of Ceramic Materials Source Book*, ASM International, Materials Park, Ohio, 1985, pp. 3–12.

90. Lucas, N., Zabel, H., Morkoc, H. and Unlu, H., "Anisotropy of Thermal Expansion of GaAs on Si(001)." (Ref. incomplete.)

91. Scheibert, K. A. and Lutz, M. A., "Novel UV-Curable Silicones for Electronic Packaging," *IEPS Proc.*, September 1989, pp. 727–741.

92. "Supertherm 2003," Tracon Technical Product Guide, Tracon, Inc., Medford, Mass.

93. Lynch, J. F., Ruderer, C. G. and Dukworth, W. H., *Engineering Properties of Selected Ceramic Materials*, American Ceramic Society, Columbus, Ohio, 1966.

94. "GaAs IC Design Course," Triquint Semiconductor, a Tektronix Co., pp. 3–6, March 1989.

95. Mosko, J., "The Mixed Dielectric Approach: Improving Speed and Density with Gore-Ply Precision Dielectric Prepreg," *IEPS Proc.*, September 1989, pp. 856–867.

96. Demarco, J. and Freifeld, B., "Elimination of Damaging Leaks in Electronic Components: A Novel Approach," *IEPS Proc.*, September 1989, pp. 779–784.

97. Uhlig, H. H., *Corrosion and Corrosion Control*, 2nd ed., Wiley, New York, 1971.

98. Burns, G., *Solid State Physics*, Academic Press, Orlando, Fla., 1985.

99. Onishi, T. and Sagami, Y. and Burkhart, A., "UV Light Transparent Epoxy Encapsulation for Epron-Cob Device," *IEPS Proc.*, September 1989, pp. 785–804.

100. *Metals Handbook*, American Society for Metals, Metals Park, Ohio, 1960.

101. Thomas, R. N., Hobgood, H. M., Ravishankar, P. S. and Braggins, T. T., "Melt Growth of Large Diameter Semiconductors: Part 1," *Solid State Technol.*, March 1990.

102. Weast, R. C. and Astle, M. J., eds., *Handbook of Chemistry and Physics*, 63rd ed., CRC Press, Boca Raton, Fla., 1982.

103. Deutchman, A. H. and Partykay, R. J., "Diamond-Film Deposition—a Gem of a Process," *Adv. Mater. Processses 135*(6) (1989).

104. McClintock, F. A. and Argon, A. S., *Mechanical Behavior of Materials*, Addison-Wesley, Reading, Mass., 1966.

105. Nielsen, O. H., "Compliance of Silicon," in *Properties of Silicon*, INSPEC, 1988.

106. *Metals Handbook*, 9th ed., Vol. 6, *Welding, Brazing and Soldering*, ASM International, Materials Park, Ohio, 1983.

107. Prasad, R. P., *Surface Mount Technology*, Van Nostrand Reinhold, New York, 1989.

108. Tanaka, J., Kajita, S. and Terasawa, M., "Mullite Ceramic for the Application to Advanced Packaging Technology," Kyocera Corporation, Japan.

109. Safranek, W. H., *The Properties of Electrodeposited Metals and Alloys*, American Elsevier, New York, 1974.

110. Coombs, C. F., *Printed Circuits Handbook*, 3rd ed., McGraw-Hill, New York, 1988.

111. Seraphim, D. P., Lasky, R. and Li, C. Y., *Principle of Electronic Packaging*, McGraw-Hill, New York, 1989.

112. Tummala, R. R. and Rymaszewski, E. J., *Microelectronic Packaging Handbook*, Van Nostrand Reinhold, New York, 1989.

113. Lee, H. and Neville, K., *Epoxy Resins—Their Applications and Technology*, McGraw-Hill, New York, 1957.

114. Johnson, R. R., "Multichip Modules; Next-Generation Packages," *IEEE Spectrum*, March 1990, pp. 34–36, 46–48.

115. Zimmerman, D. D. and Lewin, D. H., "The Fundamentals of Microjoining Processes," ISHM Technical Monograph Series 6983-003, 1983.
116. Shumay, W. C., "Corrosion in Electronics," *Adv. Mater. Processes 132*(3):73–77 (1987).
117. Brodsky, M. H., "Progress in Gallium Arsenide Semiconductors, *Sci. Am.*, February 1990, pp. 68–75.
118. Keusseyan, R. L., "Brazing to Green Tape," *Hybrid Circuit Technol.*, February 1990, pp. 9–15.
119. Kayzuo, S. and Takabatake, M., "Low-Temperature Fired Multilayer Circuit Board for High-Frequency Applications," *Hybrid Circuit Technol.*, February 1990, pp. 35–41.
120. Tyler, J. R., "Advanced Materials for Manufacturing of Microwave Power Hybrids," *Hybrid Circuit Technol.*, March 1990, pp. 25–29.
121. Reagan, J., Androde, J. and Kleve, G., "Thin Film Hybrid Technology for On-Wafer Probing of Integrated Circuits," *Hybrid Circuit Technol.*, April 1990, pp. 13–20.
122. Ham, R. E., "Prediction of Bond Wire Temperatures Using an Electronic Circuit Analogy," *Hybrid Circuit Technol.*, April 1990, pp. 53–54.
123. Shawhan, G. J. and Sutcliffe, G. R., "Plated Copper Metallization for Power Hybrid Manufacture," *Hybrid Circuit Technol.*, April 1990, pp. 37–42.
124. Lee, H. and Neville, K., *Handbook of Epoxy Resins*, McGraw-Hill, New York, 1967.
125. "New Heat Sink Material, KCW-10," Technical Bulletin No. SEMI-001, Kyocera Corporation, Japan.
126. Dasgupta, A., Bhandarkar, S., Pecht, M. and Barker, D., "Thermoelastic Properties of Woven-Fabric Composites Using Homogenization Techniques," Fifth Technical Conference on Composite Materials, East Lansing, Michigan, June 1990.

# Appendix A:
# Acronyms Used in Electronics

The ISHM list of acronyms (ISHM 1990 Industry Guide) was modified for use in this acronym list.

## Electronic and Optical Phenomena and Devices

ACEL = ac-EL

ACTFEL = ac thin-film electroluminescence

ADC = analog-to-digital converter

AEED = avalanche electron emitting diode

APC-7 = standard precision 7-mm microwave coaxial connector

APD = avalanche photodiode

ASIC = application-specific integrated circuit

BARITT diode = barrier-injected transit-time diode

BBISHE injection = band-to-band tunneling-induced substrate hot electron injection

BC-P-MOSFET = buried-channel $p$-MOSFET

BCT = ballistic collection transistor

BEMBT = ballistic-electron MBT

BiCFET = bipolar inversion-channel FET

BiCMOS = bipolar complimentary MOS

BiCMOS-II = 1.2 μm BiCMOS

BiCMOS-III = 1.0 μm BiCMOS

BiCMOS-IV = 0.8 μm BiCMOS

BiCMOS-V = 0.5 μm BiCMOS

BIENFET = bipolar-enhanced MOSFET

BOD = breakover diode

BRAQWETS = blocked-reservoir and quantum-well electron transfer structure

BTBT = band-to-band tunneling

BWO = backward wave oscillator

CCD = charge-control device

CCNR = current-controlled negative resistance

CMOS = complementary MOS

CPU = central processing unit

CPW = coplanar waveguide

CTT = composed trench transistor

DASD = direct access storage device

DBR-LD = distributed-Bragg reflector LD

DCEL = dc-EL

DDR = double drift region

DFB-LD = distributed feedback LD

DICE = drain-induced channel enlargement(P)

DIMPATT = distributed IMPATT [diode]

DMOSFET = diffused metal oxide semiconductor FET

DMT = doped-channel heteroMISFET

DQW = double quantum wave(P)

DRAM = dynamic RAM

DRO = dielectric-resonator-oscillator

DSO = dielectric-stabilized oscillator

DSP = digital signal processing

DTL = diode-transistor logic

EAROM = electrically alterable read-only memory

ECL = emitter-coupled logic

ECR = electron cyclotron resonance

ECV = electrochemical capacitance-voltage
EED = electron energy distribution
EEPROM (or $E^2$ PROM) = electrically erasable/programmable ROM
EL = electroluminescence
EOS = electrical overstress
EPAL = EPROM programmable array logic [device]
EPROM = erasable and programmable ROM
ESD = electrostatic discharge
ESDS = electrostatic discharge sensitive
FCT = field-controlled thyristor
FD-SOI = fully depleted SOI [diode]
FDM = frequency-division multiplexing
FET = field effect transistor
FET HLDD = high-dielectric LDD
FIR filter = finite impulse response filter
FRED = fast-recovery epitaxial diode
FTR = functional throughput rate
GBW = gain-bandwidth [product]
GBSC-CPW = general broadside-coupled CPW
GILD = gas immersion laser doping(P)
GOLD = gate-overlapped LDD
GCPW = ground plane CPW
GMIC = glass microwave IC
HBT = heterojunction bipolar transistor
HCI = hot carrier injection
HCMOS = high-density CMOS
HEMT = high-energy mobility transistor
HFET = heterostructure FET
HIC = hybrid integrated circuit
HIGFET = heterostructure insulated-gate FET, heterojunction insulated-gate
HMIC = hybrid microwave integrated circuit
HMOS = high-performance MOS
HS-GOLD = halo source GOLD
HSINFET = hybrid Schottky injection FET
HTRB = high-temperature, reverse biased (P)
HTSC = high temperature superconductor
HVFET = high-voltage FET
HVIC = high-voltage IC
HVTFT = high-voltage TFT
IC = integrated circuit
IGBT = insulated gate bipolar transistor
IID = impurity-induced disordering [process]
IIR filter = infinite impulse response filter
ILO = injection-locked oscillator
IMPATT = impact avalanche and transit time [diode]
I/O = input/output
IRED = infrared emitting diode

ISFET = ion-sensitive FET
ITLDD = inverse-T gate LDD (P)
JFET = junction FET
JIT = just-in-time
LATID = large-tilt-angle implanted drain(P)
LATIPS = large-tilt-angle implanted punch-through stopper(P)
LCD = liquid crystal display
LD = laser diode
LDD = lightly doped drain (P)
LDMOST = lateral double-diffused MOST
LDS = lightly doped source (P)
LED = light-emitting diode
L-HBT = lateral HBT
LIC = linear IC
LIGBT = lateral IGBT
LIGT = lateral-insulated gate transistor
LNA = low noise amplifier
LO = local oscillator
LSA = limited space-charge accumulation [diode]
LSI = large-scale integration
LUFET = line-unified FET
LVTFT = low-voltage TFT
MAGT = MOS-assisted gate-triggered thyristor
MBT = metal base transistor
MC = Monte Carlo [technique]
MCT = MOS-controlled thyristor
MDD = moderately doped drain [structure]
MECI = multiplication-enhanced channel injection (P)
MECL = Motorola emitter-coupled logic
MESFET = metal-semiconductor FET
METL = multiplication equivalent threshold lowering (P)
MFD = microelectronic functional device
MIB = multiplication-induced breakdown
MIBI = multiplication-induced bulk injection (P)
MIDFET = metal-insulator-doped-semiconductor FET
MIS = metal-insulator-semiconductor
MISFET = MIS FET
MISIM CCD = metal-insulator-semiconductor-insulator-metal charge-control device
MITATT = mixed tunneling avalanche transit time [diode]
MLC = multilayer ceramic, multilayer capacitor
MMIC = monolithic microwave IC
MNOS = metal-nitride-oxide semiconductor

MODFET = modulation-doped FET
MOS = metal-oxide semiconductor
MOSFET = MOS FET
MQW-LD = multiple quantum well LD
MS = microstrip
MSI = medium-scale integration
MSM = metal-semiconductor-metal
MSM-HEMT = MSM-HEMT
MSM-SBPD = MSM Schottky barrier photodetector(diode)
MSW = magnetostatic wave(P)
NDR = negative differential resistance
NMOS = $n$-channel MOS
NSCPW = nonsymmetrically shielded CPW
NVRAM = nonvolatile RAM
OCVD = open-circuit voltage decay [technique]
OEIC = optoelectronic IC
PEL = pixel
PIC = photonic IC
PIN diode = $p$-intrinsic-$n$ diode
PLD = programmable logic device
PLL = phase-locked loop
PMOS = $p$-channel MOS
PROM = programmable read-only memory
PTD = post-tuning drift
PSD = power semiconductor device
QWITT = quantum-well injection transit time [diode]
QW MI³SFET = quantum well metal insulator inverted-interface semiconductor FET
RAM = random-access memory
RESP = resistive Schottky barrier field plate
RESURF LDMOST = reduced surface field double-diffused lateral
RHET = resonant-tunneling hot electron transistor
RISC = reduced-instruction set computer
ROM = read-only memory
RTD = resonant-tunneling diode
RTA = rapid thermal annealing
RTT = resonant tunneling transistor
SAW = surface acoustic wave
SBD = Schottky barrier diode
SCOS = substrate collector-shorted(P)
SC-N-MOSFET = surface channel $n$-MOSFET
SEBT = selective epitaxy base transistor
SGT = surrounding gate transistor
SHJ = single heterojunction(P)
SIMMWIC = silicon monolithic millimeter wave IC

SIS = superconductor-insulator-superconductor
SJLDD = sloped-junction lightly doped drain(P)
SLIC = subscriber line interface circuit
SMMW = submillimeter wave [length]
SPEL = self-aligning polysilicon electrode [technology]
SQUID = superconducting quantum interference device
SRD = step-recovery diode
SRAM = static RAM
SSC = spread-stacked capacitor
SSDMOSFET = spread source/drain MOSFET
SSI = small-scale integration
SSIMT = suppressed sidewall injection magnetotransistor
SSWS = static-safe work station
STW = [acoustic] surface transverse wave
STL = Schottky transistor logic
TBD = triangular barrier diode
TE = transverse electric
TED = transferred-electron device
TEGFET = two-dimensional electron gas FET
TEM = transverse electromagnetic(P)
TFT = thin-film transistor
TRAPATT = trapped plasma avalanche transit time [diode]
TTL = transistor-transistor logic
TUNNET = tunnel injection transit time [diode]
TWA = traveling-wave amplifier
TWT = traveling-wave tube
2DEG-HBT = two-dimensional electron gas HBT
ULSI = ultra-large-scale integration
UMOS FCT = MOS trench gate FCT
VDNR = voltage-controlled negative resistance
VDMOS = very deep MOS
VHSIC = very high speed IC
VLCS = vertical lateral
VLSI = very large scale integration
VMOS = v-groove MOS
VSWR = voltage standing-wave ratio
VTO = varactor-tuned oscillator
W-SBA-LD = wide-stripe self-aligned bent active layer LD
WSCL = water-soluble conductive layer
XMOS = high-speed MOS
YTO = YIG-tuned oscillator
ZMR = zone-melting recrystallization

## Miscellaneous

ADP = airborne data processor
AIChE = American Institute of Chemical Engineers
ASME = American Society of Mechanical Engineers
BIU = bus interface unit
CAD = computer-aided design
CAE = computer-aided engineering
CBW = chemical biological warfare
CIM = computer-integrated manufacturing
CMS = control, monitoring, and support
COMSEC = communications security
CRT = cathode ray tube
C3I = command, control, communications, and intelligence [system]
DAM = device adaptor module
DoD = Department of Defense
ECU = environmental control unit
EMP = [nuclear burst] electromagnetic pulse
ESM = electronic surveillance measure
HDTV = high-definition television
HEMP = high-altitude EMP
IR = infrared
IEEE = Institute of Electrical and Electronics Engineers
IEPS = International Electronics Packaging Society
ISHM = International Society of Hybrid Microelectronics
JEDEC = Joint Electronics Device Engineering Council
MBB = modular building block
MBC = MBB concept
MIL-STD = military standard
NRE = nonrecurring expense
PBX = private branch exchange
PDS = power distribution system
PME = prime mission equipment
RADC = Rome Air Development Center
SMTA = Surface Mount Technology Association
SPICE = simulation program for IC emphasis
TMC3IS = transportable military C3I system
UV = ultraviolet

## Materials Processing (Semiconductors, Thin, and Thick Films)

APCVD = atmospheric process CVD
AVD = axial chemical vapor deposition

BOX = buried oxide oxidation
CMP = chemical-mechanical polishing
CVD = chemical-vapor deposition
DI = dielectric isolation
ELO = epitaxial lateral overgrowth [technique]
FIPOS = full isolation by porous oxidized silicon
FMPBL = framed-mask polybuffered LOCOS isolation [process]
FOI = framed oxide isolation
FOLD = fully overlapped LDD (P)
GSMBE = gas-source MBE
HIPOX = high-pressure [thermal] oxidation
IOP-II = improved oxide and polysilicon isolation process
IVD = inside chemical vapor deposition (O)
LEC = liquid-encapsulated Czochralski [technique]
LOCOS = local oxidation of silicon
LPCVD = low-pressure chemical vapor deposition
LPE = liquid-phase epitaxy
LTE = low-temperature epitaxial [growth]
MBE = molecular beam epitaxy
MOCVD = metal-organic chemical vapor deposition
MOVPE = metal-organic vapor phase epitaxy
NOLOCOS = nitridized-oxide local oxidized(P)
NTD = neutron-transmutation-doped(P)
NO = nitrided oxide
OMVPE = organometallic vapor-phase epitaxy
OVD = outside chemical vapor deposition
PECVD = plasma-enhanced CVD
PIVD = plasma-assisted inside chemical vapor deposition
RIE = reactive-ion etch
ROI (or ROX) = recessed oxide isolation
RONO = reoxidized nitrided oxide
RTC = rapid thermal cleaning
RTO = rapid thermal oxidation
RTP = rapid thermal processing
SCC = stress-corrosion cracking
SEEW = selective epitaxy emitter window [technology]
SEG = selective epitaxial growth
SFOX = self-aligned field oxide [cell, structure]
SICOS = sidewall base-contact structure(P)
SIMOX = separation by implanted oxygen
SODIC = soot-deposited IC
SWAMI = sidewall-masked isolation

## Testing

AES = Auger electron spectroscopy
AQL = acceptable quality level
ATE = automatic test equipment
CLA = centerline average [surface roughness]
CTE = coefficient of thermal expansion
CDPA = destructive physical analysis
DPA = destructive physical analysis
DSC = differential scanning calorimetry
DTA = differential thermal analysis
DUT = device under test
EDS = energy-dispersive spectroscopy
ENR = excess noise ratio
ESCA = electron spectroscopy for chemical analysis
FA = failure analysis
FACI = first article configuration inspection
FMEA = failure mode and effects analysis
FMECA = failure mode, effects, and criticality analysis
FR = failure rate
HAST = highly accelerated stress testing
HPLC = high-pressure liquid chromatography
IRS = infrared scan
LTPD = lot tolerance percent defective
MMAVERIC = MMIC metrology with automatic verification-in-time
MTBF = mean time between failures
MTTF = mean time to failure
NDT = nondestructive test
NIRA = near-IR reflectance analysis
NPO = negative and positive zero
NTL = low-level noise tolerance
PDA = percent defect allowable
PIND = particle impact noise detection
PPM = parts per million
OTDR = optical time domain reflectometery
RGA = residual gas analysis
RI = receiving inspection
RTD = resistance-temperature detector [probe]
SAM = scanning acoustic microscope
    = scanning Auger microprobe
SEM = scanning electron microscopy
SIMS = secondary ion mass spectrometry
SLAM = scanning laser acoustic microscope
SRP = spreading resistance profiling [technique]
SPV = surface photovoltage [technique]
STM = scanning tunneling microscopy
TCC = temperature coefficient of capacitance
TCR = temperature coefficient of resistance

TEM = transmission electron microscopy
Tg = glass transition temperature
TGA = thermal gravimetric analysis
TLC = thin-layer chromatography
TMA = thermomechanical analysis
TSM = top side metallurgy
TTF = time to failure
UUT = unit under test
WUT = wafer under test

## Packaging

APC-7 = standard precision 7-mm microwave coaxial connector
ATAB = area-array tape-automated bonding
BCW = bare-chip-and-wire [hybrid]
BLM = ball-limiting metallurgy
BSM = backside metallurgy
BTAB = bumped tape-automated bonding
C and W = chip and wire
CC = chip carrier
CCC = ceramic CC
CEEE = common electronics equipment enclosure
CERDIP = ceramic DIP
C4 = controlled-collapse chip connector
COB = chip-on-board
CSI = compliant solderless interface
DCA = direct chip attach
DIP = dual in-line [package]
DWF = dice-in-wafer form
FPC = fine-pitch CC
FRU = field replaceable unit
HCC = hermetic chip carrier
HDCM = high-density ceramic module
HDI = high-density interconnect
ILB = inner lead bond
LCC = leadless CC
LCCC = leadless ceramic CC
LDS = low-dimensional structure (of British origin)
LMCH = leadless multiple-chip hybrids
LID = leadless inverted device
LMCH = leadless multiple-chip hybrid
LRU = line replacement unit or lowest repairable unit
MC = metallized ceramic
MCC = miniature CC
MCM = multichip module
MLB = multilayer board
MLC = multilayer ceramic
OLB = outer lead bond

PAA = pad area array
PCB = printed circuit board
PCC = same as PLCC
PCI = pressure contact interconnection
PCR = plastic CC, rectangular
PES = porcelain-enamel-steel
PET = porcelain-enamel technology
PDIP = plastic DIP
PGA = pin grid array [package]
PIP = pin insertion [package]
PLCC = plastic-leaded CC
PRN = priority ranking number
POS = porcelain-on-steel
PQFP = plastic quad flat [package]
PSCR = photosensitive copper reduction process
PTF = polymer thick film
PTH = plated-through hole: pin-through hole
PWB = printed wiring board
QFB = quad flat butt-leaded [package]
QUIP = quad in-line [package]
SCM = single-chip module
SIP = single in-line [package]
SLAM = single-layer alumina metallized (P)
SLC = single-layer ceramic
SMA = surface-mounted assembly
SMMCC = standard miniature microwave co-axial connector
SMD = surface-mounted device
SOB = small-outline butt leaded [package]
SMT = surface mount technology
SOIC = small-outline IC
SOJ = small-outline J-lead [package]
SOP = small-outline [package]
SOS = silicon-on-sapphire
SOI = silicon-on-insulator
TAB = tape-automated-bonding
TAB-OB = TAB On-Board
TC = thermocompression [bond]
TCM = thermal conduction module
TS = thermosonic [bond]
TSOP = thin small outline [package]
TFSOI = thin-film SOI (P)
TO = transistor outline [package]
US = ultrasonic [bond]
VSO = very small outline [package]
WSI = wafer-scale integration
ZIF = zero insertion force
ZIP = zigzag in-line [package]

## Materials

ADP = ammonium dihydrogen phosphate
AlAs = aluminum arsenide

AlGaAs = aluminum gallium arsenide
AlInAs = aluminum indium arsenide
AlN = aluminum nitride
$Al_2O_3$ = aluminum oxide (alumina)
AlSb = aluminum antimonide
As = arsenic
Au = gold
AuGe = gold-germanium
AuSi = gold-silicon
AuSn = gold-tin
B = boron
BN = boron nitride
Ba = barium
$BaTiO_3$ = barium titanate
Be = beryllium
BeO = beryllium oxide (beryllia)
C = carbon; graphite; diamond
CDA = clean dry air
CFC = chlorinated fluorcarbon
DI = deionized water
FR4 = epoxy resin glass laminate (EIA)
Ga = gallium
GaAs = gallium arsenide
GaAsP = gallium arsenide phosphide
GaInAs = gallium indium arsenide
GaP = gallium phosphide
GaSb = gallium antimonide
Ge = germanium
$H^+$ = atomic hydrogen
$H_2$ = hydrogen molecule
$H_2O$ = water
$H_2O_2$ = hydrogen peroxide
In = indium
InP = indium phosphide
InSb = indium antimonide
IPN = interpenetrating polymer network
$Mn(NO_3)_2 \cdot 4H_2O$ = manganous nitrate
$MnO_2$ = manganese dioxide
$N_2$ = nitrogen molecule
$NO_2$ = Nitrogen peroxide
$NO_3$ = nitrogen trioxide
$N_2O$ = nitrous oxide
$O_2$ = oxygen molecule
$O_3$ = ozone molecule
$OH^-$ = hydroxyl ion
P = phosphorus
PLZT = lead-lanthanum-zirconate-titanate
PMMA = polymethyl methacrylate
PSG = phosphosilicate glass
PVC = polyvinyl chloride
PX = paraxylylene
PZT = lead-zirconate-titanate
Sb = antimony

Si = silicon
SiC = silicon carbide
SiN = silicon nitride
SiO = silicon monoxide
$SiO_2$ = silicon dioxide (silica)
    = quartz (crystalline silica)
    = fused silica (silica glass)
Sn = tin
Ta = tantalum
$Ta_2O_5$ = tantalum pentoxide
TEOS = tetraethoxysilane
Ti = titanium
TiN = titanium nitride
$TiO_2$ = titanium dioxide (titania)
$TiSi_2$ = titanium silicide
Y = yttrium
YAG = yttrium aluminum garnet
YIG = yttrium iron garnet
Zn = zinc
ZnO = zinc oxide
ZnS = zinc sulfide

## Symbols

$Z_0$ = characteristic impedance
$\alpha$ = coefficient of thermal expansion (linear)
$\sigma$ = conductivity
$T_c$ = Curie temperature
$\rho$ = density

$\epsilon$ = dielectric constant (permittivity) $\epsilon = D\text{-}E^{-1}$
$P$ = dielectric polarization $P = \epsilon_0 E - D$
$D$ = electric displacement
$E$ = electric field strength
$\phi$ = electronic work function; magnetic flux
$E_g$ = energy gap
$E_F$ = Fermi energy
$R_H$ = Hall coefficient
$\delta$ = loss angle
$\Phi$ = luminous flux
$H$ = magnetic field strength
$B$ = magnetic induction $B = \mu H$
$J$ = magnetic polarization $J = B - \mu_0 H$
$II$ = Peltier coefficient
$\mu$ = permeability
$\mu_0$ = permeability of vacuum
$\epsilon_0$ = permittivity of vacuum
$d_{mn}$ = piezoelectric coefficient
$X$ = reactance
$R$ = resistance
$S$ = Seeback coefficient
$T$ = temperature
$\lambda$ = thermal conductivity
$t$ = time
$\tau$ = time interval
$v$ = viscosity
(P) = prefix; (T) = technique

# Appendix B:
# Glossary of Hybrid Microcircuits Packaging Terms

The ISHM Glossary of Hybrid Circuit Terms (Updated Edition-Summer, 1987) by George S. Szekely was modified for use in this glossary.

**ABRASIVE TRIMMING:** Trimming a film resistor to its nominal value by notching resistor with a finely adjusted stream of an abrasive material, such as aluminum oxide, directly against the resistor surface.

**ABSOLUTE ZERO:** The lowest temperature attainable. All molecular activity is considered to cease. Its value is 0 °K ($-273.15$ °K).

**ACCEPTANCE TESTS:** Tests agreed upon by vendor and customer to determine acceptability of product.

**ACTIVATING:** A process used to sensitize a nonconductive material to facilitate electroless deposition.

**ACTIVE AREA (OF A PACKAGE):** Internal area of package bottom, usually a cavity, that actually is used for substrate attachment. The term preferably is applied to package cases of all-metal construction (as opposed to glass or ceramic).

**ACTIVE COMPONENTS:** Electronic components, such as transistors, diodes, electron tubes and thyristors, which can operate on an applied electrical signal so as to change its basic characters; e.g., rectification, amplification, switching.

**ACTIVE DEVICES:** Discrete devices such as diodes or transistors; or integrated devices, such as analog or digital circuits in monolithic or hybrid form.

**ACTIVE ELEMENT:** An element of a circuit in which an electrical input signal is converted into an output signal by the nonlinear voltage-current relationships of a semiconductor device (see active components.)

**ACTIVE SUBSTRATE:** A substrate in which active and passive circuit elements may be formed to provide discrete or integrated devices.

**ACTIVE TRIM:** Trimming of a circuit element (usually resistors) in a circuit that is electrically activated and operating to obtain a specified functional output for the circuit (see Functional trimming.)

**ADD-ON COMPONENT (OR ADD-ON DEVICE):** Discrete or integrated prepackaged or chip components that are attached to a film network to complete the circuit functions.

**ADDITIVE PATTERNING:** Processing a hybrid circuit substrate by sequentially depositing conductive, resistive, and insulative materials, each through a mask, thus defining the contours and areas of traces, pads, and elements.

**ADDITIVE PROCESS:** A process used to selectively deposit electrically conductive material on a clad or unclad base material.

**ADHESION:** The property of one material to remain attached to another; a measure of the bonding strength of the interface between film deposit and the surface which receives the deposit; the surface receiving the deposit may be another film or substrate.

**ALLOY:** A combination of two or more metallic elements in the form of a solid solution of one or more metals in another metal, or distinct phases, or components, of an alloy.

**ALUMINA:** Aluminum oxide ($Al_2O_3$), a ceramic material often used for substrates or in ceramic bottom construction packages. Most alumina substrates contain 90–99% aluminum oxide.

**AMBIENT TEMPERATURE:** Temperature of atmosphere in intimate contact with the electrical part or device.

**AMPHOTERIC:** Literally, partly one, and partly the other. Specifically, capable of reacting either as an acid or as a base. Amphoteric metallic coating materials may be applied to the backside of silicon VLSI wafers to promote adhesion when die bonding in hybrid assembly.

**ANALOG CIRCUITS:** Circuits that provide a continuous (vs. discontinuous) relationship between the input and output.

**ANGLE OF ATTACK:** The angle between the squeegee face of a thick-film printer and the plane of the screen.

**ANGLED BOND:** Bond impression of first and second bond are not in a straight line.

**ANNULAR RING:** The conductive ring surrounding a hole.

**ANNEALING:** A method of toughening certain materials and decreasing their brittleness by heating followed by slow cooling; e.g., heating of a film resistor followed by slow cooling to relieve stresses and stabilize the resistor material.

**ANODIZATION:** An electrochemical oxidation process used to change the value of thin film resistors or prepare capacitor dielectrics.

**ANTISTATIC:** Electrostatic discharge (ESD)–protective material resisting triboelectric charging; usually some form of (impregnated) plastic, identified an antistatic, with surface resistivity controlled to be less than $10^{14}$ ohms/square and more than $10^9$ ohms/square.

**ARRAY:** A group of elements or circuits arranged in rows and columns on one substrate.

**ARTWORK:** The accurately scaled configuration or pattern produced usually at an enlarged ratio, to enable the product to be made therefrom by photographic reduction to a $1:1$ working pattern; layouts and photographic films which are created to produce the working thick film screens and thin film masks.

**ARTWORK MASTER:** A dimensionally accurate imaged pattern (usually Mylar or glass) used to manufacture production masters for use in photoimaging processes.

**AS-FIRED:** Values of thick film resistors or smoothness of ceramic substrates as they come out of the firing furnace and, respectively, prior to trimming and polishing (if required).

**ASPECT RATIO:** The ratio between the length of a film resistor and its width; equal to the number of squares of the resistor.

**ASSEMBLY:** A film circuit to which discrete components have been attached. It might also show the assembly of one or more film circuits which may include several discrete components.

**ASSEMBLY DRAWING:** A drawing showing all the components and interconnections mounted or soldered to the film circuit in their proper position. It might also show the assembly of one or more film circuits which may include several discrete components.

**ATTACK ANGLE:** See Angle of attack.

**AUTOMATIC INSERTION DIP:** A dual in-line package which resembles a flat package in which leads are widened at a certain distance to allow bending to form two parallel rows to position the package at a predetermined distance above the board.

**AXIAL LEADS:** Leads coming out of the ends of a discrete component or device along the central axis rather than out the sides.

**AZEOTROPIC SYSTEM:** A system of a number of liquid compounds which has specific boiling point for a particular composition.

**BACK BONDING:** Bonding active chips to the substrate using the back of the chip, leaving the face, with its circuitry face up. The opposite of back bonding is face-down bonding.

**BACK MOUNTING:** See Back bonding.

**BACK RADIUS:** The radius of the trailing edge of a bonding tool foot.

**BACKFILL:** Filling an evacuated hybrid circuit package with a dry inert gas prior to hermetically sealing.

**BACKPLANE:** Interconnection layer for a multilayer hybrid.

**BAKE-OUT:** Subjecting an unsealed hybrid circuit package to an elevated temperature to bake out moisture and unwanted gases prior to final sealing.

**BALL BOND:** A bond formed when a ball-shaped end interconnecting wire is deformed by thermocompression against a metallized pad. The bond is also designated a nail head bond from the appearance of a flattened ball.

**BARIUM TITANATE (BaTiO$_3$):** The basic raw material used to make high-dielectric constant ceramic capacitors. Used also in high-$\epsilon$ thick film ceramic pastes.

**BARRELL PLATING:** A technique of plating large numbers of small parts in a rotating drum.

**BASE MATERIAL:** The rigid or flexible dielectric material or insulated metal sheet that forms the base for a conductive pattern.

**BATCH PROCESSING:** Manufacturing method whereby a particular process sequence operates on a large number of components simultaneously.

**BATHTUB PACKAGE:** A boxlike solid sidewall package wherein the substrate is mounted.

**BEAM LEAD:** A long structural member not supported everywhere along its length and sub-ject to the forces of flexure, one end of which is permanently attached to a chip device and the other end intended to be bonded to another material, providing an electrical interconnection or mechanical support or both.

**BEADS:** See Glass Preforms.

**BEAM LEAD DEVICE:** An active or passive chip component possessing beam leads as its primary interconnection and mechanical attachment means to a substrate.

**BELLOWS CONTACT:** A flat folded spring connector contact providing uniform spring rate over the mating contact tolerance range.

**BERYLLIA OR BERYLLIUM OXIDE (BeO):** A substrate material used where high thermal conductivity is needed.

**BIFURCATED CONTACT:** A flat length-wise-slotted contact that permits independent contact points.

**BINDERS:** Materials added to thick film compositions and unfired substrates to given sufficient strength for prefire handling.

**BLACK OXIDE:** A technique for making matched seals by forming a controlled thickness oxide on each of the mating surfaces of a package prior to sealing.

**BLANK:** A rough-dimensioned printed board cut from a panel of base material that is either unprocessed or partially processed.

**BLEEDING:** The lateral spreading or diffusion of a printed film into adjacent areas, beyond the geometric dimensions of the printing screen. This may occur during drying, or firing.

**BLENDING:** Different viscosities of the same types of materials may be blended together to achieve intermediate viscosities. This term is also applied to resistive inks that can be blended with each other to achieve intermediate resistivities.

**BLIND VIA:** An interstitial via connecting a printed board top conductive layer with one or more inner conductive layers.

**BLISTERS:** Raised parts of a conductor or resistor formed by the outgassing of the binder or vehicle during the firing cycle.

**BLOCK:** To plug up open mesh in a screen to prevent resistor or conductor pastes from being deposited in unwanted areas.

**BLOCK DIAGRAM:** A circuit diagram in which the essential units of the functional system are drawn in the form or blocks and the relationship between blocks is indicated by appropriate connecting lines.

**BLOCK-OFF:** See Block.

**BLOW HOLE:** A surface anomaly formed by outgassing as solder freezes.

**BOAT:** A container for materials to be vacuum-evaporated or fixed.

**BOILING POINT:** The temperature of a liquid at which its vapor pressure is equal to the pressure of the atmosphere surrounding the fluid.

**BOMB:** A chamber for applying various levels of positive or negative pressure on a package or component.

**BOND:** An interconnection which performs a permanent electrical and/or mechanical function.

**BOND DEFORMATION:** The change in the form of the lead produced by the bonding tool, causing plastic flow, in making the bond.

**BOND ENVELOPE:** The range of bonding parameters over which acceptable bonds may be formed.

**BOND INTERACE:** The interface between the lead and the material to which it was bonded on the substrate.

**BOND LIFT-OFF:** The failure mode whereby the bonded lead separates from the surface to which it was bonded.

**BOND-OFF:** See Bond life-off.

**BOND PAD:** See Bonding area.

**BOND SCHEDULE:** The values of the bonding machine parameters used when adjusting for bonding. For example, in ultrasonic bonding, the values of the bonding force, time, and ultrasonic power.

**BOND SEPARATION:** The distance between the attachment points of the first and second bonds of a wire bond.

**BOND SITE:** The portion of the bonding areas where the actual bonding took place (see Bonding area).

**BOND STRENGTH:** In wire bonding, the pull force at rupture of the bond interface measured in the unit gram-force.

**BOND-TO-BOND DISTANCE:** The distance measured from the bonding site on the die to the bond impression on the post, substrate land, or fingers, which must be bridged by a bonding wire or ribbon.

**BOND-TO-CHIP DISTANCE:** In beam lead bonding the distance from the heel of the bond to the component.

**BOND TOOL:** The instrument used to position the lead(s) over the desired bonding area and impart sufficient energy to the lead(s) to form a bond.

**BONDABILITY:** Those surface characteristics and conditions of cleanliness of a bonding area which must exist in order to provide a capability for successfully bonding an interconnection material by one of several methods, such as ultrasonic or thermocompression wire bonding.

**BONDING, DIE:** Attaching the semiconductor chip to the substrate, either with an epoxy, eutectic or solder alloy.

**BONDING AREA:** The area, defined by the extent of a metallization land or the top surface of the terminal, to which a lead is or is to be bonded.

**BONDING ISLAND:** Same as Bonding pad.

**BONDING PAD:** A metallized area at the end of a thin metallic strip to which a connection is to be made.

**BONDING WIRE:** Fine gold or aluminum wire for making electrical connections in hybrid circuits between various bonding pads on the semiconductor device substrate and device terminals or substrate lands.

**BOROSILICATE GLASS:** A sealing glass providing a close coefficient of expansion match between some metal leads and ceramic or glass packages.

**BOW:** A flatness deviation caused by a roughly cylindrical curvature of a board with the corners remaining in the same plane.

**BRAZE:** A joint formed by a brazing alloy (v.). To join metals with a nonferrous filler metal at temperatures above 427°C.

**BRAZING:** Also called hard soldering. Similar to soldering, it is the joining of metals with a nonferrous filler metal at temperatures above 427°C.

**BREAK LOAD:** See Bond strength.

**BREAKAWAY:** In screen printing the distance between the upper surface of the substrate and the lower surface of the screen when the screen is not deflected by the squeegee.

**BREAKDOWN VOLTAGE:** The voltage threshold beyond which there is a marked (almost infinite rate) increase in electrical current conduction.

**B-STAGED RESIN (B-STAGE):** A partially cured resin which is used in the lamination process (see also Prepreg).

**BUGGING HEIGHT:** The distance between hybrid substrate and the lower surface of the beam lead device which occurs because of deformation of beam leads during beam lead bonding.

**BULK CONDUCTANCE:** Conductance between two points of a homogeneous material.

**BURIED VIA:** An interstitial via connecting inner conductive layers of a printed board but not extending to either surface.

**BURN-IN:** The process of electrically stressing a device (usually at an elevated temperature environment) for an adequate period of time to cause failure of marginal devices.

**BURN-IN (DEBUGGING):** A reliability conditioning procedure which is a method of aging by operating the equipment under specified environmental test conditions in accordance with an established test conditioning specification. The purpose is to eliminate early failures and thus age or stabilize the operation of the equipment.

**BURN-OFF:** See Flame-off.

**BUS BAR:** A conductor usually used to distribute high current from power supplies to backplanes; also conductors for distributing power on printed boards.

**CAMBER:** A term that describes the amount of overall warpage present in a substrate, camber $= C/D$ where $C = z$-displacement, $D =$ diagonal (longest distance) of the substrate surface. The out-of-plane deflection of a flat cable or flexible laminate of specified length.

**CAPACITANCE DENSITY:** Also referred to as sheet capacity. A term used to describe the amount of capacitance available per unit area ($pF\text{-}mm^{-2}$ or $\mu F\text{-}cm^{-2}$).

**CAPACITIVE COUPLING:** Electrical interaction between two conductors due to their shared capacitance.

**CAPILLARY:** A hollow bonding tool used to guide the bonding wire and to apply pressure to the wire during the bonding cycle.

**CAPILLARY TOOL:** A tool used in bonding where the wire is fed to the bonding surface of the tool through a bore located along the long axis of the tool.

**CATALYST:** Any substance which affects the rate of chemical reaction, but which itself may be recovered unchanged to the end of the reaction.

**CENTERLINE AVERAGE (CLA):** The arithmetic average (AA) of measured deviations in a surface profile from an imaginary mean centerline located between the peaks and valleys. The rms reading for a given surface finish is about 11% higher than the AA reading.

**CENTERWIRE BREAK:** The failure mode in a wire pull test where the wire fractures at approximately midspan.

**CENTRIFUGE:** Testing the integrity of bonds in a hybrid circuit by spinning the circuit at a high rate of speed, thereby imparting a high g loading on the interconnecting wire bonds and bonded elements.

**CERAMIC:** Inorganic nonmetallic material such as alumina, beryllia, steatite, or forsterite, whose final characteristics are produced by treatment at high temperatures (and sometimes also under applied pressure), often used in microelectronics as parts of components, substrate, or package.

**CERMET:** A solid homogeneous material usually consisting of a finely divided admixture of a metal and ceramic in intimate contact. Cermet thin films are normally combinations of dielectric materials and metals.

**CHARACTERISTIC IMPEDANCE:** The voltage-to-current ratio of an electrical signal propagating through a transmission line.

**CHASE:** A vice-like tool or instrument with adjustable, flexible draw bars for prestretching wire-mesh cloth prior to installing it in a thick screen frame. The wire mesh that is to be gauged for prescribed tension.

**CHEMICAL VAPOR DEPOSITION:**
Depositing circuit elements on a substrate by chemical reduction of a vapor on contact with the substrate.

**CHESSMAN:** The disk, knob, or lever used to manually control the position of the bonding tool with respect to the substrate.

**CHIP:** The uncased and normally leadless form of an electronic component part, either passive or active, discrete or integrated.

**CHIP-AND-WIRE:** A hybrid technology employing face-up-bonded chip devices exclusively, interconnected to the substrate conventionally; e.g., by flying wires.

**CHIP CARRIER:** A special type of enclosure or package to house a semiconductor device or a hybrid microcircuit and which has metallized castellations as usually electrical terminations around its perimeter, as well as solderpads on its underside, rather than an extended lead frame or plug-in pins.

**CHIP-IN-TAPE:** See Tape Automated Bonding (TAB).

**CHISEL:** A specially shaped bonding tool in the shape of a chisel used for wedge bonding and ultrasonic bonding of aluminum or gold wires to elements or package leads.

**CHLORINATED HYDROCARBON SOLVENTS:** See Halogenated hydrocarbon solvents.

**CHOPPED BOND:** Those bonds with excessive deformation such that the strength of the bond is greatly reduced.

**CHUCK:** Portion of the bonding machine that holds the unit to be bonded.

**CIRCUIT:** The interconnection of a number of electrical elements and/or devices, performing a desired electrical functions. A circuit must contain one or more active elements (devices) in order to distinguish it from a network.

**CIRCUMFERENTIAL SEPARATION:** A crack formed around the circumference of a plated through hole or any other cylindrical interconnection made by plating or solder.

**CLAD:** A condition of one or both sides of the base material characterized by a thin layer of bonded metal foil.

**CLAMPING FORCE:** Force applied to a bonding tool to effect a bond.

**CLEAN ROOM:** A special manufacturing area where the air is filtered to remove dust particles and precuationary measures are used to keep contamination away from the unprotected circuit during processing.

**CLINCH:** A method of mechanically securing components prior to soldering, by bending that portion of the component lead that extends beyond the lip of the mounting hole, against a pad area.

**CLINCHED LEAD:** The formed portion of a component lead extending through a hole in a printed board which prevents the component from falling out during the solder process.

**COATED METAL-CORE SUBSTRATE:** A substrate consisting of a glossy, inorganic coating bonded to metal by firing at a temperature about 500°C.

**COEFFICIENT OF EXPANSION:** The fractional change in dimension of a material for a unit change in temperature.

**COEFFICIENT OF THERMAL EXPANSION:** The ratio of the change in length to the change in temperature.

**COEFFICIENT OF VISCOSITY:** A measure of the tendency of a fluid to resist shear. The unit of viscosity is the poise.

**COFIRING:** Processing thick-film conductors and resistors through the firing cycle at the same time.

**COINED:** A screen which contains the impression of a substrate because it has been subject to abuse is said to be coined. The term can also refer to a screen manufactured with a coined impression for the purpose of screening media in a special designed substrate with a cavity that standard screens cannot achieve. It is also used to describe the process by which the base of a package has been formed.

**COINED LEAD:** A cylindrical lead that has been flattened to form a ribbon lead.

**COINING:** See Coined.

**COLD SOLDER CONNECTION:** A soldered connection where the surfaces being bonded moved relative to one another while the solder was solidifying, causing an uneven solidification structure which may contain microracks. Such cold joints are usually dull and grainy in appearance.

**COLD SOLDER JOINT:** A poorly wetted, grayish-colored solder joint due to insufficient heating, solder contaminants, or inadequate pre-solder cleaning.

**COLD-WELD:** Form a hermetic seal in a metal package by welding the lid to the frame using pressure alone.

**COLLECTOR ELECTRODE:** The metallized bonding pad making ohmic contact with the collector of a transistor element.

**COMB PATTERN:** A test pattern formed on a substrate in the form of a comb.

**COMPATIBLE MATERIALS:** Materials that can be mixed or blended or brought into contact with each other with minimum reaction or separation taking place; or each material added will not degenerate the performance of the whole.

**COMPENSATION CIRCUIT:** A circuit which alters the functioning of another circuit to which applied with the goal of achieving a desired performance; temperature and frequency compensation are the most common.

**COMPENSATION NETWORK:** Same as Compensation circuit.

**COMPLEX ARRAY:** An array of integrated devices in which a large number of elements are integral to each device.

**COMPLIANT BOND:** A bond which uses an elastically and/or plastically deformable member to impart the required energy to the lead. This member is usually a thin metal foil that is expendable in the process.

**COMPLIANT MEMBER:** The elastically and/or plastically deformable medium which is used to impart the required energy to the lead(s) when forming a compliant bond.

**COMPONENT:** A diversely used term, which, dependent on context, may mean active or passive element, device, integrated or functional circuit, functional unit, or part of an operating system.

**COMPONENT DENSITY:** The number of components per unit area.

**COMPOUND (CHEMICAL):** A substance consisting of two or more elements chemically united in definite proportions by weight.

**COMPRESSION SEAL:** A seal made between an electronic package and its leads. The seal is formed as the heated metal, when cooled, shrinks around the glass insulator, thereby forming a tight joint.

**CONDITIONING:** Exposure of a test specimen to a specified environment(s) for a limited time prior to testing.

**CONDUCTIVE ADHESIVE:** An adhesive material that has metal powder added to increase electrical conductivity.

**CONDUCTIVE EPOXY:** An epoxy material (polymer resin) that has been made conductive by the addition of a metal powder, usually gold or silver.

**CONDUCTIVE FOIL:** The conductive material covering one or both sides of the base material used to form the conductive pattern.

**CONDUCTIVE PATTERN:** The design or configuration formed in the conductive foil.

**CONDUCTIVITY:** The ability of a material to conduct electricity, the reciprocal of resistivity.

**CONDUCTOR:** A single conductive path formed in the conductive foil.

**CONDUCTOR BASE SPACING:** The spacing between conductors at the surface of the base material.

**CONDUCTOR BASE WIDTH:** The width of the conductors at the surface of the base material.

**CONDUCTOR LAYER:** The entire conductive layer formed on one side of the base material.

**CONDUCTOR SPACING:** The distance between adjacent conductor edges.

**CONDUCTOR WIDTH:** The width of individual conductors in a conductive film pattern.

**CONDUCTORS:** A class of materials that conduct electricity easily, i.e., have low resistivity ($\leq 10^{-4}$ cm$^{-1}$).

**CONFIDENCE INTERVAL:** The maximum and minimum limits defining the range within which it is expected that an observation will occur in accordance with a degree of certainty as dictated by the confidence level.

**CONFIDENCE LEVEL (ONE-TAIL):** The degree of certainty, expressed as a percentage, that a given hypothesis is true within a specified limit and that the probability of its being outside this limit is proportional to the area of the tail of the probability distribution which exceeds the limit.

**CONFIDENCE LEVEL (TWO-TAIL):** The degree of certainty, expressed as a percentage, that a given hypothesis is true within a specified limits and that the probability of its being outside these limits is proportional to the area in the left and right tails of the probability distribution.

**CONFIDENCE LIMITS:** Boundaries which take sampling and other statistical fluctuations into consideration.

**CONFORMAL COATING:** A thin nonconductive coating, either plastic (e.g., poly-p-xylylene) or inorganic, applied to a circuit for environmental and/or mechanical protection.

**CONTACT ANGLE:** The angle made between the bonding material and the bonding pad.

**CONTACT AREA:** The common area between two mating conductors permitting the flow of electricity.

**CONTACT PRINTING:** A method of screen printing where the screen is almost (within a few mils) in contact with the substrate. Used for printing with metal mask.

**CONTACT RESISTANCE:** In electronic elements, such as capacitors or resistors, the apparent resistance between the terminating electrode and the body of the device. The electrical resistance between two mating conductors.

**CONTACT SPACING:** The centerline-to-centerline spacing of adjacent contact areas.

**CONTINUITY:** The uninterrupted path of current flow in an electrical circuit.

**CONTINUOUS BELT FURNACE:** A firing furnace that has a continuous belt carrying the unfired substrates through the firing cycle.

**CONTROLLING COLLAPSE:** Controlling the reduction in height of the solder balls in a flip-chip processing operation.

**COORDINATOGRAPH:** A drafting machine of great accuracy used in making original artwork for integrated circuits or microcircuits.

**COPLANER LEADS (FLAT LEADS):** Ribbon-type leads extending from the sides of the circuit package, all lying in the same plane.

**COPLANARITY:** Property of lying in the same plane.

**CORONA:** The flow of small erratic current pulses resulting from discharges in voids in a dielectric during the voltage stress; also discharge resulting from ionization of gas surrounding a conductor (frequently luminous) which occurs when the potential gradient exceeds a certain value but is not sufficient to cause sparking.

**COUPLING CAPACITOR:** A capacitor that is used to block dc signals, and to pass high-frequency signals between parts of an electronic circuit.

**COVER LAYER:** Outer layer of insulating material covering the conductive pattern on the printed board.

**CRACKING:** Breaks in the metallic and/or nonmetallic layers extending through to an underlying surface.

**CRATERING:** Defect in which portion of the chip under ultrasonic bond is torn loose by excessive amount of energy transmitted through the wire bond leaving a pit.

**CRAZING:** Fine cracks which may extend on or through layers of plastic of glass materials.

**CREEP:** The dimensional change with time of a material under load.

**CRITICAL PRESSURE:** The pressure under which a substance may exist as a gas in equilibrium with the liquid at the critical temperature.

**CRITICAL TEMPERATURE:** The temperature above which a gas cannot be liquefied by pressure alone.

**CROP MARKS:** Marks at the corners of the artwork used to locate the borders and establish the board contour.

**CROSSHATCHING:** A pattern of voids used for breaking up a conductive layer.

**CROSSOVER:** The transverse crossing of metallization paths without mutual electrical contact and achieved by the deposition of an insulating layer between the conducting paths at the area of crossing.

**CROSS TALK:** Signals from one line leaking into another nearby conductor because of capacitance or inductive coupling or both (e.g., owing to the capacitance of a thick-film crossover).

**CRYOGENICS:** The science involving phenomena at very low temperatures less than -100°C.

**CRYSTAL GROWTH:** The formation of crystals in a material over a period of time and at an established temperature.

**C-STAGED RESIN (C-STAGE):** Completed cured resin.

**CURE TIME:** The total elapsed time between the addition of a catalyst and the complete hardening of a material; also the time for hardening of premixed, frozen, or refrigerated epoxy adhesives.

**CURIE TEMPERATURE (CURIE POINT):** Critical temperature above which ferromagnetic materials lose their permanent spontaneous magnetization and ferroelectric materials lose their spontaneous polarization.

**CURING AGENT:** A material which when added to a second material activates a catalyst already present in the second material, thereby bringing about a chemical reaction, usually causing a hardening of the entire mass.

**CURLS:** Extruded material coming out from the edge of a bond.

**CURRENT CARRYING CAPACITY:** The maximum current which can be continuously carried by a circuit without causing objectable degradation of the electrical or mechanical properties.

**CUSTOM CIRCUITS:** Circuits designed to satisfy a single application requirement (hybrid or monolithic).

**CUT AND STRIP:** A method of producing artwork using a two-ply laminated plastic sheet, by cutting and stripping off the unwanted portion of the opaque layer from the translucent layer, leaving the desired art work configuration.

**CUTOFF:** The operation following the final bonding step that separates the bond from the wire magazine.

**CUTOFF SCISSORS:** The scissors on a bonder to sever the wire after bonding.

**CYCLIC STRESS:** A completed circuit subjected to stress by cycling temperature and load over a period of time to cause premature failure.

**DC VOLTAGE COEFFICIENT:** The measure of changes in the primary characteristics of a circuit element as a function of the voltage stress applied.

**DEFECT:** A characteristic which does not conform to applicable specification requirements and adversely affects or potentially affects the quality of a device.

**DEFINITION:** The sharpness of a screen printed pattern—the exactness with which a pattern is printed.

**DEGRADATION:** Change for the worse in the characteristics of an electric element because of heat, high voltage, etc. A gradual deterioration in performance as a function of time.

**DEIONIZED WATER:** Water that has been purified by removal of ionizable materials.

**DELAMINATION:** A separation between the laminated layers of base material and/or base material and conductive foil.

**DELTA [Δ] LIMIT:** The maximum change in a specified parameter reading which will permit a hybrid microcircuit to be accepted on the specified test, based on a comparison of the final measurement with a specified previous measurement. (Note: When expressed as a percentage value, it shall be calculated as a proportion of the previous measured value.)

**DENDRITIC GROWTH:** Growth of metallic filaments between conductors due to condensed moisture and electrical bias.

**DENSITY:** Weight per unit volume of a substance.

**DERATING:** The practice of subjecting parts or components to lesser electrical or mechanical stresses than they can withstand in order to increase the life expectancy of the part or component.

**DETRITUS:** Loose material, dislodged during resistor trimming but remaining in the trimmed area.

**DEVICE:** A single discrete electronic element such as a transistor or resistor, or a number of elements integrated within one die, which cannot be further reduced or divided without eliminating its stated function. Preferred usage is die or dice, bare or prepackaged.

**DEVITRIFICATION:** The action or process of devitrifing or state of being devitrified—the conversion of a glassy matter into crystalline. Contamination, e.g., grease, will accelerate devitrication.

**DEVITRIFY:** To deprive of glassy luster and transparency—to change from a vitreous to a crystalline condition.

**DEWETTING:** The condition in a soldered area in which liquid solder has not adhered intimately and has pulled back from the conductor area.

**DEW POINT:** The temperature at which liquid first condenses when a vapor is cooled.

**DICE:** The plural of die.

**DIE:** An uncased discrete or integrated device obtained from a semiconductor wafer (see Chip).

**DIE ATTACH:** The technique of mounting chips to a substrate. Methods include AuSi eutectic bonding, various solders, and conductive (and nonconductive) epoxies.

**DIE BOND:** Attachment of a die or chip to the hybrid substrate.

**DIELECTRIC:** (1) Any insulating medium which intervenes, between two conductors. (2) A material with the property that energy required to establish an electric field is recoverable in whole or in part as electric energy.

**DIELECTRIC BREAKDOWN:** A breakdown of the dielectric characteristics of a material characterized by an electrical discharge through the deteriorated dielectric or a sudden excessive increase in voltage.

**DIELECTRIC CONSTANT (OR PERMITTIVITY):** The term used to describe a material's ability to store charge when used as a capacitor dielectric. The Greek letter epsilon "$\epsilon$" is to be used (not K). The property of a dielectric which determines the electrostatic energy stored per unit volume for unit potential gradient.

**DIELECTRIC LAYER:** A layer of dielectric material between two conductor plates.

**DIELECTRIC LOSS:** The power dissipated by a dielectric as the friction of its molecules opposes the molecular motion produced by an alternating electric field. The time rate at which electrical energy is transformed into heat in a dielectric when it is subjected to a changing electric field.

**DIELECTRIC PROPERTIES:** The electrical properties of a material such as insulation resistance, breakdown voltage, etc.

**DIELECTRIC STRENGTH:** The maximum electric field that a dielectric will withstand without breaking down (physically). Expressed in volts per unit distance, such as centimeter (preferred), not inches.

**DIFFUSION:** The phenomenon of movement of matter at the atomic level from regions of high concentration to regions of low concentration. A thermal process by which minute amounts of impurities are deliberately impregnated and distributed into semiconductor material.

**DIFFUSION BOND:** See Solid-phase bond.

**DIFFUSION CONSTANT:** The relative rate at which diffusion takes place with respect to temperature.

**DIGITAL CIRCUITS:** Applied normally for switching applications where the output of the circuit normally assumes one or two states (binary operation); however, three-state operation is possible.

**DIMENSIONAL STABILITY:** A measurement of dimensional change of a material due to various environmental factors.

**DIP SOLDERING:** Soldering process for populated printed boards accomplished by dipping the exposed leads and conductive pattern in a bath of static molten solder.

**DIRECT CONTACT:** A contact made to the semiconductor die when the wire is bonded directly over the part to be electrically connected, as opposed to the expanded contact.

**DIRECT EMULSION:** Emulsion applied to a screen in a liquid form as contrasted to an emulsion that is transferred from a backing film of plastic.

**DIRECT EMULSION SCREEN:** A screen whose emulsion is applied by painting directly onto the screen, as opposed to indirect emulsion type.

**DIRECT METAL MASK:** A metal mask made by etching a pattern into a sheet of metal.

**DISCRETE:** As applied to components used in thin- and thick-film hybrid circuits; the elements that are added separately are discrete elements (or devices), as opposed to those that are made by screen printing or vacuum deposition methods as parts of the film network.

**DISCRETE COMPONENTS:** Individual components such as resistors, capacitors, and transistors.

**DISSIPATION FACTOR (ALSO LOSS TANGENT):** It is the measure of the deviation of a material from an ideal dielectric. It is the ratio of the reactance $(X_p)$ to the resistance $(R_p)$ in an equivalent parallel circuit.

**DISTRIBUTION, PROBABILITY:** A distribution which describes the probability of occurrence of the chance or event under consideration.

**DOPING:** The addition of an ionic impurity to a semiconductor to alter its conductivity in desired well-defined areas and to specified depth.

**DOUBLE-SIDED BOARD:** A single layer of base material with conductive pattern on both sides forming a printed board.

**DRAG SOLDERING:** Soldering process for populated printed boards accomplished by moving the exposed leads and conductive pattern across a bath of static molten solder.

**DRIFT:** Permanent change in value of a device parameter over a period of time because of the effects of temperature, aging, humidity, etc.

**DROSS:** Oxide and other contaminants found on the surface of molten solder.

**DRY AIR:** Air that has been circulated through a drying process to remove water molecules.

**DRY INERT ATMOSPHERE:** An inert gas such as nitrogen that has been circulated through a drying process to remove water molecules.

**DRY PRESSING:** Pressing and compacting of dry powdered materials with additives together in rigid die molds under heat and pressure to form a solid mass, usually followed by sintering as for alumina substrates.

**DRY PRINT:** The screened resistor and conductors that have gone through the drying cycle removing the solvents from the ink.

**DRYER:** A drying tube containing silica gel or a similar moisture absorbent chemical.

**DUAL IN-LINE PACKAGE (DIP):** A package having two rows of leads extending at right angles from the base and having standard spacings between leads and between rows of leads.

**DUCTILITY:** That property which permits a material to deform plastically without fracture. Property that allows a material to absorb large overloads.

**DYNAMIC PRINTING FORCE:** The fluid force which causes a pseudoplastic paste to flow through a screen mesh and wet the surface beneath. Its absolute value is a complex function of all screen printer operating parameters together with the rheological properties of the fluid being printed.

**DYNAMIC TESTING:** Testing a hybrid circuit where reactions to AC (especially high frequency) are evaluated.

**EDGE-BOARD CONTACTS:** Contacts formed in the conductive pattern of a printed board used to mate with edge-board connectors.

**EDGE-BOARD (EDGE-CARD) CONNECTOR:** Connector that makes interconnection with I/O points at the edge of a printed board.

**EDGE DEFINITION:** The reproduced pattern edge fidelity compared to the production master.

**EDGE SPACING:** The spacing between the pattern and/or components and the printed board edge.

**E-GLASS:** A low-alkali, lime alumina borosilicate glass used in the production of dielectric base materials.

**ELASTIC MODULUS:** A constant of proportionality which is indicative of the stiffness or rigidity of the material.

**ELECTRIC FIELD:** A region where there is a voltage potential, the potential level changing with distance. The strength of the field is expressed in volts per unit distance.

**ELECTRICAL ISOLATION:** Two conductors isolated from each other electrically by an insulating layer.

**ELECTRICAL PROPERTIES:** The properties of a device or material that effect its conductivity or resistivity to the flow of an electric current.

**ELECTRICALLY HOT CASE:** A hybrid circuit package that is used as part of the grounding circuit.

**ELECTRODEPOSITION:** The deposition of conductive material by applying an electric current in a plating bath.

**ELECTRODES:** The conductor or conductor lands of a hybrid circuit. Also the metallic portions of a capacitor structure.

**ELECTROLESS PLATING:** Deposit of a metallic material on a surface by chemical deposition as opposed to the use of an electric current.

**ELECTRON BEAM BONDING:** Bonding two conductors by means of heating with a stream of electrons in a vacuum.

**ELECTRONIC PACKAGING:** The technical discipline of designing a protective enclosure for an electronic circuit so that it will both survive and perform under a plurality of environmental conditions.

**ELEMENT:** A constituent unit which contributes to the operation of a hybrid microcircuit. Integral elements include deposited or screened passive circuit elements, metallization paths, and deposited or formed insulation. Discrete elements include wires or ribbon.

**ELONGATION:** The ratio of the increase in wire length at rupture, in a tensile test, to the initial length, given in percent.

**EMBEDDED:** Enclosed in a plastic material.

**EMITTER ELECTRODE:** The metallic pad making ohmic contact to the emitter area of a transistor element.

**EMULSION:** The light-sensitive material used to coat the mesh of a screen.

**ENCAPSULATE:** Sealing up or covering an element or circuit for mechanical and environmental protection.

**ENCAPSULATION:** The process of completely enclosing an article in an envelope of dielectric material.

**ENTRAPPED MATERIAL:** Gas or particles (inclusions) bound up in a solid or in an electrical package.

**ENVIRONMENT:** The physical conditions, including climate, mechanical, and electrical conditions, to which a product may be exposed during manufacture, storage, or operation.

**ENVIRONMENTAL TEST:** A test or series of tests used to determine the sum of external influences affecting the structural, mechanical,

and functional integrity of any given package or assembly.

**ETCHBACK:** Process for removing controlled amounts of dielectric material from hole sidewalls.

**ETCHED METAL MASK:** A metal mask used for screening where-in the pattern is created in a sheet of metal by the etching process.

**ETCH FACTOR:** Ratio of etch depth to lateral etch in a conductor pattern.

**ETCHING:** The chemical, or chemical and electrolytic, removal of selected conductive material to form a conductive pattern.

**EUTECTIC:** (1) A term applied to the mixture of two or more substances which has the lowest melting point. (2) An alloy or solution having its components in such proportion that the melting point is the lowest possible with those components.

**EUTECTIC ALLOY:** An alloy having the same temperature for melting and solidus.

**EXPANDED CONTACT:** A contact made to the semiconductor die where the wire bonded to an area remote from the part to be electrically connected so that a lateral interconnection path for the current is required.

**EXPONENTIAL (WEAROUT) FAILURES:** Failures that occur at an exponentially increasing rate.

**EXTERNAL LEADS:** Electronic package conductors for input and output signals, power, and ground. Leads can be either flat ribbons or round wires.

**EYELET TOOL:** A bonding tool with a square protuberance beneath the bonding tool surface which presses into the conductor and prevents the slippage between wire or conductor and tool interface. Used primarily for ribbon wire bonding.

**FACE BONDING:** The opposite of back bonding. A face bonded semiconductor chip is one that has its circuitry side facing the substrate. Flip-chip and beam lead bonding are the two common face bonding methods.

**FAILURE:** Inability of a product to meet its performance specifications.

**FAILURE ANALYSIS:** The analysis of a circuit to locate the reason for the failure of the circuit to perform to the specified level.

**FAILURE MECHANISM:** The physical or chemical process by which a device proceeds to the point of failure.

**FAILURE MODE:** The cause for rejection of any failed device as defined in terms of the specific electrical or physical requirement that it failed to meet.

**FAILURE RATE:** The rate at which devices from a given population can be expected (or were found) to fail as a function of time (e.g., %/1000 hr of operation).

**FARADAY CAGE EFFECT:** The effect created by a conductive enclosure where the electrostatic field in the interior is zero regardless of the charge (or voltage or current) on its exterior. An appropriate example of such an enclosure is the static-shielding bag capable of attenuating an outside electrostatic field so its damaging effects do not reach the ESDS items stored inside.

**FARADAY CAGE SHIELDING BAG:** Embodied in the static-shielding bag, it is a conductive semitransparent, flexible, high-strength, heat sealable container which, being an equipotential surface, prevents external electrical fields (created by any charged object or person) from causing induced charge potential differences inside. (The bag is to limit capacitive coupling of the external charge through its contents. Another form of construction for such a bag is the conductive grid bag.) Constructed, in one realization, as an outer polyester material bearing a thin metallic film with a transparent, abrasion-resistant overcoating, laminated to an inner layer of antistatic polyethylene film. The latter's function is to prevent triboelectric static charge generation by items inside the bag.

**FATIGUE:** Used to describe a failure of any structure caused by repeated application of stress over a period of time.

**FATIGUE FACTOR:** The factor causing the failure of a device under repeated stress.

**FEATHERS:** See Curls.

**FEEDTHROUGH:** A conductor through the thickness of a substrate, thereby electrically connecting both surfaces.

**FERRITE:** A powdered, compressed, and sintered magnetic material having high resistivity; cores made of sintered powders are used for ferromagnetic applications.

**FERROELECTRIC:** A crystalline dielectric that exhibits dielectric hysteresis—an electrostatic analogy to ferromagnetic materials.

**FERROMAGNETIC:** A material that has a relative permeability noticeably exceeding unity and generally exhibits hysteresis.

**FIELD TRIMMING:** Trimming of a resistor to set an output voltage, current, etc.

**FILLER:** A substance, usually dry and powdery or granular, used to thicken fluids or polymers.

**FILLET:** A concave junction formed where two surfaces meet.

**FILM:** Single or multiple layers or coatings of thin or thick material used to form various elements (resistors, capacitors, inductors) or interconnections and crossovers (conductors, insulators). Thin films are deposited by vacuum evaporation or sputtering and/or plating. Thick films are deposited by screen printing.

**FILM CONDUCTOR:** A conductor formed in situ on a substrate by depositing a conductive material by screening, plating, or evaporation techniques.

**FILM NETWORK:** An electrical network composed of thin- and/or thick-film components and interconnections deposited on a substrate.

**FINAL SEAL:** The manufacturing operation that completes the enclosure of the hybrid microcircuits so that further internal processing cannot be performed without debilling the package.

**FINE LEAK:** A leak in a sealed package less than $10^{-5}$ cm$^3$-sec$^{-1}$ at 1 atmosphere of differential air pressure.

**FIRE:** (vt) When referred to a thick-film pattern, it denotes the act of high-temperature heating in a furnace so that resistors, conductors, insulators, capacitors, etc. be transformed into their final form, generally fused.

**FIRING SENSITIVITY:** Refers to the percentage change caused in the fired film characteristics due to a change in peak firing temperature. The firing sensitivity is expressed in units of %/°C.

**FIRST ARTICLE:** A preproduction assembly used to ensure that the manufacturing process will be capable of producing acceptable product.

**FIRST BOND:** The first bond in a sequence of two or more bonds made to form a conductive connection.

**FIRST RADIUS:** The radius of the front edge of a bonding tool foot.

**FIRST SEARCH:** That period of machine cycle at which final adjustment in the location of the first bonding area (see First bond) under the tool are made prior to lowering the tool to make the first bond.

**FISSURING:** The cracking of dielectric or conductors. Often dielectrics, if incorrectly processed, will crack in the presence of conductors because of stresses occurring during firing.

**FLAG:** Support area on lead frame for die.

**FLAME-OFF:** The procedure in which the wire is severed by passing a flame across the wire, thereby melting it. The procedure is used in gold wire thermocompression bonding to form a ball for making a ball bond.

**FLAT PACK:** An integrated circuit package having its leads extended from the sides and parallel to the base.

**FLEXIBLE COATING:** A plastic coating that is still flexible after curing.

**FLEXIBLE PRINTED CIRCUIT:** Patterned printed components and board utilizing flexible base material.

**FLEXIBLE PRINTED WIRING:** Patterned printed board utilizing flexible base material.

**FLEXURAL STRENGTH:** Strength of the laminate when it is measured by bending.

**FLIP-CHIP:** A leadless monolithic structure, containing circuit elements, which is designed to electrically and mechanically interconnect to the hybrid circuit by means of an appropriate number of bumps located on its face which are covered with a conductive bonding agent.

**FLIP-CHIP MOUNTING:** A method of mounting flip chips on thick- or thin-film circuits without the need for subsequent wire bonding.

**FLOATING GROUND:** An electrical ground circuit that does not allow connection between the power and signal ground for the same circuit.

**FLOOD BAR:** A bar or other device on a screen printing device that will drag paste back to the starting point after the squeegee has made a printing stroke. The flood stroke returns the paste without pushing it through the meshes, so it does no printing, only returns the paste supply to be ready for the next print.

**FLUX:** In soldering, a material that chemically attacks surface oxides and tarnishes so that molten solder can wet the surface to be soldered.

**FLUX RESIDUE:** Particles of flux remaining on a circuit after soldering and cleaning operations.

**FOIL BURR:** A rough edge on the foil material after same type of machining operation.

**FOOT LENGTH:** The long dimension of the bonding surface of a wedge-type bonding tool.

**FOOTPRINT:** The area needed on a substrate for a component or element. Usually refers to specific geometric pattern of a chip.

**FORMING GAS:** A gas ($N_2$, with traces of $H_2$ and He) used to blanket a part being processed to prevent oxidation of the metal areas.

**FRIT:** Glass composition ground up into a powder form and used in thick film compositions as the portion of the composition that melts upon firing to give adhesion to the substrate and hold the composition together.

**FULLY ADDITIVE PROCESS:** Deposition of the entire conductive pattern thickness by electroless metal plating.

**FUNCTIONAL TRIMMING:** Trimming of a circuit element (usually resistors) on an operating circuit to set a voltage or current on the output.

**FURNACE ACTIVE ZONE:** The thermostatically controlled portion of a multizoned muffle furnace.

**FURNACE PROFILE:** A set of material-dependent temperature and belt-speed furnace parameters used to process thick film hybrid microcircuits.

**FURNACE SLAVE ZONE:** That portion of a multizoned muffle furnace where the instantaneous power supplied to the heating element is a set percentage of the power supplied to the active zone. Hence temperature control in the slave zone is not accomplished by sensing thermocouples as in the case of the active zone.

**FUSED COATING:** A metallic coating that has been melted and solidified on the base metal (usually solder or tin).

**FUSING:** Melting and cooling two or more powder materials together so that they bond together in a homogeneous mass.

**GANG BONDING:** The act of bonding a plurality of mechanical and/or electrical connections through a single act or stroke or a bonding tool.

**GAS BLANKET:** An atmosphere of inert gas, nitrogen, or forming gas flowing over a heated integrated circuit chip or a substrate that keeps the metallization from oxidizing during bonding.

**GATE CHAIN:** Large numbers of FETs connected in parallel, with common source, common drain, common gate and substrate. Gate chain is used in VLSI technology as a test structure, placed in the ''street'' (kerf) areas separating dice on a wafer, for probing; mostly for determining the levels of defects as a way of process monitoring and characterization. Benefits hybrids, using VLSI devices.

**GEL TIME:** The time (in seconds) required for prepreg resin to melt and resolidify when heated.

**GLASS BINDER:** The glass powder added to a resistor or conductor ink to bind the metallic particles together after firing.

**GLASSIVATION:** An inert, transparent, glass-like thin layer of pyrolytic insulation material that covers (passivates) the active device areas, including operating metallization on the wafer, but excluding bonding pads, bumps, and beam leads.

**GLASS PHASE:** The part of the firing cycle wherein the glass binder is in a molten phase.

**GLASS TRANSITION TEMPERATURE ($T_g$):** In polymer chemistry, the temperature below which the thermal expansion coefficient becomes nearly constant or a simple function of temperature. The temperature at which amorphous polymers change from a hard and brittle state to a soft rubbery state. Most of the material properties change rapidly and significantly at this temperature.

**GLAZE:** See Overglaze.

**GLAZED SUBSTRATE:** A glass coating on a ceramic substrate to effect a smooth and nonporous surface.

**GLOB-TOP:** A glob of encapsulant material surrounding a die in the COB (chip-on-board) assembly process. Note: the attached dice must have been pretested/inspected/passed, as rework

after final curing of the epoxy or silicone globs is virtually impossible.

**GLOSSY:** A shiny surface usually formed by the glass matrix in a conductor or resistor ink.

**GRAIN GROWTH:** The increase in the size of the crystal grains in a glass coating or other material over a period of time.

**GRAM-FORCE:** A unit of force (nominally 9.8 mN) required to support a mass of 1 gram (1 gravity unit of acceleration times one gram of mass equals 1 gram-force). Colloquially, the term gram is used for the unit.

**GREEN:** A term used in ceramic technology meaning unfired. For example, a ''green'' substrate is one that has been formed, but has not been fired.

**GROSS LEAK:** A leak in a sealed package greater than $10^{-5}$ cm³/sec at 1 atmosphere of differential air pressure.

**GROUND PLANE:** A conductive layer on a substrate or buried within a substrate that connects a number of points to one or more grounding electrodes. The conductor layer used as a common electrical circuit return, shielding and spreading heat.

**GROUND PLANE CLEARANCE:** The etched portion of the conductive material that provides clearance around a plated through hole.

**HALO EFFECT:** A glass, or epoxy, or any other semiclear material that diffracts light, halo around certain conductors. Generally, this is an undesirable effect to be avoided by changing furnace profiles or material types.

**HALOGENATED HYDROCARBON SOLVENTS:** Organic solvents containing the elements chlorine or fluorine used in cleaning substrates and completed circuits (e.g., trichloroethylene, various Freons).

**HALOING:** Fracturing or delaminating of the base material usually below the surface due to a mechanical operation.

**HAND SOLDERING:** Forming a soldered connection with solder using a hand-held soldering iron for application of the heat.

**HARD SOLDER:** Solder that has a melting point above 425°C.

**HARDNESS:** A property of solids, plastics, and viscous liquids that is indicated by their solidity and firmness; resistance of a material to indentation by an indentor of fixed shape and size under a static load or to scratching; ability of a metal to cause rebound of a small standard object dropped from a fixed height; the cohesion of the particles on the surface of a mineral as determined by its capacity to scratch another or be itself scratched. Resistance of material to plastic deformation, usually by indentation. The term may also refer to stiffness or temper or resistance to scratching, abrasion, or cutting. Indentation hardness may be measured by various hardness tests, such as Brinell, Rockwell, and Vickers.

**HEADER:** The base of a package for a (hybrid) microcircuit or a discrete semiconductor device. The bottom portion of a device package which holds the leads or pins. Also referred to as case.

**HEAT CLEAN:** The process of removing all organic material from glass cloth to approximately 343 to 371°C for a period of time ranging up to 50 hr.

**HEAT COLUMN:** The heating element in a eutectic die bonder or wire bonder used to bring the substrate up to the bonding temperature.

**HEAT FLUX:** The outward flow of heat from a heat source.

**HEAT SINK:** The supporting member to which electronic components or their substrate or their package bottom are attached. This is usually a heat conductive metal with the ability to rapidly transmit heat from the generating source (component).

**HEAT-SINKING PLANE:** A metal plane usually found in a printed board that is used to dissipate the heat from the components.

**HEAT SOAK:** Heating a circuit over a period of time to allow all parts of the package and circuit to stabilize at the same temperature.

**HEAL (OF THE BOND):** The part of the lead adjacent to the bond that has been deformed by the edge of the bonding tool used in making the bond. The back edge of the bond.

**HEEL BREAK:** The rupture of the lead at the heel of the bond.

**HEEL CRACK:** A crack across the width of the bond in the heel region.

**HERMETIC:** A description of packages that provide an absolute seal against the infusion of water to prevent degradation of the electrical components within the package. The test for

hermeticity is to observe lead rates when placed in a vacuum.

**HERMETICITY:** The ability of a package to prevent exchange of its internal gas with the external atmosphere. The figure of merit is the gaseous leak rate of the package measured in $Pa\text{-}m^3\text{-}sec^{-1}$.

**HIGH-$K$ CERAMIC:** A ceramic dielectric composition (usually $BaTiO_3$) which exhibits large dielectic constants, and nonlinear voltage and temperature response.

**HIGH-PURITY ALUMINA:** Alumina having over 99% purity of $Al_2O_3$.

**HIGH-RELIABILITY SOLDERING:** A statistically proven soldering technique that ensures a large probability of metallic joining success.

**HOLE BREAKOUT:** A condition in which the printed board drilled hole is not completely surrounded by the land pattern in the conductive pattern.

**HOLE DENSITY:** The number of holes drilled in a printed board per unit area.

**HOLE DIAMETER:** Normally refers to the diameter of the hole through the bonding tool.

**HOLE PATTERN:** The hole configuration in the printed board.

**HOLE PULL STRENGTH:** The load required to rupture the barrel of a plated through hole. The load is applied along the longitudinal axis of the hole.

**HOLE VOID:** A void exposing the base material of a plated through hole.

**HOMOGENEOUS:** Alike or uniform in composition. A thick-film composition that has settled out is not homogeneous, but after proper stirring it is. The opposite of heterogeneous.

**HORN:** Cone-shaped member which transmits ultrasonic energy from transducer to bonding tool.

**HOSTILE ENVIRONMENT:** An environment that has a degrading effect on an electronic circuit.

**HOT SPOT:** A small area on a circuit that is unable to dissipate the generated heat and therefore operates at an elevated temperature above the surrounding area.

**HOT ZONE:** The part of a continuous furnace or kiln that is held at maximum temperature.

Other zones are the preheat zone and cooling zone.

**HYBRID CIRCUIT:** A microcircuit consisting of elements which are a combination of the film circuit type and the semiconductor circuit type, or a combination of one or both of these types and may include discrete add-on components.

**HYBRID GROUP (ELECTRICALLY AND STRUCTURALLY SIMILAR CIRCUITS):** Hybrid microcircuits which are designed to perform the same type(s) of basic circuit function(s), for the same supply, bias, and single voltages and for input-output compatibility with each other under an established set of loading rules, and which are enclosed in packages of the same construction and outline.

**HYBRID INTEGRATED CIRCUIT:** A microcircuit including thick film or thin film paths and circuit elements on a supporting substrate, to which active and passive microdevices are attached, either prepackage or in uncased form as chips, usually all enclosed in a suitable package (hermetic or epoxy type). Used interchangeably with hybrid circuit and hybrid microcircuit.

**HYBRID MICROCIRCUIT:** A microcircuit that involves an insulating substrate on which are deposited networks, consisting generally of conductors, resistors, and capacitors, and to which are attached discrete semiconductor devices and/or monolithic integrated circuits and/or passive elements to form a packaged assembly. Used interchangeably with hybrid circuit and hybrid integrated circuit.

**HYBRID MICROELECTRONICS:** The entire body of electronic art which is connected with or applied to the realization of electronic systems using hybrid circuit technology.

**HYBRID MICROWAVE CIRCUIT:** See Microwave integrated circuit.

**IMBEDDED LAYER:** A conductor layer having been deposited between insulating layers.

**IMMERSION PLATING:** The process used to chemically deposit a thin metallic layer on certain base metals by partial base metal displacement.

**INACTIVE FLUX:** Flux that becomes nonconductive after being subjected to the soldering temperature.

**INCLINED PLANE FURNACE:** A resistor firing furnace having the hearth inclined so that a draft of oxidizing atmosphere will flow through the heated zones through natural convection means.

**INCLUSION:** A foreign object in either the conductive layer, plating, or base material.

**INCOMPLETE BOND:** A bond impression having dimensions less than normal size due to a portion of the bond impression being mission.

**INDIRECT EMULSION:** Screen emulsion that is transferred to the screen surface from a plastic carrier or backing material.

**INDIRECT EMULSION SCREEN:** A screen whose emulsion is a separate sheet or film of material, attached by pressing into the mesh of the screen (as opposed to the direct emulsion type).

**INERT ATMOSPHERE:** A gas atmosphere such as helium or nitrogen that is non-oxidizing or nonreducing of metals.

**INFANT MORTALITY (EARLY FAILURES):** The time regime during which hundreds of circuits are failing at a decreasing rate (usually during the first few hundred hours of operation).

**INFRARED:** The band of electromagnetic wavelengths lying between the extreme of the visible ( $= 0.75$ $\mu$m) and the shortest microwaves ( $= 1000$ $\mu$m). Warm bodies emit the radiation and bodies which absorb the radiation are warmed.

**INJECTION MOLDING:** Molding of electronic packages by injecting liquefied plastic into a mold.

**INK:** Synonymous with ''composition'' and ''paste'' when related to screenable thick-film materials, usually consisting of glass frit, metals, metal oxide, and solvents.

**INK BLENDING:** See Blending.

**IN-PROCESS:** Some step in the manufacturing operation prior to final testing.

**INSERTION LOSS:** The difference between the power received at the load before and after the insertion of apparatus at some point in the line.

**INSPECTION LOT:** A quantity of hybrid microcircuits, representing a production lot, submitted for inspection at one time to determine compliance with the requirements and acceptance criteria of the applicable procurement document. Each inspection sublot of circuits should be a group identified as having common manufacturing experience through all significant manufacturing operations.

**INSPECTION OVERLAY:** A positive inspection artwork piece created from the production master.

**INSULATION RESISTANCE:** The resistance, measured in megohms, to current flow when a potential is applied.

**INSULATORS:** A class of materials with high resistivity. Materials that do not conduct electricity. Materials with resistivity values of over $10^5$ $\Omega$-cm$^{-1}$ are generally classified as insulators.

**INTEGRATED CIRCUIT:** A microcircuit (monolithic) consisting of interconnected elements inseparably associated and formed in situ on or within a single substrate (usually silicon) to perform an electronic circuit function.

**INTERCONNECTION:** The conductive path required to achieve connection from a circuit element to the rest of the circuit.

**INTERFACE:** The boundary between dissimilar materials, such as between a film and substrate or between two films.

**INTERFACIAL BOND:** An electrical connection between the conductors on the two faces of a substrate.

**INTERLAYER CONNECTION:** A via interconnecting multiple conductive layers in a multilayer printed board.

**INTERMETALLIC BOND:** The ohmic contact made when two metal conductors are welded or fused together.

**INTERMETALLIC COMPOUND:** A compound of two or more metals that has a characteristic crystal structure that may have a definite composition corresponding to a solid solution, often refractory.

**INTERNAL LAYER:** A conductive layer internal to a multilayer printed board.

**INTERNAL VISUAL:** See Preseal visual.

**INTERSTITIAL VIA:** A plated through hole connecting two or more multilayer printed board conductive layers.

**INTRACONNECTIONS:** Connections of conductors made within a circuit on the same substrate.

**IONIC CONTAMINANT:** Any contaminant that exists as ions and, when dissolved in solution, increases electrical conductivity.

**ION MIGRATION:** The movement of free ioins within a material or across the boundary between two materials under the influence of an applied electric field.

**IONIZABLE MATERIAL:** Material that has electrons easily detracted from atoms or molecules, thus originating ions and free electrons that will reduce the electrical resistance of the material.

**JUMPER:** A direct electrical connection between two points on a film circuit. Jumpers are usually portions of bare or insulated wire mounted on the component side of the substrate.

**JUNCTION:** (1) In solid state materials, a region of transition between *p*- and *n*-type semiconductor material as in a transistor or diode. (2) A contact between two dissimilar metals or materials (e.g., in a thermocouple or rectifier). (3) A connection between two or more conductors or two or more sections of a transmission line.

**JUNCTION TEMPERATURE:** The temperature of the region of transition between the *p*- and *n*-type semiconductor material in a transistor or diode element.

**KERF:** The slit or channel cut in a resistor during trimming by laser beam or abrasive jet.

**KEY:** A device that ensures only one possible connection of two items.

**KEYING:** The use of features or additional devices in a design to ensure that there is only one possible mating method.

**KEYING SLOT:** A slot in a printed board that permits only one possible way of mating with a connector.

**KEY WAY:** A generic term describing the keying slots.

**KILN:** A high-temperature furnace used in firing ceramics.

**KINETIC THEORY OF GASES:** A theory that the minute particles in a gas moving at random with high velocities often collide with each other or with the walls that contain them.

**KIRKENDALL VOIDS:** The formation of voids by diffusion across the interface between two different materials, in the material having the greater diffusion rate into the other.

**L-CUT:** A trim notch in a film resistor that is created by the cut starting perpendicular to the resistor length and turning 90° to complete the trim parallel to the resistor axis thereby creating an L-shaped cut.

**LADDER NETWORK:** A series of film resistors with values from the highest to the lowest resistor reduced in known ratios.

**LAMINAR FLOW:** A constant directional flow of filtered air across a clean workbench. The flow is usually parallel to the surface of the bench.

**LAMINATE:** A layered sandwich of sheets of substances bonded together under heat and pressure to form a single structure.

**LAMINATE VOID:** The absence of resin in a specific location in the laminate.

**LAND:** The region of the conductive pattern used for electrical connection of components to printed boards.

**LANDLESS HOLE:** A plated through hole without any land on the outer conductive layer.

**LAND PATTERN (FOOTPRINT):** A specific conductive pattern for electrically connecting a particular component.

**LANDS:** Widened conductor areas on the major substrate used as attachment points for wire bonds or the bonding of chip devices.

**LAPPING:** Smoothing a substrate surface by moving it over a flat plate having a liquid abrasive.

**LASER BONDING:** Effecting a metal-to-metal bond of two conductors by welding the two materials together using a laser beam for a heat source.

**LATTICE STRUCTURE:** A stable arrangement of atoms and their electron-pair bonds in a crystal.

**LAYER:** One of several films in a multiple film structure on a substrate.

**LAYOUT:** The positioning of the conductors and/or resistors on artwork prior to photoreduction of the layout to obtain a working negative or positive used in screen preparation.

**LEACHING:** In soldering, the dissolving (alloying) of the material to be soldered into the molten solder.

**LEAD:** A conductive path which is usually self-supporting.

**LEAD EXTENSION:** The portion of the component lead or wire protruding beyond the solder joint.

**LEAD FRAMES:** The metallic portion of the device package that completes the electrical connection path from the die or dice and from ancillary hybrid circuit element to the outside world.

**LEAD PROJECTION:** The distance that a lead protrudes beyond the surface of the printed board.

**LEAD WIRES:** Wire conductors used for intraconnections using fine wires (Au or Al), or interconnections that include input/out leads.

**LEADLESS DEVICE:** A chip device having no input/output leads.

**LEAKAGE CURRENT:** An undesirable small stray current which flows through or across an insulator between two or more electrodes, or across a back-biased junction.

**LEGEND:** Identification marks on a printed board used for component orientation and location during the manufacturing or rework process.

**LEVELING:** A term describing the settling or smoothing out of the screen mesh marks in thick films that takes place after a pattern is screen printed.

**LIFE CHARACTERISTIC OF EQUIPMENT:** The relationship between the failure rate of the equipment and operating or test time.

**LIFE DRIFT:** The change in either absolute level or slope of a circuit element's parameter(s) under load. Rated as a percentage change from the original value per 1000 hours of life.

**LIFE TEST:** Test of a component or circuit under load over the rated life of the device.

**LIFT-OFF MARK:** Impression in bond areas left after lift-off removal of a bond.

**LINE CERTIFICATION:** Certification that a production line process sequence is under control and will produce reliable circuits in compliance with requirements of applicable mandatory documents.

**LINE DEFINITION:** A descriptive term indicating a capability of producing sharp, clean screen-printed lines. The precision of line width is determined by twice the line edge definition/

line width. A typical precision of 4% exists when the line edge definition/line width is 2%.

**LINEAR CIRCUIT:** A circuit with an output that changes in magnitude with relation to the input as defined by a constant factor (See Analog circuits).

**LINES:** Conductor runs of a film network.

**LIQUIDUS:** The line on a phase diagram above which the system has molten components. The temperature at which melting starts.

**LOAD LIFE:** The extended period of time over which a device can withstand its full power rating.

**LOOP:** The curve or arc by the wire between the attachment points at each end of a wire bond.

**LOOP HEIGHT:** A measure of the deviation of the wire loop from the straight line between the attachment points of a wire bond. Usually, it is the maximum perpendicular distance from this line to the wire loop.

**LOSS TANGENTS:** The decimal ratio of the irrecoverable to the recoverable part of the electrical energy introduced into an insulating material by the establishment of an electric field in the material.

**LOW-LOSS SUBSTRATE:** A substrate with high radio-frequency resistance and hence slight absorption of energy when used in a microwave integrated circuit.

**MAINTAINABILITY:** The probability of completing corrective and/or preventive maintenance within a set time frame at a desired confidence level under a specified environment.

**MARGIN:** The distance between a flat cable reference edge and the nearest conductor edge.

**MARKING:** The process of adding the legend to a printed board.

**MASK:** The photographic negative that serves as the master for making thick-film screens and thin-film patterns.

**MASS LAMINATION:** The simultaneous lamination of multiple C-staged sheets between B-staged sheets.

**MASS SPECTROMETER:** An instrument used to determine the lead rate of a hermetically sealed package by ionizing the gas outflow permitting an analysis of the flow rate in $cm^3\text{-}sec^{-1}$ at one atmosphere differential pressure.

**MASTER BATCH PRINCIPLE:** Blending resistor pastes to a nominal value of $\Omega$/square. The nominal value is the master control number.

**MASTER DRAWING:** The reference design document that completely describes all conductive and nonconductive patterns, holes, and other features necessary to manufacture the product.

**MASTER LAYOUT:** The original layout of a circuit.

**MATTE FINISH:** A surface finish on a material that has a grain structure and diffuses reflected light.

**MEALING:** Separating the conformal coating at specific locations on the printed board or mounted components.

**MEAN TIME BETWEEN FAILURES:** The average time of satisfactory operation of a population of equipment. It is calculated by dividing the sum of the total operating time by the total number of failures.

**MEAN TIME TO FAILURE:** The expected value or first moment of the failure probability density function for unrepairable, one-shot products.

**MEASLING:** The glass fiber and resin separation at the weave intersections. It is usually due to thermal stresses and appears as white crosses below the base material surface.

**MECHANICAL WRAPPING:** Mechanically securing a component or wire lead to a terminal prior to soldering.

**MEDIAN:** The point of a continuous random variable which divides the distribution into two equal halves such that one half of the values are greater and one half smaller than the value of the subdividing point.

**MESH SIZE:** The number of openings per inch in a screen. A 200-mesh screen has 200 openings per linear inch, 40,000 openings per square inch.

**METAL CLAD BASE MATERIAL:** Base material covered on one or both sides with metallic foil.

**METAL INCLUSION:** Metal particles embedded in a nonmetal material such as a ceramic substrate.

**METAL MASK (SCREENS):** A screen made not from wire or nylon thread but from a solid sheet of metal in which holes have been etched in the desired circuit pattern. Useful for precision and/or fine printing and for solder cream printing.

**METALLIZATION:** A film pattern (single or multilayer) of conductive material deposited on a substrate to interconnect electronic components, or the metal film on the bonding area of a substrate which becomes a part of the bond and performs both an electrical and a mechanical function.

**METAL-TO-GLASS SEAL (OR GLASS-TO-METAL SEAL):** An insulating seal made between a package lead and the metal package by forming a glass bond to oxide layers on both metal parts. In this seal, the glass has a coefficient of expansion that closely matches the metal parts.

**MICROBOND:** A bond of a small wire such as 0.001-in.-diameter gold to a conductor or to a chip device.

**MICROCIRCUIT:** A small circuit (hybrid or monolithic) having a relatively high equivalent circuit element density, which is considered as a single part on (hybrid) or with (monolithic) a single substrate to perform an electronic circuit function. (This excludes printed wiring boards, circuit card assemblies, and modules composed exclusively of discrete electronic parts.)

**MICROCIRCUIT MODULE:** An assembly of microcircuits or an assembly of microcircuits and discrete parts, designed to perform one or more electronic circuit functions and constructed such that for the purposes of specification testing, commerce, and maintenance it is considered indivisible.

**MICROCOMPONENTS:** Small discrete components such as chip transistors and capacitors.

**MICRORACKS:** A thin crack in a substrate or chip device, or in thick-film trim-kerf walls, that can only be seen under magnification and which can contribute to latent failure phenomena.

**MICRON:** An obsolete unit of length equal to a micrometer ($\mu$m).

**MICROPOSITIONER:** An instrument used in positioning a film substrate or device for bonding or trimming.

**MICROPROBE:** A small sharp-pointed probe with a positioning handle used in making temporary ohmic contact to a chip device or circuit substrate.

**MICROSTRIP:** A microwave transmission component usually on a ceramic bound together.

**MICROSTRUCTURE:** A structure composed of finely divided particles bound together.

**MICROWAVE INTEGRATED CIRCUIT:** A miniature microwave circuit usually using hybrid circuit technology to form the conductors and attach the chip devices.

**MIGRATION:** An undesirable phenomenon whereby metal ions, notably silver, are transmitted through another metal, or across an insulated surface, in the presence of moisture and an electrical potential.

**MIL:** A unit equal to 0.001 in or 0.0254 mm. Current usage is not recommended.

**MINIMUM ANNULAR RING:** The minimum dimension between the drilled hole in a printed board and the outer land edge.

**MINIMUM ELECTRICAL SPACING:** The minimum spacing required between electrical conductors to prevent dielectric breakdown or corona.

**MISLOCATED BOND:** See Off bond.

**MISREGISTRATION:** Improper alignment of successively produced features or patterns.

**MODE:** The maximum point of the frequency distribution of a continuous random variable.

**MODULE:** A generic term referring to separable units in electronic packaging.

**MODULUS OF ELASTICITY:** The ratio of stress to strain in a material that is elastically deformed.

**MOISTURE STABILITY:** The stability of a circuit under high-humidity conditions such that it will not malfunction.

**MOLECULAR WEIGHT:** The sum of the atomic weights of all atoms in a molecule.

**MONOCRYSTALLINE STRUCTURE:** The granular structure of crystals which have uniform shapes and arrangements.

**MONOLITHIC CERAMIC CAPACITOR:** A term sometimes used to indicate a multilayer ceramic capacitor.

**MONOLITHIC INTEGRATED CIRCUIT:** An integrated circuit consisting of elements formed in situ on or within a semiconductor substrate with at least one of the elements formed within the substrate.

**MOS DEVICE:** Abbreviation for a metal oxide semiconductor device.

**MOTHER BOARD (See also BACK-PLANE):** A circuit board used to interconnect smaller circuit boards called ''daughter boards.''        A printed board assembly used to interconnect electronic modules.

**MULTICHIP INTEGRATED CIR-CUIT:** An integrated circuit whose elements are formed on or within two or more semiconductor chips which are separately attached to a substrate or header.

**MULTICHIP (MICROCIRCUIT) MODULE:** A microcircuit consisting solely of active dice and passive chips which are separately attached to the major substrate and interconnected to form the circuit.

**MULTILAYER CERAMIC CAPACITOR:** A miniature ceramic capacitor manufactured by paralleling several thin layers of ceramic. The assembly is fired after the individual layers have been electroded and assembled.

**MULTILAYER CIRCUITS:** A composite circuit consisting of alternate layers of conductive circuitry and insulating materials (ceramic or dielectric compositions) bonded together with the conductive layers interconnected as required.

**MULTILAYER PRINTED BOARD:** Printed circuit or printed wiring configuration that consists of more than two conductive layers bonded together to form a multiple conductive layer assembly. The term applies to both rigid and flexible multilayer boards.

**MULTILAYER PRINTED BOARD, SEQUENTIALLY LAMINATED:** A multilayer printed board fabricated by laminating double-sided or multilayer boards together.

**MULTILAYER PRINTED CIRCUIT BOARD:** A multilayer printed board assembly manufactured from completely processed printed electrical circuits and printed conductors.

**MULTILAYER PRINTED WIRING BOARD:** A multilayer printed board assembly manufactured from completely processed

printed wiring that is used only to interconnect between components and provide I/O.

**MULTIPLE CIRCUIT LAYOUT:** Layout of an array of identical circuits on a substrate.

**NAILHEAD BOND:** See Ball bond.

**NAIL HEADING:** Flaring of the conductive inner layers around drilled holes in a multilayered board.

**NECK BREAK:** A bond breaking immediately above gold ball of a thermocompression bond.

**NEGATIVE ETCHBACK:** Etchback characterized by recessed conductor layer material relative to the surrounding base material.

**NEGATIVE IMAGE:** The reverse print of a circuit.

**NEGATIVE RESIST:** Photoresist that is polymerized by a specific wavelength of light and remains on the surface of a laminate after the unexposed areas are developed away.

**NEGATIVE TEMPERATURE COEFFICIENT:** The device changes its value in the negative direction with increased temperature.

**NOBLE METAL PASTE:** Paste materials composed partially of noble metals such as gold or ruthenium.

**NOISE:** Random small variations in voltage or current in a circuit due to the quantum nature of electronic current flow, thermal considerations, etc.

**NOMINAL RESISTANCE VALUE:** The specified resistance value of the resistor at its rated load.

**NONCONDUCTIVE EPOXY:** An epoxy material (polymer resin) either without a filler or with a ceramic powder filler added for increasing thermal conductivity and improving thixotropic properties. Nonconductive epoxy adhesives are used in chip to substrate bonds where electrical conductivity to the bottom of the chip is unnecessary or in substrate-to-package bonding.

**NONCONDUCTIVE PATTERN:** The pattern formed in the dielectric, resist, etc.

**NONFUNCTIONAL INTERFACIAL CONNECTION:** A plated through hole with one nonfunctional land.

**NONFUNCTIONAL LAND:** A land that is not connected to the conductive pattern on that layer.

**NONLINEAR DIELECTRIC:** A capacitor material that has a nonlinear capacitance voltage relationship. Titanate (usually barium titanate) ceramic capacitors (Class II) are nonlinear dielectrics. NPO and Class I capacitors are linear by definition.

**NONPOLAR SOLVENTS:** Solvents which are insufficiently ionized to be electrically conductive and which cannot dissolve polar compounds but can dissolve nonpolar compounds.

**NONWETTING:** A condition in which the solder has not adhered to all of the base metal, leaving some of the base metal exposed.

**NUGGET:** The region of recrystallized material at a bond interface which usually accompanies the melting of the materials at the interface.

**OCCLUDED CONTAMINANTS:** Contaminants that have been absorbed by a material.

**OFF BOND:** Bond that has some portion of the bond area extending off the bonding pad.

**OFF CONTACT (SCREEN PRINTING):** The opposite of contact printing in that the printer is set up with a space between the screen and the substrate and contacts the substrate only when the squeegee is cycled across the screen.

**OFF CONTACT SCREENER:** A screener machine that uses off contact printing of patterns onto substrates.

**OFFSET LAND:** A land pattern that is offset from its associated component hole.

**OHMIC CONTACT:** A contact that has linear voltage current characteristics throughout its entire operating range.

**OHMS/SQUARE:** The unit of sheet resistance or, more properly, of sheet resistivity.

**OPAQUER:** A resin additive that renders the weave of the laminate invisible to the unaided eye under either reflected or transmitted light.

**OPERATOR CERTIFICATION:** A program wherein an operator has been qualified and certified to operate a machine.

**ORGANIC FLUX:** A flux composed of rosin base and a solvent.

**ORGANIC VEHICLE:** The organic vehicle in a flux is the rosin base material.

**OUTGAS:** The release of gas from a material over a period of time.

**OUTGASSING:** Gaseous emission from a material when exposed to reduced pressure and/or heat.

**OUTGROWTH:** Increase in conductor width due beyond the conductor-resist boundary due to a plating buildup.

**OVERBONDING:** See Chopped bond.

**OVERCOAT:** A thin film of insulating material, either plastic or inorganic (e.g., glass or silicon nitride) applied over integral circuit elements for the purposes of mechanical protection and prevention of contamination.

**OVERGLAZE:** A glass coating that is grown, deposited, or secured over another element, normally for physical or electrical protection purposes.

**OVERHANG:** The amount of conductor and plating that extends beyond the conductor edge as defined at the conductor-base material interface, which may be expressed as the sum of the outgrowth and undercut.

**OVERLAP:** The contact area between dissimilar (film) materials, e.g., between a film resistor and its termination(s).

**OVERLAY:** One material applied over another material.

**OVERSPRAY:** The unwanted spreading of the abrasive material coming from the trim nozzle in a resistor trimming machine. The overspray affects the values of adjacent resistors not intended to be trimmed.

**OVERTRAVEL:** The excess downward distance of a squeegee blade would push the screen if the substrate were not in position.

**OXIDIZING ATMOSPHERE:** An air- or other oxygen-containing atmosphere in a firing furnace which oxidizes the resistor materials while they are in the molten state, thereby increasing their resistance.

**PACKAGE:** The container for an electronic component(s) with terminals to provide electrical access to the inside of the container. In addition, the container usually provides hermetic and environmental protection for, and a particular form factor to, the assembly of electronic components.

**PACKAGE CAP:** The cuplike cover that encloses the package in the final sealing operation.

**PACKAGE LID:** A flat cover plate that is used to seal a package cavity.

**PAD:** A metallized area on the surface of an active substrate as an integral portion of the conductive interconnection pattern to which bonds or test probes may be applied.

**PACKAGING DENSITY:** The amount of function per unit volume, often defined qualitatively as high, medium, or low.

**PAD GRID ARRAY:** A package that is an SMT derivative of PGA; substituting soldering pads on the underside for the customary pins.

**PANEL:** A rectangular or square section of base material containing printed boards and any required test coupons.

**PANEL PLATING:** A plating process in which the holes and surface of the panel are all plated.

**PARALLEL-GAS SOLDER:** Passing a high current through a high-resistance gap between two electrodes to remelt solder, thereby forming an electrical connection.

**PARALLEL-GAP WELD:** Passing a high current through a high-resistance gap between two electrodes that are applying force to two conductors, thereby heating the two workpieces to the welding temperature and effecting a welded connection.

**PARALLELISM (SUBSTRATE):** The degree of variation in the uniform thickness of a given substrate.

**PARASITE LOSSES:** Losses in a circuit often caused by the unintentional creation of capacitor elements in a film circuit by conductor crossovers.

**PARTIAL LIFE:** A bonded lead partially removed from the bonded area.

**PASSIVATED REGION:** Any region covered by glass, $SiO_2$, nitride, or other protective material.

**PASSIVATION:** The formation of an insulating layer directly over a circuit or circuit element to protect the surface from contaminants, moisture, or particles.

**PASSIVE COMPONENTS (ELEMENTS):** Elements (or components) such as resistors, capacitors, and inductors which do not change their basic character when an electrical signal is applied. Transistors and electron tubes are active components.

**PASSIVE NETWORK:** A circuit network of passive elements such as film resistors that are interconnected by conductors.

**PASSIVE SUBSTRATE:** A substrate that serves as a physical support and thermal sink for a film circuit that does not exhibit transistance.

**PASTE:** Synonymous with "composition" and "ink" when relating to screenable thick-film materials.

**PASTE BLENDING:** Mixing resistor pastes of different $\Omega$/square value to create a third value in between those of the two original materials.

**PASTE SOLDERING:** Finely divided particles of solder suspended in a flux paste. Used for screening application onto a film circuit and reflowed to form connections to chip components.

**PASTE TRANSFER:** The movement of a resistor, conductor, or solder paste material through a mask and deposition in a pattern onto a substrate.

**PATTERN:** The outline of a collection of circuit conductors and resistors that defines the area to be covered by the material on a film circuit substrate.

**PATTERN PLATING:** A process in which a selective conductive pattern is plated.

**PEAK FIRING TEMPERATURE:** The maximum temperatures seen by the resistor or conductor paste in the firing cycle as defined by the firing profile.

**PEEL BOND:** Similar to lift-off of the bond with the idea that the separation of the lead from the bonding surface proceeds along the interface of the metallization and substrate insulation rather than the bond-metal surface.

**PEEL STRENGTH (PEEL TEST):** A measure of adhesion between a conductor and the substrate. The test is performed by pulling or peeling the conductor off the substrate and observing the force required. Preferred unit is $g\text{-mm}^{-1}$ or $kg\text{-mm}^{-1}$ of conductor width.

**PERCENT DEFECTIVE ALLOW-ABLE (PDA):** The maximum observed percent defective which will permit the lot to be accepted after the specified 100% test.

**PERFORATED TERMINAL:** A metallic termination with a hole for inserting a lead or wire prior to soldering.

**PERIMETER SEALING AREA:** The sealing area surface on an electronic package that follows the perimeter of the package cavity and defines the area used in bonding to the lid or cap.

**PERMANENT MASK:** A masked pattern from resist that is not removed after processing.

**PERMEABILITY:** (1) The passage by diffusion (or rate of passage) of a gas, vapor, liquid, or solid through a barrier without physically or chemically affecting it. (2) The ability of a material to carry magnetism, compared to air, which has a permeability of 1.

**PHASE:** (As glassy phase or metal phase). Refers to the part or portion of materials system that is metallic or glassy in nature. A phase is a structurally homogeneous physically distinct portion of a substance or a group of substances which are in equilibrium with each other.

**PHASE DIAGRAM:** State of a metal alloy over a wide temperature range. The phase diagram is used to identify eutectic solders and their solidus/liquidus point.

**PHOTO ETCH:** The process of forming a circuit pattern in metal film by light hardening a photosensitive plastic material through a photo negative of the circuit and etching away the unprotected metal.

**PHOTOGRAPHIC REDUCTION DIMENSION:** Dimensions called out on the artwork master indicating the amount of photographic reduction required.

**PHOTORESIST:** A photosensitive plastic coating material which when exposed to UV light becomes hardened and is resistant to etching solutions. Typical use is as a mask in photochemical etching of thin films.

**PIGTAIL:** A term that describes the amount of excess wire that remains at a bond site beyond the bond. Excess pigtail refers to remnant wire in excess of three wire diameters.

**PIN DENSITY:** The number of pins per unit area on a printed board.

**PINHOLE:** Small holes occurring as imperfections which penetrate entirely through film elements, such as metallization films or dielectric films.

**PITCH:** The nominal centerline-to-centerline dimension between adjacent conductors.

**PITS:** Depressions produced in metal or ceramic surfaces by nonuniform deposition.

**PLANAR MOUNTED DEVICE LEAD:** A component lead configuration designed to sit flush on a printed board land pattern and characterized by a gull-wing configuration.

**PLASTIC:** A polymeric material, either organic (e.g., epoxy) or silicone used for conformal coating, encapsulation, or overcoating.

**PLASTIC DEVICE:** A device wherein the package, or the encapsulant material for the semiconductor die, is plastic. Such materials as epoxies, phenolics, silicones, etc. are included.

**PLASTIC ENCAPSULATION:** Environmental protection of a completed circuit by embedding it in a plastic such as epoxy or silicone.

**PLASTIC SHELL:** A thin plastic cup or box used to enclose an electronic circuit for environmental protection or used as a means to confine the plastic encapsulant used to imbed the circuit.

**PLATE FINISH:** The finish on the metal-clad base material after contact with the press plates during the lamination process.

**PLATED THROUGH HOLE:** A hole with a plated wall used for electrical interconnection between internal and/or external conductive layers.

**PLATED THROUGH HOLE STRUCTURE TEST:** Visual examination of conductors and plated through holes after the laminate structure has been dissolved away.

**PLATING:** The process used to deposit metal chemically or electrochemically on a surface.

**PLATING BAR:** A temporary electrically conductive bar used to interconnect areas to be electroplated.

**PLATING, BURNED:** Rough and dull electrodeposit plating that is caused be excessive plating current density.

**PLATING UP:** The process of electrochemical deposition of conductive material on activated base material.

**PLUG-IN-PACKAGE:** An electronic package with pins strong enough and arranged on one surface so that the package can be plugged into a test or mounting socket and removed for replacement as desired without destruction.

**POINT-TO-POINT WIRING:** An interconnecting technique wherein the connections between components are made by wires routed between connecting points.

**POISSON RATIO:** The proportionality ratio of a lateral strain to an axial strain when a material is placed in tension.

**POLARIZATION:** The elimination of inplane symmetry so that parts can be engaged in only one way.

**POLAR SOLVENTS:** Sufficiently ionized solvents that can dissolve polar solvents but cannot dissolve nonpolar compounds.

**POLYCRYSTALLINE:** A material is polycrystalline in nature if it is made of many small crystals. Alumina ceramics are polycrystalline, whereas glass substrates are amorphous.

**POLYNARY:** A material system with many basic compounds as ingredients, as thick-film resistor compositions. Binary indicates two compounds; ternary, three; etc.

**POROSITY:** The ratio of solid matter to voids in a material.

**POSITIVE:** An artwork or production master in which the transparent regions represent areas to be free of conductive material.

**POSITIVE IMAGE:** The true picture of a circuit pattern as opposed to the negative image or reversed image.

**POSITIVE RESIST:** A resist that softens when exposed to a certain wavelength of light and that is removed after exposure and developing.

**POSITIVE TEMPERATURE COEFFICIENT:** The changing of a value in the positive direction with increasing temperature.

**POST:** See Terminal.

**POSTCURING:** Heat aging of a film circuit after firing to stabilize the resistor values through stress relieving.

**POSTFIRING:** Refiring a film circuit after having gone through the firing cycle. Sometimes used to change the values of the already fired resistors.

**POSTSTRESS ELECTRICAL:** The application of an electrical load to a film circuit to stress the resistors and evaluate the resulting change in values.

**POTTING:** Encapsulating a circuit in plastic.

**POWER DENSITY:** The amount of power dissipated from a film resistor through the substrate measured in $W\text{-}cm^{-2}$.

**POWER DISSIPATION:** The dispersion of the heat generated from a film circuit when a current flows through it.

**POWER FACTOR:** The ratio of the actual power of an alternating or pulsating current as measured by an ammeter and voltmeter.

**PREFIRED:** Film conductors fired in advance of film resistors on a substrate.

**PREFORM:** An aid in soldering or in other types of adhesion functions. Preforms are generally circular- or square-shaped, punched out of thin sheets of solder, of epoxy, or of eutectic alloy. They are placed on the spot of attachment by soldering or by bonding, prior to placing the object proper there to be heated.

**PREOXIDIZED:** Resistor metal particles that have been oxidized to achieve the desired resistivity prior to formulation into a resistor ink.

**PREPREG:** Sheet material or fabric impregnated with resin and partially cured (B-stage).

**PRESEAL VISUAL:** The process of visual inspection of a completed hybrid circuit assembly for defects prior to sealing the package.

**PRESSED ALUMINA:** Aluminum oxide ceramic formed by applying pressure to the ceramic powder and a binder prior to firing in a kiln.

**PRESS-FIT CONTACT:** An electrical contact that is press fit into a hole.

**PRINT AND FIRE:** A term sometimes used to indicate steps in the thick-film process wherein the ink is printed on a substrate and is fired.

**PRINTED BOARD:** The generic term for processed printed circuit and printed wiring boards, either single or double and multilayer boards.

**PRINTED CIRCUIT BOARD:** A printed board that consists of conductive pattern which includes printed components and/or printed wiring.

**PRINTED CIRCUIT ASSEMBLY:** One or more populated printed circuit boards that perform a specific function in a system.

**PRINTED COMPONENT:** A component, such as a resistor or capacitor, which is formed in the conductive pattern.

**PRINTED CONTACT:** A contact area that is formed in the conductive pattern.

**PRINTED WIRING BOARD:** A printed board containing a conductive pattern which consists only of printed wiring.

**PRINTED WIRING ASSEMBLY:** One or more populated printed wiring boards that perform a specific function in a system.

**PRINTED WIRING LAYOUT:** A design document that depicts the printed wiring, electrical, and mechanical component sizes and locations with sufficient detail that documentation and artwork may be generated from the document.

**PRINT LAYDOWN:** Screening of the film circuit pattern onto a substrate.

**PRINTING:** Same as Print layout.

**PRINTING PARAMETERS:** The conditions that affect the screening operation such as off-contact spacing, speed and pressure of squeegee, etc.

**PROBE:** A pointed conductor used in making electrical contact to a circuit pad for testing.

**PROCURING ACTIVITY:** The organizational element (equipment manufacturer, government, contractor, subcontractor, or other responsible organization) which contracts for articles, supplies, or services and has the authority to grant waivers, deviations, or exceptions to the procurement documents.

**PRODUCT:** An item, component, device, group of devices, or system.

**PRODUCT ASSURANCE:** A technical management operation reporting to top management and devoted to the study, planning, and implementation of the required design, controls, methods, and techniques to ensure a reliable product in conformity with its applicable specification.

**PRODUCTION LOT:** Hybrid microcircuits manufactured on the same production line(s) by means of the same production techniques, materials, controls, and design. The production lot is usually date coded to permit control and trace-

ability required for maintenance of reliability programs.

**PRODUCTION MASTER:** A 1:1 scale pattern for all the layers required to manufacture a printed board base on the master drawing. The production master may have more than one pattern for making more than one printed board on a panel.

**PROPERTY:** The physical, chemical, or electrical characteristic of a given material.

**PULL STRENGTH:** The values of the pressure achieved in a test where a pulling stress is applied to determine breaking strength of a lead or bond.

**PULL TEST:** A test for bond strength of a lead, interconnecting wire, or conductor.

**PULSE SOLDERING:** Soldering a connection by melting the solder in the joint area by pulsing current through a high-resistance point applied to the joint area and the solder.

**PURGE:** To evacuate an area or volume space of all unwanted gases, moisture, or contaminants prior to backfilling with an inert gas.

**PURPLE PLAGUE:** One of several gold-aluminum compounds formed when bonding gold to aluminum and activated by reexposure to moisture and high temperature ($>340°C$). Purple plague is purplish in color and is very brittle, potentially leading to time-based failure of the bonds. Its growth is highly enhanced by the presence of silicon to form ternary compounds.

**PUSH-OFF STRENGTH:** The amount of force required to dislodge a chip device from its mounting pad by application of the force to one side of the device, parallel to the mounting surface.

**PYROLYZED (BURNED):** A material that has gained its final form by the action of heat is said to be pyrolyzed.

**Q:** The inverse ratio of the frequency band between half-power points (bandwidth) to the resonant frequency of the oscillating system. Refers to the electromechanical system of an ultrasonic bonder, or sensitivity of the mechanical resonance to changes in driving frequency.

**QUALITY ASSURANCE:** A technique or science which uses all the known methods of quality control and quality engineering to ensure the manufacture of a product of acceptable quality standards.

**QUALITY CONFORMANCE TEST CIRCUITRY:** A section of a printed board containing test coupons used to determine the board's acceptability.

**RADIOGRAPHS:** Photographs made of the interior of a sealed package by use of X-rays to expose the film.

**RANDOM FAILURES:** Circuit failures which occur randomly with the overall failure rate for the sample population being nearly constant.

**RC NETWORK:** A network composed only of resistors and capacitors.

**REACTIVE METAL:** Metals that readily form compounds.

**REAL ESTATE:** The surface area of an integrated circuit or of a substrate. The surface area required for a component or element.

**REBOND:** A second bonding attempt after a bond made on top of a removed or damaged bond or a second bond made immediately adjacent to the first bond.

**REDUCING ATMOSPHERE:** An atmosphere containing a gas such as hydrogen that will reduce the oxidation state of the subject compound.

**REDUCTION DIMENSION:** A dimension specified on enlarged scale matrices between a pair of marks which are positioned in very accurate alignment with the horizontal center locations of two manufacturing holds and their locationally coincident targets. This dimension is used to indicate and verify the exact horizontal distance required between the two targets when the matrices are photographically reduced to full size.

**REDUNDANCY:** The existence of more than one means of accomplishing a given task, where all means must fail before there is an overall failure of the system. Parallel redundancy applies to systems in which both means are working at the same time to accomplish the task and either of the systems is capable of handling the job itself in case of failure of the other system. Standby redundancy applies to a system in which there is an alternative means of accomplishing the task that is switched in by a malfunction-sensing device when the primary system fails.

**REFERENCE EDGE:** The edge of a cable or conductor used as a reference for making measurements.

**REFIRING:** Recycling a thick-film resistor through the firing cycle to change the resistor value.

**REFLOW SOLDERING:** A method of soldering involving application of solder prior to the actual joining. To solder, the parts are joined and heated, causing the solder to remelt, or reflow.

**REFRACTORY METAL:** Metal having a very high melting point, such a molybdenum.

**REGISTRATION:** The alignment of a circuit pattern on a substrate.

**REGISTRATION MARKS:** The marks used for aligning successive processing masks.

**REINFORCED PLASTIC:** Plastic having reinforcing materials such as fiberglass embedded or laminated in the cured plastic.

**RELIABILITY:** The probability that a product will perform a stated function satisfactorily at a desired confidence level for a set period of time under a specified environment.

**RELIABILITY ASSURANCE:** A technique or science which assesses the reliability of a product by means of surveillance and measurement of the factors of design and production which affect it.

**REPAIR:** An operation performed on a nonconforming part or assembly to make it functionally usable but which does not completely eliminate the nonconformance.

**RESIN RECESSION:** Void between the barrel of a plated through hole and the wall of the hole which is usually due to board exposure to high temperatures. These are usually detected by microsectioning the printed board.

**RESIN-RICH AREA:** An area of nonreinforced surface layer resin of the same composition as the base material resin.

**RESIN SMEAR:** A transfer of resin from the nonconductive layer to the exposed conductive edge surface in a drilled hole.

**RESIN-STARVED AREA:** An area of the printed board which does not have a sufficient amount of resin covering the reinforcement fabric. This area is characterized by low gloss, dry spots, or exposed fibers.

**RESIST:** A protective coating that will keep another material from attaching or coating something, as in solder resist, plating resist, or photoresist.

**RESISTANCE SOLDERING:** A soldering process in which current passing through one or more electrodes heats the soldering area.

**RESISTANCE WELD:** The joining of two conductors by heat and pressure with the heat generated by passing a high current through the two mechanically joined materials.

**RESISTIVITY:** A proportionality factor characteristic of different substances equal to the resistance that a centimeter cube of the substance offers to the passage of electricity, the current being $R = \rho L/A$, where $R$ is the resistance of a uniform conductor, $L$ its length, $A$ its cross-sectional area, and $\rho$ its resistivity. Resistivity is usually expressed in ohm-centimeters. The ability of a material to resist passage of electrical current either through its bulk or on a surface.

**RESISTOR DRIFT:** The change in resistance of a resistor through aging and usually rated as percent change per 1000 hr.

**RESISTOR GEOMETRY:** The film resistor outline.

**RESISTOR OVERLAP:** The contact area between a film resistor and a film conductor.

**RESISTOR PASTE CALIBRATION:** Characterization of a resistor paste for $\Omega$/square value, TCR, and other specified parameters by screening and firing a test pattern using the paste and recording the results.

**RESISTOR TERMINATION:** See Resistor overlap.

**RESOLUTION:** The degree of fineness or detail of a screen printed pattern (see Line definition).

**REVERSE CURRENT CLEANING:** Electrolytic cleaning using the part to be cleaned as the anode.

**REVERSE IMAGE:** The resin pattern on a printed board exposing the conductive areas to be plated.

**REVERSION:** A chemical reaction resulting in a polymerized material that degenerates to a lower polymeric state or to the original monomer.

**REWORK:** An operation performed on a nonconforming part or assembly that restores all nonconforming characteristics to the requirements in the contract, specification, drawing, or other approved product description.

**RHEOLOGY:** The science dealing with deformation and flow of matter.

**RIBBON INTERCONNECT:** A flat narrow ribbon of metal such as nickel, aluminum, or gold used to interconnect circuit elements or to connect the element to the output pins.

**RIBBON WIRE:** Metal in the form of a very flexible flat thread or slender rod or bar tending to have a rectangular cross section as opposed to a round cross section.

**RIGHT-ANGLE EDGE CONNECTOR:** A connector terminated at the edge of the board which routes the I/O perpendicular to the plane of the board.

**RIGID COATING:** A conformal coating of thermosetting plastic that has no fillers or plasticizers to keep the coating pliable.

**RISERS:** The conductive paths that run vertically from one level of conductors to another in a multilayer substrate or screen-printed film circuit.

**ROADMAP:** A pattern printed on nonconductive material used to delineate components and circuitry to facilitate board repair and servicing.

**ROSIN FLUX:** A flux having a rosin base which becomes interactive after being subjected to the soldering temperature.

**ROSIN SOLDER CONNECTION (ROSIN JOINT):** A soldered joint in which one of the parts is surrounded by an almost invisible film of insulating rosin, making the joint intermittently or continuously open even though the joint looks good.

**SAGGING OR WIRE SAG:** The failure of bonding wire to form the loop defined by the path of the bonding tool between bonds.

**SAMPLE:** A random selection of units from a lot for the purpose of evaluating the characteristics or acceptability of the lot. The sample may be either in terms of units or in terms of time.

**SAPPHIRE:** A single-crystal $Al_2O_3$ substrate material used in integrated circuits.

**SCALING:** Peeling of a film conductor or film resistor from a substrate, indicating poor adhesion.

**SCALLOP MARKS:** A screening defect which is characterized by a print having jagged edges. This condition is a result of incorrect dynamic printing pressure or insufficient emulsion thickness.

**SCAVENGING:** Same as Leaching.

**SCHEMATIC:** Diagram of a functional electronic circuit composed of symbols of all active and passive elements and their interconnecting matrix that forms the circuit.

**SCORED SUBSTRATE:** A substrate that has been scribed with a thin cut at the breaklines.

**SCRATCH:** In optical observations, a surface mark with a large length-to-width ratio.

**SCREEN:** A network of metal or fabric strands, mounted snugly on a frame, and upon which the film circuit patterns and configurations are superimposed by photographic means.

**SCREEN DEPOSITION:** The laydown of a circuit pattern on a substrate using the silk screening technique.

**SCREEN FRAME:** A metal, wood, or plastic frame that holds the silk or stainless steel screen tautly in place.

**SCREENING:** The process whereby the desired film circuit patterns and configurations are transferred to the surface of the substrate during manufacture, by forcing a material through the open areas of the screen using the wiping action of a soft squeegee.

**SCREEN PRINTING:** The process in which an image is transferred to a substrate by forcing screen-printable material through a stencil or image screen using a squeegee.

**SCRIBE COAT:** A two-layer system, sometimes used for the preparation of thick-film circuit artwork, consisting of a translucent, dimensionally stable, polyester base layer covered with a soft, opaque, strippable layer and normally processed on a coordinatograph machine (see Cut and strip).

**SCRUBBING ACTION:** Rubbing a chip device around on a bonding pad during the bonding operation to break up the oxide layer and improve wettability of the eutectic alloy used in forming the bond.

**SEARCH HEIGHT:** The height of the bonding tools above the bonding area at which final adjustments in the location of the bonding area under the tool are made prior to lowering the tool for bonding.

**SECOND BOND:** The second bond of a bond pair made to form a conductive connection.

**SECOND RADIUS:** The radius of the back edge of the bonding tool foot.

**SECOND SEARCH:** That period of machine cycle at which final adjustments in the location of the second bonding area (see Second bond) under the tool are made prior to lowering the tool for making the second wire bond.

**SELECTIVE ETCH:** Restricting the etching action on a pattern by the use of selective chemical which attack only one of the exposed materials.

**SELF-HEATING:** Generation of heat with a body by chemical action. Epoxy materials self-heat in curing due to exothermic reaction.

**SELF-PASSIVATING GLAZE:** The glassy material in a thick-film resistor that comes to the surface and seals the surface against moisture.

**SEMIADDITIVE PROCESS:** An additive conductive patterning process combining electroless metal deposition on an unclad or foil substrate with electroplating, with etching, or with both.

**SEMICONDUCTOR CARRIER:** A permanent protective structure which provides for mounting and for electrical continuity in application of a semiconductor chip to a major substrate.

**SEMICONDUCTORS:** Solid materials such as silicon that have a resistivity midway between those of a conductor and a resistor. These materials are used as substrates for semiconductor devices such as transistors, diodes, and integrated circuits.

**SERPENTINE CUT:** A trim cut in a film resistor that follows a serpentine or wiggly pattern to effectively increase the resistor length and increase resistance.

**SHADOWING, ETCHBACK:** A condition characterized by incomplete removal of dielectric material immediately next to the foil; however, acceptable etchback may have been achieved elsewhere.

**SHEAR RATE:** The relative rate of flow or movement (of viscous fluids).

**SHEAR STRENGTH:** The limiting stress of a material determined by measuring a strain resulting from applied forces that cause or tend to cause bonded contiguous parts of a body to slide relative to each other in a direction parallel to their plane of contact; the value of the force achieved when shearing stress is applied to the bond (normally parallel to the substrate) to determine the breaking load.    Strength to withstand shearing of a material.

**SHEET RESISTANCE:** The electrical resistance of a thin sheet of a material with uniform thickness as measured across opposite sides of a unit square pattern. Expressed in $\Omega$/square.

**SHELF LINE:** The maximum length of time, usually measured in months, between the date of shipment of a material to a customer and the date by which it should be used for best results.

**SHIELDING, ELECTRONIC:** A barrier designed to reduce the interaction of components and portions of circuits with magnetic fields.

**SHORT-TERM OVERLOAD:** Overload of a circuit with current or voltage for a period too short to cause breakdown of the insulation.

**SIGNAL:** An electrical impulse of specified voltage, current, polarity, and pulse width.

**SIGNAL CONDUCTOR:** An electrical conductor used to transmit a signal.

**SIGNAL PLANE:** A conductive layer designed to carry signals, as opposed to ground or fixed voltages.

**SILICON MONOXIDE:** A passivating or insulating material that is vapor deposited on selected areas of a thin-film circuit.

**SILK SCREEN:** A screen of a closely woven silk mesh stretched over a frame and used to hold an emulsion outlining a circuit pattern and used in screen printing of film circuits. Used generically to describe any screen (stainless steel or nylon) used for screen printing.

**SINGLE-SIDED BOARD:** A printed board with only one conductive layer.

**SINKING:** Shorting of one conductor to another on multilayer screen printed circuits because of a downward movement of the top conductor through the molten crossover glass.

**SINTERING:** Heating a metal powder under pressure and causing the particles to bond together in a mass. Alternately, subjecting a ceramic-powder mix to a firing cycle whereby the mix is less than completely fused and shrinks.

**SKIN EFFECT:** The increase in resistance of a conductor at microwave frequencies because of the tendency for current to concentrate at the conductor surface.

**SLICE:** A thin cross section of a crystal such as silicon that is used for semiconductor substrates.

**SLIVERS:** Plating material overhanging a conductor edge or completely detached from the rest of the plating which can cause potential electrical shorts.

**SLUMP:** A spreading of printed thick-film composition after screen printing but before drying. Too much slumping results in loss of definition.

**SLURRY:** A thick mixture of liquid and solids, the solids being in suspension in the liquid.

**SMEARED BOND:** A bond impression that has been distorted or enlarged by excess lateral movement of the bonding tool, or by the movement of the device-holding fixture.

**SNAPBACK:** The return of a screen to normal after being deflected by the squeegee moving across the screen and substrate.

**SNAP-OFF DISTANCE:** The screen printer distance setting between the bottom of the screen and top of the substrate (see Breakaway).

**SNAPSTRATE:** A scribed substrate that can be processed by gang deposition in multiples of circuits and snapped apart afterward.

**SOAK TIME:** The length of time a ceramic material (such as a substrate or thick-film composition) is held at the peak temperature of the firing cycle.

**SOFT SOLDER:** A low melting solder, generally a lead-tin alloy, with a melting point below 425°C.

**SOLDER ACCEPTANCE:** Same as Solderability.

**SOLDER BALLS:** Small solder spheres that adhere to the surface of the printed board after reflow or wave soldering. The solder balls are more predominant with an uncontrolled soldering process using solder paste.

**SOLDER BRIDGING:** A short between two or more conductors due to solder.

**SOLDER BUMPS:** The round solder balls bonded to a transistor contact area and used to make connection to a conductor by face-down bonding techniques.

**SOLDER COAT:** A solder layer applied directly to the surface conductive layer of a printed board through exposure to molten solder.

**SOLDER CONNECTION:** An electrical or mechanical connection between two metal parts formed by solder.

    * Disturbed solder connection—a cold solder joint resulting from movement of one or more of the parts during solidification.

    * Excess solder connection—a solder joint containing so much solder that the joined part contours are not visible. Also refers to solder that has flowed beyond the designated solder area.

    * Insufficient solder connection—a solder joint characterized by one or more of the parts incompletely covered by solder.

    * Overheated solder connection—a solder joint characterized by a dull, chaulky, grainy, and porous/pitted appearance which is the result of excess heat application during the solder process.

    * Preferred solder connection—a smooth, bright, and well-wetted solder connection with no exposed bare lead material and no sharp protrusions or embedded foreign material with the lead contours visible.

    * Rosin solder connection—a solder connection which has a cold solder joint appearance but is further characterized by rosin flux entrapment separating the surfaces to be joined.

**SOLDER CUP TERMINAL:** A metallic terminal with a cylindrical feature at one end used for soldering one or more leads or wires.

**SOLDER DAM:** A dielectric composition screened across a conductor to limit molten solder from spreading further onto solderable conductors.

**SOLDER FILLET:** A rounded and blended solder configuration around a component or wire lead and the land pattern.

**SOLDER GLASSES:** Glasses used in package sealing that have a low melting point and tend to wet metal and ceramic surfaces.

**SOLDER IMMERSION TEST:** A test that immerses the electronic package leads into a solder bath to check resistance to soldering temperatures.

**SOLDER LEVELING:** A process using heated gas or other material to remove excess solder.

**SOLDER PASTE:** A composition of metal (Sn/Pb) powder, flux, and other organic vehicles.

**SOLDER PLUGS:** Solder cores in plated through holes in a printed board.

**SOLDER PROJECTION:** An undesirable solder protrusion from a solidified joint.

**SOLDER RESIST:** A material used to localize and control the size of soldering areas, usually around component mounting holes. The solderable areas are defined by the solder resist matrix.

**SOLDERABILITY:** The ability of a conductor to be wetted by solder and to form a strong bond with the solder.

**SOLDERING:** The process of joining metals by fusion and solidification of an adherent alloy having a melting point below about 427°C.

**SOLDERING IRON TIP:** The end of a soldering iron that has been plated with iron, nickel, chromium, or a similar metal which is used to heat solder to reflow temperature.

**SOLDERING OIL:** An oil that is intermixed with solder in wave-soldering equipment and is used as a covering for the molten solder in still and wave solder pots to eliminate dross and reduce the solder surface tension.

**SOLDER SIDE:** The side of the board opposite the component bodies and where the leads are soldered to the conductive pattern on the printed board.

**SOLDER WICKING:** The rising of solder between individual strands of wire due to capillary action.

**SOLDERLESS WRAP:** An interconnection technique which is achieved by tightly wrapping or winding solid wire around square, rectangular, or V-shaped terminals using a special tool.

**SOLID METAL MASK:** A thin sheet of metal with an etched pattern used in contact printing of film circuits.

**SOLID-PHASE BOND:** The formation of

a bond between two parts in the absence of any liquid phase at any time prior to or during the joining process.

**SOLID STATE:** Pertaining to circuits and components using semiconductors as substrates.

**SOLID TANTALUM CHIP:** A chip or leadless capacitor whose dielectric ($Ta_2O_5$) is formed with a solid electrolyte instead of a liquid electrolyte.

**SOLIDUS:** The locus of points in a phase diagram representing the temperature, under equilibrium conditions, at which each composition in the system begins to melt during heating, or complete freezing during cooling.

**SOLUBILITY:** The ability of a substance to dissolve into a solvent.

**SOLVENT:** A material that has the ability to dissolve other materials.

**SOLVENT-RESISTANT MATERIAL:** A material that is unaffected by solvents and does not degrade when cleaned in solvents.

**SPACING:** The distance between adjacent conductor edges.

**SPAN:** The distance between the reference edges of the first and last conductors.

**SPECIFIC GRAVITY:** The density (mass/unit volume) of any material divided by that of water at a standard temperature.

**SPECIFIC HEAT:** The ratio of the heat capacities of a body and water at a reference temperature; i.e., the quantity of heat required to raise the temperature of 1 g of a substance 1°C.

**SPECIMEN:** A sample of material, device, or circuit representing the production lot removed for test.

**SPINEL:** A hard, single-crystal mineral of magnesium aluminum oxide, $MgAl_2O_4$, used as substrate in special integrated circuit structures.

**SPLAY:** A drill bit's tendency to produce off-center, out-of-round, and nonperpendicular holes.

**SPUTTERING:** The removal of atoms from a source by energetic ion bombardment, the ions supplied by a plasma. Process is used to deposit films for various thin-film applications.

**SQUASHOUT:** The deformed area of a lead which extends beyond the dimensions of the lead prior to bonding.

**SQUEEGEE:** The part of a screen printer that pushes the composition across the screen and through the mesh onto the substrate.

**STAINLESS STEEL SCREEN:** A stainless steel mesh screen stretched across a frame and used to support a circuit pattern defined by an emulsion bonded to the screen.

**STAIR-STEP PRINT:** A print which retains the pattern of the screen mesh at the line edges. This is a result of inadequate dynamic printing pressure (inserted?) on the paste or insufficient emulsion thickness coating the screen.

**STAMPED PRINTED WIRING:** Wiring produced by stamping and mounting to a dielectric base material.

**STANDARD DEVIATION:** A statistical term that helps describe the likely value of parts in a lot or batch of components in comparison with the lot's average value. Practically all of a lot will fall within 3 standard deviations of the average value if it is a normal distribution.

**STANDOFF:** A connecting post of metal bonded to a conductor and raised above the surface of the film circuit.

**STEATITE:** A ceramic consisting chiefly of a silicate of magnesium used as an insulator or circuit substrate.

**STENCIL:** A thin sheet material with a circuit pattern cut into the material. A metal mask is a stencil.

**STEP-AND-REPEAT:** A process wherein the conductor or resistor pattern is repeated many times in evenly spaced rows onto a single film or substrate.

**STEP SCALE STEP-WEDGE:** A scale of evenly spaced tones ranging from clear to black through intermediate shades of gray used as a reference in photographic reproductions.

**STEP SOLDERING:** A soldering process using solder alloys with different melting points.

**STITCH BOND:** A bond made with a capillary-type bonding tool when the wire is not formed into a ball prior to bonding.

**STRAIGHT-THROUGH LEAD:** A component lead that protrudes through a printed board and is not mechanically formed.

**STRATIFICATION:** The separation of nonvolatile components of a thick film into horizontal layers during firing, due to large differences in density of the component. It is more likely to occur with a glass containing conductor paste and under prolonged or repeated firing.

**STRAY CAPACITANCE:** Capacitance developed from adjacent conductors separated by an air dielectric or dielectric material.

**STRESS RELIEVE:** A process of reheating a film resistor to make it stress free. A portion of a component or wire lead that is sufficiently long to minimize stresses.

**STRESS-FREE MATERIAL:** The annealed or stress-relieved material in which the molecules are no longer in tension.

**STRIPLINE:** A microwave conductor on a substrate. A transmission line configuration consisting of a conductor equally spaced between two parallel ground planes.

**STYLUS:** A sharp-pointed probe used in making an electrical contact on the pad of a leadless device or a film circuit.

**SUBCARRIER SUBSTRATE:** A small substrate of a film circuit which is mounted in turn to a larger substrate.

**SUBSTRATE (OF A MICROCIRCUIT OR INTEGRATED CIRCUIT):** The supporting material upon which the elements of a hybrid microcircuit are deposited or attached or within which the elements of an integrated circuit are fabricated.

**SUBSYSTEM:** A smaller part of an electronic system which performs a part of the system function but can be removed intact and tested separately.

**SUBTRACTIVE PROCESS:** The process of selectively removing unwanted portions of a conductive layer using an etching process.

**SUPERCONDUCTIVITY:** The phenomenon of almost no electrical resistance in materials when their temperatures are reduced.

**SUPPORTED HOLE:** A hole in a printed board that has been reinforced by plating or some other technique.

**SURFACE CONDUCTANCE:** Conductance of electrons along the outer surface of a conductor.

**SURFACE DIFFUSION:** The high-temperature injection of atoms into the surface layer of a semiconductor material to form the junctions. Usually, a gaseous diffusion process.

**SURFACE FINISH:** The peaks and valleys in the surface of a substrate rated in $\mu$m-mm$^{-1}$ deviation.

**SURFACE MOUNTING:** The interconnection of components to the top conductive layer of a printed board without the use of through holes.

**SURFACE NUCLEATION:** The change in phase or state of the surface on a substrate. Applicable to the beginning of formation of evaporated thin film.

**SURFACE RESISTIVITY:** The resistance to a current flow along the surface of an insulator material.

**SURFACE TENSION:** An effect of the forces of attraction existing between the molecules of a liquid. It exists only on the boundary surface.

**SURFACE TEXTURE:** The smoothness or lack of it on the surface of a substrate.

**SURFACTANT:** A contraction for the term ''surface-active agent.''

**SWAGED LEADS:** Component leads which are mechanically flattened or swaged to the printed board during the manufacturing process.

**SWIMMING:** Lateral shifting of a thick-film conductive pattern while on molten glass, for example, upon reheating.

**SYSTEM:** A product which for most considerations is repairable. Examples include computers, radars, and power supplies.

**TACKY STATE:** A material is in a tacky state when it exhibits an adhesive bond to another surface.

**TAIL (OF THE BOND):** The free end of wire extending beyond the bond impression of a wire bond from the heel.

**TAIL PULL:** The act of removing the excess wire left when a wedge or ultrasonic bond is made.

**TANTALUM CAPACITOR:** Capacitors that utilize a thin tantalum oxide layer as the dielectric material.

**TAPE ALUMINA:** Alumina (substrates) made by tape casting of slurry into strips of green alumina of a predetermined thickness. This is followed by stamping, cutting in the green state, then firing.

**TAPE AUTOMATED BONDING (TAB):** The utilization of a metal or plastic tape material as a support and carrier of a microelectronic component in a gang bonding process.

**TAPED COMPONENTS:** Components that are attached to a tape material forming a continuous roll for use in automatic insertion or placement machines.

**TARNISH:** Chemical accretions on the surface of metals, such as sulfides and oxides. Solder fluxes have to remove tarnish in order to allow wetting.

**TEAR STRENGTH:** Measurement of the amount of force needed to tear a solid material that has been nicked on one edge and then subjected to a pulling stress. Measured in $g\text{-}cm^{-1}$.

**TEMPERATURE AGING:** Aging or stressing of a film circuit in an elevated temperature over a period of time.

**TEMPERATURE COEFFICIENT OF CAPACITANCE (TCC):** The amount of capacitance change of a capacitor with temperature, commonly expressed as the average change over a certain temperature range in ppm/°C.

**TEMPERATURE COEFFICIENT OF RESISTANCE (TCR):** The amount of resistance change of a resistor (or resistor material) with temperature, commonly expressed as the average change over a certain temperature range in ppm/°C.

**TEMPERATURE CYCLING:** An environmental test in which the film circuit is subjected to several temperature changes from a low temperature to a high temperature over a period of time.

**TEMPERATURE EXCURSION:** The extreme temperature difference seen by a film circuit under operating conditions.

**TEMPERATURE TRACKING:** The ability of a component to retrace its electrical readings going up and down the temperature scale.

**TEMPERING:** A method for heating a material followed by rapid cooling.

**TENSILE STRENGTH:** The pulling stress which has to be applied to a material to break it, usually measured in Pa.

**TENTING:** The manufacturing process for covering the surface of the printed board, usually with a dry film resist.

**TERMINAL:** A metal lead used to provide electrical access to the inside of the device package. A metallic part used for electrical interconnection.

**TEST BOARD:** A printed board that is manufactured using the same process that will be used for production quantities of product to ensure product acceptability.

**TEST COUPON:** A portion of the quality control test area used for acceptance tests or other related tests.

**TEST PATTERN:** A circuit or group of substrate elements processed on or within a substrate to act as a test site or sites for element evaluation or monitoring of fabrication processes. A pattern used for testing and/or inspection.

**TEST POINT:** A point used for access to the electrical circuit for testing purposes.

**THERMAL CONDUCTIVITY:** The amount of heat per unit time per unit area that can be conducted through a unit thickness of a material at a difference in temperature.

**THERMAL DESIGN:** The schematic heat flow path for power dissipation from within a film circuit to a heat sink.

**THERMAL DRIFT:** The drift of circuit elements from nominal value due to changes in temperature.

**THERMAL DROP:** The difference in temperature across a boundary or across a material.

**THERMAL GRADIENT:** The plot of temperature variances across the surface or the bulk thickness of a material being heated.

**THERMAL MISMATCH:** Difference of thermal coefficients of expansion of materials which are bonded together.

**THERMAL NOISE:** Noise that is generated by the random thermal motion of charged particles in an electronic device.

**THERMAL RELIEF:** Crosshatching used to minimize warping and/or blistering during the soldering process.

**THERMAL RUNAWAY:** A condition wherein the heat generated by a device causes an increase in heat generated. This spiraling rise in dissipation usually continues until a temperature is reached that results in destruction of the device.

**THERMAL SHIFT:** The permanent shift in the nominal value of a circuit element due to heating effect.

**THERMAL SHOCK:** A condition whereby

devices are subjected alternately to extreme heat and extreme cold. Used to screen out processing defects.

**THERMOCOMPRESSION BONDING:** A process involving the use of pressure and temperature to join two materials by interdiffusion across the boundary.

**THERMOPLASTIC:** A substance that becomes plastic (malleable) on being heated; a plastic material that can be repeatedly melted or softened by heat without change or properties.

**THERMOSWAGING:** Heating a pin that is inserted in a hole and upsetting the hot metal so that it swells and fills the hole, thereby forming a tight bond with the base material.

**THICK FILM:** A film deposited by screen printing processes and fired at high temperature to fuse into its final form. The basic processes of thick-film technology are screen printing and firing.

**THICK-FILM CIRCUIT:** A microcircuit in which passive components of a ceramic-metal composition are formed on a suitable substrate by screening and firing.

**THICK-FILM HYBRID CIRCUIT:** A hybrid microcircuit that has add-on components, usually chip devices added to a thick-film network to perform an electronic function.

**THICK-FILM NETWORK:** A network of thick-film resistors and/or capacitors or inductors interconnected with thick-film conductors on a ceramic substrate, formed by screening and firing.

**THICK-FILM TECHNOLOGY:** The technology whereby electrical networks or elements are formed by applying a liquid, solid, or paste coating through a screen or mask in a selective pattern onto a supporting material (substrate) and fired. Films so formed are usually 5 $\mu$m or greater in thickness.

**THIEF:** A portion of the panel pattern area or a racking device used to provide more uniform current density on plated parts.

**THIN FILM:** A thin film (usually less than 5 $\mu$m thickness) is one that is deposited onto a substrate by an accretion process such as vacuum evaporation, sputtering, chemical vapor deposition, or pyrolytic decomposition.

**THIN-FILM HYBRID CIRCUIT:** A hybrid microcircuit that has add-on components,

usually chip devices added to a thin-film network to perform an electronic function.

**THIN-FILM INTEGRATED CIRCUIT:** A functioning circuit made entirely of thin-film components. Also used to mean thin-film hybrid circuit.

**THIN-FILM NETWORK:** A resistor and/or capacitor and conductor network formed on a single substrate by vacuum evaporation and sputtering techniques.

**THIN-FILM TECHNOLOGY:** The technology whereby electronic networks or elements are formed by vacuum evaporation or sputtering films onto a supporting material (substrate). Films so formed are less than 5 and usually of the order 0.3 to 1.0 $\mu$m in thickness.

**THIN FOIL:** Metal sheet stock less than 0.0007 in. thick.

**THIXOTROPIC:** A fluid that gets less viscous as it is stirred (or moved). The term is sometimes applied to these fluids (e.g., most thick-film inks).

**THROUGH CONNECTION:** An electrical connection between different conductive layers.

**THROUGH HOLE MOUNTING:** The electrical interconnection of components with leads that insert through the holes of a printed board.

**THROWAWAY MODULE:** A functional circuit in a modular form factor that is considered expendable and will not be repaired because of its low cost.

**TINNED:** Literally, coated with tin, but commonly used to indicate coated with solder.

**TINNING:** To coat metallic surfaces with a thin layer of solder.

**TIP:** That portion of the bonding tool which deforms the wire to cause the bond impression.

**TO PACKAGE:** Abbreviation for transistor outline, established as an industry standard by JEDEC of the EIA.

**TOE:** See Tail (of the bond).

**TOOLING FEATURE:** A specific feature on a printed board or panel that facilitates alignment during the manufacturing process.

**TOOLING HOLES:** Holes in the printed board or panel that serve the same type of purpose as tooling features.

**TOP-HAT RESISTORS:** Film resistors

having a projection out one side allowing a notch to be cut into the center of the projection to form a serpentine resistor and thereby increase the resistivity.

**TOPOGRAPHY:** The surface condition of a film—bumps, craters, etc.

**TOPOLOGY:** The surface layout design study and characterization of a microcircuit. It has application chiefly in the preparation of the artwork for the layout masks used in fabrication.

**TOROIDS:** A helical winding on a ring-shaped core (doughnut-shaped coil). A popular form used for inductors.

**TORQUE TEST:** A test for determining the amount of torque required to twist off a lead or terminal.

**TRACKING:** Two similar elements on the same circuit that change values with temperature in close harmony are said to track well. Tracking of different resistors is measured in ppm/°C (difference). Tracking is also used in reference to temperature hysteresis performance and potentiometer repeatability.

**TRANSDUCER:** A device actuated by one transmission system and supplying related energy to another transmission system.

**TRANSFER MOLDING:** Molding circuit modules by transferring molten plastic into a cavity holding the circuit by using a press.

**TRANSFER SOLDERING:** A hand soldering process using ball, chip, or disk solder preforms.

**TRANSMISSION CABLE:** Two or more transmission lines forming either a flat cable or a round structure.

**TRANSMISSION LINE:** A signal-carrying line with controlled electrical characteristics used to transmit high-frequency or narrow-pulse signals.

**TRAPEZOIDAL DISTORTION:** Distortion that can occur during photoreduction. As a result, a square shape in the original master will be transformed into a trapezoid at the reduced position.

**TREATMENT TRANSFER:** The transfer of copper foil treatment to the base material which is characterized by black, brown, or red steaks after the copper has been removed.

**TRIM LINES:** Lines defining the printed board border.

**TRIM NOTCH:** The notch made in a resistor by trimming to obtain the design value (see Kerf).

**TRIMMING:** Notching a resistor by abrasive or laser means to raise the nominal resistance value.

**TRUE POSITION:** The theoretical location of a hole or feature defined by basic dimension.

**TRUE POSITION TOLERANCE:** The movement around the true position defined by a diameter and called out on the master drawing.

**TWIST:** A deformation of a rectangular sheet characterized by one corner being out of a plane defined by the other three corners.

**ULTRASONIC BONDING:** A process involving the use of ultrasonic energy and pressure to join two materials.

**ULTRASONIC CLEANING:** A method of cleaning that uses cavitation in fluids caused by applying ultrasonic vibrations to the fluid.

**ULTRASONIC POWER SUPPLY:** An electronic high-frequency generator that provides ultrasonic power to a transducer.

**UNCASED DEVICE:** A chip device.

**UNDERBONDING:** In wire bonding, insufficient deforming the wire with the bonding tool during the bonding process.

**UNDERCUT:** The distance from the outermost base conductor edge, excluding any additional cover plating, to the innermost point at the conductor-base material interface.

**UNDERDEFORMING:** Insufficient deforming of the wire by the bonding tool occurring during the bonding operation.

**UNDERGLAZE:** A glass or ceramic glaze applied to a substrate prior to the screening and firing of a resistor.

**UNIVERSAL LEAD FRAME:** Lead frame in which flag is not supported by any lead.

**UNSUPPORTED HOLE:** A hole that is not supported by any type of reinforcement.

**USEFUL LIFE:** The length of time a product functions with a failure rate which is considered to be satisfactory.

**VACUUM DEPOSITION:** Deposition of a

metal film onto a substrate in a vacuum by evaporation or sputtering techniques.

**VACUUM PICKUP:** A handling instrument with a small vacuum cup on one end used to pick up chip devices.

**VAPOR DEPOSITION (VAPOR EVAPORATION):** Same as Vacuum deposition.

**VAPOR PHASE:** The state of a compound when it is in the form of a vapor.

**VAPOR PRESSURE:** The pressure exerted by a vapor in equilibrium with the liquid phase of the same material.

**VARNISH:** A protective coating for a circuit to protect the elements from environmental damage.

**VEHICLE:** A thick-film term that refers to the organic system in the paste.

**VIA:** An opening in the dielectric layer(s) through which a riser passes, or else whose walls are made conductive.

**VIA HOLE:** A plated through hole providing electrical interconnection for two or more conductive layers but not intended to have a component lead inserted through it.

**VINTAGE CONTROL NUMBER:** An alphanumeric code related to changes made on manufactured products. The vintage control code generally follows the time part number.

**VISCOSIMETER (VISCOMETER):** A device that measures viscosity. Viscometers for thick-film compositions must be capable of measuring viscosity under conditions of varying shear rates.

**VISCOSITY:** A term used to describe the fluidity of material, or the rate of flow versus pressure. The unit of viscosity measurement is poise, more commonly centipoise. Viscosity varies inversely with temperature.

**VISCOSITY COEFFICIENT:** The coefficient of viscosity is the value of the tangential force per unit area which is necessary to maintain unit relative velocity between two parallel planes a unit distance apart.

**VISUAL EXAMINATION:** Qualitative examination of physical characteristics using the unaided eye or defined levels of magnification.

**VITREOUS:** Having the nature of glass.

**VITREOUS BINDER:** A glassy material used in a compound to bind other particles together. This takes place after melting the glass and cooling.

**VITRIFICATION:** The progressive reduction in porosity of a ceramic material as a result of heat treatment or some other process.

**VOID:** In visual inspection of solid materials, a space not filled with the specific solid material, such as a gap or opening which is an unintentional defect in the material.

**VOLTAGE GRADIENT:** The voltage drop (or change) per unit length along a resistor or other conductance path.

**VOLTAGE PLANE:** A conductor layer in a printed board that is held at a some specified voltage; it may also be used for shielding or heat sinking.

**VOLTAGE PLANE CLEARANCE:** The etched portion of the conductive material that provides clearance around a plated through hole.

**VOLTAGE RATING:** The maximum voltage which an electronic circuit can sustain to ensure long life and reliable operation.

**VOLUME RESISTIVITY:** The resistance of a material to an applied electrical voltage as a function of the configuration of the material.

**WAFER:** A slice of semiconductor crystal ingot used as a substrate for transistors, diodes, and monolithic integrated circuits.

**WARP AND WOOF:** Threads in a woven screen which cross each other at right angles.

**WARPAGE:** The distortion of a substrate from a flat plan.

**WAVES:** Any disturbance that advances through a medium with a speed determined by properties of the medium.

**WAVE SOLDERING:** A soldering process in which populated printed boards are soldered by passing them over a continuous flowing wave of molten solder.

**WAVINESS:** One or a series of elevations or depressions, or both, which are readily noticeable and which include defects such as buckles and ridges.

**WEAVE EXPOSURE:** A condition existing in the base material characterized by lack of resin covering the woven reinforcing cloth.

**WEAVE TEXTURE:** A surface condition of the base material in which woven reinforcing material is apparent even though it is completely covered by resin.

**WEDGE BOND:** A bond made with a wedge tool. The term is usually used to differentiate thermocompression wedge bonds from other thermocompression bonds. (Almost all ultrasonic bonds are wedge bonds.)

**WEDGE TOOL:** A bonding tool in the general form of a wedge with or without a wire-guide hole to position the wire under the bonding face of the tool, as opposed to a capillary-type tool.

**WETTING:** The spreading of molten solder on a metallic surface, with proper application of heat and flux.

**WHISKERS:** Needle-shaped metallic growths between conductors.

**WICKING:** The flow of solder along the strands and under the insulation of stranded lead wires.

**WIRE BOND:** A completed wire connection which includes all its constituents providing electrical continuity between the semiconductor die (pad) and a terminal. These constituents are the fine wire; metal bonding surfaces like die pad and package land; metallurgical interfaces underneath the bonded-wire deformation; underlying insulating layer (if present); and substrate.

**WIRE BONDING:** The method used to attach very fine wire to semiconductor components to interconnect these components with each other or with package leads.

**WIRE CLAMP:** A device designed to hold the wire during the cutoff operation.

**WIRE LEAD:** A length of uninsulated solid or twisted wire which may be formed to make an electrical interconnection.

**WIRE SPOOL:** The wire magazine.

**WOBBLE BOND:** A thermocompression, multicontact bond accomplished by rocking (or wobbling) a bonding tool on the beams of a beam lead device.

**WORST-CASE ANALYSIS:** The analysis of a circuit function under tolerance extremes of temperature, humidity, etc. to determine the worst possible effect on the output parameters.

**WOVEN SCREEN:** A screen mesh used for screen printing, usually made of nylon or stainless steel or possible silk.

**YIELD:** The ratio of usable components at the end of a manufacturing process to the number of components initially submitted for processing. Can be applied to any input-output stage in processing, and so must be carefully defined and understood.

**YIELD STRENGTH:** Strength at a point at which the strain begins to increase very rapidly without a corresponding increase in stress.

**ZIF CONNECTOR:** A connector used for electrical interconnection to a printed board or mating connector half that requires zero insertion force (ZIF).

**ZIF SOCKET:** A socket used to interconnect a component to a printed board that does not require any solder and is characterized by zero insertion force.

# Appendix C: Standards and Specifications for Microelectronics

## Components: Diodes
DO-207 Leadless
DO-213 Leadless
DO-214 C Bend Plastic
DO-215 GW Plastic
MIL-C-15500 Diodes and Transistors

## Active Components: Transistors
ASTM Standard F466-79
  Test Method for Small-Signal Scattering Parameters of Low-Power Transistors in the 0.2 to 2.0-GHz Frequency Range
ASTM Standard F528-81
  Method of Measurement of Common-Emitter D-C Current Gain of Junction Transistors.
ASTM Standard F570-81
  Test Methods for Transistor Collector-Emitter Saturation Voltage
ASTM Standard F616-82
  Method for Measuring MOSFET Drain Leakage Current
ASTM Standard F617-86
  Method for Measuring MOSFET Linear Threshold Voltage
ASTM Standard F632-79
  Method for Measuring Small-Signal Common Emitter Current Gain of Transistors at High Frequencies

ASTM Standard F769-84
  Method for Measurement of Transistor and Diode Leakage Currents
TO-236 ~ SOT-23
TO-243 ~ SOT-89
TO-253 ~ SOT-143
MIL-C-15500 Diodes and Transistors

## Active Components: Integrated Circuits
ASTM Standard F676-83
  Method for Measuring Unsaturated TTL Sink Current
ASTM Standard F774-82
  Guide for Analysis of Latchup Susceptibility in Bipolar Integrated Circuits

## Active Components: Semiconductor (General)
ISHM SPB 010
  Uncased Silicon Semiconductor Dice (ISHM-HMSSG)
MIL-STD-19500
  General Specifications for Semiconductor Devices

839

## Bonds and Contacts

ASTM Committee F1-07
Standard Ball-Shear Test Method
ASTM Standard E10
Test Method for Brinell Hardness of Metallic
Materials
ASTM Standard F458-84
Recommended Practice for Nondestructive
Pull Testing of Wire Bonds
ASTM Standard F459-84
Methods for Measuring Pull Strength of Mi-
croelectronic Wire Bonds
ASTM Standard F580-78
Test Method for Adhesion Strength of Bond-
able Films to Substrates in Ribbon Wire
Bonds
MIL-STD-883
Test Methods and Procedures for Microelec-
tronics

## Clean Rooms

ASTM Standard F91-70
Recommended Practice for Testing for Leaks
in the Filters Associated with Laminar Flow
Clean Rooms and Clean Work Stations by
Use of a Condensation Nuclei Detector
ASTM Standard F50-83
Test Method for Continuous Sizing and
Counting of Airborne Particles in Dust-Con-
trolled Areas Using Instruments Based upon
Light-Scattering Principles
ASTM Standard F51-68
Method for Sizing and Counting Particulate
Contaminant in and on Clean Room Garments
ASTM Standard F311-78
Practice for Processing Aerospace Liquid
Samples for Particulate Contamination Anal-
ysis Using Membrane Filters
ASTM Standard F312-69
Methods for Microscopical Sizing and Count-
ing Particles from Aerospace Fluids on Mem-
brane Filters
ASTM Standard F316-86
Test Method for Pore Size Characteristics of
Membrane Filters for Use with Aerospace
Fluids
FED-STD-209
Federal Requirement, Clean Room and Work
Station Requirements, Controlled Environ-
ment
SEMI Standards Doc.1738
Measurement of Particle Contamination Con-

tributed to the Product from the Process or
Support Tool

## Cleaning

AN SI/IPC-SC-60
Post Solder Solvent Cleaning Handbook
ANSI/IPC-SC-62
Post Solder Aqueous Cleaning Handbook

## Contaminants/Purity

ASTM Standard F57-68
Method of Concentrating and Measuring
Trace Quantities of Copper in High-Purity
Water Used in the Electronics Industry
ASTM Standard F58-68
Method of Measuring Specific Resistivity of
Electronic Grade Solvents
ASTM Standard F60-68
Methods for Detection and Enumeration of
Microbiological Contaminants in Water Used
for Processing Electron and Microelectronic
Devices
ASTM Standard F71-68
Method for Using the Morphological Key for
the Rapid Identification of Fibers for Con-
tamination Control in Electron Devices and
Microelectronics
ASTM Standard F488-79
Test Method for Total Bacterial Count in Wa-
ter Used for Processing Electron and Micro-
electronic Devices
ASTM Standard F569-78
Test Method for Filterability Index of Re-
agent-Grade Water and High-Purity Waters
for Microelectronic Device Processing
MIL-P-28809
Method for Ionic Contamination Control
SEMI Specification XXXX
Particulate Level in VLSI-Grade Gases
SEMI Guidelines XXXX
Pure Water Guidelines

## Electrostatic Discharge

DOD-STD-1686
Electrostatic Discharge Control Program for
Protection of Electrical and Electronic Parts,
Assemblies and Equipment

## Encapsulation (Including Coatings)

ASTM Standard F74-73
Recommended Practice for Determining Hy-
drolytic Stability of Plastic Encapsulants for
Electronic Devices

ASTM Standard F542-80
Test Method for Exothermic Temperature of Encapsulating Compounds for Electronic and Microelectronic Encapsulation

ASTM Standard F635-85
Guide for Selection of Tests for Material Properties of Liquid Thermoset Encapsulating Compounds for Electronic and Microelectronic Encapsulation

ASTM Standard F636-85
Guide for Selection of Tests for Material Properties of Transfer Molding Compounds for Electronic and Microelectronic Encapsulation

ASTM Standard F677-80
Test Method for Fluid and Grease Resistance of Thermoset Encapsulating Compounds Used in Electronic and Microelectronic Applications

IPC-CC-830
Quality and Performance of Electrical Insulating Compounds for PWBs Conformal Coating

MIL-STD-23586
Sealing Compound, Electrical Silicone Rubber Accelerator Required

## Hermeticity

ASTM Standard F97-72
Practices for Determining Hermeticity of Electronic Devices by Dye Penetration

ASTM Standard F98-72
Recommended Practices for Determining Hermeticity of Electron Devices by a Bubble Test

ASTM Standard F134-85
Practices for Determining Hermeticity of Electron Devices with a Helium Mass Spectrometer Leak Detector

ASTM Standard F730-81
Test Methods for Hermeticity of Electron Devices by a Weight-Gain Test

ASTM Standard F784-82
Method for Calibrating Radioisotope Hermetic Test Apparatus

ASTM Standard F785-82
Test Method for Hermeticity of Sealed Devices by a Radioisotope Test

ASTM Standard F816-83
Test Method for Combined Fine and Gross Leaks for Large Hybrid Microcircuit Packages

ASTM Standard F866-84
Method for Measuring the Package Attenuation Coefficient of a Sealed Device for Radioisotopic Hermetic Test

ASTM Standard F979-86
Test Method for Hermeticity of Hybrid Microcircuit Packages Prior to Lidding

British Telecom
M219F Specification for Type Approval of Encapsulating Resins

## Interconnections: Wire

ASTM Standard F16-67
Methods for Measuring Diameter or Thickness of Wire and Ribbon for Electronic Devices and Lamps

ASTM Standard F46-69
Practice for Room Temperature Resistivity Measurements on Thermoelectric Material

ASTM Standard F72-85
Specification for Gold Wire for Semiconductor Lead-Bonding

ASTM Standard F180-72
Test Method for Density of Fine Wire and Ribbon for Electronic Devices

ASTM Standard F205-82
Method for Measuring Diameter of Fine Wire by Weighing

ASTM Standard F487-82
Specification for Fine Aluminum–1% Silicon Wire for Semiconductor Lead-Bonding

ASTM Standard F584-82
Practice for Visual Inspection of Semiconductor Lead Bonding Wire

ASTM Standard F638-82
Specification for Fine Aluminum–1% Magnesium Wire for Semiconductor Lead-Bonding

ISHM SP 006
1% Silicon Aluminum Wire for Semiconductor Lead-Bonding (ISHM-HMSSG)

ISHM SP 007
Gold Wire for Semiconductor Lead-Bonding (ISHM-HMSSG)

## Labeling

MIL-STD-129
Marking for Shipment and Storage

MIL-STD-130
Marking, Identification of U.S. Military Property

MIL-STD-1285
Marking, Electrical and Electronic Parts

## Lead Frames

ASTM Standard F375-77
   Specification for Integrated Circuit Lead
   Frame Material
SEMI-Vol.4
   Integrated Circuit Leadframe Materials Used
   in the Production of Stamped Leadframes,
   Revision
SEMI-Vol.4
   Specifications for Stamped Leadframes of
   Plastic Molded Semiconductor Packages, Re-
   vision
SEMI-Vol.4
   Leadframe Specifications for Molded Quad
   Packages, Revision

## Leak Detection (See Also Vacuum Processing)

ASTM Standard F78-79
   Method for Calibration of Helium Leak Detectors by Use of Secondary Standards

## Measurements/Tests: General

ASTM Standard F728-81
   Practice for Preparing an Optical Microscope
   for Dimensional Measurements
ASTM Standard F889-84
   Test Method for Characterizing a Combina-
   tion of Properties of Planar Surfaces to a Tech-
   nical Level by a Vacuum Hold-Down Tech-
   nique
IEEE-218-1956
   Standard Methods of Testing Transistors
MIL-STD-202
   Test Methods for Electronic and Electrical
   Component Parts
MIL-STD-750C
   Methods, Test, Semiconductor Devices
MIL-STD-810
   Methods, Test, Environmental
MIL-STD-883
   Test Methods and Procedures for Microelec-
   tronics
MIL-C-45662
   Calibration Systems Requirements
MIL-D-105D
   Inspection by Attribute, Sampling Proce-
   dures and Tables for
MIL-T-28880
   Test Equipment, Electronic and Electrical,
   General Specifications for

SMTA Test Guide
   SMT, Design for Test Access Guidelines
SMTA Testability Design Recommendations
   SMT Design Guidelines for In-Circuit Probe-
   ability

## Microelectronics: General

EIA-JEP-95
   JEDEC Registered and Standard Outlines for
   Semiconductor Devices
JEDEC-STD-16
   Assessment of Microcircuit Outgoing Qual-
   ity Levels in Parts per Million (PPM)
MIL-STD-454
   Equipment Requirements, Standard General
MIL-STD-1313
   Terms and Definitions, Microelectronics
MIL-STD-1331
   Specifications Parameters to be Controlled,
   Microelectronics
MIL-STD-1389
   Modules, Design Requirements for Standard
MIL-E-5400
   Aerospace General Specification for Elec-
   tronic Equipment
MIL-M-28787
   Modules, General Specifications for
MIL-M-38510
   General Specifications for Microcircuits
NAVSEA 5516562
   Modules, Electronic, General Specifications

## Microelectronics: Hybrids

ANSI/IPC-HM-860
   Specification for Multilayer Hybrid Circuits
EIA-STD-RS507
   Clips, Edge, Hybrids and Chip Carrier, Di-
   mensional Characteristics
IPC-D-859
   Design Standard for Multilayer Hybrid Cir-
   cuits
ISHM-1402
   Hybrid Microcircuit Design Guide
ISHM SP 001
   Item Specification, Standard Hybrid Micro-
   electronics
ISHM SP 002
   General Specifications, Standard Hybrid Mi-
   croelectronics
ISHM STD 02
   Glossary of Terms, Standard Hybrid Micro-
   electronics

ISHM STD 03
Parameters to be Controlled, Standard Hybrid Microelectronics
MIL-STD-1772
Certification Requirements for Hybrid Microcircuit Facilities and Lines
MFSC-STD-587
Design and Quality Standard for Custom Hybrid Microcircuits
MFSC-SPEC-592
Specification for the Selection and Use of Organic Adhesives in Hybrid Microcircuits
MIL-H-38534 (Proposed)
Military Specification, Hybrid Microcircuits, General Specifications for

## Packages: General

IPC-STD-CM-78
Chip Carriers, Guidelines for Surface Mounting
MIL-STD-454
Standard General Requirements for Electronic Equipment
MIL-STD-794
Packing and Packaging of Parts and Equipment, Procedures for
MIL-STD-28787
General Specifications for Standard Electronic Modules
MIL-P-11268
Parts, Materials and Processes for Electronic Equipment

## Packages: Ceramic

ISHM SP 008
Packages
MS-002-009 CCC (JC-11)
MS-014 40 mil CCC (JC-11)
MO-056 45 mil CCC (JC-11)
MO-057 20 mil CCC (JC-11)
MO-062 25 mil CCC (JC-11)
MO-044 50 mil CJL (JC-11)
SEMI-Vol.4
Specifications for Cofired Ceramic Fine Pitch Leaded and Leadless Chip Carrier Package Constructions
SEMI-Vol.4
Specifications for Preform Integrity

## Packages: Plastic

EIAJ/EE-13
Proposed Standardization of Package Name and Code IAJ-IC-74-2-f9-III General Rules

for the Preparation of Outline Drawings of Integrated Circuits Thin Small Outline Packages
EIAJ-IC-74-2-1984
General Rules for the Preparation of Outline Drawings of Integrated Circuits Flat Dual In-Line Packages
EIAJ-IC-74-3-1983
General Rules for the Preparation of Outline Drawings of Integrated Circuits Dual In-Line Packages
EIAJ-IC-74-4-1986
General Rules for the Preparation of Outline Drawings of Integrated Quad Flat Packages
EIAJ-IC-74-5-1987
General Rules for the Preparation of Outline Drawings of Integrated Circuits Zig-Zag In-Line Packages
EIAJ-XX-XX-X-XXXX
General Rules for the Preparation of Outline Drawings of Integrated Circuits Pin Grid Array Package (Square Type)
IPC-SM-78
Technical Report on Impact of Moisture on Plastic I/C Package Cracking
MO-046 200 mil GW SOIC (JC-11)
MS-012 150 mil GW SOIC (JC-11)
MS-013 300 mil GW SOIC (JC-11)
MS-059 330 mil GW SOIC (JC-11)
MO-053 25 mil Leaded Square (JC-11)
MO-052 JL Rectangular (JC-11)
MO-047 50 mil JL Square (JC-11)
MO-071 20 mil GW CC (JC-11)
MO-041 Type E Rectangular, Leadless (JC-11)
MO-042 Type F Rectangular, Leadless (JC-11)
MO-061 0.400 in body SOJ family (JC-11)
MO-063 0.350 in body family (JC-11)
MO-065 0.300 in body family (JC-11)
SEMI-Vol.4
Specification for Measurement Method of Molded Plastic Tooling
SEMI-Vol.4
Integrated Circuit Leadframe Materials Used in the Production of Stamped Leadframes, Revision
SEMI-Vol.4
Specifications for Stamped Leadframes of Plastic Molded Semiconductor Packages, Revision
SEMI-Vol.4
Leadframe Specifications for Molded Quad Packages, Revision

## Packages: TAB

ASTM Standard F637-85
  Specification for Format, Physical Properties, and Test Methods for 19 and 35-mm Testable Tape Carrier for Perimeter Tape Carrier-Bonded Semiconductor Devices

## Passive Components: General

EIA-PDP-100
  Mechanical Outlines for Registered and Standard Electronic Parts (Passive Devices)

## Passive Components: Capacitors

ASTM Standard F752-82
  Test Method for Effective Series Resistance (ESR) and Capacitance of Multilayer Ceramic Capacitors at High Frequencies
IS-28 (EIA-s)
  Fixed Tantalum Chip Capacitor, Style 1, Protected, Standard Capacitance Range
IS-29 (EIA-s)
  Fixed Tantalum Chip Capacitor, Style 1, Protected, Extended Capacitance Range
IS-36 (EIA-s)
  Multilayer Ceramic Chip Capacitor
Bulletin CB-11(EIA-s)
  Guidelines for the Surface Mounting of Multilayer Ceramic Chip Capacitors
EIA-469-A
  Standard Test Methods for Destructive Physical Analysis of High Reliability Ceramic Monolithic Capacitors
EIA-510
  Standard Test Methods for Destructive Physical Analysis of Industrial Grade Ceramic Monolithic Capacitors
EIA-CB-11
  Guidelines for Surface Mount of MLC Capacitors
ISHM SP 004
  Capacitors, Chip, Fixed (ISHM-HMSSG)
MIL-C-55681B
  Capacitor Chip, Multilayer, Unencapsulated
MIL-C-39028
  Capacitor, Packaging of
MIL-C-39003
  Capacitors, Tantalum Electrolytic
MIL-C-55681B
  Solderability Requirements, Chip Capacitors

## Passive Components: Resistors

ASTM Standard F817-83
  Method for Characterization of Film Resistor Materials and Processes

ASTM Standard F729-81
  Test Methods for Temperature Coefficient of Resistance of Film Resistors
ASTM Standard F772-82
  Test Method for Noise Quality of Film-Type Resistors
IS-30 (EIA-s)
  Resistor, Fixed, Surface Mount
IS-34 (EIA-s)
  Resistor Networks
ISHM SP 005
  Resistors, Chip, Fixed (ISHM-HMSSG)
MIL-R-55342C
  Resistors, Chip, Solderability Requirements
MIL-R-39008
  Resistors, Fixed, Composite and Film
MIL-R-39032
  Resistors, Preparation for Delivery of
MIL-R-55342C
  Solderability Requirements, Chip Resistors

## Passive Components: Films

ASTM Standard F21-65
  Test Method for Hydrophobic Surface Films by the Atomizer Test
ASTM Standard F22-65
  Test Method for Hydrophobic Surface Films by the Water-Break Test
ASTM Standard F390-78
  Test Method for Sheet Resistance of Thin Metallic Films with a Collinear Four-Probe Array
ASTM Standard F508-77
  Recommended Practice for Specifying Thick-Film Pastes
ISHM SP 003
  Films, Thick and Thin (ISHM-HMSSG)

## Photolithography (Including Photoresist)

ASTM Standard F66-84
  Methods for Testing Photoresists Used in Microelectronic Fabrication
ASTM Standard F127-84
  Definitions of Terms Relating to Photomasking Technology for Microelectronics
ASTM Standard F518-77
  Recommended Practice for Determining the Effective Adhesion of Photoresist to Hard-Surface Photomask Blanks and Semiconductor Wafers During Etching
ASTM Standard F527-80
  Practice for Adjusting Photoresist Exposure Time

ASTM Standard F581-78
Practice for Accelerated Aging of Photoresist
ASTM Standard F583-82
Test Method for Photoresist Cleanliness-Filterability
ASTM Standard F691-80
Practice for Preparing Photoplates for Measuring Flatness Deviation
ASTM Standard F804-83
Practice for Producing Spin Coating Resist Thickness Curves
ASTM Standard F863-84
Practice for the Detection of Defects in Spin-Coated Resist
ASTM Standard F890-84
Practice for Determining Pinhole Density in Photoresist Films Used in Microelectronic Device Processing
ASTM Standard F908-85
Practice for Evaluating Safelights for Photoresists

## Plating

AMS-2422C-81
Gold Plating for Electronic and Electrical Applications
F/S QQ-N-290A
Nickel, Electrodeposited Plating
F/S QQ-S-365C
Gold, Electrodeposited Plating
MIL-C-14550
Copper Plating
MIL-C-26074C
Nickel, Electroless, Coatings
MIL-G-45204
Gold, Electrodeposited Plating
MIL-P-81728
Tin-Lead, Electrodeposited Plating
MIL-T-10727
Tin Plating, Electrodeposited or Hot Dipped
SEMI-Vol.5
Method of Viscosity Determination. Method A, Kinematic Viscosity Revision
SEMI-Vol.5
Guideline for Functional Testing of Microelectronic Resists, Revision

## Reliability, Inspection, and Quality

DoD-STD-1686
ESD Control Program, Protection, Electrical Parts Assemblies and Equipment
EIA-IS-46
Test procedure for SMC Resistance to VPS

IPC-A-610
Acceptability of Printed Board Assemblies
IPC-AI-641
User Guidelines for Automated Solder Joint Inspection Systems
IPC-AI-643
User Guidelines for Automatic Optical Inspection of Populated Packaging and Interconnection Assemblies
IPC-TR-476
(Dendritic) Metal Growth, Avoidance of
IPC-TR-579
Round Robin Evaluation for Small Diameter Plated Through Holes
IPC-SM-782-RR [1]
Round Robin Test Plan for Surface Mount Land Pattern Reliability
IPC-SM-XXX [2]
Guidelines on Accelerated Surface Mount Attachment Reliability Testing
IPC-SM-XXX [2]
Testing and Handling of Surface Mount Plastic Packages Susceptible to Moisture Induced Cracking
ISHM STD 01-01
Visual and Mechanical Inspection
ISHM STD 01-02
External Visual Inspection
ISHM STD 01-03
Internal Visual Inspection
ISHM STD 01-04
Screening Procedures
ISHM STD 01-05
Qualification and Quality Conformance Procedures
MIL-STD-105D
Inspection by Attribute, Sampling Procedures and Tables for
MIL-STD-217
Reliability Prediction of Electronic Equipment
MIL-STD-810
Methods, Test, Environmental
MIL-I-45208A
Inspection System Requirements
MIL-HDBK-217E
Reliability Prediction Methods
MIL-Q-9858
Quality Program Requirements

## Safety

BUMED 6270.3
Hazardous Air Contaminants, Personnel Exposure Limits

## Sealing

ASTM Standard F15-78
  Specification for Iron-Nickel-Cobalt Sealing Alloy
ASTM Standard F18-64
  Specification and Method for Evaluation of Glass-to-Metal Headers Used in Electron Devices
ASTM Standard F19-64
  Method for Tension and Vacuum Testing Metalized Ceramic Seals
ASTM Standard F29-78
  Specification for Dumet Wire for Glass-to-Metal Seal Applications
ASTM Standard F30-85
  Specification for Iron-Nickel Sealing Alloys
ASTM Standard F31-68
  Specification for 42% Nickel–6% Chromium-Iron Sealing Alloy
ASTM Standard F79-69
  Specification for Type 101 Sealing Glass
ASTM Standard F105-72
  Specification for Type 58 Borosilicate Sealing Glass

## Soldering

ASTM Standard F357-78
  Practice for Determining Solderability of Thick-Film Conductors
ASTM Standard F692-80
  Method of Measuring Adhesion Strength of Solderable Films to Substrates
EIA-IS-46
  Test Procedure for Resistance to Soldering (Vapor Phase Technique) for Surfaced Mount Devices
EIA-IS-49
  Solderability Test Procedures
EIA-JEDEC(Method B102)
  Surface Mount Solderability Test(JESD22B)
IPC-S-805
  Solderability Test for Component Leads and Terminations
IPC-S-815
  General Requirements for Soldering Electronic Interconnections
IPC-S-8XX
  SMT Process Guidelines and Checklist
IPC-SF-818
  General Requirements for Electronic Soldering Fluxes

IPC-SF-60
  Post Solder Solvent Cleaning Handbook
IPC-AC-62
  Post Solder Aqueous Cleaning Handbook
MIL-STD-202F (Method 208F)
  Dip and Look Test
DOD-STD-2000-1
  Soldering Technology, High Quality/High Reliability
DOD-STD-2000-2
  Solder Assembly, Component Mounting for High Quality and High Reliability
DOD-STD-2000-3
  Soldering Techniques, Criteria for High Quality and High Reliability
DOD-STD-2000-4
  General Soldering Requirement for Electrical and Electronic Equipment
DoD-STD-2000-5
  Terms and Definitions for Interconnecting and Packaging Electrical Circuits
MIL-F-14256E
  Flux, Soldering, Liquid (Resin Paste)
MIL-P-28809
  Flux Residue, Verification and Inspection
NAVY WS-6536E
  Process Specifications and Procedures for Preparing and Soldering Electronics
F/S QQ-S-571E
  Solder, Tin Alloy: Tin-Lead Alloy; and Lead Alloy
SEMI-Vol.1
  Specifications for Buffered Oxide Etchants
SEMI-Vol.1
  Specifications for Ammonium Fluoride 40% Solution
SEMI-Vol.1
  Specifications for Nitric Acid
SEMI-Vol.1
  Specifications for Dichloromethane (Methylene Chloride)
SEMI-Vol.1
  Specifications for VLSI Grade Tungsten Hexafluoride (SF6) in Cylinders (Provisional)
SEMI-Vol.1
  Specifications for Nitrogen Trifluoride (Provisional)
SEMI-Vol.1
  Particle Specification for Grade 20/0.2 Oxygen Delivered as Pipeline Gas
SEMI-Vol.1
  Particle Specification for Grade 20/0.2 Hydroxygen Delivered as Pipeline Gas

## Solvents and Reagents

F/S TT-I-735a
Alcohol, Isopropyl
F/S O-T-620C
Trichloroethane, 1,1,1, Inhibited (Methyl Chloroform)
MIL-C-81302C
Trichlorotrifluorethane Solvent Cleaning Compound
MIL-T-81533A
Trichloroethane, 1,1,1, Inhibited (Methyl Chloroform)

## Substrates: Ceramic

ASTM Standard C372-81
Test Method for Linear Thermal Expansion of Procelain Enamel and Glaze Frits and Fired Ceramic Whitewear Products by the Dilatometer Method
ASTM Standard C408-82
Test Method for Thermal Conductivity of Whitewear Ceramics
ASTM Standard C773-82
Test Method for Compressive (Crushing) Strength Whitewear Materials
ASTM Standard D150-81
Test Methods for A-C Loss Characteristics and Permittivity (Dielectric Constant) of Solid Electrical Insulating materials
ASTM Standard D2442-75
Specification for Alumina Ceramics for Electrical and Electronic Applications
ASTM Standard D1829-66
Test Method for Electrical Resistance of Ceramic Materials at Elevated Temperatures
ASTM Standard E18-79
Test Method for Rockwell Hardness and Rockwell Superficial Hardness of Metallic Materials
ASTM Standard F7-70
Specification for Aluminum Oxide Powder
ASTM Standard F44-68
Specification for Metalized Surfaces on Ceramic
ASTM Standard F77-69
Test Method for Apparent Density of Ceramics for Electron Device and Semiconductor Application
ASTM Standard F109-73
Definitions of Terms Relating to Surface Imperfections on Ceramics

ASTM Standard F356-75
Specification for Beryllia Ceramics for Electronic and Electrical Applications
ASTM Standard F394-78
Test Method for Biaxial Flexure Strength (Modulus of Rupture) of Ceramic Substrates
ASTM Standard F417-78
Test Method for Flexural Strength (Modulus of Rupture Electronic-Grade Ceramics
ASTM Standard F865-84
Test Method for Gross Camber of Ceramic Substrates for Thick Film Applications
ISHM SP 009
Substrates (ISHM-HMSSG)
NAVSEA 6228508
Board, Thick Film Multilayer Interconnection, Design Specifications
NAVSEA 6228507
Board, Thick Film Multilayer Interconnection, Performance Specifications

## Substrates: Metal

ASTM B735
Standard Test Methods for Porosity in Gold Coatings on Metal Substrates by Gas Exposures
ASTM B741
Standard Test Methods for Porosity in Gold Coatings on Metal Substrates by Paper Electrography

## Substrates: PWBs

ANSI/IPC-BP-421
General Specifications for Rigid PWB Backplanes with Press Fit Contacts
ANSI/IPC-CM-770C
Printed Board Component Mounting Guidelines
ASTM B451-86
Specification for Copper Foil, Strip, and Sheet for Printed Circuits and Carrier Tapes
IPC-CF-152
Metallic Foil Specification for Copper/Invar/Copper for Printed Wiring and Other Related Applications
IPC-D-249 1
Design Standard for Flex Single- and Double-Sided PWBs
IPC-D-319
Design Standard for Rigid 1- and 2-Sided PWBs

IPC-D-330
  Printed Wiring Design Guide
IPC-D-350
  Digital Description of PWBs
IPC-D-949
  Design Standard for Rigid Multilayer Printed
  Boards
IPC-FC-250
  Performance Specification for Single and
  Double Sided Flexible Printed Boards
IPC-MC-323
  Design Standard for Metal-Core PWBs
IPC-SD-320
  Performance Specification for Rigid Single
  and Double Sided Printed Boards
IPC-S-804
  Solderability Test Method for Printed Wiring
  Boards
IPC-SM-840
  Qualification and Performances of Permanent
  Polymer Coating (Solder Mask) for Printed
  Boards
IPC-B-29/50884
  Certification Boards, Flex and Rigid-Flex,
  Master Drawings
MIL-STD-275E
  PWBs for Electrical Equipment
MIL-STD-275
  Printed Wiring for Electronic Equipment
MIL-STD-429
  Terms and Definitions, Printed Wiring and
  PCBs
MIL-STD-1495
  Multilayer PWBs for Electrical Equipment
MIL-STD-2188
  PWBs, Flexible and Rigid Flexible, Electri-
  cal Design Requirements
MIL-I-46058C
  Coating, Insulating, PWBs
MIL-P-13949
  General Specifications for Plastic Sheet,
  Laminated, Metal Clad (for Printed Wiring
  Boards)
MIL-P-50884C
  Military Specification Printed Wiring, Flex-
  ible and Rigid Flex
MIL-P-55110
  Military Specification Printed Wiring
  Boards, General Specification for
MIL-R-9288C
  Resin, Phenolic Laminating

## Substrates: Semiconductor

ASTM Standard F26-84
  Methods for Determining the Orientation of
  a Semiconductive Single Crystal
ASTM Standard F42-72
  Test Methods for Conductivity Type of Ex-
  trinsic Semiconducting Materials
ASTM Standard F43-83
  Test Methods for Resistivity of Semiconduc-
  tor Materials
ASTM Standard F388-84
  Method for Measurement of Oxide Thickness
  on Silicon Wafers and Metallization Thick-
  ness by Multiple-Beam Interference (Tolan-
  sky Method)
ASTM Standard F533-82
  Test Method for Thickness and Thickness
  Variation of Silicon Slices
ASTM Standard F534-84
  Test Method for Bow of Silicon Slices
ASTM Standard F574-83
  Test Method for Evaluation of Polysilicon
ASTM Standard F576-84a
  Method for Measurement of Insulator Thick-
  ness and Refractive Index on Silicon Sub-
  strates by Ellipsometry
ASTM Standard F612-83
  Practice for Cleaning Surfaces of Polished
  Silicon Slices
ASTM Standard F613-82
  Test Method for Measuring Diameter of Sil-
  icon Slices and Wafers
ASTM Standard F657-80
  Method for Measuring Warp and Total Thick-
  ness Variation Silicon Wafers by a Noncontact
  Scanning Method
ASTM Standard F671-83
  Method for Measuring Flat Length on Slices
  of Electronic Materials
ASTM Standard F775-83
  Test Method for Wafer and Slice Flatness by
  Interferometric, Noncontact Technique
SEMI-Vol.3
  Silicon Wafer Specification for 200 mm Wa-
  fers, Revision
SEMI-Vol.3
  Specifications for Polycrystalline Silicon
SEMI-Vol.3
  A Decision Tree for Specifying Wafer Flat-
  ness

SEMI-Vol.3

Wafer Defect Limits Table for Polished Gallium Arsenide Wafers

SEMI-Vol.3

Specification for Ion Implantation on Gallium Arsenide Wafers

SEMI-Vol.3

Specifications for Serial Alphanumeric Marking of the Front Surface of Wafers, Revision

## Surface Mount Components (General)

EIA-PDP-100

Mechanical Outlines for Registered and Standard Electronic Parts (Passive Devices)

EIA-JEP-95

JEDEC Registered and Standard Outlines for Semiconductor Devices

EIA-RS-481

Taping of Surface Mount Components for Automatic Placement

## Vacuum Processing

ASTM Standard F798-82

Practice for Determining Gettering Rate, Sorption Capacity, and Gas Content of Nonevaporable Getters in the Molecular Flow Region

AVS-2.1

Calibration of a Leak Detectors of the Mass Spectrometer Type (1973)

AVS-2.2

Method for Vacuum Leak Calibration (1968)

AVS-2.3

Procedure for Calibrating Gas Analyzers of the Mass Spectrometer Type (1972)

AVS-3.1

Unbaked, Ungrooved Bolted Vacuum Connection Flanges, Nominal Sizes 4 in to 24 in (1965)

AVS-3.2

Flanges Bakeable to 500°C (1965)

AVS-3.3

Method for Testing Flange Seals to 500°C (1968)

AVS-3.4

Dimensions for Unbaked Flanges, Light Series (1969)

AVS-3.6

Procedure for Rating All-Metal Valves Bakeable to Above 250°C (1973)

AVS-4.1

Procedure for Measuring Speed of Oil Diffusion Pumps (1963)

AVS-4.2

Procedure for Measuring Throughput of Oil Diffusion Pumps (1963)

AVS-4.3

Procedure for Measuring Forepressure Characteristics of Oil Diffusion Pumps (1963)

AVS-4.4

Procedure for Measuring the Ultimate Pressure of Pumps Without Working Fluids (1973)

AVS-4.5

Procedure for Measuring Backstreaming of Oil Diffusion Pumps (1963)

AVS-4.6

Procedure for Measuring the Warm-Up and Cool Down Characteristics of Oil Diffusion Pumps (1964)

AVS-4.7

Procedure for Measuring the Ultimate Pressure Pumps Without Working Fluids (1973)

AVS-4.8

Procedure for Measuring Speed of Pumps Without Working Fluids (1963)

AVS-4.10

Determining the Refrigerant Consumption and Temperature Characteristics of Baffles and Traps (1968)

AVS-5.1

Measurement of Blank-Off Pressure (Permanent Gases) of Positive Displacement Mechanical Vacuum Pumps (1963)

AVS-5.2

Presentation of Pumping Speed Curves of Mechanical Pumps (1963)

AVS-5.3

Method for Measuring Pumping Speed of Mechanical Vacuum Pumps for Permanent Gases (1967)

AVS-6.2

Procedure for Calibrating Vacuum Gauges of the Thermal Conductivity Types (1969)

AVS-6.4

Procedure for Calibrating Hot Filament Ionization Gauges Against a Reference Manometer in the Range $10^{-2}$–$10^{-5}$ Torr (1969)

AVS-6.5

Procedures for the Calibration of Hot Filament Ionization Gauge Controls (1971)

AVS-7.1
Graphic Symbols in Vacuum Technology (1966)
AVS-9.1
Reporting of Outgassing Data (1964)
AVS-9.2
Reporting of Thermal Degassing Data (1969)

## Abbreviations:

Ceramic = Ceramic: CC = Chip Carrier; CCC = Ceramic Chip Carriers; F/S = Federal Specifications; GW = Gull Wing; JC-11 = JEDEC (EIA); Standards for Surface Mount (JC-11: Mechanical Standardization Committee); JL = J-Lead; MLC = Multilayer Ceramic; PLCC = Plastic Leaded Chip Carrier; SOIC = Small Outline IC; SOJ = Small Outline; XXX . . . = no. not known at time of publication.

## Organizations:

AMS
Society of Automotive Engineers, 400 Commonwealth Drive, Warrendale, PA 15096
ANSI
American Nat'l Standards Institute), 1430 Broadway, New York, NY 10018

ASTM
American Society for Testing Materials), 1916 Race Street, Philadelphia, Pa 19103
AVS
American Vacuum Society, 335 E 45th St., New York, NY 10017
EIA:JC-11
Electronic Industries Association, 2001 Eye St., N.W., Washington, DC 20006
EIAJ
Electronic Industries Association of Japan, 2-2, Marunouchi 3 Chome, Chiyoda-Ku, Tokyo 100, Japan
IPC
Institute for Interconnecting and Packaging Electronic Circuits, 7380 N. Lincoln Ave., Lincolnwood, IL 60646'
ISHM
International Society for Hybrid Microelectronics, 1861 Wiehle Ave., PO Box 2698, Reston, VA 22090
MIL;NAVSEA;DOD;F/S
Naval Publications and Forms Office, 5801 Tabor Ave, Philadelphia, PA 19120-5099, Tel:215-697-2667, Fax:215-697-5917
SEMI
Semiconductor Equipment and Materials International, 805 E. Middlefield Road, Mountain View, CA 94043

# Appendix D:
# Standards and Specifications, PCB/PWBs

## Cables and Connectors

IPC-FC-203
 Specification for Flat Cable, Round Conductor, Ground Plane

IPC-FC-210
 Performance Specification for Flat-Conductor Undercarpet Power Cable (Type FCC)

IPC-FC-213
 Performance Specification for Flat Undercarpet Telephone Cable

IPC-FC-217
 General Document for Connectors, Electric, Header/Receptable, Insulation Displacement for Use with Round Conductor Flat Cable

IPC-FC-218
 General Specification for Connectors, Electric Flat Cable Type

IPC-FC-219
 Environmentally Sealed Flat Cable Connectors for Use in Aerospace Applications

IPC-FC-220
 Specification for Flat Cable, Flat Conductor, Unshielded

IPC-FC-221
 Specification for Flat-Copper Conductors for Flat Cables

IPC-FC-222
 Specification for Flat Cable, Round Conductor, Unshielded

IPC-FC-231
 Flexible Bare Dielectrics for Use in Flexible Printed Wiring

IPC-FC-232
 Adhesive Coated Dielectric Films for Use as Cover Sheets for Flexible Printed Wiring

## Cleaning

IPC-AC-62
 Post Solder Aqueous Cleaning Handbook

IPC-SC-60
 Post Solder Solvent Cleaning Handbook

## Design, Guidelines for

ANSI-ASQC-Z-1.15
 Generic Guidelines for Quality Systems

DOD-D-1000
 Drawings, Engineering and Associated Lists

EIA-RS-208
 Definition & Register Printed Wiring

IPC-241B
 Flexible Metal-Clad Dielectrics for Use in Fabrication of Flexible Printed Wiring

IPC-A-600C
 Acceptability of Printed Boards

IPC-A-610
 Acceptability of Printed Assemblies

IPC-BP-421
General Specification for Rigid Printed Board
Backplanes with Press-Fit Contacts
IPC-CF-150
Copper Foil for Printed Wiring Applications
IPC-D-249
Design Standard for Flexible Single and Dou-
ble-Sided Printed Boards
IPC-D-300G
Printed Board Dimensions and Tolerances
IPC-D-319
Design Standards for Rigid Single and Dou-
ble-Sided Printed Boards
IPC-D-322
Guidelines for Selecting Printed Wiring
Board Sizes Using Standard Panel Sizes
IPC-D-330
Printed Wiring Design Guide
IPC-D-859
Design Standard for Thick Film, Multilayer
Hybrid Circuits
IPC-D-949
Design Standard for Rigid Multilayer Printed
Boards
IPC-DW-425A
Design and End Product Requiremednts for
Discrete Wiring Boards
IPC-DW-426
Specifications for Assembly of Discrete Wir-
ing
IPC-FC-233
Specification for Flexible Adhesive Bonding
Films
IPC-FC-250A
Specification for Single- and Double-Sided
Flexible Printed Wiring
IPC-HM-860
Specification for Multilayer Hybrid Circuits
IPC-MC-323
Design Standard for Metal Core Printed
Boards
IPC-MC-324
Performance Specification for Metal Core
Boards
IPC-ML-949
Design Standard for Rigid Multilayer Printed
Boards
IPC-ML-950C
Performance Specification for Rigid Multi-
layer Printed Boards
IPC-PMC-90
General Requirements for Implementation of
Statistical Process Control

IPC-R-700
Suggested Guidelines for Modifications, Re-
work, and Repair of Printed Boards and As-
semblies
IPC-SD-320B
Performance Specification for Rigid Single-
and Double-Sided Printed Boards
JEDEC-95
JEDEC Registered and Standard Outlines for
Solid State Products
MIL-I-45208
Inspection System Requirements
MIL-P-55110
Printed Wiring Boards, General Specification
for
MIL-S-19500
Semiconductor Devices, General Specifica-
tion
MIL-STD-105
Sampling Procedures and Tables for Inspec-
tion by Attributes
MIL-STD-129
Marking for Shipment and Storage
MIL-STD-275
Printed Wiring for Electronic Equipment
MIL-STD-414/ANSI Z-1.9
Sampling Procedures and Tables for Inspec-
tion by Variables for Percent Defective
MIL-/STD-454
Standard General Requirements for Elec-
tronic Equipment
RS-167
Type Designation for Receiver-Type Tube
Sockets
RS-428
Type Designation System for Microelec-
tronic Devices
RS-471
Symbol and Label for Electrostatic Sensitive
Devices
TR-TSY-000078
Generic Physical Design Requirements for
Telecommunications Products and Equip-
ment

## Drawings and Documentation
IPC-100043-3
Master Drawing: IPC Multilayer Qualifica-
tion Specimen
IPC-D-300
Printed Wiring Board Dimensions and Tol-
erances

IPC-D-325
End Product Documentation for Printed Boards
IPC-D-350C
Printed Board Description in Digital Form
IPC-D-351
Printed Board Drawings in Digital Form
IPC-D-352
Electronic Design Data Description for Printed Boards in Digital Form
IPC-D-354
Library Format Description for Printed Boards in Digital Form
IPC-NC-349

## Insulation
IPC-CC-830
Electrical Insulating Compounds for Printed Board Assemblies
MIL-C-14550
Copper Plating (Electrodeposited)
MIL-G-45204
Gold Plating, Electrodeposited
MIL-I-7444
Insulation Sleeving, Electrical, Flexible
MIL-I-23053
Insulation Sleeving, Electrical, Heat Shrinkable
MIL-I-46058
Insulating Compound Electrical (for Coating Printed Circuit Assays)
MIL-T-10727
Tin Plating, Electrodeposited or Hot Oil Dipped, for Ferrous and Nonferrous Metals

## Laminates
IPC-L-108B
Specification for Thin Metal Clad Base Materials for Multilayer Printed Boards
IPC-L-109
Specification for Glass Cloth, Resin Preimpregnated (B Stage) for High Temperature Multilayer Printed Boards
IPC-L-109B
Specification for Resin Preimpregnated Fabric (Prepreg) for Multilayer Printed Boards
IPC-L-112
Standard for Foil Clad, Polymeric, Composite Laminate
IPC-L-115
Specification for Thin Laminates for Extended High Temperature Multilayer Printed Board (Polyimide)

IPC-L-115B
Specification for Rigid Metal-Clad Base Materials for Printed Boards
IPC-L-125
Specification for Plastic Sheet, Laminated, Metal-Clad for High Frequency Interconnections (Microwave) Printed Boards
IPC-L-125A
Specification for Plastic Substrates, Clad or Unclad, for High Speed/High Frequency Interconnections
IPC-L-130
Specifications for Thin Laminates, Metal Clad Primarily for General-Purpose Multilayer Printed Boards
MIL-P-13949
General Specifications for Plastic Sheet, Laminated Metal Clad (for Printed Wiring)

## Mounting, Component
IPC-CN-78
Guidelines for Surface Mounting
IPC-CM-770
Component Mounting Guidelines for Printed Boards
IPC-SM-782
Surface Mount Land Patterns (Configurations and Design Rules)
MIL-S-12883
Sockets and Accessories for Plug-in Electrical Components
MIL-S-83502
Sockets, Plug-in Electronic Component Round Style
RS-192
Holder Outlines and Pin Connections for Quartz Crystal Units
RS-296
Lead Taping, Axial Components
RS-367
Dimensional and Electrical Characteristics Defining Receiver-Type Sockets
RS-415
Dimensional and Electrical Characteristic Defining Dual-Line-Type Sockets
RS-468
Lead Taping, Radial Components
RS-481
Lead Taping, Leadless Components
RS-488
Sockets, Individual Lead-Types (for Electrical and Electronic Components)

## Production

IPC-AM-361
   Specification for Rigid Substrates for Additive Process Printed Boards
IPC-AM-372
   Electrodes Copper Film for Additive Printed Boards
IPC-CF-148
   Resin Coated Metal for Printed Boards
IPC-CF-152
   Metallic Foil Specification for Copper Invar Copper (CIC) or Printed Wiring and Other Related Applications
IPC-DR-570
   General Specification for $\frac{1}{8}$ Inch Diameter Shank Carbide Drills for Printed Boards
IPC-EG-140
   Specification for Finished Fabric Woven from "E" Glass for Printed Boards
MIL-P-11268
   Parts, Materials, and Processes Used in Electronic Equipment
MIL-P-28809
   Printed Wiring Assemblies
MIL-P-46843
   Printed Wiring Assemblies, Production of
MIL-P-66110
   Printed Wiring Boards
MIL-P-81728
   Plating, Tin-Lead (Electrodeposited)

## Soldering

IPC-AJ-820
   Assembly and Joining Handbook
IPC-S-815
   General Requirements for Soldering Electrical Connections and Printed Board Assemblies
IPC-SF-818
   General Requirements for Electronic Soldering Fluxes
IPC-SF-819
   General Requirements and Test Methods for Electronic Grade Solder Paste
IPC-SM-839
   Pre and Post-Solder Mask Application Cleaning Guidelines

IPC-SM-840B
   Qualification and Performance of Permanent Polymer Coating (Solder Mask) for Printed Boards
IPC-SP-819
   General Requirements and Test Methods for Electronic Grade Solder Paste
MIL-F-14256
   Flux, Soldering, Liquid (Rosin Base)
MIL-S-45743
   Soldering, Manual Type, High Reliability, Electrical & Electronic Equipment
MIL-S-46844
   Solder Bath Soldering of Printed Wiring Assemblies
MIL-S-46880
   Soldering of Metallic Ribbon Lead Material to Solder Terminal, Process for Reflow
QQ-S-571
   Soldering, Tin Alloy, Tin-Lead Alloy and Lead Alloy

## Testing

IPC-HF-318
   Microwave End Product Board Inspection and Test
IPC-S-804
   Solderability Test Methods for Printed Wiring Boards
IPC-S-805
   Solderability Tests for Component Leads and Terminations
IPC-SM-840
   Qualification and Performance of Permanent Polymer Coating (Solder Mask) for Printed Boards
IPC-TF-870
   Qualification and Performance of Polymer Thick Film Printed Boards
IPC-TM-Test Methods Manual
IPC-QL-653
   Qualification of Facilities That Inspect/Test Printed Boards, Components and Materials
MIL-STD-202
   Test Methods for Electronic and Electrical Component Parts
MIL-STD-1344
   Test Methods for Electrical Connectors

*(For organization addresses, please see the end of Appendix C, page 850.)*

# Appendix E: Unit Conversion Tables

**Table E.1**   Thermal Conductivity

|  | watt/m-°C | watt/cm-°C | watt/in-°C | Btu/hr-ft-°F | cal/scc-cm-°C |
|---|---|---|---|---|---|
| 1 watt/m-°C | 1.0 | 0.01 | 0.0254 | 0.578 | 0.002389 |
| 1 watt/cm-°C | 100 | 1.0 | 2.540 | 57.8 | 0.2389 |
| 1 watt/in-°C | 39.37 | 0.3937 | 1.0 | 22.75 | 0.09405 |
| 1 Btu/hr-ft-°F | 1.730 | 0.01730 | −0.0440 | 1.0 | $4.234 \times 10^{-3}$ |
| 1 cal/scc-cm-°C | 418.6 | 4.186 | 10.63 | 241.9 | 1.0 |

**Table E.2**   Specific Heat (Entropy)

|  | watt-sec/kg-°C | Btu/lb$_m$-°F | cal/g-°C |
|---|---|---|---|
| 1 watt-sec/kg-°C | 1.0 | $2.388 \times 10^{-4}$ | 1.0007 |
| 1 Btu/lb$_m$-°F | 4186 | 1.0 | $2.390 \times 10^{-4}$ |
| 1 cal/g-°C | 4184 | 0.9993 | 1.0 |

$$1 \frac{J}{kg - K} = \frac{1 \text{ watt-sec}}{kg - °C}$$

**Table E.3**   Stress/Pressure

|  | Pascal | lb/ft$^2$ | lb/in.$^2$ | Inches of water | Atmospheres |
|---|---|---|---|---|---|
| 1 pascal | 1.0 | 0.021 | $1.45 \times 10^{-4}$ | $4.02 \times 10^{-3}$ | $9.8760 \times 10^{-6}$ |
| 1 lb$_f$/ft$^2$ | 47.88 | 1.0 | $6.944 \times 10^{-3}$ | 0.1926 | $4.725 \times 10^{-4}$ |
| 1 lb$_f$/in.$^2$ | $6.895 \times 10^3$ | 144.0 | 1.0 | 27.73 | 0.06805 |
| 1 in.-H$_2$O | $2.49 \times 10^2$ | 5.193 | 0.03606 | 1.0 | $2.454 \times 10^{-3}$ |
| 1 atmosphere | $1.013 \times 10^5$ | 2116.0 | 14.7 | 407.5 | 1.0 |

**Table E.4**  Mass

|            | kg                     | g                     | Slug                    | lb                      | oz                      |
|------------|------------------------|-----------------------|-------------------------|-------------------------|-------------------------|
| 1 kilogram | 1                      | 1000                  | $6.852 \times 10^{-2}$  | 2.205                   | 35.27                   |
| 1 gram     | 0.001                  | 1                     | $6.852 \times 10^{-5}$  | $2.205 \times 10^{-3}$  | $3.527 \times 10^{-2}$  |
| 1 slug     | 14.59                  | $1.459 \times 10^{4}$ | 1                       | 32.17                   | 514.8                   |
| 1 pound (avoirdupois) | 0.4536      | 453.6                 | $3.108 \times 10^{-2}$  | 1                       | 16                      |
| 1 ounce    | $2.835 \times 10^{-2}$ | 28.35                 | $1.943 \times 10^{-3}$  | $6.250 \times 10^{-2}$  | 1                       |

**Table E.5**  Force

|                   | N          | $kg_f$                 | Dyne                   | lb                      |
|-------------------|------------|------------------------|------------------------|-------------------------|
| 1 Newton          | 1          | 0.1020                 | $10^{5}$               | 0.2248                  |
| 1 kilogram-force  | 9.807      | 1                      | $9.807 \times 10^{5}$  | 2.205                   |
| 1 dyne            | $10^{-5}$  | $1.020 \times 10^{-6}$ | 1                      | $2.248 \times 10^{-6}$  |
| 1 pound           | 4.448      | 0.4536                 | $4.448 \times 10^{5}$  | 1                       |

**Table E.6**  Length

|              | Meter                  | cm     | ft                     | in.                     |
|--------------|------------------------|--------|------------------------|-------------------------|
| 1 meter      | 1                      | 100    | 3.281                  | 39.37                   |
| 1 centimeter | $10^{-2}$              | 1      | $3.281 \times 10^{-2}$ | 0.3937                  |
| 1 foot       | 0.3048                 | 30.48  | 1                      | 12                      |
| 1 inch       | $2.540 \times 10^{-2}$ | 2.540  | $8.333 \times 10^{-2}$ | 1                       |

**Table E.7**  Heat Flux

|                   | watts/m$^2$           | watts/cm$^2$           | watts/in.$^2$          | Btu/hr-ft$^2$          | Btu/hr-in.$^2$         | Btu/sec-in.$^2$         |
|-------------------|-----------------------|------------------------|------------------------|------------------------|------------------------|-------------------------|
| 1 watt/m$^2$      | 1.0                   | $10^{-4}$              | $6.45 \times 10^{-4}$  | 0.3172                 | $2.203 \times 10^{-3}$ | $6.12 \times 10^{-7}$   |
| 1 watt/cm$^2$     | $10^{4}$              | 1.0                    | 6.452                  | 3172                   | 22.03                  | $6.12 \times 10^{-3}$   |
| 1 watt/in.$^2$    | 1550                  | 0.155                  | 1.0                    | 491.67                 | 3.415                  | $9.486 \times 10^{-4}$  |
| 1 Btu/hr-ft$^2$   | 3.1525                | $3.1525 \times 10^{-4}$ | $2.034 \times 10^{-3}$ | 1.0                   | $6.944 \times 10^{-3}$ | $1.929 \times 10^{-6}$  |
| 1 But/hr-in.$^2$  | 453.9                 | 0.04539                | 0.2928                 | 144                    | 1.0                    | $2.778 \times 10^{-4}$  |
| 1 Btu/sec-in.$^2$ | $1.634 \times 10^{6}$ | 163.4                  | 1054                   | $5.184 \times 10^{5}$  | 3600                   | 1.0                     |

**Table E.8**  Density

|             | kg/m$^3$              | gm/cm$^3$              | lb/ft$^3$              | lb/in.$^3$              |
|-------------|-----------------------|------------------------|------------------------|-------------------------|
| 1 kg/m$^3$  | 1                     | 0.001                  | $6.243 \times 10^{-2}$ | $3.613 \times 10^{-5}$  |
| 1 gram/cm$^3$ | 1000                | 1                      | 62.43                  | $3.613 \times 10^{-2}$  |
| 1 lb/ft$^3$ | 16.02                 | $1.602 \times 10^{-2}$ | 1                      | $5.787 \times 10^{-4}$  |
| 1 lb/in.$^3$ | $2.768 \times 10^{4}$ | 27.68                 | 1728                   | 1                       |

**Table E.9** Temperature

|      | °C     | °F  | K      | R   |
|------|--------|-----|--------|-----|
| 1°C  | 1.0    | 1.8 | 1.0    | 1.8 |
| 1°F  | 0.5556 | 1.0 | 0.5556 | 1.0 |
| 1 K  | 1.0    | 1.8 | 1.0    | 1.8 |
| 1 R  | 0.5556 | 1.0 | 0.5556 | 1.0 |

$$T_{°C} = (T_{°F} - 32)/1.8$$
$$T_K = T_{°C} + 273.16$$
$$T_R = T_{°F} + 459.67$$

**Table E.10** Heat (Energy, Work)

|                        | kW-hr                    | watt-sec             | ft-lb                    | Btu                      | cal                      | erg                      |
|------------------------|--------------------------|----------------------|--------------------------|--------------------------|--------------------------|--------------------------|
| 1 kilowatt-hour        | 1                        | $3.6 \times 10^6$    | $2.655 \times 10^6$      | 3413                     | $8.601 \times 10^5$      | $3.6 \times 10^{13}$     |
| 1 watt-sec (1 joule)   | $2.778 \times 10^{-7}$   | 1                    | 0.7376                   | $9.481 \times 10^{-4}$   | 0.2389                   | $10^7$                   |
| 1 foot-pound           | $3.766 \times 10^{-7}$   | 1.356                | 1                        | $1.285 \times 10^{-3}$   | 0.3239                   | $1.356 \times 10^7$      |
| 1 Btu                  | $2.930 \times 10^{-4}$   | 1055                 | 777.9                    | 1                        | 252.0                    | $1.055 \times 10^{10}$   |
| 1 calorie              | $1.163 \times 10^{-6}$   | 4.186                | 3.007                    | $3.968 \times 10^{-3}$   | 1                        | $4.186 \times 10^7$      |
| 1 erg                  | $2.778 \times 10^{-14}$  | $10^{-7}$            | $7.376 \times 10^{-8}$   | $9.481 \times 10^{-11}$  | $2.389 \times 10^{-8}$   | 1                        |

**Table E.11** Heat Rate (Power)

|              | kW                     | watt                   | ft-lb/min            | ft-lb/sec              | Btu/hr                 | cal/sec                |
|--------------|------------------------|------------------------|----------------------|------------------------|------------------------|------------------------|
| 1 kilowatt   | 1                      | 1000                   | $4.425 \times 10^4$  | 737.6                  | 3413                   | 238.9                  |
| 1 watt       | 0.001                  | 1                      | 44.25                | 0.73776                | 3.413                  | 0.2389                 |
| 1 ft-lb/min  | $2.260 \times 10^{-5}$ | $2.260 \times 10^{-2}$ | 1                    | $1.667 \times 10^{-2}$ | $7.713 \times 10^{-2}$ | $5.399 \times 10^{-3}$ |
| 1 ft-lb/sec  | $1.356 \times 10^{-3}$ | 1.356                  | 60                   | 1                      | 4.628                  | 0.3239                 |
| Btu-hr       | $2.930 \times 10^{-4}$ | 0.2930                 | 12.97                | 0.2161                 | 1                      | $7.000 \times 10^{-2}$ |
| 1 calorie/sec| $4.186 \times 10^{-3}$ | 4.186                  | $1.852 \times 10^2$  | 3.087                  | 14.29                  | 1                      |

**Table E.12** Area

|                      | $m^2$                   | $cm^2$                 | cir mil               | $ft^2$                 | $in.^2$                |
|----------------------|-------------------------|------------------------|-----------------------|------------------------|------------------------|
| 1 square meter       | 1                       | $10^4$                 | $1.974 \times 10^9$   | 10.76                  | 1550                   |
| 1 square centimeter  | $10^{-4}$               | 1                      | $1.974 \times 10^5$   | $1.076 \times 10^{-3}$ | 0.155                  |
| 1 circular mil       | $5.067 \times 10^{-10}$ | $5.067 \times 10^{-6}$ | 1                     | $5.454 \times 10^{-9}$ | $7.854 \times 10^{-7}$ |
| 1 square root        | $9.290 \times 10^{-2}$  | 929.0                  | $1.833 \times 10^8$   | 1                      | 144                    |
| 1 square inch        | $6.452 \times 10^{-4}$  | 6.452                  | $1.273 \times 10^6$   | $6.944 \times 10^{-3}$ | 1                      |

**Table E.13**  Electric Resistivity

|                                | ohm-m | ohm-cm | $\mu$ohm-cm | statohm-cm | ohm-circ mil/ft |
|--------------------------------|-------|--------|-------------|------------|-----------------|
| 1 ohm-meter                    | 1 | 100 | $10^8$ | $1.113 \times 10^{-10}$ | $6.015 \times 10^8$ |
| 1 ohm-centimeter               | 0.01 | 1 | $10^6$ | $1.113 \times 10^{-12}$ | $6.015 \times 10^6$ |
| 1 micro-ohm-centimeter         | $10^{-8}$ | $10^{-6}$ | 1 | $1.113 \times 10^{-18}$ | 6.015 |
| 1 statohm-centimeter (1 esu)   | $8.987 \times 10^9$ | $8.987 \times 10^{11}$ | $8.987 \times 10^{17}$ | 1 | $5.406 \times 10^{18}$ |
| 1 ohm-circular mil per foot    | $1.662 \times 10^{-9}$ | $1.662 \times 10^{-7}$ | 0.1662 | $1.850 \times 10^{-19}$ | 1 |

**Table E.14**  Temperature Conversion

Look up the quantity (°C or °F) in the "Number" column and read across the desired column for the temperature conversion; e.g., 500°F to 260°C.

| Number | °C | °F | Number | °C | °F | Number | °C | °F |
|--------|------|--------|--------|-------|--------|--------|-------|--------|
| − 200 | − 128.9 | − 328.0 | 310 | 154.4 | 590.0 | 820 | 437.8 | 1508.0 |
| − 190 | − 123.3 | − 310.0 | 320 | 160.0 | 608.0 | 830 | 443.3 | 1526.0 |
| − 180 | − 117.8 | − 292.0 | 330 | 165.6 | 626.0 | 840 | 448.9 | 1544.0 |
| − 170 | − 112.2 | − 274.0 | 340 | 171.1 | 644.0 | 850 | 454.4 | 1562.0 |
| − 160 | − 106.7 | − 256.0 | 350 | 176.7 | 662.0 | 860 | 460.0 | 1580.0 |
| − 150 | − 101.1 | − 238.0 | 360 | 182.2 | 680.0 | 870 | 465.6 | 1598.0 |
| − 140 | − 95.6 | − 220.0 | 370 | 187.8 | 698.0 | 880 | 471.1 | 1616.0 |
| − 130 | − 90.0 | − 202.0 | 380 | 193.3 | 716.0 | 890 | 476.7 | 1634.0 |
| − 120 | − 84.4 | − 184.0 | 390 | 198.9 | 734.0 | 900 | 482.2 | 1652.0 |
| − 110 | − 78.9 | − 166.0 | 400 | 204.4 | 752.0 | 910 | 487.8 | 1670.0 |
| − 100 | − 73.3 | − 148.0 | 410 | 210.0 | 770.0 | 920 | 493.3 | 1688.0 |
| − 90 | − 67.8 | − 130.0 | 420 | 215.6 | 788.0 | 930 | 498.9 | 1706.0 |
| − 80 | − 62.2 | − 112.0 | 430 | 221.1 | 806.0 | 940 | 504.4 | 1724.0 |
| − 70 | − 56.7 | − 94.0 | 440 | 226.7 | 824.0 | 950 | 510.0 | 1742.0 |
| − 60 | − 51.1 | − 76.0 | 450 | 232.2 | 842.0 | 960 | 515.6 | 1760.0 |
| − 50 | − 45.6 | − 58.0 | 460 | 237.8 | 860.0 | 970 | 521.1 | 1778.0 |
| − 40 | − 40.0 | − 40.0 | 470 | 243.3 | 878.0 | 980 | 526.7 | 1796.0 |
| − 30 | − 34.4 | − 22.0 | 480 | 248.9 | 896.0 | 990 | 532.2 | 1814.0 |
| − 20 | − 28.9 | − 4.0 | 490 | 254.4 | 914.0 | 1000 | 537.8 | 1832.0 |
| − 10 | − 23.3 | 14.0 | 500 | 260.0 | 932.0 | 1010 | 543.3 | 1850.0 |
| 0 | − 17.8 | 32.0 | 510 | 265.6 | 950.0 | 1020 | 548.9 | 1868.0 |
| 10 | − 12.2 | 50.0 | 520 | 271.1 | 968.0 | 1030 | 554.4 | 1886.0 |
| 20 | − 6.7 | 68.0 | 530 | 276.7 | 986.0 | 1040 | 560.0 | 1904.0 |
| 30 | − 1.1 | 86.0 | 540 | 282.2 | 1004.0 | 1050 | 565.6 | 1922.0 |
| 40 | 4.4 | 104.0 | 550 | 287.8 | 1022.0 | 1060 | 571.1 | 1940.0 |
| 50 | 10.0 | 122.0 | 560 | 293.3 | 1040.0 | 1070 | 576.7 | 1958.0 |
| 60 | 15.6 | 140.0 | 570 | 298.9 | 1058.0 | 1080 | 582.2 | 1976.0 |
| 70 | 21.1 | 158.0 | 580 | 304.4 | 1076.0 | 1090 | 587.8 | 1994.0 |
| 80 | 26.7 | 176.0 | 590 | 310.0 | 1094.0 | 1100 | 593.3 | 2012.0 |
| 90 | 32.2 | 194.0 | 600 | 315.6 | 1112.0 | 1110 | 598.9 | 2030.0 |
| 100 | 37.8 | 212.0 | 610 | 321.1 | 1130.0 | 1120 | 604.4 | 2048.0 |
| 110 | 43.3 | 230.0 | 620 | 326.7 | 1148.0 | 1130 | 610.0 | 2066.0 |

**Table E.14   (Continued)**

| Number | °C | °F | Number | °C | °F | Number | °C | °F |
|--------|------|-------|--------|-------|--------|--------|-------|--------|
| 120 | 48.9 | 248.0 | 630 | 332.2 | 1166.0 | 1140 | 615.6 | 2084.0 |
| 130 | 54.4 | 266.0 | 640 | 337.8 | 1184.0 | 1150 | 621.1 | 2102.0 |
| 140 | 60.0 | 284.0 | 650 | 343.3 | 1202.0 | 1160 | 626.7 | 2120.0 |
| 150 | 65.6 | 302.0 | 660 | 348.9 | 1220.0 | 1170 | 632.2 | 2138.0 |
| 160 | 71.1 | 320.0 | 670 | 354.4 | 1238.0 | 1180 | 637.8 | 2156.0 |
| 170 | 76.7 | 338.0 | 680 | 360.0 | 1256.0 | 1190 | 643.3 | 2174.0 |
| 180 | 82.2 | 356.0 | 690 | 365.6 | 1274.0 | 1200 | 648.9 | 2192.0 |
| 190 | 87.8 | 374.0 | 700 | 371.1 | 1292.0 | 1210 | 654.4 | 2210.0 |
| 200 | 93.3 | 392.0 | 710 | 376.7 | 1310.0 | 1220 | 660.0 | 2228.0 |
| 210 | 98.9 | 410.0 | 720 | 382.2 | 1328.0 | 1230 | 665.6 | 2246.0 |
| 220 | 104.4 | 428.0 | 730 | 387.8 | 1346.0 | 1240 | 671.1 | 2264.0 |
| 230 | 110.0 | 446.0 | 740 | 393.3 | 1364.0 | 1250 | 676.7 | 2282.0 |
| 240 | 115.6 | 464.0 | 750 | 398.9 | 1382.0 | 1260 | 682.2 | 2300.0 |
| 250 | 121.1 | 482.0 | 760 | 404.4 | 1400.0 | 1270 | 687.8 | 2318.0 |
| 260 | 126.7 | 500.0 | 770 | 410.0 | 1418.0 | 1280 | 693.3 | 2336.0 |
| 270 | 132.2 | 518.0 | 780 | 415.6 | 1436.0 | 1290 | 698.9 | 2354.0 |
| 280 | 137.8 | 536.0 | 790 | 421.1 | 1454.0 | 1300 | 704.4 | 2372.0 |
| 290 | 143.3 | 554.0 | 800 | 426.7 | 1472.0 | | | |
| 300 | 148.9 | 572.0 | 810 | 432.2 | 1490.0 | | | |

$$°C = 5/9(°F - 32) \qquad °F = 9/5(°C) + 32$$

# Index